Chemistry of Multiphase Atmospheric Systems

NATO ASI Series

Advanced Science Institutes Series

A series presenting the results of activities sponsored by the NATO Science Committee, which aims at the dissemination of advanced scientific and technological knowledge, with a view to strengthening links between scientific communities.

The Series is published by an international board of publishers in conjunction with the NATO Scientific Affairs Division

A Life Sciences **B Physics**	Plenum Publishing Corporation London and New York
C Mathematical and Physical Sciences	D. Reidel Publishing Company Dordrecht, Boston and Lancaster
D Behavioural and Social Sciences **E Applied Sciences**	Martinus Nijhoff Publishers Boston, The Hague, Dordrecht and Lancaster
F Computer and Systems Sciences **G Ecological Sciences**	Springer-Verlag Berlin Heidelberg New York Tokyo

Series G: Ecological Sciences Vol. 6

Chemistry of Multiphase Atmospheric Systems

Edited by

Wolfgang Jaeschke

Center for Environmental Research, University of Frankfurt, FRG

Springer-Verlag Berlin Heidelberg New York Tokyo

Published in cooperation with NATO Scientific Affairs Division

Proceedings of the NATO Advanced Study Institute on Chemistry of Multiphase Atmospheric Systems held on the Island of Corfu, Greece, September 26 – October 8, 1983

ISBN 3-540-15647-X Springer-Verlag Berlin Heidelberg New York Tokyo
ISBN 0-387-15647-X Springer-Verlag New York Heidelberg Berlin Tokyo

Library of Congress Cataloging in Publication Data. NATO Advanced Study Institute on Chemistry of Multiphase Atmospheric Systems (1983 : Kerkyra, Corfu) Chemistry of multiphase atmospheric systems. (NATO ASI series. Series G, Ecological sciences ; vol. 6) "Proceedings of the NATO Advanced Study Institute on Chemistry of Multiphase Atmospheric Systems held on the Island of Corfu, Greece, September 26 – October 8, 1983"—P. Includes index. 1. Atmospheric chemistry—Congresses. I. Jaeschke, W. (Wolfgang) II. Title. III. Series.
QC879.6.N38 1983 551.5 85-27883
ISBN 0-387-15647-X (U.S.)

Printed in Germany

Printing: Beltz Offsetdruck, Hemsbach; Bookbinding: J. Schäffer OHG, Grünstadt
2131/3140-543210

TABLE OF CONTENTS

Part III

ORIGIN, DISTRIBUTION AND REMOVAL OF ATMOSPHERIC TRACE COMPOUNDS IN THE PRESENCE OF AIRBORNE PARTICLES AND LIQUID WATER

Part IV

CHEMICAL CONVERSIONS IN MULTIPHASE ATMOSPHERIC SYSTEMS AND THEIR PHYSICO CHEMICAL INVESTIGATION

Part V

THE ROLE OF MULTIPHASE ATMOSPHERIC CHEMISTRY IN SOURCE-RECEPTOR CONSIDERATIONS

PREFACE

Rapidly increasing interest in the problems of air pollution and source-receptor relationships has led to a significant expansion of knowledge in the field of atmospheric chemistry. In general the chemistry of atmospheric trace constituents is governed by the oxygen content of the atmosphere. Upon entering the atmosphere in a more or less reduced state, trace substances are oxidized via various pathways and the generated products are often precursors of acidic compounds. Beside oxidation processes occurring in the gas phase, gaseous compounds are often converted into solid aerosol particles.

The various steps within gas-to-particle conversion are constantly interacting with condensation processes, which are caused by the tropospheric water content. Thus in addition to the gaseous state, a liquid and solid state exists within the troposphere. The solid phase consists of atmospheric conversion products or fly ash and mineral dust. The liquid phase consists of water, conversion products and soluble compounds. The chemistry occurring within this system is often referred to as hydrogeneous chemistry. The chemist interprets this term, however, more strictly as reactions which occur only at an interphase between phases. This, however, is not always what happens in the atmosphere. There are indeed heterogeneous processes such as reactions occurring on the surface of dry aerosol particles. But apart from these, we must focus as well on reactions in the homogeneous phase, which are single steps of consecutive reactions running through various phases. Therefore, instead of heterogeneous chemistry, the interaction of all chemical processes occurring in and between the three phases of the atmosphere should better be entitled 'Multiphase Atmospheric Chemistry'.

In recent years this kind of chemistry was given great scholarly attention, and slogans like 'acid precipitation', 'dirty cloud' or 'killer fog' indicate that these phenomena are of a high public interest. Therefore scientists of the Atmospheric Sciences Research Center of the State University of New York at Albany and scientists of the Center for Environmental Research of the University of Frankfurt came up with the idea of the common organization of an 'Advanced Study Institute' on the topic of 'Multiphase Atmospheric Chemistry'.

The contents of this book are lectures and research papers presented at an 'Advanced Study Institute' entitled 'Chemistry of Multiphase Atmospheric Systems' held from September 26th to October 8th, 1983, on the island of Corfu, Greece. The institute was mainly sponsored by the 'NATO Scientific and Environmental Affairs Division', and we want to express our gratitude to this institution for the important assistance and cooperation we received.

The sequence of the chapters subdivided into five parts reflects the tutorial system of the program offered during the Corfu institute. In an introductory part the analytical and experimental methodology is described, which is necessary for the investigation of chemical compounds and chemical conversion processes appearing in cloud droplets, precipitation and aerosol particles. It is followed by chapters on field studies of cloud, fog and precipitation. This part is introduced by a chapter with ten basic statements on cloud and aerosol physics.

The third part is mainly concerned with the origin, the distribution and the removal of the most important atmospheric trace compounds in the presence of airborne particles and liquid water. With regard to acid deposition the most important compounds considered here are free protones, nitric and sulphuric acid, hydrocarbons and oxydants. This consideration leads into the forth part of the book in which physico-chemical investigations of conversion processes in 'Multiphase Atmospheric Systems' are treated. This includes photochemistry, absorbtion and diffusion processes, equilibrium distribution of compounds between the phases, the connection of pH-values with ion balance, a compilation of measured S^{IV} to S^{VI} conversion rates and the role of transition metals as potential catalysts of oxidation processes. Finally, there is a group of six chapters dealing with the role of 'Multiphase Atmospheric Chemistry' in source receptor considerations.

The advanced study institute held in Corfu in 1983 was an ideal opportunity to compile this interdisciplinary survey of problems and results gained in the last few years. Although the chapters contain much current research material, the main objective is to give the reader a basic understanding of the chemistry and physics of cloud and precipitation and aerosol science. The institute in 1983 and its herewith published proceedings can be seen as a continuation of a former NATO advance study institute entitled 'Chemistry of the Unpolluted and Polluted Troposphere' which was conducted in 1981 also on the island of Corfu in Greece. Based on experiences with the edition of proceedings of this type of institute it was decided that this time all chapters should undergo a careful review in order to avoid overlapping or underlapping between the different chapters. Moreover, the final versions of all papers were typed by the staff of the editor, so that all chapters appear in a common phenotype. The disadvantage of this strenuous editing is a substantial delay of the publication date due to an intensive reiteration of texts between the authors and the editor. Within this context I want to thank all authors and the publisher for their patience. My thanks also go to the members of my staff, especially to Mr. Duelfer, who brought all chapters into a common form by the virtuous handling of his word processing system.

I do hope that the content of this book is a substantial contribution to the understanding of the chemistry of 'Multiphase Atmospheric Systems' which have received great scientific attention in recent years. The transport and conversion of chemical compounds occurring within the 'Multiphase Atmospheric Systems' are the link between pollutant emission sources and aquatic and terrestrial receptor areas of dry and wet deposition. Effects of acidic deposition like forest disease or aquatic damage are only touched upon in the last chapters of this book. There is still much work to be done on the research of acid deposition effects, and another conference on this topic may be meaningful in coming years.

Frankfurt a.M., August 1985

Wolfgang Jaeschke

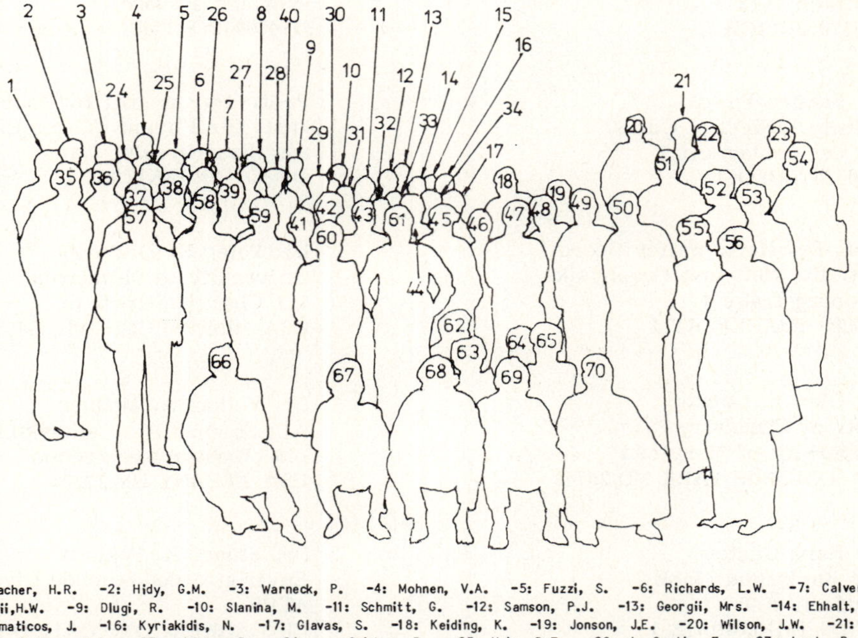

-1: Pruppacher, H.R. -2: Hidy, G.M. -3: Warneck, P. -4: Mohnen, V.A. -5: Fuzzi, S. -6: Richards, L.W. -7: Calvert, J.G. -8: Georgii,H.W. -9: Dlugi, R. -10: Slanina, M. -11: Schmitt, G. -12: Samson, P.J. -13: Georgii, Mrs. -14: Ehhalt, D.H. -15: Pneumaticos, J. -16: Kyriakidis, N. -17: Glavas, S. -18: Keiding, K. -19: Jonson, J.E. -20: Wilson, J.W. -21: Stedman, D.H. -22: Delmas, R.J. -23: Winkler, P. -24: van Grieken, R. -25: Huie, R.E. -26: de Santis, F. -27: Jacob, P. -28: Scherer, B. -29: Hamrud, M.H. -30: McMurry, P.H. -31: Kinz, L. -32: Newiger, M. -33: Penkett, S.A. -34: Vaughan, G. -35: Schwartz, S.E. -36: Altwicker, E.R. -37: Beltz, N. -38: Meixner, F. -39: Mallant, R.K. -40: Sethi, D.S. -41: Gordon, G.E. -42: Bahnemann, D. -43: Schaefer, L.U. -44: Schneider, W. -45: Kohlmaier, G.H. -46: Duvignac, P. -47: Spoelstra, H. -48: Kelm, H. -49: Slanina, J. -50: Jaeschke, W. -51: Hoffmann, M.R. -52: Schiff, H.I. -53: Schiff, D. -54: Herrmann, J. -55: Quin, Y. -56: Ottar, B. -57: Klug, W. -58: Buseck, P. -59: Berresheim, H. -60: Campell, H. -61: Diederen, H. -62: Mugler, H. -63: Alvim Ferras, M. -64: Wurzinger, C. -65: Tavares, T.N. -66: Platt, U. -67: Neftel, A. -68: Meyrahn, H. -69: van Eldik, R. -70: During, G.

LIST OF AUTHORS AND PARTICIPANTS

Authors

Dr. Jack. G. Calvert
NCAR
P.O. Box 3000
USA BOULDER CO,80307

Dr. William L. Chameides
Georgia Inst.of Technology
School of Geophysical Scien.
USA ATLANTA GA,30332

Dr. Robert J. Delmas
Lab. de Glaciologie
P.O. Box 68
F-38402 SAINT MARTIN D'HERES

Prof. Dr. Dieter H. Ehhalt
KFA Jülich, Inst. 3
Postfach 1913
D-5170 JÜLICH 1

Dr. Sandro Fuzzi
Istituto FISBAT - C.N.R.
Via de' Castagnoli 1
I-40126 BOLOGNA

Prof. Dr. Hans-Walter Georgii
Inst. für Met. und Geophysik
Feldbergstraße 47
D-6000 FRANKFURT 1

Dr. Glen E. Gordon
Dept. of Chemistry
University of Maryland
USA COLLEGE PARK MD,20742

Dr. Hans Güsten
KFZ Karlsruhe GmbH
Postfach 3640
D-7500 KARLSRUHE 1

Dr. George M. Hidy
Energy Res. & Technology
2625 Townsgate Road, Sui.360
USA WESTLAKE VILLAGE CA,91361

Prof. Dr. Peter V. Hobbs
Atmosp. Science Dept. AK-40
University of Washington
USA SEATTLE WA,98195

Dr. Wolfgang Jaeschke
Zentrum für Umweltforschung
Robert-Mayer-Straße 7-9
D-6000 FRANKFURT 1

Dr. Robert G. Keesee
152 Davey Laboratory
Pennsylvania State Univers.
USA UNIVERSITY PARK PA,16802

Prof. Dr. Dieter Klockow
Institut für Chemie
Postfach 50 0500
D-4600 DORTMUND 50

Prof. Dr. Gundolf H. Kohlmaier
Inst. für Physik.+Theor.Chem.
Am Niederurseler Hang
D-6000 FRANKFURT 1

Dr. Peter H. McMurry
University of Minnesota
111 Church Street, SE
USA MINNEAPOLIS MN,55455

Dr. Volker A. Mohnen
Atm.Science Res.Center/SUNYA
1400 Washington Avenue
USA ALBANY NY,12205

Dr. Stuart A. Penkett
Environ.& Med.Science Div.
AERE
GB HARWELL,OX. OX11 ORA

Dr. Ulrich Platt
KFA - Inst. f. Chemie 3
Postfach 1913
D-5170 JÜLICH 1

Prof. Dr. Hans R. Pruppacher
Institut für Meteorologie
Postfach 3980
D-6500 MAINZ

Prof. Perry J. Samson
The University of Michigan
2455 Hayward
USA ANN ARBOR MI,48109

Dr. Stephen E. Schwartz
Brookhaven National Lab.
Dept. of Applied Sciene
USA UPTON, NEW YORK NY,11973

Dr. Jakob Slanina
ENC Research center
P.O. Box 1
NL-1755 ZG PETTEN

Dr. Donald H. Stedman
Dept. of Chemistry
University of Denver
USA DENVER CO,80208

Dr. Rudi Van Eldik
Inst. für Physik. Chemie
Am Niederurseler Hang
D-6000 FRANKFURT

Prof. Dr. Rene Van Grieken
Dept. of Chemistry (U.I.A.)
Universiteitsplein 1
B-2610 ANTWERP-WILRIJK

Prof. Dr. Peter Warneck
MPI für Chemie
Saarstr.23
D-6500 MAINZ 31

Dr. Peter Winkler
Deutscher Wetterdienst
Strahlungsobs. Frahmredder
D-2000 HAMBURG 65

Participants

Dr. Elmar R. Altwicker
Dept.of Chem.Engng.& Env.Eng.
RPI
USA TROY NY,12181

Dr. Detlef Bahnemann
Hahn-Meitner-Inst./Kernfor.
Glienicker Straße 100
D-1000 BERLIN 39

Dr. Ian Barnes
Physikalische Chemie
Postfach 10 01 27
D-5600 WUPPERTAL 1

Norbert Beltz
Zentrum für Umweltforschung
Robert-Mayer-Str. 9-7
D-6000 FRANKFURT 1

Dr. Harald Berresheim
Dept. of Oceanography
Florida State University
USA TELLAHASSEE FL,32306

Dr. Andrew Blommers
GEOSENS B.V.
P.O. Box 12067
NL-3004 GB ROTTERDAM

Dr. Brüll
Parkweg 32
NL-6212 MAASTRICHT

Dr. Peter R. Buseck
Dept. of Chemistry
Arizona State University
USA TEMPE AZ,85281

Prof. Dr. Knut Bächmann
TH Darmstadt
Hochschulstraße 4
D-6100 DARMSTADT 1

Dr. Hilary Campbell
BP-Britannic House
Moor Lane
GB LONDON EC2Y 9BU

Dr. Franco De Santis
Inst.Inquinamento Atm.C.N.R.
Via Salaria km.29,300-C.P.10
I-00016 MONTEROTONDO STAZIONE

Hubert Diederen
TNO Research Institute
P.O. Box 214
NL-2600 AE DELFT

Dr. Ralph Dlugi
KFZ Karlsruhe GmbH LAF 1
Postfach 3640
D-7500 KARLSRUHE 1

Dr. Gustav During
Försvarets Forskningsanstalt
P.O. Box 27322
S-102 54 STOCKHOLM

Philippe Duvignac
Comm.a l'Energie Atomique
LA BARP
F-33830 BELIN-BELIET

Dr. Karl-Heinz Gericke
Inst.f. Physikalische Chemie
Niederurseler Hang
D-6000 FRANKFURT

Prof. Dr. Lois T. Gidel
Georgia Inst.of Technology
School of Geophysical Scien.
USA ATLANTA GA,30332

Dr. Sotiris Glavas
260 Alex. Ypsilanton St.
GR PATRAI

Mats Hubert Hamrud
Dept. of Meteorology
Arrhenius Laboratory
S-106 91 STOCKHOLM

Jürgen Herrmann
Zentrum für Umweltforschung
Robert-Mayer-Str. 7-9
D-6000 FRANKFURT

Prof. Dr. Klaus Heumann
Inst. für Anorg. Chemie
Universitätsstraße 31
D-8400 REGENSBURG

Dr. Michael Robert Hoffmann
CALTECH
1201 E.California Blv.138-78
USA PASADENA CA,91125

Dr. Robert E. Huie
Chemical Kinetics Division
National Bureau of Standards
USA WASHINGTON DC,20234

Peter Jacob
Universität Dortmund
Postfach 50 0500
D-4600 DORTMUND 50

Jan Eiof Jonson
Furubraatvn. 14 a
N OSLO 11

Kristian Keiding
Air Pollution Laboratory
RISOE National Laboratory
DK-4000 ROSKILDE

Prof. Dr. Hartwig Kelm
Inst. für Physik. Chemie
Am Niederurseler Hang
D-6000 FRANKFURT

Dr. C. S. Kiang
Georgia Inst. of Technology
School of Geophysical Scien.
USA ATLANTA GA,30332

Lucia Kinz
KFA Jülich
Postfach 1913
D-5170 JÜLICH 1

Prof. Dr. Werner Klug
Inst. für Meteorologie
Hochschulstraße 1
D-6100 DARMSTADT

Dr. Chiang Hai Kuo
Mississippi State University
P.O.Box 52
USA MISS STATE MS,39762

Dr. Nicolas Kyriakidis
Menelaou 34
Ano Petralona
GR ATHENS

Maria da Concei. M. Alvim Ferraz
Centro de Engenharia Quimica
Rua dos Bragas
P-4099 PORTO - CODEX

Dr. Ronald Karel Mallant
ECN Research Center
P.O. Box 1
NL-1755 ZG PETTEN

Horst Mayrahn
MPI für Chemie/Luftchemie
Postfach 3060
D-6500 MAINZ

Dr. Eugene McLaren
Dept. of Chemistry
1400 Washington Avenue
USA ALBANY NY,12222

Dr. Franz Meixner
KFA Jülich, JCH 3
Postfach 1913
D-5170 JÜLICH 1

John Mugler
NASA Langley Res.C., M/S 401
Mail Stop 401-8
USA HAMPTON VA,23665

S. Feda Mutlu
Gazi University
Hosdere cad. yayla ap. II/2
TR-ANKARA YUKARI AYRANCI

Dr. Albrecht Neftel
MPI für Chemie
Postfach 3060
D-6500 MAINZ

Madeleine Newiger
MPI für Meteorologie
Bundesstraße 55
D-2000 HAMBURG 13

Dr. Brynjulf Ottar
Norwegian Inst. for Air Res.
P.O. Box 130
N-2001 LILLESTRÖM

John Pneumaticos
Atmospheric Physics Lab.
University of Patras
GR PATRAI

Doz. Dr. Hans Puxbaum
TU Wien
Getreidemarkt 9/151
A-1060 WIEN

Dr. Yu Qin
Dept. of Geophysics
Peking University
TJ BEIJING 100871

Dr. L. Willard Richards
Sonoma Technology Inc.
3402 Mendocino Avenue
USA SANTA ROSA CA,95401

Dr. Charles B. Richardson
P.O.Box 636
USA UPTON, LONG ISLAND NY,11973

Dr. Ram Sant
Dept. of Applied Mathematics
11, College Street
GB ST.ANDREWS,FIFE KY16 9SS

Bernhard Scherer
Inst.f. Geophysik. Wissensch.
Thielallee 50
D-1000 BERLIN 33

Prof. Dr. H. I. Schiff
Faculty of Science
4700 Keele Street
CDN DOWNSVIEW, ONT. M3J 1P3

Günther Schmitt
Inst.f.Meteorologie u.Geoph.
Feldbergstraße 47
D-6000 FRANKFURT 1

Wolfgang Schneider
MPI für Chemie, Abt. Luftch.
Postfach 3060
D-6500 MAINZ

Dr. Lutz U. Schäfer
MPI für Chemie
Saarstraße 23
D-6500 MAINZ 1

Dr. Dhanwant S. Sethi
Dept. of Chemistry
University of Bridgeport
USA BRIDGEPORT CT,06601

Rama Shankar Shrivastava
P.O. Box 50867
Z LUSAKA

Hendrik Spoelstra
P.O.Box 12067
NL-3004 GB ROTTERDAM

Dr. Turan Söylemez
Charlotten Straße 99
D-4330 MÜHLHEIM-RUHR

Prof. Tania N. Tavares
Mal. Floreano 9 apt.1701
Canela 4000
BR SALVADOR/BAHIA

Dr. Phil. Geraint Vaughan
Meteorol. Office
London Road
GB BRACKNELL BERKS. RG12 2SZ

Erich Weber
Regierungsdirektor
Weberstraße 29
D-5300 BONN 1

James Charles Wilson
125 Mechanical Eng.
111 Church Street S.E.
USA MINNEAPOLIS MI,55455

Dr. Jerre W. Wilson
United States Military Acad.
Science Research Laboratory
USA WEST POINT NY,10996

William E. Wilson, jr.
U.S. Environm.Prot.Agency
MD-84
USA RES. TRIANGLE PARK NC,27711

Christel Wurzinger
MPI/Airchemistry
Saarstraße 23
D-6500 MAINZ

Part I

ANALYTICAL AND EXPERIMENTAL METHODOLOGY

MULTIPHASE ATMOSPHERIC CHEMISTRY

Wolfgang Jaeschke
Zentrum fuer Umweltforschung
J.W. Goethe-Universitaet, D-6000 Frankfurt
F.R.G.

ABSTRACT

The various reactions in the atmosphere are constantly interacting with condensation processes. Since these processes lead to the formation of fogs and clouds, in a first part a phenomenology of the aqueous phase in the atmosphere is presented. In a second part the sampling and analyzing methodology applicable in the three phases of Multiphase Atmospheric Systems (MAS) are briefly discussed. In a final part some experiments and theoretical considerations with respect to the removal of SO_2 in MAS are discribed. The investigations were performed in a dynamic flow reactor, where multiphase systems relevant to the atmosphere could be simulated.

INTRODUCTION

The atmosphere is a gaseous medium in which water appears in the three states of aggregation. Chemical reactions occuring in this Multiphase Atmospheric System (MAS) are largely governed by the dynamics of clouds and fog. These dynamics are quantitatively described by the so-called microphysics of clouds. The issues of these microphysics should be emphasized briefly by the following statements:

- The presence of a particular phase depends on temperature, humidity and the presence of condensation nuclei.
- Visible results of these phenomena are fogs and clouds.
- Clouds are dynamic structures, in which the water substance passes through the three states of aggregation possibly up to hundreds of times.
- The distribution of trace gases and particles is influenced by the dynamics of clouds and fogs depending on their individual solubilities.
- In principle the physics of these processes can be described in dynamic mathematical

NATO ASI Series, Vol. G6
Chemistry of Multiphase Atmospheric Systems
Edited by W. Jaeschke

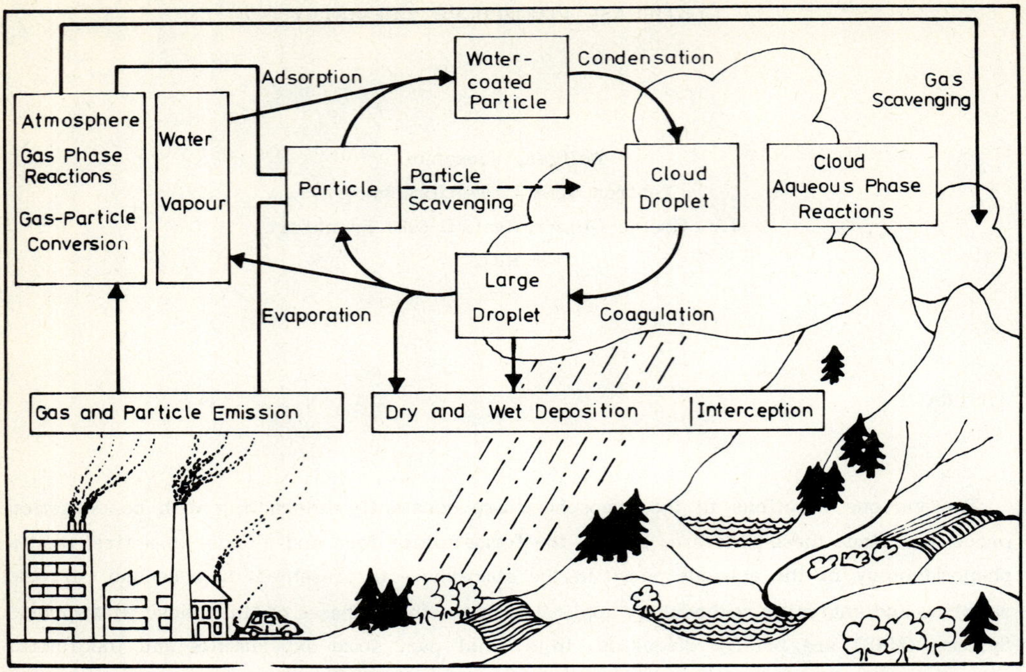

Fig. 1 Scheme of the Multiphase Atmospheric System (MAS) and its interaction with the environment.

models (cloud models).

Chemical trace substances present in the multiphase atmospheric system undergo chemical transformations while the watersubstance is repeatedly passing through the cycle of the three states of aggregation (Fig. 1). The chemical transformations occuring within these cycles can be summarized briefly by the following statements:

- in the oxygen-rich atmosphere the transformations result in an increase in the oxidation state of the transformed substances.
- An increase in solubility is usually associated with the increase in oxidation state, which leads to an enrichment of the oxidation products in the aqueous phase.
- In so far as the oxidation products are acidic anhydrides, their solution results in an increase in the concentration of free protons (decreasing pH-values).
- The rate of chemical reactions in clouds depends on the chemical constitution of the multiphase system and from the droplet size spectrum.

PHENOMENOLOGY OF MAS

For the study of chemical reactions in MAS particles, cloud- and fog-droplets must be collected and chemically analyzed. During these measurements informations about the meteorological conditions and the resulting microphysical properties of clouds and fogs must also be gained. Most important parameters are the liquid water content ranging from 10^{-2} to 10 g water / m^2 of air and the size distribution of the droplets in the range from 10^{-2} to 100 µm diameters. According to the weather situation, various types of fogs and clouds can appear:

1. Fog types

- Advective fog.

  This fog forms when warm, moist air flows horizontally over water or land surfaces with lower temperatures. The air is cooled by turbulent heat exchange with the surface, so that the dew point is reached and condensation takes place. It occurs primarily under weather conditions in late fall and winter (November fog).

- Radiation fog.

  Formation of radiation fog is caused by long wave heat radiation from the underlying surface into a clear atmosphere (negative heat balance for the surface). The moist air lying over the surface of the earth is thus cooled to its dewpoint. This occurs mostly during the evening or night in fall and winter. It is often called ground fog, and is always coupled with a temperature inversion.

- Evaporation fog.

  This fog forms by evaporation of a water surface into cooler air, or by warm rain, which falls through a colder air layer. It exists only in unstable air layers. The so-called sea-smoke and smoke of forests after summer rain showers are fogs of this type.

- Frontal fog.

  This type of fog forms when the warm, moist air associated with a warm front moves over and is mixed with colder underlying air.

- Mountain fog.

  Adiabatic cooling occurs through the lifting of an air volume up a mountain slope, so that clouds are formed in the uprising air. If they cover the mountain, they are perceived as fog by an observer on the mountain. This fog is a special case, and represents a transition to cloud formation.

2. Cloud Types.

The ten main cloud families can be classified into two principal kinds, based on the manner in which they are usually seen to be formed:

- Convective clouds.

  Cumulus and cumulonimbus belong to this type. The necessary and often violent vertical movement of the air (several meters per second) originates through driving forces. An air parcel, warmed by solar radiation near the surface of the earth, rises and expands under lower pressure. It is thus adiabatically cooled at a rate of about 1 degree Celsius per 100 meter rise. A relative humidity of 100 % is finally reached at the condensation level and, with the availability of suitable condensation nuclei, condensation and associated cloud formation takes place.

- Stratus clouds.

  The remaining 8 cloud families belong to this type and they should be named for the purpose of completeness: Cirrus, Cirrocumulus, Cirrostratus, Altocumulus and Altostratus, and finally Stratocumulus, Stratus and Nimbostratus.

In the histogram of Fig. 2 an example of the chemical composition of fog water is given.

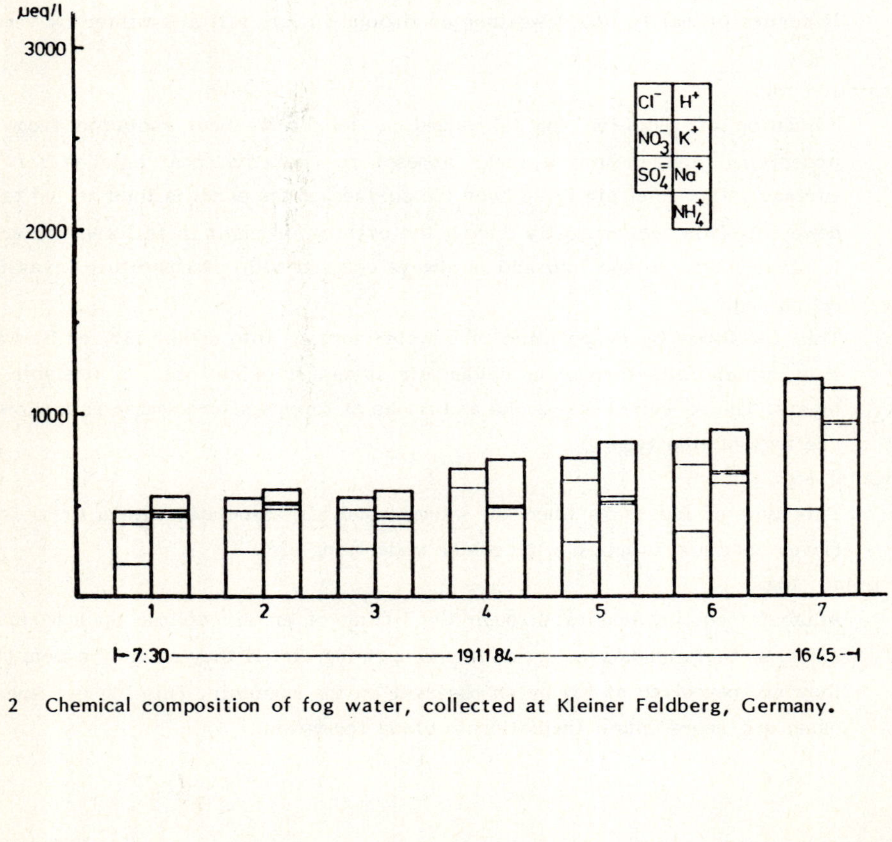

Fig. 2 Chemical composition of fog water, collected at Kleiner Feldberg, Germany.

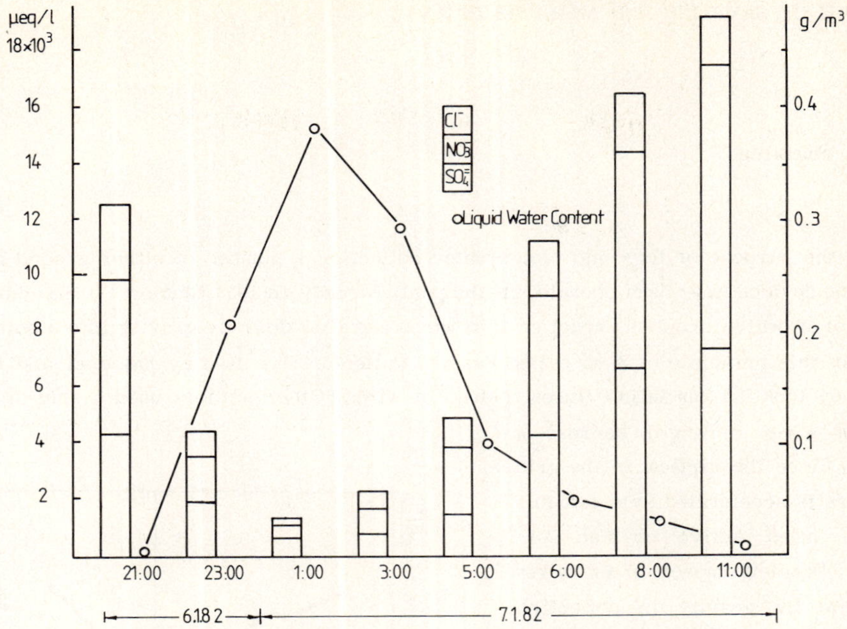

Fig. 3 Chemical composition of fog water measured together with the liquid water content in the Po Valley, Italy by S. Fuzzi et al. [1].

The samples were collected at a mountain station on Kleiner Feldberg, Taunus, Germany, during an advective fog episode. One notices the increase in total concentration of trace materials during the course of the measuring day. This probably results from a decreasing liquid water content of the fog during the day, as a result of evaporation. This example points out that a microphysical characterization of the MAS is the prerequisite for interpretation of ion concentrations in fog and cloud water. The minimum requirement is the knowledge of the liquid water content.

In the diagram of Fig. 3 the trace substance content of radiation fog from the Po Valley in Northern Italy is presented together with the measured liquid water content. From these data reported by Fuzzi et al. [1] the negative correlation between the measured anion concentration and liquid water content can easily be recognized.

With respect to particles it should be emphasized that chemical trace compounds are not homogeneously distributed in the different size ranges of the particles. The existing so called external mixing, which means that certain elements preferrably occur in different size ranges, can be seen in Fig. 4. The particles were sampled with a six stage impactor during the 57th cruise of the German research vessel 'Meteor' over the North Atlantic between Madeira and the Azores.

METHODS OF SAMPLING AND ANALYSIS IN MAS

1. Sampling

For the purpose of fog- and cloud-water collection a number of either ground based or airborne devices have been operated in the past. Already in 1953 Grunow [2] described a fog collector in which droplets impact on thin wires and flow down by gravity into a bottle. More recently this principle of a so called Passive Collector was used by Falconer and Falconer [3]. They took 0.4 mm teflon fibres instead of wires. Other authors used a grid of vertical 0.4 mm wires arranged accross a tunnel. Since the airflow in the grid samplers is accelerated by a fan this type is called 'Active' or 'Fan Collector'. Droplets impact on the wires and flow by gravity in collection bottles at the base of the grid. Another principle is the 'Rotating Arm Collector' in which fog droplets impact on rotating slotted tubes and are collected by centrifugal force in small vials mounted at the ends of the tube. According to the respective method which provides the relative speed between droplet and collection surface the several kinds of collectors may be devided into three categories: passive collectors, fan collectors and rotating collectors. The wide variety of efforts made to collect fog water can be seen in the cited literature [4] - [14].

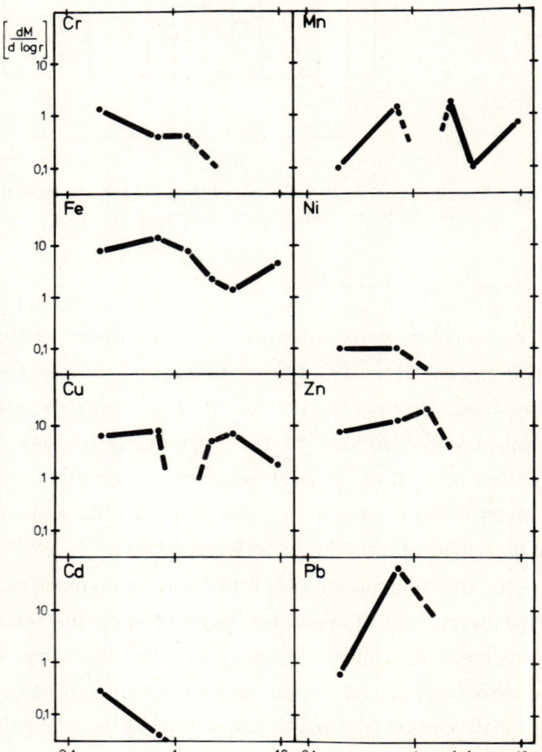

Fig. 4 Mass distribution of heavy metals in the aerosol over the Atlantic Ocean. The particles were accumulated on plates of a six stage impactor from 7.22. to 7.29.1983 between the Madeira Islands and the Azores.

Passive collectors are only applicable at minimum wind speeds of about 3 m s^{-1} [15]. The most prominent version of this type was developed and tested at the Atmospheric Sciences Research Center (ASRC), State University of New York at

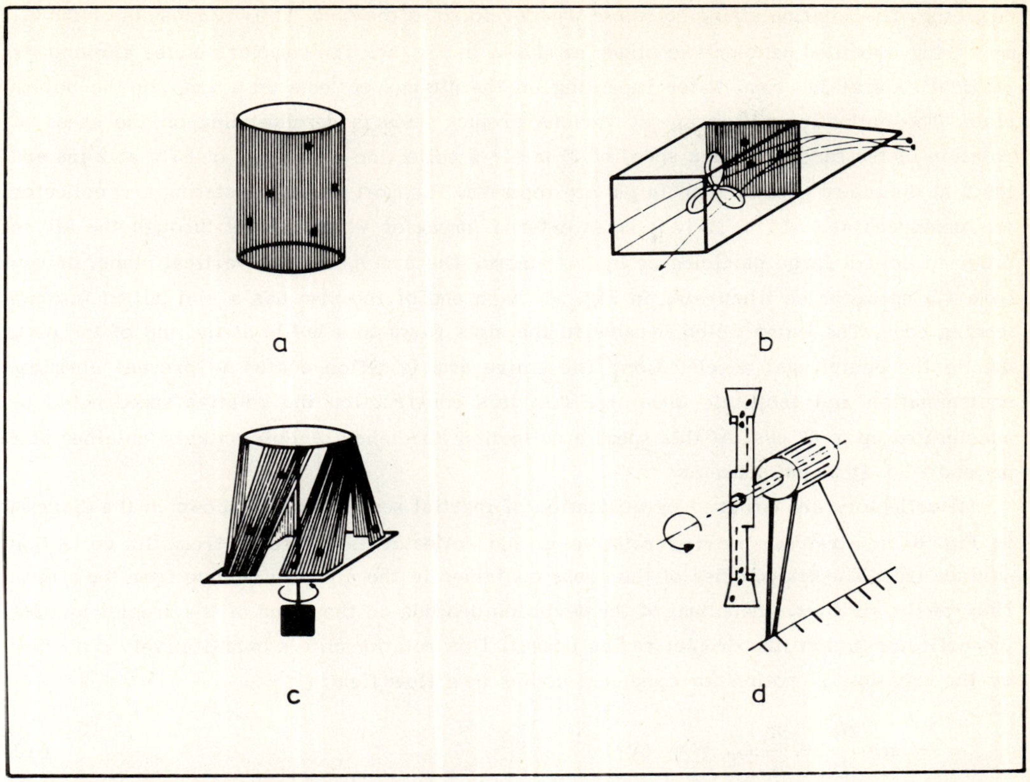

Fig. 5 Principles of fog water collectors
a) Passive Collector
b) Fan Collector
c) Rotation String Collector
d) Rotating Arm Collector

Albany [16]. Teflon filaments (0.4 mm diameter) are strained between two polypropylene discs every 3 mm around the periphery (see Fig. 5a). Because the device is entirely passive, collection rate depends on the liquid water content of the cloud or fog and on changes of efficiency at very low or very high wind speeds. Sampling times are normally a few minutes to collect enough water (5 ml) for all analyses intended.

In a fan collector the fog- or cloud-water containing air is drawn through mashes made out of teflon or polypropylene filaments by the aid of an electric fan. Coalesced droplets drain into a collection bottle at the bottom of the mash (Fig. 5b). A recent version of this collector type is mounted on a wind vane such that it is automatically brought into the main wind direction. The sampling efficiency is strongly dependent on the size of the droplets. At a flow rate of 1 m^3 per minute collection efficiencies of 91 % at three μm and 99 % at 12 μm diameter are given in the literature [17].

Main versions of rotating collectors are a rotating string collector and a rotating arm collector. The rotating string collector was developed at the ASRC [18]. It consists of 150 0.1 mm strings mounted between two plates as shown in Fig. 5c. The sampler rotates arround its vertical axis at 100 rpm. Water impacting on the strings collects in a trap in the bottom plate. The collection efficiency at various droplet sizes is dependending on the speed of rotation of the sampler. For a speed of 25 m s^{-1} a collection efficiency of 53 % at 2 µm and 100 % at diameters greater than 10 µm are reported. The most efficient rotating arm collector was developed at CALTEC [19]. It is an external impactor which sweeps through the air in order to collect large particles or hydrometeors. The arm spins in a vertical plane, driven by a 1.5 hp motor as illustrated in Fig. 5d. Each end of the arm has a slot milled into its leading edge. The water which impacts in the slots flows to a bottle at the end of the arm, due to the centrifugal acceleration. The entire arm is teflon-coated to prevent chemical contamination and facilitate cleaning. With this construction the relative speed could be accelerated up to 47 m/s. At this speed a collection efficiency of 70 % could be obtained at 8 um and 80 % at 20 µm diameter.

All collectors are designed by utilization of inertial seperation. As shown in the diagram in Fig. 6 an airstream moving relative to the collector is deflected from the collection surface [20]. The trajectories of the drops contained in the air then deviate from the stream lines of the air flow. The extent of the deviation depends on the speed of the droplet towards the collector and on the droplet radius (mass). This relationship is quantitatively described by the equation of motion for spherical bodies in a flow field:

$$m_T \frac{du}{dt} = \frac{\pi}{8} d_T^2 \rho C_T (V-u)^2 \qquad (1)$$

where: m_T = droplet mass
u = droplet speed
V = air speed
ρ = density of the air
d_T = droplet diameter
C_T = drag coefficient

The right side of the equation represents the resistance force arising from the air, while the left side represents the accelerating force on the droplet. This equation forms the basis for the theoretical investigations of Langmuir and Blodgett [21] for the calculation of droplet separation efficiencies. Results of such calculations are sumarized in Table 1. The approximate impaction efficiency for 0.4 mm diameter cylindrical rods over a range of wind speeds is shown. To arrive at optimal collection properties, the design of a collector should initially based on such theoreti-

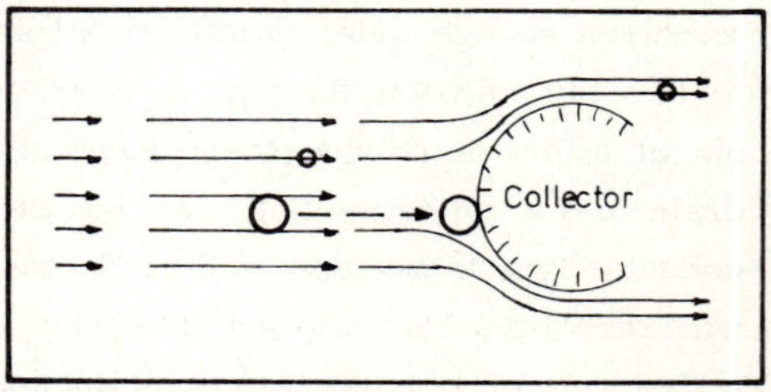

Fig. 6 Scheme of inertial separation of a droplet (after Fuchs [20]).

Table 1 Approximate impaction efficiency for 0.4 mm diameter cylindrical rods over a range of wind speeds

Droplet Diameter [µm]	Wind Speed [knots]					
	5	10	20	30	40	50
2	.03	.10	.25	.33	.43	.53
6	.43	.63	.77	.84	.88	.91
10	.71	.83	.92	.94	.98	.99
14	.83	.92	.98	.99	1	1
18	.90	.95	1	1	1	1
22	.93	.99	1	1	1	1
26	.95	1	1	1	1	1
30	.98	1	1	1	1	1

cal calculations from hydrodynamic laws. Then its practical application could be tested by field measurements and preferably by intercomparison exercises. Such comparison measurements have been carried out with various USA collectors in June 1983 at Henninger flats in the St. Gabriel Mountains north of Pasadena, California [22]. More recently similar tests were carried out at the White Face Mountain Station of the New York State University or in Europe at the field station of San Piedro Capofiume (Po-Valley close to Bologna, Italy) or at the mountain station Kleiner Feldberg close to Frankfurt/Germany. In all cases the following comparison criteria had been considered:

- Amount of water collected per unit time (collection rate).
- Collection rate as a function of liquid water content of the cloud or fog, as determined by the method of hot-wire liquid evaporation or gravimetrically by high volume filtration.
- Determination of water collected in specific size ranges, with the droplet size spectrum measured by laser light scattering (Knollenberg Probe)[23].

The tests and comparisons showed, that up to now no equipment or procedure could be recommended as an optimum one, however, the exchange of informations during the experiments yielded a better understanding of the problems inherent in the sampling of cloud- and fog-water [24] - [32].

In order to collect cloud water besides ground based collectors cloud water samplers have been developed which can be used in an aircraft. The separation of cloud water drops is also achieved by use of the principle of inertia. In this case the relative speed of the drops to the collection surface is determined by the aircraft speed. In order to prevent vaporization of the drops on the collection surface at the high relative speed (> 60 m s^{-1}), rods with a slot facing in flight direction are used. An air-stream free cusion of gas exists in the slot, so

Fig. 7 ASRC Cloudwater Collector
1. Teflon rods, 1 cm diameter;
2. Slods, 2 mm wide 4 mm deep;
3. Teflon tubing;
4. Collector Bottle.

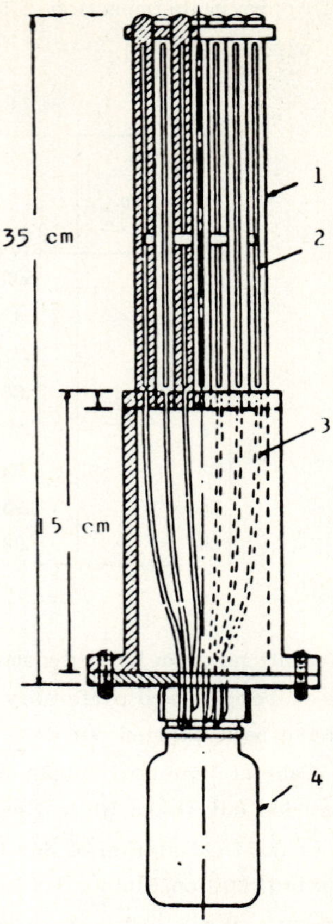

that the captured drops can coagulate and flow down into a collection vessel without vaporization. A most efficient cloud water collector based on these principles was developed by the ASRC in Albany [16] (Fig. 7). The separation effiency of the cloud water collector is a function of drop diameter at various aircraft speeds (Fig. 8). The ASRC cloud water collector is used by the Center of Environmental Research of the University of Frankfurt in a "Partenavia Victor" as the research aircraft (Fig. 9). Before the installation, the collector is stored in a container which is filled with distilled water. Only during the measuring flight it is taken from this analytically pure environment and placed into its working position through a shaft in the aircraft door. At a speed of 120 knots (62 m s^{-1} = 222 Km/h) 2 to 10 ml of cloud water can be collected within 15 min depending on the liquid water content of the cloud.

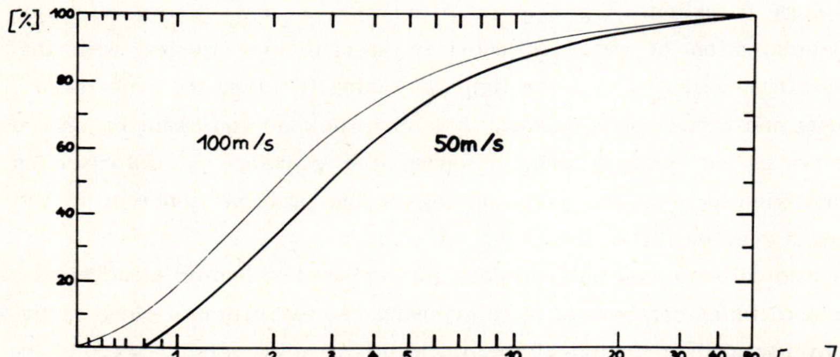

Fig. 8 Sampling efficiency of the ASRC cloud water collector as a function of drop diameter and aircraft speed.

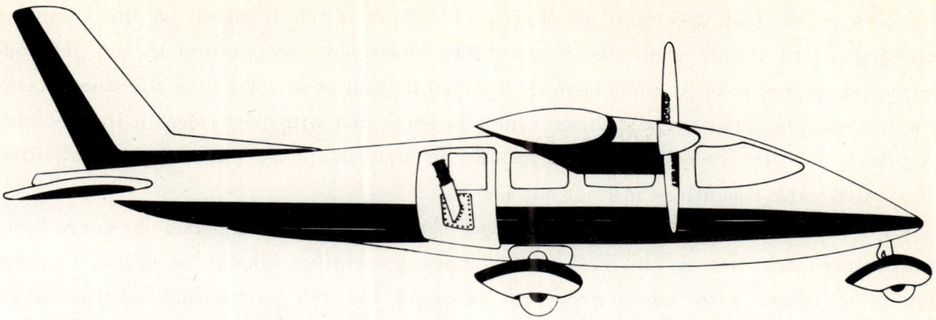

Fig. 9 Position of the cloudwater collector at the research aircraft (Partenavia Victor) used by the University of Frankfurt.

Analysis

a. Liquid Phase

Commercial apparatus and procedures can be used to determine the pH-value and the concentration of anions, cations and heavy metals in fog and cloud water. The scheme in Figure 10 summarizes procedures used for the analysis of the aqueous phase. For the analysis of H_2O_2 and sulfite special procedures have been developed. For the detection of

Sampling	Detection Method	Compounds
Rain Collector, Fog Water Sampler, Cloud Water Sampler	Ion Chromatography	$SO_4^=$, NO_3^-, Cl^-, NH_4^+, Na^+, Ca^+, Mg^{++}, Formaldehyde
	Ion Selective Electrode	$SO_4^=$, NO_3^-, Cl^-, NH_4^+, Avail. H^+
	Atom Absorption Spectroscopy	Transition Metals
	Chemiluminescence ($KMnO_4$)	S^{IV} (Complexed and Free)
	Chemiluminescence (Oxalicester)	H_2O_2
	Fluorescence (Peroxidase)	H_2O_2
Org. Solvent Extraction	Gas Chromatography	Organics (Aliphatic)
	High Pressure Liquid Chromatography	Organics (Aromatic)

Fig. 10 Procedures used for the analysis in the atmospheric aqueous phase.

H_2O_2 Zika et al. [33] developed an enzymatic method which is based on the fluorescence quenching effect which occurs when peroxides react with scopoletine in the presence of horseradish peroxidase. Another method reported by Kok et al. [34] uses the fluorescence of the p-hydroxyphenylacetic acid dimer which is generated with peroxides in the presence of peroxidase. For the detection of sulfite in the aqueous phase upto now the photometric method with para rosaniline is most common.

Own developments for the detection of H_2O_2 and sulfite in the aqueous phase are based on chemiluminescence effects [35] [36] and a short description should be given. The sample solution is mixed with the appropriate reagents for the particular chemiluminescence analysis reaction in a cuvette, from which all light is excluded. The mixing is accomplished by means of an air-pressure driven system (Fig. 11). The light signal generated is monitored by a photomultiplier with attached photon counter. The light output is a function of the sample concentration.

The chemistry of the H_2O_2 reaction is only effective in the organic phase [35]. Therefore the aqueous H_2O_2 sample must be put into an organic solvent. H_2O_2 reacts with the oxalic acid ester of trichlorophenol to form dioxetandion and trichlorophenol [37].

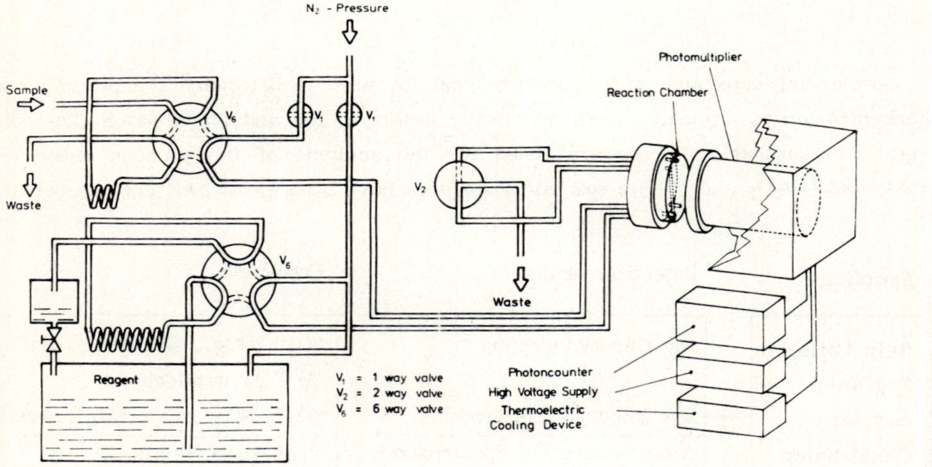

Fig. 11 Scheme of the chemiluminescence set up used for the detection of either H_2O_2 or sulfite.

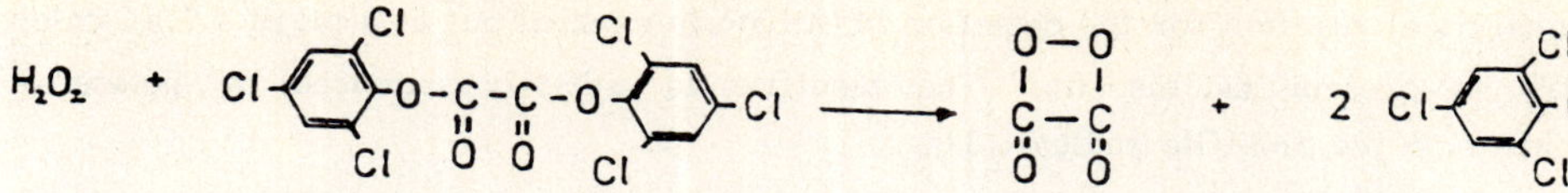

(2)

The dioxetandion decomposes spontaneously into two excited molecules of carbon dioxide in the form of excited triplett states.

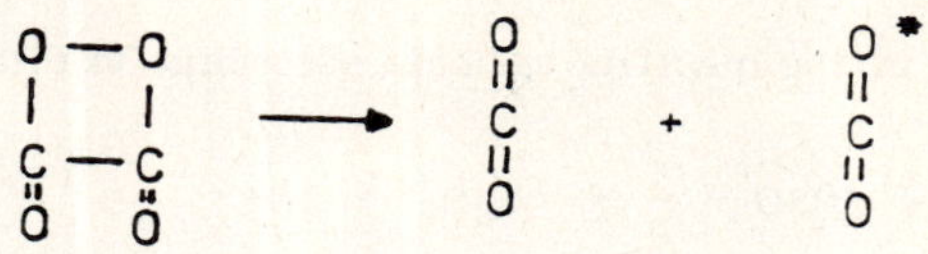
(3)

The triplett energy of the excited CO_2 is transferred to perylene.

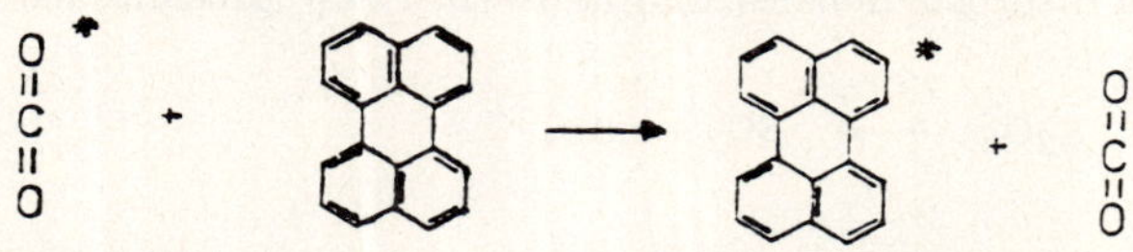
(4)

The perylene causes the chemiluminescence by its conversion from the excited triplett state into the ground state.

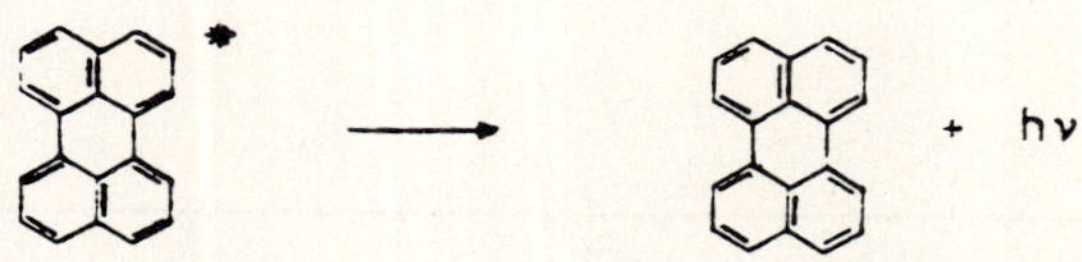

(5)

More details of this method are described in the chapter of D. Klockow et al. in this book. The detection limit for H_2O_2 in the aqueous phase is 1 ng ml^{-1} (Fig. 12).

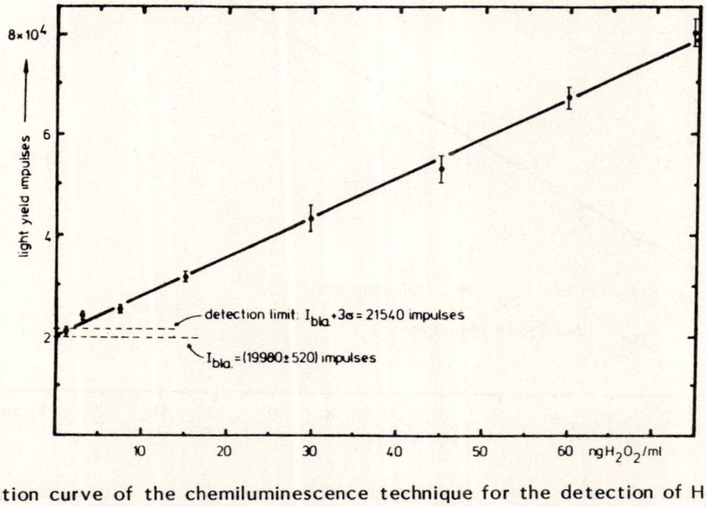

Fig. 12 Calibration curve of the chemiluminescence technique for the detection of H_2O_2, detection limit 1 ng/ml.

The analytical reaction for the detection of sulfite is carried out at a pH of 2.7 at which the S(IV) is present as bisulfite. The bisulfite is initially oxydised by potassium permanganate to the bisulfite radical [38].

$$HSO_3^- + MnO_4^- \longrightarrow HSO_3^- + MnO_4^= \qquad (6)$$

A small portion of the bisulfite radicals recombine to dithionate.

$$2HSO_3^- \longrightarrow S_2O_6^= + 2H^+ \qquad (7)$$

The dithionate ion disproportionates into an excited SO_2 molecule and a sulfate ion.

$$S_2O_6^= \longrightarrow SO_2 + SO_4^= \qquad (8)$$

The deactivation of the excited SO_2 molecule causes the chemiluminescence. The detection limit for the bisulfite reaction is 0.5 ng/ml. The calibration curve is shown in Fig. 13.

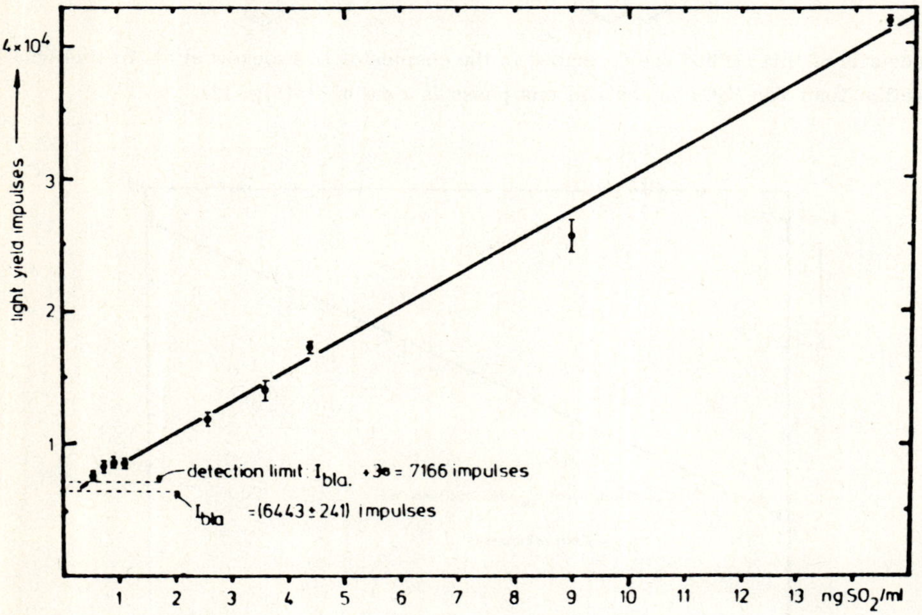

Fig. 13 Calibration curve of the chemiluminescence technique for the detection of bisulfite, detection limit 0,5 ng/ml.

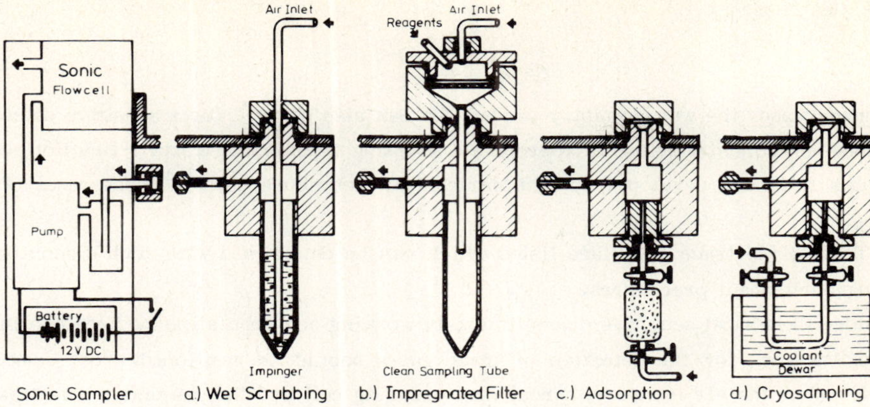

Fig. 14 Multifunction sampling apparatus for enrichment techniques.

Table 2

Discontinuously Working Accumulation Methods					
Tracegas	Accumulation	Device	Indentification	Detection	Range
CO_2	$Ba(OH)_2$	Impinger	$BaCO_3$	Nephelometry	percent
CO	Pd^{++}	Impinger	Pd	Nephelometry	ppm
CH_4	Cryosampling	Cold Trap	Gaschromatography	FID	ppb
Hydro Carbon	Cryosampling	Cold Trap	Gaschromatography	FID	ppt
F-Hydro Carbon	Cryosampling	Cold Trap	Gaschromatography	ECD	ppt
Aldehyde	Bisulfito	Impinger	Chromotr. Acid Deriv.	Photometer	ppb
NH_3	Sulfuric Acid	Filter	Indophenolblue	Photometry	ppb
NH_3	Oxalic Acid	Denuder	Ionchromatography	Conductometry	ppt
NH_3	Citric Acid	Denuder	Ionchromatography	Conductometry	ppt
NH_3	Tungstic Acid	Denuder	NH_3 NO conversion	Chemiluminescense	ppb
N_2O	Adsorption	Molec Sieve	Gaschromatography	ECD	ppb
NO_2	Sulfuric Acid	Impinger	Azo Dye	Photometry	ppb
PAN	Adsorption	Molec Sieve	Gaschromatography	ECD	ppt
HNO_3	Sodium Fluoride	Denuder	Ionchromatography	Conductometry	ppt
O_3	Iodine	Impinger	J_2 - bleaching	Photometry	ppm
H_2S, RSH	Adsorption	Goldwool	Gaschromatography	FPD, HECD	ppt
CS_2, COS	Cryosampling	Cold Trap	Gaschromatography	FPD, HECD	ppt
SO_2	TCM	Impinger	Rosanilin Derivate	Photometry	ppb
SO_2	TCM	Filter	$KMnO_4$ Reduction	Chemiluminescense	ppt
HCl	$HgCl_2$	Tube	Diphenylcarbazon	Photometry	ppt

b. Gas Phase

In many cases the wet chemistry procedures can also be used for a sensitive detection of gas-phase components, by use of appropriate sampling techniques. A multi-function sampling apparatus has been developed for the enrichment techniques used in these cases (Fig. 14) [39].

In Table 2 the trace gases are listed which can be determined with such discontinuously working enrichment procedures.

Besides these most sensitive discontinuously working accummulation methods continuously working methods for the detection of trace gases should be mentioned. Most common are methods where the air is sucked through a measuring cell and the gas under investigation is detected by a specific physical or chemical effect. Compounds and methods together with the detection limits are summarized in the scheme in Fig 15. Since several years remote sensing techniques have also been applied for the detection of atmospheric trace gases. Well known is the correlation spectroscopy where the radiation of the sun is used as light source and the absorbtion caused by certain trace gases in the free atmosphere is measured by a spectroscope [40]. This method is applicable only at relative high concentrations as they occur in plumes of industrial stacks. In order to localize these plumes the spectroscope is mostly installed in a mobile laboratory.

Besides this the tunable laser spectroscopy is applied in the free atmosphere for instance for the detection of OH-radicals [41]. A compilation of the compounds detectable with remote sensing techniques is given in the scheme of Fig. 16.

c. Solid phase (particles)

Particles acting as condensation nuclei in the MAS are dissolved and their material appears in the liquid phase. The theoretical treatment of this phenomen will be described in the next section of this chapter. Besides this, unsoluble or inactive particles may remain between the water droplets as interstitial particles. Their sampling and separation is still in development. Most common techniques for the sampling of particles are so-called denuders, or filters and impactors. Since the subsequent analysis is performed in liquid extractions, in principle the same detection methods as applied in the aqueous phase can be used. For the analysis of individual particles the so-called LAMMA-technique is applied. An intensive description of this very important method will be found in the chapter of R. van Grieken in this book. A compilation of the most common methods used for the analysis of airborne particles is given in the scheme of Fig. 17.

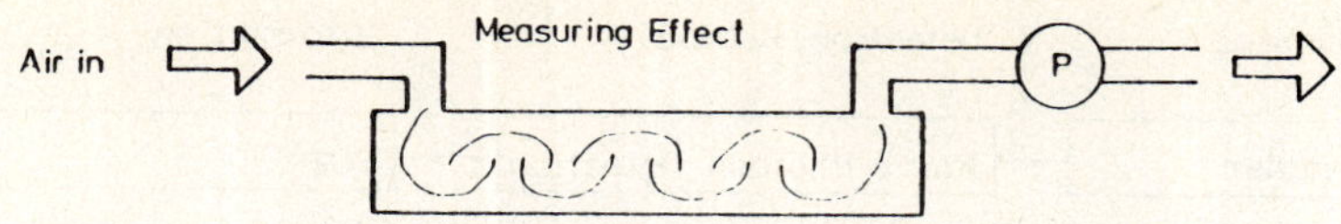

Method \ Compound	SO_2	NH_3	N_2O	NO	HNO_2	NO_2	HNO_3	NO_3	PAN	O_3	OH	H_2O_2
Laser Induced Fluorescence	3×10^{-3}	10	–	1	–	10^{-3}	–	10^{-3}	–	–	10^6	3×10^{-2}
Furier Transform IR-Spectrosc.	–	2	4	10^{-3}	5	–	4	–	5	–	–	–
Spectroscopy (UV-IR) and Fluoresc.	2	–	–	–	–	–	–	–	–	–	–	–
Chemiluminescence	–	10^{-2}	–	2×10^{-3}	–	1×10^{-2}	0,1	–	–	0,1	–	–
Flamephotometry	0,4	–	–	–	–	–	–	–	–	–	–	–
Coulometry (wet chemistry)	0,15	–	–	–	–	–	–	–	–	1	–	–
Conductometry (wet chemistry)	10^3	–	–	–	–	–	–	–	–	–	–	–
Detection Limit	ppb	ppb	ppb	ppb	ppb	ppb	ppb	ppb	ppb	ppb	Molec·cm^{-3}	ppb

Fig. 15 Continuously working methods for the detection of trace gases.

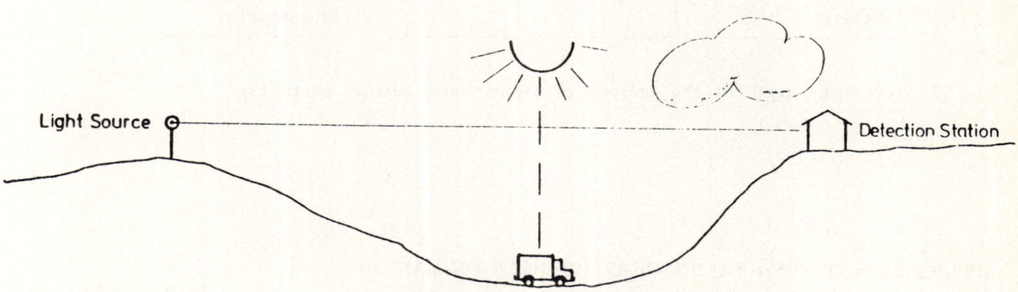

Method \ Compound	SO_2	NH_3	N_2O	NO	HNO_2	NO_2	HNO_3	NO_3	PAN	OH	H_2O_2
Correlation Spectroscopy	0,2	0,2	–	0,4	0,02	0,1	–	5	–	$3{,}7\times10^5$	–
Tunable Laser Absorption Spectr.	0,3	1	3	0,3	–	0,3	0,3	1	0,3	$3{,}7\times10^5$	1
Detection Limit	ppb	ppb	ppb	ppb	ppb	ppb	ppb	ppb	ppb	Molec·cm^{-3}	ppb

Fig. 16 Remote sensing techniques for the detection of trace gases.

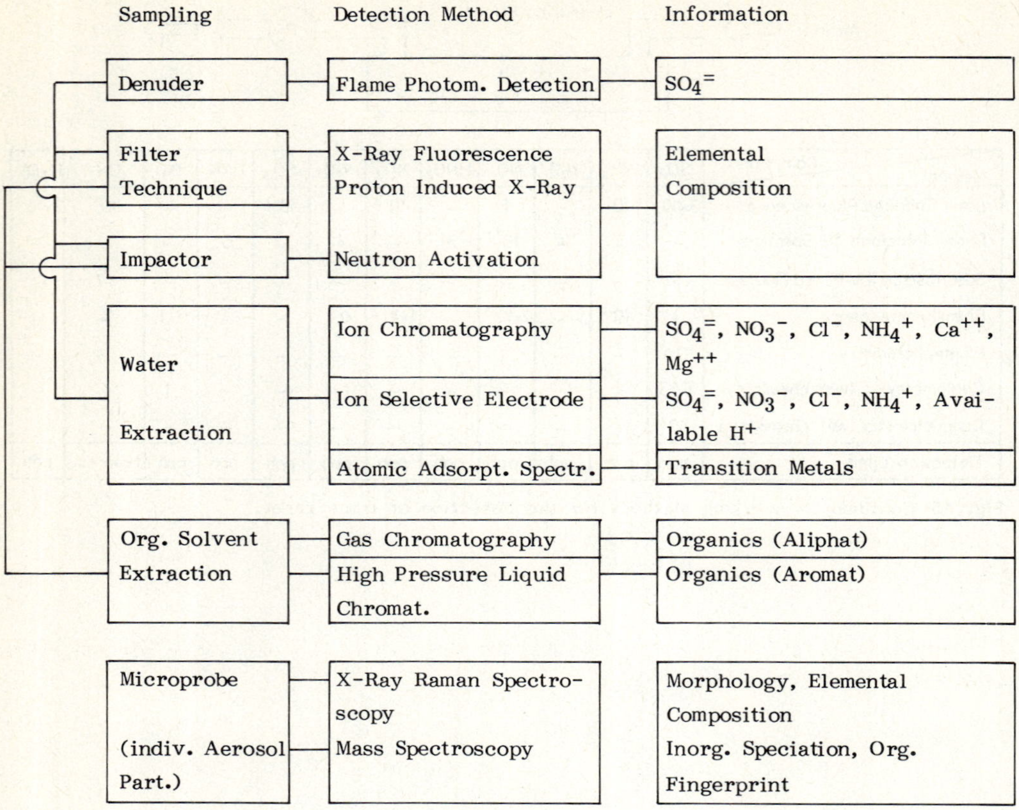

Fig. 17 Procedures used for the analysis of atmospheric aerosol particles.

APPLICATION OF PHYSICOCHEMICAL METHODS ON MAS

Based on results gained with the sampling and analysing methods, described in the previous part of that chapter, physico chemical investigations can be performed in order to obtain rate laws of chemical conversion processes occuring in MAS. These investigations may either take place during field experiments in the real world or during laboratory experiments under simulation of environmental conditions. Prominent examples of this kind of laboratory experiments are investigations of the chemical removal of SO_2 in MAS. Therefore this system was chosen to discuss some problems, which arise when physico chemical methodology is applied to MAS.

1. Theory

Since the troposphere can generally be regarded as a dynamic multiphase system coupled with a vast number of interrelated chemical reactions, with regard to the chemical removal of atmospheric SO_2 three main processes can be distinguished: 1. homogeneous reactions in the gas phase including photochemical oxidation (one-phase processes), 2. heterogeneous reactions on the surface of dry aerosol particles (two-phase processes), and 3. homogeneous or heterogeneous reactions in the aqueous phase of wetted aerosol particles or in droplets containing soluble and/or insoluble substances (two- or three-phase processes).

A large number of investigations have indicated that absorption and oxidation in the aqueous phase is in general the dominant chemical removal process for atmospheric SO_2 (Middleton et al.[42]; Moeller [43]). Gas phase reactions are relatively unimportant except under smog or dry clean air conditions combined with a high sunlight intensity (Calvert and Stockwell [44]). Adsorption and oxidation of SO_2 on aerosol particles produces a fast poisoning of their surfaces and is therefore mostly a short-lived process which may be important only in the vicinity of strong particle sources (Baldwin [45]; Harrison and Pio [46]).

When gaseous SO_2 is absorbed in an aqueous solution it undergoes the following equilibria:

$$SO_2(g) + H_2O \underset{}{\overset{H_{SO_2}}{\rightleftharpoons}} SO_2 \cdot H_2O \qquad (9)$$

$$SO_2 \cdot H_2O \overset{K_1}{\rightleftharpoons} H^+ + HSO_3^- \qquad (10)$$

$$HSO_3^- \overset{K_2}{\rightleftharpoons} H^+ + SO_3^= \qquad (11)$$

H_{SO_2}, K_1, K_2 = equilibrium constants
(g) = index for gas phase

Eq. (9) describes the phase equilibrium between the SO_2 concentration in the gas phase and in the aqueous phase according to Henry's law. Eq. (10) and (11) represent the dissociation equilibria which are established among the dissolved sulfur (IV) species depending on temperature and pH of the solution. The above equations thus describe reversible physical and chemical pathways including only S(IV) species, i.e. the oxidation state of sulfur is not changed by these processes.

In the atmosphere SO_2 is reversibly absorbed by aqueous suspensions i.e. fog-, cloud- and rain-droplets or by wetted aerosol particles. As indicated in Figure 18 the S(IV) species dissolved in the aqueous phase can be irreversibly oxidized to S(VI) species. The partial consumption of S(IV) due to this reaction provokes a reestablishment of the equilibria (9) -

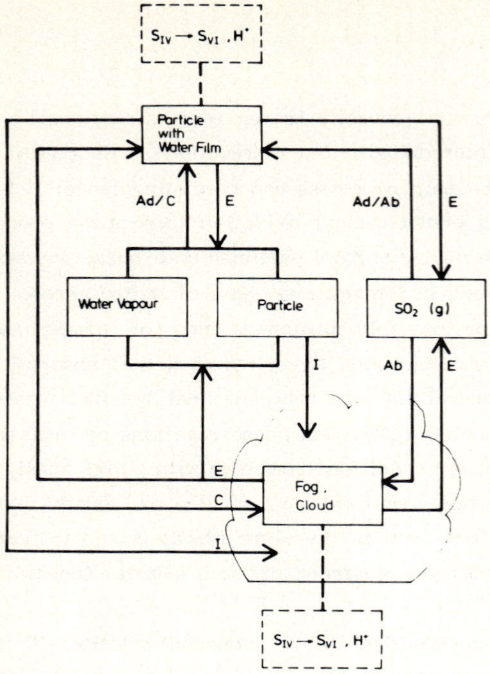

Fig. 18 Physico-chemical removal of SO_2(g) in an atmospheric multiphase system (surface reactions on dry particles and washout processes are omited)
C = condensation,
E = evaporation,
I = incorporation,
Ad = adsorption,
Ab = absorption;
---- irreversible oxidation in the liquid phase.

(11) including the absorption of additional SO_2 molecules from the gas phase. Thus gaseous SO_2 is definitely removed from the atmosphere by the aqueous phase oxidation of S(IV) to S(VI).

Aqueous solutions in atmospheric aerosol systems normally have a pH < 7. Under these conditions, mainly sulfuric acid is formed by the above process. Thus, if there are not enough cationic substances present to neutralize the H^+-activity produced by the dissociation of H_2SO_4 and by the absorption equilibria given in eq. (10) and (11) the aqueous phase may become highly acidified. In this way the chemical removal of SO_2 by aqueous phase processes is partially responsable for the formation of acid fog or acid rain in the atmosphere.

Numerous efforts in laboratory research have been undertaken in recent years to study several competitive aqueous phase oxidation mechanisms which may be important for the production of sulfuric acid under atmospheric conditions. These studies suggest that in atmospheric aerosol systems S(IV) is predominately oxidized by 1. ozone, 2. hydrogen peroxide, and/or 3. molecular oxygen in the presence of catalytically active transition metal complexes (Moller [43]; Martin [47]).

According to Graedel and Weschler (1981) the catalytic removal process is preferentially promoted under atmospheric conditions by Fe(III), Mn(II), and Cu(II) complexes dissolved in the liquid phase of such particles. In several experiments we have thus used different types of artificial aerosols containing one of these species. The overall catalytic reaction can be summarized as follows:

$$SO_2(aq) + H_2O + \frac{1}{2} O_2 \xrightarrow{Me^{n+}} [H_2SO_4 \rightleftharpoons HSO_4^- + H^+] \qquad (12)$$

$SO_2(aq)$ represents the dissolved S(IV) species $SO_2 \cdot H_2O$, HSO_3^- and $SO_3^=$ while Me^{n+} symbolizes the corresponding metal ion complex. The above reaction can be described by a chain mechanism including free radical intermediates (Franck and Haber [49]; Backstrom [50]). Alternative reaction pathways involving O-bonded sulfito species have recently been proposed by van Eldik [51].

The catalytic process is presumably only of minor importance in cloud or rain droplets compared to H_2O_2 or O_3 chemistry (Martin [47]; Penkett [52]). It is however possible that metal ion complexes contribute significantly to the generation of H_2O_2 in the presence of U.V. light or to its decomposition by a Fenton reaction. Likewise metal ion complexes are assumed to participate in the aqueous phase oxidation of S(IV) by ozone (Harrison et al. [53]; Hoffmann and Jacob [54]). Furthermore, in wetted aerosol particles these metal compounds can be concentrated in solutions of several molal since the amount of liquid water in such particles is about three to five orders of magnitude lower than in fog or cloud droplets (Ho et al. [55]). Thus, if the rate of the aqueous S(IV) oxidation increases with the activities of the dissolved metal compounds it may become significantly high in such concentrated solutions and a low pH value may consequently be produced. Since these wetted aerosol particles can be incorporated into larger droplets they may be very effective in the acidification of dilute fog or cloud water. Furthermore they can act directly as acidic cloud condensation nuclei. These physical processes of incorporation or condensation are likewise included in Figure 18.

In laboratory research it is very difficult to study these mechanisms simultaneously under controlled conditions. Furthermore the rather complex microphysical processes coupled with SO_2 chemistry in atmospheric multiphase systems are often completely neglected in such studies. In general, the different reaction mechanisms are investigated separately in very simplified experiments. With the severe limitations inherent in such experiments one may consequently draw wrong conclusions by extrapolating the laboratory data to atmospheric conditions (Hoffmann [56]).

Experimental studies of the chemical removal of SO_2 and the aqueous phase oxidation of

S(IV) are mostly based on one of the following techniques:

- studies in bulk solutions in equilibrium with gaseous SO_2 and/or other S(IV) species (e.g. Holt et al. [57]; Martin and Damschen [58]; Harrison et al. [53]; Maahs [59]; Martin [47]; Ibusuki and Barnes [60]),
- static exposure of a definite number of particles or droplets to a mixture of SO_2 gas and clean humid air (e.g. van den Heuvel and Mason [61]; Barrie and Georgii [62]; Judeikis et al. [63]; Clarke and Williams [64]; Dlugi [65]),
- dynamic mixing of SO_2 with a definite steady-state number of airborne particles or droplets (e.g. Matteson et al. [66]; Cheng et al. [67]; Cains and Carabine [68]; Kaplan et al. [69]; Crump et al. [70]; Dlugi [65]).

In view of the restrictions mentioned above, bulk phase studies may not be suitable to describe chemical processes occurring in real atmospheric aerosol systems particularly if the SO_2 removal by aqueous phase oxidation depends in any way on the microphysics of the corresponding system. The equivalence of the chemical removal of SO_2 over a bulk solution and in an aerosol system has not yet been proved by adequate experiments. Furthermore, with regard to the static exposure studies mentioned above it must be questioned in this context if the data obtained for the SO_2 removal rate can be linearly extrapolated to any number density of particles or droplets different from that produced in the experiments, i.e. to any population density of a corresponding aerosol system in the atmosphere.

The importance of different microphysical conditions by measuring the SO_2 removal rate in artificial aerosol systems as a function of the particle number density, and the relative humidity. In our opinion atmospheric conditions can best be approximated in experiments using a dynamic multiphase reaction system. Compared to bulk or static exposure techniques a dynamic method offers the advantage of performing measurements in a lower SO_2 concentration range (< 1 ppm). Furthermore deposition losses to the wall of the reactor are rather low even at high relative humidities. In previous dynamic flow studies, however, high SO_2 mixing ratios in the ppm-range have been used except in the work of Kaplan et al. [69]. Since Martin [47] has demonstrated that the reaction order of the S(IV) oxidation in the aqueous phase may be a function of the S(IV) or SO_2 gas phase concentration itself these studies may be relevant only for the SO_2 chemistry in plumes. Also the data obtained by Kaplan et al. [69] are rather few and have been criticized by Crump et al. [70]. The only catalytic substance that has been extensively investigated was $MnSO_4$. In newer studies three different Mn(II) and Cu(II) aerosols have been used (Berresheim [71]). Thereby the potential influence of different anions on the SO_2 removal rate could be studied. Some authors have already emphasized the possibility of such an influence, especially with regard to Cl^- which is supposed to support the catalytic oxidation of S(IV) in the aqueous phase (Bassett and Parker [72]; Abel [73]; Cheng et al. [67]; Barrie and Georgii [62]; Clarke and Radojevic [74]). In view of relevant coordination chemistry, this effect can probably be explained by several ligand substitution mechanisms.

The temporal removal of SO_2 from the gas phase due to a chemical sink mechanism can be described by the SO_2 removal rate R which is defined as

$$R = - d(SO_2)_g/dt \tag{13}$$

In order to express the reaction kinetics of this removal process we can first assume a general relationship between R and the SO_2 (g) concentration itself:

$$R = - d(SO_2)_g/dt = k_1 \ (SO_2)_g^d \tag{14}$$

In this equation the exponent d represents the reaction order with regard to $(SO_2)_g$. The pre-factor k_1 includes the relationship between R and all remaining parameters relevant for this process which have yet to be identified. The removal process investigated in this study is characterized by the steady-state absorption and oxidation of SO_2 in the aqueous phase of heavy metal aerosols. Since this is a multiphase process we consider k_1 to include different parameters relative to the aerosol as well as to the aqueous phase. We have assumed the following parameters to be important in this process:

a) particle number density N_p, metal concentration C_{Me} (aerosol phase in Fig. 19)

b) relative humidity rH, mean molal salt activity $a(s,\pm)$, H^+-activity $a(H^+)$ (aqueous phase in Fig. 19).

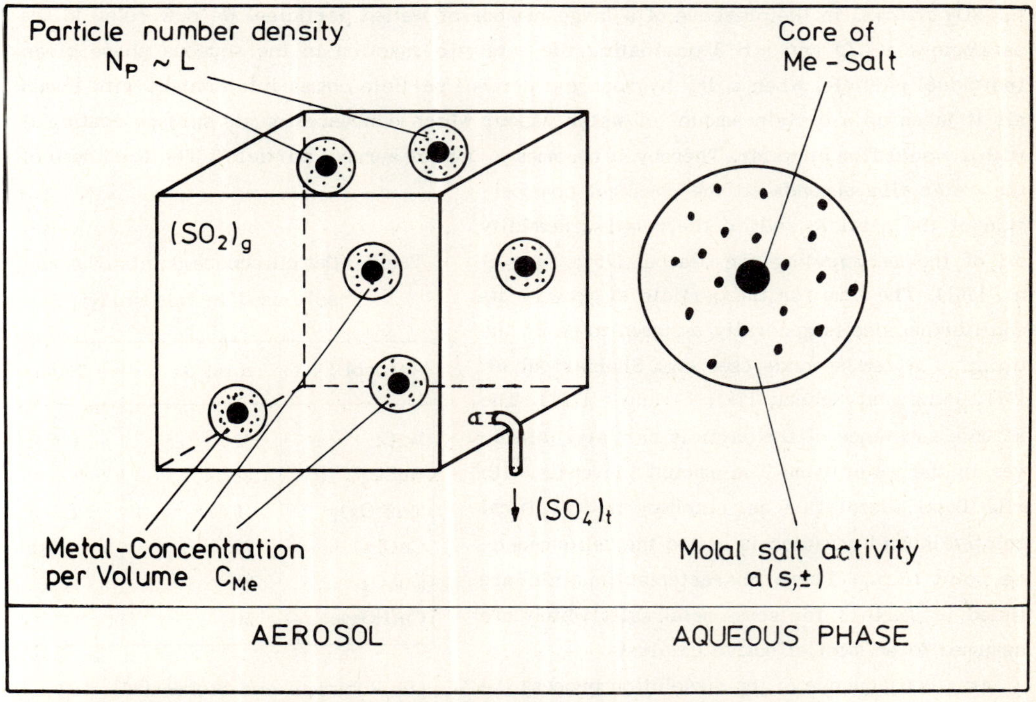

Fig. 19 Illustration of the most important variables in a multiphase system.

The particle number density N_p is defined as the number of particles per unit volume of air. With respect to the aerosol phase of the observed system an increase of N_p should also cause an increase of the SO_2 removal rate R since then more SO_2 molecules can be absorbed and oxidized in the same time period. We must thus define k_1 as a function of N_p. The relationship between k_1 (or R) and N_p has not yet been investigated with respect to wetted metal aerosols, and it remains to be determined whether or not it is linear. Therefore we may first assume a general non-linear relationship as follows:

$$k_1 = k_2 \, C_{Me}^{n} \qquad (n = const.) \tag{15}$$

Here k_2 represents a new prefactor which is introduced into eq. (14). The particle number density N_p is replaced in eq. (15) by the metal concentration C_{Me}, i.e. the total mass of reactive metal per unit volume of air. This is an admissible replacement since C_{Me} is a function of N_p. On the other hand it is better to use C_{Me} to represent the particle number density because it also incorporates chemical concentration information which is otherwise omitted. Also, the particle number is reintroduced at a later step through integration of the size distribution function for the aerosol.

Up to this point only the aerosol phase parameters have been considered with respect to the SO_2 removal in the presence of a large number of wetted particles. We now refer to the parameters $a(s,\pm)$ and $a(H^+)$ dominating the catalytic reaction in the aqueous phase of an individual particle. When a dry hygroscopic aerosol particle comes into contact with humid air it takes up a certain amount of water vapour which condenses on its surface coating it with a liquid film of water. Thereby it becomes a "wetted aerosol particle". The thickness of the water film depends on the chemical composition of the particle and on the relative humidity rH of the surrounding air (Haenel [75]; Tang et al. [76]). The time for the particle to grow to its equilibrium size is generally assumed to be on the order of a few seconds (see e.g. Shariari et al. [77]; Bunz and Schock [78]; Straubel [79]). The soluble substance of the particle partially dissolves in the water film. The amount increases with rH. Total dissolution is reached at a critical relative humidity which is called the deliquescence point (d.p.). These characteristic points are listed in Table 3 for some metal salts which are assumed to be most effective catalysts.

Table 3 Deliquescence points of aerosols used in this study

Aerosol	d.p., %	Ref.
$MnCl_2$	57	1
$MnSO_4$	86	1
$Mn(NO_3)_2$	3	2
$CuCl_2$	72 (67)	1 (3)
$CuSO_4$	98	1, 4
$Cu(NO_3)_2$	50	1

(1) Robinson and Stokes [80]

(2) Gmelin [91]

(3) Rockland [92]

(4) Low [93]

As a consequence of the dissolution process the liquid film covering each salt particle consists of a saturated salt solution until the d.p. is reached. Above this point a further uptake of water vapour dilutes the solution leading to a decrease of the mean molal activity $a(s,\pm)$ of the respective salt dissolved in the aqueous phase. This parameter is by definition the product of the mean molal activity coefficient $\gamma^{\pm}$ and the molal concentration <Me> of the dissolved metal compound:

$$a(s,\pm) = \gamma^{\pm} \langle Me\rangle \quad (16)$$

The molality of the metal compound can be calculated from the following relation:

$$\langle Me\rangle = C_{Me} / L M \quad (17)$$

where M is the molecular weight of the metal and L is the liquid water content of the aerosol system, i.e. the total amount of liquid water per unit volume of air. C_{Me} and L are given in ug m^{-3} and mg m^{-3}, respectively. The determination of L will be described in a later section (see eq. (22)). In case of a polydisperse aerosol eq. (17) gives the average value for the metal molality in an individual particle solution:

$$\langle Me\rangle = \frac{1}{N_p} \int_r N_p(r) \langle Me\rangle(r)\, dr \quad (18)$$

r = radius of a wetted particle

Values for the activity coefficient $\gamma^{\pm}$ are tabulated in standard books (e.g. Robinson and Stokes [80]).

The catalytic oxidation of $SO_2(aq)$ taking place in each particle solution is a function of $a(s,\pm)$. Therefore, at steady- state conditions the SO_2 removal rate R is also a function of $a(s,\pm)$ which can be described by an exponential term implied in eq. (15):

$$k_1 = k_3\, C_{Me}^{n}\, a(s,\pm)^{b} \quad (19)$$

k_3 = 3rd proportionality factor

b = reaction order with respect to $a(s,\pm)$

According to eq. (10), (11), and (12) we may similarly assume an influence of the pH of the individual particle solutions on the SO_2 removal rate. Thus eq. (15) can further be specified by an expression for the H^+-activity $a(H^+)$:

$$k_1 = k_4\, C_{Me}^{n}\, a(s,\pm)^{b}\, a(H^+)^{c} \quad (20)$$

k_4 = 4th proportionality factor

= const. at a given temperature

If this expression is introduced into eq. (14) we obtain the final basic equation for the experimental conditions investigated in MAS:

$$R = - d(SO_2)_g/dt = k_4\, C_{Me}^{n}\, a(s,\pm)^{b}\, a(H^+)^{c}\, (SO_2)_g^{d} \quad (21)$$

2. Experimental examples

Based on the theory developed in the previous section the SO_2 removal rate should be measured as a function of $(SO_2)_g$, N_p (represented by C_{Me}), and the parameters $a(s,\pm)$, $a(H^+)$ which are associated with the relative humidity rH, in order to determine the constant k_4 and the exponents n, b, c, and d in eq. (21) for different types of heavy metal aerosols.

Fig. 20 shows an experimental set up used in our laboratory for this kind of measurements. It consists mainly of a clean air generation system, a thermostated humidifier, a SO_2 source, a Collison Nebulizer as aerosol generation system, a vertical tubular flow

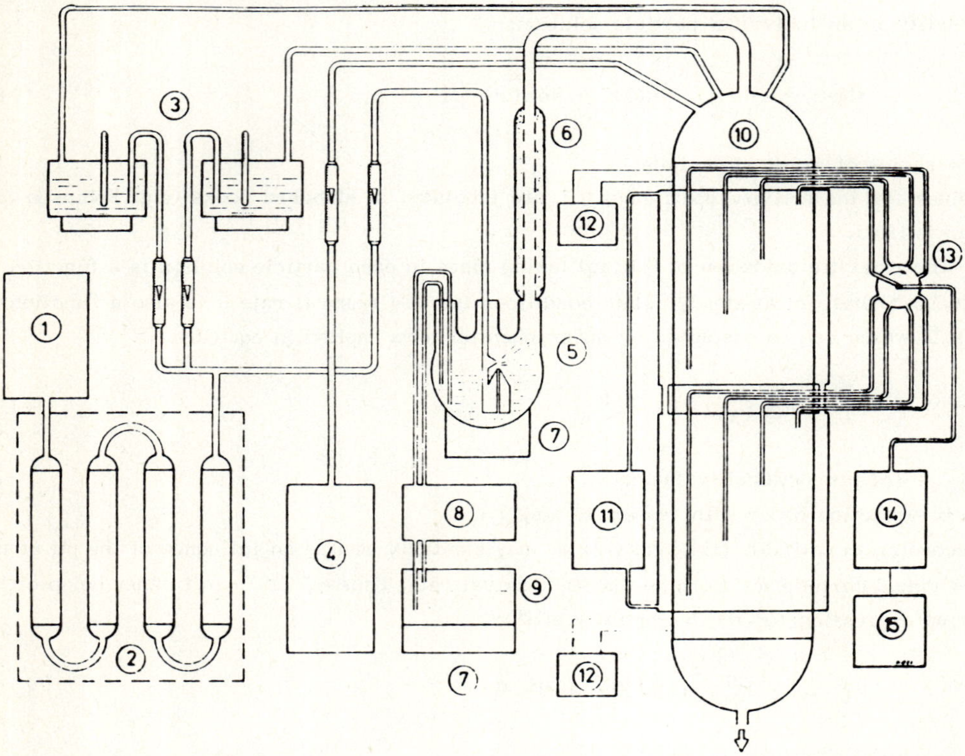

Fig. 20 Scheme of the experimental set up, which was used for kinetic studies of the SO_2 removal in simulated multiphase systems.
1. Compressor, 2. Air cleaning system, 3. Humidifier, 4. SO_2 source (permeation tube), 5. Collison Nebulizer [90], 6. Diffusion dryer, 7. Stirrers, 8. Peristaltic Pump, 9. Stock solution, 10. Tubular flow reactor, 11. Thermostated water cycle, 12. Humidity sensors, 13. six way valve, 14. Coulometric SO_2 detector, 15. Recorder.

Fig. 21 Size distribution of dry $MnCl_2$ aerosol as a function of the molarity of the reservoir solution: ● = 1.82×10^{-2} M, x = 10^{-3} M, □ = 10^{-4} M, o = 10^{-5} M.

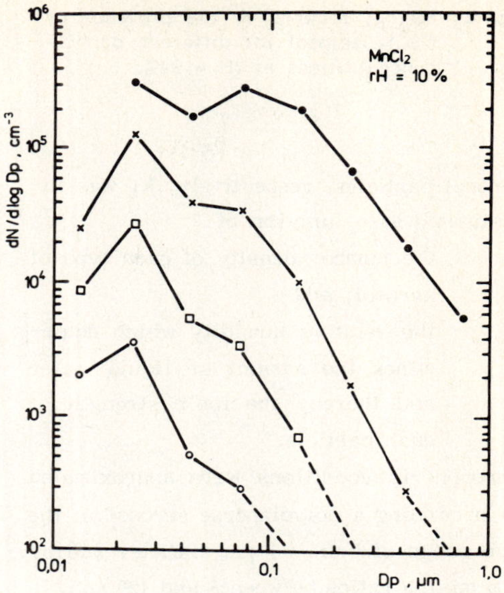

reactor, and a coulometric SO_2 detector. A more detailed description of the apparatus has been given in previous publications (Berresheim and Jaeschke [81]). The gas/aerosol mixture is continuously moved down to the outlet of the reactor by an airstream. During this transport SO_2 molecules are removed from the gas phase due to the catalytic oxidation of S(IV) in the aqueous phase of the aerosol particles. This consumption of SO_2 produces a steady-state SO_2 concentration gradient between the inlet and the outlet of the reactor which is measured at different sampling points. The SO_2 removal rate R can be obtained from this gradient and the flow rate of the gas/aerosol mixture.

The size distribution of the dry aerosol is measured at all three PTFE joints of the reactor with a TSI Electrical Aerosol Analyzer (EAA) 3030. As an example Figure 21 shows different size distributions of $MnCl_2$ aerosol as a function of the $MnCl_2$ concentration in the reservoir solution. A linear relationship is suggested by these data. Based on such curves the respective size distributions at humid conditions can be calculated (Haenel [75]). This is shown in Figure 22 for $MnSO_4$ and $CuCl_2$ aerosol. In separate experiments using $MnCl_2$, $MnSO_4$, $Mn(NO_3)_2$, $CuCl_2$, $CuSO_4$, and $Cu(NO_3)_2$

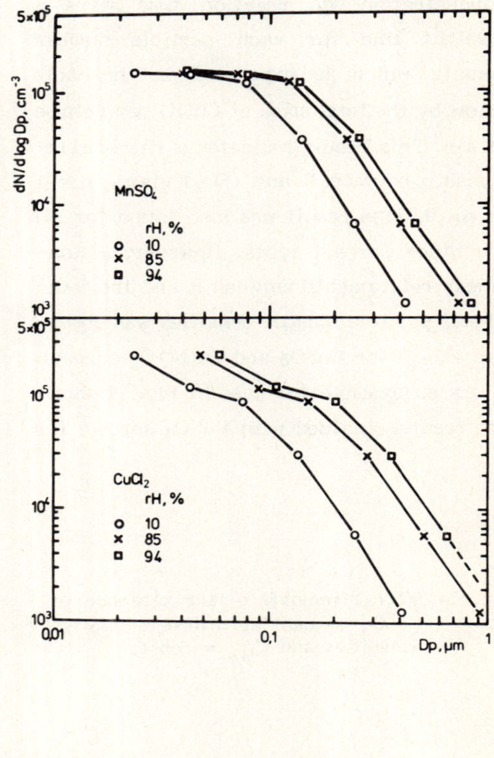

Fig. 22 Size distribution of $MnSO_4$ and $CuCl_2$ aerosol as a function of the relative humidity, calculated after Haenel (1976) for rH = 85% and 94%; reservoir solution concentration: 10^{-3} M.

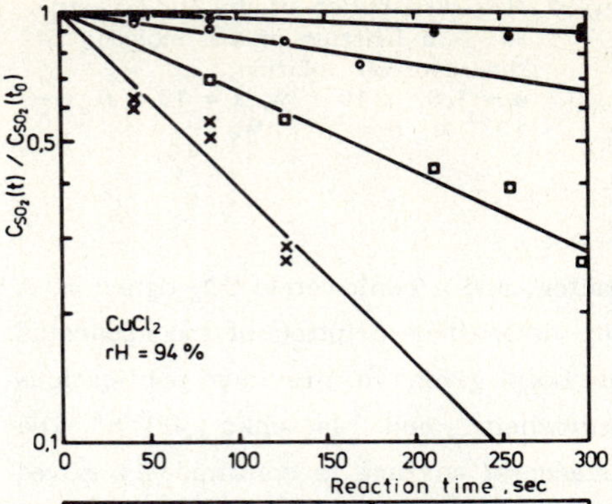

Symbol	C_{Cu} , µg m^{-3}	k_1 , 10^{-3}sec^{-1}
●	86	0,43
○	173	1,42
□	432	4,57
×	696	9,45

Fig. 23 SO_2(g) removal in the presence of $CuCl_2$ aerosol for different particle concentrations at rH = 94%.

aerosol particles, respectively, k_1 was investigated as a function of

a) the number density of each type of aerosol, and

b) the relative humidity which determines the amount of liquid water and thereby the ionic strength of each particle.

Atmospheric conditions were approximated by producing a polydisperse aerosol in the size range of 0.01 - 1 µm diameter and by SO_2 mixing ratios between 5 and 120 ppb. In case b) the relative humidity was varied from 73% to 94% while the ionic strength of the particles was always > 1 M. The temperature in the reactor was kept constant at 25°C.

Fig. 23 shows an example of the results obtained in case a). $CuCl_2$ aerosol was mixed with SO_2 at a constant relative humidity of 94%. The semi-logarithmic plot of the relative SO_2 concentration vs. reaction time gives a straight line for each particle number density which is expressed in the table below by the total mass of Cu(II) per volume of air. This result indicates a linear relationship between R and (SO_2) giving n = 1 in eq. 9. This result was also found for all the other aerosol types. However, a nonlinear relationship between R and the individual particle number densities was observed except for $MnSO_4$ and $Cu(NO_3)_2$ aerosol.

As an example for case b) Fig. 24 shows the results obtained with $MnSO_4$ aerosol for

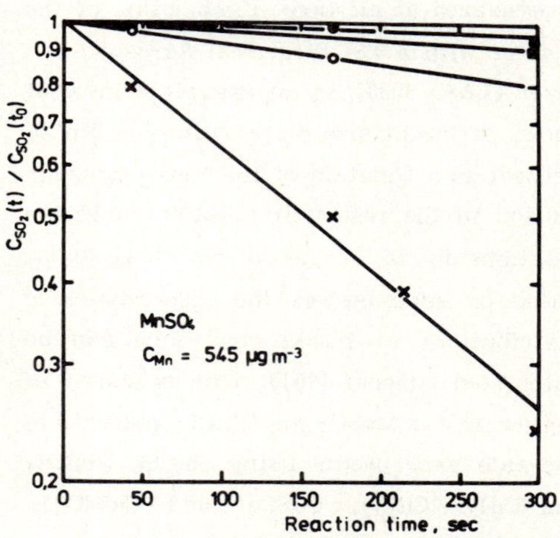

Symbol	rH, %	[Mn], mol kg^{-1}	pH	k_1, 10^{-4}sec^{-1}
●	73	5.50	2.1	2.03
□	85	4.20	2.1	2.88
○	89	3.60	2.2	7.0
×	94	2.62	2.6	42.8

Fig. 24 SO_2(g) removal in the presence of $MnSO_4$ aerosol at different relative humidities and C_{Me} = const.

a fixed particle number density and different relative humidities. At dry conditions (10% r.h.) none of the aerosols investigated in this study produced a detectable SO_2 concentration gradient in the reactor. However, above 70% r.h. a significant SO_2 removal rate was measured which showed a specific non-linear relationship to the corresponding relative humidity for each type of aerosol.

As can be seen from eq. (21) the SO_2 removal rate R is a function of the aqueous phase parameters $a(s,\pm)$ and $a(H^+)$. Previous studies of the S(IV) oxidation in diluted salt solutions or in cloud droplets with $a(s,\pm) < 10^{-2}$ M have shown a reciprocal linear relationship between R and $a(H^+)$ (Hegg and Hobbs [82]; Jacob and Hoffmann [83]). In this study the same relationship is assumed with regard to the highly concentrated particle solutions produced in the reactor. Thus the exponent c in eq. (21) obtains a value of c = -1. With this assumption eq. (20) can be transformed into the following expression:

$$k_1 \; a(H^+) \; / \; C^n_{Me} = k_4 \; a(s,\pm)^b \qquad (22)$$

Since the data for all parameters on the left side of eq. (22) as well as for $a(s,\pm)$ were known from our experiments the values for k_4 and b could be obtained by a regression analysis.

Fractional numbers were found in general for the exponent b which represents the order of the SO_2 removal rate with regard to $a(s,\pm)$. This typically indicates that the catalytic S(IV) oxidation occurring in the aqueous phase of the aerosol particles proceeds via a complex reaction pathway.

With the results of both parts of the experiments involving heavy metal aerosols the SO_2 removal rate can now explicitly be described by the following expression:

$$R = - \, d(SO_2)_g/dt = k_4 \; C^n_{Me} \; a(s,\pm)^b \; a(H^+)^{-1} \; (SO_2)_g \qquad (23)$$

3. Applications to the Atmosphere

Equation 23 can be used to calculate R for different particle or metal concentrations and different relative humidities. This is shown in Figure 25 for $Mn(NO_3)_2$ aerosol. In this diagram the experimental results have been extrapolated to Mn(II) concentrations which are typical for the atmospheric background aerosol (Charlson et al. [84]). With respect to the SO_2 removal rate at these conditions the following order of efficiency was found for the individual aerosols: $Mn(NO_3)_2 > MnCl_2 > Cu(NO_3)_2 > MnSO_4 > CuSO_4 > CuCl_2$.

The results prove that not only fog- or cloud-droplets containing strong oxidants like H_2O_2 or O_3 but also wetted aerosol particles may act as very efficient sinks for atmospheric SO_2. This is clearly shown by the following example: If we assume typical atmospheric fog conditions with L = 0.1 g m^{-3} and pH (droplets) = 4 an SO_2 removal rate of 0.53 - 5.3 % h^{-1} for 0.1 - 1 ppm H_2O_2 can be calculated using the data of Martin and Damschen [58]. Assuming on the other hand an ensemble of wetted aerosol particles at rH = 94 % containing $Mn(NO_3)_2$ with C_{Mn} = 55 ng m^{-3}, L = 0.9 ug m^{-3} and pH (wetted particles) = 1.5 we obtain an SO_2 removal rate of 2.5 % h^{-1} using eq. (23).

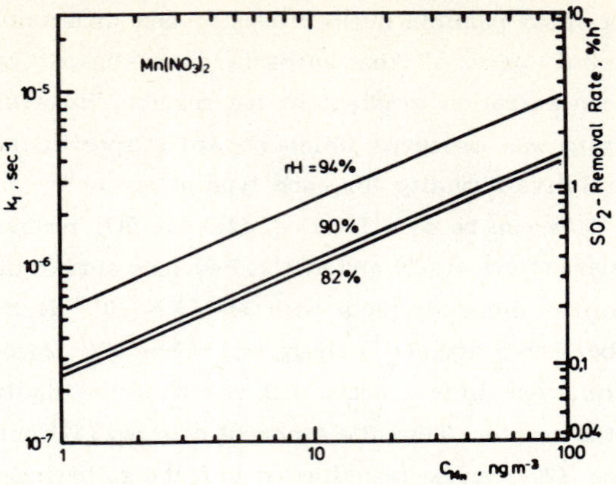

Fig. 25 SO_2 removal rate in the pressence of $Mn(NO_3)_2$ aerosol at atmospheric background concentrations of Mn(II) and different relative humidities.

Furthermore these measurements point out that the catalytic oxidation of $SO_2(aq)$ is most efficient in fog droplets with a size range of D = 2 - 20 µm. This is tentatively demonstrated in Figure 13 which shows a semi-quantitative plot of the SO_2 removal rate vs. the condensational growth of monodisperse particles, represented by the ratio m_w/m_0 where m_w is the liquid water mass and m_0 the mass of the particles at dry conditions. Instead of this ratio one can also choose the particle radius r as abscissa. In Figure 26 this parameter is quantified with regard to a classification of the atmospheric aerosol. The particle number density and the initial SO_2 gas phase concentration are assumed to be constant. The SO_2 removal rate shall be discussed here only as a function of the relative humidity. The SO_2 removal rate strongly increases with the relative humidity above the deliquescence points of the individual particles. At the d.p. each particle consists of a saturated salt solution which is generally characterized by an inhomogeneous mixture of the dissolved substances. The diffusion processes occurring in such a salt droplet may partially be supported by ion pairs or larger cluster molecules (Cussler [85]). Consequently no steady-state conditions are established in the liquid phase and the $SO_2(aq)$ oxidation is inhomogeneously distributed over the whole volume of the droplet. The degree of this inhomogeneity is characterized by the activity coefficient $\gamma^{\pm}$ of the respective solution. Furthermore the diffusivity of the individual S(IV) components is reduced in a concentrated solution (Eriksen [86]). Based on the data of Ratcliff and Holdcroft [87] the following order of diffusivity reduction for 1 M solutions of the salts used in this study could be calculated: $MnSO_4$ > $CuSO_4$ > $FeCl_3$ > $Mn(NO_3)_2$ > $Cu(NO_3)_2$ > $MnCl_2$ > $CuCl_2$ > NaCl. This shows that the aqueous phase diffusion of $SO_2(aq)$ was especially restricted in the sulfate particles which had the highest d.p. values (Table 3). Under such conditions mass transport limitations as described above may determine the rate of the aqueous phase reaction. However if the relative humidity exceeds the respective deliquescence point the concentrated solutions are diluted by a further uptake of water vapour. The individual activity coefficients of the solutions will pass through a minimum

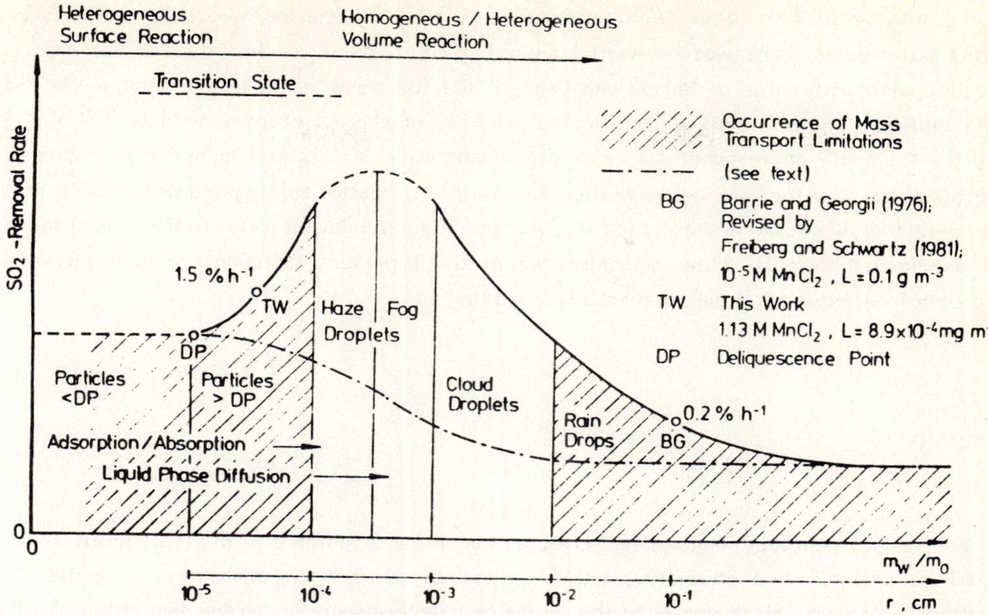

Fig. 26 Qualitative draft of the SO_2 removal rate as a function of the condensational growth of atmospheric aerosol particles containing catalytic substances (see text for further explanations).

and increase up to $\gamma^{\pm} = 1$ at relative humidities close to rH = 100 %. Then $a(s,\pm)$ becomes identical to the metal molality <Me> (ideal solution). The aqueous phase reaction rate increases since the above mass transport limitations disappear. Simultaneously the pH of each solution increases and more SO_2 molecules are absorbed and oxidized when the wetted particles are changing into growing droplets.

If these droplets continue to grow by condensation the dilution of the catalytic substance in the droplets finally begins to dominate the aqueous phase oxidation process. Thus the SO_2 removal rate will decrease until the dilution process has reached a point where the catalytic reaction is determined by mass transport processes of the S(IV) molecules in the droplets (Schwartz and Freiberg [88] [89]. Consequently a maximum of the removal rate must exist somewhere in the size range between small haze and large fog droplets (e.g. r = 1 - 10 μm). This is qualitatively indicated in Figure 26 by the dashed connection line. The chain-dotted curve in this Figure illustrates the decrease of the SO_2 removal rate in the absence of any mass transport limitations, starting at the deliquescence point if R were only a function of the metal molality <Me>. When the relative humidity decreases below the d.p. the aqueous phase parameters $a(s,\pm)$ and $a(H^+)$ remain constant. The amount of liquid water is then very small and the SO_2 removal is dominated by the adsorption and oxidation of undissociated SO_2 molecules on the surface of the particles. Thus eq. (21) implies a quasi-constant removal rate below the corresponding deliquescence point.

In previous investigations on the removal of SO_2 by multiphase processes mostly large dilute salt droplets were used as reactive aerosol species. As shown in Figure 26 our results are compared with those of Barrie and Georgii [62] for aerosols containing $MnCl_2$. The BG data represent the SO_2 removal rate in the presence of aerosol droplets with 10^{-5} M $MnCl_2$ and $L = 0.1$ g m^{-3} corresponding to a metal concentration of $C_{Mn} = 55$ ng m^{-3} with regard to the gas phase. For the same concentration represented by wetted $MnCl_2$ particles at rH = 94 % we obtaine an SO_2 removal rate which was 7.5 times higher than the rate observed in droplets by Barrie and Georgii. This emphasizes again the important influence of microphysical properties on rates of chemical reactions occuring in MAS.

SUMMARY

In this introductory chapter an attempt was made to point out what difficulties and problems arrise, when chemistry, either analytical or physico chemistry is applied to Multiphase Atmospheric Systems. In the following chapters many of the touched problems will be treated more intensively by experts actively engaged in the respective topics.

REFERENCES

(1) Fuzzi, S., Orsi, G., Mariotti, M.; Radiation fog liquid water acidity at a field station in the Po Valley; J. Aerosol Sci. 14; 1983; pp. 135-138

(2) Grunow, J.; Kritische Nebelfroststudien; Arch. Met. Geoph. Biokl. Ser. B.4; 1953, pp. 389-419

(3) Falconer, R. E., Falconer, D. D.; Determination of cloud water acidity at a mountain observatory in the Adirondack Mountains of New York State; J. Geophys. Res. 85; 1980, C12; pp. 7465 - 7476

(4) Mossop, S. C.; Intercomparison of instruments used of cloud drop concentration and size distribution; J. Clim. Appl. Meteorol. 22, 1983, 3; pp. 419 - 428

(5) Baumgardner, D.; An analysis and comparaison of five water droplet instruments; J. Clim. Appl. Meteorol. 22, 1983, 5; pp. 891 - 910

(6) Radford, W. H.; An instrument for sampling and measuring liquid fog water; Pap. Phys. Oceanogr. Meteorol. 6, 1938, 4; pp. 19 - 31

(7) Fukutomi, T., Tabaka, T.; Two simple self-recording apparatus for liquid water content in natural fogs; Low Temp. Sci. 4, 1948, p. 43

(8) Tabaka, T., Fujioka, T., Matsumara, N.; On a recording fog meteorological studies on fogs; T. Hori, Ed.; Tanne Trading Co, Ltd, Sapporo, Japan; 1953; pp. 169 - 173

(9) Houghton, H.; On the chemical composition of fog and cloud water; J. Meteorol. 12, 1955; pp. 355 - 357

(10) Mrose, H.; Measurements of pH, and chemical analysis of rain-, snow-, and fog-water; Tellus 18, 1966; pp. 266 - 270

(11) Okita, T.; Concentration of sulfate and other organic materials in fog and cloud water and in aerosol; J. Meteorol. Soc. Jpn. 46, 1968; pp. 120 - 126

(12) Petrenchuk, O. P., Drozdora, V. M.; On the chemical composition of cloud water; Tellus 18, 1966; pp. 280 - 286

(13) Lazrus, A. L., Baynton, H. W., Lodge, J. P.; Trace constitaents in oceanic cloud water and their origin; Tellus 22, 1970; pp. 106 - 114

(14) Oddie, B. C. V.; The chemical composition of precipitation at cloud level; Quart. J. Roy. Meteorol. Soc. 18, 1962; pp. 535 - 538

(15) Castillo, R. A., Jiusto, J. E., McLaren, E.; The pH and ionoc composition of stratiform cloud water; Atmos. Envirnon. 17, 1983, 8; pp. 1497 - 1505

(16) Mohnen, V., Wilson, J., Kadlecek, S.; Acid precipitation in the north eastern United States; in: Chemsitry of the unpolluted and polluted troposphere; H. W. Georgii and W. Jaeschke, Ed.; Reidel Pub. Comp.; 1982; pp. 395 - 423

(17) Brewer, R. L., Gordon, R. J., Shepard, L. S.; Chemistry of mist and fog from the Los Angeles urban area; Atmos. Environ. 17, 1983, 11; pp. 2267 - 2270

(18) Proceedings of EPA/NSF workshop on the atmospheric deposition and its impact on high elevation montain forest systems; Albany, N. Y.; 1984

(19) Jacob, D. J., Flagan, R. C., Waldman, J. M., Hoffmann, M. R.; Design and calibration of a rotating arm collector for ambient fog sampling, precipitation scavenging, dry deposition, and resuspension; Elsevier; 1983; pp. 125 - 134

(20) Fuchs, N. A.; The mechanics of aerosols; Pergamon Press, Oxford; 1964; p. 408

(21) Langmuir, J., Blodgett, K. B.; U.S. Army Air Forces Techn. Rep. No. 5418; 1946

(22) Hering, S.V., Blumenthal, D.L.; Fog sampler incomparison study; Sonoma Technilogy Inc. (STI); Dez. 1983

(23) Knollenberg, R. G.; Technique for probing cloud microstructure; New York; 1981; pp. 15 - 91

(24) Petrenchuk, O. P., Dozdora, V. M.; On the chemical composition of cloud water; Tellus 18, 1966; pp. 280 - 286

(25) Lazrus, A. L., Baynton, H. W., Lodge, J. P.; Trace constituents in oceanic cloud water and their origin; Tellus 22, 1970; pp. 106 - 114

(26) Falconer, R. E., Schaefer, K. J.; A new plane model cloud water. Projectcirrus final report; General Electric Research Lab. Report No. 140; 1948

(27) Oddie, B. C. K.; The chemical composition of precipitation at cloud level; Quart. J. Roy. Meteorol. Soc. 88, 1962; pp. 535 - 538

(28) Fricke, W., Georgii, H. W., Gravenhorst, V. G.; Application of a new sampling device for cloud water analysis. In some problems of cloud physics (collector papers); Gidrometeoizdat, Leningrad; 1978; pp. 200 - 213

(29) Scolt, W. O.; The pH of cloud water and the production of sulfate; Atmos. Environ. 12, 1978; pp. 917 - 924

(30) Kallend, A. S.; The fate of atmospheric emissions along plume trajectories over the north sea. First annual report to EPRI, Mars 1979; Report No. RD/L/R 1998; Central Electricity Research Lab., Leatherhead, Surrey, England; 165; 1979

(31) Winters, W., Hogan, A., Mohnen, V., Barnard, S.; ASRC- airborne cloud water collection system; ASRC-SUNY Pub. No. 728, Report to NSF; Atmos. Res. Section

(32) Walters, P. T., Moore, M. J., Webb, A. H.; A separator for obtaining samples of cloud water in aircraft; Atmos. Environ. 17, 1983, 6; pp. 1083 - 1091

(33) Zika, R.G., Satzman, E.S.; Interactions of ozone and hydrogen peroxide in water: Implications for analysis of H_2O_2 in air; Geophys. Res. Lett. 9; 1982; pp. 231-234

(34) Kok, G.L., Wilson, K.; A derivatization technique for the determination of peroxides in precipitation; Brookhaven Nat. Lab. No. 51757 19; 1984; pp. 1-8

(35) Stauff, J., Jaeschke, W., Schloegl, G.; Chemilumineszenz des Dioxetandions; Z. f. Naturforschung 27b; 1972; pp. 1434-1435

(36) Jaeschke, W., New methods for the analysis of SO_2 and H_2S in remote areas and their application in the atmosphere; Atmos. Environ. 12, 1978, p. 715

(37) Stauff, J., Jaeschke, W., Schloegl, G., CO_2 Nachleuchten und Chemilumineszenz der Oxalester-Oxidation. Vergleich der Spektren und der Kinetik; Z. f. phys. Chemie, N.F. 99; 1976; pp. 37-58

(38) Jaeschke, W., Stauff, J., Die Chemilumineszenz der SO_2 - Oxidation und ihre Anwendung in der Chemie der Atmosphaere; Ber. Buns. Ges. 82; 1978; pp. 1180-1184

(39) Jaeschke, W., Problems and methods of trace gas measurements in the troposphere. In: Chemistry of the unpolluted and polluted troposphere; Georgii, H. W., and Jaeschke, W., Eds.; D. Reidel Pub. Comp.; 1982; pp. 3-38

(40) Moffat, A.J., Millan, M.M., The applications of optical correlation technique to the remote sensing of SO_2 plumes using sky light; Atmos. Environ. 5; 19xx, pp. 677-690

(41) Perner, D., Huebler, G., Experimental detection of OH in the troposphere, in: Chemistry of the unpolluted and polluted troposphere, H.W. Georgii and W. Jaeschke, Eds.; D. Reidel Publ. Comp., 1982, pp. 267-294

(42) Middleton, P., Kiang, C.S. and Mohnen, V.A., Theoretical estimates of the relative importance of various urban sulfate aerosol production mechanisms, Atmos. Environ. 14, 1980, 463-472.

(43) Moller, D., Kinetic model of atmospheric SO_2 oxidation based on published data, Atmos. Environ. 14, 1980, 1067-1076.

(44) Calvert, J.G. and Stockwell, W.R., Mechanism and rates of the gas-phase oxidations of sulfur dioxide and nitrogen oxides in the atmosphere, in: J.G. Calvert (ed.), SO_2, NO and NO_2 Oxidation Mechanisms: Atmospheric Considerations, Ann Arbor Science, Boston, 1984, pp. 1-62.

(45) Baldwin, A.C., Reactions of gases on prototype aerosol surfaces, in: D.R. Schryer (ed.), Heterogeneous Atmospheric Chemistry, Geophysical Monograph 26, AGU, Washington, 1982, pp. 99-102.

(46) Harrison, R.M. and Pio, C.A., Kinetics of SO_2 oxidation over carbonaceous particles in the presence of H_2O, NO_2, NH_3 and O_3, Atmos. Environ. 17, 1983, pp. 1261-1275.

(47) Martin, L.R., Kinetic studies of sulfite oxidation in aqueous solution, in: J.G. Calvert (ed.), SO_2, NO and NO_2 Oxidation Mechanisms: Atmospheric Considerations, Ann Arbor Science, Boston, 1984, pp. 63-100.

(48) Graedel, T.E. and Weschler, C.J., Chemistry within aqueous atmospheric aerosols and raindrops, Rev. Geophys. Space Phys. 19, 1981, pp. 505-539.

(49) Franck, J. and Haber, F., Zur Theorie der Katalyse durch Schwermetallionen in waessriger Loesung und insbesondere zur Autoxydation der Sulfitloesungen, Sitzungsber. Preuss. Akad. Wiss., Phys.-Math. Kl. 13, 1931, pp. 250-256.

(50) Baeckstroem, H.L.J., Der Kettenmechanismus bei der Autoxydation von Natriumsulfitloesungen, Z. Phys. Chemie 25, 1934, pp. 122-138.

(51) van Eldik, R., Transition metals as potential catalysts in atmospheric oxidation processes, 1985, this book.

(52) Penkett, S.A., Laboratory studies of the multiphase S_{IV} - S_{VI} conversion rate, 1985, this book.

(53) Harrison, H., Larson, T.V. and Monkman, C.S., Aqueous phase oxidation of sulfites by ozone in the presence of iron and manganese, Atmos. Environ. 16, 1982, pp. 1039-1041.

(54) Hoffmann, M.R. and Jacob, D.J., Kinetics and mechanisms of the catalytic oxidation of dissolved sulfur dioxide in aqueous solution: An application to nighttime fog water chemistry, in: J.G. Calvert (ed.), SO_2, NO and NO_2 Oxidation Mechanisms: Atmospheric Considerations, Ann Arbor Science, Boston, 1984, pp. 101-172.

(55) Ho, W., Hidy, G.M. and Govan, R.M., Microwave measurements of the liquid water content of atmospheric aerosols, J. Appl. Met. 13, 1974, pp. 871-879.

(56) Hoffmann, M.R., Are the results obtained in controlled kinetic studies under laboratory conditions applicable to a wide variety of field situations? - The real world versus the idealized batch or continuous flow reactor at standard conditions, Paper presented on the Workshop on Research Priorities and Criteria to Establish Factors which Govern Precipitation Chemistry, Lake Placid, 1982.

(57) Holt, B.D., Kumar, R. and Cunningham, P.T., Oxygen-18 study of the aqueous-phase oxidation of sulfur dioxide, Atmos. Environ. 15, 1981, pp. 557-566.

(58) Martin, L.R. and Damschen, D.E., Aqueous oxidation of sulfur dioxide by hydrogen peroxide at low pH, Atmos. Environ. 15, 1981, pp. 1615-1621.

(59) Maahs, H.G., Measurements of the oxidation rate of sulfur (IV) by ozone in aqueous solution and their relevance to SO_2 conversion in nonurban tropospheric clouds, Atmos. Environ. 17, 1983, pp. 341-345.

(60) Ibusuki, T. and Barnes, H.M., Manganese (II) catalyzed sulfur dioxide oxidation in aqueous solution at environmental concentrations, Atmos. Environ. 18, 1984, pp. 145-151.

(61) van den Heuvel, A.P. and Mason, B.J., The formation of ammonium sulphate in water droplets exposed to sulphur dioxide and ammonia, Quart. J. Roy. Met. Soc. 89, 1963, pp. 271-275.

(62) Barrie, L.A. and Georgii, H.W., An experimental investigation of the absorption of sulphur dioxide by water drops containing heavy metal ions, Atmos. Environ. 10, 1976, pp. 743-749.

(63) Judeikis, H.S., Stewart, T.B. and Wren, A.G., Laboratory studies of heterogeneous reactions of SO_2, Atmos. Environ. 12, 1978, pp. 1633-1641.

(64) Clarke, A.G. and Williams, P.T., The oxidation of sulphur dioxide in electrolyte droplets, Atmos. Environ. 17, 1983, pp. 607-615.

(65) Dlugi, R., SO_2-oxidation in aerosol particles and droplets, J. Aerosol Science 14, 1983, pp. 292-297.

(66) Matteson, M.J., Stober, W. and Luther, H., Kinetics of the oxidation of sulfur dioxide by aerosols of manganese sulfate, Ind. Eng. Chem. Fundam. 8, 1969, pp. 677-687.

(67) Cheng, R.T., Corn, M. and Frohliger, J.O., Contribution to the reaction kinetics of water soluble aerosols and SO_2 in air at ppm concentrations, Atmos. Environ. 5, 1971, pp. 987-1008.

(68) Cains, P.W. and Carabine, M.D., Oxidation of sulphur dioxide in aerosol droplets, catalysed by manganous sulphate, J. Chem. Soc. Faraday Trans. I 11, 1978, pp. 2689-2702.

(69) Kaplan, D.J., Himmelblau, D.M. and Kanaoka, C., Oxidation of sulfur dioxide in aqueous ammonium sulfate aerosols containing manganese as a catalyst, Atmos. Environ. 15, 1981, pp. 763-773.

(70) Crump, J.G., Flagan, R.C. and Seinfeld, J.H., An experimental study of the oxidation of sulfur dioxide in aqueous manganese sulfate aerosols, Atmos. Environ. 17, 1983, pp. 1277-1289.

(71) Berresheim, H., Reaktionskinetische Untersuchungen zum Abbau von SO_2 in Aerosolpartikeln und Wolkentroepfchen, Ph. D. Thesis, 1984, University of Frankfurt.

(72) Bassett, W. and Parker, W.G., The oxidation of sulphurous acid, J. Chem. Soc. 47, 1951, pp. 1540-1560.

(73) Abel, E., Zur Anionkatalyse der Autoxydation, Monatsh. f. Chemie 87, 1956, pp. 113-119.

(74) Clarke, A.G. and Radojevic, M., Chloride ion effects on the aqueous oxidation of SO_2, Atmos. Environ. 17, 1983, pp. 617-624.

(75) Haenel, G., The properties of atmospheric aerosol particles as functions of the relative humidity at thermodyamic equilibrium with the surrounding moist air, in: H.E. Landsberg and J. van Mieghem (eds.), Advances in Geophysics, Academic Press, N.Y., 1976, pp. 74-188.

(76) Tang, I.N., Munkelwitz, H.R. and Davies, J.G., Aerosol growth studies - II. Preparation and growth measurements of monodisperse salt aerosols, J. Aerosol Science 8, 1977, pp. 149-159.

(77) Shariari, S., Sarmiento, A.N. and Goodrich, F.C., The kinetics of growth of an aerosol in a flow reactor - I. Experimental, in: G.M. Hidy (ed.), Aerosols and Atmospheric Chemistry, Academic Press, N.Y., 1972, pp. 67-73.

(78) Bunz, H. and Schock, W., Beeinflussung der Partikelform durch Wechselwirkung zwischen Aerosol und gesaettigtem Wasserdampf, in: W. Stober and D. Hochrainer (eds.), Proceedings of the 8th Conference on Aerosols in Science, Medicine and Technology: Physical and Chemical Properties of Aerosols, Gesellschaft f. Aerosolforschung, Schmallenberg, FRG, 1980, pp. 316-322.

(79) Straubel, H., Observation of chemical reactions on single particles, in: W. Stober and D. Hochrainer (eds.), Proceedings of the 8th Conference on Aerosols in Science, Medicine and Technology: Physical and Chemical Properties of Aerosols, Gesellschaft f. Aerosolforschung, Schmallenberg, FRG, 1980, pp. 347-350.

(80) Robinson, R.A. and Stokes, R.H., Electrolyte Solutions, Butterworths, London, 1970.

(81) Berresheim, H., Jaeschke, W., Laboratory studies of the multiphase SO_2-oxidation. Proc.: Gas-liquid Chemistry of Natural Water; Brookhaven Nat. Lab., 1984, pp. 15/1-15/8

(82) Hegg, D.A. and Hobbs, P.V., Oxidation of sulfur dioxide in aqueous systems with particular reference to the atmosphere, Atmos. Environ. 12, 1978, pp. 242-253.

(83) Jacob, D.J. and Hoffmann, M.R., The chemistry of nighttime urban fogs, in: H.R. Pruppacher, R.G. Semonin and W.G.N. Slinn (eds.), Precipitation Scavenging, Dry Deposition, and Resuspension, Vol. 1, Elsevier, N.Y., 1983, pp. 149-157.

(84) Davies, C.W., Ion Association, Butterworths, London, 1962.

(85) Cussler, E.L., Cluster diffusion in liquids, AIChE J. 26, 1980, pp. 43-51.

(86) Eriksen, T.E., Diffusion studies in aqueous solutions of sulfur dioxide, Chem. Eng. Sci. 24, 1969, pp. 273-278.

(87) Ratcliff, G.A. and Holdcroft, J.G., Diffusivities of gases in aqueous electrolyte solutions, Trans. Instn. Chem. Engrs. 41, 1963, pp. 315- 319.

(88) Schwartz, S.E. and Freiberg, J.E., Mass transport limitation to the rate of reaction of gases in liquid droplets: Application to oxidation of SO_2 in aqueous solutions, Atmos. Environ. 15, 1981, pp. 1129-1144.

(89) Freiberg, J.E. and Schwartz, S.E., Oxidation of SO_2 in aqueous droplets: Mass transport limitation in laboratory studies and the ambient atmosphere, Atmos. Environ. 15, 1981, pp. 1145-1154.

(90) May, K.R., The Collison nebulizer: description, performance and application, J. Aerosol Science 4, 1973, pp. 235-243.

(91) Gmelin Handbuch der Anorganischen Chemie, Springer, Berlin, 1958, 60 (B1), pp. 169, 270, 526. 1975, 56 (C3), p. 282. 1976, 56 (C6), pp. 99, 108.

(92) Rockland, L.B., Saturated salt solutions for static control of relative humidity between 5 ^{o}C and 40 ^{o}C, Analyt. Chem. 32, 1960, pp. 1375-1376.

(93) Low, R.D.H., A theoretical study of nineteen condensation nuclei, J. Rech. Atmos. 4, 1969, pp. 65-78.

LABORATORY AND FIELD STUDIES OF GAS-AEROSOL REACTIONS

Peter H. McMurry, J. Charles Wilson, and Daniel J. Rader
Particle Technology Laboratory, Department of Mechanical Engineering
111 Church Street Southeast, University of Minnesota
Minneapolis, MN 55455
U.S.A.

ABSTRACT

In this paper research that pertains to reactions between gases and aerosol particles in the atmosphere is discussed. The paper is divided into two sections. In the first section methods that have been used to investigate the significance of secondary aerosol formation by liquid phase reactions in the atmosphere are outlined. This discussion emphasizes the evidence for liquid phase chemical reactions in non-cloud aerosol systems. It is shown that there is some data which suggest that liquid phase chemical reactions of dissolved gases (e.g., sulfur dioxide) may occur in non-cloud aerosol systems.

In the second section experimental techniques that are being used in our laboratory to study reactions between various gases and submicron aerosol droplets are discussed. The key analytical tool in these experiments is the differential mobility analyzer (DMA) which is used both to produce the monodisperse aerosol particles that are fed into the reactors, as well as to measure particle size changes associated with the chemical reactions. With this technique size changes of less than 1% for particles smaller than 0.1 μm can be measured. Flow reactor studies designed to measure the rate of reaction between ammonia gas and sulfuric acid aerosol have been completed, and techniques and results of that study are summarized. Batch reactor studies of aqueous phase oxidation of sulfur dioxide are currently underway. Preliminary work shows that particle growth rates comparable to those that have been measured in the atmosphere can be measured with this technique.

INTRODUCTION

Reactive gases and aerosol particles normally coexist in the atmosphere. Aerosol particle concentrations can range from hundreds to millions per cubic centimeter of air, and have characteristic sizes ranging from about 1 nm to 100 μm. Chemical interactions between gases

NATO ASI Series, Vol. G6
Chemistry of Multiphase Atmospheric Systems
Edited by W. Jaeschke

and aerosol particles may be significant when the characteristic lifetime of a gas phase species is long compared to the characteristic time for collision with an aerosol particle. Characteristic collision times for a gas phase molecule with an aerosol particle tend to be greater than about 10^{-4} seconds.

There are a variety of ways in which gas-aerosol interactions may affect atmospheric chemistry. Aerosols may serve as sinks for reactive gas phase species. Gases can also be absorbed by liquid airborne droplets, and may undergo further chemical reactions within the droplets. Aerosol particles can also serve as sites for condensation of low vapor pressure vapors that are formed by gas phase chemical reactions.

It is useful to distinguish between liquid phase reactions of dissolved gases that involve cloud droplets and those that involve non-cloud liquid droplets. Cloud droplets tend to be larger than about 5 μm in diameter, and consist predominantly of water with relatively low concentrations of other species. In contrast, non-cloud aerosol droplets are typically in the submicrometer size range, and usually consist of highly concentrated aqueous solutions; ionic strengths exceeding 10 molal are not uncommon (Stelson and Seinfeld, 1983). Therefore, the effects of ionic strength must be accounted for if the chemical equilibria (solubilities, dissociations, activities) and reaction kinetics are to be modelled. Because most studies of reaction kinetics in aqueous solutions are done in more dilute solutions, it is not straightforward to apply this work to non-cloud aerosols. Another significant difference between cloud and non-cloud aerosols is liquid water content. The concentration of liquid water in clouds is typically 4 orders of magnitude greater than for non-cloud aerosol systems.

In this paper, research that deals with two aspects of gas-aerosol interactions is discussed. In both cases the focus will be on reactions that take place in non-cloud aerosol systems. First, the evidence for the formation of aerosol sulfur by liquid phase chemical reactions in non-cloud ambient aerosol is discussed. Then, a new experimental technique for studying gas-aerosol reactions in the laboratory is described. This technique has been applied in measurements of the mass accomodation coefficient for collisions between ammonia gas and sulfuric acid aerosol, and results of these measurements are presented.

LIQUID PHASE CHEMISTRY IN NON-CLOUD AMBIENT AEROSOLS

A great deal of attention has recently been directed towards studies of rates and mechanisms of chemical reactions that may be important in cloud and fog droplets. There is some evidence, however, which suggests that chemical reactions in non-cloud aerosol systems may also lead to the formation of secondary sulfur-containing aerosols. In this section the evidence for such processes is presented.

Gillani and Wilson (1983) presented evidence for the formation of secondary aerosol sulfur by liquid phase reactions involving non-cloud aerosols. Their conclusions were based on analysis of data that were acquired in power plant plumes during the MISTT (Labadie power plant, St. Louis, MO, July 1976) and the STATE (Cumberland and Johnsonville power plants, TN, 1978) field studies. A total of 10 days of data was analyzed (Gillani et al., 1981). They found that on days with relative humidity less than 75%, measured conversion rates were consistent with an empirical rate expression given by:

$$SO_2 \text{ conversion rate } (\%\text{hr}^{-1}) = 0.03\ R\ \Delta z_p\ O_3$$

where R is the total solar radiation in kWm^{-2}, Δz_p is the vertical spread of the plume in meters, and O_3 is ozone concentration in ppm. Conversion rates on days with relative humidities exceeding 75% were found to be greater than expected on the basis of this empirical rate expression, and this difference was ascribed to liquid phase chemistry.

In an effort to distinguish between cloud and non-cloud liquid phase conversion of sulfur dioxide, Gillani and Wilson (1983) correlated the trajectories of air parcels with cloud locations. It was argued that if the cloud base was higher than the top of the mixing layer, then air parcels that were transported through the mixing layer could not have had contact with clouds. Evidence for liquid phase reactions was found for six of the ten days that were analyzed. Of these six days, in-cloud processes were impossible on one day, and the observed liquid phase contribution to secondary sulfur aerosol formation was attributed to non-cloud liquid phase chemistry. On the remaining five days, both cloud and non-cloud processes may have contributed to the observed liquid phase conversion.

McMurry and Wilson (1983) also attempted to distinguish between gas and liquid phase contributions to oxidation of sulfur dioxide in ambient air. In this work, the "growth law" technique was applied. Growth law analyses are based on measurements of aerosol size distributions that are growing by gas-to-particle conversion. It is necessary to have size distribution measurements in an air parcel at (at least) two times. With these data, diameter growth rates that are necessary to grow the aerosol from the initial to the final size distribution are calculated. These diameter growth rates generally vary with particles size, and this size dependence contains information about chemical mechanisms of aerosol formation. When sulfur dioxide is oxidized by gas phase reactions, the low vapor pressure products tend to condense on aerosol particles. For condensational growth, growth rates tend to decrease with increasing particle size for particles larger than about 0.1 μm. In contrast, diameter growth rates for liquid aerosol droplets that are growing by liquid phase chemical reactions tend to increase with increasing particle size.

The data that McMurry and Wilson analyzed were obtained in a variety of power plant plumes (5 episodes) and in the Columbus, OH, urban plume (8 episodes). They found that the diameter dependence of measured diameter growth rates could usually be fit with a theoretical growth law that included simultaneous growth by condensation and droplet phase reactions. A comparison of data for diameter growth rates and theoretical growth law of this type is shown in Fig. 1. The data that are shown in Fig. 1 were acquired in the urban plume of Columbus, OH. Although these calculations are based on measurements of aerosol size distributions, independent measurements of aerosol sulfur distributions with respect to size support the argument that sulfur contributed to the growth of these aerosols in all particles size ranges. Therefore, it is concluded that both gas and liquid phase reactions contribute to the formation of secondary sulfur-containing aerosols in the atmosphere. It was found that liquid phase reactions were the predominant reaction pathway for relative humidities greater than 75% (see Fig. 2).

In the work of McMurry and Wilson (1983), no effort was made to distinguish between liquid phase reactions that took place in non-cloud aerosols, and those that took place in clouds. More recently, Wilson has applied the criteria of Gillani and Wilson (1983) to the

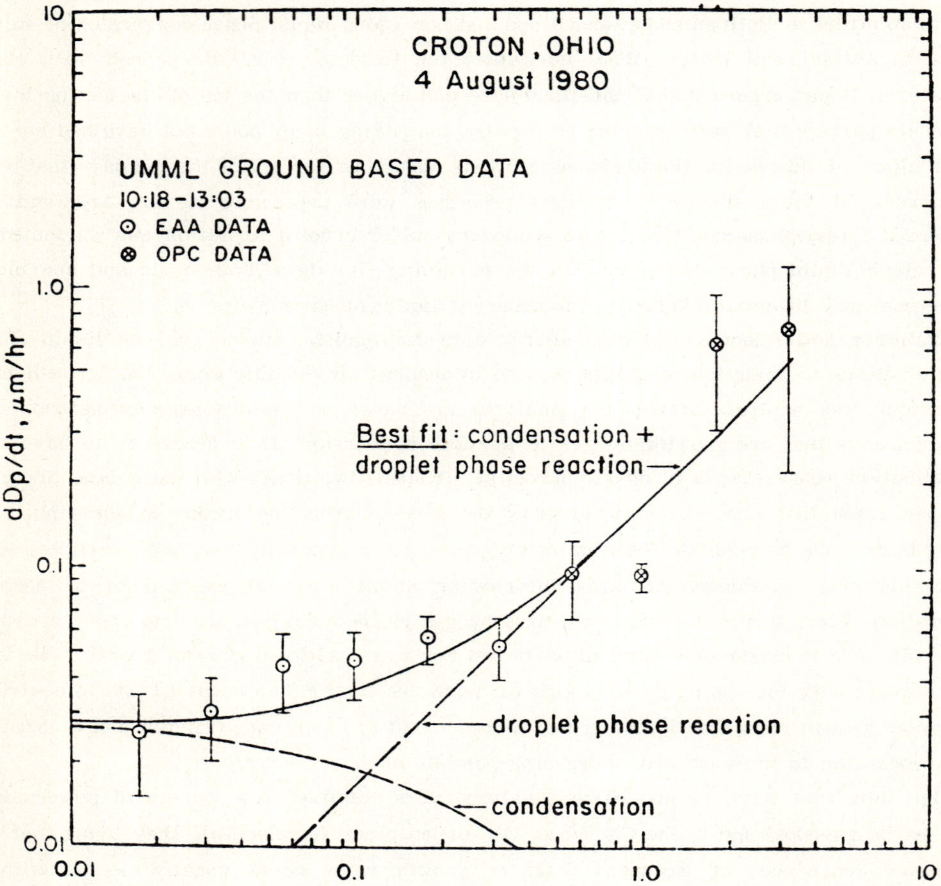

Figure 1. Comparison of data with a theoretical growth law that accounts for simultaneous particle growth by condensation and liquid phase reactions. (McMurry and Wilson, 1983).

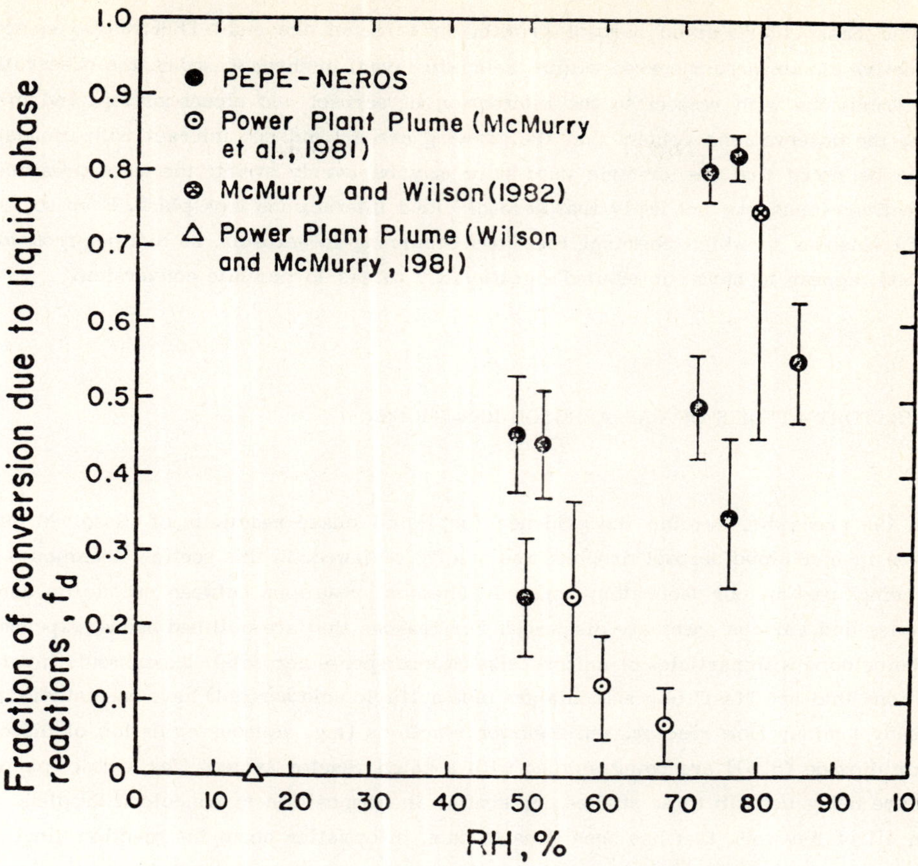

Figure 2. Fraction of the total secondary aerosol formation due to liquid phase reactions as a function of relative humidity. Note that for relative humidities exceeding about 75% liquid phase reactions tend to be predominant. (McMurry and Wilson, 1983).

growth law results, to try and determine whether or not chemical reactions in clouds might have played a role in observed liquid phase contributions to secondary aerosol formation. At this time these analyses have been done only for the data that were acquired in the Columbus urban plume during the PEPE/NEROS study (July, August, 1980). In order to determine if clouds might have had an influence on the evolution of the aerosol, soundings of temperature and dewpoint as a function of height and hourly observations of cloud type, cover and height were compared. (These data were acquired and provided by the National Weather Service). If convective clouds were observed within the mixed layer during the time that the aerosol was growing, it was concluded that processes occurring in clouds may have influenced the growth of aerosol. If the mixed layer was below observed clouds, it was concluded that cloud-aerosol interactions did not occur and that chemical conversion in aerosol particles was the cause of observed aerosol growth. In three of the eight PEPE/NEROS episodes, convective clouds were observed within the mixed layer; in three episodes, the observations were ambiguous with respect to the interaction of aerosol and clouds of fog; and in two cases, the observations suggest that the growing aerosol did not interact with clouds. It should be noted that the criteria used here may be overly strict: the mere presence of convective clouds may not imply that aerosol-cloud interactions took place. Even so, there remain episodes in which chemical reactions occurring in aerosols, as distinct from cloud droplets, appear to have contributed significantly to gas-to-particle conversion.

LABORATORY STUDIES OF GAS-AEROSOL REACTIONS

In the preceeding section the evidence for liquid phase reactions of dissolved sulfur dioxide in non-cloud aerosol droplets was briefly reviewed. In this section, techniques that are being used in our laboratory to study chemical reactions between submicron aerosol particles and various gases are discussed. For reasons that are outlined below, experiments are being done with particles of uniform size (monodisperse aerosols). Experiments involving reactions that are "fast" (e.g ammonia gas plus sulfuric acid aerosol) have been studied with a steady laminar flow reactor, while slower reactions (e.g. aqueous oxidation of dissolved sulfur dioxide (SO_2)) are being studied with a batch reactor system. The aerosol particles that are being used in these studies are similar in composition to non-cloud droplets.

In all of the work that has been done to date, information about the reaction kinetics is inferred from measurements of particle size changes associated with chemical reactions. In some cases, such changes occur directly as a result of the reaction. For example, if a molecule containing sulfur IV is oxidized in an aqueous aerosol droplet, another molecule will be absorbed from the gas phase in order to maintain Henry's Law equilibrium. When this occurs, additional water molecules may also need to be absorbed. As a result of these absorption processes, the droplet will grow (and may also change in composition). In other cases, the relationship between particle size change and chemical kinetics is less direct. In the case of NH_3 + sulfuric acid aerosols, for example, particles may actually give up water molecules as the aerosol is neutralized. If the humidity in the vicinity of the reacting droplet is sufficiently low, the droplet will eventually crystallize. It is possible that in such a case the resulting crystal will be smaller in volume than the original droplet. Rather than measuring size changes directly, the product aerosols in the ammonia-sulfuric acid study were humidified, and their composition was inferred from measurements of their deliquescent behavior.

Studies of these types have obvious limitations. For example, it is not possible to determine any direct information about the reaction products or intermediaries. However, it is also true that aerosol particles in the atmosphere do grow as a result of gas-to-particle conversion processes (McMurry and Wilson, 1983). During episodes of haze formation in the Eastern United States, aerosol droplets have been observed to double in diameter during the course of a single day (McMurry and Wilson, 1983). With the techniques described in this paper, size changes of less than 1% (corresponding to one molecular monolayer) can be measured for particles smaller than 0.1 μm. Therefore, if chemical kinetics studies suggest that a particular reaction is a significant pathway for secondary aerosol formation, the proposed scheme can be tested directly to see if it does indeed lead to aerosol growth. Furthermore, these techniques can be done at concentration levels (for aerosols and reacting gases) that are comparable to those found in the atmosphere. For example, for the studies of the ammonia-sulfuric acid reaction rate (McMurry et al., 1983) accurate data were obtained with aerosol concentrations as low as 1 ng/m^3. This is several orders of magnitude below concentrations that are typically required for chemical analysis.

In the following sections, experimental techniques that are being used in our laboratory to study gas-aerosol reactions are outlined. The capacity of these techniques to resolve changes in particles size is first discussed. The laminar flow and batch reactor studies are then briefly discussed. The emphasis will be on techniques rather than a discussion of results.

Instrumentation and Size Change Resolving Power

The differential mobility analyzer (DMA) (Liu and Pui, 1974; Knutson and Whitby, 1975) is the key instrument in these studies of aerosol chemistry. The DMA is used both to generate monodisperse aerosol particles that are fed into the reactors, as well as to measure particle size changes associated with chemical reaction. The use of the DMA for generating monodisperse aerosols was discussed in the above references. In this paper we focus on measuring changes in particle size.

A schematic of a typical experiment is shown in Fig. 3. A DMA is first used to generate monodisperse aerosols. These aerosols are then fed to a reactor (either a steady flow reactor or a batch reactor). The product aerosols are returned to a DMA for measuring size changes associated with any reactions that took place within the reactor. For steady flow experiments at least two DMA's are required: one for generating the aerosols, and a second for measurements on the product aerosols. Only one DMA is required for batch reactor studies. In this case, the DMA is first used to fill the reactor with a monodisperse aerosol, and then used to sample the aerosol from the reactor as the reaction progresses.

The size distribution of the aerosol that is produced when using a DMA to generate monodisperse aerosols has been described theoretically (Knutson and Whitby, 1975). Aerosols that are produced by the DMA are not perfectly monodisperse, but rather have a distribution of sizes that is determined by the flowrates through the DMA, and by the geometry of the instrument. A good approximation to this theoretically predicted distribution is the triangular "monodisperse" aerosol curve shown in Fig. 3. This distribution is quite narrow, so for practical purposes the reacting aerosol can be considered to be monodisperse. However, it is necessary to know the size distribution of these (nearly) monodisperse particles in order to maximize resolution in measurements of particle size changes.

In using a DMA to measure size changes for monodisperse reacting aerosols, the monodisperse aerosol is first passed through a bipolar charger (a "neutralizer"). At the exit from the charger, most of the particles have zero charge. A small fraction of the particles will have a single charge, and if experiments are being done with particles smaller than 0.1 µm (where this technique works best), a negligibly small fraction will be multiply changed. Only charged particles can be manipulated and measured with the DMA. Therefore, measurements of size changes are based on the measured mobility distribution of these charged particles. These mobility distributions are measured by determining the relationship between concentration at the monodisperse aerosol exit from the DMA and voltage applied to

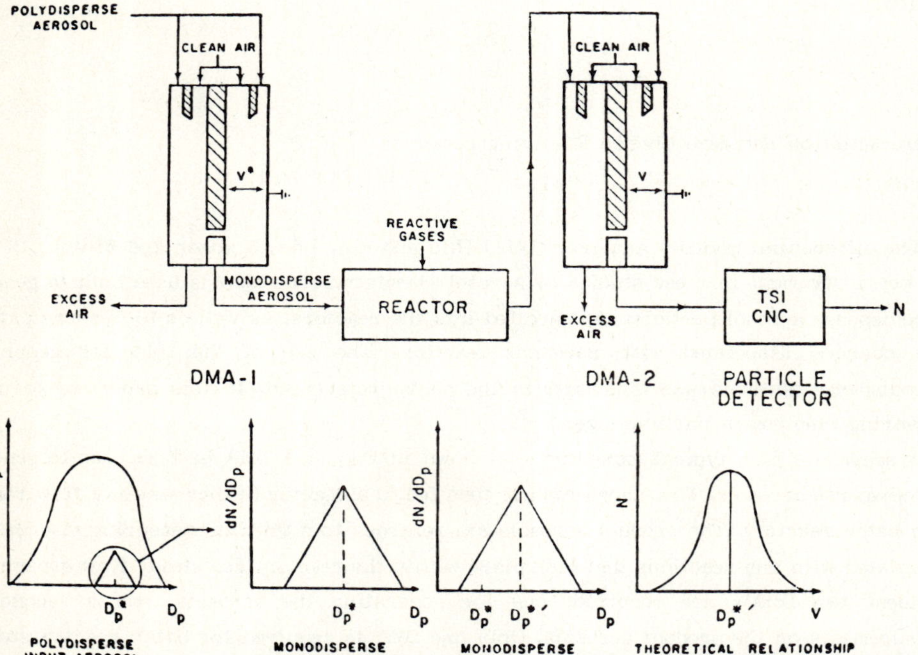

Figure 3. A schematic diagram of the approach used in studying chemical reactions involving monodisperse aerosol particles. Monodisperse aerosol is first produced with a differential mobility analyzer (DMA1) by selecting particles of a known size from any polydisperse aerosol. These particles are then reacted with various gases in a batch or steady laminar flow reactor. Size changes associated with the reaction are then measured, again with a differential mobility analyzer (DMA2).

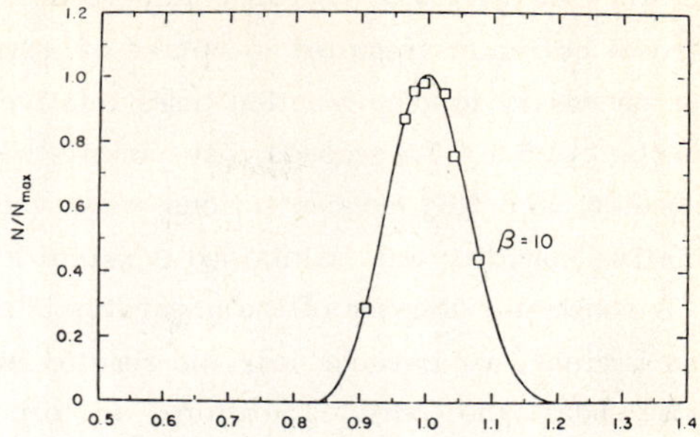

Figure 4. Concentration-voltage relationship at the monodisperse aerosol exit of DMA2 (see Fig. 3). These results apply when the input aerosol to DMA2 was monodisperse, and was produced by DMA1. The solid line shows the theoretical relationship between concentration and collecting rod voltage; several data points are also shown. Note that the theory and data are in excellent agreement. By fitting the theoretical curve to the data with a least squares technique, it is possible to determine the mean size of the reacted monodisperse aerosols with excellent precision.

the rod. As voltage is increased, the electrical force that attracts charged particles towards the center electrode on the DMA increases, thereby allowing larger particles (i.e. particles with smaller electrical mobilities) to penetrate through the exit slit on the center rod.

In recent work, Rader (1985) has done a theoretical analysis to calculate the expected relationship between number concentration leaving the DMA and voltage applied to the center rod for the case when a DMA-generated "monodisperse" aerosol is passed into the aerosol inlet of the DMA. Theory predicts that the concentration-voltage relationship should resemble a Gaussian distribution in shape (see Fig. 3). The theory has been compared with data that were obtained when a monodisperse aerosol generated with one DMA was measured with a second DMA as outlined above. Results of this comparison are shown in Fig. 4. Note that the theoretical relationship between the number concentration and the applied voltage is in excellent agreement with the data.

In future work, particle size changes will be determined by fitting the theoretical transfer function shown graphically in Figs. 3 and 4 to the data for concentration versus voltage. With this technique, the mean particle size of the monodisperse aerosol can be determined with excellent precision. Experiments are currently under way to determine the ultimate sensitivity of the technique, but it is known that diameter changes of less than 1% can easily be resolved with this technique. The key challenges to the experimentalist are to maintain flows at known and constant values, and also to determine number concentrations at the DMA exit with good accuracy.

Flow Reactor Studies

A series of three DMA's was used in a steady laminar flow reactor to measure the reaction rate between NH_3 gas and H_2SO_4 aerosol particles. Details of this study were reported by

McMurry et al. (1983). Because this reaction is fast (it was found that 20-50% of the NH_3-aerosol collisions resulted in uptake of ammonia molecules, depending on particle size), it was necessary to keep reaction times relatively short. In this study, reaction times were kept constant at 5.0 ± 0.5 seconds. Experiments were with aerosol particles from 0.03 to 0.2 μm in diameter, and NH_3 concentrations were varied from 13 to 63 ppb. In all experiments the relative humidity was maintained constant at 6%.

A schematic diagram of the apparatus is shown in Fig. 5. Because the flow in the reactor was laminar, air parcels near the reactor walls tended to have longer residence times than those near the core. Therefore, in order to ensure that reaction times were well characterized, a sample probe was installed along the centerline, and only 10% of the total flow was sampled through this probe.

The rate of reaction was determined by measuring hygroscopic properties of the product aerosols. The product aerosols were exposed to moist air within DMA3. Tang has shown that H_2SO_4 particles that are between 0 and 50% neutralized by NH_3 are hygroscopic at all humidities. Particles that are 50-75% neutralized deliquesce at 39% RH, the deliquescence point for NH_4HSO_4. Particles that are greater than 75% but less than 100% neutralized should deliquesce at 69%, the deliquescence point for letovicite $(((NH_4)_3H(SO_4)_2))$. These transitions are sharp functions for both composition and relative humidity.

It was reasoned that according to the laws of mass transfer, small particles should become

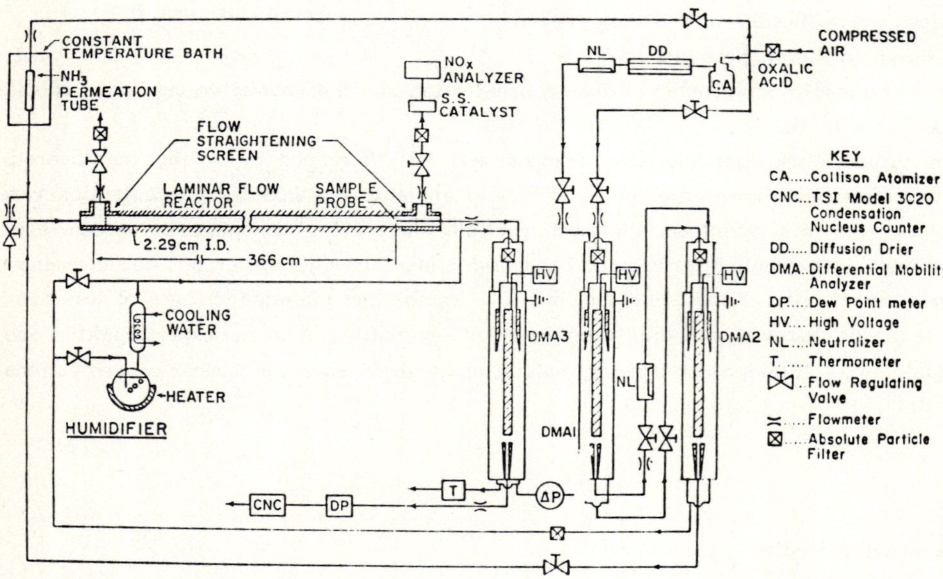

Figure 5. Schematic diagram of the steady laminar flow reactor system that was used in studying the reaction between ammonia gas and monodisperse sulfuric acid aerosols. In this case, two differential mobility analyzers were used to produce the monodisperse particles, and one was used to measure the hygroscopic behaviour of the product aerosols.

neutralized more quickly than larger particles. Therefore, experiments at a given reaction time and NH_3 concentration were done by titrating with respect to particle size (which is easy to vary experimentally). The smallest particle that was hygroscopic at relative humidities between 39 and 69% was judged to be 50% neutralized, and the smallest particle that was hygroscopic at relative humidities between 69 and 79% (79% is the point at which ammonium sulfate deliquesces) was assumed to be letovicite. Since both reaction time, ammonia concentration, and percent neutralization were known, it was possible to use the laws of mass transfer to determine the fraction of ammonia-aerosol collisions that resulted in chemical reaction. These results, along with results of experiments by Huntzicker (1980) and Dydek (1982) are shown in Fig. 6. The three data sets are consistent at 0.2 μm. Experimental techniques that were used by the other investigators were limited to particles larger than 0.2 μm, and the present study was limited to particles smaller than 0.2 μm. The reason for the minimum in the reaction efficiency at 0.2 μm is unknown.

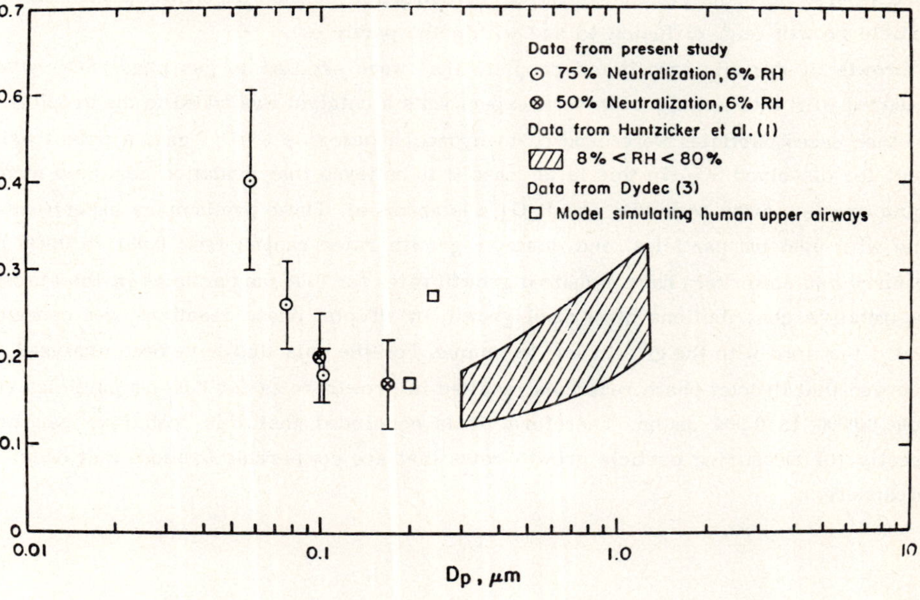

Figure 6. The fraction of ammonia gas-aerosol collisions than resulted in chemical reaction, α_r, as a function of particle size.

Batch Reactor Studies

When chemical reactions are "slow" (timescales of minutes to hours) a steady flow reactor is not practical. Instead, the gases and aerosol particles can be contained in a reactor and sampled periodically to measure particle size. The use of monodisperse aerosol particles in such batch reactor studies has significant advantages over other techniques that have been used to study the liquid phase conversion of dissolved gases in aerosol droplets. First, since conclusions are based on measurements of particle size changes, results are unaffected by wall deposition. This is important because wall deposition rates of submicron particles in such a reactor are likely to be on the order of 30-90% hr^{-1} (McMurry and Rader, 1985), much greater than likely rates of aerosol growth by liquid phase conversion. Also, since only a relatively small volume of air is required for a single measurement of particle size, numerous measurements can be made without seriously depleting the air in a reactor of modest size (in our work collapsible teflon film reactors that are about 250 liters in volume are being used). Therefore, particle sizes can be monitored for timescales on the order of several hours. This ensures that experiments can be done over timescales that are of interest in the atmosphere. Therefore, if a given reaction produces significant particle growth during the course of a day in the atmosphere, it will be possible to study this process under well controlled laboratory conditions with this technique. Finally, experiments with monodisperse aerosols permit systematic investigation of size-dependent phenomena during particle growth (e.g. diffusion to and within the particles).

Growth of monodisperse liquid droplets that were exposed to gas phase SO_2 have been measured with this technique. In some experiments a catalyst was added to the droplets, while in other cases particles were exposed to hydrogen peroxide (H_2O_2) gas, a potent oxidizing agent for dissolved SO_2. In this latter case it is believed that oxidation occurred within the liquid droplets after some SO_2 and H_2O_2 was absorbed. These preliminary experiments were done with 0.06 um particles, and diameter growth rates ranged from 0.001 to 0.039 μm/hr. McMurry and coworkers have measured growth rates for 0.06 μm particles in the atmosphere, The relative contributions of aerosol growth by droplet phase reactions and condensation were determined with the growth law technique. For the data that have been analyzed, it was observed that droplet phase reactions resulted in growth rates for 0.06 μm particles ranging from 0.0006 to 0.004 μm/hr. Therefore, it is concluded that this technique can be used directly for measuring particle growth rates that are comparable to those that occur in the atmosphere.

CONCLUSIONS

Liquid phase chemical reactions that lead to the formation of secondary aerosols may occur in clouds or in non-cloud liquid aerosol droplets. Two methods of data analysis have been applied to several data sets to show that secondary sulfur-containing aerosols can be formed by liquid phase reactions in non-cloud aerosol droplets. The evidence is not sufficiently strong to determine the overall importance of such reactions; similar analyses need to be done under a wider variety of meteorological conditions. Still, the finding is significant, and it demonstrates that the chemistry of non-cloud liquid aerosol systems must be considered.

Techniques that are being used for studying gas-aerosol reactions in this laboratory have been briefly discussed. The differential mobility analyzer (DMA) is the key to these studies. This instrument is used both to generate the monodisperse aerosol particles that are being used in the experiments as well as to measure size changes in these particles associated with chemical reactions.

The DMA is best suited for work with particles that are smaller than about 0.2 μm. It becomes increasingly difficult to generate monodisperse aerosols as particles sizes increase beyond this because particles tend to become multiply charged. So long as there is only one charge on particles, there is a unique relationship between the voltage applied to the DMA collector rod and the diameter of particles leaving the DMA. When multiple charging becomes important, however, particles of several sizes will be produced for each collector rod voltage setting.

The reaction between ammonia gas and sulfuric acid aerosols was studied with this technique. Because this reaction is fast, experiments were done in a steady laminar flow reactor. Product aerosols were sampled from the core of the tubular reactor to ensure that reaction times were well defined. It was found that between 20 and 50% of the ammonia-aerosol collisions resulted in chemical reaction.

Experiments to study growth of liquid aerosol droplets exposed to SO_2 are currently underway. Because these reactions are slow (timescales of hours), experiments are being done with a batch reactor system. Preliminary data show that particle growth rates similar to those of significance in the atmosphere can be measured with this technique.

ACKNOWLEDGEMENTS

We thank Ms. Rebecca Calby, Mesomet, Inc., for helpful discussions of meteorological data. The growth law research has been funded by the United States Environmental Protection Agency through cooperative agreement CR806801-01 to the University of Minnesota. It has not been subjected to the Agency's required peer and policy review, and therefore does not necessarily reflect the views of the Agency, and no official endorsement should be inferred. The laboratory research is being supported by Grant ATM-8113156 from the United States National Science Foundation. The work has not been reviewed by this agency and no official endorsement should be inferred.

REFERENCES

Dydek S.T. (1981), Neutralization and Size Changes of Sulfuric Acid Mist Particles in a Model of the Human Upper Airways, Ph.D. Dissertation, University of North Carolina, Chapel Hill, NC.

Gillani N.V., Kohli S., and Wilson W.E. (1981), Gas-to-particle conversion of sulfur in power plant plumes-I. parametrization of the conversion rate for dry, moderately polluted ambient conditions., Atmos. Environ., 15, pp 2293-2312.

Gillani N.V. and Wilson W.E. (1983), Gas-to-particle conversion of sulfur in power plant plumes-II. observations of liquid-phase conversions., Atmos. Environ., 17, pp 1739-1752.

Huntzicker J.J., Cary R.A., Chaur-Sun Ling (1980), Neutralization of sulfuric acid aerosol by ammonia. Environ. Sci. Technol., 14, pp 819-824.

Knutson E.O. and Whitby K.T. (1975), Aerosol classification by electric mobility: apparatus, theory and applications., J. Aerosol Sci., 6: pp 443-451.

Liu B.Y.H. and Pui D.Y.H. (1974), A submicron aerosol standard and the primary, absolute calibration of the condensation nuclei counter., J. Colloid Interface Sci., 47, pp 155-171.

McMurry P.H. and D.J. Rader (1985), Aerosol wall losses in electrically charged chambers., Aerosol Sci. Technol. I, in press.

McMurry P.H. and Wilson J.C. (1983), Droplet phase (heterogeneous) and gas phase (homogeneous) contributions to secondary ambient aerosol formation as functions of relative humidity., J. Geophysical Res., 88, pp 5101-5108.

McMurry P.H., Takano H. and Anderson G.R. (1983) Study of the ammonia (gas)-sulfuric acid (aerosol) reaction rate., Environ. Sci. Technol. 17, pp 347-352.

Rader D.J. (1985) Ph.D. dissertation research, Particle Technology Laboratory, Department of Mechanical Engineering, University of Minnesota, Minneapolis, MN 55455.

Stelson A.W. and Seinfeld J.H. (1982), Thermodynamic prediction of the water activity, NH_4NO_3 dissociation constant, density and refractive index for the NH_4NO_3-$(NH_4)_2SO_4$-H_2O system at 25 C., Atmos. Environ., 16, pp 2507-2514.

CHARACTERIZATION OF AIR PARTICULATE MATTER BY X-RAY, RAMAN AND MASS SPECTROMETRIC TECHNIQUES

R. Van Grieken and F. Adams
Department of Chemistry
University of Antwerp (U.I.A.)
B - 2610 Antwerp-Wilrijk
Belgium

ABSTRACT

In recent years, X-ray emission spectrometry with X-ray- tube, radioisotope or charged-particle excitation, and with energy- or wavelength-dispersive detection, has been extensively applied to the direct multi-element analysis of aerosol loaded filters or aerosol deposits from impactors. Spark source mass spectrometry has been invoked for truly panoramic analysis of large aliquots of air particulate material.

The electron microprobe with X-ray detection attachment, the new laser microprobe mass analysis instrument and the Raman microprobe are now being used to study the morphology, elemental composition, inorganic speciation and organic fingerprint of individual aerosol particles or of particle coatings.

Of all these techniques, the principles, instrumentation and methodological advantages and limitations are discussed and their applicability in the field of aerosol research is illustrated.

INTRODUCTION

The elemental analysis of bulk or size-fractionated air particulate matter represents an unique problem. The total amount of sample material is usually very small but often may samples have to be analysed and many elements in widely differing concentration and atomic number are of interest. Therefore, emphasis should be placed on techniques that combine high speed, low cost, advantageous sensitivity and multi-element character. Not only is rapid and multi-element analysis necessary to handle the large sample loads often encountered, but

NATO ASI Series, Vol. G6
Chemistry of Multiphase Atmospheric Systems
Edited by W. Jaeschke

also these characteristics help to reveal significant correlations between different pollutant components. In addition, the ability to measure more elements than just a few pollutants often greatly increases the usefulness of the results.

As is the case in many other fields of application, atomic absorption spectrometry has been used most often for the analysis of air particulate material. However, this is a single element technique, and like the now popular inductively-coupled plasma emission spectrometry, it suffers from the need to dissolve the particulate matter prior to the analysis step. Other multi-element spectrometric methods like DC-arc emission spectrometry suffer from limitations in sensitivity and precision. Therefore instrumental neutron activation analysis has been used extensively. A recent literature survey lists not less than 225 references (Fox and Jeffries, 1981). A drawback of neutron activation is that toxicologically important elements such as lead and cadmium are insensitive. X-ray spectrometry (XRS) offers an attractive alternative because it allows a satisfactory sensitivity and provides reliable results for some 10-20 elements simultaneously. The difficulties inherent to XRS, namely in the data analysis and matrix absorption correction, have gradually been solved by computer analysis. Spark source mass spectrometry (SSMS) has also been invoked in the literature on particulate matter analysis, albeit less frequently, mainly because it offers unique truly multi-element analysis.

Analytical results on microgram or milligram quantities of air particulate material are statistically representative for a very large number of airborne particles. In principle the most definitive information can only be extracted by the analysis of a large number of individual particles. Indeed, with such data it is possible to obtain, in a direct and straightforward way, information about the sources of pollution and their relative strength, about the changes in the aerosol as it travels from its source, and on the potential effects of the particles. The techniques that have been used most frequently for single particle analysis are electron probe micro analysis (EPMA), the new technique of laser microprobe mass analysis (LAMMA), and the Raman microprobe analyser or molecular optical laser examiner (MOLE).

For the surface analysis of particles, electron spectroscopy (X-ray photoelectron spectroscopy and Auger electron spectrometry) and secondary ion mass spectrometry (SIMS) have been used.

In the present report, the principles, instrumentation, advantages and limitations, and applicability to air particulate matter, will be discussed for the techniques based on X-ray, Raman and mass spectrometry, namely XRS and SSMS for bulk samples and EPMA, LAMMA, MOLE and SIMS for single particles. Most of the illustrative examples will be chosen from applications that have been studied at the authors' laboratory.

X-RAY EMISSION SPECTROMETRY METHODS

Principle

X-ray spectrometry has been in use as an analytical technique for several decades. Its principles are very simple. When atoms of the sample to be analysed are irradiated with particles or electromagnetic radiation of sufficient energy, inner shell ionization can occur. A higher level electron will fill up the vacancy, while the difference in energy between both levels is released as a characteristic X-ray (or an Auger electron is emitted). When the vacancies in the inner electron shells are created in a controlled way and the energy and intensity of the resulting X-rays are measured, quantitative analysis becomes obvious. A more thorough description of the X-ray fluorescence can be found in several handbooks such as those of Birks (1969), Jenkins (1974), Bertin (1975).

Instrumentation

The various XRS-techniques can be subdivided according to the detection and excitation type. The older, more conventional XRS-instruments are of the wavelength-dispersive (WD) detection type. The characteristic X-rays emitted by the sample are angularly dispersed, by an analysing crystal or pseudocrystal, according to Bragg's law:

$$n\lambda = 2d \sin \Theta$$

where n is an integer, λ is the radiation wavelength, d is the interlattice spacing of the crystal and O is the angle of incidence of the radiation. Usually, the radiation of a particular wavelength is detected at a certain angle by either a gas flow proportional counter (for long wavelengths) or a NaI(Tl) scintillation counter (for the short wavelength range). Various crystals are commercially available; usually one with a d-value such that $20° < \Theta < 80°$ is selected. Lately in WD-XRS instruments, the choice of parameters, crystal change and sample change have been automated thoroughly, so that automated sequential analysis may be performed. Also, multichannel instruments are available with a dozen or more crystals and detectors located around the sample.

Since Bowman et al. (1966) reported that the resolution of solid-state Si(Li) detectors had become sufficient to resolve X-ray peaks of adjacent elements in the periodic table, a radically new XRS technique has emerged, namely the energy-dispersive (ED) type. This design eliminated the need for an analysing crystal and for moving parts in the instrument; the undispersed characteristic X-ray beam impinges on a liquid nitrogen cooled Si(Li) detector that produces pulses proportional to the individual X-ray photon energies. Alternative detectors for ED-XRS which have drawn much attention recently are the high-purity Ge detector, which is more efficient than Si(Li) at high energies, and the HgI_2 detector which does not require liquid-nitrogen cooling and is thus suitable for portable instruments, but offers an unfavourable resolution of 300-400 eV at 5.9 keV. The advantages of WD- and ED-XRS are compared in Table I.

Excitation by electrons was applied already by von Hevesy (1932) and it is revived in modern electron microprobe instruments (see below) which offer an excellent spatial resolution, but rather unfavourable relative detection limits.

Excitation by electromagnetic radiation can be achieved by conventional X-ray tubes that emit Bremsstrahlung spectra; which is usually the case in WD-XRS instruments. In ED-XRS, X-ray tubes are often combined with appropriately chosen secondary fluorescers and filters in order to reduce the continuum Bremsstrahlung background; the radiation from a high power tube is then focused on a secondary target material (usually a pure metal foil) which in turn emits a high flux of its characteristic X-rays onto the sample. Since optimal sensitivities are achieved when the energy of the exciting radiation is slightly above the absorption edge for the element(s) to be determined, usually various secondary targets are available. In the authors' laboratory, secondary excitation by Mo, has mostly been applied as an optimal compromise for wide elemental range and low detection limits, e.g. in the case of air particulate analysis. In some cases pulsed X-ray tubes have been used (Jaklevic et al.,

Table I: Comparison of the advantages of wavelength- and energy-dispersive detection in XRS

Advantages of wavelength-dispersion XRS	Advantages of energy-dispersive XRS
High resolution; little peak overlap; simple peak evaluation	Simultaneous analysis, faster multi-element screening;
Simple background correction for quantitative analysis	Less chances of missing unexpected pollutants
Advantageous detection of low-Z elements, down to B in some cases	High detection efficiency; can be combined with low-power X-ray tubes and low beam current; little sample damage

1977) for ED-XRS, or transmission X-ray tubes, in which the high energy continuum background is reduced while the primary intensity is lower due to the cooling requirements.

The use of heavy charged particles has found much interest since the pioneering work of Johansson et al. (1970); the use of accelerator produced particles in XRS is referred to as particle-excited X-ray emission analysis or PIXE. The particles are usually 2-4 MeV protons or alpha particles produced by a Van de Graaff or cyclotron accelerator. Such expensive machines are usually not specifically installed for the purpose of PIXE-analysis, but once installed, it is easy to include X-ray equipment in these setups. Since proton beams produce little Bremsstrahlung background, high particle fluxes are available and protons have a high cross-section for X-ray production, a very advantageous detection limit can usually be reached. Because proton beams can easily be focused, the detection limits are more advantageous, relative to those obtained with tube-excited XRS, especially for small-size targets, such as particulate deposits from cascade impactors. Particularly when numerous targets are to be analysed, as can be the case in extensive air pollution surveys, extending the counting time within practical limits in conventional XRS cannot nearly result in a compensation for the lower detection limits in PIXE. Table II gives a comparison of detection limits obtained with PIXE and with secondary target XRS for impactor samples.

Table II: Detection limits for ED-XRS and PIXE analysis of aerosol impactor samples. Comparison with typical environmental concentrations of air particulate material (Van Espen et al., 1981).

Element	Detection limit (ng)			Typical atmospheric concentration (ng/stage)	
	ED-XRS with		PIXE		
	Ti- excitation	Mo- excitation		Polluted	Remote
Al	100	-	20	4000	2000
S	20	400	9	10000	6
Cl	20	100	8	3000	3
Ti	-	20	2	700	100
Mn	-	7	0.8	300	20
Cu	-	4	0.6	200	1
As	-	3	0.8	20	0.05
Br	-	2	0.6	200	0.06
Sr	-	3	2	-	9
Pb	-	5	2	1300	0.3

In spite of their 10^6-10^7 times lower primary intensity, radio isotopes emitting gamma- or X-radiation, such as ^{55}Fe, ^{109}Cd, ^{241}Am, ^{57}Co, ^{153}Gd, have also been used as exciation sources in ED-XRS for air particulate material, because of their low cost, ability to excite high-energy X-rays and portability (La Brecque, 1981).

Comparisons of detection limits of various X-ray techniques for air particulate matter have been given in several publications, e.g. Gilfrich et al. (1973), Jaklevic et al. (1976).

The choice of the excitation-detection system in XRS is certainly not a trivial one, and will depend on the number of elements to be determined, the desired detection limits and precision, the number of targets to be analysed, the equipment already available and the price of the equipment to be purchased. Gilfrich et al., (1973) have performed a comparison of several instrument types. In general, if one is to routinely determine a dozen or more elements in many air particulate samples, energy-dispersive spectrometers should seriously be considered, and, if possible, PIXE. With fewer elements, wavelength-dispersive instruments can compare favourable, if automated.

XRS-methodology for air particulate analysis

Spectrum evaluation

In WD-XRS it is usually sufficient, in view of the high resolution, and little overlap, to measure the background area between peaks and to subtract this average background to obtain net peak intensity. In ED-XRS severe overlap of X-ray peaks can occur. The K_α X-rays of an element with atomic number Z often overlap with the K_β -rays of an element with atomic number Z-1, e.g. Co-K_α and Fe-K_β, and K-lines of an element sometimes overlap with L-lines of another element, e.g. As-K and Pb-L. Rhodes (1979) has reviewed 71 references for the 1968-1978 period concerning spectrum acquisition and analysis. When utilizing a least-square fitting routine, the experimental spectrum is fitted to a linear combination of pure element spectra. There are many mathematical ways to accomplish this, as discussed by Arinc et al. (1977). For example Schamber (1977) has applied a simple digital filter operator (digital cross correlation function) in order to make the linear least-square method independent of an explicit model for the continuum background. Since linear least-squares techniques are based upon the assumption that peak profiles are exactly known, significant errors can occur if the spectrometer system used cannot maintain sufficiently constant peak positions, widths and overall shape. This difficulty has stimulated the design of programs based on non-linear least-squares fitting procedures, which are able to account for resolution and gain shifts, at the expense, however, of using more complex mathematics and requiring longer calculation times and/or more powerful computers.

Van Espen et al. (1977a, b, 1979) have described a Fortran program called "Analytical X-ray analysis by Iterative Least-squares" (AXIL) for routine analysis of ED-XRS spectra from secondary target excitation and from PIXE. In this version the number of counts at every energy is given by adding the contributions from the background, from a Gaussian part of each line for each element, and from an empirically determined non-Gaussian part which is proportional to the area of the main peak. Fig. 1 illustrates for an air particulate spectrum excited by a Mo secondary target, how well this routine can evaluate extremely complex parts of ED-XRS spectra, with extreme spectral overlap.

Matrix effects

Emerging characteristic X-rays, particularly those from low-Z elements, can severely be attenuated by the air particulate material itself, and the impinging exciting electromagnetic radiation or particle beam can be attenuated or slowed down to some extent. The effect of secondary enhancement by the characteristic radiation of a major element, which is often important in bulk analysis, can mostly be neglected for thin layers of air particulate

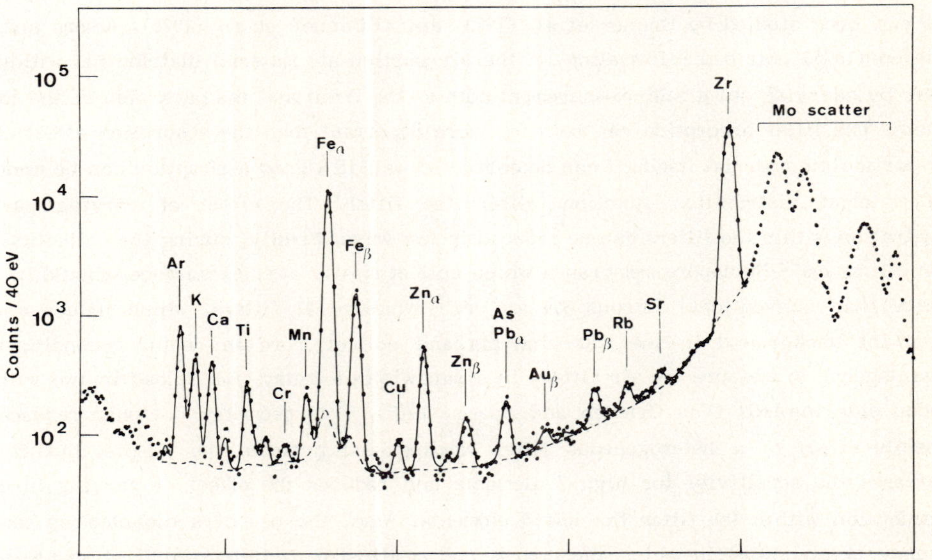

Fig. 1: Fit to a complex molybdenum secondary excitation spectrum. The dashed line is the background, while the full line is the fitted sum of the background and the response functions of all the elements. The Zr-signal is from an external reference wire.

material. In aerosol analysis, however, additional problems arise because of the particulate nature of the material. Moreover, the absorption by the filter, into which the particles penetrate to a certain extent when the collection is done by filtration, require an additional correction, while the lateral heterogeneity of the aerosol deposit can represent an additional problem when cascade impactors are used for collection. An extended literature is available on these subjects, as was recently reviewed by Van Grieken and La Brecque (1985), and discussed in a general paper on XRS-analysis of filter-collected aerosol particles by Dzubay and Rickel (1978). In practice the effects are only significant for low-Z elements, e.g. below Fe, when the K-radiation is observed.

Conventional procedures for matrix effect corrections, mainly for bulk materials, can be found in all reference works on XRS. Generally, when the mass-per-unit-area of the material is known (e.g. from weighing) and its composition is known or roughly constant, a correction for the attenuation in the material itself can be carried out. The X-ray absorption can be calculated from literature data (McMaster et al., 1969) for an assumed composition, or be determined by transmission measurements (Van Dyck et al., 1980). Also they can be derived from the ratio and intensity of the coherent and incoherent scatter peaks which are always prominent in secondary fluorescer ED-XRS spectra (Van Dyck and Van Grieken, 1980). However, for low-Z elements like Na, Mg, Al and Si, the matrix absorption effect is so large that reliable XRF-analyses are not possible. For the particle size correction, Criss (1976) has employed a simple equation, which requires the knowledge of the particle size or particle size distribution.

The effect of X-ray absorption by the air filter material, which is sometimes neglected, has e.g. been studied by Bonner et al. (1973) and O'Connor et al. (1975). Adams and Van Grieken (1975) obtained information on the air particulate material distribution within the filter by carrying out a XRS-measurement both at the front and the back side of the loaded filters. The filter absorption can be much more important than the absorption effect of the air particulate material itself; it can be corrected well if a good assumption can be made for the material distribution function within the filter. The effect of varying particle penetration within the filters can be reduced in two ways. Firstly, during the collection step thin 1.1 mg cm^{-2} Nuclepore membranes which collect mostly at their surface, should be used rather than conventional fibrous 8.5 mg cm^{-2} Whatman-41 filters, which induce a more important background in the XRS-analysis and collect more in depth. Secondly, it is advantageous to measure the air filters in a sandwich geometry, i.e. folded in two with the loaded side inwards (Van Grieken and Adams, 1976). This geometry not only reduces the possible effect of a heterogeneous filter loading and protects the sample, but it also increases the sensitivity for high-Z elements and reduces the effect of varying in-depth distribution within the filter for low-Z elements. Also, the use of a dichotomous sampler (Dzubay et al., 1976) in which the coarse and small-size particles are first separated by inertia and then collected separately on e.g. Nuclepore filters, greatly simplifies the correction for self absorption.

The lateral heterogeneity of a cascade impactor deposit does usually not represent a problem in PIXE where a homogeneous proton beam larger than the deposit area can easily be obtained (Johansson et al., 1975). In e.g. secondary-target ED-XRS however, the excitation detection efficiency is also heterogeneous over the aerosol deposit area, thus necessitating the use of empirical scaling factors which differ for every stage of the impactor and are determined by comparison with other analysis techniques (Van Espen et al., 1981).

Calibration

There are basically two possibilities for calibration: one is based on the use of suitable calibration standards for every element to be determined, the other relies on the use of fundamental physical constants. Both methods have advantages and drawbacks, and for both the radiation absorption in the sample may become significant for samples or standards of finite thickness. The standardisation procedure for tube-excited ED-XRS with secondary target geometry, adopted in our laboratory, relies on:

(1) use of a minimum number of thin uniform calibrated deposits,

(2) the knowledge of a consistent set of fundamental physical parameters and

(3) the assumption that the excitation spectrum is sufficiently "monochromatic" in the secondary target set-up, so that excitation by the remaining background continuum of the tube can be neglected.

Straightforward calculation in this manner provides calibration data with an accuracy comparable to that obtained experimentally with commercial standard samples or standards prepared by a carefully optimized procedure in the laboratory. A full account was given by Van Espen and Adams (1981).

Commercially available standards consisting of pure elements or simple inorganic compounds, vacuum evaporated on a 4 µm Mylar backing can be purchased from Micromatter Co, Seattle, Wa. and Columbia Scientific Instruments, Austin, Tx., USA. The accuracy of the weight per unit area is specified to be better than 5%. Filter standards prepared from aerosols, created by blown pulverised materials, are also commercially available from Columbia Scientific Instruments. Another type of standards consists of multi-element thin film standards obtained by mixing a solution of different elements with a short-lived radioactive tracer and a water soluble polymer; a film is produced by spreading the liquid mixture evenly on glass plates (Billiet et al., 1980).

XRS-applications in aerosol analysis

Although the analysis of air particulate material, particularly for low-Z elements, is not a trivial task, as appeared above, most problems can be overcome by using modern sampling methods and computer programs for performing the necessary corrections. The widespread utilization of XRS for aerosol analysis is evidence to the novice in the technique that, with careful use, XRS is a very viable method. More than 500 XRS-applications to air pollution analysis have been published in the last few years (Van Grieken and La Brecque, 1985). A complete review of all applications is hence not possible here. Only a few selected XRS-applications in air particulate studies can be discussed.

In view of its simplicity, the use of tube-excited ED- and WD-XRS seems, in general, preferable when a limited number of targets is to be analysed, particularly in the case of loaded filters. These methods are used as routine air quality monitoring tools by many environmental agencies.

The exciting flux in a radioisotope excitation is usually not sufficient to offer a great sensitivity, but still, on loaded filters, a suite of elements can often be assessed simultaneously especially in polluted air (Panadellis and Katselis, 1978) and the use of portable instruments for on-site air pollution measurements becomes easy: e.g. Rhodes et al. (1983) have described the application of two such portable instruments in industrial hygiene problems.

Because of the possibilities of collimating the exciting particle beam and achieving some spatial resolution, PIXE has been used in combination with an automatic time-sequence filter sampler called "streaker", developed by Nelson et al. (1979). This device holds a 200 m x 5 mm strip of Nuclepore filter. A 2 mm x 5 mm orifice located against the filter strip pulls air through the filter, while moving continously along the filter (at 1 mm/hr). A time resolution of 2 h can thus straightforwardly be obtained by PIXE.

More frequently PIXE has been used on impactor deposits of aerosols because of its impressively low detection limits for small-size samples as was already illustrated in Table II. Because of the high sensitivity, deposits from small air volumes can be analysed, allowing the use of battery operated pumps, hence of sampling in very remote environments. The present authors have been involved in PIXE applications to study aerosols from remote continental (Adams et al., 1983), desert (Annegarn et al., 1983) and marine (Johansson et al., 1976) environments.

Quite spectacularly and successfully, PIXE has been applied to analyse size-fractionated aerosol samples collected from a nearly metal-free sailboat made predominantly of wood and polyester (Maenhaut et al., 1981 and 1983). Sampling was done by drawing air with battery powered miniature pumps, through two parallel 1 l/min cascade impactors, suspended from the mast at 8 m above the sea surface. The sampling period was two days continuously for every sample set. The sailboat track was from Europe to West-Africa, through the Panama Canal to Polynesia; it took 15 month all together. Many conclusions could be drawn, e.g.:

- the precision of PIXE for impactor deposits is such that the ratio of two subsequent analyses on the same targets, averaged over the 10-15 detected elements, was 1.00 to 1.02, with a standard deviation per measurement of 3 to 11%.
- the reliability of the sampling step and the accuracy of the PIXE-analyses are such that the concentrations obtained from the two parallel impactors, usually deviated by less than 20%.
- the detection limit values for a pure marine aerosol and added for 3 impactor stages (corresponding to the 0.25 - 2 µm aerodynamic size range) analysed by 5 µC of 2.4 MeV protons are e.g., in ng m^{-3}, for V: 0.13, Cr: 0.09, Mn: 0.07, Ni: 0.04, Cu: 0.05, Zn: 0.3, As: 0.2, Se: 0.15, and Pb: 0.9.
- the crustal elements were found to decrease significantly with distance to the continent. Between the Galapagos Island and Polynesia, a maximum of 0.07 µg m^{-3} of crustal material was found with average Si and Fe levels of only 4.8 and 3.3 ng m^{-3}, respectively.
- for this purely marine air, the excess submicrometer S decreases to a mean of 45 ng m^{-3}, and the excess fine K to 5 ng m^{-3}, suggesting predominant continental sources for these components.
- under truly marine conditions several concentration ratios of the major seawater elements were significantly different from those in bulk seawater: Ca, Sr and S were enriched relative to K and Na by up to a factor of 2 for 1-4 µm diameter particles.
- chlorine and bromine were only slightly depleted relative to bulk seawater in the pure marine aerosols (e.g. only 5 to 10% for Cl).
- all heavy metals were below the detection limits in pure marine air, e.g. below 0.05 ng Cu m^{-3} and 0.3 ng Zn m^{-3}, ruling out enrichment effects of more than 1000-fold in the sea-air transfer process.

SPARK SOURCE MASS SPECTROMETRY (SSMS)

Principle

Inorganic analysis by SSMS is typically performed by drawing a high-voltage radio-frequency spark between two electrodes which contain the sample material. The resulting atomic and molecular ions of various charges are accelerated into a mass spectrometer for ion separation according to the m/e-ratio and for detection with an ion sensitive emulsion (photoplates) or by an electric detection system. Several reviews on the principles, instrumentation, drawbacks and advantages of SSMS are available in the literature (Ahearn, 1982; Cornides, 1978; Morrison, 1980; Adams, 1982).

Instrumentation

The basic components are normally a r.f. spark discharge, and a double focussing mass spectrometer. The r.f. spark process provides a repetitive series of single high-voltage breakdowns in which the ions are generated. It makes use of an oscillating potential difference of 20-100 kV at ca. 1 MHz, which is generated in short pulse trains of variable length (20-200μs) and with a repetition frequency of 1-10^4 s^{-1}. The spark ion source was originally selected because of its prolific production efficiency of singly charged atomic species with comparable ionization efficiency, and therefore similar sensitivity, for all elements. However, the r.f spark is a discontinuous and quite erratic ion source; the ionized species are formed with a considerable energy spread, typically 1 keV or more, and several other ionized species complicate the measurement of the singly charged atomic species (M^+): multiply charged ions (M^{n+}), cluster ions (M_n^+) and complex (molecular) ions (e.g. MO^+, MC^+).

These features require the use of a double focussing mass spectrometer of high mass resolution with simultaneous integrating detection of the entire mass spectrum, which is performed with a spectrograph of the Mattauch-Herzog design. After acceleration (20-30 kV), part of the energy distribution is selected (200-600 eV) with the electrostatic sector. These ions then enter the magnetic field and follow a path with a radius of curvature proportional to the square root of ion mass. They are brought to focus along a flat plane onto the photoplate.

Taking into account the low dynamic range of the photographic plate, one analysis consists of a series of sequential exposures. Typically, 15 exposures are used to cover the entire concentration range. Apart from its poor dynamic range, photoplate detection suffers from poor measurement precision and difficulties in extracting quantitative spectral information. Densitometry and calibration of the emulsion is necessary to achieve a precision of 5-20%. There has been an increasing tendency to use computer controlled microdensitometers. Fully implemented systems (see, e.g. Vanderborght and Van Grieken, 1978; Pilate and Adams, 1980) provide for the following steps: identification of the mass lines, densitometry on the exposure stages, determination of the calibration curve and the calculation of concentrations. Reading and processing of a single plate may require several hours.

For some applications electrical detection with an electron multiplier is preferred. Mass resolution with electrical detection is in general considerably worse than photoplate detection because the slit widths need to be increased to obtain stable measurement conditions. Moreover, electrical detection is more sensitive to fluctuations in the spark process and to sample heterogeneities. Advantages are the increased precision with peak switching compared with the photoplate detection mode, and the speed with which data become available. These factors and poor ion statistics limit the precision of mass scanning to worse than 30% relative standard deviation. Significant improvements are possible by repetitive scanning and signal averaging.

Promising recent advantages include the use of an inductively coupled plasma source, hallow cathode discharge or laser ionization, and of position-sensitive electro-optical ion detectors.

Methodology

SSMS is now well established for the determination of trace and minor constituents in many different types of material (semiconductors, metals, environmental, geological and biological samples).

The merits of SSMS include:

(i) simultaneous detection capability of nearly all elements and the availability of isotopic information;

(ii) extremely wide concentration range: major, minor and trace constituents down to concentration levels of ca. 10 ng g^{-1} can be detected;

At present the method has however, failed to establish itself as a routine analytical technique, principally because the instrumentation is present only in a few well equipped research laboratories. In fact, only JEOL Ltd., Tokyo, manufactures a SSMS instrument at present. Also statements are frequently made in the literature about the insufficient accuracy of the results for many purposes, and the cost of the apparatus and its maintenance.

If sufficient quantities are available the sample preparation of air particulate for SSMS can consist of directly mixing the material with a conducting powder at a 50-80% concentration, and compressing to electrodes. High purity graphite is mostly preferred as a matrix. Aerosol-loaded filters are ashed, e.g. in a low temperature ashing unit, redissolved in acid, and mixed with graphite (Pilate and Adams, 1981).

Calibration and matrix correction factors are usually determined by analysing the matrix material doped with known amounts of the elements to be determined, and by using relative sensitive coefficients of the elements relative to a matrix element or an element whose concentration is known in the sample to be analysed. Relative sensitivity coefficients are now available for ca. 50 elements in graphite (Vanderborght and Van Grieken, 1979), with a precision of ca. 20%. They do however depend on the instrument and the conditions used.

SSMS-applications in aerosol analysis

In view of all its limitations, SSMS has only seldom been proposed for aerosol analysis (Brown and Vossen, 1970; Pilate and Adams, 1981) and, to our knowledge, not been invoked on a large-scale and routine basis. Its major advantage, namely the capability for truly panoramic and sensitive multi-element determinations, imply that in e.g. typical municipal incinerator samples, some 30 elements can be assessed simultaneously, above the typical detection limits which are often around 0.4 ppm for a 400 nC plate exposure and typical sample preparation conditions (Pilate and Adams, 1981).

ELECTRON MICROSCOPY AND ELECTRON PROBE MICRO-ANALYSIS

The analytical tools that have most often been used for individual microparticle analysis are the electron beam instruments: the electron probe micro-analyser (EPMA), the scanning electron microscope (SEM) and the transmission electron microscope (TEM), equipped with wavelength- or energy-dispersive X-ray spectrometers (WD and ED). In these instruments the sample is bombarded with a focused, high energy electron beam and the resulting characteristic X-rays are examined, together with the abundant concomitant Bremsstrahlung radiation. In general, these techniques allow individual particle analyses for elements above C at concentration levels above 500 ppm or elemental quantities above 10^{-15} g i.e. the relative detection limits are much less advantageous than for XRS (see above), or for LAMMA (see below). Simultaneously however, the electron beam techniques provide detailed information on the morphology and size of the particles by using the secondary electron signals and also on the topography and average composition via the backscattered electron signal.

Several reviews on the capabilities of electron beam analysis are available (e.g. Andersen, 1975; Chandler, 1977; Goldstein et al., 1981). Although EPMA, SEM and TEM provide similar data, the EPMA is optimized and superior for micro-analysis and analysis of particles larger than 0.5 µm, due to differences in design (a more stable electron beam, a more intense electron flux, higher spectrometer resolution and an optical system and sample stage allowing more precise positioning of the specimen). The SEM with ED- or WD-attachment is, however, the instrument of choice for the morphologic study of particles and for the analysis of particles smaller than 0.5 µm in diameter.

In fact, the term electron microscope (SEM or TEM) has long been used to describe electron beam instruments whose primary function is electron imaging of samples, while the term electron microprobe (EPMA) has been used to describe instruments fitted with X-ray spectrometers, whose primary function is the X-ray emission analysis of microvolumes in a sample. Formerly the design characteristics of the SEM/TEM and EMPA prevented their effective use for each other's primary function, but recent advances have led to the decrease of the distinction between the two instrument types.

Fig. 2 represents setup of the EPMA-instrument JXA-733 Superprobe of JEOL Inc., which has been used recently in our laboratory for particle analyses. It includes a column with an electron gun, column liner tube, gun alignment coils, gun airlock value, condenser lens, magnetic shield, objective aperture and probe current detector and an optical microscope, a secondary electron detector, a backscattered electron detector, two double crystal spectrometers for WD X-ray detection, a Si(Li)-detector for LED X-ray spectrometry, a cathodeluminescence detector and a transmission stage. Modern systems are highly automated. The computer control system can drive the crystal spectrometer, exchange the crystals, control the ED-counting of X-rays and evaluate the spectra, and drive the specimen stage.

Particle analysis can be performed with a very high degree of automation, e.g. using an ED-setup like the Tracor TN-2000 system with PRC program ("Particle Recognition and Characterization"). Particles on a substrate such as a Nuclepore membrane are automatically detected, sized and analysed. The electron beam scans the selected target field. A particle is "detected" when the digitalized backscattered electron signal exceeds a chosen threshold. The path length during which the signal exceeds the threshold is interpreted as one diameter of the particle. The mean diameter of the particle is the average result of 16 differently

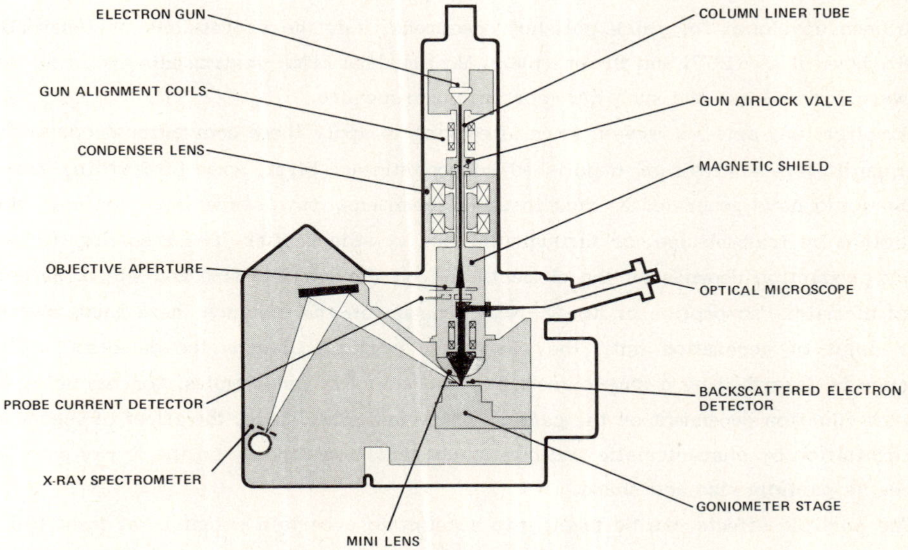

Fig. 2: Schematic representation of an electron microprobe

oriented diameter measurements. The projected area, the circumference and a shape factor for the particle are calculated. The electron beam is then positioned in the centre of the particle and the ED X-ray spectrum is recorded for a preset time and the beam moves on to a next particle. Hundreds of particles can be sized and analysed overnight. Such a system offers unique and nearly unexploited possibilities for air particulate studies.

EPMA-Methodology

The first step, namely the characteristic X-ray peak evaluation, is nearly identical in EPMA and bulk XRS, in both the WD- and ED-mode. As in XRS, determining the relation between X-ray intensities and elemental concentrations poses severe problems in EPMA. The intensity is significantly affected by both instrumental and physical factors.

Effects of instrumental variables, such as the X-ray detection efficiency and the electron beam potential and current, can be minimized by ratios of peaks of the measured elements in the sample to those in standards of known composition.

The physical factors are dependent on the sample composition, are commonly called ZAF-corrections (Z = atomic number, A = absorption, F = fluorescence) and include: the dependence of the efficiency of specific X-ray production by the various elements (Z-factor), the amount of absorption by the sample of the element's characteristic X-rays (A-factor) and the amount of secondary X-ray fluorescence, generated by the primary characteristic X-rays and the Bremsstrahlung radiation (F-factor). Several ZAF-corrections have been developed for thick polished specimens, e.g. the expressions of Duncumb-Reed (1968), Love et al. (1978) and Brown (1082). Most modern EPMA-instruments are provided with software for at least one such ZAF-correction procedure.

Complications arise, however, when attempting to apply these conventional corrections to the quantitative analysis of unpolished microparticles. First, some bombarding electrons, which would have generated X-rays in thick specimens, may escape from particles, through the bottom by transmission, or through the side by side-scatter. The resulting decrease in X-ray production depends on the elements and on the particle size and shape. Second, the effect of matrix absorption on the X-rays depends on the distance these must travel from their point of generation until they leave the particle towards the detector. While this distance is described by a simple geometric function for flat samples, for particles it is a complex function dependent on the particle shape and size. Third, the effect of secondary X-ray excitation by characteristic and continuum radiation depends on the X-ray path length, hence the particle size and shape.

The particle effects can be taken into account to a certain extent in at least two ways. First a recent approach (Small et al., 1979; Statham, 1979) is based on the fact that the ratio of the characteristic X-ray intensity to the continuous Bremsstrahlung X-rays of similar energy (peak-to-background ratio) is nearly identical in particles and bulk material of the same composition over a wide range of particle size. The practical applicability of this approach is still under investigation (Small et al., 1980). Second, in the modified ZAF-correction scheme of Armstrong and Buseck (1975), X-ray attenuation corrections are calculated for particles of an assumed type and geometric shape (e.g. rectangular prism, tetragonal prism, cylinder, triangular prism, square pyramid, hemisphere, sphere). The prerequisite of knowing the particle type is a less severe problem when this correction is used in combination with automatic systems for identifying and sizing particles, like the PRC-program (Van Dyck et al., 1983).

As in most other microanalysis methods, the need to bring the specimens into vacuum will create problems when liquid or volatile particles are of interest.

Applications

Because of the limitations mentioned above, electron beam techniques have not been used much for quantitative analysis of airborne particles. In our laboratory, the applications of TEM and EPMA to aerosols have been mainly along three lines, of which we give a typical example.

- Assessment of asbestos pollution by TEM

An accurate identification of asbestos fibers and a determination of the fiber concentration is required. The problem is compounded by the existence of different varieties (chrysotile, crocidolite, amosite, tremolite, anthophyllite) and also by the occurrence of non-asbestos fibers (e.g. glass fibers). Asbestos is present in the respirable fraction of the urban aerosol in concentrations of the order of a few ng m^{-3}, although it may be present at much higher concentration levels in or near factories or mines. Diameters range between about 20 nm and several hundred nm and lengths from ca. 100 nm to several µm. Presently the only successful approach for their measurement in the particulate fraction of the air or water samples consists in their identification with TEM or STEM. Single particle quantitative X-ray analysis or at least the measurement of the ratio of elemental intensities is required for a more or less safe identification. There is an uncertainty of identification and a non-negligible possibility for a confusion between asbestos minerals and others which happen to have a similar silicate composition. Therefore, morphological characterization and crystal structure as obtained by selected area electron diffraction (SAED) are often considered mandatory (Bloch et al., 1983).

- TEM as a source tracing tool in an antimony smelter

Air particulate material was collected near an antimony and lead refining plant. The morphology seen by TEM, elemental analysis performed by TEM with ED X-ray attachment and the results of SAED showed that four distinct crystallographig Sb containing particles were present, namely senarmontite (cubic Sb_2O_3), a super-structure of senarmontite, valentinite (orthorombic Sb_2O_3 needles) and Sb_2O_5 (crystalline or amorphous).

Certain specific particles were identified as typical for given emissions:

- valentinite and the superstructure of senarmontite are specifically produced in the convertors and refinery ovens,
- transparent senarmontite is observed only in the particulate material emitted through the stacks by a refinery oven,
- aggregates of Sb_2O_5 which contain a high lead concentration are typical for blast furnace fugitive or stack emissions,
- Zn, Cl and As-containing Sb_2O_5 particles are indicators for stack emissions of a refinery oven,
- lead containing Sb_2O_5 single spheres are abundant for convertor emission (fugitive or stack), at least if they are not accompanied by Sb_2O_5-aggregates; otherwise they could also be produced by a blast furnace,
- pure Sb_2O_5 spheres are indicative for the fugitive emissions of a refinery oven.

The manual analysis of particles with EPMA and TEM may be a very time-consuming process. Computer based automatic image analysis methods for analytical electron microscopy, such as the PRC-program discussed above, allow unaided rapid analysis of size, shape and composition of a number of particles in sequence. Particles can also be automatically classified.

- EPMA screening of remote and polluted aerosols

Using a cascade impactor, particles were collected at many sites which ranged from truly remote to heavily polluted, from purely marine to continental. The preliminary results of EPMA and the corresponding secondary electron pictures showed the following (Storms, 1982):

- large particle marine aerosols are cubic in shape, as expected for halite, and contain the major seawater-elements;

- unexpected complications such as pile-up of large seasalt particles in the sampling device can arise under the sampling conditions used routinely in combination with PIXE-analysis, implying differential recrystallisation of seawater according to the Van 't Hoff scheme after the sampling step, and leading to erroneous matrix corrections in the PIXE-technique;
- large particle continental aerosols contain mainly Al, Si, Mg, Ca and Fe as detectable elements;
- the quantification procedures commercially available in EPMA-software (conventional ZAF-corrections) are not adequate for accurate analysis for particles with variable shape;
- submicrometer particles, at all the examined sampling sites, both remote and polluted, marine and continental, are predominantly spherical, suggesting condensation from the gas phase as a source process, and they contain only S (and sometimes some Na, Cl and Ca) as detectable elements, with most of the mass made up by low-Z elements such as N and O. This is probably indicative of NH_4HSO_4, $(NH_4)_2SO_4$ and/or organic S-compounds formed from natural or pollution S-compounds. In areas with rich vegetation, K occurs concomitantly with S in the small spherical particles, probably due to natural or anthropogenic vegetation fires;
- at polluted continental sites, S and Si are often detected simultaneously in small spherical particles.

LASER MICROPROBE MASS ANALYSIS

Principle and instrumentation

Due primarily to the recent advent of a commercially available instrument, considerable interest is currently being focussed on laser microprobe mass analysis (LAMMA).

The LAMMA®-500 instrument (Leybold Heraeus, Cologne, F.R. Germany) is shown in Fig. 3. A detailed description is given in the literature (Hillenkamp and Kaufmann, 1981). The system is equipped with two lasers. The first of these is a low-powered (2 mW) visible (λ = 613 nm) He-Ne search laser that can be focussed to a spot for illuminating the analytical region. The second is a Q-switched (pulse duration = 15 ns) neodymium-YAG laser whose output is quadrupled in frequency to provide a high power density (10^{10}-10^{11} W cm^{-2} for a 0.5 μm diameter spot) at 265 nm for vaporization and ionization. The power of this laser beam can be attenuated to 1% with of a 15 step optical filter sequence.

The optical paths of the two lasers are collinear so they may be focused onto the

Fig. 3: Schematic representation of LAMMA-instrument

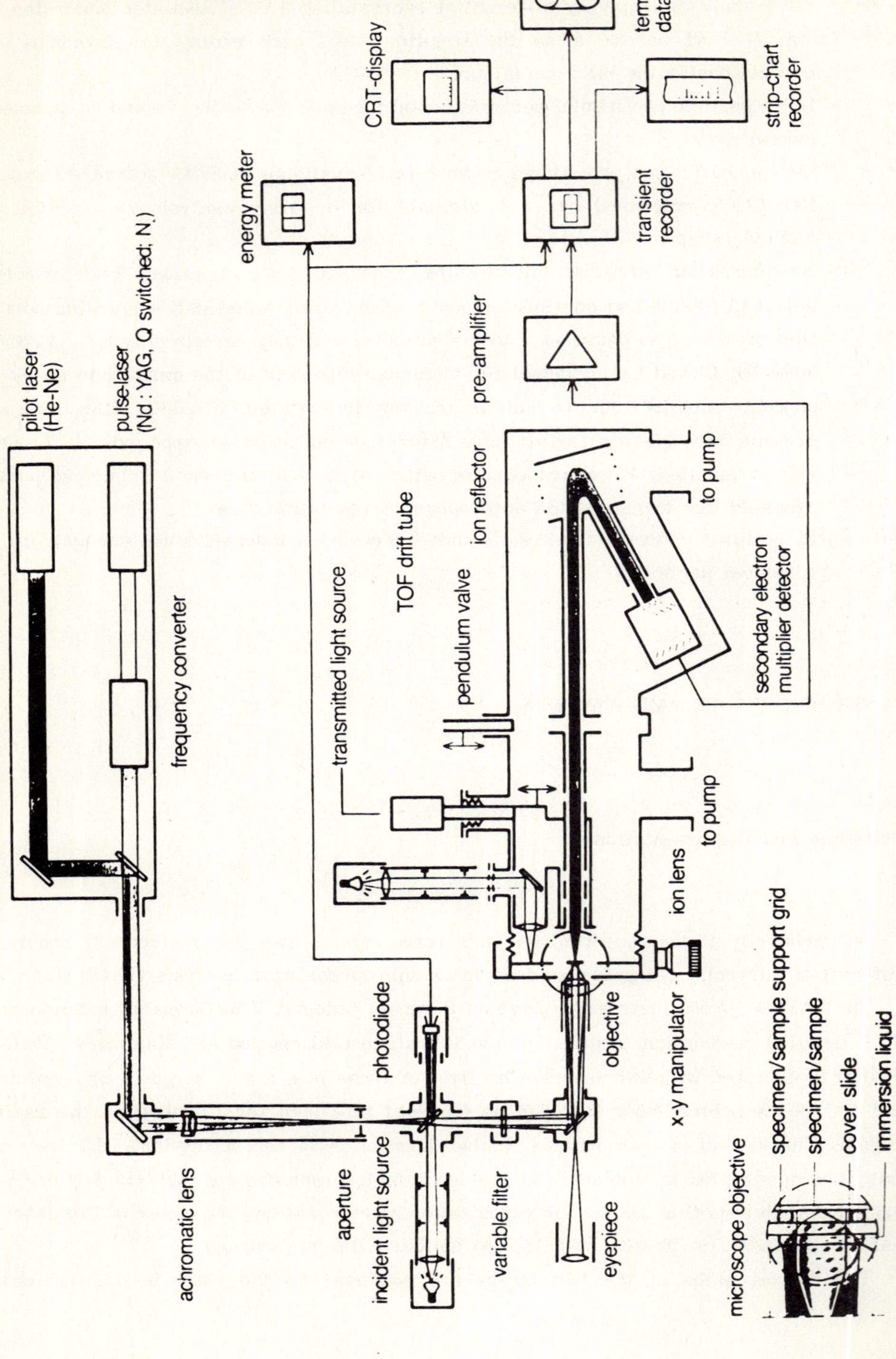

analytical region by an optical microscope fitted with UV-transparent glycerol immersion lenses. The sample may be viewed with a binocular microscope using top or trans-illumination, in either phase contrast or interference contrast modes, via the same optical path as followed by the lasers. All optics prior to the sample are external to the vacuum system, and magnifications in the range 100X to 1250X are available.

The sample is mounted in vacuum on a movable x-y stage and viewed through a thin quartz vacuum seal. Once the analytical region is chosen, the transmitted light condenser lens (which is located in the vacuum system) is pneumatically exchanged for ion optics. Since ions are extracted for mass analysis at 180° from the path of the incident laser beam, it is necessary for the ionization laser to pass through the analytical region, which can be not thicker than a few um.

Activation of the excitation laser vaporizes analyte material to produce a microplasma that consists of neutral fragments together with elementary, molecular, and fragment ions, having predominantly unit charge. The ions are extracted into the drift tube of a 1.8 m TOF mass spectrometer with high extraction and transmission efficiency. The remaining energy spread is largely compensated for by an ion reflector in the drift tube.

The mass spectrum obtained from a laser pulse is detected by an open Cu-Be secondary electron multiplier after additional acceleration to increase ion-to-electron conversion efficiency and to decrease the mass discrimination of the detector. The analog signal is digitized for storage in a 2048-channel 100-MHz transient recorder having 8-bit precision and a time resolution of 10 ns. Spectra can be displayed on an oscilloscope, recorded as an analog trace, or transferred to a computer for storage or data treatment. Mass spectra of either positive or negative ions may be recorded simply by switching the polarity of the electric fields. The power of each laser pulse is monitored and displayed directly for reference purposes.

Recently, a new instrumental version, namely the LAMMA®-1000 has become available, which operates in a 90° geometry, so that bulk specimens and particles on a thicker backing can be examined.

Methodology

With respect to the special requirements for particle analysis it has been claimed that LAMMA offers a number of advantages compared to other techniques (Denoyer et al., 1982), including:

- As far as atomic composition is concerned, all elements of the periodic table can be detected, many of them at relative detection limits in the ppm or even sub-ppm range and with absolute detection limits between 2 x 10^{-18} and 2 x 10^{-20} g (Kaufmann and Wieser, 1980). This allows a high sensitivity for the detection of many trace contaminants in single particles.

- For the chemical characterization of aerosol particles with the aim to establish their origin, history, and ecological significance, element analysis is usually not sufficient. In this regard one needs as much information as possible about the molecular constituents (both organic and inorganic). One of the unique features of LAMMA is that one obtains also organic fingerprint spectra and some indication of inorganic speciation.
- If microprobe techniques are applied to single particle analysis, usually the problem of distribution statistics is raised. Since with the LAMMA instrument the time needed for the analysis is limited only by data processing or writing the recorded mass spectra (usually 30-60 s), a large number of analyses can be executed in a reasonable period of time.
- In many studies one wants to know particle morphology (size and shape), the distribution of particles over the support or the exact position of individual particles chosen for analysis, either as independent information or in correlation with the analytical data. This requires a suitable imaging technique. In the LAMMA instrument one obtains the structural information which a light microscope can deliver. If the structural elements of the particle or the whole particle itself are below the spatial resolution of the optical microscope, there is still the possibility to make a SEM picture of the particles to be analyzed prior to the LAMMA analysis.
- A very interesting feature is that, when a low laser power of about 0.2 µJ is applied to particles larger than 1 um, LAMMA can be used in the "desorption" mode so that only the surface material of a particle is evaporated and ionized, and subsequent layers of one particle can be assessed.

There are also some limitations and disadvantages of the LAMMA technique either based on physical or instrumental conditions (Denoyer et al., 1982):

- LAMMA analysis is essentially destructive. This, for instance, excludes the possibility of recording the spectra of positive and negative ions from the same particle. Furthermore, optimizing the operational conditions of the instrument (e.g. adjusting the optimal power density) usually requires some test shots with the unavoidable loss of a corresponding number of particles.
- With every laser-induced evaporation of an individual particle, part of the supporting film is also evaporated. This material, of course, contributes to the recorded LAMMA-spectrum and inevitably creates some kind of a "background". The ratio of these background signals to mass signals originating from the particle itself depends on the size of the particle, the thickness of the supporting foil, and the chemical nature of both materials. Since formvar, collodium or carbon foils can be made as thin as 100 Å, there is usually no background problem encountered with particles in the range down to about 0.5 µm in size. If particles become smaller or if trace constituents are to be measured the foil-induced background can no longer be neglected.

- The need to bring the specimen into vacuum will create problems when the analysis of liquid or volatile particles is of interest.
- Precision or reproducibility of LAMMA spectra is often poor. From an instrumental standpoint, one cause of poor reproducibility is the variation in power density delivered to the sample. This is determined in part by the reproducibility of the focus of the laser beam. For a uniform sample presented in a thin section it is possible to reproduce the focused spot diameter. However, for irregularly shaped samples (e.g. particles) it is difficult to achieve a reproducible focus.

As in the case of other microbeam analytical techniques, the accuracy of LAMMA depends on the ability to obtain authentic standards, the ability to define the analytical volume, and the intrinsic reproducibility. As the ion production process depends upon the chemical, thermal, and optical properties of the sample, appropriate standards must closely simulate the sample matrix. Indeed, much more theoretical understanding of the processes of ion production by laser irradiation is required before even semiquantitative analyses are routinely possible.

Applications to air particulate analysis

The first report on the applicability of LAMMA in aerosol studies were published by P. Wieser and co-workers (1980).

Since then, several applications have been explored in our laboratory.

- Identification of asbestos types and detection of adsorbed organics.

Systematic LAMMA measurements on some 30 different asbestos samples of widely differing origin have shown that the characteristic positive and negative fingerprint mass spectra provide enough information on elemental composition to distinguish between the asbestos types, while also the molecular ions are of considerable help. Also in the laser desorption mode, LAMMA can readily and quite specifically detect organic impurities adsorbed on single particles, as shown in Fig. 4.

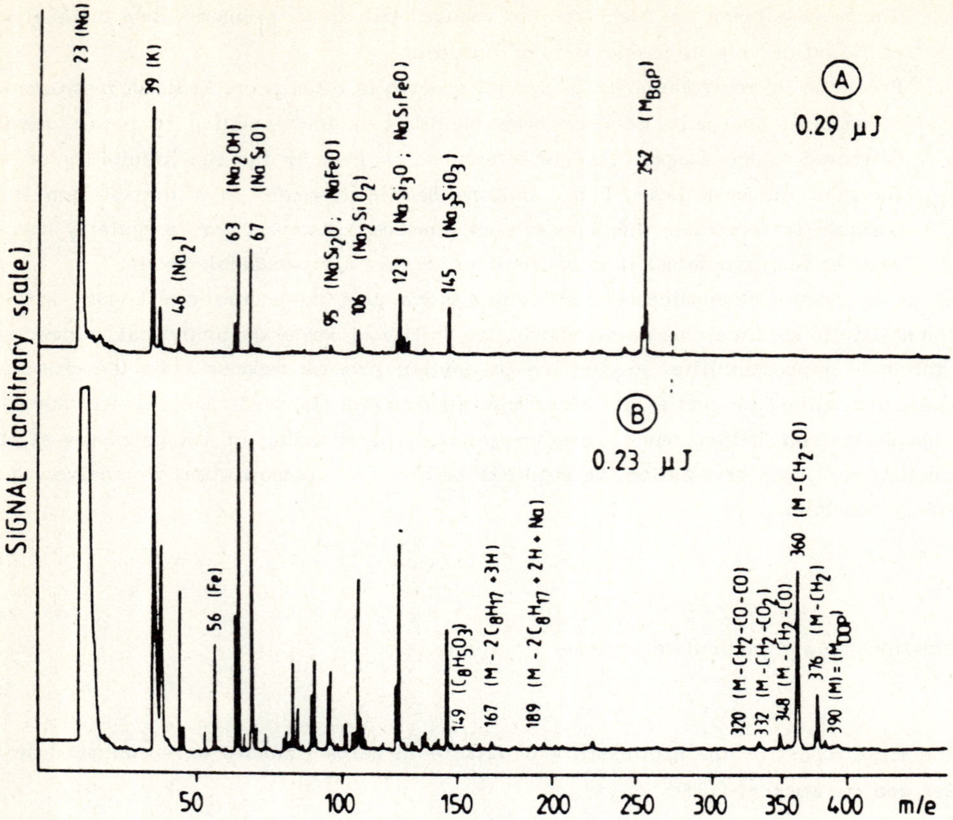

Fig. 4: Positive LAMMA spectra of (A) benzo(a)pyrene doped crocidolite (50 ppm) and of (B) di-(2-ethylhexyl)phtalate (1000 ppm) as surface contaminants on crocidolite.

- Identifications of sulfur species in single particles

LAMMA is capable of identifying sulfur compounds in micrometric particles and of distinguishing e.g. between Na_2SO_4, Na_2SO_3 and $Na_2S_2O_3$. The negative ion mode spectra are dominated by SO_n^- ions; their pattern with respect to the fragment valence can be modelled but it is identical for the three compounds and does not allow identification. This is also true for the $Na_2O_n^+$ clusters in the positive ion mode spectra. The ratio of the strong $Na_3SO_3^+$ and $Na_3SO_4^+$ mass peaks, which is not critically affected by the laser power, does offer the possibility of unambiguous identification that can be confirmed via the Na_3S^+ and $Na_2SO_2^+$ intensities (Bruynseels and Van Grieken, 1984).

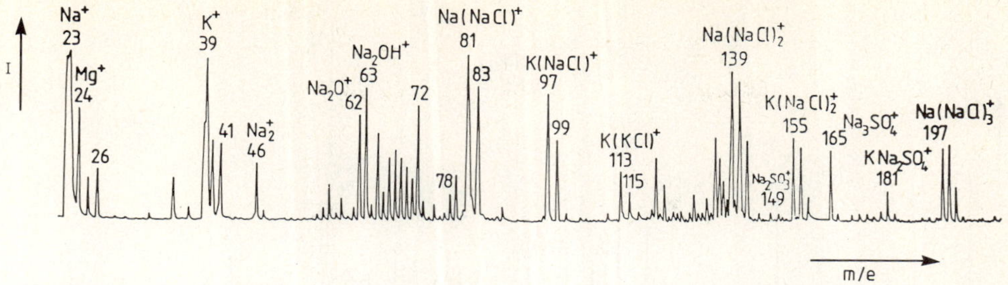

Fig. 5: Positive LAMMA mass-spectrum of an aerosol particle from a purely marine environment

- Screening of remote and polluted aerosols

LAMMA was applied to cascade impactor deposits of particles collected at remote, purely marine sites, more continental remote sites and stations in highly polluted areas. The results led to the following observations:

* On large particle-size marine particles only signals from the expected major seawater ions and their cluster compounds can be detected, as shown in Fig. 5. No direct evidence of trace metal enrichment during ocean/atmosphere transfer can be found.
* In all examined large atmospheric aerosol particles, sulfur is present as sulfate.
* In smaller particles, an excess S is always found and it is enriched at the surface, as was evident from subsequent shots at low laser power. The S-excess increases with increasing anthropogenic character of the environment. The mass peaks at m/e = 18 (NH_4^+) and m/e = 97 (HSO_4^-) are the major mass peaks, suggesting that these aerosols contain mainly NH_4HSO_4 or $(NH_4)_2SO_4$.
* In seasalt particles of diameter 1-2 µm collected at more continental locations, an enrichment with $NaNO_3$ was discovered, probably formed by a heterogeneous gas phase/salt particle interaction. The number of these $NaNO_3$ rich particles increases and their size range widens with increasing pollution character.
* With increasing distance inland from the sea, terrestrial indicators such as positive ions from V, Ti, Fe and Pb and negative ions such as SiO_n^- become more predominant, as well as negative ion species indicative of hydrocarbons.
* In a heavily polluted industrial area with petrochemical, fertilizer and industrial plants some 30 km downwind from the ocean, not less than 8 distinct aerosol types can be distinguished by LAMMA:

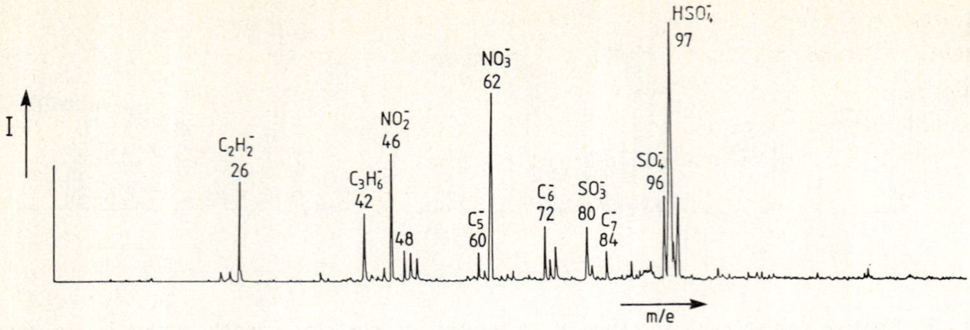

Fig. 6: Negative LAMMA mass-spectrum of an aerosol particle from a polluted industrial environment

- in the large size-ranges:

(1) sea salt particles

(2) sea salt particles with $NaNO_3$ enrichment

- in the small size range:

(3) HSO_4^- rich particles with organic backbone fragmentation (C_n^-),

(4) HN_4^+, Na, Ca, V, Fe-rich particles

(5) Na, Mg, Al, Si, Ca, Ti, Fe, Ba, Sr-rich particles

(6) particles containing HSO_4^- and NO_n^- anions and the organic backbone fragmentation pattern (see Fig. 6)

(7) particles containing phosphorus and nitrogen anions (PO_n^- and NO_n^-)

(8) particles with NH_4^+, Na, K, V, Fe and mass peaks at m/e = 30, 44, 86, 100 which can be ascribed to organic amines.

RAMAN MICROPROBE SPECTROMETRY

Raman spectroscopy is possible for microscopical samples since the availibility of lasers as sources for excitation. There are actually two types of instruments; the Raman microprobe developed at the National Bureau of Standards, Washington, U.S.A. (Etz and Blaha, 1980) and the commercially available Raman microprobe/microscope MOLE (molecular optical laser examiner), developed at CNRS in France by Dhamelincourt et al., (1979). A second generation instrument, MICROMARS, was introduced commercially recently (Delhaye et al., 1982). Both instruments are adapted for the analysis of micrometer size particles.

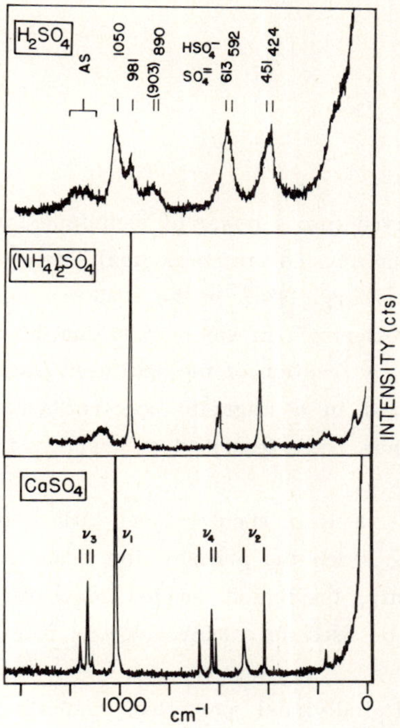

Fig. 7: Raman microprobe spectra of sulfate species in liquid and solid microparticles.
(i) Spectrum of a microdroplet (30 μm diameter) of sulfuric acid. Indicated are the characteristic Raman shifts (cm^{-1}) of HSO_4^- and SO_4^{2-} ions in the aqueous solution phase.
(ii) Spectrum of a ~8 μm particle of crystalline ammonium sulfate.
(iii) Spectrum of a ~5 μm particle of anhydrite, The ν_i designations indicate the frequencies of the fundamental internal vibrational modes of the sulfate ion in $CaSO_4$.

Raman spectroscopy of light is an inefficient process in which a fraction of the scattered light undergoes wavelength shifts with a spectrum characteristic of molecular vibrations. The spectral patterns are indicative of selected molecular subgroups. In Raman spectroscopy of bulk samples, the energy shift of each peak is unique and the intensity is proportional to the concentration of the molecule. This is not so for an ensemble of microscopical objects where the Raman scattering method becomes considerably more complicated by variations of the scattering cross section (Stafford et al., 1977) and morphology dependent spectra (Owenetal, 1982). Moreover there is a non-negligible possibility for heating of the sample under the laser illumination conditions applied. In order to avoid catastrophic sample heating with chemical or physical transformations, the microscopical particle must be in good thermal contact with a good thermal conductor such as sapphire. It is a common practice to place isolated microparticles onto a sapphire disk. Some particles however, are too fragile to be analyzed this way.

In contrast with the fast measurements with the LAMMA instrument, Raman microprobe measurements are inherently slow, of the order of hours for smaller particles. On the other hand, this technique does not require putting the sample in vacuum.

There are several exploratory reports

on the use of the Raman microprobe for air particulate analysis e.g. in the analysis of Mount St. Helens ash (Delhaye et al., 1982), particulate contamination studies (Dhamelincourt et al., 1979, 1982) and environmental pollution analysis (Blaha et al., 1978, Cunningham et al., 1979). The principal inorganic species which can be identified are common minerals (e.g. silicates), oxides, carbonates and sulfates (see Fig. 7). The chemical state of species can be identified. For example the Raman microprobe spectra of impactor collected South Pole aerosols indicate that they consist mainly of $(NH_4)_2SO_4$. Also, the method allows the detection of polycyclic hydrocarbons. An excellent review has been given by Etz and Blaha (1980).

SECONDARY ION MASS SPECTROMETRY

Secondary ion mass spectrometry (SIMS) has evolved into a powerful technique for the analysis of a wide range of specimens of interest in the physical and biological sciences and in various technological applications (Hugh, 1975; Newbury, 1978). In the SIMS technique, a beam of energetic ions bombard a target, causing "sputtering", in which individual atoms or clusters of atoms lying near the surface are ejected. A fraction of the sputtered particles, the secondary ions, are collected and mass analyzed in a magnetic spectrometer or a quadrupole mass filter. Instrumentation for SIMS has been described extensively in the literature (Newbury, 1978; Czanderna, 1975).

Because the secondary ion signal originates from a depth of about 1-2 nm, SIMS is one of the major surface analysis techniques. In addition, since the primary ion beam can be focused to form a small probe on the sample, restricting the region sampled to an area with dimensions in the micrometer range, SIMS can also be used as a microanalysis technique, with a minimum sampling volume of 0.01 μm^3.

SIMS offers several advantages for study of individual particles, including the capabilities for detection of all elements, isotope ratio measurements, compound detection, depth profiling, and ion imaging of specific constituents.

For the most part, SIMS applications have involved qualitative analyses as quantitative work, although possible in principle, is far from straightforward. The reasons for this include:

- Difficulties are encountered in establishing even semi-quantitative depth scales. Depth calibration is normally obtained by calibrating the rate of removal of surface material. This can be done with sample standard materials, but with environmental particles there are considerable difficulties in matching the surface composition of a standard with that of a poorly defined particle.
- The method is subject to several types of interferences and artifacts, e.g., spectral interferences from multiply charged and molecular ions, which may obscure measurements even at high mass resolution.

- Volatilisation losses and migration of sample species may occur under the influence of the primary ion beam. Also, the poor electrical conductivity of the samples gives rise to problems.

Potential applications of SIMS to the study of atmospheric particles have been discussed by Newbury (1980). The method has been applied to semi-quantitative studies of the surface of coal fly ash and automobile exhaust particles in conjunction with several other microanalytical and surface sensitive techniques including EPMA, AES and ESCA. The results have been of considerable importance from the standpoint of environmental chemistry and demonstrated the need for more quantitative determinations of both particle surface concentrations and elemental depth profiles.

In our laboratory quantitative results have been achieved (Van Craen et al., 1982) by applying a matrix-correction procedure which corrects changes of composition with depth. E.g. the variation of concentration with erosion time for Na and Co in a typical steel-dust sample has been investigated. Comparison of the interior regions of the sample with SIMS and X-ray fluorescence proved the accuracy of the depth profiles.

Finally, Table III compares the characteristics of the microanalysis techniques considered above.

Table III: Analytical characteristics of some microanalytical techniques

	EMPA	LAMMA	MOLE	SIMS
Resolution				
lateral, µm	1	0.5	0.5-1	1
depth, µm	1	?	?	5×10^{-4}
Elements	B-U	all	molecules	all
Isotopes	no	yes		yes
Molecular	some	yes	yes	yes
Abs. sensitivity (g)	10^{-14}-10^{-16}	10^{-18}-10^{-20}	10^{-8}	10^{-19}
Rel. sensitivity (ppm)	100		10^{4}-10^{5}	1

ACKNOWLEDGEMENT

This work was supported in part by the Belgian Ministry of Science Policy through grant 80-85/10, and by NATO through grant MS-20.

REFERENCES

Adams F.; (1982); Phil. Trans. R. Soc. Lond. A305; 509.

Adams F. C. and Van Grieken R. E.; (1975); Anal. Chem. 47; 1767.

Adams F., Van Espen P. and Maenhaut W.; (1983); Atmosph. Environ. 17; 1521.

Ahearn A. J. (ed.); (1972); Trace Analysis by Mass Spectrometry; Academic Press, New York.

Andersen C. A.; (1973); Microprobe Analysis; John Wiley & Sons, New York.

Annegarn H. J., Van Grieken R. E., Bibby D. M. and Von Blottnitz F.; (1983); Atmosph. Environ. 10; 2045.

Arinc F., Wielopolski L. and Gardner R. P.; (1977); X-Ray Fluorescence Analysis of Environmental Samples; Dzubay T. S. (ed.); Ann Arbor Science, Ann Arbor, MI, 227.

Armstrong J. and Buseck P.; (1975), Anal. Chem. 47; 2118.

Bertin E. P.; (1975); Principles and Practice of X-Ray Spectrometric Analysis; 2nd ed.; Plenum Press, New York.

Billiet J., Dams, R. and Hoste J.; (1980); X-Ray Spectrom. 9; 206.

Birks, L. S.; (1969); X-Ray Spectrochemical Analysis; 2nd ed.; Interscience, New York.

Blaha J. J., Rosasco G. J. and Etz E. S.; (1978); Appl. Spectrosc. 32; 292.

Bloch P., Vanderborght B., Adams F. and Van Landuyt J.; (1983); Intern. J. Environ. Anal. Chem. 14; 257.

Bonner N. A., Bazan F. and Camp D. C.; (1973); Elemental Analysis of Air Filter Samples Using X-Ray Fluorescence; Report UCRL-5/1388; Lawrence Radiation Lab., University of California, Livermore, CA.

Bowman H. H., Hyde E. K., Thompson S. G. and Jared R. C.; (1966); Science 151; 562.

Brown R. and Vossen P. G. T.; (1970); Anal. Chem. 42; 1820.

Brown J.; (1982); Microbeam Analysis-1982; D. Newbury (ed.); San Francisco Press, San Francisco, CA, 151.

Bruynseels F. J. and Van Grieken R. E.; (1984); Anal. Chem.; 56; 871.

Chandler J. A.; (1977); X-Ray Microanalysis in the Electron Microscope; North-Holland Publ. Comp., Amsterdam.

Cornides I.; (1978); Analysis of Airborne Particles by Physical Methods; Malissa H. (ed.); CRC Press, Palm Beach, FL, 73.

Criss J. W.; (1976); Anal. Chem. 48; 179.

Cunningham W. C., Etz E. S. and Zoller W. H.; (1979); Microbeam Analysis-1979; Heinrich K. F. J. (ed.); San Francisco Press, San Francisco, CA, 148.

Czanderna A. W.; (1979); Methods of Surface Analysis; Elsevier Publ. Co., Amsterdam.

Delhaye M., Bridoux M., Dhamelincourt P., Barbillat J., Da Silva E. and Roussel B.; (1982); Microbeam Analysis-1982; Heinrich K. F. J. (ed.); San Francisco Press, San Francisco, CA; 275.

Denoyer E., Van Grieken R., Adams F. and Natusch D. F. S.; (1982); Anal. Chem. 54; 26A.

Dhamelincourt P.; (1979); Microbeam Analysis-1979; Heinrich K. F. J. (ed.); San Francisco Press, San Francisco, Ca; 155.

Dhamelincourt P., Wallart F., Leclerq M., N'Guyen A. T. and Landon D. O.; (1979); Anal. Chem. 51; 414 A.

Duncumb P. and Reed S.; (1968); Quantitative Electron Probe Microanalysis; Heinrich K. F. J. (ed.); NBS Special Publ., 298; 135.

Dzubay T. C. and Rickel D. G.; (1978); Electron Microscopy and X-Ray Applications; Ann Arbor Science, Ann Arbor, MI, 3.

Dzubay T. G., Stevens R. K. and Peterson C. M.; (1976); X-Ray-Fluorescence Methods for Analysis of Environmental Samples; Dzubay T. G. (ed.); Ann Arbor Science, Ann Arbor, MI, 95.

Etz E. S. and Blaha J. J.; (1980); Characterization of Particles; NBS Special Publ. 533; 153.

Fox D. L. and Jeffries H.; (1981); Anal. Chem. 53; 1 R.

Gilfrich J. V., Burkhalter P. G. and Birks L. S.; (1973); Anal. chem. 45; 2002.

Goldstein J. I., Newbury D. E., Echlin P., Joy D. C., Fiori C. and Lifshin E.; (1981); Scanning Electron Microscopy and X-Ray Microanalysis; Plenum Press, New York.

Hillenkamp F. and Kaufmann R.; (1981); Sessions of LAMMA Symposium held in Duesseldorf, F.R.G., Oct. 8-10, 1980; Z. Anal. Chem. 308; 193-320.

Jaklevic J. M. and Walter R. C.; (1976); X-Ray-Fluorescence Methods for Analysis of Environmental Sampling; Dzubay T. S. (ed.); Ann Arbor Science, Ann Arbor, MI; 63.

Jaklevic J. M., Goulding E. S. and Landin D. A.; (1972); IEEE Tran Nucl. Sci. NS19; 392.

Jenkins R.; (1974); An Introduction to X-Ray-Spectrometry; Heyden & Son, London.

Johansson T. B., Akselsson R. and Johansson S. A. E.; (1970); Nucl. Instr. Methods 84; 141.

Johansson T. B., Van Grieken R. E., Nelson J. W. and Winchester J. W.; (1975); Anal. Chem. 47; 855.

Johansson T. B., Van Grieken R. E. and Winchester J. W.; (1976); J. Geophys. Res. 81; 1039.

Kaufmann R. and Wieser P.; (1980); Characterization of Particles; NBS Special Publ., 533; 199.

LaBrecque J. J.; (1981); Progress in Analytical Spectrometry 4; 191.

Love G., Cox M. and Scott V.; (1978); J. Physics D: Apll. Phys. 11; 7.

Maenhaut W., Selen A., Van Espen P., Van Grieken R. and Winchester J. W.; (1981); Nucl. Instr. Methods; 181, 399.

Maenhaut W., Raemdonck H., Selen A., Van Grieken R. and Winchester J. W.; (1983); J. Geophys. Res. 88; 5353.

McHugh J. A.; (1975); Methods of Surface Analysis; Czanderna A. W. (ed.); Elsevier Publ. Co., Amsterdam; 223.

McMaster W. H., Delgrande N., Mallett J. H. and Hubell J. H.; (1969); Compilation of X-Ray Cross Sections; Report UCRL-50174; Lawrence Radiation Lab., California University, Livermore, CA.

Morrison G. H.; (1980); in: Techn. Rep. Ser. no 197; IAEA, Vienna; 201.

Nelson J. W., Jensen B., De Saedeleer G. G., Akselsson K. R. and Winchester J. W.; (1976); Adv. X-Ray Anal. 19; Kendall/Hunt, Dubuque, Ia; 403.

Newbury D. E.; (1978); Environmental Pollutants Detection and Measurement; Toribara T. Y. et al. (eds.); Plenum Press, New York; 317.

Newbury D. E.; (1982); Characterization of Particles; NBS Special Publ. 533; 139.

O'Connor B. H., Kerrigan G. C., Thomas W. W. and Gasseng R.; (1975); X-Ray Spectrom. 4; 190.

Owen J. F., Barber P. W. and Chang R. K.; (1982); Microbeam Analysis-1982; Heinrich K. F. J. (ed.); San Francisco Press, San Francisco, Ca; 255.

Paradellis T. and Katselis V.; (1978); J. Radioanal. Chem. 42; 153.

Pilate A. and Adams F.; (1980); Anal. Chim. Acta 122; 57.

Pilate A. and Adams F.; (1981); Z. Anal. Chem. 309; 295.

Rhodes J. R.; (1979); Practical Aspects of Energy-Dispersive X-Ray Emission Spectrometers; (IAEA-216); Vienna, Austria.

Rhodes J. R., Stout J. A., Schindler J. S. and Piorek S.; (1983); Portable X-Ray Survey Meters for In-situ Trace Element Monitoring of Air Particulates; ASTM Special Technical Publ. 786; ASTM, Philadelphia, PA.

Schamber F. H.; (1977); X-Ray-Fluorescence Analysis of Environmental Samples; Dzubay T. S. (ed.); Ann Arbor Science, Ann Arbor, MI; 241.

Small J., Heinrich K., Newbury D. and Mykleburst R.; (1979); Scanning Electron Microscopy; SEM Inc., AMF O'Hare, IL; 801.

Small J. A., Heinrich K. F. J., Newbury D. E., Mykleburst R. C. and Fiori C. E.; (1980); Characterization of Particles; NBS Special Publ. 533; 29.

Stafford R. G., Chang R. K. and Kindlman P. J.; (1977); Methods and Standards for Environmental Measurements; NBS Special Publ. 464; 659.

Statham P.; (1979); Microbeam Analysis-1979; Newbury D. (ed.); San Francisco Press, San Francisco, CA; 247.

Storms H.; (1982); M. Sc. Thesis; University of Antwerp.

Van Craen M., Natusch D. F. S. and Adams F.; (1982); Anal. Chem. 54; 1786.

Vanderborght B. and Van Grieken R.; (1978); Anal. Chim. Acta 103; 223.

Vanderborght B. and Van Grieken R.; (1979); Talanta 26; 461.

Van Dyck P. M. and Van Grieken R. E.; (1980); Anal. Chem. 52; 1859.

Van Dyck P., Markowicz A. and Van Grieken R.; (1980); X-Ray Spectrom. 9; 70.

Van Dyck P., Storms H. and Van Grieken R.; (1984); J. de Physique 45; C2; 781.

Van Espen P., Nullens H. and Adams F.; (1977a); Nucl. Instr. Methods 142; 243.

Van Espen P., Nullens H. and Adams F.; (1977b); Nucl. Instr. Methods 145; 579.

Van Espen P., Nullens H. and Maenhaut W.; (1979); Microbeam Analysis-1979; Newbury D. E. (ed.); San Francisco Press, San Francisco, CA; 283.

Van Espen P., Adams F. and Maenhaut W.; (1981); Bull Soc. Chim. Belg. 90; 305-31.

Van Espen P. and Adams F.; (1981); X-Ray Spectrom. 10; 64.

Van Grieken R. E. and Adams F. C.; (1976); X-Ray Spectrom. 5; 61.

Van Grieken R. E. and LaBrecque J. J.; (1985); Trace Analysis 4; Lawrence J. F. (ed.); Academic Press Inc., New York, in press.

Van Hevesy G.; (1982); Chemical Analysis by X-Ray and its Application; McGraw-Hill, New York.

Wieser P., Wuerster R. and Seiler H.; (1980); Atmos. Environ. 14; 485.

STANDARDIZED TECHNIQUES FOR THE COLLECTION AND ANALYSIS OF PRECIPITATION.

J. Slanina
The Netherlands Energy Research Foundation (ECN)
Petten
The Netherlands

ABSTRACT

Sampling of precipitation is complicated by a number of factors: -the sample must be representative so dry deposition must be excluded and sampling of the whole precipitation event must take place. -Contamination and absorption of chemical species must be avoided. -The sample must be kept intact until the moment of chemical analysis.

A survey of the existing wet-only samplers is given and their characteristics are discussed.

The wide choice of available analytical techniques offers sufficient adequate solutions for the analysis of precipitation samples. The merits of different methods, such as Continuous Flow Analysis, Ion-chromatography, Ion Selective Electrodes, Photometric methods, Atomic Absorption Spectrometry, Inductive Coupled Plasma Emission Spectrometry, Polarographic methods and Radio-analytical techniques, are discussed in regard of sample load, precision and accuracy. Problems are encountered in the analysis of instable species such as H_2O_2 and sulfite, organic compounds, and some trace elements such as As, Sb, Se and B.

In many cases only indirect statistical methods, such as regression analysis or cluster analysis, can extract the desired information out of the available data. Different sampling strategies are required for studies on monitoring deposition, detection of trends or source identification.

An optimisation of the complete procedures of sampling, analysis and data evaluation is necessary to obtain useful results.

INTRODUCTION

Research in precipitation chemistry is not a new subject, a few investigations were even performed in the nineteenth century, but more recent findings of ecological damages,

NATO ASI Series, Vol. G6
Chemistry of Multiphase Atmospheric Systems
Edited by W. Jaeschke

contributed to the deposition of acidic and toxic substances, have triggered an enormously increasing interest.

The acidification of lakes in Scandinavia and the U.S. gave the first impulse, (Sandefjord 1980), the large-scale deterioration of woods in Germany provides a second one (Lindau, 1983).

Nowadays the chemical composition of precipitation is investigated for many aims:

- To monitor the deposition of certain compounds in a region by precipitation.
- To detect trends in the composition of precipitation as a function of time in order to make predictions for the future situation.
- To investigate the source regions of compounds in precipitation in order to be able to take effective abatement measures.
- Special studies, e.g. the measurement of wet removal of pollutants present in plumes of large sources.

The first requirement for sampling and analysis of precipitation is that the result is representative of the chemical composition of the precipitation which is studied. This seems a rather trivial remark, but many problems are encountered:

- Sampling must be representative, which also means that dry deposition must be excluded, contamination and absorption is to be avoided. The sample must be kept intact until chemical analysis is performed.
- Sample pretreatment and analysis must avoid artifacts and guarantee sufficient accurate results.

Furthermore, sampling strategy is a very important parameter, it is essential that a database is obtained which is suitable for the data-evaluation and interpretation, to provide the answers to the questions posed.

Indirect statistical methods such as regression analysis, cluster-analysis or factor analysis must often be applied to obtain useful information, and the precision of a database and the number of outlying results influence the results of these statistical techniques.

So the whole set of procedures for sampling, analysis and interpretation must be optimized for good results.

Unfortunately no standard procedures exist today which ensure adequate results for all types of research in precipitation chemistry.

SAMPLING

The sampling of precipitation, in liquid or solid form, presents a number of problems.

In open samplers a mixture of wet and dry deposited materials is sampled. The contribution of dry deposition can be large in areas where high concentrations of precursors and their reaction products are encountered in the atmosphere. Soederlund (1982) estimates by comparing the results of open and wet-only samplers that dry deposition is responsible for 30% of the calcium and 11% of the nitrate found in open samplers in Sweden. The contribution of dry deposition for other bulk elements is neglegible. In the Netherlands dry deposition accounts for 10 to 20% for the concentration of bulk elements found in open samplers (Slanina et al., 1979). But 30-90% of heavy metal concentrations found near Sudbury, Canada, in open samplers are due to dry deposition (Jeffries and Snijder, 1981). Furthermore contamination of samplers by bird-droppings, insects etc. is less severe in the case of wet-only sampling.

So wet-only samplers are generally to be preferred for precipitation chemistry research. Such a wet-only sampler must fulfill a number of requirements:

- The sampler must be equipped with a tightly fitted cover, which prevents dry deposition.
- All parts of the sampler, exposed to the sampled precipitation must be constructed from inert material, e.g. polyethylene, propylene or teflon if inorganic components are measured and glass or metal for organic components. This does not only include the sample funnel but all parts that are placed so close to the open parts of the sampler that splash-off can contribute to contamination.
- The cover should not, in open position, disturb the wind flow pattern around the sample funnel too much, otherwise the sampling could be dependent on the wind direction and droplet size.
- The sample must be kept intact inside the sample bottle which means that evaporation must be prevented (most easily done by the use of a sample funnel connected with a tube of small diameter to the sample bottle). The sample must be protected from sunlight to prevent growth of algae. Some authors report that instant cooling to 4°C or even freezing is necessary (MAP3S 1977, Mueller 1982), but investigations at ECN indicate that little or no change is observed in the first 10 days after onset of the sample collection. Experiments based on monthly samplings in the Netherlands however showed that in some cases quite drastic changes in NH_4^+ and NO_3^- concentrations occured during the sampling period (Ridder 1982). Comparison of the results of daily, weekly and biweekly sampling intervals in Florida indicated that little deterioration was observed in 14 days (Madsen 1982).

  In Scandinavia no problems regarding stability were found in the case of monthly sampling periodes (Granat 1974).

  A study in Illinois (U.S.) also indicated that at least for a period of up to 15 days stability is guaranteed in the case of wet-only samplers. Open samplers gave more problems (Peden and Skowron, 1978). In summing up all experiments regarding sample stability under field conditions, it seems that no problems are to be expected at sampling periods of one week or shorter, but local conditions could cause problems in this respect.
- The delay between the onset of a precipitation event and opening of the sampler must be kept as small as possible. Sequential sampling of precipitation events has led to

the conclusion that in some cases concentrations of bulk elements in the first 0.5 mm of a precipitation event can be 50 times the average concentration found for the whole event (Asman 1980, Dawson 1978). In consequence detectors with a large surface (at least 10 to 15% of the sampling surface of the sampler) must be employed.

- Sampling of solid precipitation is in principle possible if the funnel of the sampler is heated. But representative sampling of solid precipitation is seen as very difficult problem.

  Samplers equipped with a bucket instead of a funnel are more suitable (Granat 1975b), but even in this case many problems are encountered because snow is easily blown out or into the samplers in adverse circumstances.

- If official rain gauges are employed in the vicinity of precipitation chemistry samplers to provide an accurate measure for the amount of precipitation, the sample efficiency of the sampler can differ from 100%. But if deviations of 5% or more are encountered, problems can arise. If the efficiency is more than 100%, extensive splash off occurs which can lead to contamination, if the efficiency is below 100%, opening of the cover may be delayed or the flow pattern around the sampler is disturbed in open position.

- Sampling of mist or dew is generally to be avoided. The simplest solution is to heat the detector surface at low power, so no water film is formed on the detector surface except in case of precipitation. If one considers all desiderata mentioned above, it is not surprising that the construction of adequate samplers has been a difficult problem. A limited number of commercial samplers are available mainly in the U.S. and Canada. A number of these collectors have been tested in comparative experiments for bulk elements: The Misco, Wong, Erni (all three wet-only collectors), Parelco Oy, HASL (fig. 1), (both "wet-dry" collectors) were compared by Galloway and Likens (1976, 1978), who concluded that only the HASL collector functioned satisfactorily. The Sangamo (patterned after the HASL-collector), Aerchem. and Battelle (fig. 2) (used in the MAP3S program) collectors were tested by De Pena et al. (1980), with the result that the Battelle and Aerchem. collectors gave comparable results but the Sangamo collector deviated, and that the collection efficiency of the three collectors was 1.08, 0.95 and 0.98 compared to a standard rain-gauge.

  A number of institutes have built their own samplers, mainly because they were not convinced of the merits of the commercial ones especially for the analysis of trace elements. A, by no means complete, survey is given below.

  - The Battelle sampler is used in the MAP3S experiment (MAP3S 1977). This sampler cools the collected precipitation to 4°C (fig. 2).
  - K.F.A. Juelich used a collector, largely constructed from polyethylene, in programs directed to sampling of bulk and trace elements (fig. 3) (Nuernberg et al., 1979).
  - The wet-only sampler developed by Granat is employed in the monitoring

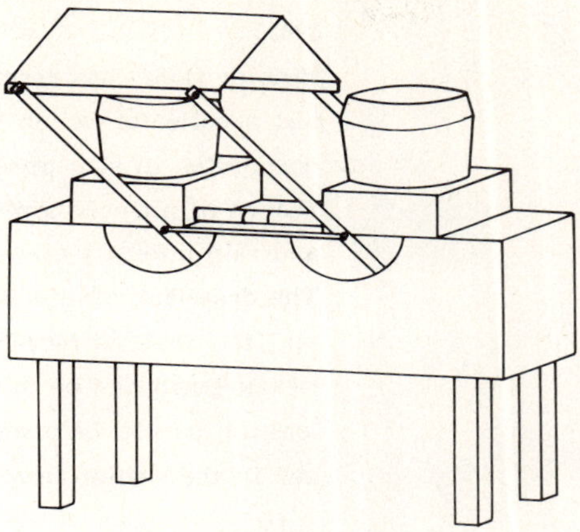

Fig. 1 HASL Collector

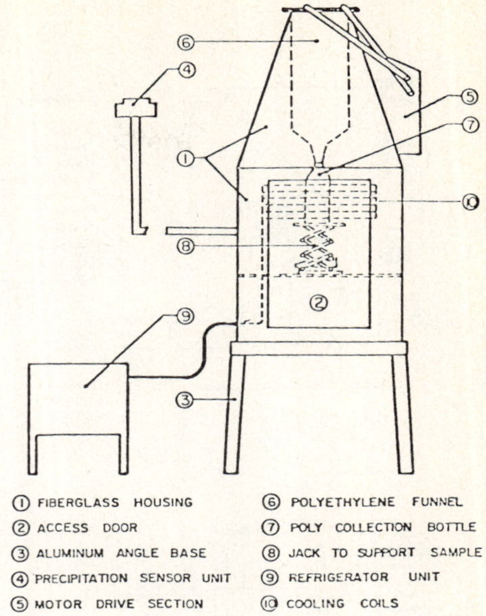

Fig. 2 Battelle Collector

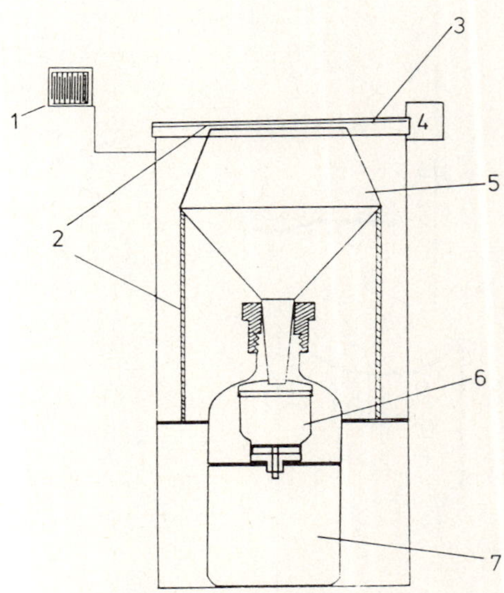

Fig. 3 KFA Juelich Collector

Fig. 4 Collector Met. Institute University of Stockholm

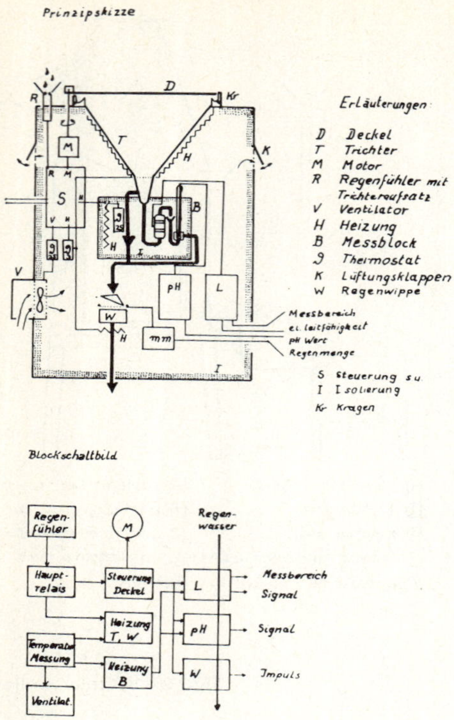

Fig. 5 Collector Deutscher Wetterdienst

programs of the Meteorological Institute of Stockholm University (figure 4) (Soederlund and Granat 1982).

- Winkler (1982) has described a collector which incorporates in situ measurement of pH conductivity and rain intensity (fig. 5). The drawback of this set-up is that direct pH measurement can be problematic, as will be discussed in the section devoted to analytical problems, and that each extra handling of sampled precipitation enhances the possibilities of contamination.
- ECN employes a collector equipped with a 7 position sample changer and adapted to computer control to facilitate field work (fig. 6) (Slanina 1983).

Most of these collectors, as far as known, have not been tested in comparative studies. These studies can provide the necessary information which is needed to develop

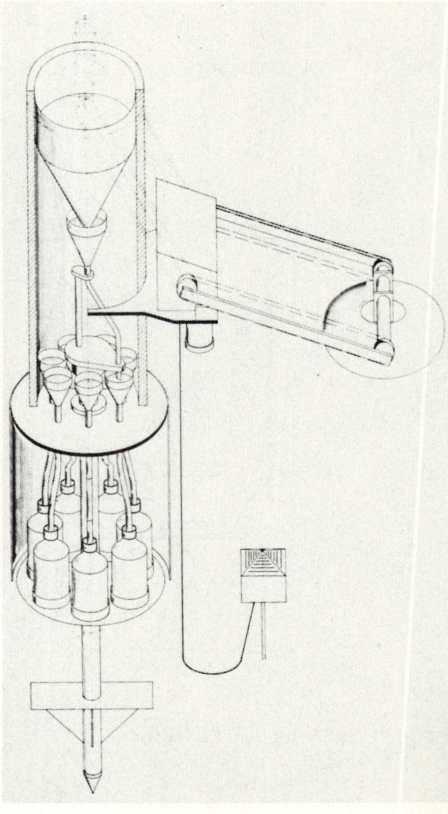

Fig. 6 ECN Collector

generally accepted collector types.

At ECN another approach has been chosen to test our wet-only collectors. Groups of 8 identical wet-only and 8 open samplers of the same shape as the wet-only collectors are placed on one location and sampled daily. Comparison of the mean amount of precipitation and the standard deviation of both groups is indicative whether the wet-only samplers are functioning properly. Our latest wet-only sampler sampled the same amount (100%) found for the open collectors.

If sampling, sample pretreatment and analytical procedures are correct it is generally observed that the standard deviation calculated from the results of 8 identical collectors is constant or goes down at increasing concentrations. If the contrary is found, contamination, a severe problem especially for trace components, occurs during sampling or analysis. Figure 7 gives the results for SO_4^{2-} for open and wet-only samplers, indicating that no contamination has taken place.

In case of copper (fig. 8) it was clear that a copper contamination was present in our former type wet-only collectors. Properly functioning wet-only collectors should give results with a small standard deviation as can be seen in table 1, giving the results of 8 wet-only collectors for three 24 hours sampling periods at Lelystad in the Netherlands.

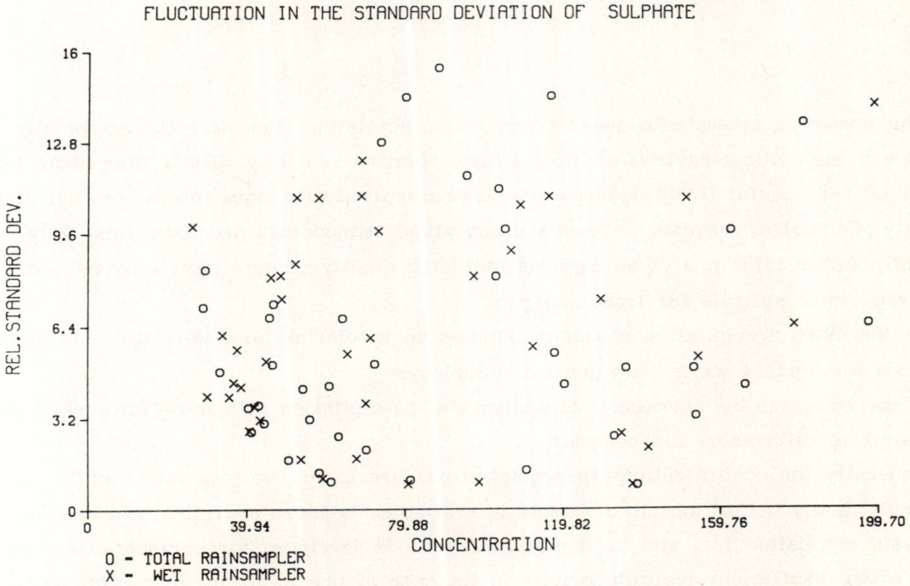

Fig. 7 Variation in SO_4 data

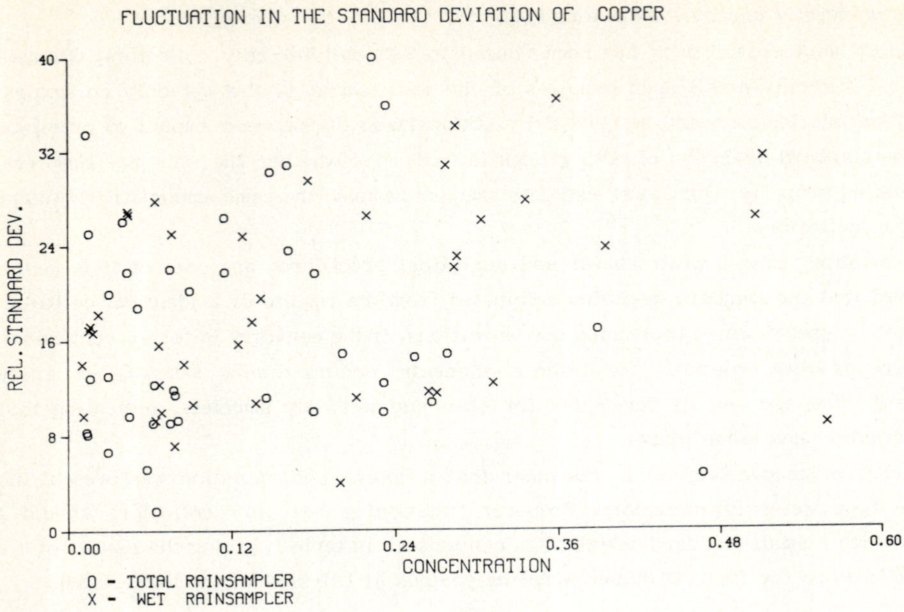

Fig. 8

ANALYSIS

The numerous attempts to detect trends in precipitation chemistry (Likens et al., 1979, 1981) and the critical reviews of these efforts (Hansen and Hidy, 1982), make clear that it would be very useful if old databases could be compared with more recent ones but that the quality of the older sampling methods and analytical methodology precludes this. Only fairly recently (after 1975) analytical methods including quality control have acquired sufficient accuracy to be suitable for trend analysis.

In the older precipitation chemistry studies no special methods were used but standard methods for surface water were generally employed.

If one compares the parameters of analyses of precipitation with those for surface water, two striking differences are observed:

Generally the concentrations in precipitations are lower (up to a factor of 50, e.g. for chloride) and a limited amount of sample is available. So micro analytical methods for trace analysis are called for, and the traditional methods developed for surface water analysis gave often insufficient accurate results in the case of precipitation. In recent years more appropriate methods have been developed.

Table 1 Results for 3 sample periods of 8 wet-only collectors

Date	Sampler	H ppb	NH_4 ppm	Na ppm	K ppm	Ca ppm	Mg ppm	SO_4 ppm	NO_3 ppm	Cl ppm	Rain gram	Ion balance
1 - 6	1	25	1.75	0.09	0.07	0.42	0.02	4.19	2.90	0.29	267.30	1.05
	2	22	1.84	0.11	0.09	0.42	0.05	4.19	3.02	0.34	273.40	1.07
	3	21	1.90	0.11	0.09	0.45	0.03	4.20	3.19	0.39	264.90	1.06
	4	24	1.72	0.08	0.05	0.44	0.04	4.16	3.02	0.31	268.20	1.03
	5	24	1.70	0.10	0.07	0.43	0.04	4.12	2.99	0.30	272.60	1.04
	6	22	1.78	0.09	0.07	0.42	0.04	4.22	2.93	0.30	277.40	1.04
	7	25	1.70	0.08	0.07	0.45	0.04	4.19	2.98	0.32	271.90	1.04
	8	23	1.84	0.07	0.06	0.42	0.04	4.17	2.98	0.33	266.90	1.07
	S in %	6	4	16	19	3	24	1	3	10	1	
2 - 6	1	32	1.66	0.16	0.19	0.21	0.04	4.41	2.54	0.43	488.60	1.03
	2	31	1.59	0.19	0.21	0.20	0.04	4.28	2.44	0.44	529.30	1.03
	3	30	1.59	0.18	0.23	0.20	0.04	4.29	2.44	0.42	538.10	1.04
	4	32	1.57	0.15	0.22	0.21	0.04	4.31	2.46	0.42	537.80	1.02
	5	32	1.61	0.14	0.20	0.21	0.03	4.31	2.50	0.41	550.40	1.03
	6	32	1.59	0.18	0.26	0.25	0.04	4.35	2.47	0.42	550.00	1.06
	7	32	1.60	0.17	0.23	0.21	0.04	4.36	2.50	0.43	496.30	1.03
	8	30	1.63	0.14	0.19	0.19	0.02	4.35	2.53	0.44	504.50	0.99
	S in %	3	2	12	11	8	21	1	2	2	5	
18 - 5	1	30	2.01	0.70	0.31	0.71	0.14	5.20	3.96	1.61	59.50	1.04
	2	28	2.05	0.95	0.40	0.81	0.18	5.54	4.22	2.02	57.50	1.03
	3	31	1.94	0.63	0.17	0.58	0.14	4.94	3.76	1.50	62.80	1.02
	4	35	2.01	0.70	0.18	0.67	0.14	5.17	4.02	1.57	57.40	1.04
	5	35	1.90	0.70	0.16	0.66	0.14	5.06	3.86	1.52	60.90	1.04
	6	33	1.99	0.72	0.19	0.67	0.14	5.20	3.91	1.60	60.40	1.04
	7	33	1.90	0.73	0.21	0.70	0.14	5.08	3.91	1.60	60.60	1.04
	8	37	1.96	0.70	0.18	0.67	0.14	5.24	4.10	1.60	57.30	1.02
	S in %	19	3	13	30	9	10	3	4	10	3	

Determination of anions in precipitation

Spectrophotometry has long been the basis of many methods for the analysis of anions. Chloride is measured by means of its reaction with mercuric thiocyanate followed by the formation of ferric thiocyanate which is measured at 480 nm. The detection limit is 0.2 ppm (Zall et al., 1956). All other halogens and compounds which can form insoluble mercuric salts interfere. Nitrate can be measured after the reduction to nitrite by cadmium by means of α

coupling reaction involving sulphonic amide and N-1-naphtylethylene diamine. In manual procedures the reduction of NO_2^- can be quite problematic, but in an auto-analyzer version this problem is less servere (Technicon). Direct measurement of NO_3^- at 210 nm, preceded by removal of interfering organic substances by activated carbon has also been applied (Slanina et al., 1980a). The wellknown thorin method for sulphate in which an excess of Ba^{2+} ions was added followed by measurement of the decrease in Ba^{2+} ions by formation of $BaSO_4$ is problematic because of many interferences (PO_4^{3-}, Cl^- and trace heavy metals) (ASTM, 1976). Another version in which arsenazo III is used at low pH is better (Reijnders et al., 1979), but the results of both methods at SO_4^{2-} concentrations under 1 ppm are erratic.

No real alternative is available for the wellknown phosphate determination by means of the ammonium phosphor molybdate complex. The method is quite sensitive (0.01 ppm) and in case of precipitation sufficiently selective (APHA 1976). Most spectrophotometric methods can be automized either in an Auto-analyzer version (Goulden, 1978), or by means of flow Injection Analysis (Ruzicka and Hansen, 1981, Slanina et al., 1980a). In the auto-analyzer or related systems the flowsystem is interrupted by airbubbles to diminish broadening of the sample plug, in the FIA-system the sample is injected into the flowsystem without any airsegmentation. Especially the latter system is fast, up to 300 determinations per hour can be performed.

Some years ago Ion Selective Electrodes were seen as the solution for many problems in the field of analysis of surface waters and precipitation. In general they were a deception, either because their selectivity was insufficient (e.g. in case of the NO_3^--electrode) or because they presented grave troubles in maintenance (e.g. the NH_3 electrode). Other electrodes, such as the halogenide ISE, were in no way superior to the classical photometric methods. There are two notable exceptions: the glass electrode for H^+ and the LaF_3 electrode for F^- measurements. A problem in the case of the H^+ electrode is the varying but generally low concentrations of electrolytes present in precipitation which result in variations in liquid junction- and streaming potentials in weakly stirred or flow cells.

The best remedy is the addition of 0.005 to 0.01 M KNO_3. Direct pH measurements without addition of electrolytes can be off as much as 0.2 pH in the pH range of 5 to 7 (Winkel et al., 1974, Tyree, 1981).

In the case of the fluoride electrode, the slow response at the low concentrations encountered in precipitation is a problem. But both electrodes can very well be used as detectors in Flow Injection Systems, and, with the proper precautions, low detection limits (1 ppb for F^-), good accuracy and a high through-put (more than 100 samples per hour) can be achieved (Slanina et al., 1980b).

Nuclear analytical methods have proven their usefulness in this field. Neutron activation has been applied to the determination of chloride and bromide quite successfully (Slanina et al., 1979, Schutijser et al., 1978). Another example is the use of isotope dilution for the determination of sulphate by means of the addition of labelled sulphate to the sample, precipitation of the sulphate by a substoichiometric amount of barium, followed by determination of the sulphate activity in the supernatant solution (Klockow et al., 1974).

Of course, these methods have the drawback that one has to have a nuclear reactor, or in the latter case a laboratory equipped for radiochemistry at hand.

The largest contribution to the analytical chemistry of precipitation has been made, in my opinion, by Small, Stevens and Baumann (1975) in developing ion-chromatography. In this method a sample is injected on an anion exchange column with low exchange capacity using carbonate or a weak organic acid as eluent. The anions are separated in this way. Next the eluent containing the different anions is pumped through a column filled with cation exchanger in the H^+ form. The eluent is converted to CO_2 or a weakly dissociated organic acid giving a low background for conductivity detection, while the anions are converted to their respective acids, which give in case of F^-, Cl^-, NO_2^-, NO_3^-, PO_4^{3-}, SO_3^{2-}, Br^- and SO_4^{2-} a strong conductivity signal. The detection limit is about 0.1 ppm or lower for the above mentioned anions if a sample of 0.5 ml is injected. The selectivity of the anion exchange columns is limited, so one either has the choice to analyze at a rate of 1 sample per 12 to 25 minutes, or to apply, as is done modern HPLC very often, multiple detection systems and post column reactions. If the latter approach is chosen with e.g. UV-detection for NO_2^- and NO_3^- (Slanina et al., 1981), amperometric detection for the halogens (Rocklin and Johnson 1982), ion-selective electrodes for F^- and halogens or simply photometric postcolumn reactions for sulphite and chloride, a very versatile, accurate and fast system can be assembled which can measure up to 8 anions, including the bulk species Cl^-, NO_3^- and SO_4^{2-}, at a rate of 15 samples per hour. The selectivity is extraordinary good and the detection limits are in the order of 0.01 to 0.03 ppm.

Ion-chromatography has also been used for the determination of organic acids in precipitation. But in view of the very low concentrations, gas-chromatography seems to be more suitable for this task (Klockow et al., 1978).

The determination of SO_3^{2-} and NO_2^- poses special problems. Their stability in precipitation is quite limited so stabilizers must be present in the sample-bottles right at the beginning of the sampling period. Both mercuric tetrachloride and formaldehyde solutions have been successfully applied to stabilize SO_3^-, followed by either a determination by IC (Dasgupta et al., 1980) or photometry (Hales et al., 1971).

Determination of cations in precipitation.

The determination of H^+ by means of a glass electrode has already been described in the last paragraph.

For ammonium several good spectrophotometric methods are available. The widely used Berthelot reaction based on formation of indophenol by ammonia with hypochlorite and phenol, gives excellent results (APHA 1976). It has been automated by means of auto-analyzer and Flow Injection Analysis (Ruzicka and Hansen 1981, Slanina et al., 1980a).

The role of flame emission spectrometry in the determination of the bulk cations Na^+, K^+, Mg^{2+} and Ca^{2+} has been taken over by atomic absorption spectrometry. Good standard methods are availabe, with sufficient accuracy and quite low detection limits. In most cases Zn and Fe can be measured by straight forward flame technique, the detection limit is, in our experience, given by contamination occuring in sample pretreatment procedures, which will be discussed later.

The graphite furnace technique, using a hollow graphite rod in which the sample (10-50 µl) is injected and subsequently heated to 3000°C, is for many elements a good approach. The determination of Pb, Cd and Cu is straight forward.

Refractory metals such as Al, Mn, Cr can be measured, but the high furnace temperatures (2500-3000°C) which are required reduce the lifetime of the graphite furnaces. Indirect AAS procedures, combining sample extraction and concentration are available. The detection limits are low (Kinrade and Van Loon 1974), but we found the reliability of these methods insufficient.

The use of hydride generation for As, Sb and Se by means of sodium borohydride followed by AAS measurement of the formed hydrides in quartzcell is, in our experience problematic. A sample of 50 to 100 ml is required to achieve meaningful detection limits. Commercial apparatuses accept samples of 10 to 20 ml, resulting in detection limits for As and Se of 0.1 and 0.25 ppb in practice, which is too high in many cases (Robbins and Caruso 1979). The interference of Cu^{2+} and Fe^{3+} can be a problem. The reduction of Hg by Sn^{2+} followed by measurement in a cell is very sensitive (Oda and Ingle, 1981), but special measures, such as the addition of strong oxidants to bring Hg in the Hg^{2+} form are necessary to obtain representative samples.

Recently, Induction Coupled Plasma Emission Spectrometry, has come of age. The sample is aspirated into a plasma torch at temperatures up to 10.000 K, and the emission of the elements of interest are measured (Greenfield 1980). Though ICP is not free from matrix effects, as was claimed in the past, the technique is very useful because many refractory elements (e.g. Ba, Sr, Mn, Cr, Al, Fe, V, Co) can be measured fairly easy in environmental samples and precipitation (Floyd et al., 1980, Broekaert et al., 1982). ICP is also suitable for the determination of B and P.

Polarographic methods are used for the determination of Zn, Cd, Pb and Cu (Nuernberg et al., 1976), but also for As, Sb, Se and Te (Ebhart and Umland 1982). The availability of modern instrumentation for anodic stripping voltammetry has contributed to the popularity of this technique. The advantages are that the detection limits are sufficiently low for precipitation analysis, that 2 to 4 elements can be measured simultaneously and that information about the speciation of the elements can be obtained. Problems can arise because of the interference by intermetallic complexes and the influence of organic compounds forming complexes with the measured species or films blocking the transfer of ions to and from the measuring electrodes (Schulze and Frenzel 1983). Careful sample pretreatment is therefore needed and in our experience the reliability of polarographic techniques is inferior compared to AAS and ICP.

Neutron activation Analysis is quite useful for the determination of e.g. V, Al, Mn, Co, but has only be applied in limited number of occasions.

To summarize the analytical possibilities to determine (trace) metals in precipitation one could state that in general adequate methods are available, except in the case of As, Sb, Se and perhaps B, which are however important species.

The most important problems are encountered in sampling and pretreatment procedures:

- Contamination by parts of the collector, such as funnels and sample-bottles and by splash-off of the lid and walls, is a very difficult problem and imposes rigid restrictions in collector construction.
- Contamination during pretreatment procedures in the laboratory is difficult to prevent. At ECN all handling of precipitation is done in laminar flow clean benches. This contamination problem is also observed in the case of NH_4^+, Cl^- and NO_3^-, substances which are abundantly present in laboratory air.

Determination of H_2O_2 and organic substances.

The discovery that H_2O_2 could play an important role in the oxidation of SO_2 to SO_4^{2-} has started many attempts to measure the concentration of H_2O_2 in precipitation. The first results obtained by Kok (1980) were surprising. Later work indicated that the luminol method (basd on the chemiluminescence observed on mixing luminol, a Cu^{2+} solution and sample in alkaline conditions) used in this study was prone to interference e.g. by heavy metals. Better methods are available at the moment (see Klockow and Jacob, this volume). Cationexchangers have been used to get rid of the heavy metal interference (Roemer and Slanina 1982). Methods based on the use of enzymes such as peroxidase are more selective, and even capable of making a distinction between H_2O_2 and organic peroxides (Ten Brink 1983). Sampling is a severe problem: The stability of H_2O_2 in precipitation samples is very limited and polyethylene destroys H_2O_2 rather rapidly, a fact confirmed in studies at ECN. Glass seems to be the best material for sampling H_2O_2 but it is feared that H_2O_2 production can occur easily if water, in contact with glass, is exposed to light.

The determination of organic substances has not yet been persued with a lot of interest. Gaschromatographic methods have been described for the determination of e.g. carbonic and organosulphonic acids (Klockow et al., 1978). The content in precipitation seems to be quite low (less than 100 ppb). Investigations by means of GC and GC-MS of alkalines and fatty acids in rainwater have turned out that their concentration is below 10 ppb. Recently some programs were started in the Netherlands to determine the concentrations of poly-aromatic hydrocarbons by means of HPLC and GC-MS, the first results indicate that the concentrations are below 1000 ppt (Noord, van, 1983).

SAMPLING STRATEGY AND THE INTERPRETATION OF THE RESULTS

As mentioned in the introduction, the sample strategy is a function of the goals of an investigation. The subject of measurement strategy in direct relation to the subsequent interpretation of the obtained data base, has not yet been studied intensively, and so it is not possible to indicate standard solutions for each problem, but the problems and possible solutions will be discussed for a number of important fields in precipitation chemistry research.

1. Measurement of the wet deposition for a certain area.

Remarkedly few studies have been published about network design for wet deposition measurements, even though many monitoring networks have been in operation during the last 5 to 10 years. Granat (1975a), Barrie (1982), Haszpra (1980, 1981) and Horvath (1980) come to the conclusion, based on Scandinavian and U.S. data, that the uncertainty in deposition measurements are a function of the sample period, sampler density and of the compound which is measured. In case of annual deposition estimates obtained by monthly sampling, an average distance of 50 km between samplers would typically result in a standard deviaton of ca. 10% for SO_4^{2-} and NO_3^-, 20% for NH_4^+, and 40% for Mg^{2+}. This uncertainty goes up as the period for which the deposition is calculated, is smaller. In order to arrive at the same uncertainty for monthly deposition values, the sampler density must be six times higher (figures 9 and 10).

The data available to calculate the necessary collector densities for shorter periods are insufficient.

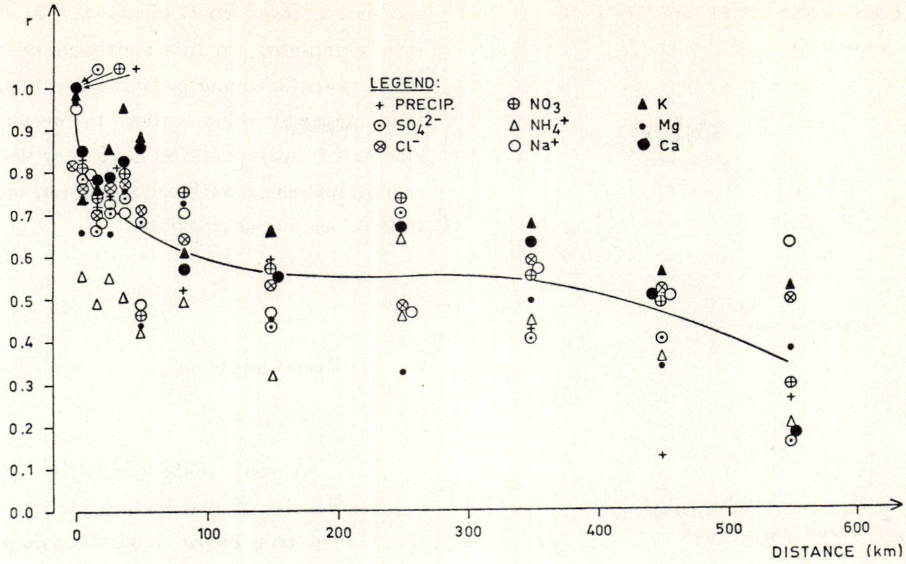

Fig. 9 Correlation coefficient as a function of separation between stations (after Granat)

In deposition measurements around large single sources in Scandinavia (Granat 1975a) and in the Netherlands (Slanina et al., 1982) variations of a factor 3 to 6 in deposition were observed over distances of 50 km for single events (figure 11). The aerial variation in concentration was smaller in both cases (ca. 50%). So the best method to measure deposition on a small time-scale (a week or less) would be to calculate the area variability of the concentrations, obtained by chemical precipitation collectors in a low density network, and use for deposition calculations measurements of the amount of rainfall by means of ordinary standard raingauges. This procedure is only possible if a high correlation between the concentration in adjacent collectors is observed. In our experiments 60 samplers were placed on an arc segment of 25 km,

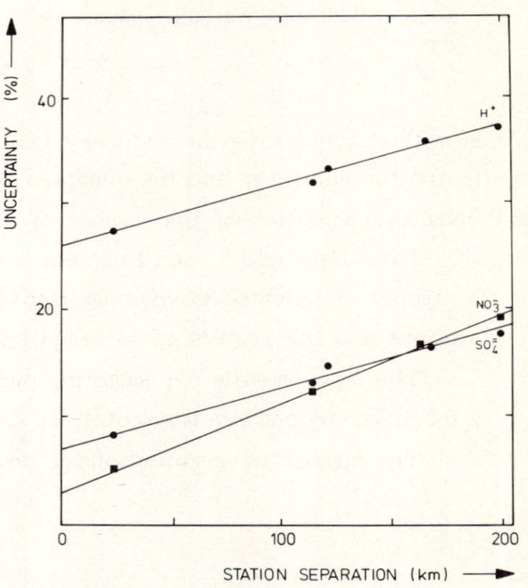

Fig. 10 The total uncertainty in annual precipitation concentration in eastern U.S. as a function of network spacing (after Barrie)

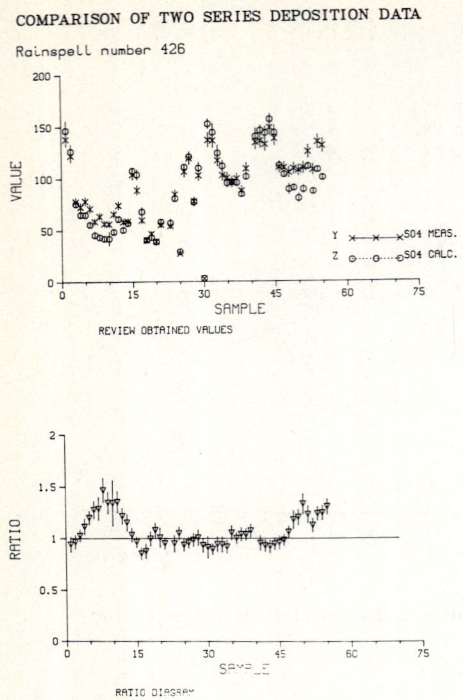

Fig. 11 Deposition of SO_4 around a point source. Plume washout measurements

and correlation coefficients up to 0.98 were calculated, so this approach is feasible, given adequate sampling and analytical procedures in order to keep the number of outliers which have very detrimental influence on the correlation coefficient, as low as possible.

2. Trend analysis

A small scale variability in the results of one station of 4 to 10% relative on an annual basis will obscure a trend in the chemical composition of precipitation caused by e.g. a 4% increase of emissions to such an extend that it will take 9 years of data to estimate this trend with 25% accuracy (Granat, 1975a). So it is important to have data which are as accurate as possible for one location. An obvious method to improve the accuracy is to use more than 1 collector at one site. We have investigated the accuracy and the standard deviation of the concentrations of bulk and trace compounds as a function of the number of collectors used.

8 wet only and 8 open samplers were placed on one spot (150 x 200 meter) near the airport of Lelystad where only light planes are allowed. The sampling period was 24 hours and the results of 44 sampling periods were used in this study.

The eight results per sampling period can be rearranged as if they were obtained by different combinations of 1, 2, 3, 4, 5, 7 and 8 samplers.

The number of combinations is dependent on how many samplers are investigated.

Number of samplers	Number of combinations
1	8
2	28
3	56
4	70
5	56
6	28
7	8
8	1

The mean value and standard deviation of any combination of 1 to 8 samplers are computed per sample period.

The results of each combination are tested for Gaussian distribution and a Dixon test (a = 0.05) is applied to detect outliers. All samples with 3 or more outlying results are deleted. The outlier test is, of course, less powerful at a lower number of

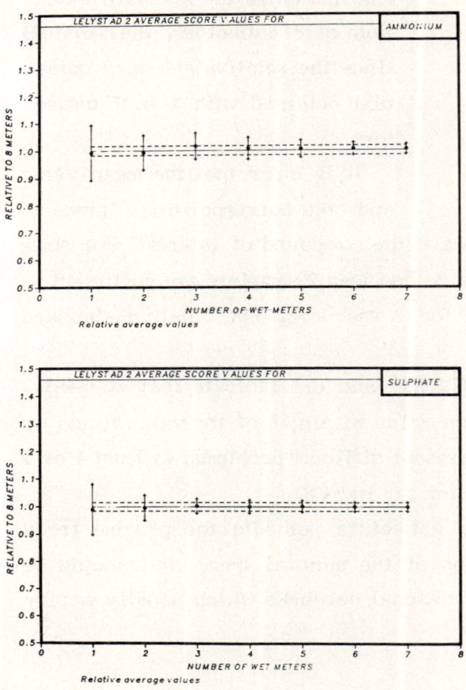

Fig. 12 Mean value and standard deviation as a function of number of samplers for NH_4^+ and sO_4^{2-}

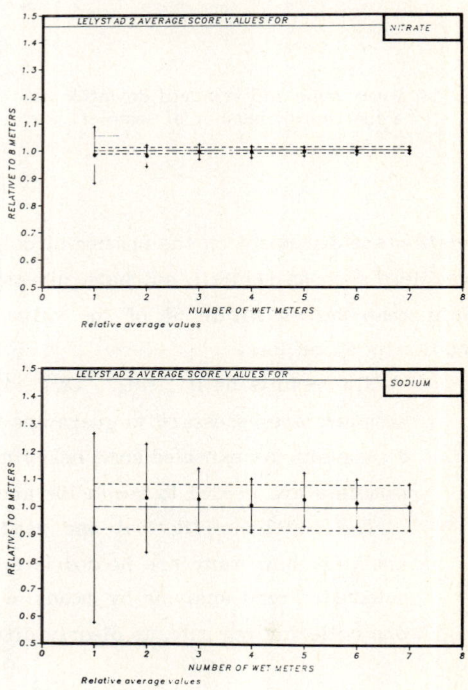

Fig. 13 Mean value and standard deviation as a function of number of samplers for NO_3^- and Na^+

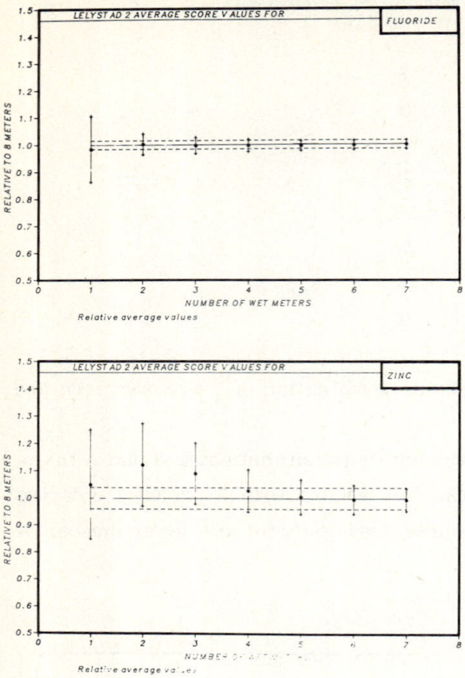

Fig. 14 Mean value and standard deviation as a function of number of samplers for F^- and Zn

samplers and practically meaningless if combinations of less than 4 samplers are tested.

The overall mean value and the standard deviation over the whole experimental period is obtained by pooling the results of each combination of samples per sampling period.

An extensive statistical treatment is given elsewhere (Slanina 1983). In figures 12, 13 and 14 the mean concentration found by all combinations of 1 to 7 collectors is given as the ratio of the mean value found for 8 samplers and the mean value obtained with different number of collectors. The dashed line indicates the standard deviation of 8 collectors, the vertical lines the relative standard deviation obtained with 1 to 7 collectors.

It is clear that the mean value and the corresponding standard deviation are dependent on the number of collectors and the compound of interest. For some parameters such as rainfall, sulphate, nitrate, fluoride and lead 2 samplers are sufficient to obtain mean values within 5% of the value found for 8 samplers. The relative standard deviation is 5% or less.

The results of H^+, NH_4^+, Ca^{2+}, Mg^{2+}, K^+, Na^+ and Cl^- indicate that at least 3 samplers are necessary to guarantee the mean value within 5% of the value found by 8 samplers. As expected some heavy metals present difficult problems, at least 4 or 5 samplers are needed to reach 10% accuracy for Zn and Cd.

The conclusion is that one needs more collectors per site for precise trend analysis; how many are needed is a function of the minimal trend that should be detected. Trend analysis by means of the traditional networks which usually employ one collector per site, is clearly difficult.

3. Source-Receptor relations

Generally, source-receptor relations for the chemical composition of precipitation are investigated by statistical analysis, such as factor analysis, cluster analysis and multiple regression analysis. The statistical methods are very sensitive to the number of outlying results in data bases and the standard deviation of the results. The use of more collectors per site is often necessary to obtain significant results by means of this techniques.

As an example the relative distribution of H^+, NH_4^+, Ca^{2+}, Mg^{2+}, K^+, and Na^+ as counter ions to SO_4^{2-}, NO_3^-, and Cl^- is calculated for two experiments in figure 15.

This distribution was calculated by means of multiple regression analysis, outlined elsewhere (Slanina et al., 1982). The sample periods are divided according to origin of the airmasses in marine and continental. DK and LS refer to the location of the experiments in the north-western part and the centre of The Netherlands respectively.

A few conclusions:

- The distribution of counter ions is different for sulphate compared to

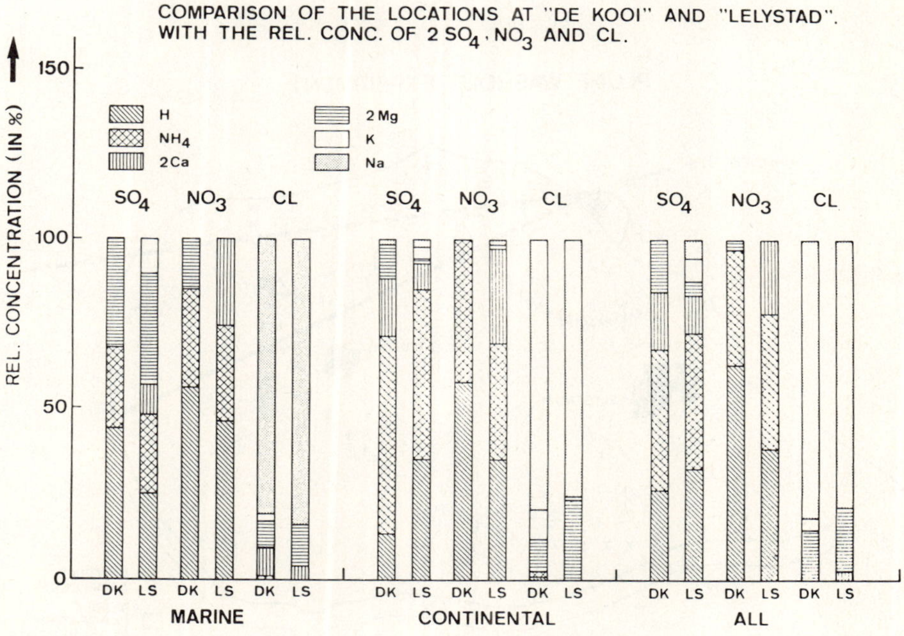

Fig. 15 Distribution of H^+, NH_4^+, Ca^{2+}, Mg^{2+}, K^+ and Na^+ as counterions to SO_4^{2-} and NO_3^- in precipitation

nitrate, which indicates that the history of the both species is different and that at least one of them has not been formed in the water phase e.g. by cloud chemistry.

- The distribution of counter ions is dependent on the location. But in both locations the contribution of HNO_3 to total nitrate exceeds the fraction of H_2SO_4 of total sulphate.
- The counter ions found for Cl^- reflect the marine origin in all cases. Little or no HCl and NH_4Cl are found indicating that HCl does not contribute significantly to the acidity in precipitation.

4. Special studies around large single sources.

In a series of experiments, near the Flevo power station, conducted by ECN, IMOU and KEMA, the wash-out of SO_2, NO_x, SO_4^{2-}, and NO_3^- out of the plume of the power station by precipitation was investigated. The set-up of the experiment is given in figure 16. Event or daily sampling was employed. The variation in the background deposition was so large (as mentioned above) that it was very difficult to detect the

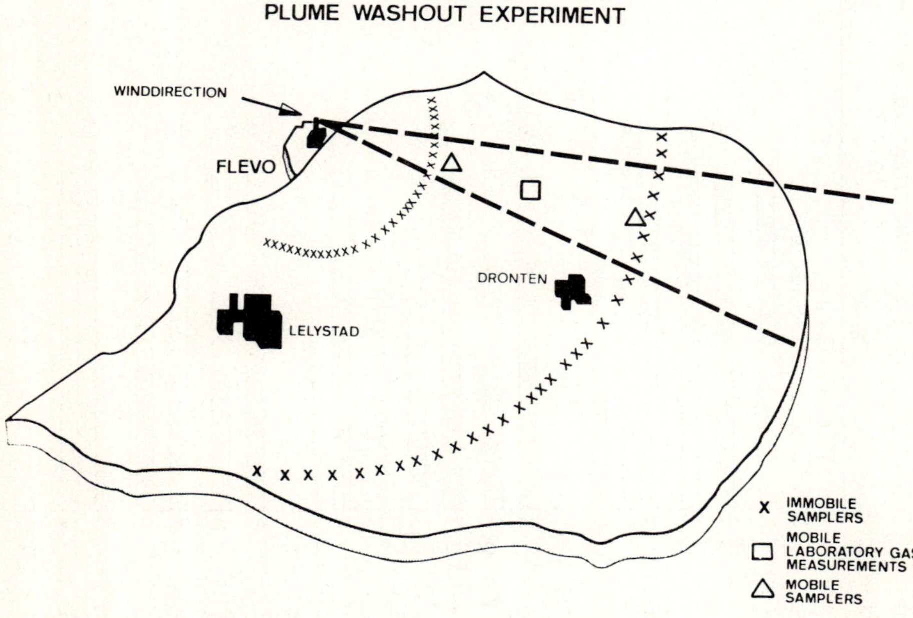

Fig. 16

contribution of the plume. But if meteorological parameters (amount and rate of rainfall, droplet size etc.) are the main cause of this variation, high correlation for all compounds between the collectors should be present, as indeed was observed.

For each sampler a sulphate concentration can be calculated based on the correlation with other species $(SO(i))_{calc}$ = A.C(i) + B; C_i = concentration of species, not present in the plume.

The mean of these calculated sulphate concentrations is then computed taking into account the "goodness of fit".

For background samples $\frac{(SO_4)_{measured}}{(SO_4)_{calculated}}$ varied between 0.9 and 1.1, indicating that:

- the data base probably does not contain many outliers, which is an indirect way to validate the data base,
- the contribution of the background to sulphate deposition under a plume can be calculated.

An extra contribution of the plume was indeed observed in good accordance with the prevailing wind direction (figure 14) in a number of experiments and a wash-out coefficient for SO_4^{2-} of 1% per hour over 20 km was calculated, which corresponds well with recent data obtained elsewhere (Dana and Parinos 1982).

The extra contribution of the Flevo power plant (emitting 6 tons of SO_2 per hour) to the wet deposition of SO_4^{2-} can be calculated using this wash-out coefficient, and is about 5% of the "natural" background north-east of the power plant. The extra H^+ deposition is maximal 15% of the background.

Wash-out of NO_x was investigated in the same way, and detected in two experiments (fig. 17).

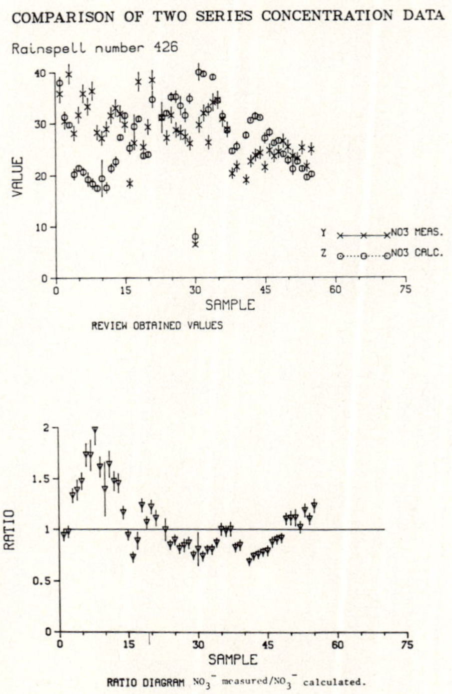

Fig. 17 NO_3^- conc. around a point source. Plume washout measurements

Statistical evaluation has indicated that the background corrections by means of multiple regression are very dependent on the number of samplers used; if less than 30 samples are employed, the plume can be missed altogether (Slanina et al., 1982).

CONCLUSION

To obtain meaningful results it is necessary to optimize the sampling strategy, sampling and analytical methodology for each object of precipitation research.

Good statistical design of each study is essential to its success.

REFERENCES

APHA, ASSA and WPCF, Standard Method for the Examination of Water and Waste Water, Washington, 1976.

Asman W.A.H., Draft, Construction and Operation of a Sequential Rain Sampler., J. Wat. Air Soil Poll. 13, 1980, pp 235-245.

ASTM Standards, Part 31, 1976, American Society for Testing and Materials.

Barrie L.A., Interpretation of Network Data: Spatial Temporal Variations, Proc. Symp. on Monitoring, Ottawa, Nat. Res. Council Canada, 1982.

Broekaert, J.A.C., Wopenka B. and Puxbaum H., Inductively Coupled Plasma Optical Emission Spectrometry for the Analysis of Aerosol Samples Collected by Cascade Impactors, Anal. Chem. 54, 1982, pp 2174-2179.

Dana M.T. and Parinos A.A.N., Scavenging measurements near large point source: a review and evaluation, Procr. 4th Intern. Conf. on Precipitation Scavenging, Dry Deposition and Resuspension, Pruppacher H.R., Semonin R.G., Slinn W.G.N., eds., Elsevier, 1982, pp 335-349.

Dasgupta P.K., DeCesare K. and Ullrey J.C., Determination of Atmospheric Sulphur Dioxide without tetrachloromercurate (II) and the Mechanism of the Schiff Reaction, Anal. Chem. 52, 1980, pp 1912-1922.

Dawson G.A., Ionic Composition of Rain During Sixteen Convective Showers, Atmos. Environ. 12, 1978, pp 1991-1999.

De Pena R.G., De Pena J.A. and Bowersok V.C., Precipitation Collectors Intercomparison Study, Report, Department of Meteorology, Pennsylvania State University, 1980.

Ebhart K.B., Umland F., Untersuchung zur Verbesserung der simultanen Voltammetrischen Bestimmung von Selen und Tellur durch "Cathodic-Stripping", Fres. Z. Anal. Chem. 310, 1982, pp 406-409.

Floyd M.A., Fassel V.A. and Silva A.P.,Computer-Controlled Scanning Monochromator for the Determination of 50 Elements in Geochemical and Environmental Samples by Inductively Coupled Plasma-Atomic Emission Spectroscopy, Anal. Chem. 52, 1980, pp 2168-2173.

Galloway J.N. and Likens G.E., Calibration of Collection Procedures for the Determination of Precipitation Chemistry, J. Wat. Air Soil Poll. 6, 1976, pp 241-258.

Galloway J.N. and Likens G.E., The Collection of Precipitation for Chemical Analysis, Tellus 30, 1978, pp 71-82.

Goulden P.D., Environmental Pollution Analysis, Heijden, London, 1978, pp 71-119.

Granat L., On the Deposition of Chemical Substances by Precipitation, WMO report no. 368, 1974, pp 71-78.

Granat L., Principles in Network Design for Precipitation Chemistry Measurement, Proc. First Specialty Symposium on Atmospheric Contribution to the Chemistry of Lake Waters, Intern. Assoc. Great Lakes Res., 1975a, pp 42-55.

Granat L., On the Variability of Rain Water Composition and Errors in Estimates of Areal Wet Deposition, Report AC-30, 1975b, Department of Meteorology, University of Stockholm.

Greenfield S., Plasma Spectroscopy Comes of Age, Analyst. 105, 1980, pp 1022-1044.

Hales J.M., Thorp J.M. and Wolf M.A., Field Investigation of Sulphur Dioxide Washout from the Plume of a large Coal-Fired Power Plant by Natural Precipitation, NTIS Report PB 203129, 1971.

Hansen D.A. and Hidy G.M., Review of Questions Regarding Rain Acidity, Atm. Environ. 16, 1982, pp 2107-2126.

Haszpra L., On the Planning of Precipitation Chemistry Networks I, Idojaras, 84, 1980, pp 121-129.

Haszpra L., On the Planning of Precipitation Chemistry Networks II, Idojaras, 85, 1981, pp 309-316.

Horvath L., Effect of the Sampling Period and Method on the Measured Rate of Wet Deposition of Inorganic Compounds, Idojaras, 84, 1980, pp 130-134.

Jeffries D.S. and Snyder W.R., Atmospheric Deposition of Heavy Metals in Central Ontario, J. Wat. Air Soil Poll. 15, 1981, pp 127-152.

Kinrade J.D. and Van Loon J.C., Solvent Extraction for Use with Flame Atomic Absorption Spectrometry, Anal. Chem. 46, 1974, pp 1894-1898.

Klockow D., Denzinger H. and Roenicke G., Anwendung der substoichiometrischen Isotopenverduennungsanalyse auf die Bestimmung von atmosphaerischem Sulfat und Chlorid in "Background"-Luft, Chemie-Ing.-Technik 46, 1974, p 831.

Klockow D., Bayer W. and Faigle W., Die Gaschromatographische Bestimmung von Spuren niederer Carbon- und Sulfonsaeuren in waessrigen Loesungen, Fres. Z. Anal. Chem. 292, 1978, pp 385-390.

Kok G.L., Measurement of Hydrogen Peroxide in Rainwater, Atmos. Environ. 14, 1980, pp 653-656.

Likens G.E., Wright R.F., Galloway J.N., Butler T.J., Acid Rain, Scient. Am. 24, 1979, pp 43-51.

Likens G.E., Butler T.J., Recent Acidification of Precipitation in North America, Atm. Environ. 15, 1981, pp 1103-1109.

Lindau, Conference: Saure Niederschlaege - Ursachen und Wirkung, 1983, Lindau.

Madsen B.C., An Evaluation of Sampling Interval Length on the Chemical Composition of Wet-Only Deposition, Atmos. Environ. 16, 1982, pp 2515-2519.

The MAP 3S Precipitation Chemistry Network, First Periodic Summary Report, PNL-2402, 1977, Battelle Pacific Northwest Laboratories.

Mueller K.P., Aheimer G. and Gravenhorst G., The Influence of Immediate Freezing on the Chemical Composition of Rain Samples, in: Deposition of Atmospheric Pollutants, H.W. Georgii and J. Pankrath, eds., D. Reidel, 1982, pp 125-132.

Noord P. van, Preleminary Result RID Project (personal comm.), 1983.

Nuernberg H.W., Valenta P. and Nguyen V.D., Ein neuer Weg zur Messung toxischer Metalle im Regen. KFA Juelich, Jahresbericht 1979.

Nuernberg H.W., Valenta P., Mart L., Raspor B. and Sipos L., Applications of Polarography and Voltammetry to Marine and Aquatic Chemistry, Fres. Z. Anal. Chem. 282, 1976, pp 357-367.

Oda C.E. and Ingle J.G., Anal. Chem. 54, 1981, pp 2030-2033.

Peden M.E. and Skowron L.M., Ionic Stability of Precipitation Samples, Atm. Environ. 12, 1978, pp 2343-2349.

Reijnders H.F.R., van Staden J.J. and Grieping B., Batchwise Photometric Determination of Sulphate in Water Samples, Fres. Z. Anal. Chem. 298, 1979, pp 156-157.

Robbins W.R. and Caruso J.A., Development of Hydride Generation Methods for Atomic Spectroscopic Analysis, Anal. Chem. 51, 1979, pp 889A-899A.

Ridder T., K.N.M.I., unpublished results, 1982.

Rocklin R.D. and Johnson E.L., Determination of Cyanide, Sulfide, Iodide and Bromide by Ion Chromatography with Electrochemical Detection, Anal. Chem. 55, 1982, pp 4-7.

Roemer F.G. and Slanina J., Unpublished results, 1982.

Ruzicka J. and Hansen E.H., Flow Injection Analysis, J. Wiley and Sons, New York, 1981.

Sandefjord, Conference: Ecological impact of acid precipitation, 1980.

Schulze G. and Frenzel W., Stoereinfluesse durch Kupfer bei der pontentiometrischen Stripping-Analyse von Zink und Cadmium, Fres. Z. Anal. Chem. 314, 1983, pp 459-462.

Schutijser P., Maenhout W. and Dams R., Instrumental Neutron Activation Analysis of Dry Atmospheric Fall-out and Rainwater, Anal. Chim. Acta 100, 1978, pp 75-85.

Slanina J., Moels J.J., Baard J.H., van der Sloot H.A., van Raaphorst J.G. and Asman W., Collection and Analysis of Rainwater: Experimental Problems and Interpretation of the Results, Int. J. Environ. Anal. Chem. 7, 1979, pp 161-176.

Slanina J., Bakker F., Bruyn-Hes A. and Moels J.J., A Computer Controlled Multichannel Continuous Flow Analysis System Applied to the Measurement of Nitrate, Chloride and Ammonium Ions in Small Samples of Rain Water, Anal. Chim. Acta 113, 1980a, pp 331-342.

Slanina J., Lingerak W.A. and Bakker F., The Use of Ion-Selective Electrodes in Manual and Computer-Controlled Flow Injection Analysis, Anal. Chim. Acta 117, 1980b, pp 91-98.

Slanina J., Bakker F.P., Jongejan P.A.C., van Lamoen L. and Moels J.J., Fast Determination of Anions by Computerized Ion Chromatography Coupled with Selective Detectors, Anal. Chim. Acta 130, 1981, pp 1-8.

Slanina J., Roemer F.G. and Asman W.A.H., Investigation of the Source Regions for Acid Deposition in the Netherlands. Proceedings of the Symposium on Acid Deposition, Berlin, Sept. 1982, pp 141-148, S. Beilke and A.J. Elshout (eds.), commission of the European Communities.

Slanina J., Collection and Analysis of Precipitation, Methods, Data Evaluation and Interpretation. Proceedings of the Symposium: Saure Niederschlaege - Ursachen und Wirkung, Lindau 1983, In Press.

Small M., Stevens T.S. and Bauman W.C., Novel Ion Exchange Chromatographic Method Using Conductometric Detection, Anal. Chem. 47, 1975, pp 1801-1809.

Soederlund R., and Granat L., On the Difference in Chemical composition of Precipitation Collected in Bulk and Wet-only Collectors, Report CM-57, 1982, Department of Meteorology, University of Stockholm.

Technicon Instrument Corporation, Industrial Method no. 100-70 W.

Ten Brink H.M., Private Communication, 1983.

Tyree S.Y. jr., Rain water Acidity Measurement Problems. Atmos. Environ. 5, 1981, pp 67-60.

van den Winkel P., Mertens J. and Massat D.L., Streaming Potentials in Automatic Potentiometric Systems, Anal. Chem. 46, 1974, pp 1765-1768.

Winkler P., Deposition of Acid in Precipitation. In: Deposition of Atmospheric Pollutants, H.W. Georgii and J. Pankrath (eds.), D. Reidel Comp., 1982, pp 67-76.

Zall D.M., Fischer D. and Garner M.D., Photometric Determination of Chlorides in Water, Anal. Chem. 28, 1956, pp 1665-1668.

THE PEROXYOXALATE CHEMILUMINESCENCE AND ITS APPLICATION TO THE DETERMINATION OF HYDROGEN PEROXIDE IN PRECIPITATION.

D. Klockow and P. Jacob
Department of Chemistry
Univeristy of Dortmund
D-4600 Dortmund 50
F.R.G.

ABSTRACT

It is well known that during oxidation of oxalic acid esters by hydrogen peroxide an intense chemiluminescence radiation is emitted. This phenomenon has been used already several times for the determination of hydrogen peroxide in different matrices down to very small concentrations.

This contribution deals with the adaptation of the peroxyoxalate chemiluminescence method to precipitation analysis. Instrumentation, reagents and analytical procedures are given as well as results of extensive interference studies,

Furthermore data obtained in field measurements of hydrogen peroxide in precipitation and surface water is presented, and possible correlations are discussed.

INTRODUCTION

In recent years aqueous atmospheric chemistry has received increased attention, in particular in connection with acid rain formation. Research in this field was concentrated especially on liquid phase sulfur dioxide conversion, and an important step forward was the finding that the powerful oxidant hydrogen peroxide plays a leading role in this process (Penkett et al., 1979, Moeller, 1980, Martin and Damschen, 1981, Shaw and Rodhe, 1982, Kunen et al., 1983). It is interesting to note that hydrogen peroxide has been detected in different forms of precipitation (rain, snow, graupel) already in the last century (Struve, 1969, Schoene, 1874) so that the results obtained nowadays might be compared with those reported more than hundred years ago.

Since about 1972 hydrogen peroxide has been measured in rain and cloudwater by several

NATO ASI Series, Vol. G6
Chemistry of Multiphase Atmospheric Systems
Edited by W. Jaeschke

groups, mainly by employing the chemiluminescent oxidation of luminol (Kok, 1980, Bufalini et al., 1979, Heikes et al., 1982). The application of this reaction to the determination of hydrogen peroxide in the gaseous state has been tried, but due to serious sampling artifacts has not been found to produce reliable data (Zika and Saltzman, 1982).

Recently two alternative methods have been proposed as substitutes for the luminol reaction: the enzyme (peroxidase) catalyzed oxidation of p-hydroxyphenyl acetic acid (Kadlecek et al., 1982) and of scopoletin (Zika et al., 1982b, Zika and Saltzman, 1982), which leads to a fluorescent product or to fluorescence quenching, respectively.

Another useful chemiluminescent reaction of hydrogen peroxide is the well known oxidation of oxalic acid esters with dioxetandione as a hypothetic intermediate (Rauhut et al., 1967, Stauff et al., 1976, Scott et al., 1980, Sherman et al., 1978, Shaw, 1980, Shoemaker and Birks, 1981). Decay of the latter in the presence of a suitable fluorescer leads to an intense light emission. We have evaluated the potentialities of this reaction and after several modifications are able to use it for reliable and sensitive determinations of hydrogen peroxide in aqueous samples.

EXPERIMENTAL

Reagents and solutions

- Bis(trichlorophenyl)oxalate (TCPO), prepared according to Mohan and Turro (1974).
- Perylene, analytical grade.
- Acetone, analytical grade. Purified by distillation over $CaCl_2$.
- Reagent solution: 150 mg of TCPO and 25 mg of perylene dissolved in 1 l acetone. Handled only in the dark and stored in brown glass bottles in order to reduce background signal.
- Borate buffer solution A, pH 7.8: 0.1 M H_3BO_3, 0.1 M KCl, 0.005 M NaOH; used in the flow system.
- Borate buffer B, pH 7.8: 0.5 M H_3BO_3, 0.5 M KCl, 0.025 M NaOH; used for sample conditioning.
- Hydrogen peroxide: 30% aqueous solution, analytical grade.
- Hydrogen peroxide stock solution: 2×10^{-2} M, standardized by titration with permanganate solution.
- Hydrogen peroxide standard solutions for calibration: always freshly prepared by proper dilution of the stock solution with deionized distilled water.

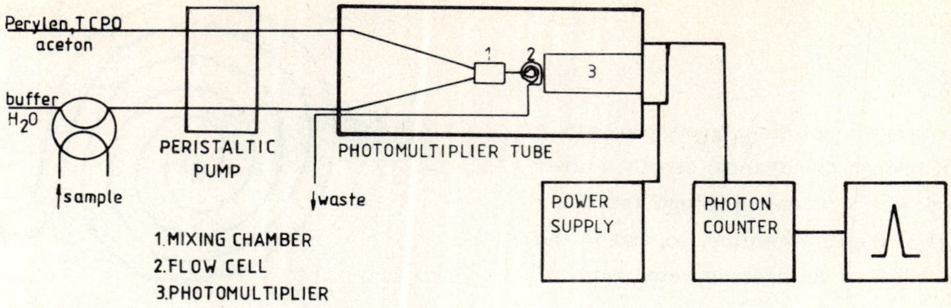

Figure 1: Scheme of the flow system.

Apparatus

Flow and detection system (Figure 1) consisting of:

- a 500 µl injection valve (Serva, Heidelberg)
- a peristaltic pump (Ismatec, Zuerich/Switzerland) with Viton^R and silicon tubing (1 mm and 2 mm, resp., inner diameter)
- a light tight housing for a photomultiplier tube (EMI 9862B/350) fitted with a tap water cooling coil and containing a mixing chamber (Figure 2) and a flowcell (Figure 3), placed in front of the window of the photomultiplier tube.
- a high voltage power supply (456H, Ortec Ltd.)
- a photon counter (5Cl, Ortec Ltd.)
- a pen recorder (5M 8251, Philips).

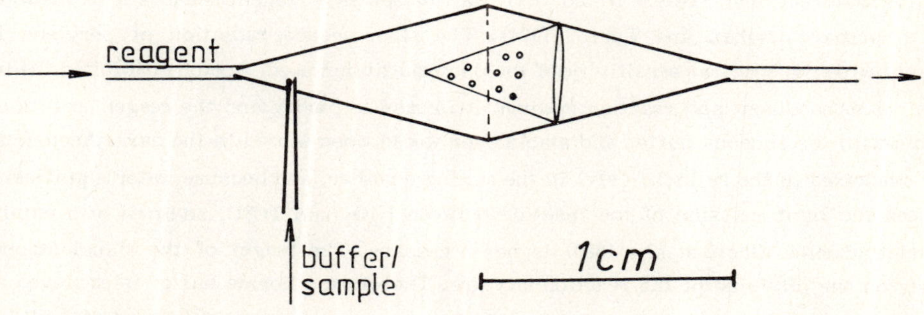

Figure 2: Mixing chamber.

PROCEDURE

The reagent solution is pumped with 3 ml/min through one channel and the buffer solution with 1 ml/min through the other channel of the flow reactor, so that in the mixing chamber an acetone:water ratio of 3:1 is obtained.

The PM tube is operated at 2000 V. The intensity of the emitted light is recorded as a signal analogous to the number of counts accumulated within a preselected counting time of typically 10 seconds.

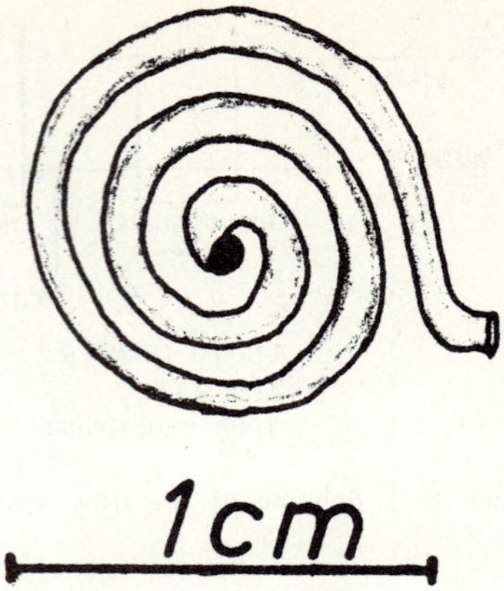

Figure 3: Flow cell.

Except at very low H_2O_2 concentrations, where a sample volume of 1 ml is necessary for analysis, 30 to 100 µl of rain are made up to 1 ml using deionized distilled water and mixed with 0.2 ml of borate buffer B. An aliquot of 500 µl of this mixture is injected into the buffer flow (borate buffer A) and the light emission measured as described above. Each determination is accomplished in about 3 minutes.

Calibration is generally carried out by employing a standard addition technique. By this means an influence on the signal caused by too high acidities of the aqueous samples (pH < 3.5) is eliminated.

RESULTS AND DISCUSSION

Of the different oxalic esters tested TCPO was chosen as a reagent because it is stable and easy to prepare (Mohan and Turro, 1974). The fluorescence radiation of perylene fits perfectly with the spectral sensitivity of the photomultiplier used. Acetone exhibited the best properties as a solvent and reaction medium. It is easy to purify and the reagent solution is miscible with the aqueous buffer and stable over weeks when stored in the dark. Acetone and water are mixed in the ratio 3:1 (v/v) in the mixing chamber, and because water significantly enhances the light emission of the reaction between H_2O_2 and TCPO, addition of a catalyst like triethylamine (Scott et al., 1980) is not necessary. The height of the signal strongly depends on the pH value of the reaction mixture. Therefore a borate buffer is employed and the samples are adjusted to the working pH before they are injected into the buffer flow.

Under the experimental conditions given a linear relationship exists between the measured

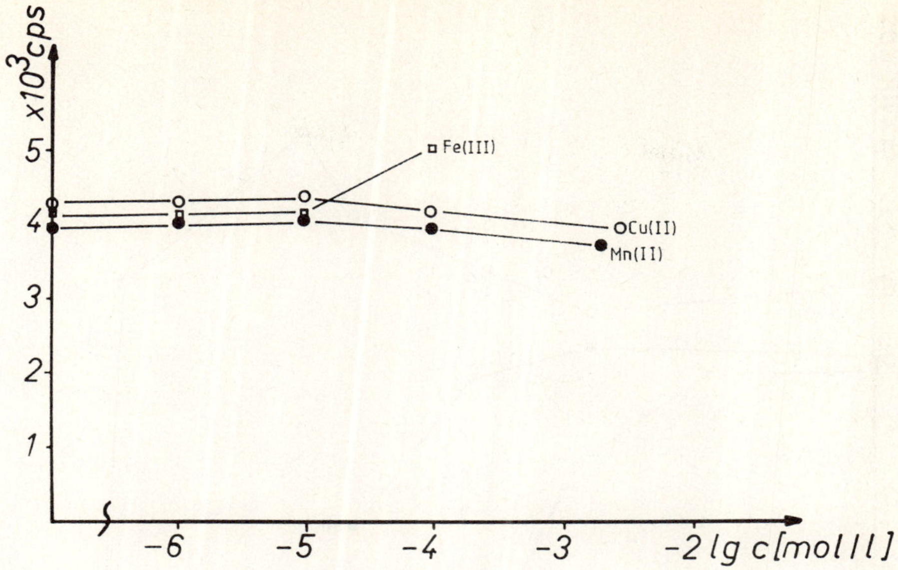

Figure 4: Influence of metal species on the signal at 5 ppb ($=1.5 \times 10^{-7}$ mol/l) of hydrogen peroxide.

light intensity and the H_2O_2 concentration in the range between the detection limit of 0.3 ppb (w/w) and 30 ppm (w/w). This means the dynamic range of the method covers 5 orders of magnitude. The relative standard deviation at a concentration of 2 ppb H_2O_2 is 4.1% (95% confidence level).

INTERFERENCE STUDIES

The influence on the observed light emission of those species was investigated, which seem to be relevant for aqueous phase atmospheric chemistry:

- A thousandfold excess of Mn(II) and Cu(II), and a hundredfold excess of Fe(III) does not effect the photomultiplier signal at a H_2O_2 concentration of 5 ppb ($= 1.5 \times 10^{-7}$ mol/l) (see Figure 4). This is in contrast to the metal ion catalyzed method in which these components exhibit a distinct influence on the reaction rate (Ibusuki, 1983). Elemental iron (in suspended form) as well as Fe(II) and Cu(I) interfere so far as they produce H_2O_2 by reduction of dissolved oxygen (Kaden et al., 1960).
- In concentrations above 5×10^{-5} mol/l Ca^{2+} causes slight positive and negative

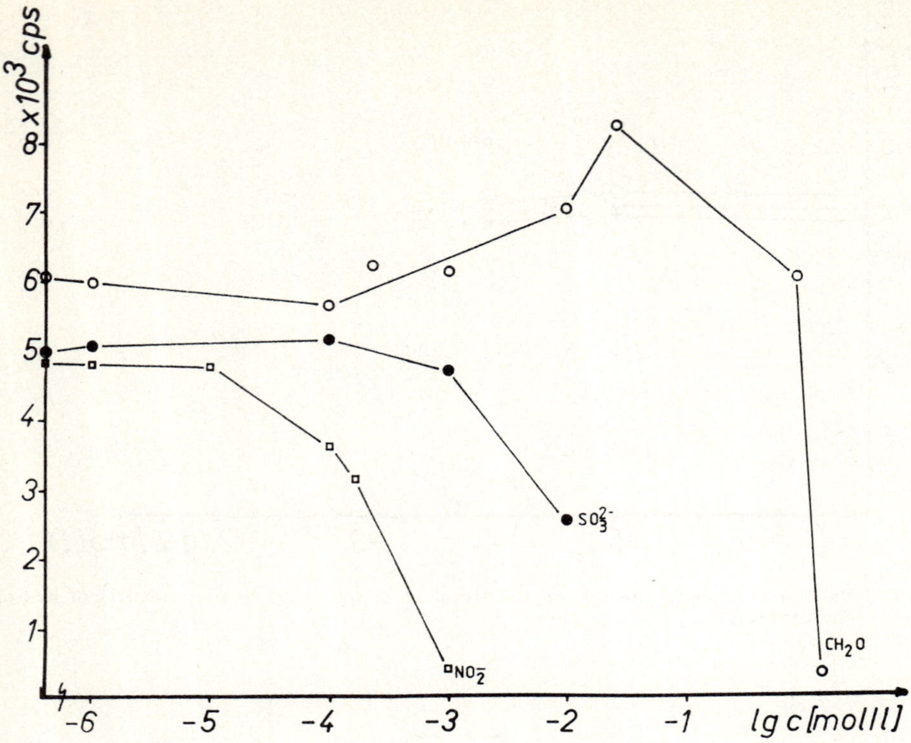

Figure 5: Influence of NO_2^-, HSO_3^- and formaldehyde on the signal at 5 ppb (=1.5 x 10^{-7} mol/l) of hydrogen peroxide.

changes of the signal, most probably due to its interaction with the borate buffer.

- A signal depression has been observed in presence of nitrite (> 10^{-5} mol/l) and sulfite (> 10^{-4} mol/l) (see Figure 5). This effect is a consequence of the reaction of these anions with H_2O_2 - possibly including other species - rather than a influence on the TCPO oxidation. The interference caused by formaldehyde (Figure 5) is of no relevance for precipitation analysis because it is not significant below about 10^{-3} mol/l.
- Hydrogen sulfide in concentrations higher than 10^{-4} mol/l enhances the light emission (see Figure 6). This might be the result of H_2O_2 formation through autoxidation of sulfide or of the start of an additional chemiluminescence reaction between H_2O_2 and H_2S (Burguera and Townshend, 1980).
- The possible influence of organic hydroperoxides was investigated by carrying out measurements in the presence of tert.-butylhydroperoxide. As can be seen from

Figure 6 a distinct positive effect is observed not before a concentration of 10^{-5} mol/l of the hydroperoxide is used. This finding compares well with the selectivity of the H_2O_2/TCPO reaction as reported by Rauhut et al. (1967).

- Dissolved ozone can contribute to the emission of light only at rather high concentrations, most probably by formation of H_2O_2 in aqueous solution according to a process like (Hoigne and Bader, 1975):

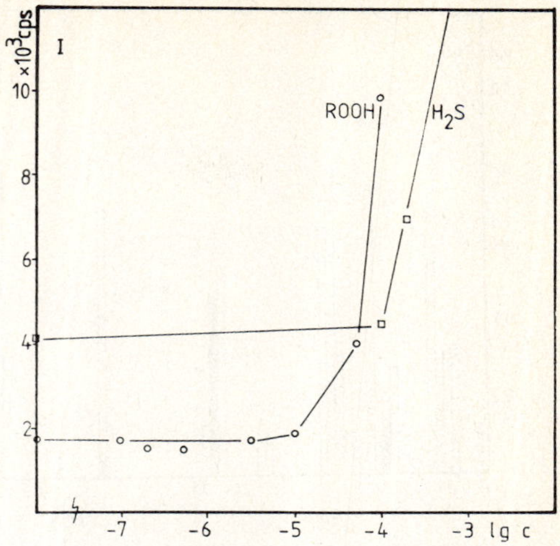

Figure 6: Influence of t-butylhydroperoxide (ROOH) and H_2S on the signal at 2 ppb (= 6 x 10^{-8} mol/l) and 5 ppb (1.5 x 10^{-7} mol/l), respectively, of hydrogen peroxide.

$$4O_3 + 2OH^- \longrightarrow 2OH + 2O_2^- + 4O_2$$

$$2O_2^- + 2H^+ \longrightarrow 2HO_2$$

$$HO_2 + HO_2 \longrightarrow H_2O_2 + O_2$$

$$4O_3 + 2H_2O \longrightarrow H_2O_2 + 5O_2 + 2OH$$

Such a process has been made responsible for the generation of artifact H_2O_2 during collection of airborne H_2O_2 in wash bottles (Zika and Saltzman, 1982). In our case a solution saturated with ozone at 20°C and about 0.1 atmosphere partial pressure (= 4 x 10^{-5} mol/l) (Graedel and Weschler, 1981) exhibited a signal corresponding to 2.3 x 10^{-6} mol/l H_2O_2. A linear extrapolation to atmospheric background conditions, i.e. 30 ppb (v/v) ozone would lead to a signal equivalent to 0.7 x 10^{-12} mol/l H_2O_2. This concentration is several orders of magnitude below the detection limit of our method (10^{-8} mol/l).

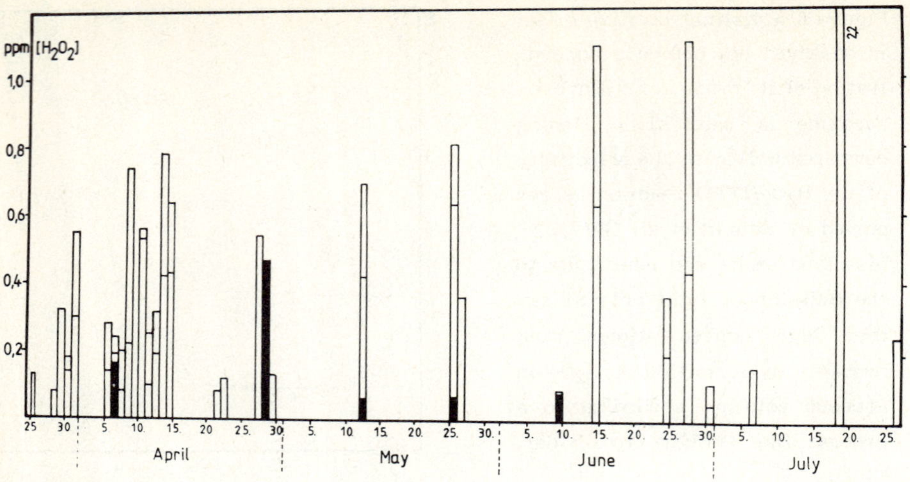

Figure 7: Hydrogen peroxide concentrations in rainwater collected at Dortmund between March and July 1983. Full columns: Results of rain collected during night. Crossline in the open columns: Mean concentration of several samples of a rain event during daytime.

ANALYSIS OF RAIN

The samples were collected at a site on the campus of the University of Dortmund. The rain collector consisted of a borosilicate glass funnel with a diameter of 19 cm and a polyethylene tube (150 cm long), which delivered the rainwater directly into the field laboratory. Daily rinsing of the funnel with deionized distilled water made sure that only wet deposition was collected.

Some of the results obtained are shown in Figure 7. Though at present a detailed interpretation is difficult the following two statements can be made:

- Precipitation collected during daytime contains more H_2O_2 (range: 2 ppb - 2.2 ppm, mean value: 260 ppb) than the one collected during night (range 0 - 440 ppb, mean value: 80 ppb).
- During the winter months (Dec. 82, Jan. 83, Jan. 84) the hydrogen peroxide concentrations are lower (range 0 - 290 ppb, mean value: 30 ppb) than during summer (May - August 83; range 25 - 2200 ppb mean value: 340 ppb).

These findings are not surprising if one accepts that hydrogen peroxide is a product of photochemical processes in the gas phase as well as in the aqueous phase (Chameides and Davis, 1982) and is easily transfered from the gas to the aqueous phase because of its high solubility (Martin and Damschen, 1981).

Fractionating sampling and analysis during rain events gave the following results:

- In 70% of the events the concentration of hydrogen peroxide decreased with increasing duration of the rain.

- Sometimes sudden changes in the H_2O_2 level occured, even at night (see Figure 8).
- In the most cases a negative correlation existed between H_2O_2 concentration and rainfall rate (mm/hr). But a substantial number of exceptions, i.e. positive correlations, have been found.

Interesting observations could be made following the changes in H_2O_2 concentrations in rainwater samples stored at diffuse light and 20°C in polypropylene vials:

- In many cases a H_2O_2 degradation occured the rate of which is significantly higher in winter than in summer (see Figure 9). This phenomenon might be explained by different HSO_3^-:H_2O_2 concentration ratios found in winter and summer (see Table 1).
- Sometimes the stored rainwater samples exhibited a sharp increase in H_2O_2 concentration. During a period of 10 to 30 minutes after collection in the field H_2O_2 was produced in these samples so that two times or even three times the initial H_2O_2 concentration was accumulated (Figure 10). A similar effect can be created by adding deionized distilled water to dry deposition collected in a borosilicate glass beaker (Figure 11). Both observations lead to the assumption that particulate matter, suspended or being dissolved in the solutions, plays a role in aqueous phase H_2O_2 generation (Jacob and Hoffmann, 1983).

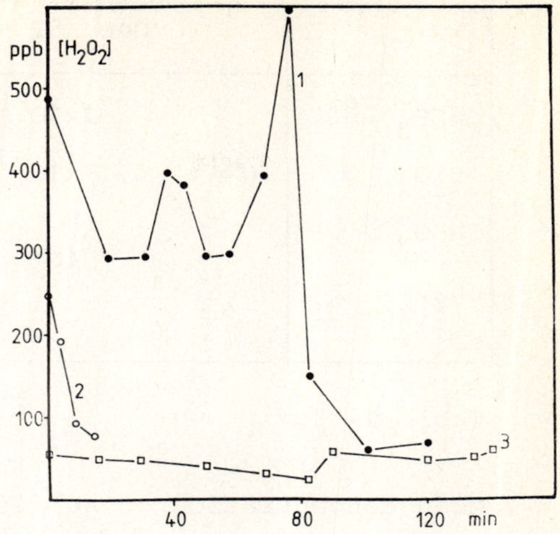

Figure 8: Hydrogen peroxide concentration during precipitation events.
1. April 4, 1983, 9.40-11.40,
2. March 30, 1983, 16.50-17.05,
3. April 12/13,1983,22.00-0.30.

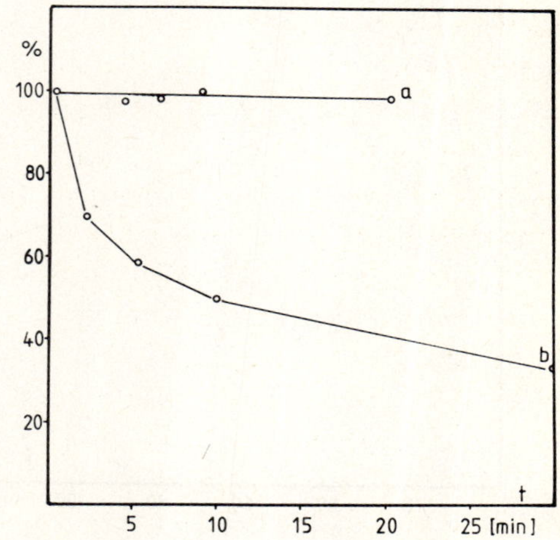

Figure 9: Hydrogen peroxide decay in stored rainwater.
a) May 25, 1983; 100% = 125 ppb H_2O_2, (pH 3.5)
b) Jan. 1, 1983; 100% = 3.7 ppb H_2O_2 (pH 3.3)

	Dec. 82 - Jan. 83	Apr. 83-June 83
HSO_3^- a)	$3x10^{-5}$mol/l	$5x10^{-6}$mol/l
H_2O_2	$9x10^{-7}$mol/l	10^{-5}mol/l
$\frac{HSO_3^-}{H_2O_2}$	40	0.5

Table 1

Mean concentration of sulfite and hydrogen peroxide and sulfite to hydrogen peroxide concentration ratios during different periods of the year.
a) Measurements made by A. Bambauer (1983).

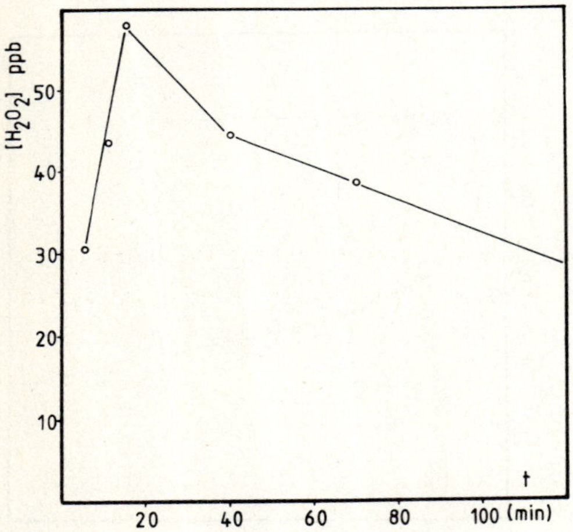

Figure 10: Increase in hydrogen peroxide concentration in stored rainwater at Jan. 20, 1984.

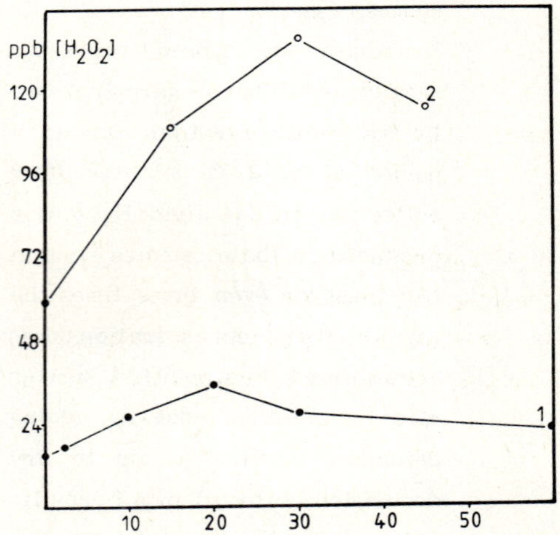

Figure 11: Hydrogen peroxide growth in water mixed with dry deposition matter. 1. Dry deposition collected within 1 day. 2. Dry deposition collected within 3 days.

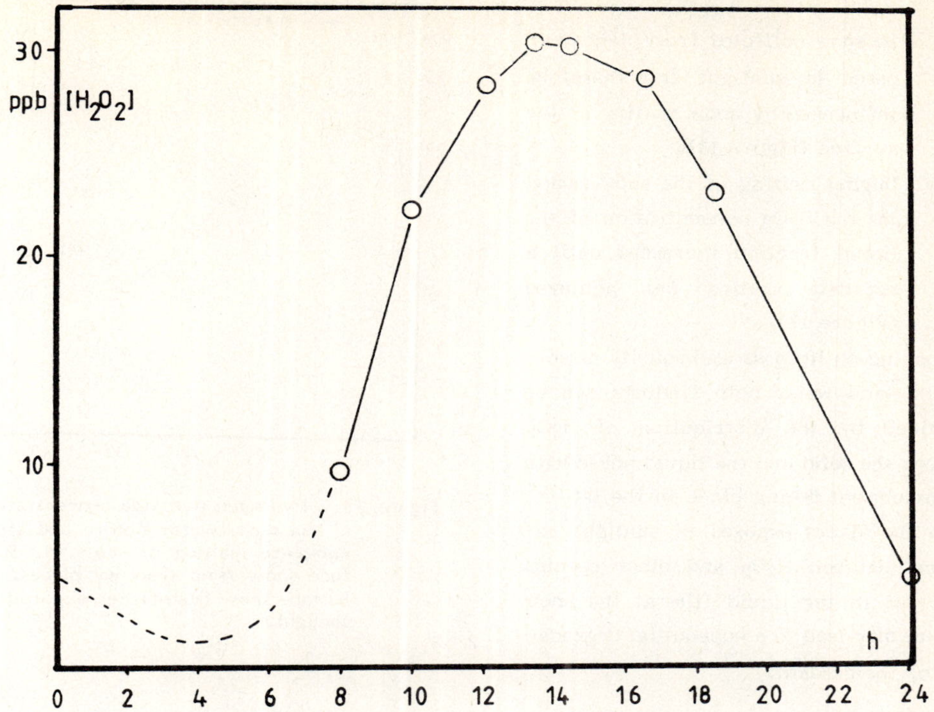

Figure 12: Diurnal variation of the hydrogen peroxide concentration in the surface layer (1 cm) of a pond at a clear day (July 7, 1983).

ANALYSIS OF SURFACE WATER

The general importance of H_2O_2 production in surface waters exposed to sunlight has recently been reported by Cooper and Zika (1983). We have made almost the same observations as these authors employing the peroxyoxalate chemiluminescence for H_2O_2 determination. One example of the results is given in Figure 12.

ANALYSIS OF SNOW

During a short field investigation of snow at the campus of the University of Dortmund in January 1983 the following results were obtained:

- The H_2O_2 level was significantly higher in snow collected from shadowy places than in snow collected from places exposed to sunlight and therefore influenced by snow melting at the surface (Figure 13).

- During melting of the snow samples the H_2O_2 concentration of the liquid fraction increased until a complete melting was achieved (Figure 13).

Looking on H_2O_2 as an impurity dissolved in solid water both findings can be explained by the distribution of H_2O_2 between the solid and the liquid phase with an enrichment taking place in the latter.

In the places exposed to sunlight reactions between H_2O_2 and other species dissolved in the liquid film at the snow surface may lead to a substantial degradation of the oxidant.

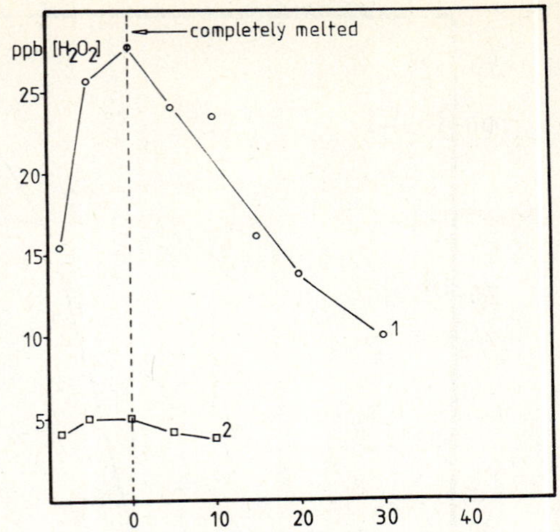

Figure 13: Hydrogen peroxide concentration in snow water before and after complete melting of snow. 1. Surface snow from shadowy places. 2. Surface snow from places exposed to sunlight.

CONCLUSIONS

The reaction between hydrogen peroxide and an oxalic acid ester has been used to establish a selective and sensitive chemiluminescence technique for the determination of traces of hydrogen peroxide in aqueous samples. The method has been applied to the investigation of rain, snow and surface water and has been found to give reliable and reproducible results also under field conditions. The method requires only small sample amounts and is fast enough to allow time resolved analyses. At present emphasis is placed upon the development of simpler instrumentation for routine field work, the investigation of snow, graupel and cloudwater, and upon the detection of hydrogen peroxide in the gas phase. Results obtained with polar ice samples will be reported elsewhere (Neftel et al., 1984).

ACKNOWLEDGEMENT

The financial support of this work by the Minister of Science and Research of the State of Nordrhein-Westfalen is gratefully acknowledged.

REFERENCES

Bambauer A. (1983), Untersuchungen zum Auftreten und Reaktionsverhalten von Sulfit, Nitrit und Formaldehyd in Niederschlaegen, Diplomarbeit, Dortmund, p. 45

Bufalini J.J., Lancaster H.T., Namie G.R. and Gay B.W. (1979), Hydrogen peroxide formation from the photooxidation of formaldehyde and its presence in rainwater. J. Environ. Health A14(2), pp. 135-141.

Burguera J.L. and Townshend A. (1980), Determination of ng/ml levels of sulfide by a chemiluminescence reaction. Talanta 27, pp. 309.

Chameides W.L. and Davis D.D. (1982), The free radical chemistry of cloud droplets and its impact upon the composition of rain. J. Geophys. Res. 87, pp. 4863-4877.

Cooper W.J. and Zika R.G. (1983), Photochemical formation of hydrogen peroxide in surface and ground waters exposed to sunlight. Science 220, pp. 711-712.

Graedel T.E. and Weschler C.J. (1981), Chemistry within Aqueous atmospheric aerosols and raindrops. Review of Geophysics and Space Physics 19, pp. 505-539.

Heikes B.G., Lazrus A.L., Kok G.L., Kunen S.M., Gantrud B.W., Gitlin S.N. and Sperry P.D. (1982), Evidence for aqueous phase hydrogen peroxide synthesis in the troposphere. J. Geophys. Res. 87, pp. 3045-3051.

Hoigne J. and Bader H. (1975), Role of hydroxyl radicals as oxidizing intermediate. Science 190, pp. 782-784.

Ibusuki T. (1983), Influence of trace metal ions on the determination of hydrogen peroxide in rainwater by using a chemiluminescence technique. Atmospheric Environment 17, pp. 393-396.

Jacob D.J. and Hoffmann M.R. (1983), A dynamic model for the production of H^+, NO_3^- and SO_4^{2-} in urban fog. J. Geophys. Res. 88, pp. 6611-6621.

Kaden Th., Walz D. and Fallab S. (1960), Reaktivitaet von Koordinationsverbindungen IV. ueber die Autoxidation von Eisen(II)-Salz-Loesungen. Helv. Chim. Acta 43, pp. 1639-1645.

Kadlecek J., McLaren S., Camarota N., Mohnen V. and Wilson J. (1982), Cloud water chemistry at Whiteface Mountain, 4th Int. Conference on Precipitation Scavenging, Dry Deposition and Resuspension. Santa Monica, California, 11.2.-12.3.1982.

Kok G.L. (1980), Measurements of hydrogen peroxide in rainwater. Atmospheric Environment 14, pp. 653-656.

Kunen S.M., Lazrus A.L., Kok G.L. and Heikes B.G. (1983), Aqueous oxidation of SO_2 by hydrogen peroxide. J. Geophys. Res. 88, pp. 3671-3674.

Martin L.R. and Damschen D.E. (1981), Aqueous oxidation of sulfur dioxide by hydrogen peroxide at low pH. Atmospheric Environment 15, pp. 1615-1621.

Moeller D. (1980), Kinetic model of atmospheric SO_2 oxidation based on published data. Atmospheric Environment 14, pp. 1067-1076.

Mohan A.G. and Turro N.J. (1974), A facile and effective chemiluminescent demonstration experiment. Chem. Education 51, pp. 528-529.

Neftel A., Oeschger H., Jacob P. and Klockow D. (1984), Measurements of hydrogen peroxide in polar ice samples, Nature, in press.

Penkett S.A., Jones B.M.R., Brice K.A. and Eggleton A.E.J. (1979), The importance of atmospheric ozon and hydrogen peroxide in cloud- and rainwater. Atmospheric Environment 13, pp. 123-137.

Rauhut M.M, Bollyky L.J., Roberts B.G., Loy M., Whitman R.H., Iannotta A.V., Semsel A.M. and Clarke R.A. (1967), Chemiluminescence from reactions of electronegatively substituted ary oxalates with hydrogen peroxide and fluorescent compounds. J. Am. Chem. Soc. 89, pp. 6515-6522.

Schoene Em. (1874), Ueber das atmosphaerische Wasserstoffhyperoxyd. Ber. 7, pp. 1693-1708.

Scott G. and Seitz W.R. (1980), Improved determination of hydrogen peroxide by measurement of peroxyoxalate chemiluminescence. Anal. Chim. Acta 115, pp. 221-228.

Shaw F. (1980), Development and construction of an analyser for the determination of the hydrogen peroxide content of natural water using a chemiluminescence-reaction. Analyst 105, pp. 11-17.

Shaw, R. W., and Rodhe, H. (1982), Non-photochemical oxidation in regionally polluted air during winter, Atmospheric Environment 16, pp. 2879-2888.

Sherman P.A., Holzbecher J. and Ryan D.E. (1978), Analytical application of peroxyoxalate chemiluminescence. Anal. Chim. Acta 97, pp. 21-27.

Shoemaker B. and Birks J.W. (1981), Generalized chemiluminescence spray cell for liquid chromatography detection selective detection using $O_2(^1\ _g)$. J. Chromatog . 209, pp. 251-263.

Stauff J., Jaeschke W. and Schloegl G. (1976), CO_2-Nachleuchten und Chemielumineszenz der Oxalester-Oxidation. Z. phys. Chem. Neue Folge 99, pp. 37-58.

Struve H. (1869), Ueber die Gegenwart von Wasserstoffperoxyd in der Luft. Z. anal. Chem. 8, pp. 315-321.

Zika R.G. and Saltzman E.S. (1982), Interaction of ozone and hydrogen peroxide in water: implication for analysis of H_2O_2 in air. Geophys. Res. Lett. 9, pp. 231-234.

Zika R., Saltzman E., Chameides W.L. and Davis D.D. (1982b), Hydrogen peroxide levels in rainwater collected in South Florida and the Bahama Islands. J. Geophys. Res. 87, pp. 5015-5017.

Part II

FIELD STUDIES OF CLOUDS, FOGS AND PRECIPITATION

THE ROLE OF CLOUDPHYSICS IN ATMOSPHERIC MULTIPHASE SYSTEMS: TEN BASIC STATEMENTS

H.R. Pruppacher
Meteorologisches Institut
Johannes Gutenberg Universitaet
6500 Mainz
F.R.G.

ABSTRACT

The role of cloud physics in atmospheric multiphase systems is summarized below in the form of ten basic statements: 1: In the atmosphere, the phase change of water vapor to water drops is a heterogeneous process involving aerosol particles (AP). 2: The phase change from water vapor to drops involves "preferred" AP of specific characteristics. 3: The water soluble portion of atmospheric aerosol particles, picked up by nucleation or impaction scavenging, goes into solution inside the cloud and rain drops, while the water insoluble portion remains suspended in particulate from inside the drops. 4: Most clouds in the atmosphere evaporate again after they had formed. 5: The phase change in the atmosphere from water vapor to ice crystals is a heterogeneous process involving AP. 6: In the atmosphere, the phase change to ice involves "preferred" AP of specific characteristics. 7: The atmosphere contains aerosol particles of a wide range of sizes and number concentrations which are related to each other in a characteristic manner, analogously for all air masses. 8: Atmospheric clouds and precipitation contain hydrometeors of a wide range of sizes and number concentrations which are related to each other in a manner depending characteristically on: (1) the aerosol of the air mass in which the cloud formed, (2) the dynamics of the cloud, (3) the microphysical processes resulting from interaction between the cloud particles, and (4) on whether the hydrometeors consist of water particles or ice particles. 9: Atmospheric clouds contain hydrometeors of a wide range of shapes. These decisively affect the flow field around them, their fall velocity and their fall mode. 10: The development of precipitation particles is a result of four basic mechanisms: (1) Diffusional growth of ice crystals surrounded by a water saturated atmosphere filled with cloud drops; (2) Stochastic growth of snow crystals colliding with and sticking to other snow crystals to form snow flakes; (3) Semi-continuous growth of snow crystals by collision with supercooled drops to form graupel and hailstones; (4) Stochastic growth of cloud drops colliding and coalescing with other cloud drops to form rain drops.

NATO ASI Series, Vol. G6
Chemistry of Multiphase Atmospheric Systems
Edited by W. Jaeschke

STATEMENT 1: IN THE ATMOSPHERE, THE PHASE CHANGE OF WATER VAPOR TO WATER DROPS IS A HETEROGENEOUS PROCESS INVOLVING AEROSOL PARTICLES (AP).

Experiment and theory show that in the absence of aerosol particles, the supersaturation needed to produce drops is of order several hundred percent, whereas in atmospheric clouds the supersaturation never exceeds a few percent and mostly remains below 1% (Fig. 1). Indeed, theoretical studies of the condensation process involving mixed AP, which consisted of water soluble and water insoluble materials in various proportions, and originated in continental, maritime and urban air masses, show that

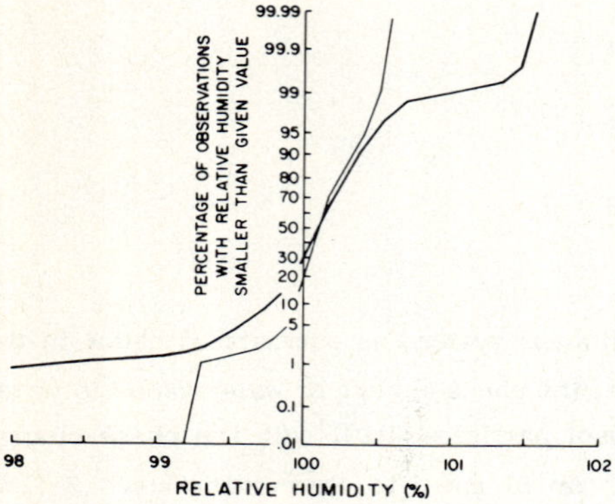

Fig. 1. Percentage of observations with relative humidity less than given value for all samples (heavyline) and for samples taken within 300 m of cloud base (thin line). (From Warner, 1968; by courtesy of J. de Rech. Atm. and the author).

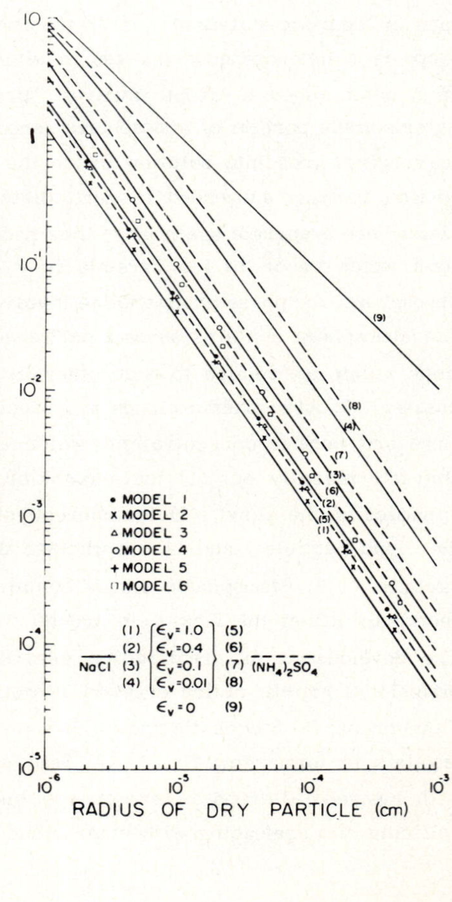

Fig. 2. Critical supersaturation for activating aerosol particles composed of various volume proportions ε_v of NaCl. $(NH_4)_2SO_4$ and a water insoluble substance as a function of particle size. Comparison with natural aerosols is made: Model 1: urban aerosol, Summer 1966, Mainz; Model 2: sea spray aerosol; Model 3: maritime aerosol over Atlantic, April 1969; Model 4: maritime aerosol over Atlantic, with Sahara dust, April 1969; Model 5: urban aerosol, March 1970, Mainz; Model 6: continental background aerosol, Summer 1970, Hohenpeissenberg. Based on date of Haenel, 1976. (From Pruppacher and Klett, 1978; by courtesy of D. Reidel Publ. Co. and the authors).

Fig. 3. Dependence of the sulfate mass in aerosol particles as a function of their radius (% of sulfate mass in particles of size < r_*.) (From Georgii et al, 1971; by courtesy of the authors).

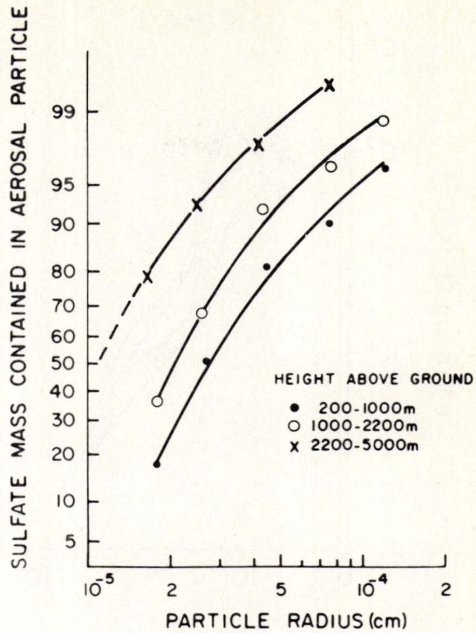

except for the very smallest particles of the Aitken size range, activation to drops generally takes place at a supersaturation below one percent (Fig. 2).

The need for cloud drops to be initiated by AP implies that AP will become removed from the aerosol population by becoming incorporated in the cloud particles through "nucleation scavenging". The importance of nucleation scavenging to the removal of sulfate- and mixed sulfate particles was advocated among others by Hegg and Hobbs (1982). They argued that most of the sulfate particle mass lies in the 0.1 μm to 2.0 μm size range as shown by Georgii (1972) (Fig. 3). In this size range impaction scavenging is relatively small. On the other hand we notice from Fig. 2 that typical mixed AP in this size range may be activated at supersatu-

Date	P_{SO_2}	Cloud Water pH	Cloud Type	Cloud† Supersaturation %	$[SO_4^=]$ in Air ①	$[SO_4^=]$ in Cloud Water per m^3 of Air ②	F ②/①	F_C*
4/7/82	--	5.3	Cu	--	4.9±1.0 μg m^{-3}	0.56±0.28 μg m^{-3}	0.11±0.06	0
9/22/81	10	4.8	Cu	0.71	3.2±0.3	4.5 ±1.2	1.41±0.40	1.1
9/24/81	--	4.6	Cu	0.97	9.65±9.68	5.36±3.73	0.56±0.68	0.36
9/29/81	10	4.1	Cu	1.70	3.06±1.1	3.5±1.3	1.14±0.59	0.85
9/29/81	10	4.1	Cu	1.70	3.06±1.1	2.6±0.6	0.85±0.36	0.56
9/29/81	10	3.7	Cu	1.56	2.15±0.8	2.42±0.6	1.13±0.29	0.71
10/9/81	4	5.4	Cu	1.54	1.91±0.33	2.32±0.47	1.21±0.32	0.74
10/9/81	4	5.8	Cu	1.10	2.41±1.12	1.93±0.56	0.80±0.44	0.43
4/22/82	--	3.7	St	--	8.0±0.9	3.24±0.3	0.41±0.06	0.29
8/12/82	12	3.7	Cu	--	0.23±0.02	1.21±0.36	5.26±1.63	1.35
8/12/82	12	3.9	Cu	--	1.28±0.13	1.02±0.30	0.80±0.25	0.09
10/1/82	--	3.6	Sc	--	1.73±0.35	0.55±0.17	0.32±0.12	0
10/4/82	8	5.2	Sc	--	0.62±0.12	1.98±0.2	3.19±0.70	1.74

†Determined by comparison of cloud base CCN spectra with cloud droplet number concentrations. $\bar{F}_c = 0.63$

*Scavenging efficiency after correction for the effect of in-solution sulfate production.

Table 1. In cloud sulfate scavenging efficiency F. (From Hegg and Hobbs, 1982; by courtesy of the authors).

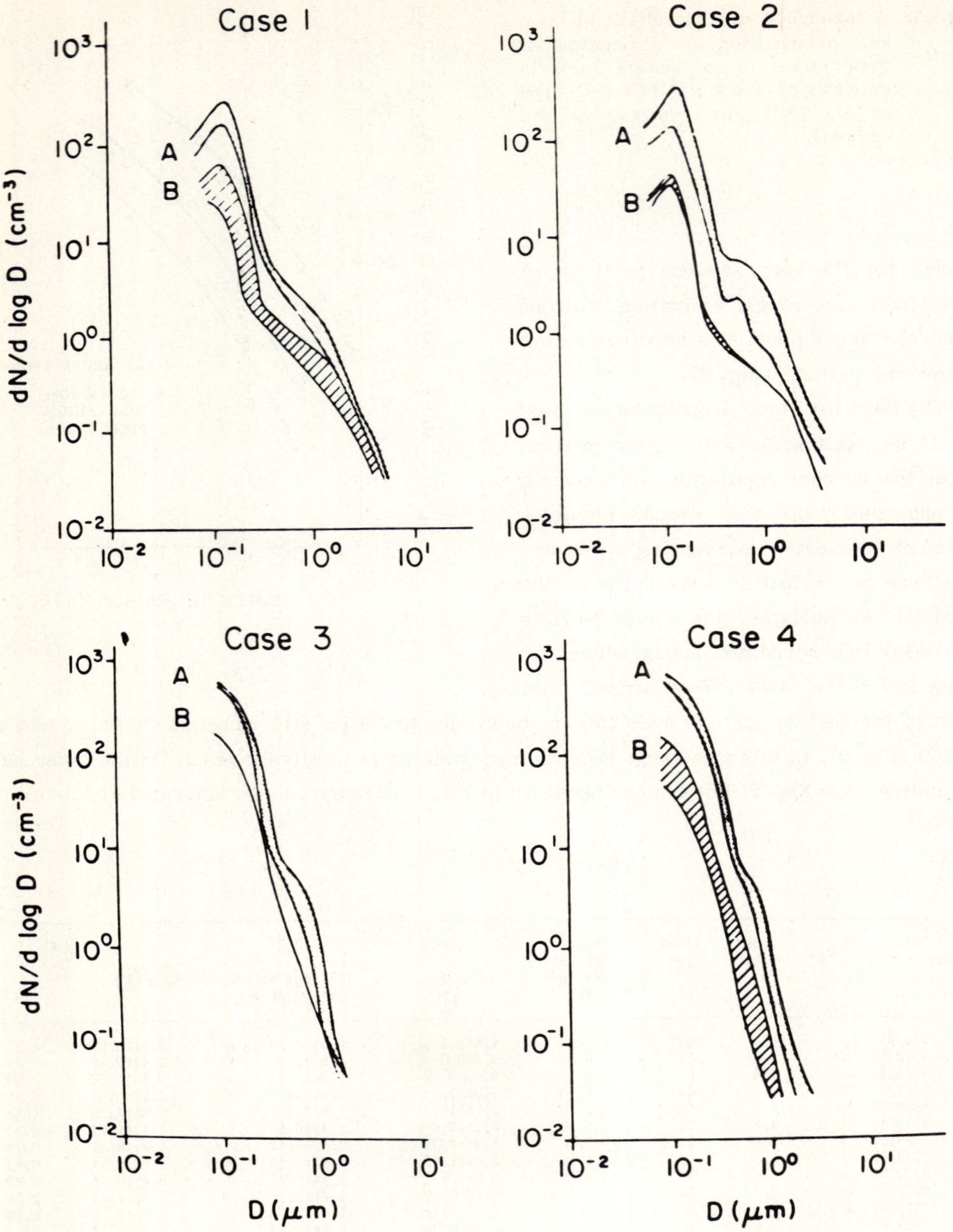

Fig. 4. Aerosol size distribution measured below cloud base (A) and a few hundred meters above cloud base (B) in cumulus clouds (cases 1 and 2) and in stratocumulus clouds (cases 3 and 4); shaded areas indicate range of measurements, (From Radke, 1982; by courtesy of the author).

rations of less than 0.5 %, achievable by modest updrafts. Consistent with this, the field observations of Hegg and Hobbs show from a comparison of the sulfate mass mixing ratio in air to the sulfate mass mixing ration in the cloud water (multiplied by the liquid water content of the cloud) that the sulfate in air picked up by drops via nucleation scavenging may on the average account for over 60 % of the sulfate mass in the cloud water (Table 1). The remaining sulfate was attributed to in situ absorption of SO_2. Radke (1982) came to a similar conclusion from field measurements of the cloud interstitial aerosol (CIA) just above the cloud base inside of convective clouds, as seen in Fig. 4. Comparison of the CIA concentrations with the concentration of AP just below cloud base showed that the loss of AP across the cloud base may be as much as 80 to 90 % in the 0.1 µm size range. Since this decrease approximately equalled the peak concentration of the cloud drops in the cloud, Radke concluded that the observed decrease in aerosol population is to a large degree the result of nucleation scavenging.

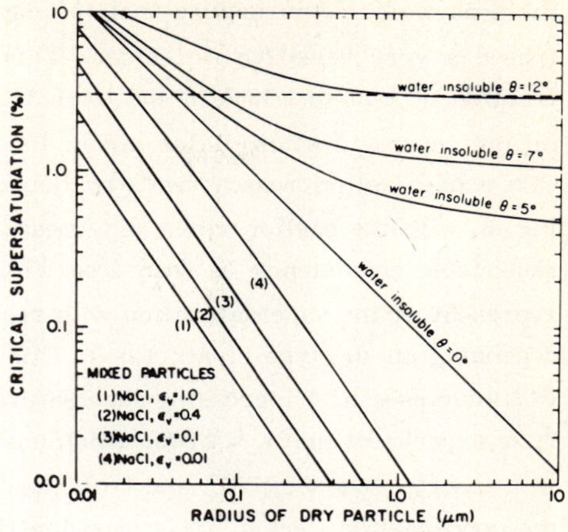

Fig. 5. Critical supersaturation for water nucleation on a spherical, water insoluble substrate of radius r_N as a function of wetting angle Θ for water, and critical supersaturation for activation of mixed aerosol particles as a function of volume fraction ε_v of NaCl in particle, for 0°C. (From Pruppacher and Klett, 1978; by courtesy of D. Reidel Publ. Co. and the authors).

It should be added here, however, that the results of Hegg and Hobbs and of Radke apply to mixed AP of considerable soluble mass fraction. Obviously AP with none or very low soluble material must to a large degree be remove via impaction scavenging.

STATEMENT 2: THE PHASE CHANGE FROM WATER VAPOR TO DROPS INVOLVES "PREFERRED" AP OF SPECIFIC CHARACTERISTICS.

These "preferred" AP of cloud condensation nuclei (CCN) either are completely water soluble or consist of a mixture of water soluble and water insoluble materials in various proportions. The activation to drops by particles which are completely water insoluble depends on their wettability, characterized by the wetting angle Θ, which should, for drop formation, be as small as possible. We note from Fig. 5 that water soluble and mixed particles require a supersaturation for activation which is considerably lower than for water insoluble particles. This implies that the capture of AP by drops via nucleation scavenging is indeed a very selective mechanism. This selectivity is expressed by the fact that only a fraction of 1/10 to 1/1000 of the total AP population serves as CCN, this fraction depending on the supersaturation of the air in the aerosols environment. Comparison of some recent CCN concentration measurements by Hobbs et al., (1978) for the US High Planes, as shown in Fig. 6, with the earlier world wide measurements of Twomey and Wojciechowski (1969) show reasonable consistency as seen from Fig. 7. Generally, one finds that $N_{CCN} = Cs_{wv}^k$ (s_{wv} representing the supersaturation with respect to water and the parameters C and k varying depending on the type of air mass). This relating implies that the number of AP serving as CCN increases with increasing supersaturation. According to Jiusto and Lala (1981) one finds from experiment that $k \approx 2\beta/3$, which has its theoretical justification since from condensation theory $s_{wv} \propto r_N^{-3/2}$, and from the Junge power law for AP, $N_{AP} \propto r_N^{-\beta}$. As expected, the condensation mechanism is importantly determined by the CCN spectrum which adequately determines the drop spectrum in clouds. This earlier finding of Twomey and Warner (1967) was re-emphasized by Fitzgerald (1974) and more recently by Hindman et al., (1977). The former showed that the broadness of the drop size distribution is not significantly affected by the composition of the CCN, while the total number density increases as the percentage of the insolube nuclei decreases. The latter emphasized the fact that high concentrations of large ($0.1 \leq r_N \leq 1$ µm) and giant ($r_N > 1$ µm) CCN lead to drop size spectra which contain

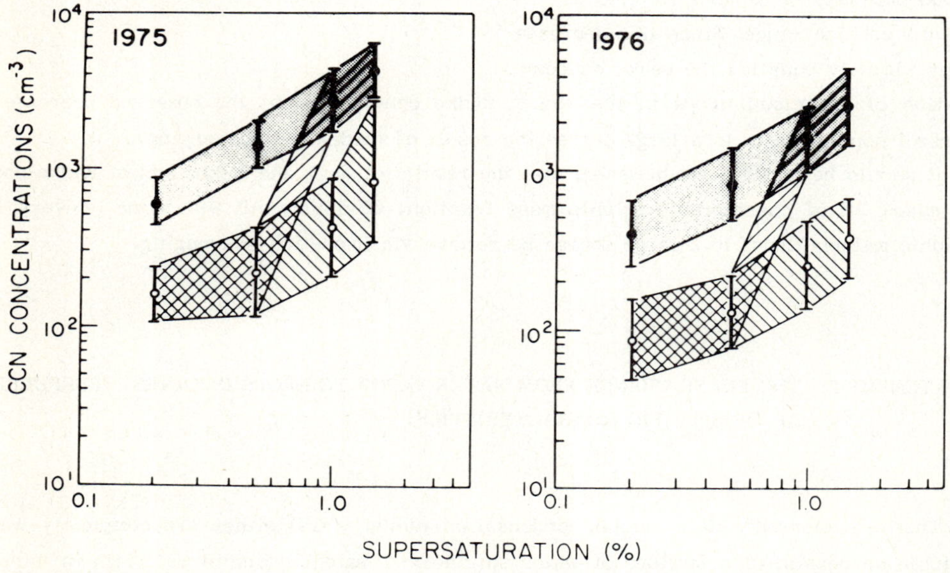

Fig. 6. Cloud condensation nucleus spectra measured over the US High Planes during Summer 1975 and 1976; shaded: aged continental; hatched left to right: maritime; hatched right to left: transitional. (From Hobbs et al, 1978; by courtesy of the authors).

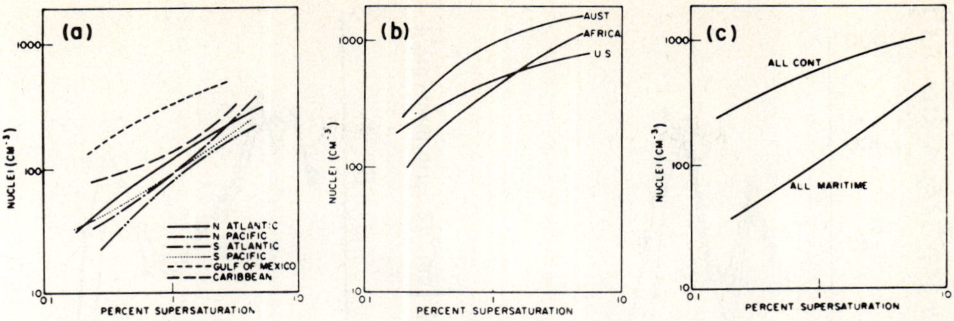

Fig. 7. Median world wide concentration of CCN as a function of supersaturation required for activation; (a) in air over ocean, (b) in air over continents, (c) all observations. (From Pruppacher and Klett, 1978; by courtesy of D. Reidel Publ. Co. and the authors).

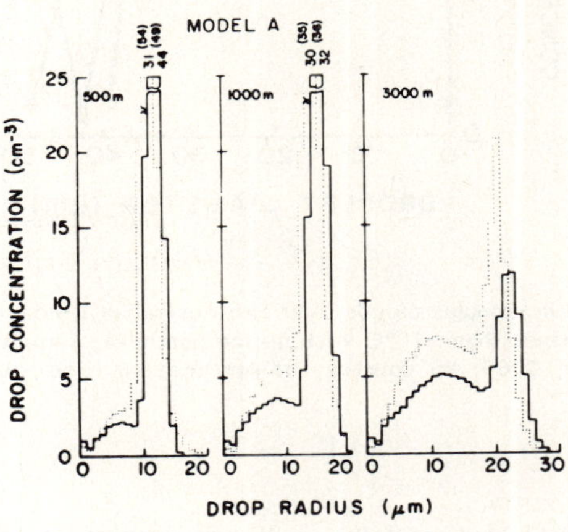

8

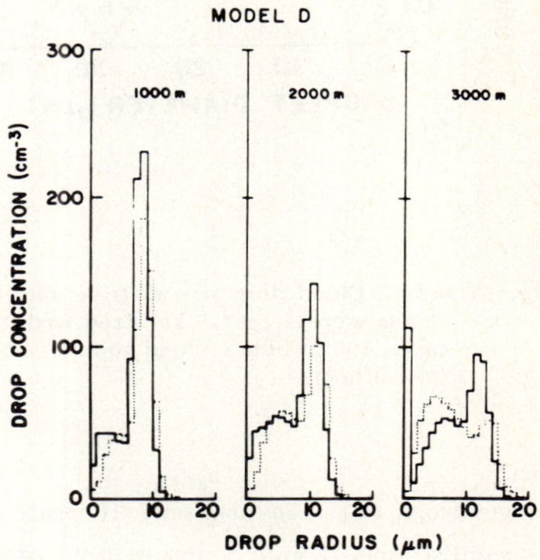

9

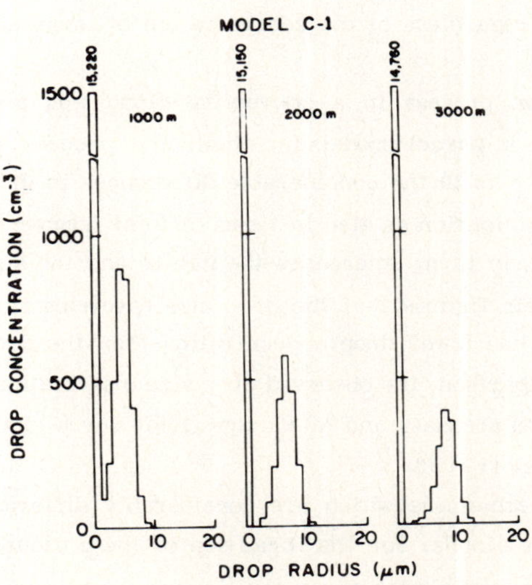

10

Fig. 8-10 Histogram representation of drop size distribution at selected levels above cloud base computed by a parcel model with entrainment; Fig. 8: for aerosol Model A (maritime air mass over Atlantic, Summer 1969; Fig. 9: for aerosol model B (continental background aerosol: Hohenpeissenberg, FRG, 1000 MSL, Summer 1970; Fig. 10: for aerosol model C-1, urban aerosol, Mainz, FRG, Summer 1966. (From Lee et al, 1980; by courtesy of Amer. Meteor. Soc. and the autors).

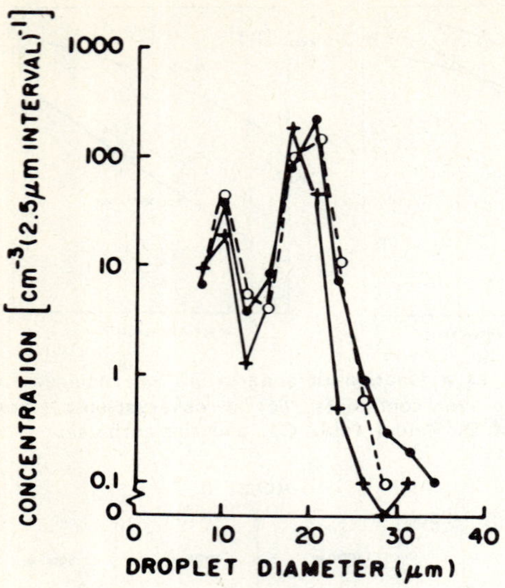

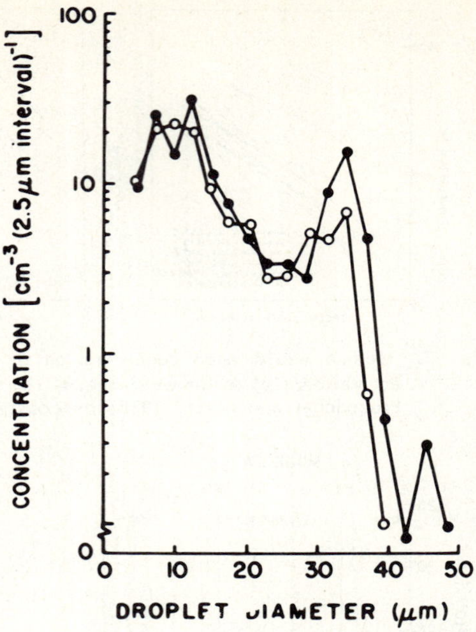

Fig. 11 and 12 Cloud drop size distribution observed in cumulus clouds over the Australian E-coast; clouds were 1 to 1.5 km deep with tops warmer than -10°C with no ice particles; samples taken 300 m below cloud top. (From Warner, 1969; by courtesy of Am. Meteor. Soc. and the author).

large drops (a_d > 30 μm) only if small CCN of ($0.03 \leq r_N \leq 0.1$ μm) were present in low concentrations. If high concentrations of such small CCN were present, the drop size spectra contained few if any large (a_d > 30 μm) drops, regardless of the concentration of large and giant nuclei.

The selective nature of the drop nucleation process in a convective cloud was also demonstrated by Lee et al., (1980) who used a parcel model in which the process of entrainment was included. We notice from Figs. 8 to 10 the considerable differences in drop spectra both in terms of the width of the size distribution as also in terms of total number of drops, depending on the type of air mass which, in turn, determines the nature and the size distribution of the AP. Thus, we note from their figures that the drop size spectrum in a maritime air mass is considerably broader and has lower drop concentration than the drop size spectrum in a continental air mass. For comparison, the observed drop size distributions of Warner (1969) for clouds formed in a modified air mass and of Eldrige (1957) for a cloud formed in a polluted air mass are given in Figs. 11 - 13.

Fog and low level stratus are formed by mechanisms which are considerably different than those which are responsible for convective clouds. For the formation of these clouds, slow, isobaric radiative cooling allows the AP to grow almost under equilibrium conditions at

Fig. 13. Cloud drop size spectra inside a strato-cumulus cloud at top of Mt. Washington, New Hampshire (based on data of Eldrige, 1957). (From Lee et al, 1980; by courtesy of Amer. Meteor. Soc. and the authors).

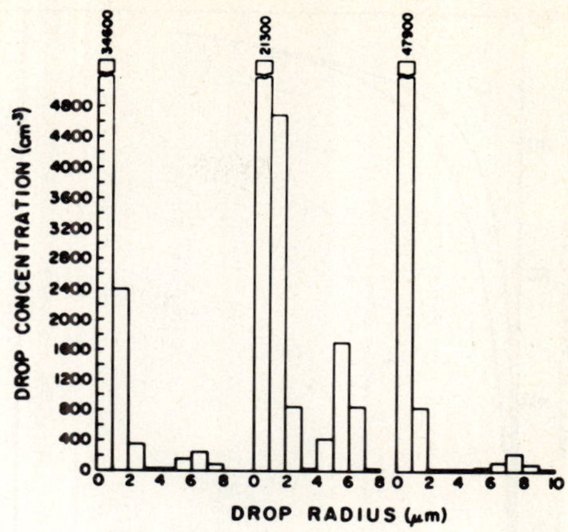

very low or even zero supersaturations. The selective nature of a deposit of AP to respond in growth to relative humidities below 100% under equilibrium conditions is demonstrated by Fig. 14 and 15 for a maritime and an urban aerosol. We notice that the maritime aerosol picks up considerably more water mass now at any relative humidity than an urban aerosol. Appropriately, fog condensation nuclei/FCN) are those AP which can be activated at supersaturations between 0.01 and 0.1%, Examples for FCN spectra are given in Fig. 16 derived from field experiments of Hudson (1980).

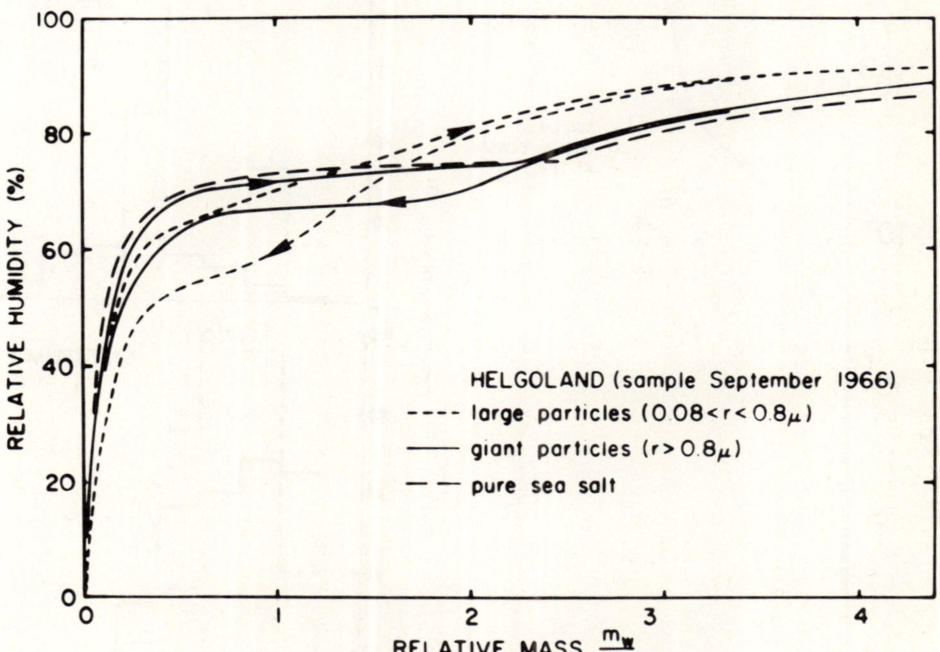

Fig. 14. Equilibrium growth of maritime aerosol deposits as a function of relative humidity (m_w = mass of water taken up by a dry deposit of mass m_0). (From Winkler, 1967; by courtesy of the author).

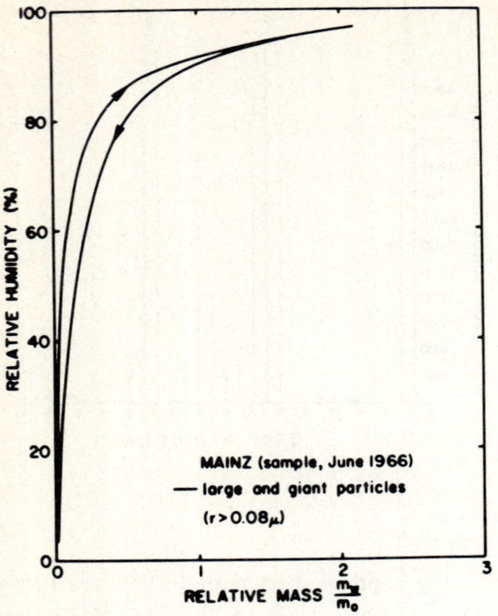

Fig. 15. Equilibrium growth of an urban aerosol deposit as a function of relative humidity (symbols same as Fig. 14). (From Winkler, 1967; by courtesy of the author).

Fig. 16. Typical fog condensation nucleus (FCN) spectra at the locations (number per unit volume as a function of supersaturation.) (From Hudson, 1980; by courtesy of Amer. Meteor. Soc., and the author).

Fig. 17. Drop size distribution in fogs at height of 2 m above ground, after 4 hours of model integration for masses of AP; 1×10^{-5} g $\leq m_s \leq 1 \times 10^{-12}$ g; spectrum B: low concentration of nuclei; spectrum E: high concentration of nuclei; cont. line: with radiative cooling; dashed line: without radiative cooling; (From Brown, 1980; by courtesy of Roy. Meteor. Soc. and the author).

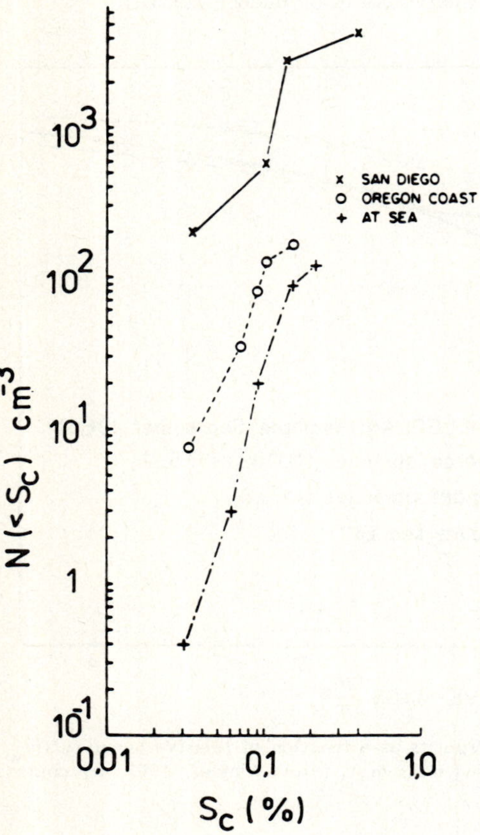

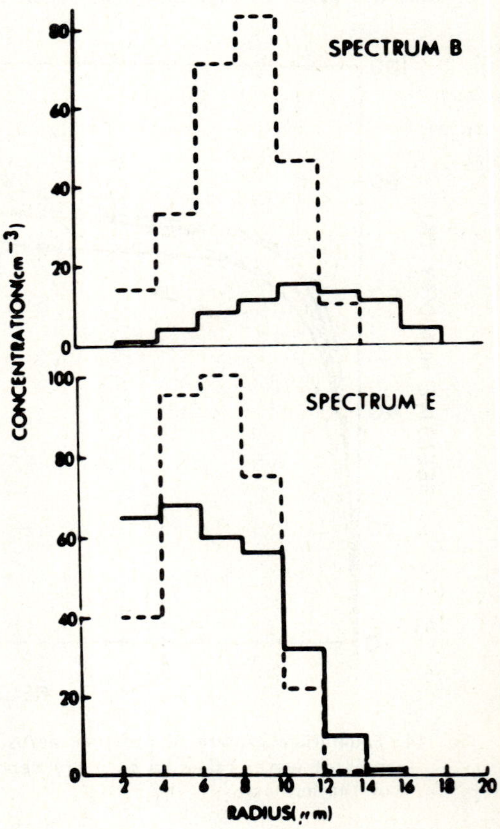

In Fig. 17 an illustration is given for the effect of radiative cooling on the drop spectrum formed by condensation in a fog model which includes turbulent heat and vapor exchange (Roach, 1976; Brown, 1980). The figure shows that for relatively low AP concentrations (spectrum B) radiative cooling lowers the maximum supersaturation by lowering the critical supersaturation of the larger AP. Thus, large nuclei are activated before the air becomes supersaturated. As the nuclei grow, vapor depletion inhibits the increase of the super-saturation. Since the radiative cooling hardly affects the critical supersaturation of small AP they are left un-activated, resulting in an overall lowering of the total drop concentration due to radiative cooling. For high nucleus concentrations (Spectrum E) the drop spectrum is much less affected by the radiative cooling effect.

STATEMENT 3: THE WATER SOLUBLE PORTION OF ATMOSPHERIC AEROSOL PARTICLES, PICKED UP BY NUCLEATION OR IMPACTION SCAVENGING, GOES INTO SOLUTION INSIDE THE CLOUD AND RAIN DROPS, WHILE THE WATER INSOLUBLE PORTION REMAINS SUSPENDED IN PARTICULATE FROM INSIDE THE DROPS.

Generally, aerosol particles contain more than 50% of their mass in form of water soluble inorganic or organic material. This is illustrated in Fig. 18, which in addition shows that a rural aerosol contains usually a larger fraction of water soluble material than an urban aerosol.

This finding implies that cloud and rain water represents an aqueous solution of salts, acids and bases, and water soluble organics, with the insoluble particulate matter in

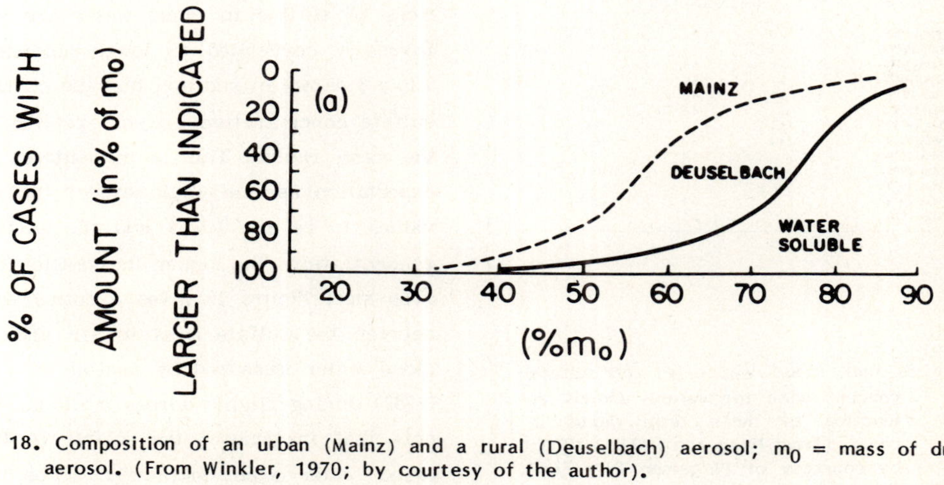

Fig. 18. Composition of an urban (Mainz) and a rural (Deuselbach) aerosol; m_0 = mass of dry aerosol. (From Winkler, 1970; by courtesy of the author).

Location	N	Date	pH	Concentration (μeq/liter)						Concentration (μeq/liter)				Concentration (mg/liter)	
				H^+	Na^+	K^+	NH_4^+	Ca^{2+}	Mg^{2+}	F^-	Cl^-	NO_3^-	SO_4^{2-}	Fe	CH_2O
Pasadena	4	11/15/81	5.25-4.74	5.6-55	12-496	4-39	370	19-360	7-153	120	56-280	130-930	62-380	0.094-2.1	
Pasadena	4	11/23/81	4.85-2.92	14-1,200	320-500	33-53	1,290-2,380	140-530	89-360	180-410	480-730	1,220-3,520	481-944	0.92-1.77	3.1-3.5
Lennox	8	12/7/81	5.78-2.55	2-2,820	28-480	6-160	1,120-4,060	44-4,350	17-1,380	115-395	111-1,110	820-4,560	540-2,090	0.34-24	4.6-12.8
Lennox	3	12/17/81	2.81-2.52	1,550-3,020	80-166	19-40	950-1,570	73-190	43-99	180-500	90-197	2,070-3,690	610-1,970	1.02-2.08	
Bakersfield*	3	1/14/82	3.07-2.90	850-1,260	151-1,220	39-224	5,370-10,520	165-1,326	20-151	126-242	203-592	3,140-5,140	2,250-5,000		6.1-14.4
Pasadena*	1	1/17/82	2.25	5,625	2,180	500	7,870	2,050	1,190	637	676	12,000	5,060		
Los Angeles rain (volume-weighted means at nine sites)		1978-79	5.4-4.4	4-39	4-37	0.24-4.9	1-36	3.9-17	1.7-11		5-52	11-44	7-56	0.004-0.17	
Coastal California fogs		9-10/76			78-944	8-26	1-578	9-102	17-175		96-1,230	23-234	52-490		
Whiteface Mountain		7-8/80	4.2-3.2	63-630	1-55	1-6	4-310			11-54	1-15	7-190	40-800		
Fog and clouds, U.S.S.R.		1961-64	5.3-4.7	5-20	30-104	15-44	33-100	20-50	17-83		59-177	2-13	13-185		

Table 2. Composition of bulk fog water of 6 fogs in Southern California; comparison is made with composition of rain and of clouds and fogs at other locations. (From Waldman et al, 1982; by courtesy of the Am. Assoc. for the Advancement of Sci. and the authors).

Date of sample	Mean radius of cloud droplets (μm)	pH	Mass concentration of sulfur (expressed as SO_4 in mg l^{-1})
Los Angeles (a) *Stratus clouds*			
10 January 1980	407	4.6	16.5
10 January 1980	10	5.3	1.5
11 January 1980	120	5.3	6.0
10 January 1980	39	5.5	11.1
10 January 1980	1110	5.5	15.3
8 January 1980	11	5.5	19.8
8 January 1980		5.7	21.9
8 January 1980	13	6.6	19.5
8 January 1980	26	6.8	20.4
12 January 1980	57	6.8	16.2
Western Washington (a) *Stratus clouds*			
30 April 1979	243	3.7	17.8
30 April 1979		3.8	17.9
30 April 1979	105	3.9	12.0
30 April 1979	16	3.9	16.0
3 August 1979	32	3.9	6.4
3 August 1979	32	3.9	4.8
12 July 1979		4.1	19.7
12 July 1979		4.8	14.9
12 July 1979	20	5.2	13.0
12 July 1979	12	5.6	10.9
(b) *Wave clouds*			
27 April 1979	10.0	4.2	13.2
13 July 1979	3.7	4.3	4.4
24 April 1979	5.4	4.4	7.3
24 April 1979	6.1	4.4	5.5
13 July 1979	3.6	4.5	3.0
3 May 1979		4.6	5.8
3 May 1979	3.1	4.8	9.4
15 May 1979	3.4	5.9	6.3
15 May 1979		5.9	4.8
15 May 1979		5.9	10.0
15 May 1979		6.0	2.5
10 May 1979		6.2	15.7
10 May 1979		6.5	14.7

Table 3. Bulk cloud water pH and sulfate concentration for various clouds as function of mean drop radius in cloud. (From Hegg and Hobbs, 1981; by courtesy of Pergamon Press and the authors).

suspension.

In Table 2 an example is given for typical concentrations of various compounds dissolved in cloud rain and fog water together with the acidity (pH = -log $[H^+]$, where $[H^+]$ is the concentration of the hydrogen ions). We note that in clouds, fogs and rain the concentration of dissolved materials varies greatly. Highest concentrations of dissolved materials and lowest values for pH were found in urban air-influenced fogs with long times of exposure to the pollution in the air. Hegg and Hobbs (1982) among others show in Table 3 that higher concentrations of sulfate in cloud water are not inversely correlated to lower acidities. Table 3 shows in addition that the pH and sulfate concentration are not related to the mean radius. This is in contrast to expectation as one would expect the pH values to be the lower and the sulfate concentrations the higher the smaller the drop size. Figure 19 gives a comparison between the sulfate mass in air and in cloud water measured by Leaitch et al., (1982) during flights across a cloud. We note that in general the sulfate in the cloud water is not completely accounted

ALT. (M.)	PRE (mb)	TEMP (°C)	AC* (#/cm³)	SO_2 (µg/m³)	DC* (#/cm³)	MVD* (µm)	Ave LWC* (g/m³)	Dissolved $SO_4^=$ (µg/ml)	Dissolved NO_3^- (µg/ml)	Air Equiv.** $SO_4^=$ (µg/m³)	Air Equiv.** NO_3^- (µg/m³)	pH	In-Cloud AC*
1530	840	7.0	600	0	715	12.7	0.61	15.4	20.4	9.4	12.5	5.0	450
1220	872	9.5	1000	4	728	11.9	0.47	16.4	29.2	7.7	13.7	4.1	480
920	902	11.	1200	8	511	10.9	0.26	24.2	40.1	6.3	10.5	3.5	580
800	916	11.	940	6	514	9.5	0.11	48.6	62.9	5.2	6.7	-	500
735	925	12.	950	6	-	-	-	-	-	6-9	6-10	-	
290	976	15	1400	6						8-13	9-14		

Fig. 19. Composition of bulk cloud water at different locations in a non-precipitating convective cloud. DC = drop concentration, MVD = mean drop volume diameter, LWC = liquid water content, AC = aerosol particle concentration. (From Leaitch et al, 1982; by courtesy of the authors).

Fig. 20. Concentration of SO_2 absorbed by 2.1 mm water drop as a function of exposure time to SO_2 for 8°C and 570 ppb SO_2; C: distilled water, B: $1:^{-5}$ m $MnCl_2$ + 10^{-5} m $FeCl_2$; A: 10^{-5} m $MnCl_2$ + 10^{-5} m $FeCl_2$. (From Barrie and Georgii, 1976; by courtesy of Pergamon Press and the authors).

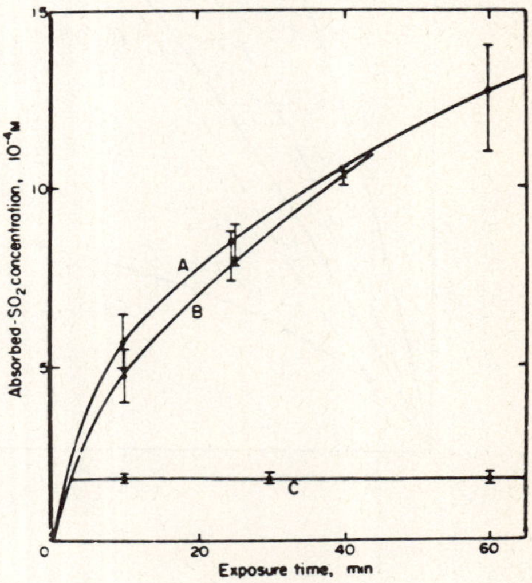

for by the sulfate in the air surrounding the drops, suggesting in accordance to Hegg and Hobbs (1982) (see Table 1), that substantial amounts of sulfate is formed in situ as a result of SO_2 gas absorption and subsequent oxidation inside the cloud water.

The presence of water insoluble material inside the cloud water provides an

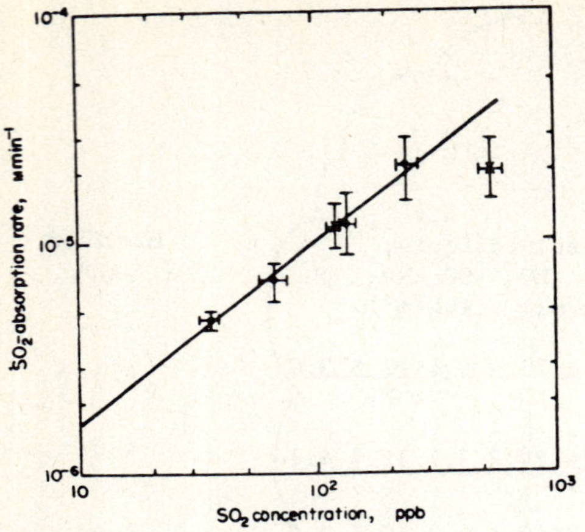

Fig. 21. Rate of SO_2 absorption by 2.1 mm water drop as a function of SO_2 gas concentration in drop environment; drop contained 10^{-5} m $MnCl_2$; dots: 25°C, crosses: 8°C. (From Barrie and Georgii, 1976; by courtesy of Pergamon Press and the authors).

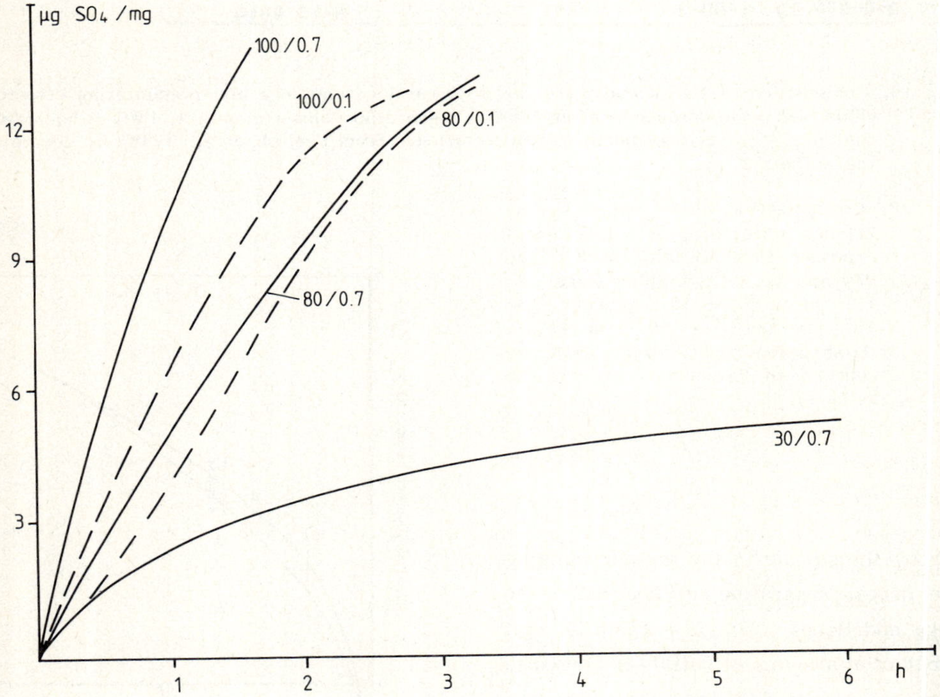

Fig. 22. Rate of sulfate formation during 6 hours (h) on Mangan acetate aerosol (µg SO_4 per mg aerosol) as a function of relative humidity (100%, 80%, 30%) and SO_2 gas concentration (0.1, 0.7 ppm SO_2). (From Runca-Koeberich, 1979; by courtesy of the authors).

opportunity for heterogeneous reactions in solution. Little is known on reactions catalyzed by the surface of solid particles suspended inside drops. On the other hand, Barrie and Georgii (1976) showed that manganese and other metalions inside water drops efficiently catalyze the oxidation of absorbed SO_2 to sulfate, raising the amount of absorbed SO_2 significantly above that absorbed by pure water, the absorption rate increasing with increasing SO_2 concentration in air (Figs. 20, 21). Sulfate also is efficiently catalyzed at the surface of solid particles such as manganese acetate, where the production of sulfate was shown to rapidly increases with increasing relative humidity, i.e. with progressing deliquescence of the salt (Runca-Koeberich, 1979). In particular, we notice from Figs. 22 and 23 that the amount of sulfate formed at the surface of these particles increase with increasing relative humidity and with increasing time of exposure, and is, for the same relative humidity, larger for a size distribution with small AP than for a distribution where the small AP are absent (Fig. 24). The sulfate formation catalyzed on the surface of solid water insoluble aerosol particles on which SO_2 was adsorbed was studied by Dlugi (1982). Thus, the uptake capacity (the saturation value) for SO_2 on ash particles is given in Fig 25.

Although one notices from Figs. 23 and 25 that considerable time is involved in the process of SO_2 adsorption on solid AP, it must be remembered that even in the lower troposphere the life time of AP ranges for AP of $10^{-2} \leq r \leq 10\ \mu m$ between 1 day and 1 week, as seen in Fig. 26 (Jaenicke, 1978). It is therefore quite likely that adsorption processed which take place at the surface of aerosol particles are also contributing to the reduction of the concentration of trace gases in the atmosphere, whereby smaller AP appear to contribute more than larger ones.

STATEMENT 4: MOST CLOUDS IN THE ATMOSPHERE EVAPORATE AGAIN AFTER THEY HAD FORMED.

On a global average, the process of cloud formation and re-evaporation repeats itself 10 to 25 times before a cloud system develops which is capable to precipitate (see Pruppacher in Plate, 1982). This implies on the one hand that the atmospheric aerosol is becoming "processed" through its involvement in cloud forming-evaporation cycles. Unfortunately, no studies have been reported in literature on the effect of this processing on AP. On the other hand statement 4 implies that, on a global average, most natural and anthropogenic pollutants, incorporated in clouds by nucleation scavenging, impaction scavenging of particulates, and by absorption scavenging of gases, become airborne again. Gaseous material may desorb. Soluble compounds may crystallize out and become airborne as particulates, some in their original chemical form and some as new compounds after having involved in variety of chemical reaction. Water insoluble particles may become airborne as mixed particles, being coated and having coagulated with crystallized soluble substances. No

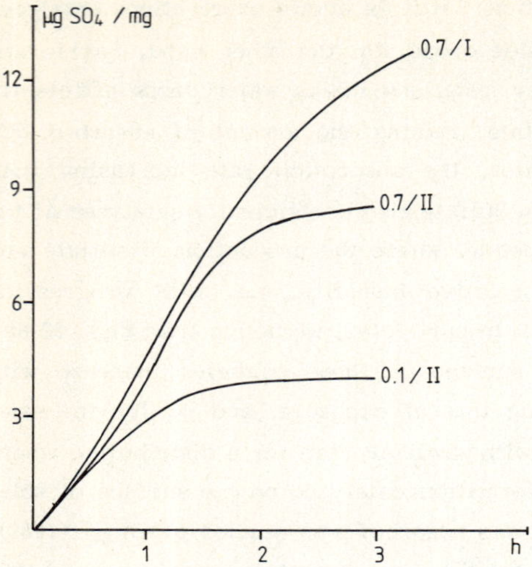

Fig. 23. Rate of sulfate formation during 3 hours (h) on mangan acetat (µg SO_4 per mg aerosol) for 80% relative humidity for two aerosol particle size distributions (I and II). (From Runca-Koeberich, 1979; by courtesy of the author).

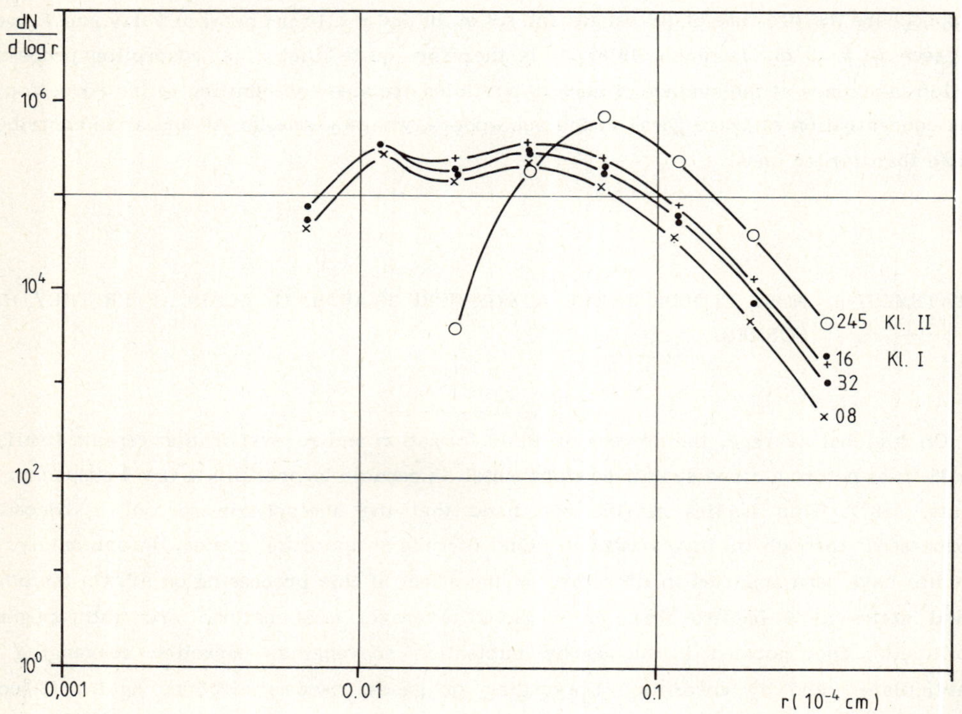

Fig. 24. Size distribution I and II of aerosol used for results in Fig. 22 and 23. (From Runca-Koeberich, 1979, by courtesy of the author).

data are available on the efficiency of desorption of trace gases from cloud and rain drops under atmospheric conditions. However, some measurements of the production of water soluble and mixed CCN by evaporation of wave clouds have been made by Hegg et al., (1980). Thus, Fig. 27 shows that the concentration of CCN is larger downwind than upwind of clouds, clearly demonstrating that clouds can be a source of AP releasing them into the air when evaporating. Comparison of the downwind and upwind size spectrum of CCN illustrated by Fig. 28 shows that the particle production is most pronounced for particles in the 1 µm diameter range.

Computation and laboratory measurements show that not all drop sizes are equally likely to evaporate. Thus, Fig. 29 shows that in air of slightly below saturation (90% relative humidity), drops of less than 100 µm do not survive an isobaric-isothermal excursion distance of as little as 300 meters (Beard & Pruppacher, 1971). Although larger drops are less susceptible to total evaporation, they nevertheless suffer considerable mass loss and may evaporate before reaching the ground depending on their original size and on the relative humidity of the air between cloud base and the ground, as shown in Fig. 30.

Relative humidities less than 100% are not only confined to outside of a cloud but are often also found inside a cloud, in particular near the cloud top and cloud edges where drier air from the cloud environment is entrained. This fact already is implied in Fig. 1. A more detailed study of the distribution of the relative humidity in two clouds is reproduced in Fig 31 a, b where the flight observations of Hoffer (1960) are reported. We note that apart of supersaturated regions which contribute to cloud drop growth, other portions of the clouds

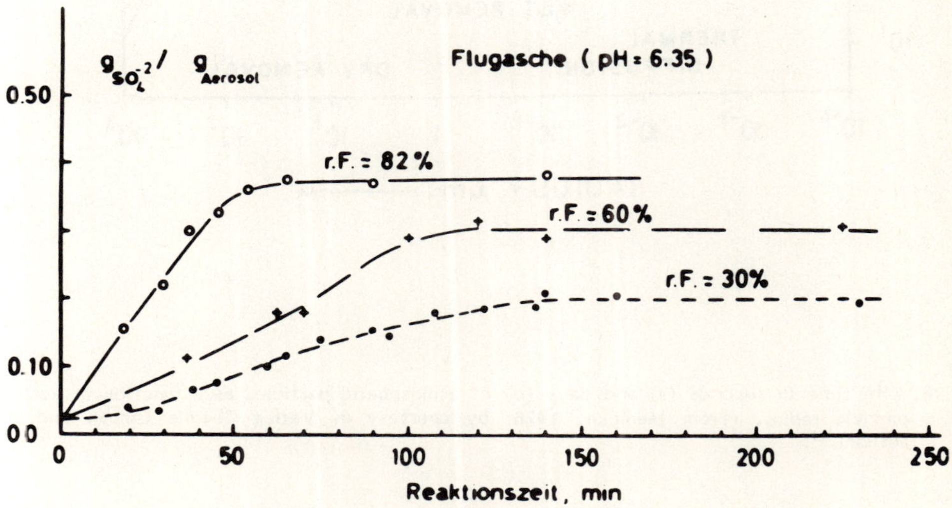

Fig. 25. Rate of sulfate formations on coal-ash particles during 250 minutes as a function of relative humidities (r.F.). (From Dlugi, 1982; by courtesy of the author).

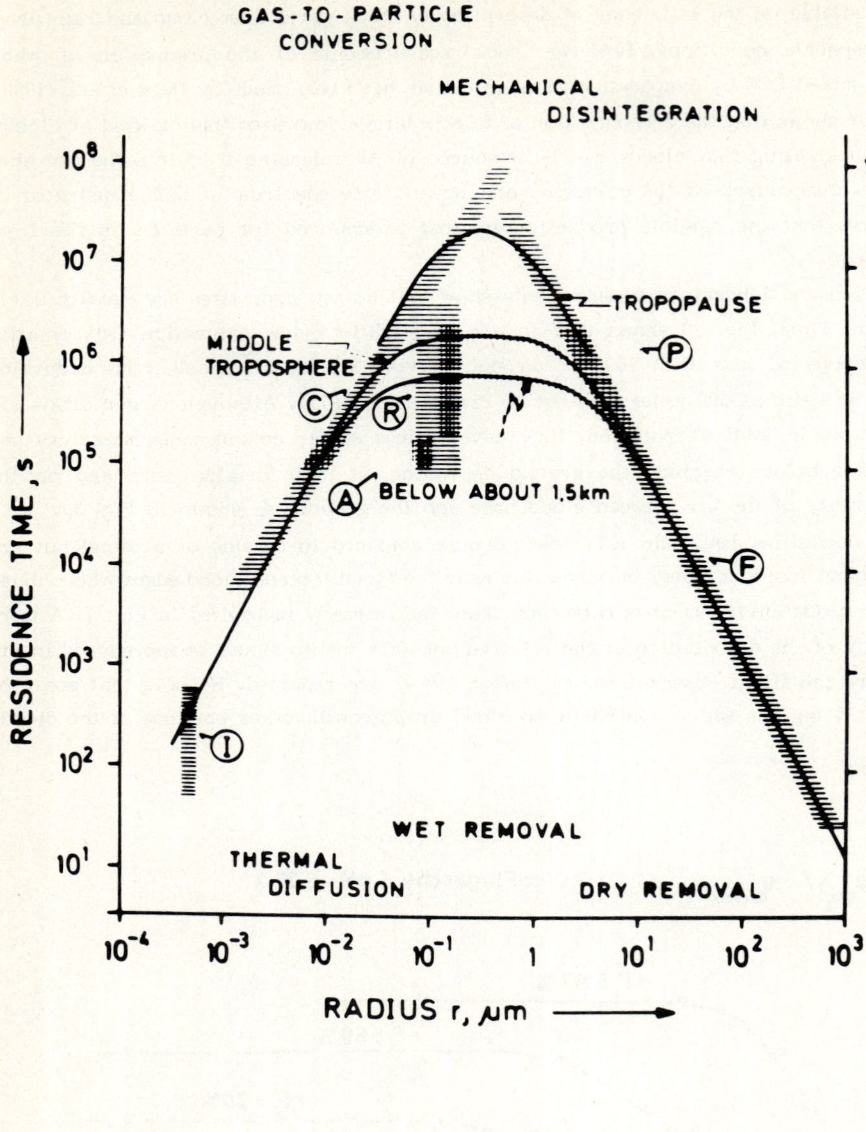

Fig. 26. Life time in seconds (s) and days (d) of atmospheric particles as a function of aerosol particle radius. (From Jaenicke, 1978; by courtesy of Verlag Chemie GmbH. and the author).

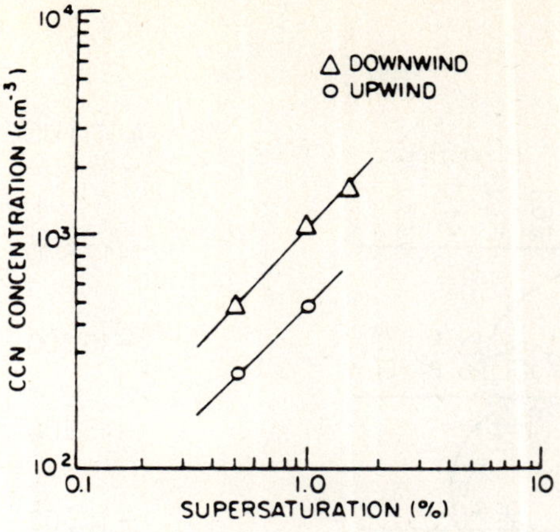

Fig. 27. CCN spectra measured upwind (circles) and down wind (triangles) of a wave cloud. (From Hett et al, 1980; by courtesy of the authors).

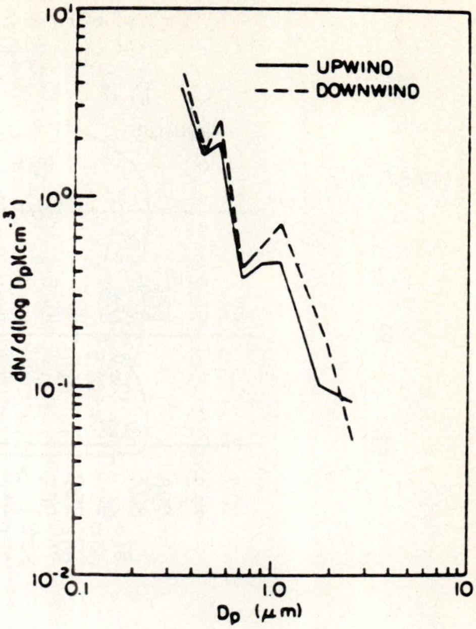

Fig. 28. AP spectra measured upwind (continuous line) and downwind (dashed line) of same wave cloud as in Fig. 27. (From Hegg et al, 1980; by courtesy of the authors).

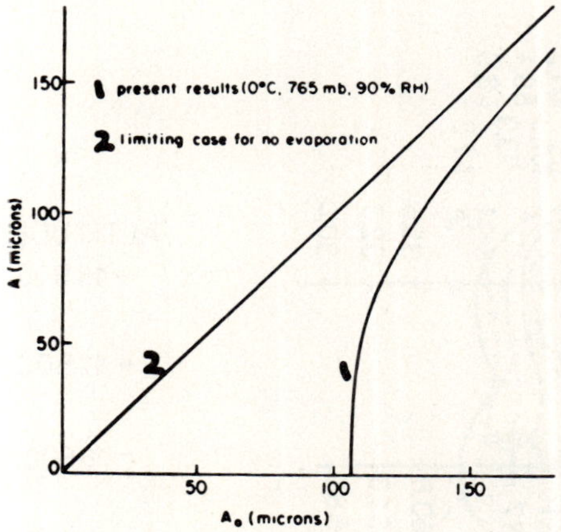

Fig. 29. Final size A of various drops of initial size A_o after the home mode an isothermal, isobaric excursion of 300 meters into air of 0°C, 765 mbar and 90% relative humidity. (From Beard and Pruppacher, 1971; by courtesy of the Am. Meteor. Soc., and the authors).

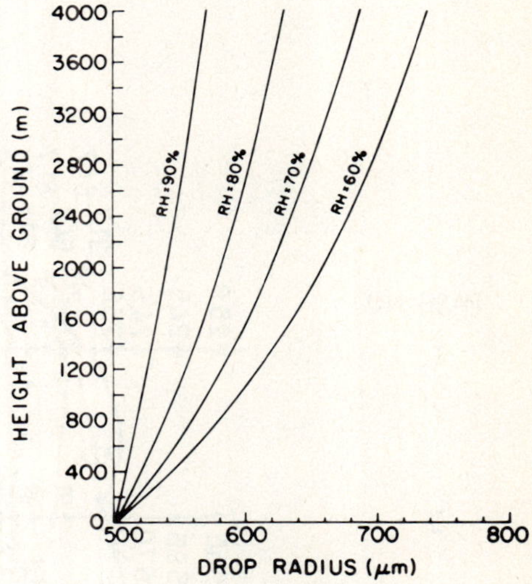

Fig. 30. Variation of the size of a rain drop falling from cloud base to ground through clean air (NACA Standard Atmosphere) as a function of cloud base height and relative humidity of the air. (From Pruppacher and Rasmussen, 1979; by courtesy of Am. Meteor. Soc. and the authors).

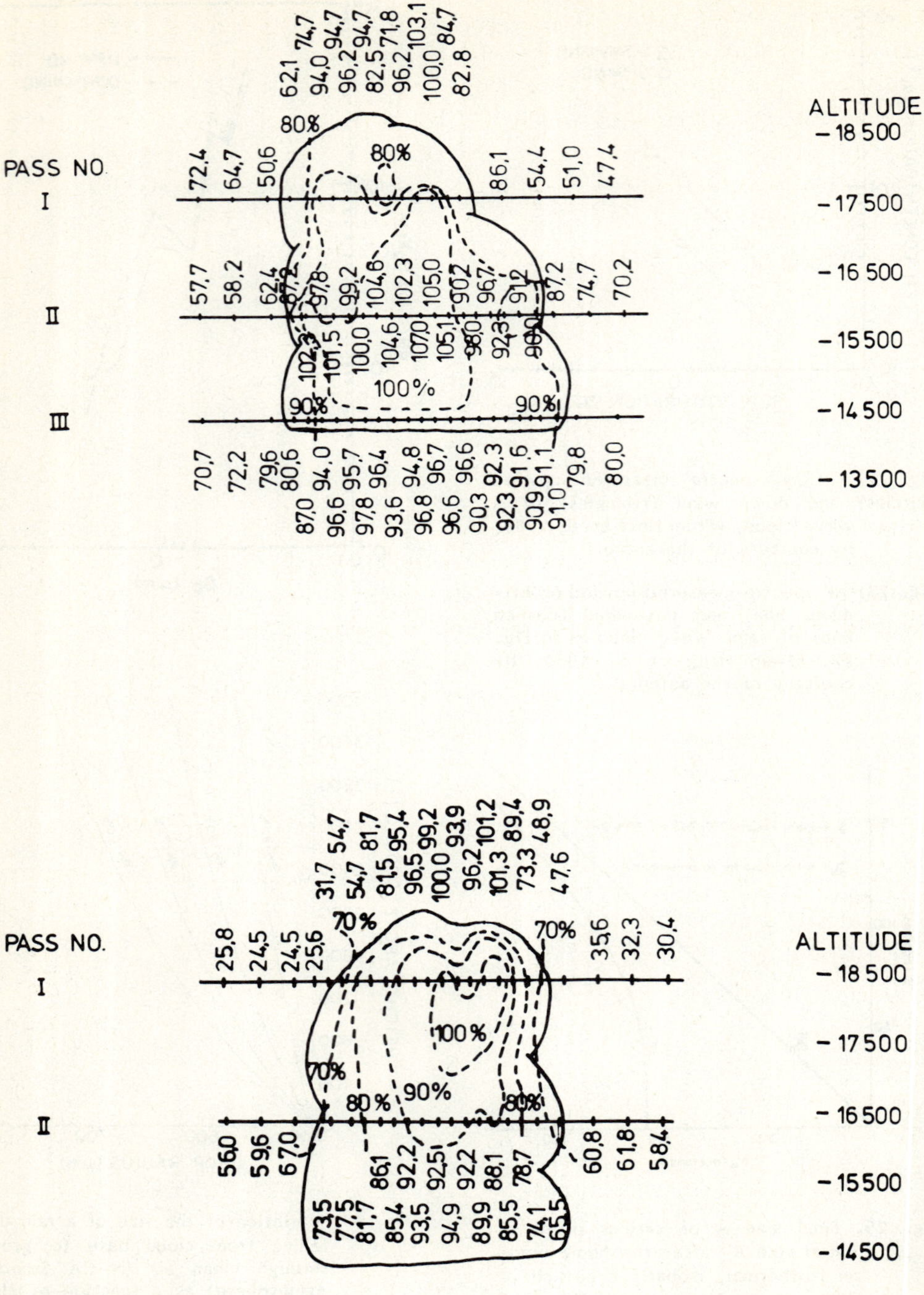

Fig. 31, a, b. Distribution of relative humidity in two convective non-precipitating clouds. (From Hoffer, 1960; by courtesy of the author).

exhibit relative humidities considerably below 100%. Although these relative humidities affect mainly the size of the smaller cloud drops they are sufficiently low to affect the scavenging of CIA by all the cloud drops. This scavenging is accomplished via a netphoretic force which is directed towards an evaporating drop with a surface temperature which is colder than that of the drops's environment. For such a condition, the thermophoretic force is directed towards the evaporating drop overpowering for all particles of radius less than about 2 μm the diffusiophoretic force, directed away from the evaporating drop. The same effect applies to raindrops falling in subsaturated air from cloud base to ground.

STATEMENT 5: THE PHASE CHANGE IN THE ATMOSPHERE FROM WATER VAPOR TO ICE CRYSTALS IS A HETEROGENEOUS PROCESS INVOLVING AP.

Experiment and theory show that without any AP the supersaturation for ice formation would need to be several hundred percent and the temperature would need to be below -30°C. In actuality, ice crystals form on AP at or slightly below water saturation (i.e. at a few percent ice supersaturation) and at temperatures as warm as a few degrees below 0°C. The temperature dependence of the occurance of ice in atmospheric clouds is illustrated in Fig. 32 which shows that with decreasing temperature the percentage of supercooled clouds in the

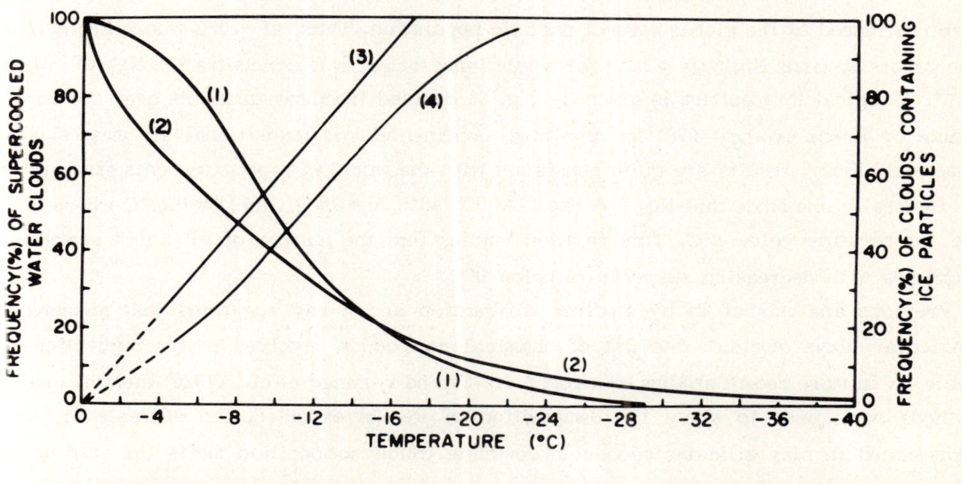

Fig. 32. Variation of the frequency of supercooled clouds and of clouds containing ice crystals. Curves 1 and 2 pertain to ordinate at left; curves 3 and 4 pertain to ordinate at right; (1) Clouds over Germany, (2) clouds over European territory of Russia; (3) clouds over Tasmania, (4) clouds over Minnesota. (From Pruppacher and Klett, 1978; by courtesy of D. Reidel Publ. Co. and the authors).

atmosphere rapidly decreases.

The need for AP to form atmospheric ice particles implies that AP become removed from the aerosol population as a result ice nucleation scavenging. Direct photographic evidence for AP serving as ice forming nuclei, and as such being located at the center of the ice crystal formed, was given by Rucklidge (1965).

Apart from ice crystal formation via deposition of water vapor onto ice forming nuclei, IC may be formed by the condensation - freezing mechanism which involves the activation of an AP to a drop by condensation and the subsequent freezing of the drops by means of the water insoluble compounds inside the drop. Ice crystals may, thirdly, be formed by contact nucleation whereby AP initiate freezing of existing supercooled drops by collisional contact with them. Nucleation by deposition and by condensation freezing remove AP by necleation scavenging while contact nucleation removes the AP by impaction scavenging.

STATEMENT 6: IN THE ATMOSPHERE, THE PHASE CHANGE TO ICE INVOLVES "PREFERRED" AP OF SPECIFIC CHARACTERISTICS.

These "preferred" AP of ice forming nuclei (IN) are generally water insoluble, belong to the large or giant size category, exhibit at their surface crystallographic characteristics in terms of bond arrangements similar to that in ice itself, exhibit at their surface ionic or preferredly hydrogen bonds, and contain at their surface "surface-active-sites".

This selectivity is expressed by the fact that the concentration of IN is a very small fraction indeed of the global aerosol particle population. Thus, at -20°C where $N_{IN} \approx 1$/liter one generally finds $N_{IN}/N_{AP} \approx 10^{-6}$ for a total aerosol particle concentration N_{AP} of only 10^3 cm^{-3}. A typical IN spectrum is given by Fig. 33 derived from measurements over the US High Planes by Hobbs et al., (1978) for maritime, continental and transitional air masses. These somewhat "local" results are quite consistent with the world wide measurements given in Fig. 34. Generally one finds that $N_{IN} = A \exp(\beta \Delta T)$, with $A \approx 10^{-5}$/liter, $\beta \approx 0.6$/°C where ΔT is the temperature below 0°C. This relation implies that the number of AP which serve as IN increases with decreasing temperature below 0°C.

Previous analysis of IN by electron diffraction and x-ray suggested that silicates, in particular clays dominate the list of chemical compounds involved in ice nucleation (see Table 4). In more recent studies by Grosch (1978) and Parungo et al., (1976) element analysis methods were used to study the composition of IN. As expected, the elements Si, Al, K represented in clay silicates appear to dominate. Such composition meets the insolubility, crystallographic, and chemical bond requirement, as silicates are water insoluble and exhibit on certain crystallographic faces a pseudo-hexagonal symmetry of the hydrogen bonds which emanate from the OH-sites at the surface.

Georgii (1982) and Rosinski et al., (1980) studied the size distribution of IN. Their results

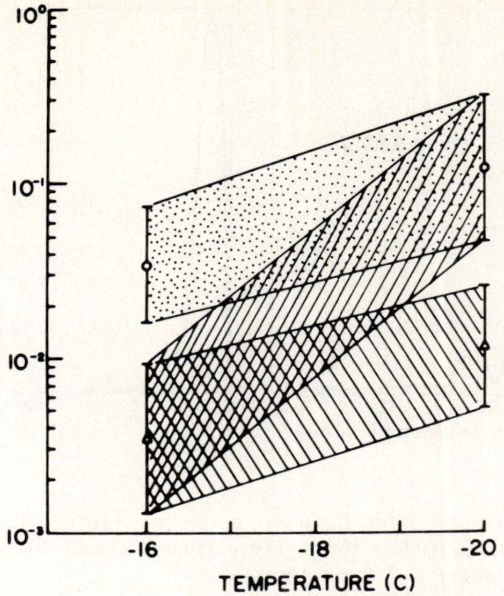

Fig. 33. Ice nucleus spectrum measured over the US High Planes during Summer of 1975 and 76; dotted: continental background air mass, hatched from left to right: clean air mass, hatched from right to left: intermediate air mass. (From Hobbs et al, 1978; by courtesy of the authors).

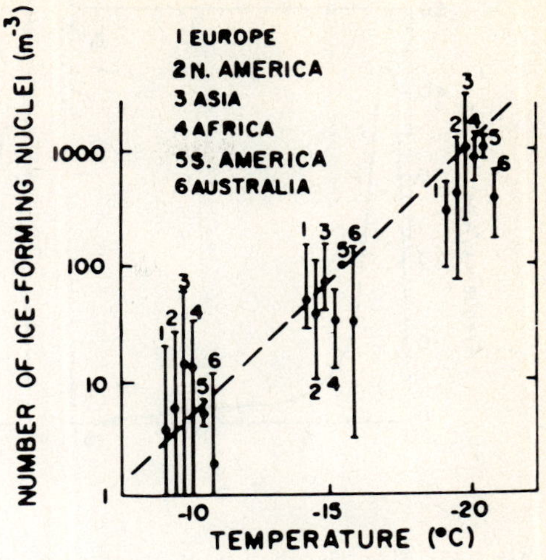

Fig. 34. Range of median number concentration of ice forming nuclei as a function of temperature for various geographic locations (44 stations). Dashed line: $N_{IN} = 10^{-5} \exp(0.6\ \Delta t)$, with $\Delta T = T_0 - T$, $T_0 = 273$ K. (From Pruppacher and Klett, 1978; by courtesy of D. Reidel Publ. Co. and the authors).

Composition of center particle	Hokkaido (Japan)		Honshu (Japan)		Michigan (U.S.A.)		Missouri (U.S.A.)		Thule (Greenland)		Amundsen-Scott (South Pole)	
	Number	%	Number	%	Number	%	Number	%	Number	%	Number	%
Clay mineral	176	57	46	88	235	87	70	28	302	84	55	59.1
Hygroscopic particle	57	19	0	0	2	1	5	2	2	1	19	20.4
Combustion product	26	8	2	4	6	2	7	3	0	0	-	-
Micro-organism	3	1	0	0	0	0	3	1	0	0	-	-
Unidentified material	30	10	4	8	25	9	100	40	39	11	5	5.4
Not observed	15	5	0	0	3	1	65	26	13	4	14	15.1
Total	307	100	52	100	271	100	250	100	356	100	93	100
Reference	Kumai (1961)						Rucklidge (1965)		Kumai and Francis (1962b)		Kumai (1976)	

Table 4. Composition of aerosol particles in central portion of snow crystals. (From Pruppacher and Klett, 1978; by courtesy of D. Reidel Publ. Co. and the authors).

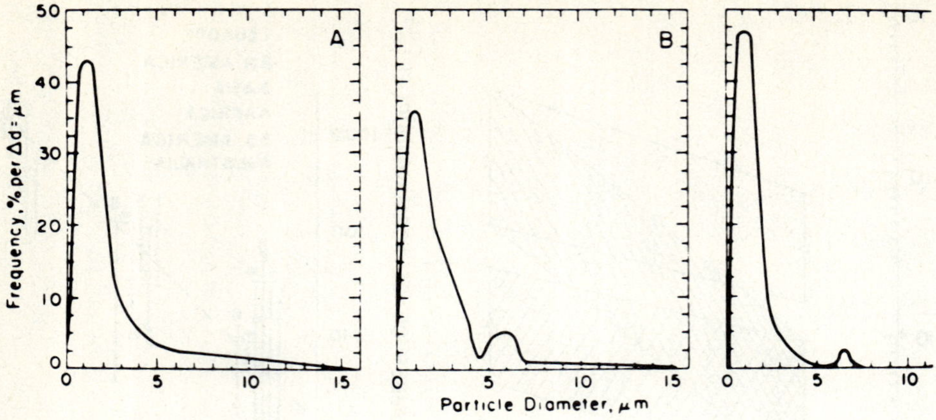

Fig. 35. Size distribution of ice forming nuclei in air over Colorado active at -20°C; A: all particles; B: inorganic particles; C: organic particles; total: 278 particles. (From Rosinski et al, 1980; by courtesy of Gebr. Borntraeger and the authors).

are summarized in Figs. 35 and 36 which show that the size distribution of IN has a pronounced mode near a diameter of 2 um, indicating that most IN belong to the category of giant AP.

The requirement for "active sites" on the surface of an AP which is to serve as IN was demonstrated by numerous scientists (see Pruppacher and Klett, 1978). This requirement is illustrated by Fig. 37 by the experiment of Roberts and Hallett (1968). This figure shows for clay minerals that the ice forming efficiency for a particular silicate dramatically increased with decreasing temperature, despite the fact that the size and the chemical and crystallographic characteristics of the AP studied was kept constant. Four different types of "active sites" have thus far been identi-

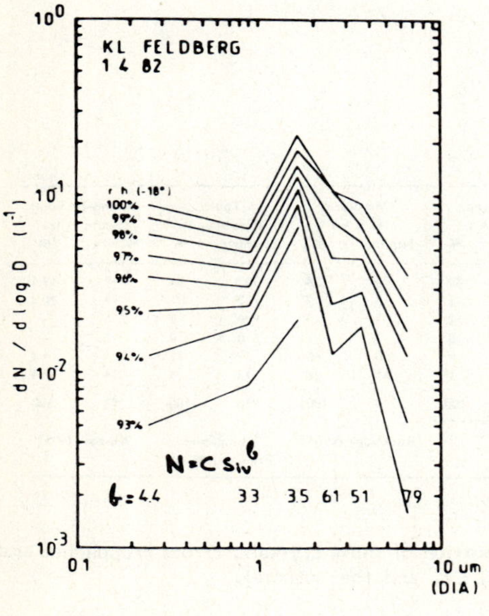

Fig. 36. Size distribution of ice forming nuclei in air over kl. Feldberg (FRG) at different relative humidity of the air. (From Stein and Georgii, 1982; by courtesy of Gebr. Borntraeger and the authors).

fied: (1) morphological sites in form of steps on crystallographic lattice dislocations, (2) chemical sites in terms of hydrophilic atoms surrounded by a hydrophobic crystallographic matrix, (3) electric sites in terms of electric surface fields of oriented electric dipole groups, and (4) electronic sites in terms of atomic electron donor or receptor states.

Analogously to the behaviour of CCN, one finds that $N_{IN} = Cs_{iv}^{k}$ (where s_{iv} is the supersaturation with respect to ice). This relation implies that at any given temperature below 0°C the number of IN increases with increasing supersaturation in the environment. In addition to the requirement for sufficient supersaturation, Anderson and Hallett (1976) pointed out that the time of exposure to a particular supersaturation decisively determines whether a particular AP can serve as IN. This nucleation "time lag" is illustrated by Fig. 38 a, b which shows for CuS and AgI that the time lag increases as the supersaturation decreases.

An additional constraint on an AP to serve as IN results from the fact that between 0°C and critical temperature below 0°C water saturation is needed to exist in the aerosol particle's environment, first activating the AP to a drop and subsequently freezing it. At temperatures below this critical temperature humidities below water saturation (but still above ice saturation) are sufficient for activation of an AP to ice. This result is illustrated by Fig. 39 derived from the experiments of Schaller and Fukuta (1979).

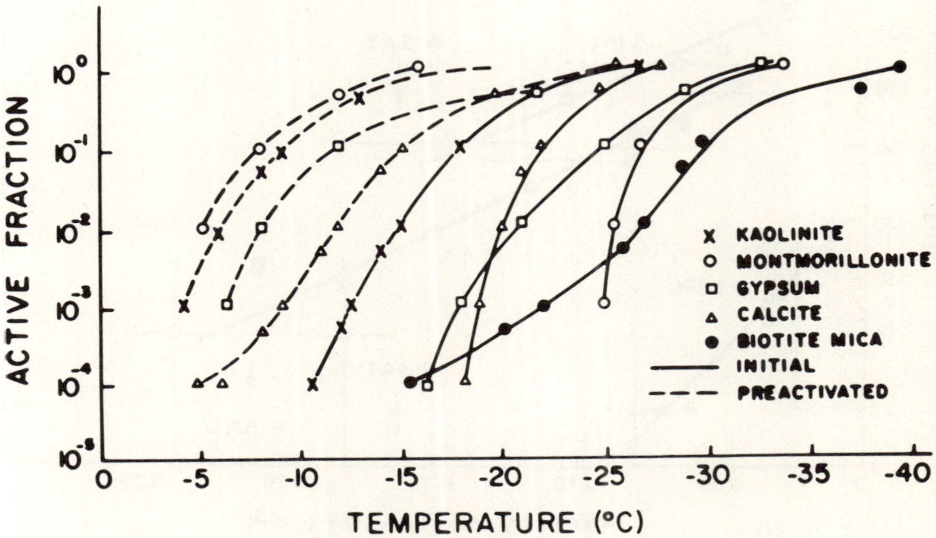

Fig. 37. Variation of the fraction of aerosol particles acting as ice forming nuclei as a function of temperature for various silicates. (From Roberts and Hallett, 1968; by courtesy of Royal Meteor. Soc. and the authors.)

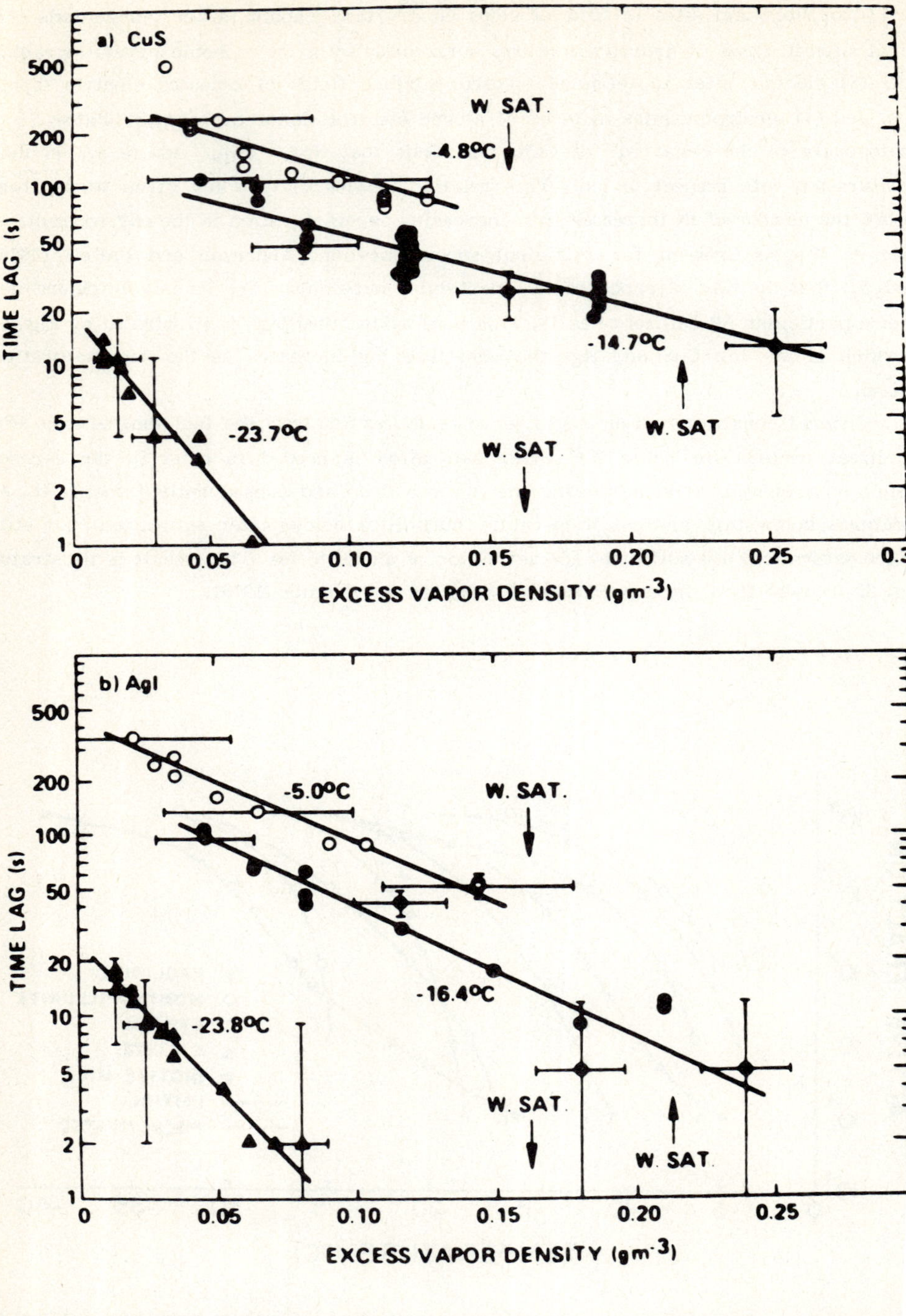

Fig. 38 a, b. Time lag for ice nucleation on surfaces of: a: covellite, b: silveriodide. (From Anderson and Hallett, 1976; by courtesy of Amer. Meteor. Soc. and the authors).

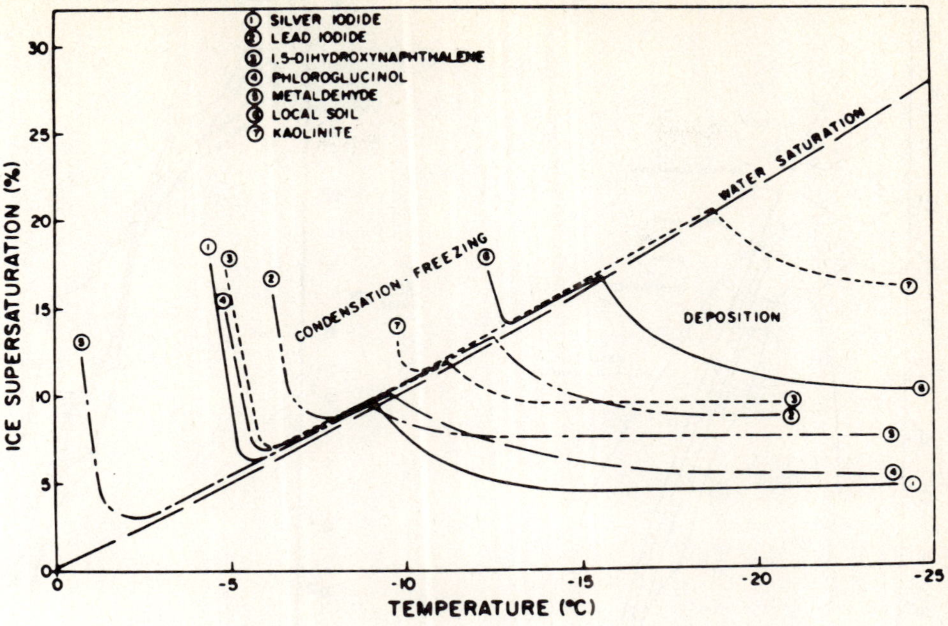

Fig. 39. Limiting curves for different aerosol particles to serve condensation-freezing or deposition ice forming nuclei as a function of temperature and ice supersaturation. (From Schaller and Fukuta, 1979; by courtesy of the Am. Meteor. Soc. and the authors).

STATEMENT 7: THE ATMOSPHERE CONTAINS AEROSOL PARTICLES OF A WIDE RANGE OF SIZES AND NUMBER CONCENTRATIONS WHICH ARE RELATED TO EACH OTHER IN A CHARACTERISTIC MANNER, ANALOGOUSLY FOR ALL AIR MASSES.

This is demonstrated by Figs. 40 to 43 where the number distributions are given for a model maritime, continental and urban aerosol (Jaenicke, 1982), for a silicate aerosol over desert area (d'Almeida and Schuetz, 1983), for a sulfate aerosol (Georgii, 1971), and for sodium containing AP (Radke et al., 1976). Generally, these distributions show an increase in number concentration with decreasing AP size approximately according to a power law of the type $-dN/d\log r = n^*(r)$, where $n^*(r) = C/r^{\alpha}$, C and α being parameters which depend on the airmass characteristics. Generally, these distributions have a concentration maximum near $r = 0.1$ µm. At smaller sizes, i.e. in the Aitken particle size range, minima and maxima may follow, caused by an intricate interplay between gas-to-particle-conversion and AP coagulation. When the number-size distributions are transformed to volume-size distributions apparent minor deviations in the power law for a number distribution become signi-

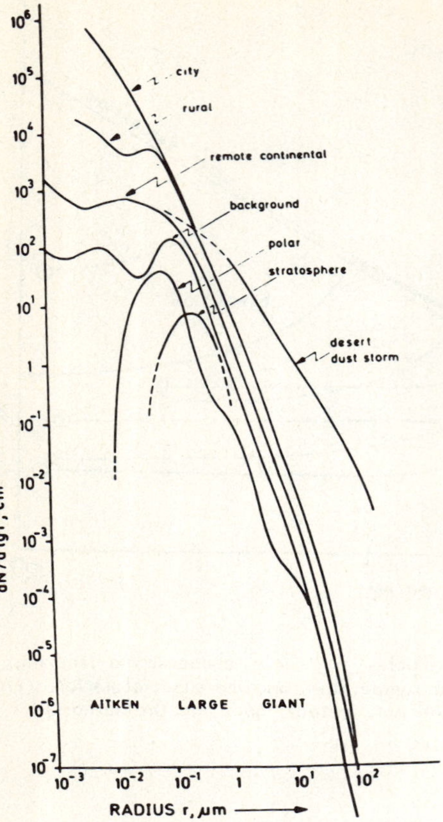

Fig. 40. Model aerosol particle number size distributions for various air masses. (From Jaenicke, 1978; by courtesy of the author).

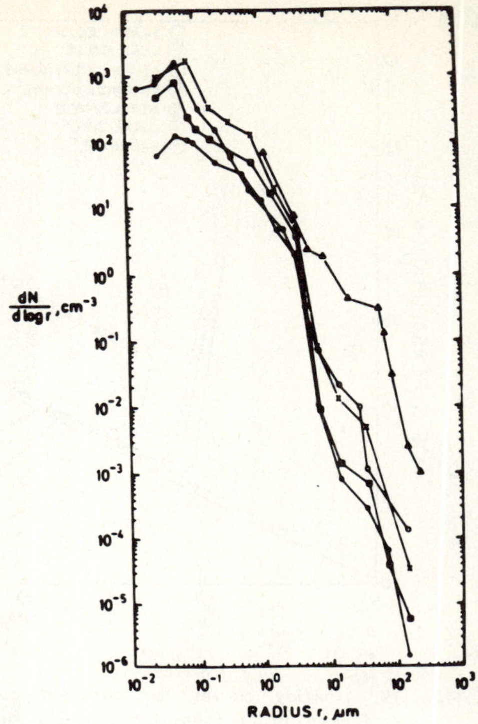

Fig. 41. Aerosol particle number size distribution for airborne silicates over Sahara desert. (From d'Almeida and Schuetz, 1983; by courtesy of Am. Meteor. Soc. and the authors).

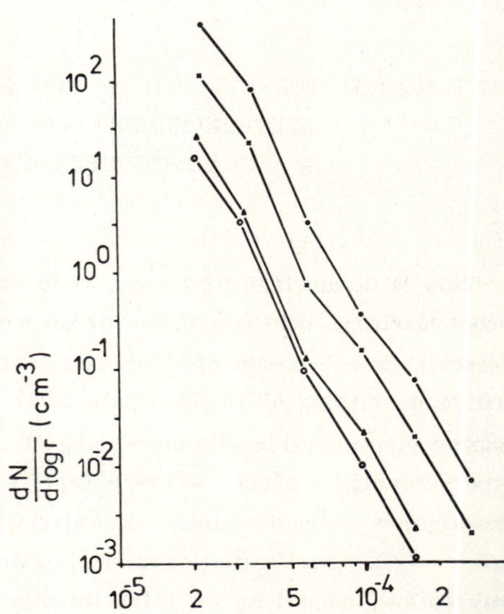

Fig. 42. Size distributions of sulfate aerosol particles at various altitudes. (From Georgii et al, 1971; by courtesy of the authors).

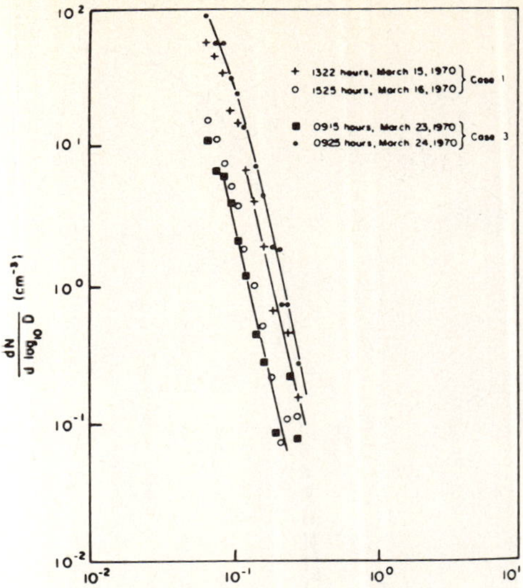

Fig. 43. Size distribution of sodium containing aerosol particles. (From Radke et al, 1976; by courtesy of Amer. Meteor. Soc. and the authors).

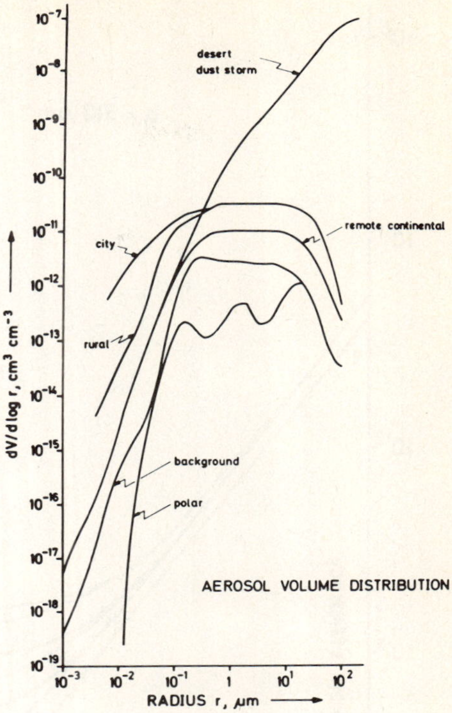

Fig. 44. Model aerosol particle volume size distribution for various air masses. (From Jaenicke, 1982; by courtesy of the author).

ficant modes, as originally pointed out by Whitby (1978) and shown in Fig. 44 by Jaenicke (1982).

Theory and experiment show that the AP size importantly determines the efficiency with which AP are removed from the air and incorporated in cloud and rain water not only by nucleation scavenging but also by impaction-scavenging. Four mechanisms control impaction scavenging: Brownian convective diffusion, thermo- and diffusiophoresis, and inertial impaction controlled by the hydrodynamic flow around the collector and by gravity. Superimposed on these four mechanisms are the effects of electric charges on the AP and on the collector, external electric fields, and turbulence in the air. Brownian motion appears to increasingly dominate the impaction scavenging for AP of $r \lesssim 0.1$ µm while inertial impaction appears to increasingly dominate impaction scavenging for AP of $r > 1.0$ µm. For drops as scavengers an efficiency minimum appears for $0.1 \leq r \leq 1.0$ µm, termed the "Greenfield gap" which for planar ice crystals as scavengers is shifted to $0.01 \leq r \leq 0.1$ µm. Both, phoretic and electric forces affect the scavenging of AP in this size range and tend to "bridge" the gap increasingly with decreasing relative humidity in the air and with increasing electric charges on the collector and the AP. This behavior is illustrated by Figs. 45 and 46. Although the actual "debth"of the Greenfield gap is still considerably debated considering

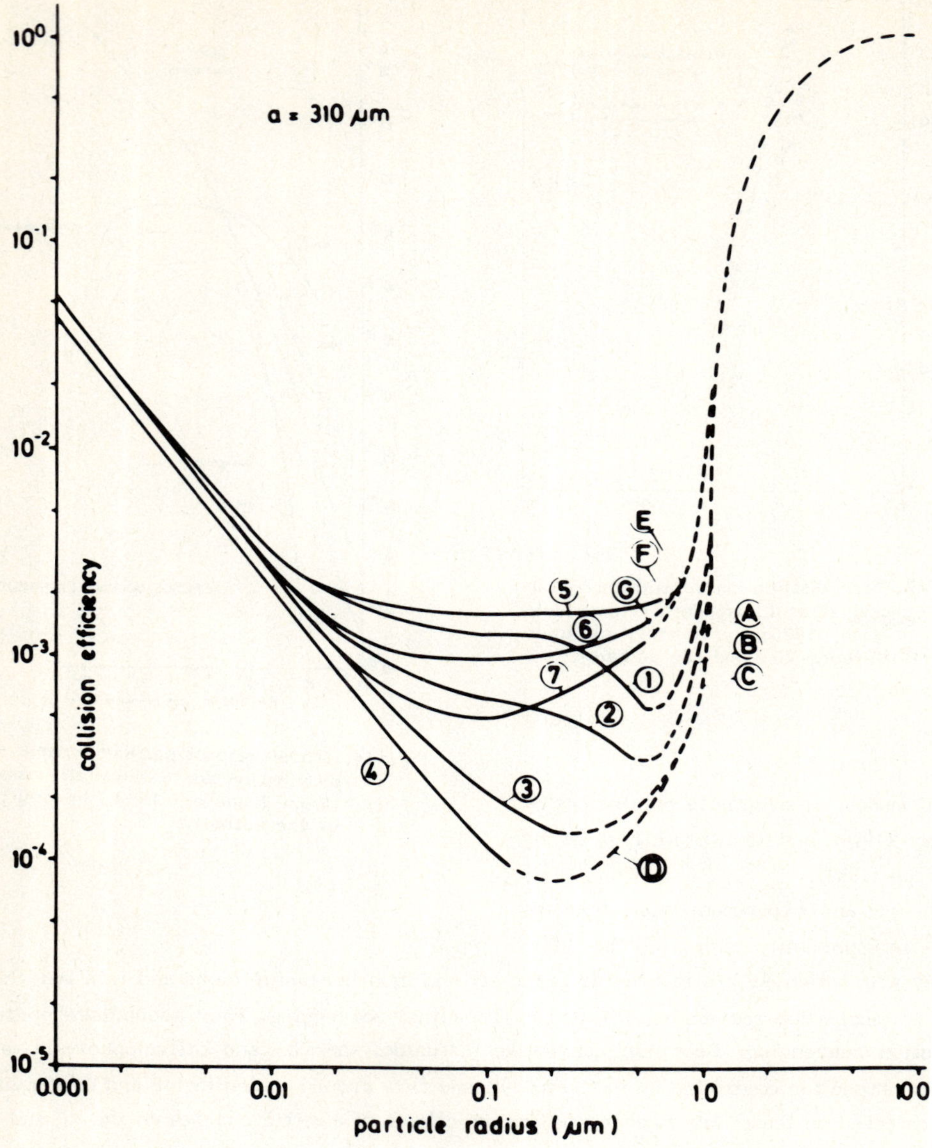

Fig. 45. Efficiency with which a water drop of 310 um collects aerosol particles as a function of aerosol particle size; curves 4 and D, 3 and C, 2 and B, 1 and A: for 100%, 95%, 75%, 50%, relative humidity and no electric charge; curves 7 and G, 6 and F, 5 and E: for 95%, 75%, 50% relative humidity and electric charge $q_r = Q_r/r^2$, $q_a = Q_a/a^2$ with $q_r = -2.0$ esu, $q_a = +2.0$ esu; for 20°C and 900 mb. (From Wang et al, 1978; with courtesy of Amer. Meteor. Soc. and the authors).

the field observations of Radke et al., (1980) which show a relatively shallow gap the fact remains that not only nucleation-scavenging but also impaction-scavenging is a very selective process, the latter favouring the capture of the very large and the very small AP and of intermediate size AP only for low relative humidities and high electric charges. We recall, however, that for drops nucleation scavenging of AP is favoured precisely in the "gap-region" of impaction scavenging mechanisms together may indeed "bridge" the gap completely. This characteristic applying to drops does not apply to the scavenging behavior of planar ice crystals as for them the impaction scavenging gap lies between 0.01 and 0.1 μm AP radius while ice nucleation scavenging favours AP of r > 1 μm.

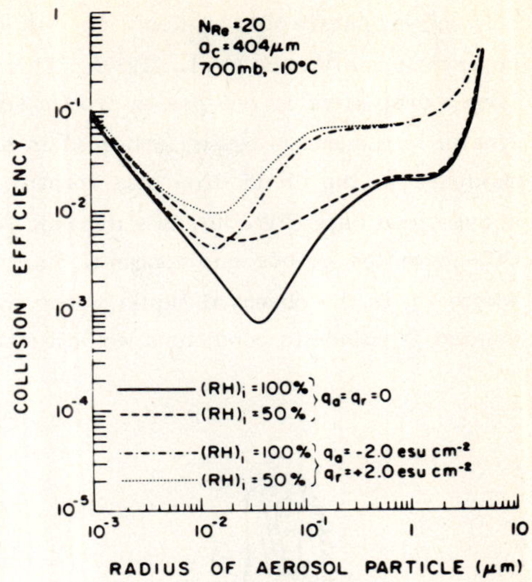

Fig. 46. Efficiency with which a planar ice crystal of 404 μm radius collects aerosol particles as a function of aerosol particle size. (From Martin et al, 1980; by courtesy of Amer. Meteor. Soc. and the authors).

STATEMENT 8: ATMOSPHERIC CLOUDS AND PRECIPITATION CONTAIN HYDROMETEORS OF A WIDE RANGE OD SIZES AND NUMBER CONCENTRATIONS WHICH ARE RELATED TO EACH OTHER IN A MANNER DEPENDING CHARACTERISTICALLY ON: (1) THE AEROSOL OF THE AIR MASS IN WHICH THE CLOUD FORMED, (2) THE DYNAMICS OF THE CLOUD, (3) THE MICROPHYSICAL PROCESSES RESULTING FROM INTERACTION BETWEEN THE CLOUD PARTICLES, AND (4) ON WHETHER THE HYDROMETEORS CONSIST OF WATER PARTICLES OR ICE PARTICLES.

These four controlling factors cause the clouds to exhibit noticeable space variations of the drop number concentration, of the drop size, and of the liquid water content, as exhibited in Fig. 47, for a non-precipitating convective cloud observed by Mei-Yuan (1963). The effect of air mass in the drop size spectrum is illustrated in Fig. 48 by the observations of Battan and Reitan (1957), showing that with increasing maritime character the drop spectrum widens to larger sizes while the total concentration becomes less. The effects of the convective nature of a cloud on the drop size spectrum is exhibited in Fig. 49 by observations of Hobbs et al., (1980). This figure shows that the drop spectrum shifts toward larger drop sizes as one passes from a small developing cumulus to a cumulus complex and finally to a cumulus system embedded in a stratiform cloud. Also the often observed double maximum in the cloud drop size spectra, increasingly pronounced with increasing height above cloud base (Fig. 50), is a dynamic feature caused by the entrainment of drier air and CCN from the clouds environment. Such entrainment is reflected by the ratio $W_L/W_{L,ad}$, where W_L is the observed liquid water content in the cloud and $W_{L,ad}$ is the liquid water content if adiabatic conditions would exist in the cloud. We note from Fig. 51 that generally

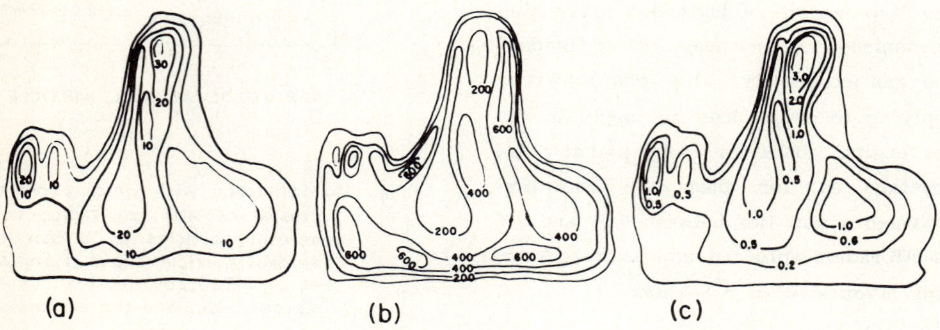

Fig. 47. Distribution of: (a) drop diameter, (b) drop concentration, (c) liquid water content, inside an non-precipitating cloud. (From Pruppacher and Klett, 1978; by courtesy of D. Reidel Publ. Co. and the authors).

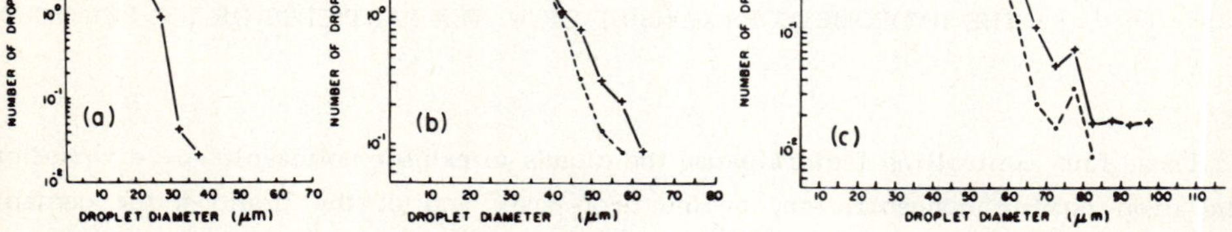

Fig. 48. Mean drop size distribution inside various types of clouds: (a) fair weather cumulus, (b) cumulus congested over central US, (c) cumulus over Gulf of Mexico. (From Battan and Reitan, 1957; by courtesy of Pergamon Press and the authors).

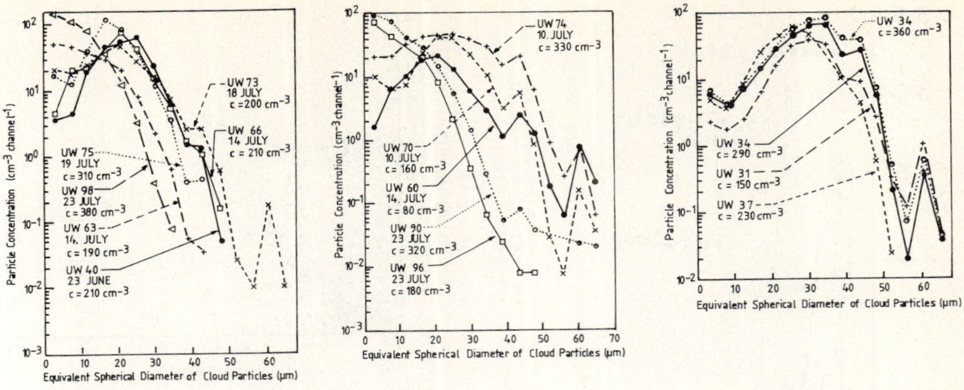

Fig. 49. Mean cloud particle size distribution inside various types of clouds; from left to right: developing small cumulus, cumulus complexes, cumulus embedded in stratoform cloud. (From Hobbs et al, 1980; by courtesy of Amer. Meteor. Soc. and the authors).

$(W_L/W_{L,ad}) < 1$, due to entrainment. However, in certain clouds $(W_L/W_{L,ad})$ may become superadiabatic and assume values considerably larger than unity due to accumulation zones as shown by Soulage et al., (1980) for cumulonimbus clouds (see Fig. 52).

Cloud drop size distribution without a double maximum may adequately be described by a modified gamma distribution of the form $n(a) = Aa^{\beta} \exp(-Ba^{\gamma})$, where A, B, β and γ are positive parameters, and a is the dropradius. Often: $\beta = 2$, $\gamma = 1$, $A = 1{,}45 \times 10^{-6}(W_L/\rho_W a^{-6})$, and $B = 3/\bar{a}$, $\bar{a}$ being the mean dropradius, as suggested in Borovikov et al., (1963). Unfortunately, drop size spectra with double maxima have no single analytic description. On the other hand, microphysical processes, such as drop collision, coalescence and break-up, produce rain drop size distributions of rather simple, exponential form of the

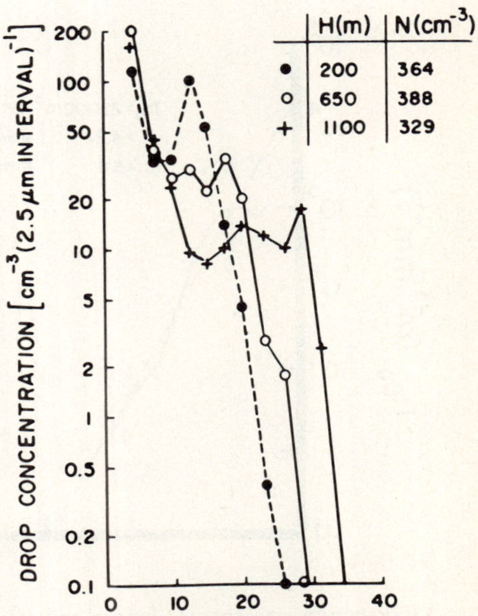

Fig. 50. Variation of drop size distribution with height above cloud base in convective cloud. (From Warner, 1969; by courtesy of Amer. Meteor. Soc. and the author).

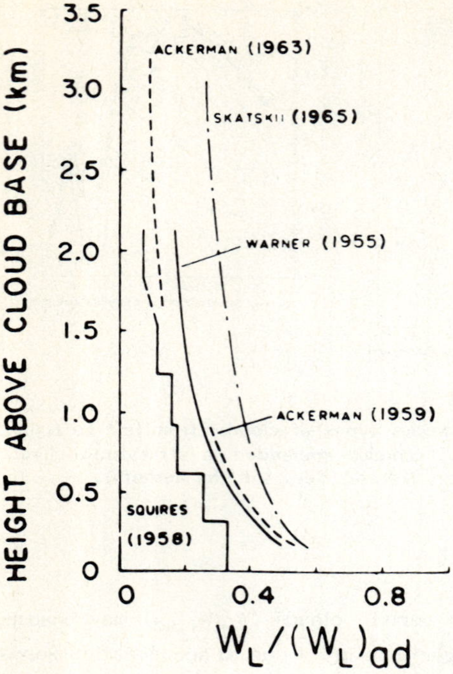

Fig. 51. Variations of $W_L/W_{L,ad}$ with height above cloud base (W_L = actual liquid water content in cloud, $W_{L,ad}$ = liquid water content for adiatic cooling). (From Warner, 1970; by courtesy of Amer. Meteor. Soc. and the author).

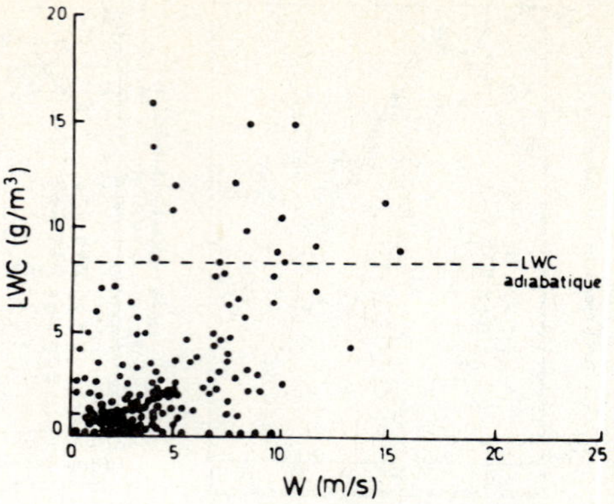

Fig. 52. Variation of the liquid water content with mean updraft speed in cumulonimbus clouds. (From Soulage et al, 1980; by courtesy of J. de Rech. Atm. and the authors).

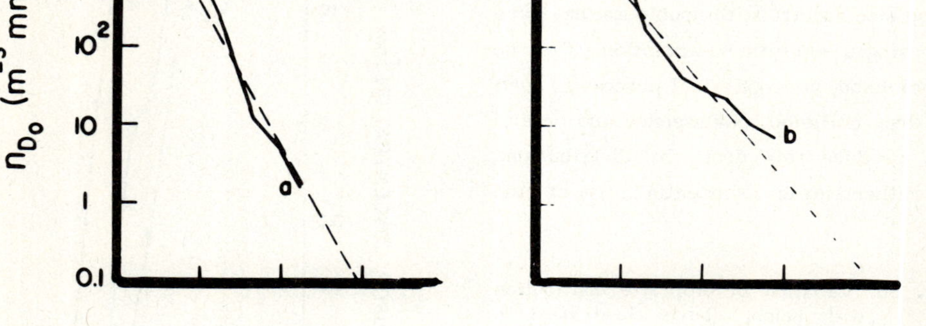

Fig. 53. Raindrop size spectra before and after n_0-jump during a particular rainfall at station in S. Switzerland. (From Waldvogel, 1974; by courtesy of Amer. Meteor. Soc. and the author).

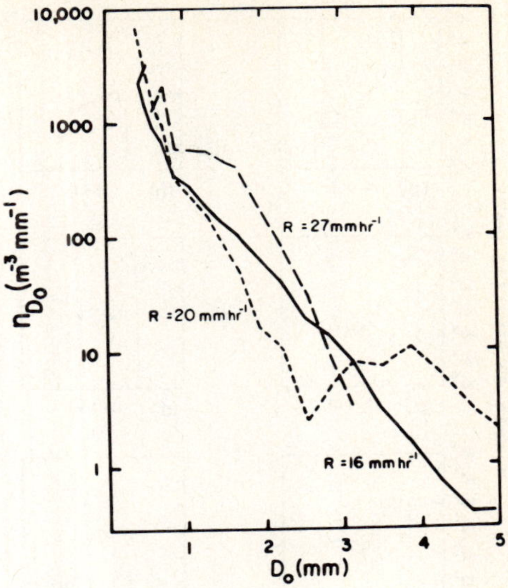

Fig. 54. Typical raindrop size distributions measured at different occasion in S. Switzerland. (From Joos et al, 1968; by courtesy of Amer. Meteor Soc. and the authors).

type $n(D_o) = n_0 \exp(-\Lambda D_o$, where D_o is the equivalent drop radius of the rain drop and n_0 and Λ are parameters which depend on the rain rate and the microphysical process in clouds, the latter being illustrated by Fig. 53. This figure shows an exponential raindrop size distribution with a jump in the value of and n_o during a single precipitation event, as observed by Waldvogel (1974). The fact that raindrop spectra often are not exponential is examplified in Fig. 54. Six typical model distributions given in Fig. 55 were suggested by Czerwinski and Pfisterer (1972) based on 10'650 raindrop spectra observed at different geographic locations.

Fogs have generally small drop concentrations for radii a > 1 μm, a narrow size spectrum and low liquid water contents, although some maritime fogs may have relatively broad size

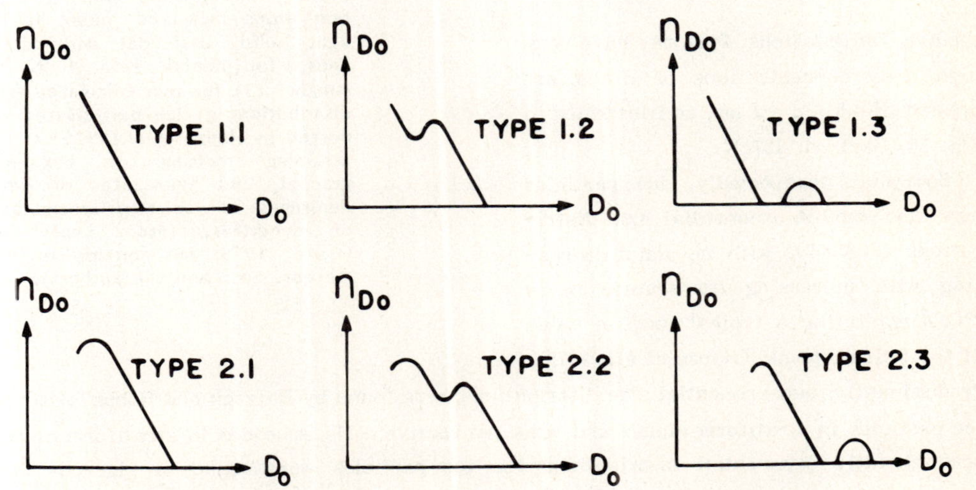

Fig. 55. Schematic representation of raindrop spectra found from 10'650 spectra observed at different locations. (From Czerwinski and Pfisterer, 1972; by courtesy of J. de Rech. Atmos. and the authors).

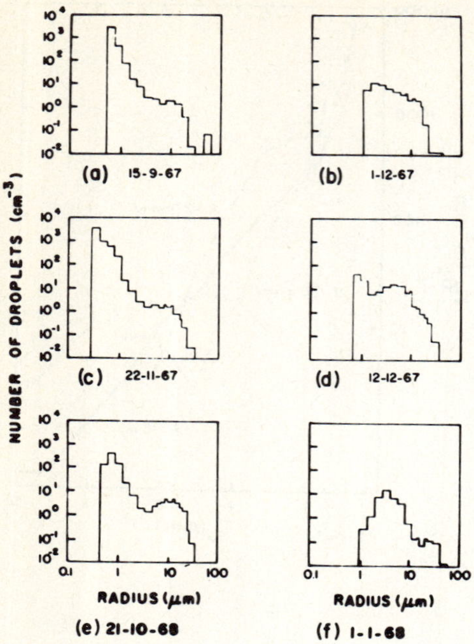

Fig. 56. Drop size distribution in fogs. (From Garland, 1971; by courtesy of Roy. Meteor. Soc. and the author).

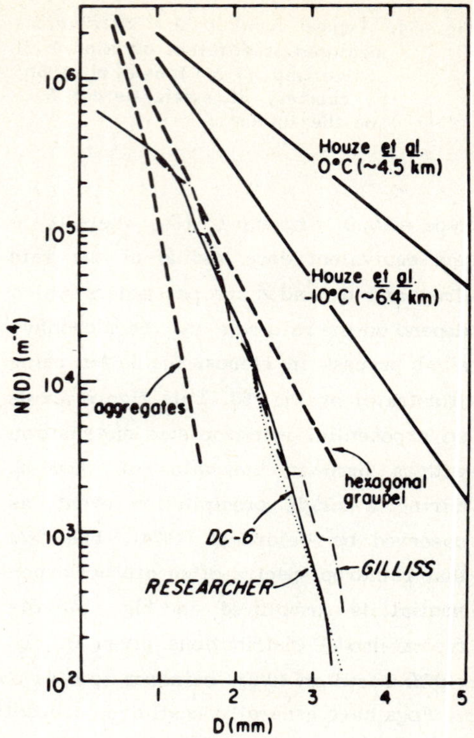

Fig. 57. Size distribution: (1) for ice particles as reported by Houze et al (1979), heavy solid lines; (2) for rain drops measured during GATE, light solid, dash dot and dotted lines, for rainfall rata 1 to 10 mm/hr; (3) for rain calculated from distributions of ice particulates reported by Houze et al (1979) at 0°C assuming melting of hexagonal graupel, and aggregates of rimed denisities or radiating assemblages of dendrites. (From Leary and Houze, 1979; by courtesy of Am. Meteor. Soc. and the authors).

spectra. On occasions, fogs may have very high drop concentrations of drops are present of $0.5 \leq a \leq 1$ μm, as illustrated in Fig. 56 (Garland, 1971).

Somewhat unexpectedly, ice particles may also exhibit exponential size distributions (Fig. 57), with n_0 and Λ decreasing with increasing temperature below 0°C, Λ exhibiting a typical jump in value at the melting level (Houze et al., 1979). Predominantly, subexponential size distributions were found by Herzegh and Hobbs (1980) for ice particles in stratiform clouds and weak convective cells embedded in stratiform clouds, predominantly exponential distributions for ice particles were found in vigerous cells embedded in stratiform clouds, and predominantly super-exponential distributions were found for ice particles in deep convective cells. Graupel and snowflakes exhibit exponential distribution if the number concentration is plotted against the equivalent drop diameter of the melted ice particles, as found by Yagi and Uyeda (1980) in confirmation of the much earlier studies of Gunn and Marshall (1958). Exponential size spectra were observed also for hailstones (Federer and Waldvogel, 1975) (see Fig. 58).

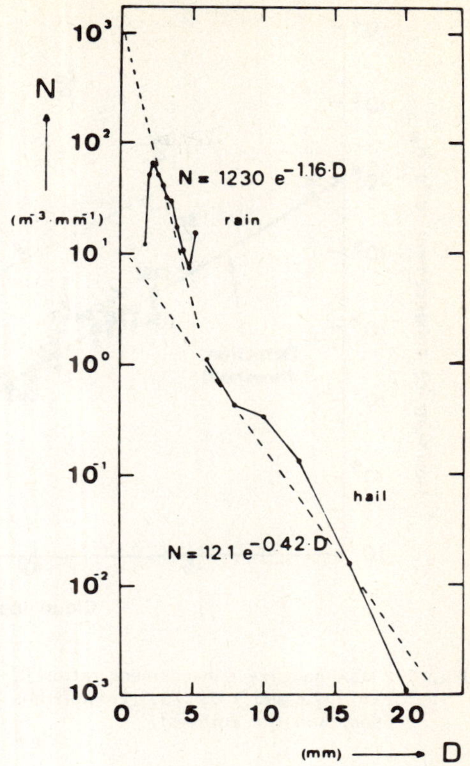

Fig. 58. Mean hail and rain spectra over a hailswath; solid lines: observed data; dashed lines: corresponding exponential distribution. (From Federer and Waldvogel, 1975; by courtesy of Amer. Meteor. Soc. and the authors).

In many clouds a comparison of the total number of ice particles with the number of ice forming nuclei reveals a considerable enhancement of the ice crystal concentration up to a factor of 10^4 near -5°C decreasing to 10° near -25°C, as shown in Fig. 59. Actual ice particle concentrations as a function of cloud top temperature and in-situ cloud temperature is given in Figs. 60 and 61. A comparison of the IN spectrum with the minimum ice crystal concentration suggests that regions of clouds either contain very few ice particles (< 0.1/liter), or ice particles in concentration at least as large as those required by the IN population. Many different mechanisms have over the years been suggested for explaining this ice particle enhancement- or ice multiplication process. Two mechanisms appear to stand out (1) the fracturing or delicate dendritic snow crystals, usually very abundant in snow carrying clouds, and (2) the ice splinter ejection found to take place during riming of larger ice particles. This splintering process critically depends on the efficiency with which the ejected ice splinters, grown by vapor diffusion to small columnar crystals, are captured by the larger, still unfrozen water drops. This efficiency has recently been computed by Lew and Pruppacher (1983). Their results are summarized in Fig. 62 which gives this efficiency for a collector drop of 416 um radius collecting columnar ice crystals of various lengths, and length to diameter ratios, colliding in various orientations with the drop. We note that the efficiency rapidly decreases with increasing crystal length, particularly for length to diameter ratios (L/D) > 3.

From the definition of the scavenging coefficient for drops $\Lambda \equiv (-1/n_p)\ (dn_p/dt) = \int_0^\infty K(r,a)\ n_d(a,t)\ da$, (where a and r are the drop radius and AP radius, respectively, n_d and n_p are number concentration of drops and AP, respectively, t is time, and K(r,a) is the collection kernel being a function of the efficiency E(r,a) with which a particular AP is

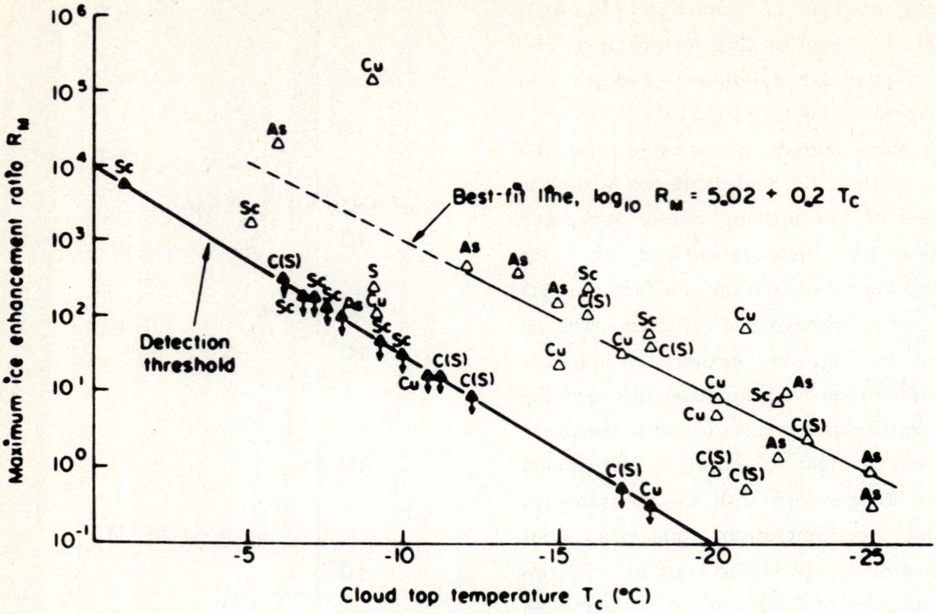

Fig. 59. Maximum ice enhancement ratios R_M for flights over Cascade Mts (Wash.) during winters 1971/72 and 1972/73. (From Hobbs and Atkinson, 1976; by courtesy of Amer. Meteor. Soc. and the authors).

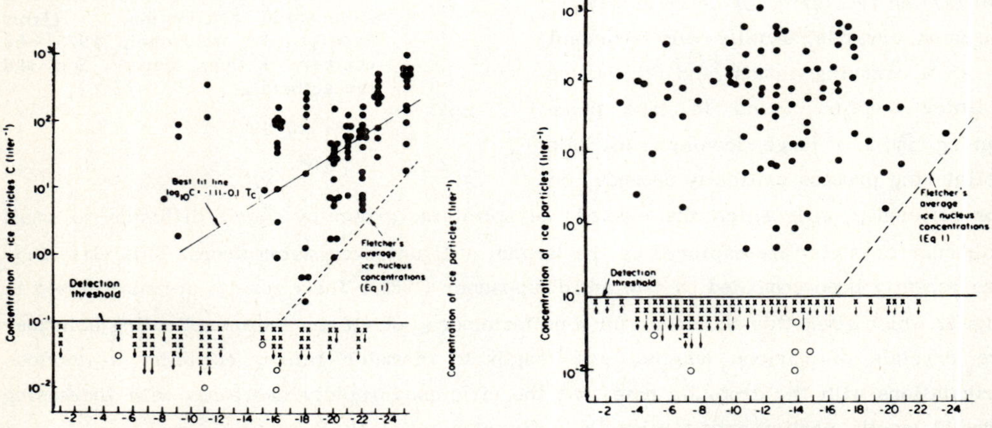

Fig. 60, 61. Ice crystal concentrations in clouds as indicated in Fig. 59, as function of cloud top temperature (Fig. 60), and as a function of in situ cloud temperature (Fig. 61). (From Hobbs and Atkinson, 1976; by courtesy of Amer. Meteor. Soc. and the authors).

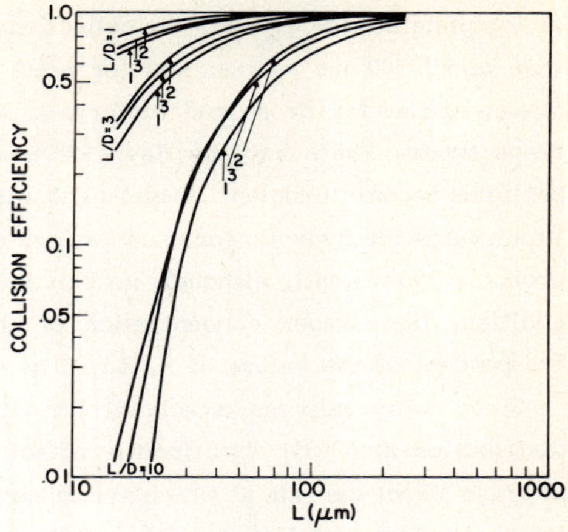

Fig. 62. Efficiency with which a cloud drop of 416 μm radius collects columnar ice crystals of various length (L) to diameter (D) ratios as a function of L, for 3 different orientations of columns with respect to drop. (From Lew and Pruppacher, 1983; by courtesy of Amer. Meteor. Soc. and the authors).

captured by a particular drop, and a function of the relative velocity between drop and AP), one notices that the drop size distribution decisively determines the rate at which AP are being removed from the atmosphere by impaction scavenging. An analogous statement applies to the removal of AP by ice hydrometers.

Having given in Figs. 45 and 46 the variation of the collection efficiency E for a drop of given size collecting by impaction scavenging AP of various radii, it is of interest to display the variation of E for a given r with the radius of the collector. This relationship is shown in Figs. 63 and 64 based on data of Wang et al., (1978) and of Martin et al., (1980). We note in particular that for drops as well

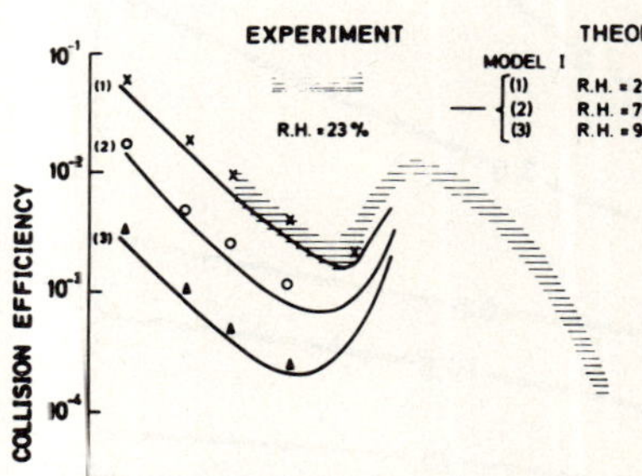

Fig. 63. Efficiency with which aerosol particles of 0.25 μm radius are captured by cloud drops and rain drops of various sizes; comparison of experiment and theory. (From Wang et al, 1978; by courtesy of Amer. Meteor. Soc. and the authors).

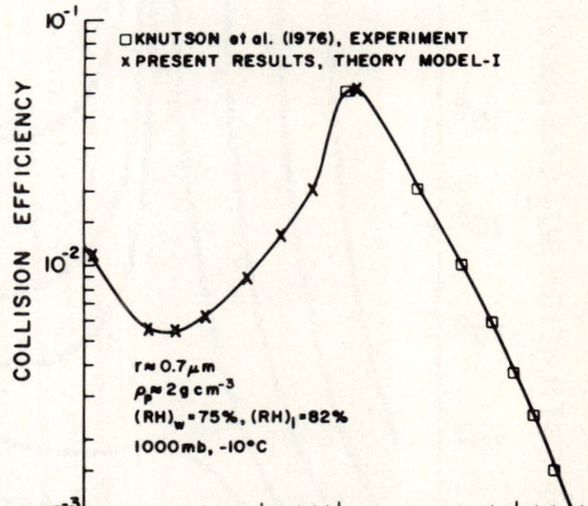

Fig. 64. Efficiency with which aerosol particles of 0.7 μm radius are captured by planar ice crystals of various sizes; comparison of theory and observation. (From Martin et al, 1980; by courtesy of Amer. Meteor. Soc. and the authors).

as for platelike ice crystals the collection efficiency rapidly decreases for drop sizes larger than about 500 um radius, and for plate type ice crystals larger than about 1000 um in diameter, i.e. for drops and ice crystal Reynolds numbers larger than about 300 and 100, respectively. These are the Reynolds numbers at which the hydrometeors wake of these particles become turbulent. A turbulent wake may also be the reason why the efficiency, with which large ice hydrometeors such as graupel and hailstones remove AP by direct pickup, is probably very small, although no detailed measurements exist to verify this statement. In addition, the number concentration of these large hydrometeors is very small so that independent of the values of K, Λ will be small.

Since water acts as a solvent for numerous gases in the atmosphere the drop size distribution also will significantly affect the rate at which atmospheric AP are scavenged but also affect the rate at which atmospheric trace gases are scavenged. Thus we notice from the theoretical model of Walceck and Pruppacher (1984) that the time for drops to reach 63% of saturation is a strong function of drop size and gas concentration in the environment, as shown in Fig. 65. For drops of 1.5 mm radius and typical environmental gas concentrations, this time is about 110 seconds. Considering the fall velocity of these drops this time represents a fall distance of about 1000 meters. This is a large distance considering the

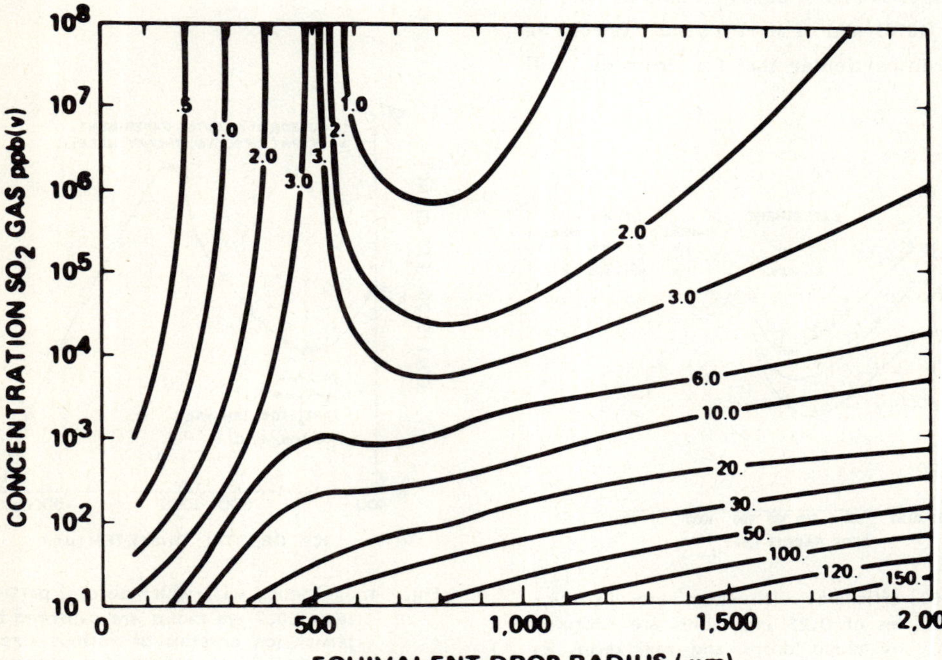

Fig. 65. Time (seconds) for drop to reach 63% of saturation as a function of drop size and SO_2 concentration in air surrounding drop. (From Walceck and Pruppacher, 1984; by courtesy of D. Reidel Publ. Co., and the authors).

vertical extent and the rapid variations of the concentration found in typical pollution layers. The work of Walceck and Pruppacher also shows that the amount of sulfur taken up by drops falling through a Gaussian plume of SO_2 is an important function of drop size, oxidation rate in the drop, and amount of rain having fallen through the layer. These parameters also appear to control the "wash-down" i.e. the lowering of the plume. Both findings are illustrated by Figs. 66 to 69 in which absorption as well as desorption of SO_2 is considered as the drop falls through the SO_2 plume. Thus, we note from Fig. 66 that e.g. 300 um radius drops tend to absorb SO_2 in the upper portion of the plume fairly rapidly, almost equilibrating with the environmental gas concentration. However, they appear to lose the taken-up SO_2 almost as rapidly to the air below the plume, arriving at the ground with relatively little sulfur. On the other hand, large drops of e.g. 2000 um radius fall rapidly and as a result spend little time in the pollution layer also transporting little sulfur to the ground. There appears to be an optimum drop size of between 1.0 and 1.5 mm radius at which a drop just spends sufficient time in the plume and loses by desorption still relatively little to arrive at the ground with maximum sulfur. Fig. 67 and 68 show for a typical raindrop spectrum the total amount of sulfur and the pH inside drops of different sizes as they arrive at the ground. We note the important role the oxidation rate plays in determining the amount of sulfur scavenged, and in determining the pH inside the drops. While for very small oxidation rates drops between 1000 to 1500 µm appear to be the most efficient scavengers, for very fast oxidation rates the scavenging efficiency seems to be inversely proportional to the drop size. Thus, the smaller drops, which have a larger surface area to volume ratio scavenge the most sulfur and are the most acidic. Fig. 69 demonstrates the effect of plume washdown for a typical rain drop spectrum as a function of amount of rain fallen through the plume for no oxidation in rain water. We note that significantly progressive lowering of the pollution plume with increasing rain amount having fallen through the plume.

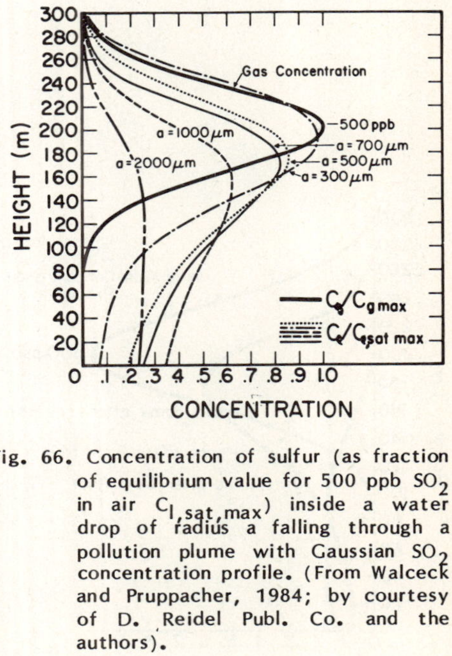

Fig. 66. Concentration of sulfur (as fraction of equilibrium value for 500 ppb SO_2 in air $C_{l,sat,max}$) inside a water drop of radius a falling through a pollution plume with Gaussian SO_2 concentration profile. (From Walceck and Pruppacher, 1984; by courtesy of D. Reidel Publ. Co. and the authors).

Due to the relatively slow growth and long life time of fogs, the small size of their drops and their corresponding very small fall velocities, fogs may be quite efficient in the uptake of trace gases. It is therefore not astonishing that fog water often has lower pH-values than rain or cloud water. However, thus far no theoretical model has been formulated which

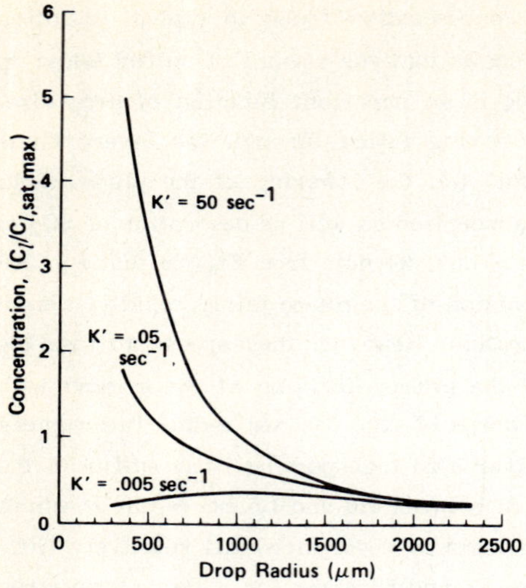

Fig. 67. Concentration of sulfur ($C_l(C_{l,sat,max})$) inside drops of various radii after they have fallen through pollution plume specified in Fig. 65; for various oxidation rates K', where: $[S^{VI}] = K'[S^{IV}]$. (From Walceck and Pruppacher, 1984; by courtesy of D. Reidel Publ. Co. and the authors).

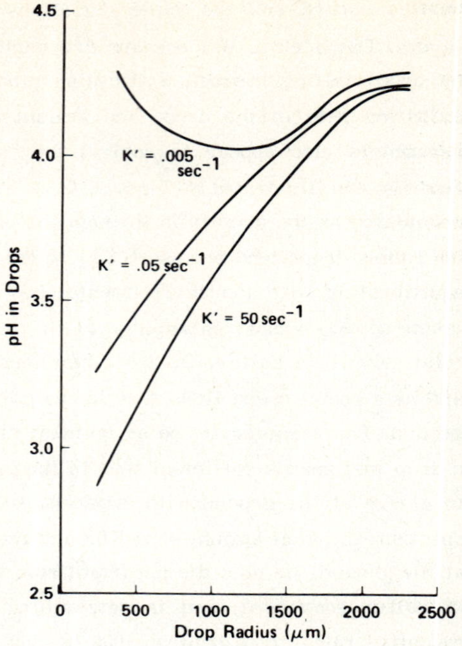

Fig. 68. Acidity (pH) of drops of various radii after they had fallen through pollution plume specified in Fig. 65; for various oxidation rates, specified in Fig. 66. (From Walceck and Pruppacher, 1984; by courtesy of D. Reidel Publ. Co. and the authors).

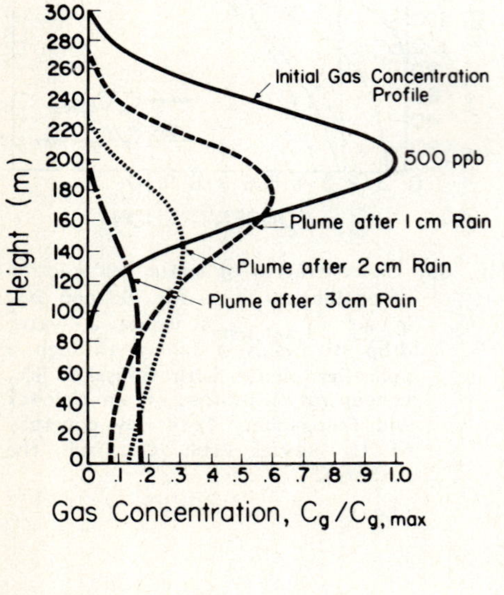

Fig. 69. Variation of SO_2 concentration profile in pollution plume specified in Fig. 65 after various amounts of rain had fallen through; for a typical atmospheric raindrop size distribution; K' = 0 (no oxidation). (From Walceck and Pruppacher, 1984; by courtesy of D. Reidel Publ. Co. and the authors).

could describe the uptake of trace gases by fogs and stratiform clouds.

It is likely that also snowcrystals take up trace gases such as SO_2 or NO_2, especially at temperatures only slightly below 0°C. Such a behaviour is expected due to the presence of a surface layer of highly mobile water molecules having a thickness of about 10 molecular layers near -5°C, decreasing to 1 molecular layer near -25°C. However, unfortunately no measurements have yet been made to determine the uptake of gases by such a mobile layer, although several studies presented at recent conferences indicate that snow crystals are not only efficient scavengers of AP, but also of trace gases.

STATEMENT 9: ATMOSPHERIC CLOUDS CONTAIN HYDROMETEORS OF A WIDE RANGE OF SHAPES. THESE DECISIVELY AFFECT THE FLOW FIELD AROUND THEM, THEIR FALL VELOCITY AND THEIR FALL MODE.

The flow fields outside and inside a spherical water drop of Reynolds number 100, i.e. a radius of about 310 µm, and of an idealized planar ice crystal of Reynolds number 20, i.e. a radius of 404 µm are given in Fig. 70 a, b and 71, respectively. In these figures results are given of numerical solutions of the complete, steady state Navier-Stokes equation of motion solved by Le Clair et al., (1972) for waterspheres in air, and by Pitter et al., (1973) for idealized ice plates in air.

It is obvious that the flow of air outside a hydrometeor importantly affects the trajectory

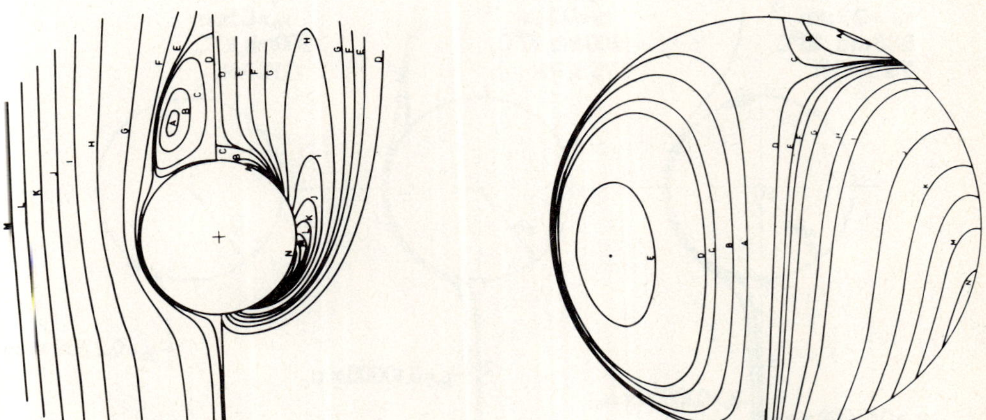

Fig. 70. a, b, Streamlines (left hand side) and lines of equal vorticity (right hand side) for airflow outside (a) and for circulation inside a drop of Reynolds number: 100 (drop radius 310 µm); from numerical solution of complete steady state Navier-Stokes equation of motion. (From LeClair et al, 1972; by courtesy of Amer. Meteor. Soc. and the authors).

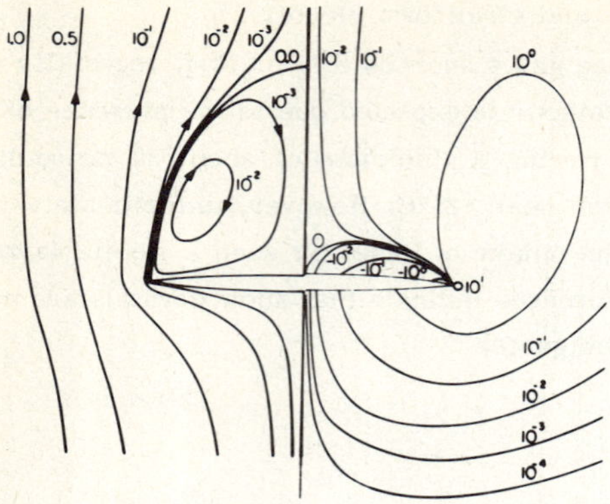

Fig. 71. Streamlines (left hand side) and lines of equal vorticity (right hand side) for air flow outside a planar ice crystal of Reynolds number: 20 (crystal radius: 404 μm).

of any AP to be collected by it. This is illustrated in Fig. 72 a, b, c for AP trajectories around falling drop as determined by Lew and Pruppacher (1983) and in Fig. 73 a, b for trajectories around falling ice crystal plates as determined by Martin et al., (1980): We note in particular that AP may be captured by drops anywhere on their surface and by the eddy located on the downstream side of the drops. On the other hand, AP are preferentially captured near the edges of planar ice crystals, a finding experimentally verified by Prodi (1976). Similar results were reported by Murakami et al., (1981) who found up to 10^4 AP attached to the surface of snow crystals as shown in Fig. 74.

The absorption of trace gases is affected by the flow field outside as well as inside a falling water drop. The rate at which gases are enhancedly transferred into a drop is

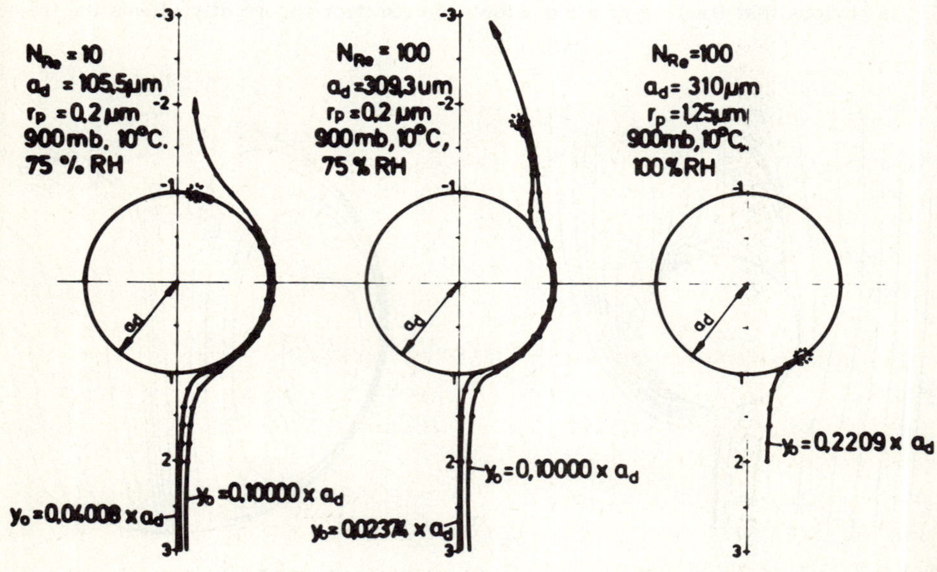

Fig. 72. Trajectories of aerosol particles of various sizes around drops of various sizes (From Pruppacher and Lew, 1983; by courtesy of the authors).

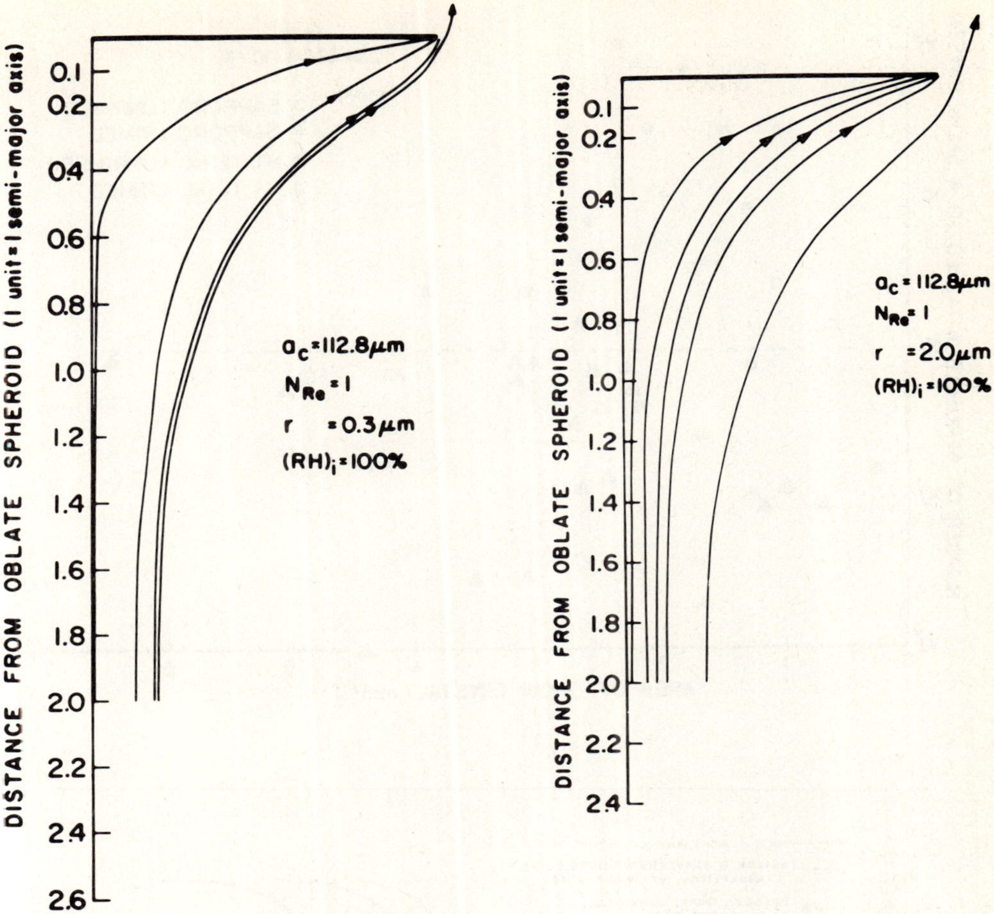

Fig. 73. Trajectories of aerosol particles of various sizes around planar snowcrystal of 112.8 μm radius. (From Martin et al, 1980; by courtesy of Am. Meteor. Soc. and the authors).

examplified in Fig. 75 (Baboolal et al., 1981). We notice that gas transfer is larger for a moving internally circulating drop than for a stagnant drop.

Since the fall velocity of AP is considerably smaller than that of hydrometeors it usually may be neglected so that the relative velocity between a hydrometeor and an AP to be captured is determined by the velocity of the hydrometeor alone. These velocities are well known for a wide range of hydrometeor masses and shapes, but generally apply to a specific temperature and pressure level in the atmosphere. This problem was recently circumvented by Beard (1980) who, from an analysis of a wide range of drag data for differently shaped bodies concluded that for all the pertinent hydrometeors in the atmosphere their terminal fall velocity $V_{\infty,0}$ at the level of measurement could be extrapolated to any other level by

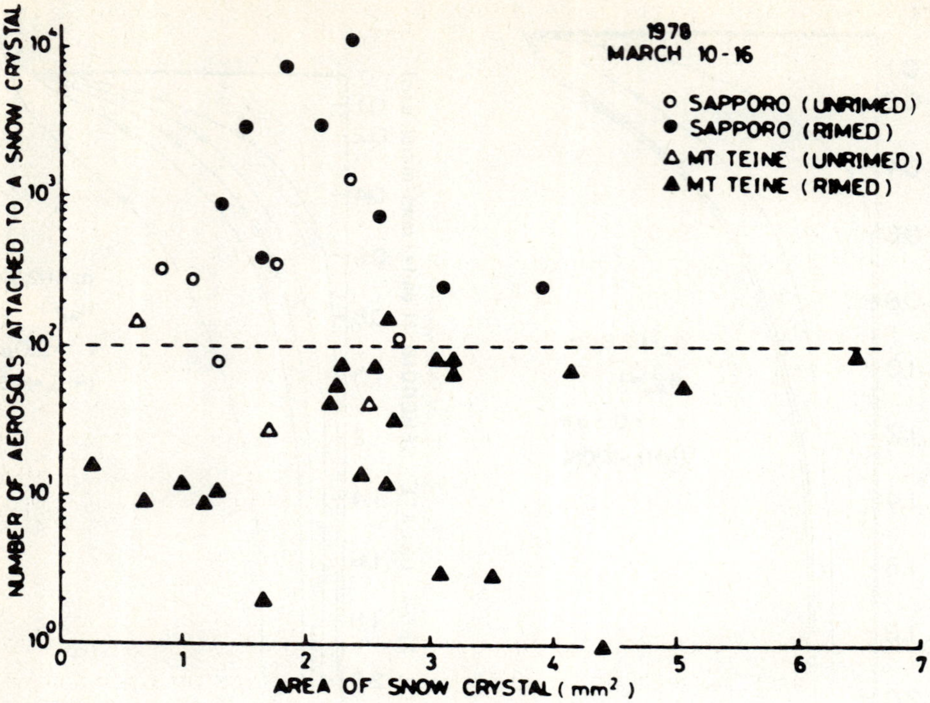
1978
MARCH 10-16
○ SAPPORO (UNRIMED)
● SAPPORO (RIMED)
△ MT TEINE (UNRIMED)
▲ MT TEINE (RIMED)
NUMBER OF AEROSOLS ATTACHED TO A SNOW CRYSTAL
10^0
10^1
10^2
10^3
10^4
0
1
2
3
4
5
6
7
AREA OF SNOW CRYSTAL (mm^2)

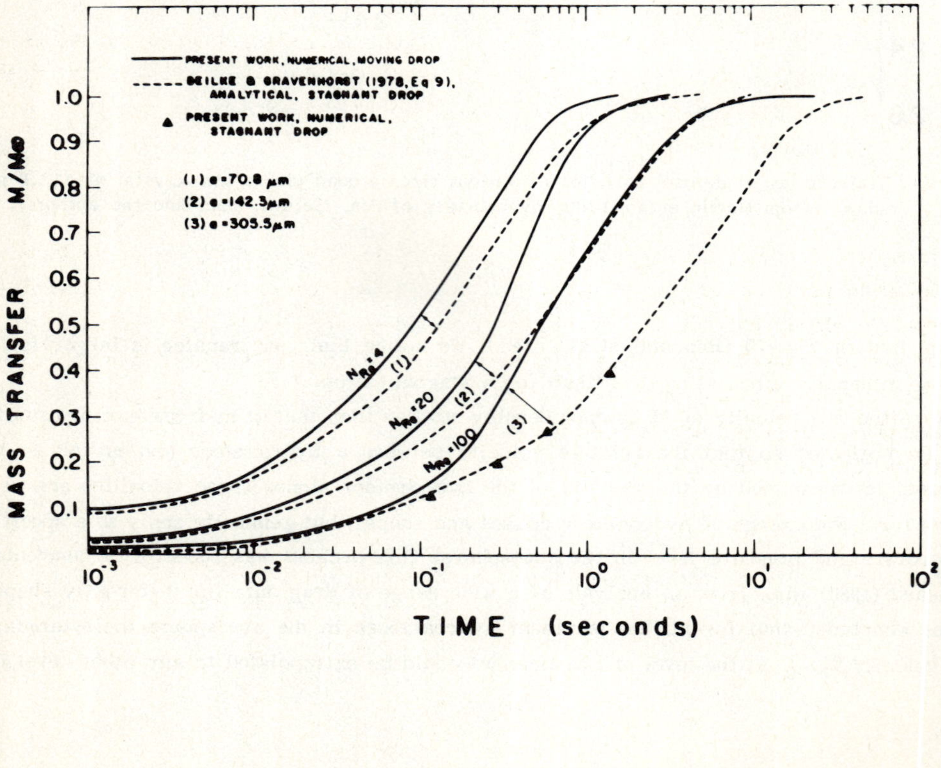
PRESENT WORK, NUMERICAL, MOVING DROP
BEILKE & GRAVENHORST (1978, Eq 9), ANALYTICAL, STAGNANT DROP
PRESENT WORK, NUMERICAL, STAGNANT DROP
(1) e = 70.8 μm
(2) e = 142.3 μm
(3) e = 305.5 μm
N_{Re} = 4
N_{Re} = 20
N_{Re} = 100
(1)
(2)
(3)
MASS TRANSFER M/M∞
0.1
0.2
0.3
0.4
0.5
0.6
0.7
0.8
0.9
1.0
10^{-3}
10^{-2}
10^{-1}
10^0
10^1
10^2
TIME (seconds)

Fig. 74. Number of aerosol particles of diameter larger than 0.4 μm on individual snow crystals of various surface area. (From Murakami et al, 1981; by courtesy of Japan Meteor. Soc. and the authors).

setting $V_\infty = f\, V_{\infty,o}$, where for $N_{Re,o} \leq 0.2$: $f = f_0$, for $N_{Re,o} > 1000$: $f = f_\infty$, and for the between Reynolds number range $f = f_0 + [(f_\infty - f_0)\,(1.61 + \ln N_{Re,o})] / 8.52$, with $f_0\ \eta / \eta_0$, $f_\infty = (\rho_0/\rho)^{1/2}$, η and ρ being the dynamic viscosity and density of air, respectively, and $N_{Re,0}$, being the Reynolds number of the hydrometeor at the level of measurement.

The fall mode of a hydrometeor determines the actual air volume swept out during its fall. A study of the main fall modes of atmospheric hydrometeors shows that straight, undeviated fall is rare and confined only to low Reynolds numbers. Little is known of how this increased volume swept out by a hydrometeor which falls in a manner differently than straight, affects its capacity to scavenge AP and gases, and for that matter, how it affects the growth of the hydrometeors themselves.

STATEMENT 10: THE DEVELOPMENT OF PRECIPITATION PARTICLES IS A RESULT OF FOUR BASIC MECHANISMS: (1) DIFFUSIONAL GROWTH OF ICE CRYSTALS SURROUNDED BY A WATER SATURATED ATMOSPHERE FILLED WITH CLOUD DROPS; (2) STOCHASTIC GROWTH OF SNOW CRYSTALS COLLIDING WITH AND STICKING TO OTHER SNOW CRYSTALS TO FORM SNOW FLAKES; (3) SEMI-CONTINUOUS GROWTH OF SNOW CRYSTALS BY COLLISION WITH SUPERCOOLED DROPS TO FORM GRAUPEL AND HAILSTONES; (4) STOCHASTIC GROWTH OF CLOUD DROPS COLLIDING AND COALESCING WITH OTHER CLOUD DROPS TO FORM RAIN DROPS.

Atmospheric clouds meet the conditions necessary for the growth of cloud particles to precipitation particles only with great difficulty. Thus the precipitation efficiency of atmospheric clouds is usually very low: on a world average basis only about 10% of all clouds are in a state to precipitate (see Pruppacher in Plate, 1982). This suggests that the time determining step for wet deposition of pollutants is not the incorporation of these pollutants into the cloud particles or the incorporation of the pollutants into precipitation, but is the formation of the precipitation itself.

Fig. 75. Mass M of sulfur transferred into drops of different sizes in terms of mass M_∞ transferred under equilibrium conditions; effect of flow inside and outside drop on mass transfer. (From Baboolal et al, 1981; by courtesy of Am. Meteor. Soc. and the authors).

Diffusional growth of snow crystals in a watersaturated environment filled with drops inhibits the scavenging of AP by snow crystals as the thermophoretic force is directed away from a growing crystal. On the other hand, the vapor pressure gradient set up between the surface of the ice crystals and drops, forces the drops to evaporate, and thus the drops exhibit phoretic scavenging eventhough the air surrounding the drops is at slightly above 100% relative humidity.

The growth of snow crystals by stochastic collision with other crystals was for a long time considered a very inefficient process as solid bodies are "supposed" to bounce-off from each other. However, recent work of Passarelli (1978), and of Passarelli and Srivastava (1979) shows that the growth rate to snow flakes is significant not only near 0°C due to a highly mobile layer of water molecules at the surface of ice, thus improving the sticking efficiency of collided crystals, but also due to the wide spectrum of fall speeds for crystals of the same mass, and due to the capability of the very abundant dentritic crystals to interlock. In Fig. 76 the evolution of a snow flake spectrum is presented in terms of the size distribution developed after 1000 seconds of growth. As each of the snow crystals is acquiring by itself a certain amount of aerosol material through nucleation scavenging and impaction scavenging, and a certain amount of gas through absorption into the surface film, the scavenged pollution material becomes part of the snow flakes formed by stochastic growth, each snowflake continuing to scavenge by itself.

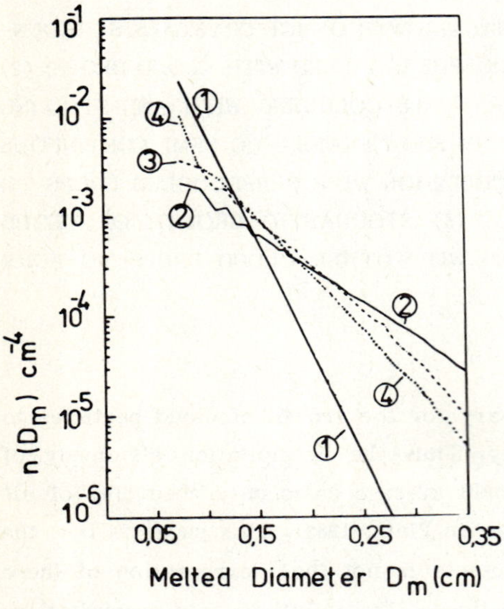

Fig. 76. Stochastic evaluation of size spectrum of snow crystals, in terms of their melted diameters, after 1000 seconds of model evolution. (1) Initial Spectrum. (2) to (4) Final spectrum for different collection kernels. (From Passarelli and Srivastava, 1979; by courtesy of Am. Meteor. Soc. and the authors).

The growth of graupel particles results from a non-stochastic, semi-continuous acquisition of supercooled drops by snow crystals due to the fact that the riming ice centers are small in number concentration and thus sufficiently far from each other so that they do not stochastically interact. Initially, riming is quite uniform around columnar crystals but preferrentially at the rim of planar crystals (Pitter, 1974); (Schlamp et al, 1975). Subsequently the frozen-on drops increasingly cover the whole crystal leading to a porous, conical, spherical or irregular shaped graupel. Fig. 77 a, b, c shows the evolution of a graupel spectrum growing from a given ice crystal spectrum. We note the effect of this growth on reducing the collision-coalescence growth of the cloud drops feeding the graupel. During this growth each snow crystal provides to the graupel the pollution mass it has acquired during its own life time. Through the acquisition of drops new pollution material is added to the graupel by means of the pollution acquired by the drops themselves. Due to the low number concentration of graupel most of the scavenged pollution mass is due to that acquired by the cloud drops rather than due to what the original snow crystal of the graupel itself has picked up.

The growth of cloud drops by stochastic collision and coalescence with other drops is well investigated and quite well understood, except that the coalescence efficiency of two colliding drops is still only very poorly known. On the one hand this efficiency is controlled by the air gap between the two approaching water surfaces during a collision, this gap being bridged depending on the impact velocity, impact angle, and sizes of the drops involved. On the other hand, partially coalescing drops may break up again in a variety of modes, forming one larger drop fragment and several smaller fragments. Both of these

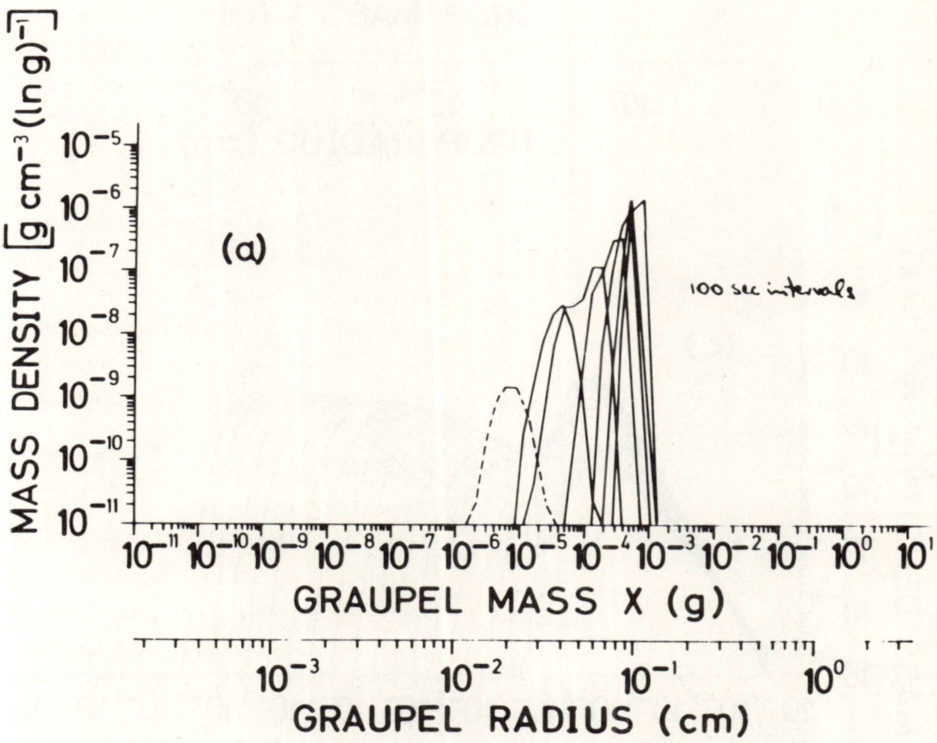

Fig. 77. a, b, c. Evolution of a size spectrum of graupel particles growing by riming (a) and simultaneously stochastic evolution of cloud drop spectrum (b); stochastic evolution of cloud drop spectrum for no ice crystals present in clouds (c). (From Beheng, 1978; by courtesy of Am. Meteor. Soc. and the author).

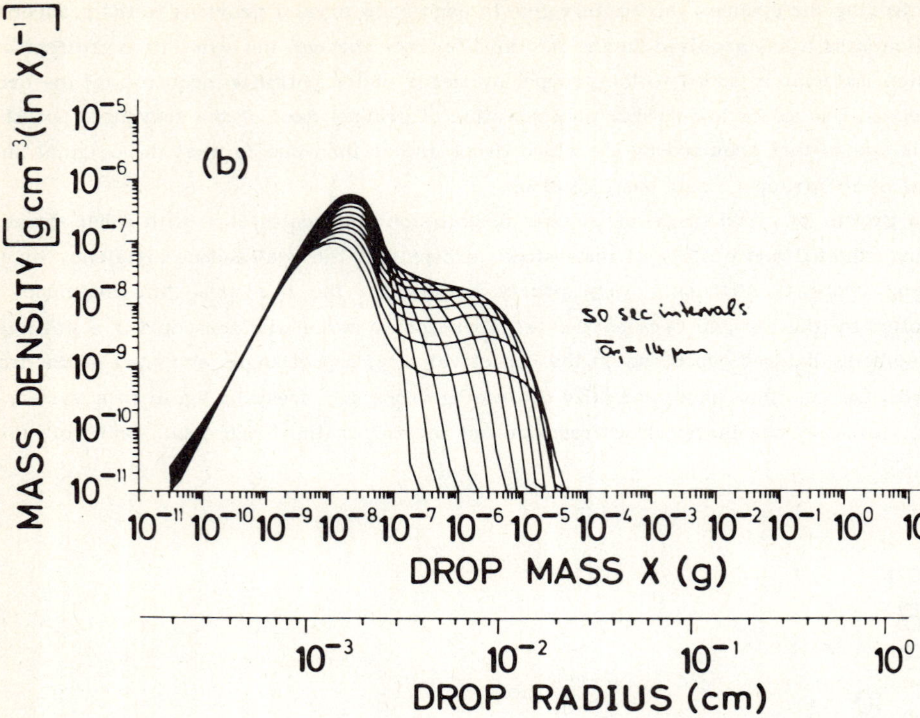
MASS DENSITY [g cm^-3 (ln X)^-1]
(b)
50 sec intervals
ā₁ = 14 μm
DROP MASS X (g)
DROP RADIUS (cm)

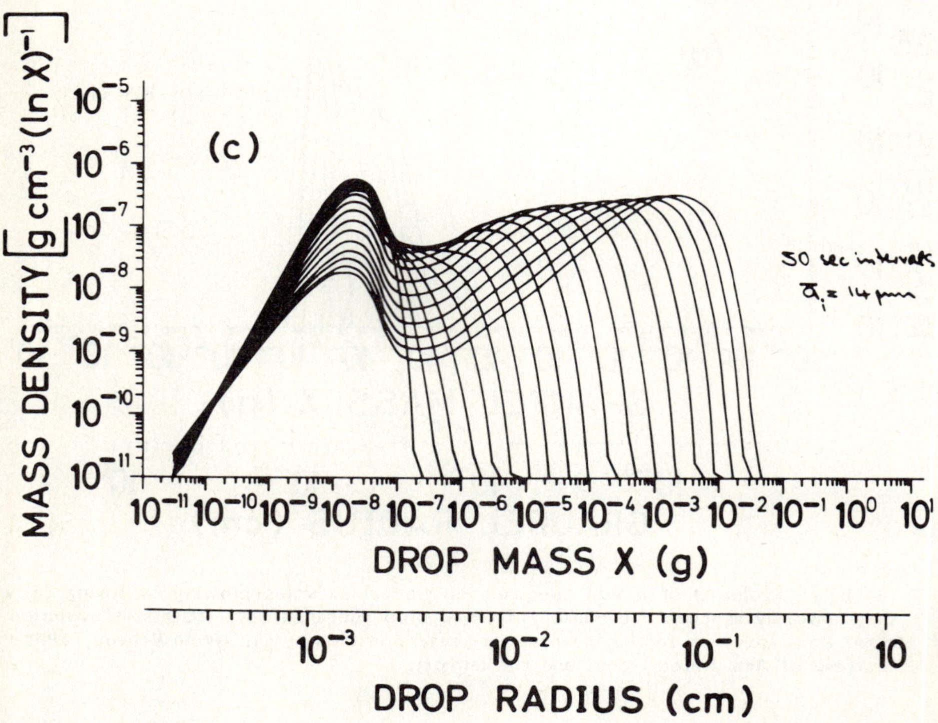
MASS DENSITY [g cm^-3 (ln X)^-1]
(c)
50 sec intervals
ā₁ = 14 μm
DROP MASS X (g)
DROP RADIUS (cm)

mechanisms are difficult to quantitatively describe. Disregarding break-up and assuming incorrectly a coalescence efficiency of unity one finds according to Berry (1973) for different initial drop spectra drop size evolutions displayed in Fig. 78. One notices that also this process is characterized by a selective nature, providing a "run-away" group of "favoured" drops which carry with them most of the water mass in the cloud.

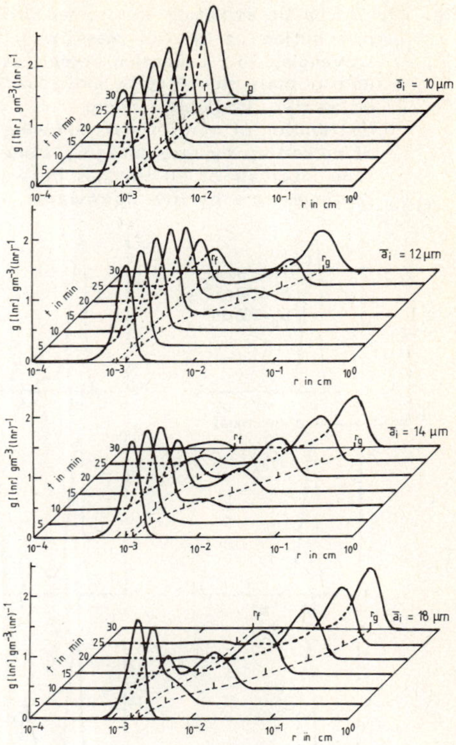

Fig. 78. Stochastic evolution of cloud drop spectra with varying initial mean drop size. (From Berry, 1983; by courtesy of the author).

In a recent study by Flossmann und Pruppacher (1983) the importance of this "run-away" water mass to AP scavenging has been illustrated by a model which takes into account the stochastic redistribution of the AP mass picked up due to impaction and nucleation scavenging by cloud drops which themselves undergo a continuous size evolution by the stochastic collision coalescence mechanism (Fig. 79 a). If one incorrectly would have assumed that AP mass enters each drop size category only by what this size category scavenges alone we find the results given in Fig. 79 d from which we notice that the AP mass remains associated with the small drop size categories, despite the fact that during the scavenging time the main water mass has moved to the large drop size categories (Fig. 79 a).

If, however, one correctly includes the stochastic re-distribution of the AP while the stochastic drop growth goes on, one finds as shown in Fig. 79 b, that the main AP mass is associated with the main water mass, also for the case of nucleation scavenging on (as here assumed) water insoluble but completely wettable AP (Fig. 79 c).

This result implies that if a cloud does precipitate at all it will allow not only the main water mass but also the main AP mass to be carried to the ground. This emphasizes the previously made statement that the bottle neck to scavenging lies in the formation of precipitation.

Fig. 79. Stochastic evolution of assumed drop size spectrum (a), and corresponding stochastic re-distribution of aerosol mass W_{AP} picked up by individual cloud drops by impaction scavenging (b); collection kernels as given by Wang et al; corresponding re-distribution of aerosol mass picked up by individual cloud drops by impaction scavenging and condensation on water insoluble, completely wettable aerosol particles (c); (d) corresponding distribution of aerosol mass allowing no re-distribution in drops but only is due to amount of picked up by size category alone; for 95% relative humidity and AP of r = 0.3 μm for time intervals of 50 seconds between distributions. Evolution begins with left row and continues row by row downwards. (From Flossmann and Pruppacher).

a)

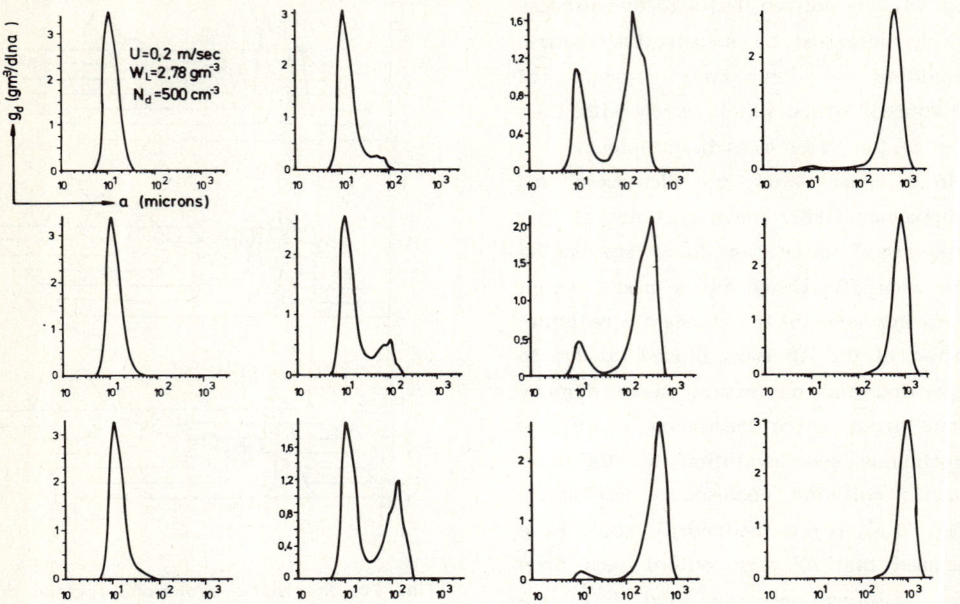

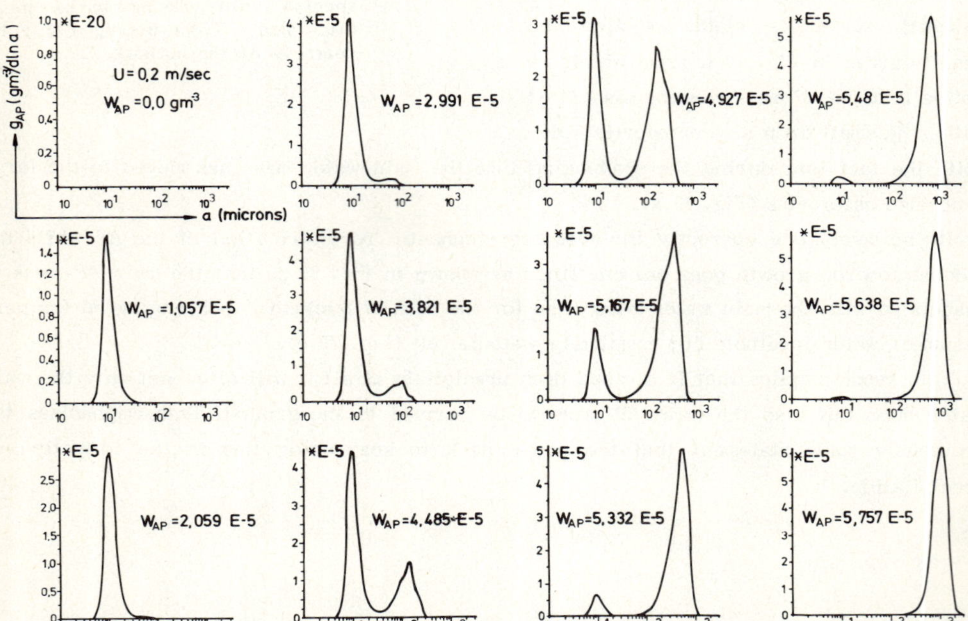

b)

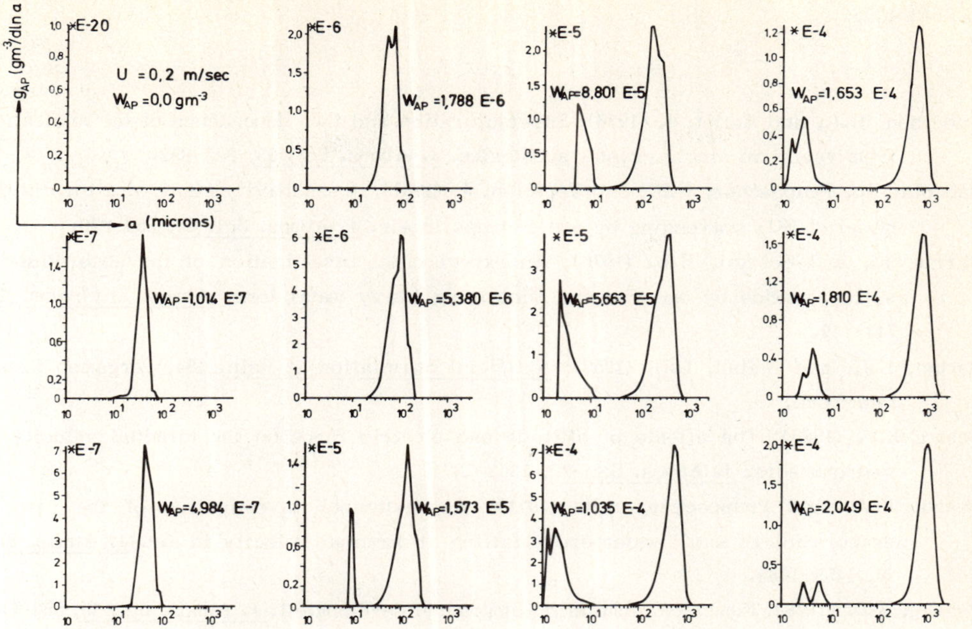
g_{AP} (gm^{-3}/dln a)
a (microns)
U = 0,2 m/sec
W_{AP} =0,0 gm^{-3}
*E-20
*E-6
W_{AP} =1,788 E-6
*E-5
W_{AP}=8,801 E-5
*E-4
W_{AP}=1,653 E-4
*E-7
W_{AP}=1,014 E-7
*E-6
W_{AP}=5,380 E-6
*E-5
W_{AP}=5,663 E-5
*E-4
W_{AP}=1,810 E-4
*E-7
W_{AP}=4,984 E-7
*E-5
W_{AP}=1,573 E-5
*E-4
W_{AP}=1,035 E-4
*E-4
W_{AP}=2,049 E-4

c

d

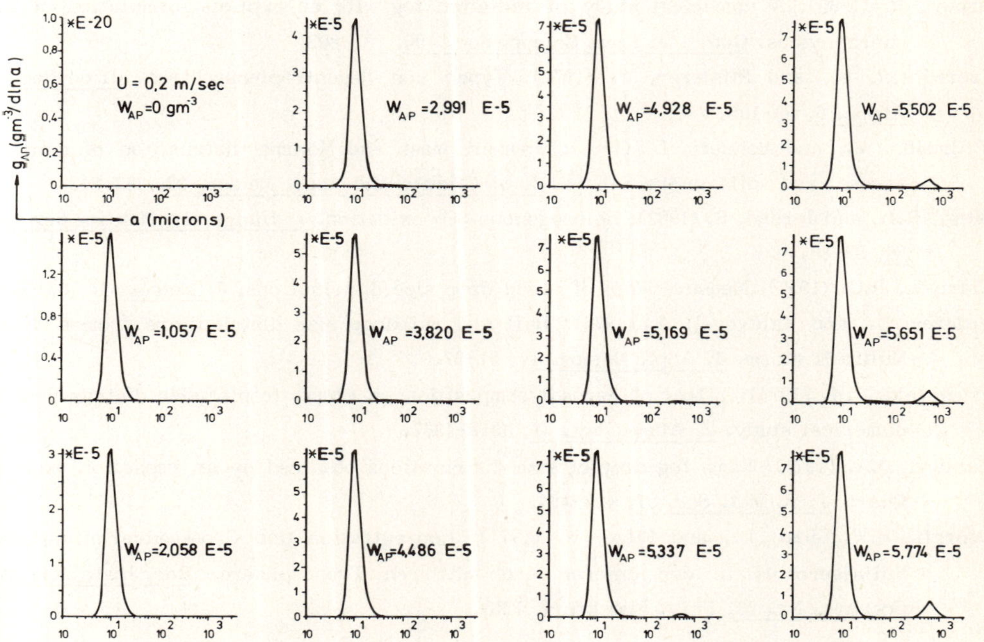
g_{AP} (gm^{-3}/dln a)
a (microns)
U = 0,2 m/sec
W_{AP} =0 gm^{-3}
*E-20
*E-5
W_{AP} =2,991 E-5
W_{AP} =4,928 E-5
W_{AP} =5,502 E-5
W_{AP} =1,057 E-5
W_{AP} =3,820 E-5
W_{AP} =5,169 E-5
W_{AP} =5,651 E-5
W_{AP} =2,058 E-5
W_{AP} =4,486 E-5
W_{AP} =5,337 E-5
W_{AP} =5,774 E-5

REFERENCES

Anderson, B.J., and Hallet, J. (1976): Supersaturation and time dependence of ice nucleation from vapor on single crystal substrates. J. Atmos. Sci. 33, 822-832.

Baboolal, L., Pruppacher, H.R., and Topalian, J. (1981): A sensitivity study of a theoretical model of SO_2 scavenging by water drops in air. J. Atmos. Sci. 38, 856-870.

Barrie, L., and Georgii, H.W. (1976): An experimental investigation of the absorption of sulfur dioxide by water drops containing heavy metal ions. Atmos. Environm. 10, 743-749.

Battan, L.J., and Reitan, C.H. (1957): Artificial Stimulation of Rain, 184, Pergamon Press, New York.

Beard, K.V. (1980): The effects of altitude and electric force on the terminal velocity of hydrometeors. J. Atmos. Sci. 37, 1363-1374.

Beard, K.V., and Pruppacher, H.R. (1971): A windtunnel investigation of the rate of evaporation of small water drops falling at terminal velocity in air. J. Atmos. Sci. 28, 1455-1464.

Beheng, K.D. (1978): Numerical simulation of graupel development. J. Atmos. Sci. 35, 683-689.

Berry, E.X. (1973): Final Rept. to NSF on grant GH 213350. Desert. Res. Inst. Reno, Nevada.

Borovikov (1963): Cloud Physics, Israel Progr. for Sci. Transl., U.S. Dept. of Commerce, Techn. Series, Washington, D.C.

Brown, R. (1980): A numerical study of radiation fog with an explicit formulation of the microphysics. Quart. J. Roy. Meteor. Soc. 106, 781-802.

Czerwinski, N., and Pfisterer, W. (1972): Typen von Regentropfenspektren. J. de Rech. Atmos. 6, 89-105.

d'Almeida, G., and Schuetz, L. (1983): Number, Mass and Volume distribution of mineral aerosol and soils in the Sahara. J. of Climate and Appl. Meteor. 22, 233-243.

Dlugi, R.J., and Jordan, S. (1982): Heterogenous SO_2 oxidation. J. Hungarian Meteor. Service 86, 82-88.

Eldrigde, R.G. (1957): Measurements of cloud drop size distributions. J. Meteor. 14, 55-59.

Federer, B., and Waldvogel, A. (1975): Hail and raindrop size distributions from a Swiss Multicell storm. J. Appl. Meteor. 14, 91-97.

Fitzgerald, J.W. (1974): Effect of aerosol composition on cloud droplet size distribution: a numerical study. J. Atmos. Sci. 31, 1358-1367.

Garland, J.A. (1971): Some fog droplet size distributions obtained by an impaction method. Quart. J. R. Met. Soc. 97, 483-494.

Georgii, H.W., Jost, D., and Fitze, W. (1971): Konzentration und Groessenverteilung des Sulfataerosols in der unteren und mittleren Troposphaere. Ber. Inst. Meteor. Geophys. Nr. 23, Univ. Frankfurt, FRG.

Gori, E.G., and Joss, J. (1980): Shape of raindrop size distributions simultaneously observed at three altitudes. Preprints, Cloud Phys. Conference, Clermont-Ferrand, France, 149-152.

Grosch, M.L. (1978): Reaktionen von SO_2 an Aerosolpartikeln unter atmosphaerischen Bedingungen, Ber. Inst. Meteor. Geophys. Nr. 36, Univ. Frankfurt, FRG.

Grover, S. (1980): A numerical investigation of the efficiency with which aerosol particles collide with drops. Ph.D. Thesis, Dept. Atmos. Sci., University of California, Los Angeles, California.

Gunn, R., and Marshall, J.S. (1958): The distribution with size of aggregate snowflakes. J. Meteor. 15, 452-461.

Hegg, D.A., and Hobbs, P.V. (1982): Cloud water chemistry and the production of sulfate in clouds. Atmos. Environ. 15, 1597-1604.

Hegg, D.A., and Hobbs, P.V. (1983): Precipitation Scavenging, Dry Deposition and Resuspension, Vol. I, 79-89. Elsevier Publ. Co., New York.

Hegg, D.A., Hobbs, P.V., and Radke, L.F. (1980): A preliminary study of cloud chemistry. Preprints, Cloud Phys. Conference at Clermont-Ferrand, France, 7-10.

Herzegh, P.H., and Hobbs, P.V. (1980): Observations of snow size spectra in frontal clouds, Preprint, Cloud Phys. Conference, Clermont-Ferrand, France, 201-204.

Hindman, E.E., Hobbs, P.V., and Radke, L.F. (1977): Cloud condensation nucleus size distributions and their effects on cloud droplet distributions. J. Atmos. Sci. 34, 951-956.

Hobbs, P.V. (1978): Res. Rept. XIII, Univ. of Washington, Dept. Atmospheric Sciences, Cloud Physics Group.

Hobbs, P.V. (1978): Res. Rept. VI, Cloud Phys. Group, Dept. Atmos. Sci., University of Washington, Wash.

Hobbs, P.V., and Atkinson, D.G. (1976): The concentration of ice particles in orographic clouds and cyclonic storms over the Cascade Mountains. J. Atmos. Sci. 33, 1362-1374.

Hobbs, P.V., Politovich, M.K., and Radke, L.F. (1980): The structure of summer convective clouds in Eastern Montana. I. J. App. Meteor. 19, 645-663.

Hoffer, T.E. (1960): A laboratory investigation of droplet freezing. Techn. Note Nr. 22, Univ. of Chicago, Chicago, Ill.

Houze, R.H., Hobbs, P.V., Herzegh, P.H., and Parsons, D.B. (1979): Size distributions of precipitation particles in frontal clouds. J. Atmos. Sci. 36, 156-162.

Hudson, J.G. (1980): Microphysics of coastal fog and stratus. Preprints, Cloud Phys. Conf., Clermont-Ferrand, France, 205-208.

Jaenicke, R. (1978): Ueber die Dynamik atmosphaerischer Aitkenteilchen. Ber. Bunsengesellschaft, Ser. Phys. Chem. 82, 1198-1202.

Jaenicke, R. (1982): Chemistry of the unpolluted and polluted Troposphere, 341-373, D. Reidel Publ. Co., Dordrecht, Holland

Jiusto., J.E., and Lala, G.G. (1981): CCN-Supersaturation Spectra Slopes. J. de Rech. Atmos. 15, 303-311.

Joss, J., Thams, J.C., and Waldvogel, A. (1968): The variations of rain drop size distributions at Locarno. Preprint, Cloud Phys. Conference, Toronto, 369-373.

Knight, N. (1981): The climatology of hailstone embryos. J. Appl. Meteor. 20, 750-755.

Leaitch, W.R. (1983): Precipitation Scavenging, Dry Deposition and Resuspension, Vol. I.53-69, Elsevier Publ. Co., New York.

Leary, C.A., and Houze, R.A. (1979): Melting and evaporation of hydrometeors in precipitation from anvil clouds. J. Atmos. Sci. 35, 669-679.

LeClair, B.P., Hamielec, A.E., Pruppacher, H.P., and Hall, W.D. (1972): A theoretical and experimental study of the internal circulation in water drops falling at terminal velocity. J. Atmos. Sci. 29, 728-740.

Lee, I.Y., Haenel, G., and Pruppacher, H.R. (1980): A numerical determination of the evolution of cloud drop spectra due to condensation on natural aerosol particles. J. Atmos. Sci. 37, 1839-1853.

Lew, J., and Pruppacher, H.R. (1983): A theoretical determination of the capture efficiency of small columnar ice crystals by large cloud drops. J. Atmos. Sci. 40, 139-145.

Martin, J., Wang, P.K., and Pruppacher, H.R. (1980): A theoretical determination of the efficiency with which aerosol particles are collected by simple ice crystal plates. J. Atmos. Sci. 37, 1628-1638.

Mei-Yuan, H. (1963): Microstructure of cumulus clouds. Izv. Akad. Nauk. SSSR, Ser. Geofiz. No. 2, 362-376.

Murakani, H., Hiramatsu, C., and Magono, C. (1981): Observations of aerosol scavenging by falling snow crystals at two sites of different heights. J. Meteor. Soc. Japan, 59, 763-771.

Parungo, P., Ackermann, E., and Proulx, H. (1976): Natural ice nuclei. J. Rech. Atmos. 10, 45-60.

Passarelli, R.E. (1978): Techn. Note, Nr. 52, Dept. of Geophys. Sci., Univ. of Chicago, Chicago.

Passarelli, R.E., and Srivastava, R.C. (1979): A new aspect of snow flake aggregation theory. J. Atmos. Sci. 36, 484-493.

Pitter, R.L. (1977): A re-examination of riming on thin ice plates. J. Atmos. Sci. 34, 684-685.

Pitter, R.L., and Pruppacher, H.R. (1974): A numerical investigation of collision efficiencies of simple ice plates colliding with supercooled drops. J. Atmos. Sci. 31, 551-559.

Pitter, R.L., Pruppacher, H.R., and Hamielec, H.E. (1973): A numerical study of viscous flow past a thin oblate spheroid at low and intermediate Reynolds numbers. J. Atmos. Sci. 30, 125-134.

Prodi, F. (1976): Scavenging of aerosol particles by growing ice crystals. Preprints, Cloud Phys. Conference, Boulder, Colorado, 70-75.

Pruppacher, H.R., and Klett, J.D. (1978): Microphysics of Clouds. D. Reidel Publ. Co., Dordrecht, Holland.

Pruppacher, H.R., and Lew. J. (1983): private communication.

Pruppacher, H.R., and Rasmussen, R. (1979): A windtunnel investigation of the rate of evaporation of large water drops falling at terminal velocity in air. J. Atmos. Sci. 36, 1255-1260.

Radke, L. (1983): Precipitation Scavenging, Dry Deposition and Pesuspension, Vol. I, 71-78. Elsevier Publ. Co., New York.

Radke, L.F., Hobbs, P.V., and Eltgroth, M.W. (1980): Scavenging of aerosol particles by precipitation. J. Appl. Meteor. 19, 715-722.

Radke, L.F., Hobbs, P.V., and Pinnons, J.E. (1976): Observations of cloud condensation nuclei, sodium containing particles, ice nuclei, and the light scattering coefficient near Barrow, Alaska. J. Atmos. Sci. 15, 982-995.

Roach, W.T. (1976): On the effect of radiative exchange on the growth by condensation of a cloud or fog droplet. Quart. J. Roy. Meteor. Soc. 102, 361-372.

Roberts, P., and Hallett, J. (1968): A laboratory study of ice nucleating properties of some mineral particles. Quart. J. Roy. Meteor. Soc. 94, 25-34.

Rosinski, J., Yamate, G., and Parungo, F. (1980): Size distribution of inorganic and organic ice-forming nuclei present in downdrafts of convective storms. Meteor. Rundschau 33, 97-106.

Rucklidge, J. (1965): The examination by electron microscope of ice crystal nuclei from cloud chamber experiments. J. Atmos. Sci. 22, 301-308.

Runca-Koeberich, D.R. (1979): Ein Beitrag zur Konstitution atmosphaerischer eisbildender Kerne. Ber. Inst. Meteor. Geophys. Nr. 37, Univ. Frankfurt, FRG.

Schaller, R.C., and Fukuta, N. (1979): On the hydrodynamic behavior of supercooled water drops interacting with columnar ice crystals. J. Atmos. Sci. 36, 1788-1802.

Schlamp, R.J., and Pruppacher, H.R. (1977): Pure and Appl. Geophys. 115, 805-843.

Schlamp, R.J., Pruppacher, H.R., and Hamielec, A.E. (1975): A numerical investigation of the efficiency with which simple columnar ice crystals collide with supercooled water drops. J. Atmos. Sci. 32, 2330-2337.

Soulange, R.G., Andraud, G., and Isaka, A. (1980): Etude d'un cas de zone d'accumulation d'eau dans un cumulonimbus tropical. J. de Rech. Atmos. 14, 477-486.

Stein, D., and Georgii, H.W. (1983): Messung der Groessenverteilung atmosphaerischer Gefrierkerne unter Beruecksichtigung der Feuchte. Meteor. Rundschau 36, 127-132.

Twomey, S., and Wojciechowski, T.A. (1969): Observation of the geographical variation of cloud nuclei. J. Atmos. Sci. 26, 684-688.

Twomey, S., and Warner, K. (1967): Comparison of measurements of cloud droplets and cloud nuclei. J. Atmos. Sci. 24, 702-793.

Walceck, C., and Pruppacher, H.R. (1984): On the scavenging of SO_2 by cloud and rain drops. III. J. Atmos. Chem. in press.

Waldman, J.M., Munger, J.W., Jacob, D.J., Flagan, R.C., Morgan, J.J., and Hoffmann, M.R. (1982): Chemical composition of acid fog. Science 218, 677-679.

Waldvogel, A. (1974): The No-jump of raindrop spectra. J. Atmos. Sci. 31, 1067-1078.

Wang, P.K., Grover, S.N., and Pruppacher, H.R. (1978): On the effect of electric charges on the scavenging of aerosol particles by cloud and small rain drops. J. Atmos. Sci. 35, 1735-1743.

Warner, J. (1968): The supersaturation in natural clouds. J. de Rech. Atmos. 3, 233-237.

Warner, J. (1969): The microstructure of cumulus clouds, I. J. Atmos. Sci. 26, 1049-1059.

Warner, J. (1970): On steady-state one dimensional models of cumulus convection. J. Atmos. Sci. 27, 1035-1040.

Whitby K.T. (1978): The physical characteristics of sulfur aerosols. Atmos. Environ. 12, 135-159.

Winkler, P. (1967): Wachstum natuerlicher Aerosolteilchenproben. Diploma Thesis. Meteor. Inst., University of Mainz, FRG.

Yagi, T., and Uyeda, H. (1980): Different size distributions of snow based on meteorological situations. Preprints, Cloud Physics Conference, Clermont-Ferrand, France, 231-234.

FIELD STUDIES OF CLOUD CHEMISTRY AND THE RELATIVE IMPORTANCE OF VARIOUS MECHANISMS OF THE INCORPORATION OF SULFATE AND NITRATE INTO CLOUD WATER

Peter V. Hobbs
Cloud and Aerosol Research Group
Atmospheric Science Department
University of Washington
Seattle WA 98195
U.S.A.

ABSTRACT

A review is given of cloud chemistry studies carried out over the past 20 years by the Cloud and Aerosol Research Group at the University of Washington.

Work in the early 1960's included attempts to explain laboratory observations of the formation of sulfate in the SO_2-NH_3-H_2O(liquid) system. In the late 1960's and early 1970's field evidence was obtained indicating that atmospheric aerosol can be modified when they are processed by clouds. An interactive cloud physics-cloud chemistry model showed that this modification could be due to the production of sulfate within the cloud droplets. This hypothesis was verified in 1979 by measuring sulfate concentrations in the air flowing into and out of wave clouds. The measured sulfate production in clouds was 0.9 ± 0.4 µg m^{-3}. The in-cloud production rates averaged over the first few minutes ranged from 0 to 1500 $\pm$ 800% h^{-1}. These data were used to derive an empirical relationship for the rate of sulfate production in wave clouds.

In the first half of the 1980's an extensive data set was collected on the scavenging and production of sulfate and nitrate by convective and layered clouds. The total scavenging coefficient (F_s) of sulfate by clouds ranged from 0.1 to 11; values of $F_s > 1$ were attributed to the production of sulfate within cloud droplets. The mean value of the in-cloud sulfate production was 1.0 ± 0.3 ug m^{-3}. The total scavenging coefficient for nitrate ranged from ~ 0 to 38 ± 25. Conservative estimates of in-cloud nitrate production yielded 0.3 to 1.7 µg m^{-3}.

Also described are recent modeling studies of aqueous-phase oxidation mechanisms and the relative contributions of nucleation scavenging, aqueous-phase chemical reactions, and below-cloud scavenging to the sulfate content of precipitation from convective clouds in polluted and clean air.

The paper concludes with "a look to the future" in cloud chemistry studies.

NATO ASI Series, Vol. G6
Chemistry of Multiphase Atmospheric Systems
Edited by W. Jaeschke

1. INTRODUCTORY REMARKS

The rather grandiose title of this paper might better be replaced by "A Cloud Physicist's View of the Acid Rain Problem: Some Personal Reflections". For, the fact is, that I have deliberately eschewed a formal review paper in favor of an informal discussion of my own studies in cloud chemistry.

I first became aware and interested in the subject of chemical processes in clouds in 1962, when I was working on my Ph.D. degree in B.J. Mason's cloud physics group at Imperial College, London. My own research at the time was quite unrelated to cloud chemistry, but Tony van den Heuval, who was working in the same group, was obtaining some laboratory measurements on the rates of formation of ammonium sulfate in water drops exposed to air containing SO_2 and NH_3 (van den Heuval and Mason, 1963). The indications from this and previous laboratory studies (Junge and Ryan, 1958; Cadle and Robbins, 1960) were that appreciable ammonium sulfate might be formed through the oxidation of SO_2 in cloud and fog droplets. In fact, the oxidation of SO_2 in aqueous solutions (uncatalyzed and catalyzed) is a classical process that has been studied extensively in the laboratory for over eighty years. When one reflects that there is still much uncertainty as to the rate of even the uncatalyzed aqueous-phase oxidation of SO_2 by oxygen in the laboratory, it is apparent that cloud chemists are faced with extremely complex tasks.

2. EARLY STUDIES OF CLOUD CHEMISTRY AT THE UNIVERSITY OF WASHINGTON

In 1963 I moved from England to the Atmospheric Sciences Department at the University of Washington where I founded the Cloud and Aerosol Research Group. During the period 1963-69 the works of my group was concerned with a combination of laboratory, theoretical and (ground-based) field studies.

As far as cloud chemistry was concerned, my first goal was to try to explain the formation of sulfate in the SO_2-NH_3-liquid H_2O system in a more rigorous manner than had been done previously. By considering the chemical equilibria involved in this system, and by assuming that the rate limiting mechanism in the production of sulfate is the oxidation of the sulfite ion, the "catalytic" action of ammonia in the reaction was explained and predictions were made of the rates of formation of sulfate under various conditions (Scott and Hobbs, 1967). Of course, it is now known that this in only one of several possible means by which sulfate may be produced in aqueous solutions.

Our first indications of the possible importance of chemical reactions in natural clouds came in 1968 during the course of ground-based field measurements in the Olympic Mountains of Washington State (Radke and Hobbs, 1969). We observed that the concentrations of cloud condensation nuclei (CCN) active at a given supersaturation sometimes increased by several hundred percent when cumulus clouds were evaporating upwind. I suggested that these increases were due to the formation of sulfate within the cloud droplets, which, when the droplets evaporated, was deposited onto the original CCN. The deposition of sulfate onto the original population of CCN, would make each CCN activate at a lower supersaturation than that at which it activated prior to being involved in the formation of the cloud. Since the concentration of CCN in the air increases with increasing supersaturation, the effect of the added sulfate would be to increase the concentration of CCN measured at a fixed supersaturation. Similar observations of increases in CCN at a given supersaturation in the air from evaporating clouds were made by Saxena et al. (1970) and Dinger et al. (1970).

3. FIELD MEASUREMENTS AND NUMERICAL MODELING OF THE ENHANCEMENT OF CLOUD CONDENSATION NUCLEI BY CHEMICAL REACTIONS IN CLOUDS

By 1970 I had decided that future progress in understanding cloud processes would be largely dependent on obtaining in situ measurements from aircraft. Consequently, my group acquired an aircraft (a Douglas B-23) and began to equip it for atmospheric research. One of the first tasks we undertook with this aircraft was to test my hypothesis concerning sulfate production in clouds and the enhancement of the activity of CCN. To do this we decided to make measurements on a relatively simple type of cloud called a wave cloud. Wave clouds form when air is forced to rise over a hill or mountain. If, during this rise, the air is lifted above its lifting condensation level, a wave cloud will form above the hill. Due to oscillations in the streamlines downwind of the mountain, a series of leewave clouds may also form. The attractive features of wave clouds from our point of view was that they remain relatively stationary and they have a well defined in-flow region (the upwind edge of the cloud) and an outflow region (the downwind edge). It should be noted that although the cloud as a whole remains relatively fixed with respect to the mountain, air and cloud droplets are continually moving through the cloud, so that droplets that form on CCN grow as they are carried up in the wave and diminish in size as they move down in the wave. Finally, in the outflow region of the cloud, only CCN are left. However, these CCN will differ from those that enter the cloud to the extent that any material produced by chemical reactions within the cloud droplets will be added to the substances comprising the original CCN.

Our preliminary field measurements, obtained in 1970, consisted of measuring the concentrations of CCN active at a fixed supersaturation in the air entering and leaving wave clouds that formed over Mt. Rainier, Washington. In three such case studies it was found that the concentrations of CCN at the measured supersaturation in the air leaving the wave clouds were significantly higher than those entering the clouds (Radke, 1970) These results have since been confirmed by more detailed observations in which both CCN and the spectra of particle sizes have been measured in air parcels entering and leaving wave clouds (Hegg et al., 1980).

In 1974 we published the results of a numerical study of the formation of sulfate and the enhancement of CCN by wave clouds (Easter and Hobbs, 1974). Although this model considered the production of sulfate in the cloud droplets by the Scott-Hobbs mechanism alone, it is of interest because, to my knowledge, it was the first interactive cloud physics-cloud chemistry model and it provided some useful insights into the relative importance of cloud physics and chemical processes and cloud-environment interactions.

In the Easter-Hobbs model a parcel of air is followed as it moves along a sinusoidal path that simulates a streamline through a wave cloud. As the parcel rises above the lifting condensation level, droplets form on 75 prescribed CCN sizes. Each CCN is composed of an insoluble, chemically inert substance and an appropriate amount of ammonium sulfate; activation supersaturations ranged from 0.04 to 1.52%. As the air parcel rises the droplets grow by condensation to remain in equilibrium with the ambient vapor pressure and, as the air parcel descends, the droplets evaporate. Physical parameters of importance include the vertical velocity of the air at the inflow edge of the cloud, the maximum vertical displacement of the air parcel, and the period of time that the air parcel remains above the condensation level. The vertical velocity of the air determines the peak supersaturation of the air and therefore how many CCN are activated to form cloud droplets. The maximum vertical displacement of the air determines the maximum liquid water content which, together with the number of activated CCN, determine the maximum droplet size. The mass of sulfate produced in a droplet is determined by the time that it remains above the condensation level and the time-averaged size of the droplet.

As the droplets form and grow in the modeled wave cloud, ambient SO_2 and NH_3 dissolve in the droplets and the dissolved SO_2 is oxidized to sulfate ions according to the first-order rate equation:

$$\frac{d[SO_4^{=}]}{dt} = K_o[SO_3^{=}] \qquad (1)$$

In view of uncertainties in the value of K_o, calculations were carried out for a range of values covering those measured in early laboratory measurements. As SO_2 and NH_3 dissolve in the cloud droplets they are depleted in the ambient air. It is assumed that the transfer of these gases to and within the cloud droplets is through gas and liquid phase diffusion. Even so, the diffusion processes are very fast compared to the assumed rate of sulfate oxidation.

Results from the Easter-Hobbs model were presented primarily in the form of CCN spectra (ie concentration of CCN versus supersaturation) for the air entering and leaving wave clouds. The model results showed that the CCN spectrum is an extremely sensitive indicator of sulfate production. For a comparatively high value of K_o (0.032 s^{-1} at neutrality and

25°C), and initial ambient concentrations of SO_2 and NH_3 of 1 and 3 ppb, respectively, the sulfate produced in the cloud was 0.3 μg m^{-3}, all the CCN had sufficient sulfate added to them to be active at 0.25% supersaturation, and the number of CCN active at 0.5% supersaturation was predicted to increase by 84% after cloud processing for ~ 4 mins by a wave cloud. For a lower oxidation rate (K_o = 0.0017 s^{-1} at 25°C) there was ~ 30% enhancement in the concentration of CCN active at 0.5% supersaturation.

The Easter-Hobbs model also revealed that when an isolated air parcel moves through a wave cloud the amount of SO_2 and NH_3 in the air can be significantly depleted by dissolving in the cloud droplets; this depletion can restrict the total amount of sulfate produced in solution. Also, the model calculations showed that the sulfate production is fairly sensitive to the assumed initial ammonia concentration, which controls the pH of the solution and hence the reaction rate. Finally, a crude generalization of these calculations to global scale suggested that the production of sulfate in cloud water by the Scott-Hobbs mechanism alone was comparable to estimates that have been made of the world-wide production of sulfate in the atmosphere. Later in this paper we will review subsequent work in which the Easter-Hobbs model was used to explore some other possible SO_2 oxidation mechanisms in clouds.

4. PRODUCTION OF SULFATE IN CLOUDS

In 1979 we initiated a series of airborne measurements designed to measure the rates of sulfate production in wave clouds. A summary of this work is given below. Full details may be found in Hegg and Hobbs (1981, 1982).

Particulate sulfate productions in wave clouds over mountains in the Western portion of Washington State were determined by exposing filters from the aircraft to the air entering and leaving clouds. The masses of sulfate on the filters were subsequently determined by ion exchange chromatography, and the rates of production of sulfate in the clouds were determined from the sulfate concentrations measured upwind and downwind of the cloud and the transit time through the cloud (= horizontal length of cloud/wind speed). The derived sulfate productions were then analyzed as a function of various other measured variables, such as cloud liquid water content, cloud water pH, SO_2 and O_3 concentrations, and temperature. A major uncertainty in the rate calculations was the transit times in the clouds. While the wind speeds, determined by <u>in situ</u> Doppler radar measurements aboard the B-23 aircraft, were generally accurate to ± 20%, the length of the cloud was sometimes known to only ± 50%. This resulted in uncertainties in the transit time which occasionally exceeded ± 70%.

A total of twenty-eight case studies were obtained (for tabulation of the complete data set see Hegg and Hobbs, 1982). Of these, sixteen showed sulfate production in the wave clouds that differed significantly from zero. The mean value of the amount of sulfate produced in the wave clouds was $0.9 \pm 0.4\ \mu g\ m^{-3}$. The sulfate production rates ranged from 0 to $1500 \pm 800\%\ h^{-1}$. Even the lower limit of the highest rate ($700\%\ h^{-1}$) is far greater than had been previously suggested for clouds. However, it must be remembered that these (average) rates were measured for in-cloud transit times of only a few minutes; as we will see later, the average production rates are expected to decline over longer transit times.

Employing a more typical sulfate production rate from our measurements of $50 \pm 20\%\ h^{-1}$, and assuming that the rate is given by Eq. (1) we obtain a value of $K_o = 0.3 \pm 0.1\ s^{-1}$ at a pH of ~ 5. This value for K_o lies toward the upper end of the range of rates measured in the laboratory (see, for example, Hegg and Hobbs, 1978).

To derive an empirical relationship for the rate of sulfate production in cloud water, we performed a multiple regression of the measured sulfate production rates onto an expression of the form:

$$\frac{d[SO_4^=]}{dt} = K_1 [H^+]^{\alpha} [SO_3^=] \exp(-\beta/T) \qquad (2)$$

This yielded the following values:

$K_1 = 3.3 \times 10^5 \pm 6.2 \times 10^5$ liter$^{1.1}$ s^{-1} mole$^{-1.1}$

$\alpha = 1.1 \pm 0.1$

$\beta = 3.5 \times 10^2 \pm 3.3 \times 10^2$ K

While the uncertainties in the values of K_1 and β are high, indicating that Eqn (2) does not describe the data set completely, some important conclusions can be drawn from this regression analysis. For example, the activation energy E_A ($= \beta R$) is only 2.9 ± 2.7 kJ mole^{-1}, which suggests the reaction has little temperature dependence. (This was also indicated by the low correlation coefficient of 0.2 between sulfate production and temperature.) Thus, highly temperature-sensitive mechanisms, such as Fe or Mn catalyzed aerobic oxidation, can be excluded as the dominant sulfate production mechanism in this case. (NB. The measurements were made at altitudes of ~ 4 km and in relatively clean air.) The value for the explicit hydrogen ion dependence (α), which was determined with good accuracy, is that expected if the sulfate is produced in solution via O_3 oxidation. This mechanism receives further support from the fact that the correlation between the sulfate production rates and $[O_3]$ was 0.66, although this relatively high correlation is highly dependent on the data associated with the two highest values of O_3.

The most significant variance in the regression is associated with the value of K_1. The uncertainty associated with the derived value of this coefficient is $\sim 200\%$ and indicates that the scaling of the rate data, as opposed to trends associated with $[H^+]$ and $[SO_3^=]$, is highly uncertain.

Explicit hydrogen ion dependencies and activation energies for various sulfate production mechanisms derived from laboratory experiments are summarized in Table 1 where they can be compared with the values derived from our measurements in wave clouds. It can be seen that no single mechanism can fully account for the values observed in natural clouds. However, the sulfate production rates that we measured are relatively close to those expected from oxidation by H_2O_2 (Penkett et al., 1979) if H_2O_2 is present in air in concentrations of $\sim$0.3 ppb. It is conceivable that variability in the ambient concentrations of H_2O_2 is partially responsible for the differences between our observed values of α and E_A and those predicted for the H_2O_2 rate, and also for the large uncertainty in K_o. However, it would be equally plausible to postulate either various combinations of the mechanisms listed in Table 1 or an entirely new mechanism as being responsible for the observed sulfate production rates.

Finally, we consider the utility of Eq (2) for predicting sulfate production rates in clouds. Insertion of the derived values of K_1, α and β into Eq(2) yields a value for K_o of 0.3 $\pm$ 0.7 s^{-1} at pH = 5 and T = 298°K. However, the large uncertainty in this derived value for K_o is misleading as to the utility of Eq (2) in predicting sulfate production rates in clouds as a function of cloud water pH. For example, if Eq (2) is used to derive values of K_o as a function of pH, it is found that the values obtained provide reasonable estimates of the measured rate coefficients (see Fig. 2 in Hegg and Hobbs, 1982).

Mechanism	α	E_A (= βR) (kJ $mole^{-1}$)	Reference
Catalyzed (Mn or Fe) aerobic	1	100	Brimblecombe and Spedding (1974) Barrie and Georgii (1976)
Catalyzed (Mn + Fe) aerobic	0.6	Negligible	Barrie and Georgii (1976)
Uncatalyzed aerobic	0.5	1-10	Larson et al (1978)
Ozone	1	10-20	Penkett (1972) Larson et al (1978)
Hydrogen peroxide	1.7	30-40	Penkett et al (1979)
Total rate measured in natural clouds	1.1 ± 0.1	2.9 ± 2.7	Hegg and Hobbs (1982)

TABLE 1: Comparisons of $[H^+]$ and temperature dependencies for sulfate production in solution over a pH range of 3-6. The Parameters α and β are defined by Equation (2). From Hegg and Hobbs (1982).

5. MEASUREMENTS OF THE SCAVENGING AND PRODUCTION OF SULFATE AND NITRATE BY CONVECTIVE AND LAYERED CLOUDS

Atmospheric aerosol and trace gases may be scavenged by clouds and precipitation in several ways. When water condenses onto CCN to form a cloud, the materials comprising the CCN are said to be removed from the air by nucleation scavenging. Gases may be removed from the air by dissolving into cloud drops where, as we have seen, they may subsequently be transformed by aqueous-phase reactions. When a drop leaves a cloud as a raindrop, collection of atmospheric aerosol by impaction becomes important; we will refer to this as below-cloud scavenging.

The mechanism that is most likely of greatest importance for the incorporation of sulfate into hydrometeors is nucleation scavenging. This is because nearly all of the sulfate mass in the atmosphere resides in particles from 0.1-1 μm radius, and particles in this size range are relatively immune to scavenging by impaction, diffusion or phoretic effects. The situation with nitrate is somewhat different because, unlike sulfate, a substantial fraction of atmospheric nitrate is in the gas phase. Thus the scavenging of gases should play an important role in the incorporation of nitrate into hydrometeors.

Recently, my research group has embarked on a series of airborne measurements designed to derive information on the efficiencies with which cloud drops scavenge sulfate and nitrate (Hegg et al., 1984). Measurements have been obtained in cumulus, stratocumulus and stratus clouds. The procedure is to obtain measurements of droplet and interstitial particle size distributions, liquid water content, temperature, vertical air velocity and the chemical composition of the cloud water just above the cloud base and measurements of particle size and CCN spectra, vertical air velocity, temperature and the chemical nature of the particles and gases just below cloud base. We then define the scavenging coefficient of sulfate (F_S) as the ratio of the sulfate concentration in the cloud water to the sulfate concentration in clear air just below cloud base.

Derived values of F_S ranged from 0.1 to 11 without of 35 values of F_S being greater than unity. Values of $F_S > 1$ are attributed to production of sulfate within the drops. The contribution of nucleation scavenging to F_S was estimated by assuming that the bulk of the mass of sulfate resided in the size range 0.1 to 1.0 μm radius, and the scavenging coefficient of the latter (F_a) by the clouds was derived from the measurements of the particles size distributions just below cloud base and interstitially within the cloud. The value of F_a derived in this way was 0.7 ± 0.2 (from 14 measurements).

The mass of sulfate produced in each cloud was determined from the relationship

$$[SO_4^{=}]_P = [SO_4^{=}]_C - F_a \times [SO_4^{=}]_A \qquad (3)$$

where the subscripts P, C and A refer to the sulfate concentrations (per unit volume of air) produced in solution, present in the cloud water, and present in the clear air, respectively.

The mean derived value of $[SO_4^=]_P$ was 1.0 $\pm$ 0.3 μg m^{-3} (compared to 0.9 $\pm$ 0.4 μg m^{-3} derived more directly from our wave cloud measurements).

In the case of nitrate, the total scavenging coefficient (FN) was defined as the nitrate concentration in the cloud water divided by the nitrate (particulate + gas) concentration in the clear air just below cloud base. The values of F_N ranged from 0.4 $\pm$ 0.3 to 38 $\pm$ 25, with 7 out of 27 values for F_N being significantly greater than unity (due, presumably to aqueous-phase nitrate production). Ultra-conservative estimates of the nitrate production in cloud yielded values ranging from 0.3 to 1.7 μg m^{-3}. These values are in reasonable agreement with the calculations of Heikes and Thompson (1983) in which various reactions of inorganic nitrogen compounds (e.g. NO_2 and N_2O_5) are assumed to take place interstitially between cloud drops and the products of these reaction (e.g. N_2O_5) then absorb into cloud drops to produce nitrates.

6. RECENT MODELING STUDIES

In support of our recent observational studies of cloud chemistry we have carried out some further numerical modeling studies, a brief description of which is given below.

Using the Easter-Hobbs model, Hobbs (1979) described the changes in the acidity of a cloud droplet activated early in the life history of a cloud that is formed in a rising parcel of air. During the initial growth stage, when water is condensing onto a soluble CCN, the dissolution of the CCN can make the droplet relatively acidic. However, the acidity is rapidly diluted (in ~100s) as further water condenses onto the droplet. As the supersaturation in the rising parcel of air reaches a peak value, relatively large numbers of additional CCN are activated and the cloud droplets increase in mass rather rapidly. This causes a sudden decrease in the concentration of ammonia in the air as it is absorbed by the growing droplets. Consequently, ammonia desorbs from droplets that were activated early in the life history of the cloud (and which had come to equilibrium with higher ambient NH_3 concentrations); this, in turn, causes a sharp decrease in the pH of the drop. Further growth of the droplet by condensation causes a slow rise in its pH, which is offset in part by any aqueous-phase production of acidic substances. Due to these time-varying effects, the average pH of cloud water is not always representative of the pH at which SO_2 oxidation occurs within a droplet, and oxidation may occur more rapidly in some cloud droplets than in others (see also Hegg and Hobbs, 1979).

Hegg and Hobbs (1979) updated the Easter-Hobbs cloud model by employing more recent estimates of ambient trace gas concentrations and the rate of uncatalyzed oxidation of aqueous SO_2 by dissolved oxygen. The rate expression of Larson et al. (1978) produced significant sulfate under most atmospheric conditions and Brimblecombe and Spedding's (1974) rate expression produced significant sulfate under commonly occuring atmospheric conditions. However, Beilke et al's. (1975) rate expression produced significant sulfate only under conditions that are rarely observed in clouds. The effects of ozone oxidation on sulfate production were also investigated using Larson et al's. (1978) rate expression. This had little effect on those runs where Larson et al's. oxygen rate expression was employed, but it had a substantial effect on those runs where Brimblecombe and Spedding oxygen rate law was used and the trace gas concentrations inputted into the model were such that the cloud water pH was relatively low.

These model calculations also showed that the cloud water pH is relatively sensitive to the CCN spectrum but the amount of sulfate produced in solution is relatively insensitive to this parameter. For CCN spectra characteristic of regions with high sulfate, proportionally fewer CCN were activated and thus a lower percentage of the ambient sulfate was incorporated into the cloud water.

Finally, the model calculations of Hegg and Hobbs (1979) showed that droplets that form on CCN which are activated at low supersaturation generally have a lower pH than those that from on CCN activated at higher supersaturations. This is due to the generally larger quantities of soluble salts in the CCN activated at lower supersaturations.

Hegg and Hobbs (1981) explored (in a relatively crude fashion) some possibilities for sulfate production in clouds by comparing predictions from the Easter-Hobbs model, using various oxidation mechanisms and rates, with our measurements in wave clouds. It was found that various combinations of O_2, O_3 and H_2O_2 oxidation rates provided reasonable agreements with the amounts of sulfate that was observed to be produced in clouds after 5 min. Also, using Penkett et al's. (1979) H_2O_2 oxidation rate, and assuming H_2O_2 concentrations in the air of 3 ppb, oxidation by H_2O_2 alone could explain the observed sulfate production. The measured pH dependence of the sulfate production was intermediate between those predicted by laboratory experiments for the O_2 + O_3 and the H_2O_2 dominated oxidation mechanisms. On this basis, it was suggested that the sulfate production might have been the result of a combination of O_2, O_3 and H_2O_2 oxidation within cloud droplets.

Finally, and importantly, both the observations and the model indicate that aqueous-phase sulfate production may be self limiting. This is due to the strong pH dependence of several of the proposed mechanisms for sulfate production in clouds and the progressive lowering of the pH of the droplets as sulfate is produced. Thus, the high SO_2 conversion rates measured during the short ($\sim$ few minutes) transit times in wave clouds should not be extrapolated to much longer time periods.

More recently, we have combined field measurements and model calculations to estimate the relative contributions of nucleation scavenging, aqueous-phase chemical reactions in cloud droplets, and below-cloud scavenging to the sulfate content of precipitation from convective clouds (Hegg, 1983). The results indicate that variations in cloud physics parameters (such as liquid water content and the presence of ice) can have substantial impact on the relative contributions of these three sulfate inclusion mechanisms. Variations in below-cloud scavenging efficiencies also produce marked changes in the sulfate content of the precipitation. In the case of all-liquid, convective clouds forming in polluted air (such at the North-eastern United States) the contributions to the sulfate content of the precipitations were: 25% from nucleation scavenging, ~45% from aqueous-phase production and 30% from below-cloud scavenging. The corresponding percentages for the cloud forming in clean oceanic air (e.g. in the Southern Oceans) were ~65%, ~19% and ~25%. In the polluted case, the SO_2 concentrations in the ambient air were high enough to allow aqueous-phase production to dominate the sulfate content of the precipitation. However, in the clean air case, the lower concentrations of sulfur gas reduced the aqueous-phase production and the majority of the sulfate in the precipitation was from nucleation scavenging.

7. CONCLUDING REMARKS AND A LOOK TO THE FUTURE

It is interesting to note that the early 1960s marked the pinnacle of laboratory studies of cloud physics. At that time it was commonly believed that many cloud processes could be simulated and studied through relatively simple experiments in the laboratory. Subsequently, the cloud physics community shifted progressively to the view that few (if any) of the complex conditions and processes that occur in natural clouds can be accurately reproduced in the laboratory. This viewpoint hastened the development, in the late 1960s and 1970s, of the means for studying many aspects of cloud and precipitation physics through in situ measurements and by remote sensing. During the same period, the ability developed to model numerically many aspects of cloud and precipitation processes. Finally, and most importantly, cloud physics is now at the point where the details provided by field measurements are commensurate with those provided by numerical models. Hence, field measurements can be used to provide data both for initializing numerical models and to check model predictions. Even so, some 50 years after the initiation of modern cloud physics research, many aspects of the physics of clouds are still only poorly understood.

Cloud chemistry is a very recent offspring of cloud physics. Like any young progeny it can learn much from its parent. Thus, in common with physical processes, it is unlikely that the chemical processes that occur in clouds will be accurately reproduced in the laboratory or that they will be predicted theoretically in their full complexity. Of course, both laboratory and theoretical studies can play important roles in elucidating certain isolated aspects of cloud chemistry, as they have done in cloud physics, but measurements in real clouds must form the basis of cloud chemistry studies. Similarly, although ground-based studies of clouds (e.g. from mountains) can provide some useful information they are not adequate substitutes for studies of clouds in the free atmosphere.

Also, like a child, cloud chemistry must learn to walk before it can run. Thus, measurements should be carried out on relatively simple cloud systems before tackling progressively more complex system. There is a pressing need for more sophisticated instrumentation for chemical measurements in the atmosphere and in clouds. At present, many of the instruments used for airborne measurements are adaptations of laboratory instruments and are inadequate for field studies. Techniques for the remote sensing and measurement of the chemistry of clouds and their environments are urgently needed.

In common with most other branches of the atmospheric sciences, cloud chemistry is very much an interdisciplinary subject. In particular, as I have tried to emphasize in this paper, chemical processes in clouds interact strongly with physical processes. Interactions on various scales undoubtedly occur, but have scarcely been considered to date. Clouds are such an ubiquitous feature of the earth's atmosphere that they are likely to play a role in many aspects of atmospheric chemistry on local, regional and global scales.

Sulfate (and more recently nitrate) cloud chemistry alone have received any serious attention. However, clouds are probably involved at some stage in the lifecycles of many inorganic and organic chemicals in the atmosphere, including, presumably, some micro-organisms.

The next decade should be an exciting one for cloud chemistry.

ACKNOWLEDGEMENTS

I thank my colleagues Drs. Dean Hegg and Lawrence Radke for their considerable contributions and help. This paper was written while I was on study leave in Europe at the Meteorological Office Radar Research Laboratory, Royal Signals and Radar Establishment, Malvern, England and at the Fraunhofer- Institut fuer Atmosphaerische Umweltforschung, Garmisch-Partenkirchen, FRG, where I was in receipt of a Senior US Scientist Award from the Alexander von Humboldt Foundation. This work was supported in part by Grants ATM81-12089 and ATM-8419000 from the US National Science Foundation.

APPENDIX

THE UNIVERSITY OF WASHINGTON'S AIRBORNE RESEARCH FACILITY

The University of Washington's research aircraft is a Douglas B-23 (this aircraft has recently been replaced by a Convair C-131 A). This is a twin-engine, propeller-drive aircraft that is large enough to carry substantial instrumentation and a crew of six, but not too large to be maintained and used efficiently by an active university research group. The B-23 has a sampling speed of 60-70 m s^{-1} and a stall speed of $\sim$ 30 m s^{-1}. These low speeds reduce problems associated with sampling aerosol and cloud particles from an aircraft.

The instrumentation aboard the aircraft (as of September 1983) is listed in Table A.1 where it is sub-divided under the following broad headings: navigational and flight characteristics, meteorological, cloud physics, aerosol, cloud and atmospheric chemistry and data processing and display.

Of particular interest here is the system for measuring the size spectra of aerosol in the air and interstitially between cloud droplets. Aerosol in the size range 0.01-45 um are measured using several instruments that are all fed from a common batch sampler of air. The batch sampler consists of a stainless steel cylinder 90 liter volume and 1.5 m high which has a freely-floating piston (Fig. A.1). Electric valves control the filling and emptying of ambient air samples into and from the cylinder with ambient air and closing the air inlet valve. Since the piston offers negligible resistance to the in-rushing air (the pressure above the piston is reduced), sampling of particles is close to isokinetic.

After the cylinder is full, air from its base passes into the various instruments shown in Fig. A.1 (and described in Table A.1). The instruments shown on the right hand side of Fig. A.1 size the particles after any water on them has been evaporated by passage through a diffusion drier. Hence, these instruments provide the size spectra of the dried particles from 0.01 to 11 µm. The Royco 245, on the other hand, measures particles in the size range 2 to 45 um without any significant drying.

When measurements are made in clouds, the particles that enter the diffusion drier, and are subsequently sized, are primarily cloud interstitial particles. This is because the largest cloud drops are removed by impaction at the first bend in the air inlet sampling tube, and droplets > 10 µm are largely removed by sedimentation in the cylinder and associated pipes to the sampling devices.

Fig. A.1 High-volume, isokinetic batch sampler aboard the University of Washington's research aircraft.

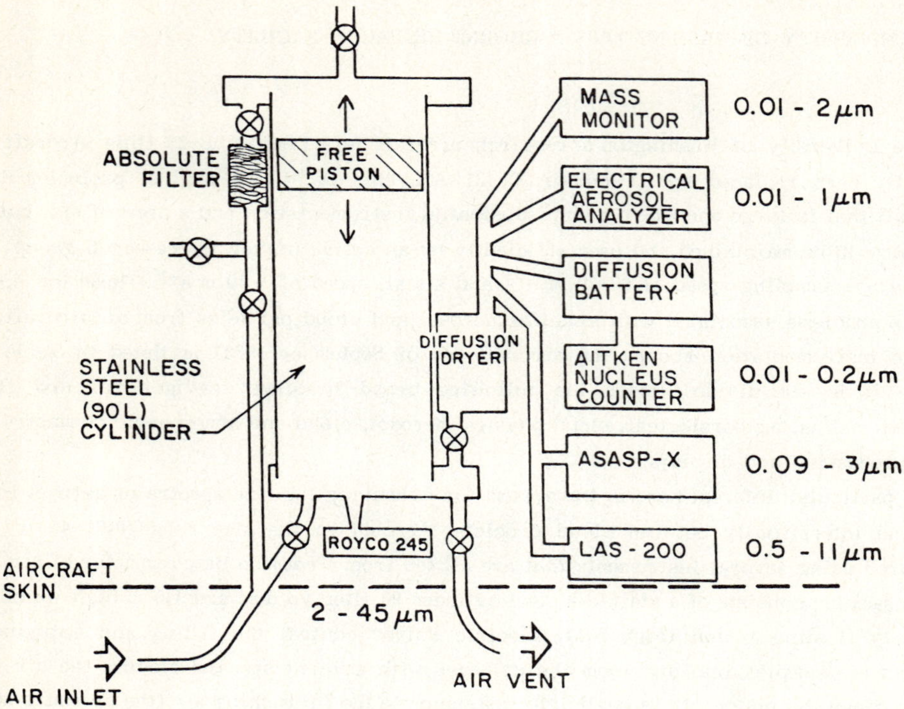

TABLE A.1 - INSTRUMENTATION ABOARD THE UNIVERSITY OF WASHINGTON'S B-23 AIRCRAFT

(a) Navigational and Flight Characteristics

Parameter	Instrument Type	Manufacturer	Range (and error)
Latitude, longitude and ground speed	VLF Omega navigator	Litton	0 to 300 m s^{-1} (± 1 m s^{-1} ground speed and ± 1° drift angle)
Ground speed and drift angle	Doppler navigator	Bendix Model DRA-12	0 to 6 km altitude
True airspeed	Variable capacitance	Rosemont Model 831 BA	0 to 230 m s^{-1} (< 0.2%)
Heading	Gyrocompass	Sperry Model C-2	0 to 360° (< 2%)
Pressure altitude	Variable capacitance	Rosemont Model 830 BA	150 to 1060 mb (< 0.2%)
Altitude above terrain	Radar altimeter	AN/APN22	0 to 6 km (< 5%)
Aircraft position and course plotter	Works off DME and VOR (to be integrated with VLF Omega system)	In-house	18 km (1 km)
Angle of attack	Potentiometer	Rosemont Model 861	± 23° (< 0.5°)

TABLE A.1 (contd)

(b) Meteorological

Parameter	Instrument Type	Manufacturer	Range (and error)
Total air temperature	Platinum wire resistance	Rosemont Model 102CY2CG + 414 L Bridge	-70 to 30°C (< 0.1°C)
Static air temperature	Computed	In-house	-70 to 30°C (< 0.5°C)
Dew point	Dew condensation	Cambridge Systems Model TH73-244	-40 to 50°C (< 1°C)
Air turbulence	Differential	Meteorology Research, Inc. Model 1120	0 to 10 $cm^{2/3}$ s^{-1} (< 10%)
Weather radar	5 cm gyro-stabilized	Radio Corp. of America, AVQ-10	100 km
Total solar radiation (up and down)	Eppley thermopile	Eppley Lab. Model PSP	0 to 1400 W m^{-2} (~ 1%)
Ultraviolet radiation	Barrier-layer photo-electric cell	Eppley Lab. Inc. Model 14042	0 to 70 W m^{-2} (< 5%)
Photographs	35 mm time-lapse camera	Automax Model GS-2D-111	1 s to 10 min

TABLE A.1 (contd)

(c) Cloud Physics

Parameter	Instrument Type	Manufacturer	Range (and error)*
Spectra of concentrations of cloud condensation nuclei	Four vertical thermal diffusion chambers	In-house	0 to 5000 cm^{-3} (< 10%) Simultaneous measurements at 0.2, 0.5, 1.0, 1.5% supersaturation
Liquid water content	Hot wire resistance	Johnson-Williams	0 to 2 and 0 to 6 g m^{-3}
Size spectrum of cloud particles	Forward light-scattering	Particle Measuring Systems Model FSSP	2 to 47 μm
Size spectrum of cloud particles	Diode occulation	Particle Measuring Systems Model OAP-200X	20 to 300 μm
Size spectrum of precipitation particles	Diode occulation	Particle Measuring Systems Model OAP-200Y	300 to 4500 μm
Images of cloud particles	Diode occulation imaging	Particle Measuring Systems Model OAP-2D-C	Resolution 25 μm
Images of precipitation particles	Diode imaging	Measuring Systems Model OAP-2D-P	Resolution 200 μm
Ice particle concentrations	Optical polarization technique	In-house	0 to 1000 ℓ^{-1} (detects particles > 50 μm)
Electric field	Rotary field mill	Meteorology Research Inc. Model 611	0 to 100 kV m^{-1} (< 10%)
Ion conductivity	Gerdian	In-house	0 to 4 x 10^{-14} ohm^{-1} m^{-1}

*All particle sizes refer to maximum particle dimensions.

TABLE A.1 (Contd)

(d) Aerosol

Parameter	Instrument Type	Manufacturer	Range (and error)*
Number concentration of particles	Light transmission	General Electric Model CNC II	10^2 to 10^6 cm^{-3} (particles >0.005 μm)
Number concentration of particles	Rapid expansion	Gardner	$2x10^2$ to 10^7 cm^{-3}
Mass concentration of particles	Electrostatic deposition onto matched oscillators	Thermal Systems, Inc., Model 3205	0.1 to 3000 μg m^{-3} (<0.2 μg m^{-3})
Sizes and types of particles	Direct impaction	Glass slides	5 to 100 μm
Size spectrum of particles	Electric aerosol analyzer	Thermal Systems, Inc., Model 3030	0.0032 to 1.0 μm
Size spectrum of particles	90° light-scattering	Particle Measuring System (LAS-200)	0.5 to 11 μm
Size spectrum of particles	Forward light-scattering	Royco 245 (in-house modified)	1.5 to 40 μm
Size spectrum of	Diffusion battery	Thermal Systems, Inc., Model 3040 (with in-house automatic valves and sequencing)	0.01 to 0.2 μm
Size spectrum of particles	35-120° light-scattering	Particle Measuring Systems, Model ASASP-X	0.09 to 3.0 μm (< 0.007 μm)
Size spectrum of particles	Forward light-scattering	Particle Measuring Systems, Model FSSP	2 to 47 μm
Size-segregated concentrations of particles	Cascade impactor	Sierra Instruments Inc.	0.1 to 3 μm (6 size fractions)
Light-scattering coefficient	Integrating nephelometer	Meteorology Res. Inc., Model 1567 (modified for increased stability and better response time)	0 to $1.0x10^{-4}$ m^{-1} or 0 to $2.5x10^{-3}$ m^{-1}
Number concentration of sodium-containing particles	Flame spectrometer	In-house	0 to 10,000 ℓ^{-1} (<1%)

* All particle sizes refer to maximum particle dimensions.

TABLE A.1 (Contd)

(e) Cloud and Atmospheric Chemistry

Parameter	Instrument Type	Manufacturer	Range (and error)
Cloud water samples	Impaction on slotted rods	In-house (based on ASRC design)	Reported to collect cloud droplets > 5 μm with an efficiency > 70%
Particulate sulfur $SO_4^=$, NO_3^-, Cl^-, Na^+, K^+, NH_4^+	Teflon filters CSI and Dionex XRF spectroscopy and ion exchange chromatography	In-house	0.1 to50 μg m^{-3} (for 500 liter air sample)
SO_2	Pulsed fluorescence	Teco SP43	1.0 ppb to 5 ppm
Total gaseous sulfur	FPD flame photometric detector	Meloy Model 285	0.5 ppb to 1 ppm
Ozone	Chemiluminescence (C_2H_4)	Monitor Labs Model 8410 A	0 to 5 ppm (< 7 ppb)
NH_3, NO, NO_2, NO_x, HNO_3	Chemiluminescence (O_3)	Modified Monitor Labs Model 8840	0 to 5 ppm (~ 1 ppb)
HNO_3	Nylon filters with teflon pre-filter followed by ion chromatography and/or tungstic acid denuder tubes followed by chemiluminescent detection	Dionex/Monitor Labs	Variable
Total hydrocarbons	Gas chromatograph	Analytical Instruments Inc.	>0.5 ppm
H_2O_2 (liquid phase)	Chemiluminescent reaction with luminol	In-house	>10 ppb (liquid phase)

TABLE A.1 (Contd)

(f) Date Processing and Display

Parameter	Instrument Type	Manufacturer	Range (and error)
Time	Time code generator	Systron Donner Model 8220	h, min, s (1:10^5)
Time	Radio WWV	Gertsch RHF 1	min
Ground communication	FM transceiver	Motorola	200 km
Inflight data processing	Mini-computer	Computer Automation LSI-III	-
Inflight color graphics	Micro-computer	Cepple II	-
Recording (digital)	Micro-computer directed cartridge recorder	3M	-
Recording (digital)	Floppy disk	Calcomp Model 140C	-
Recording (analog voice transcription)	Cassette recorder	Radio Shack Model 3C	-
Digital printout	Impact printer	-	-
Analog ship charts	6-channel hi-speed link recorder	Brush Model 260	-

REFERENCES

Barrie, L.A., and Georgii, H.W. 1976: An experimental investigation of the absorption of sulfur dioxide by water drops containing heavy metal ions. Atmospheric Environment 10, 743-749.

Beilke, S., Lamb, D. and Muller, J. 1975: On the uncatalyzed oxidation of atmospheric SO_2 by oxygen in aqueous systems. Atmospheric Environment 9, 1083-1090.

Brimblecombe, P. and Spedding, D. 1974: The catalytic oxidation of micropolar aqueous sulfur dioxide - I Oxidation of dilute solution of iron. Atmospheric Environment 8, 937-945.

Cadle, R.D., and Robbins, R.C. 1960: Kinetics of atmospheric chemical reactions involving aerosols. Discussions Faraday Soc. 30, 155-161.

Dinger, J.E., Howell, H.B., and Wojciechowski, T.A., 1970: On the source and composition of cloud nuclei in a subsident air mass over the North Atlantic. J. Atmos. Sci. 27, 791-797.

Easter, R.C., and Hobbs, P.V., 1974: The formation of sulfates and the enhancement of cloud condensation nuclei in clouds. J. Atmos. Sci. 31, 1586-1594.

Hegg, D.A., 1983: The sources of sulfate in precipitation. I. parameterization scheme and physical sensitivities. J. Geophys. Res. 88, 1369-1374.

Hegg, D.A. and Hobbs, P.V., 1978: Oxidation of sulfur dioxide in aqueous systems with particular reference to the atmosphere. Atmospheric Environment 12, 241-253.

Hegg, D.A., and Hobbs, P.V., 1979: The homogeneous oxidation of sulfur dioxide in cloud droplets. Atmospheric Environment 13, 981-987.

Hegg, D.A., and Hobbs, P.V., 1981: Cloud water chemistry and the production of sulfates in clouds. Atmospheric Environment 15, 1592-1604.

Hegg, D.A., and Hobbs, P.V., 1982: Measurements of sulfate production in natural clouds. Atmospheric Environment 16, 2663-2668.

Hegg, D.A. and Hobbs, P.V., and Radke, L.F., 1980: Observations of the modification of cloud condensation nuclei in wave clouds. J. Rech. Atmos. 14, 217-222.

Hegg. D.A., Hobbs, P.V., and Radke, L.F., 1984: Measurements of the scavenging of sulfate and nitrate in clouds. Atmospheric Environment, 18, 1939-1946.

Heikes, B.G. and Thompson, A.M., 1983: Effects of heterogeneous processes on NO_3, HONO and HNO_3 chemistry in the troposphere. J. Geophys. Res. (in press).

Hobbs, P.V., 1979: A reassessment of the mechanisms responsible for the sulfur content of acid rain. Proceedings of the Advisory Workshop to Identify Research Needs on the Formation of Acid Precipitation, Electric Power Research Institute, Palo Alto, California, Res. Rpt. EA-1074, pp. 2.31-2.37.

Junge, C., and Ryan, T.G., 1958: Study of SO_2 oxidation in solution and its rate in atmospheric chemistry. Quart. J. Roy. Meteor. Soc. 84, 46-55.

Larson, T.V., Horike, N., and Harrison, H., 1978: Oxidation of sulfur dioxide by oxygen and ozone in aqueous solution: a kinetic study with significance to atmospheric rate processes. Atmospheric Environment 12, 1597-1611.

Penkett, S.A., 1972: Oxidation of SO_2 and other atmospheric gases by ozone in aqueous solution. Nature 240, 105-106.

Penkett, S.A., Jones, B.M., Brice, A. and Eggleton, A.E., 1979: The importance of atmospheric ozone and hydrogen peroxide in oxidising sulfur dioxide in clouds and rainwater. Atmospheric Environment 13, 123-137.

Radke, L.F., 1970: Field and laboratory measurements with an improved automatic cloud condensation nucleus chamber. Preprints of Papers presented at the Amer. Meteor. Soc. Conf. on Cloud Physics, Fort Collins, CO, August 24-27, 7-8.

Radke, L.F., and Hobbs, P.V., 1969: Measurement of cloud condensation nuclei, light scattering coefficient, sodium-containing particles, and Aitken nuclei in the Olympic Mountains of Washington. J. Atmos. Sci. 26, 281-288.

Saxena, V.K., Burford, J.N., and Kassner, J.L., 1970: Operation of a thermal diffusion chamber for measurements on cloud condensation nuclei. J. Atmos. Sci. 27, 73-80.

Scott, W.S., and Hobbs, P.V., 1967: The formation of sulfate in water droplets. J. Atmos. Sci. 24, 54-57.

Van den Heuval, A.P., and Mason, B.J., 1963: The formation of ammonium sulfate in water droplets exposed to gaseous sulfur dioxide and ammonia. Quart. J. Roy. Meteor. Soc. 89, 271-275.

RADIATION FOG CHEMISTRY AND MICROPHYSICS

Sandro Fuzzi
Istituto FISBAT - C.N.R.
Via De'Castagnoli 1, 40126 Bologna
Italy

ABSTRACT

Acid precipitations have become one of the major environmental issues in recent years. Wet deposition occurs also by fog: in fact, some areas of the world experience fog with a frequency well in excess to that of rain. Moreover, recent experiments have shown that fog water can be highly acidic in polluted areas, so that wet acid deposition can be important to the environment. It has also been suggested that moisture deposited by fog on various surfaces exposed to the atmosphere (plants, buildings etc.) could enhance the effect of previously dry-deposited substances.

Presented here are some results of radiation fog water samplings and chemical analysis at two different locations: Po Valley (Northern Italy) and Albany, New York (Northeastern United States). Both climatological and pollution conditions are very different in the two places. The link between microphysics and chemistry of fog water is particularly examined and some suggestions are given about the sampling improvements necessary for a better understanding of fog chemistry.

INTRODUCTION

The subject of chemical reactions in the atmospheric dispersed liquid phase has been extensively studied in the last decade. The growing environmental problem of wet acid deposition has provided a further stimulus to these studies. Precipitation sampling networks have been organized and expanded in Europe, United States and Canada and several studies on cloud chemistry have been carried out, both by aircraft measurement (Hegg and Hobbs, 1981, 1982; Parungo et al., 1982; Lazrus et al., 1983; Richards et al., 1983) and ground based instrumentation (Falconer and Falconer, 1980; Falconer and Kadlecek, 1980; Castillo et al., 1983).

NATO ASI Series, Vol. G6
Chemistry of Multiphase Atmospheric Systems
Edited by W. Jaeschke

Fog constitutes a particular meteorological phenomenon which can be included within the same context, in so far as scavenging of gas and particles, chemical reactions and subsequent deposition of material take place.

One of the characteristics of fog is that it is usually of long duration and can persist for hours, or even days. Fog occurrence is highly variable, in different regions of the world, depending on climatological conditions (see Jiusto (1981) for a complete reference on fog properties). In Fig. 1 and 2, the frequencies of dense fog, in days per year, are reported for continental United States and Northern Italy, where some experiments have recently been carried out, to which reference will be made later. The figures show that certain areas experience fog occurrence close to 100 days per year, much more than rain frequency. Although not a great deal of moisture is associated with fog, the amount of water deposited can be important to the environment, especially if we take into account the fog "drip" interception by vegetation (Wisniewski, 1982).

Fog has recently been referred to as an "occult precipitation", due to the fact that water deposition is not detected by the normal raingauges. Dollard et al. (1983) estimated, from measurements made in England, that deposition due to fog, mist and cloud interception (occult precipitations) could account, on a yearly basis, for approx. 20% of the wet deposition by precipitations.

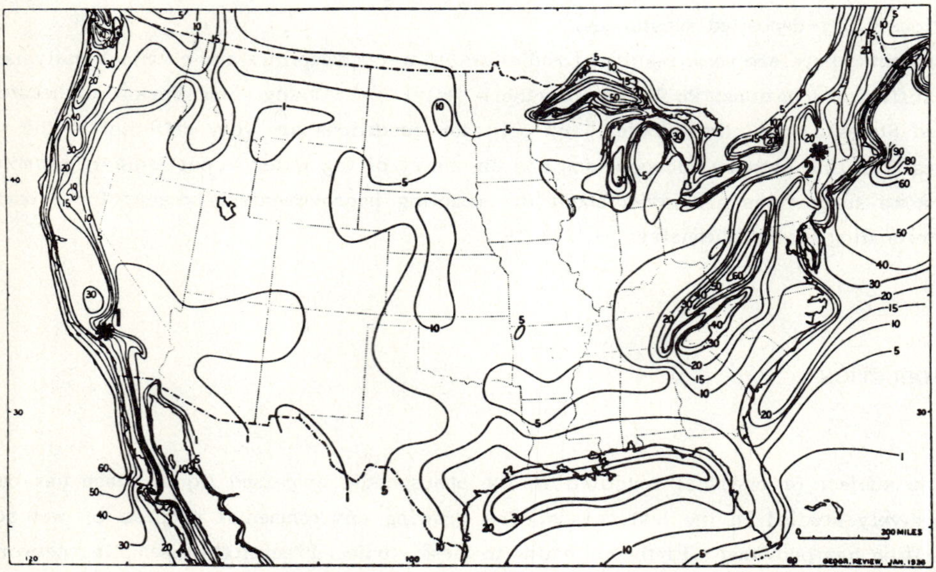

Figure 1: Distribution of fog in continental United States published by Stone (1936). Isopleths represent the mean annual number of days with dense fog (visibility less than 300 m). The two asterisks indicate the areas where experiments on fog chemistry were recently carried out.
1) Waldman et al. (1982); Munger et al. (1983)
2) Fuzzi et al. (1983 b)

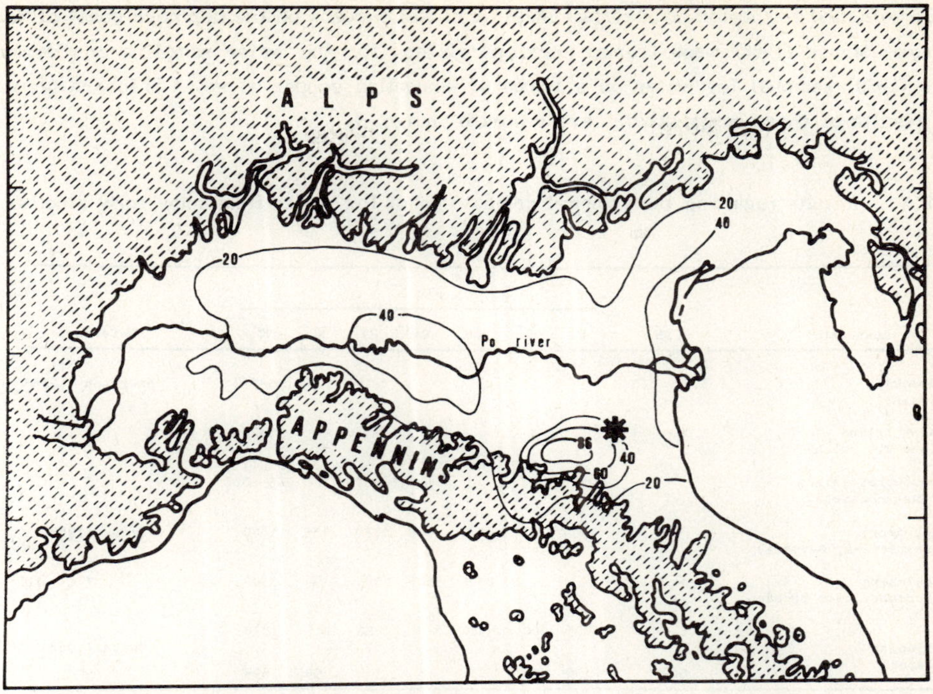

Figure 2: Distribution of fog in the Po Valley (Northern Italy) reported by Hesse (1954). Isopleths represent the mean annual number of days with fog (visibility less than 1000 m). The asterisk indicates the location where experiments on fog chemistry have been carried out (Fuzzi et al., 1983 a). The shaded area refers to elevations greater than 500 m.

Moreover, an indirect effect of water deposited by fog is that of enhancing the effects of acid substances previously dry-deposited on surfaces exposed to the atmosphere, as depicted in Fig. 3 (Wisniewski, 1982).

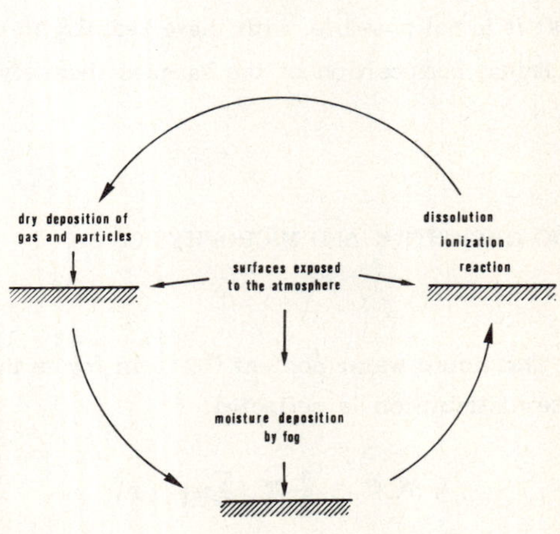

Figure 3: Schematic representation of interaction between wet deposition by fog and dry deposition.

EARLY STUDIES ON
FOG CHEMICAL COMPOSITION

Until recently, only a few studies were available, which gave an idea of the fog water chemical composition: Houghton

(1955) collected fog water on stainless steel or nickel screens on the northern coast of North America; Mrose (1966) used framed pervious clothes for fog water sampling in Germany; Okita (1968) sampled fog water on a frame of enamelled copper wires at Shiobara (Japan). The results of these early studies are reported in tab. 1.

Table 1: Results regarding the chemical composition of fog water from some early studies.

Location	pH	µeq l^{-1}							Reference
		Na^+	K^+	NH_4^+	Ca^{++}	Cl^-	NO_3^-	$SO_4^=$	
Brooklin (Maine)	3.5 - 6.3					0-140		95-770	Houghton (1955)
Kent Island (New Brunswich)	7.0 - 7.4					10-980		75-275	"
Nantucket Island (Massachussets)	-					650-5755		285-2600	"
Kap Arona* (Baltic Sea, Germany)	3.8	1500	245	2330	750	1740	455	1860	Mrose (1966)
Collmberg* (Germany, near Dresden)	4.2			2095	3175	585	190	3300	"
Shiobara (Japan)	5.9	805	220	580		620	215	1790	Okita (1968)
	5.9	190	125	235		335	55	695	

*mean values

The pH and ionic composition range of the reported measurements is quite large and this is accounted for by the different climatological and pollution conditions of the different sampling locations. Nevertheless all these fog collections represent bulk water samples, so that it is not possible, with these results, to interpret the various mechanisms leading to the chemical composition of the samples themselves.

FOG CHEMISTRY AND MICROPHYSICS

The liquid water content (LWC) in fog is the macroscopical parameter in which the droplet size distribution is reflected:

$$LWC = \frac{4}{3}\pi \sum n_i \quad r_i^3$$

where n_i is the number of droplets having radii r_i.

LWC values in fog may vary from a few mg m^{-3} to a few tenths of g m^{-3}. Peak values rarely exceed 0.5 g m^{-3} (see, for example, Fig. 8). Also LWC is inversely proportional to visibility in fog, although no absolutely fixed relationship exists between the two parameters (Jiusto, 1981).

As fog condensation nuclei start growing at the onset of fog, LWC increases following the trend of supersaturation, which controls the size of droplets and the number of activated nuclei. Then LWC decreases as fog dissipates.

Bearing in mind this simplified picture of the LWC trend during fog evolution, one could predict the occurrence of dilution/concentration effect of the solution droplets, following the growth/evaporation of the droplets themselves, as macroscopically reflected by LWC variations.

Recently, Waldman et al. (1982) and Munger et al. (1983) performed some fog water samplings at several locations in Southern California; Fuzzi et al. (1983 a; b) collected fog water samples in the Po Valley (Northern Italy) and at Albany (NY) in the American Northeast (sampling locations shown in Figs. 1 and 2). All these studies attempted to take into account the link between fog water chemistry and fog microphysics. In all cases, fog water samples

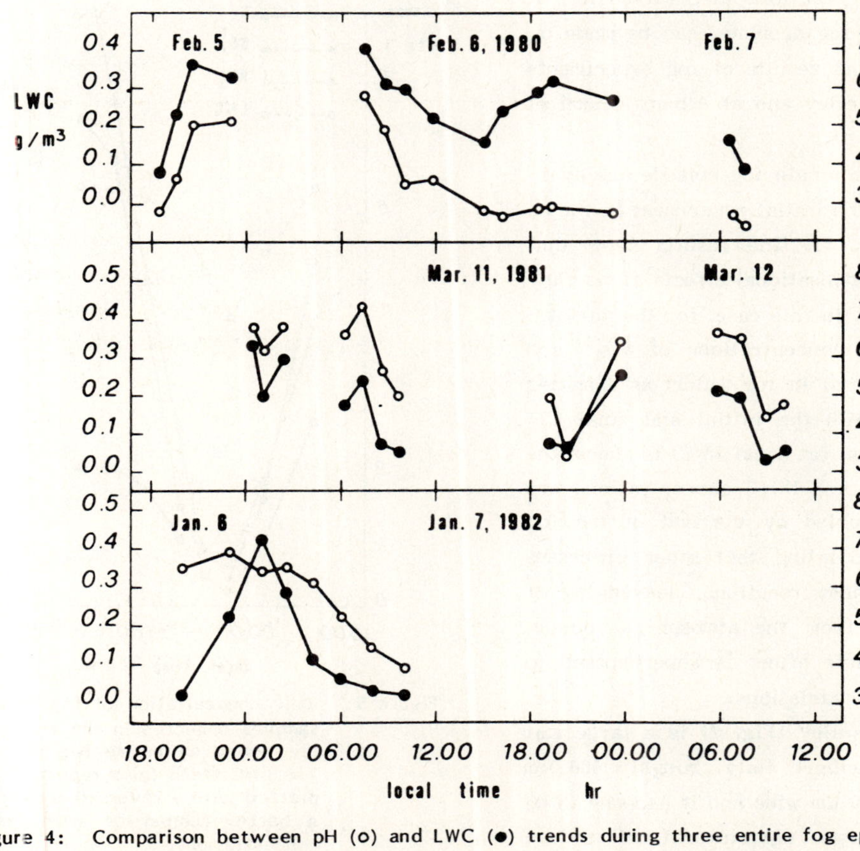

Figure 4: Comparison between pH (o) and LWC (•) trends during three entire fog episodes.

were collected sequentially, from the onset of fog until dissipation; the pH and chemical composition of the various samples were then measured. Although Waldman et al. (1982) and Munger et al. (1983) did not measure fog LWC during their samplings, the ionic concentration of sequential samples shows the characteristic U-shaped trend, which is an indication of the dilution/concentration effect of the fog droplet solution, caused by the microphysical evolution of the fog itself. Fuzzi et al. (1983 a) showed the same effect by comparing the trends of both pH and LWC during three fog events (Fig. 4). Although the two parameters are not analytically correlated (pH, on the other hand, is a logarithmic function), the effect of the microphysical evolution on fog water chemical composition is again evident.

Besides LWC variations, the initial composition of fog condensation nuclei and gas and particle scavenging affect the chemistry of the solution droplets, as shown by Jacob and Hoffmann (1983) who elaborated a model of fog H^+, NO_3^- and $SO_4^=$ production in an urban atmosphere.

A first experimental assessment as regards the relative importance of the microphysical evolution in fog, with respect to chemical reactions and scavenging of material by fog droplets, can be made by examining the results of fog experiments in the Po Valley and at Albany (Fuzzi et al., 1983 a; b).

Data of a single fog episode are available from the initial experiment in the Po Valley (Fig. 5). The results show that dilution/condensational effects alone cannot account, in this case, for the variations in the concentrations of $SO_4^=$ and NO_3^- measured in fog water: as a matter of fact, while the initial and final Cl^- concentration (at equal LWC) is almost the same, $SO_4^=$ and NO_3^- concentrations are roughly doubled by the end of the fog episode, indicating that other processes (aqueous phase reactions, scavenging of substances from the atmosphere) occur, and that their effect is superimposed to that of LWC variations.

The Po Valley (Fig. 2) is a large flat area in Northern Italy, roughly 400 km long and 100 km wide and it happens to be both the largest agricultural and indus-

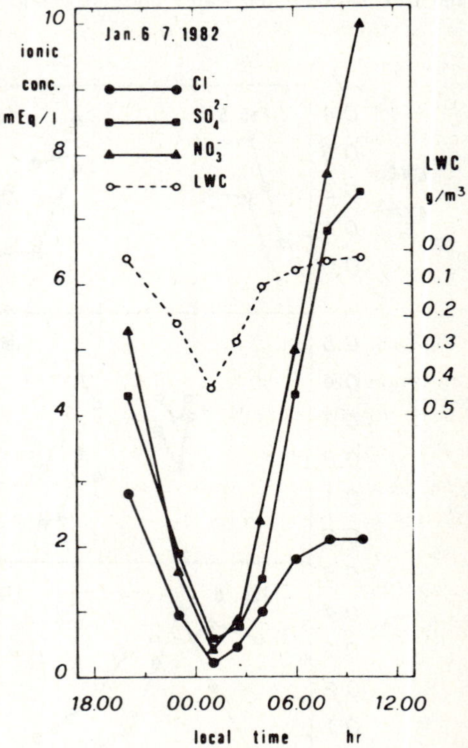

Figure 5: Ionic concentration of fog water samples collected in the Po Valley during an entire fog event. The LWC trend (also reported) is plotted with a reversed scale for a better comparison with ionic concentrations.

trial area of Italy. The atmospheric circulation within the valley is affected by the high elevations of the Alps and Appennins, which border the valley on three sides (Gandino, 1976; Tampieri et al., 1981). Nocturnal surface temperature inversions are quite common, especially in fall-winter season (Tomasi, 1983); radiation fogs are a common phenomenon and conditions of only a few meters visibility often occur (Hesse, 1954).

The pollution situation (Regione Emilia Romagna, 1974) and the fact that radiation fog forms in conditions of strong temperature inversion, when the airborne material is trapped in the lower levels of the atmosphere, account for the measured acidity and $SO_4^=$ and NO_3^- concentrations, which are only slightly lower than those reported for Los Angeles fogs.

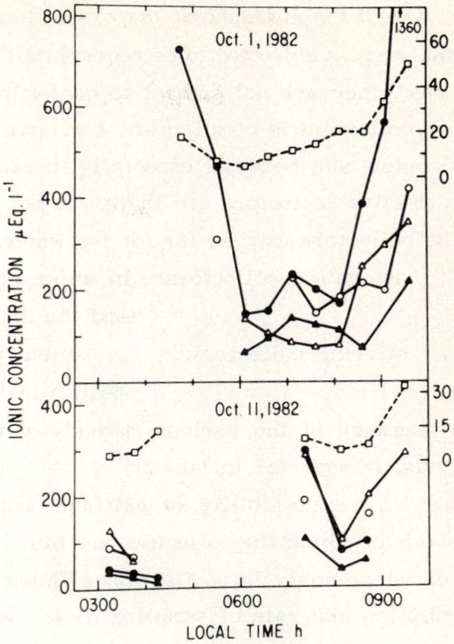

Figure 6: Ionic concentration of fog water samples collected during two individual fog events at Albany (NY). 1/LWC trend is also shown for comparison with ionic concentrations. $SO_4^=$ (●), NO_3^- (▲), Ca^{++} (○), NH_4^+ (△), 1/LWC (□).

A different situation is found at Albany (NY), in the American Northeast (see Fig. 1), where other radiation fog measurements were taken (Fuzzi et al., 1983 b). The sampling site, Albany County Airport, is typical for a suburban area, with only light industrial activity, residential complexes and some vegetable farming. Relatively low levels of pollution are reported in this area (New York State Department of Conservation, 1981). The concentration of the principal ions in the fog water at this location (Fig. 6) can be roughly accounted for by the chemical composition of the fog condensation nuclei and LWC evolution; scavenging of soil particles could account for the Ca^{++} content. The $SO_4^=$ and NO_3^- concentrations are more than one order of magnitude less than those measured in the Po Valley fogs. The pH of the samples is close to the CO_2 - H_2O atmospheric equilibrium. All these data are an indication that LWC evolution is the leading factor in controlling fog water chemistry at Albany.

FOG WATER SAMPLING FOR CHEMICAL ANALYSIS

In order to completely understand the chemical and physical processes which control the chemistry of the atmospheric dispersed liquid phase (in this specific case fog), it is essential for the samples collected to be representative of the actual atmospheric system and that, once collected, they are not subject to contamination or creation of artifacts.

Fog collection is complicated, compared to that of cloud, because of the smaller size of the droplets and because, especially in radiation fogs, windspeed is very low (less than 1 m s^{-1}). Passive collectors are therefore not suitable in this case.

The collectors used so far for fog water sampling can be divided into two main categories:

a) aspiration collectors: in which air containing droplets is accelerated by aspiration and the droplets are impacted onto the collection surface;

b) stirring collectors: in which the sampling device is accelerated in the fog medium and droplets are collected on it.

A summary of the various methods of fog water collection, for the purpose of chemical analysis, is reported in tab. 2.

Besides the possibility of artifact creation, due to the chemical characteristics of the materials of which the collectors are built, two main problems arise when sampling fog water for chemical analysis: collection efficiency of the samplings with respect to the fog size distribution and rate of sampling of fog water.

Collector	operating principle	collection efficiency	Reference(s)
Screen collector	impaction on metal or plastic screen by aspiration	70% d > 5 μm 90% d > 10 μm	Houghton (1955)
		n.r.	Okita (1968)
		50% d > 4 μm 100% d > 10 μm	Fuzzi et al. (1983b)
Pervious clothes collector	passive collector (only with high windspeed)	---	Mrose (1966)
Impinger	air aspiration through an orifice capture by inertial impaction	100% d > 10 μm	Fuzzi et al. (1983a)
Rotating arm collector	inertial impaction by centrifugal force in a slot of the rotating arm	high eff. d > 8 μm	Jacob et al. (1982)
Rectangular jet impactor	air aspiration and impaction on slowly rotating cylinders	0 d < 2 μm 100% d > 5 μm	Katz (1980)

n.r. = not reported

Table 2: List of the collectors used in fog water sampling for chemical analysis and their reported collection efficiency.

Collection efficiency of the samplings

Fog microstructure evolves from the onset, when droplets start growing on preexisting condensation nuclei, and the size spectrum follows the trend of supersaturation during the evolution of the system. Even considering only the simple model of condensational dilution and evaporative concentration, it appears quite likely that different size droplets could also have different chemical compositions (see for example Levine and Schwartz (1982)). Therefore, collectors designed for the purpose of fog water chemical analysis should efficiently sample the sizes of droplets most relevant to the total LWC. An example of size spectrum evolution at different stages of fog is given in Fig. 7. The histograms represent the volume percent distribution of droplets, measured by means of a PMS FSSP 100, integrated over 12 min period, in order to suppress the short term variations. Only the droplets with r > 2 μm were considered, which contribute more than 95% of LWC in dense fog. The solid line represents the percentage of the total volume of water (LWC) as a function of size. The contribution of smaller droplets is variable at different stages of fog evolution, but at the onset of fog and in the dissipation stage, when the effect of condensational growth and evaporative shrinking of droplets is most relevant, they contribute to a considerable fraction of the total LWC. From the data on collection efficiency reported in tab. 2, it should

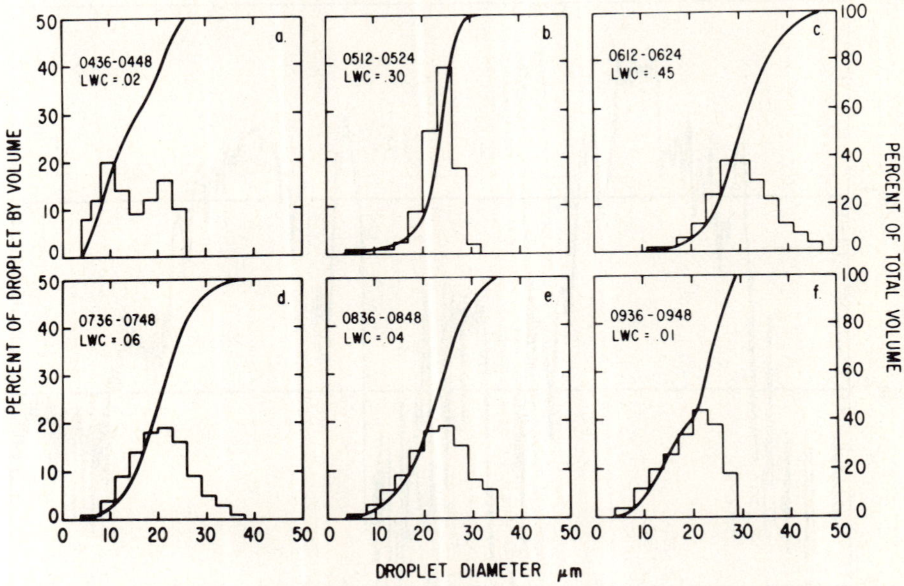

Figure 7: Volume percent distribution of fog droplets taken at different times during a fog epsisode (Albany - October 1, 1982). The solid line represents the integral percent volume as a function of droplet size. Data were obtained by means of a PMS FSSP 100 and integrated over 12 min. period. The droplet size interval taken into account is 4 - 47 μm diameter.

be noted that not all devices used for fog sampling achieve a sufficiently high collection efficiency for smaller droplets. This could imply a non-representative sample from the chemical point of view. It would be quite interesting to be able to measure the chemical composition of droplets per size class, in order to quantify the above stated effect; such a measurement, however, has not yet been made, due to the obvious difficulty in obtaining sufficient water for chemical analysis.

Rate of sampling for fog water

When the samplings of fog water are performed for the evaluation of chemical reaction rates in the dispersed liquid phase, it is, of course, necessary to have available samples with the shortest possible time resolution, in order to carefully follow the evolution of the reactions under investigation. On the other hand, fogs are characterized by lower LWC than clouds, as previously mentioned; 10 - 20 cc of water are necessary in order to perform the usual chemical analysis, which implies an average sampling time of the order of 30 - 60 min with the collectors currently in use. Fig. 8 illustrates the temporal trend of LWC during a

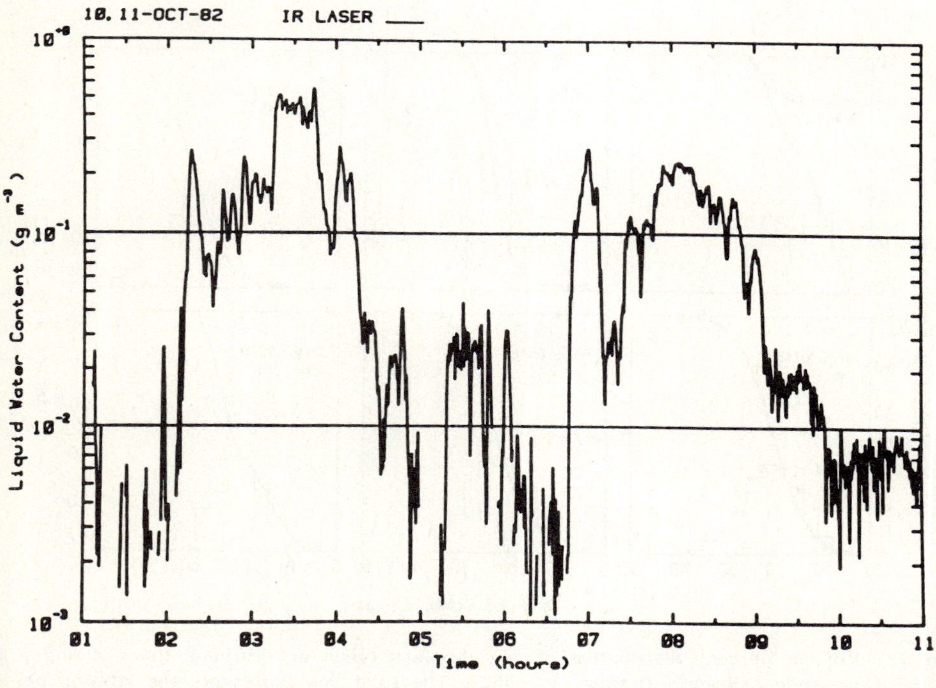

Figure 8: LWC temporal trend during a fog episode (Albany - October 10-11, 1982). Data were obtained by means of an IR laser transmissometer (after Jiusto and Lala, 1983).

typical fog event, measured by means of an IR transmissometer (Jiusto and Lala, 1983). It can be seen from the figure that the time basis of LWC variations is very short and variations of more than one order of magnitude may occur in a 30 min period (see for example the period 0700-0730 hours), which imply microphysical and therefore chemical variations.

It is to be hoped that in the near future collectors will be made available, which result in a better compromise between collection efficiency and collection rate.

CONCLUSIONS

The studies reported herein show:

a) in certain areas of the world, wet deposition by fog can be important to the environment, due to both the long duration of this meteorological phenomenon and its frequency;

b) fog chemical composition is extremely variable, as a function of the climatological and pollution situation of the locations under investigation;

c) fog chemical composition is controlled by:
   i) scavenging processes;
   ii) chemical reactions in the dispersed phase;
   iii) microphysical evolution;

   the relative importance of these different processes cannot be defined a priori, but can only be assessed by local studies;

d) it is very difficult to design fog collectors, for the purpose of chemical analysis, which combine a good collection efficiency for the smaller droplets, with a good sampling rate, in such a way as to obtain representative samples on a short time basis.

As a further conclusion it could be suggested that radiation fog is an ideal natural laboratory for the empirical rate evaluation of chemical reactions in the atmospheric dispersed liquid phase. In fact, due to the high stability of the atmosphere during radiation fog, it seems reasonable to assume a closed box approximation, as far as air mass motion is concerned, so that a mass balance based chemical study can be performed. Furthermore, a ground based network of instruments is, of course, more convenient than aircraft-carried instrumentation.

ACKNOWLEDGEMENTS

Part of the studies reported herein were performed while on leave at the Atmospheric Sciences Research Center, State University of New York at Albany. Drs. R.A. Castillo, J.E. Jiusto (I deeply regret the death of Jim Jiusto) and G.G. Lala are gratefully acknowledged for their collaboration in this work. Thanks are extended to Dr. V.A. Mohnen for his support during my stay at ASRC.

REFERENCES

Castillo R.A., Jiusto J.E. and McLaren E. (1983), The pH and ionic composition of stratiform cloud water. Atmos. Environ. 17, pp. 1997-1505.

Dollard G.J., Unsworth M.H. and Harve M.J. (1983), Pollutant transfer in upland regions by occult precipitation. Nature 302, pp. 241-243.

Falconer P.D. and Kadlecek J.A. (1980), Hurricane chemistry. Weatherwise 33, pp. 162-167.

Falconer R.E. and Falconer P.D. (1980), Determination of cloud water acidity at a mountain observatory in the Adirondack Mountains of New York State. J. Geophys. Res. 85, pp. 7465-7470.

Fuzzi S., Orsi G. and Mariotti M. (1983a), Radiation fog liquid water acidity at a field station in the Po Valley. J. Aerosol Sci. 14, pp. 135-138.

Fuzzi S., Castillo R.A., Jiusto J.E. and Lala G.G. (1983 b), Chemical composition of radiation fog water at Albany, New York, and its relationship to fog microphysics. submitted to J. Geophys. Res.

Gandino C. (1976), The influence of the Alps on the diurnal winds. Riv. Ital. Geofis. Sci. Affini 3, pp. 150-152.

Hegg D.A. and Hobbs. P.V. (1981), Cloud water chemistry and the production of sulfates in clouds. Atmos. Environ. 15, pp. 1597-1604.

Hegg D.A. and Hobbs P.V. (1982), Measurements of sulfate production in natural clouds. Atmos. Environ. 16, pp. 2663-2668.

Hesse W. (1954), Nebel and Sichtverhaeltnisse in Ober- und Mittelitalien. Pageoph 27, pp. 179-189.

Houghton H.G. (1955), On the chemical composition of fog and cloud water. J. Met. 12, pp. 355-357.

Jacob D.J., Flagan R.C., Waldman J.M. and Hoffmann M.R. (1982), Design and calibration of rotating-arm collectors for ambient fog sampling. 4th International Conference on Precipitation Scavenging, Dry Deposition and Resuspension, Nov. 29 - Dec. 3, Santa Monica, California.

Jacob D.J. and Hoffmann M.R. (1983), A dynamic model for the production of H^+, NO_3^- and $SO_4^=$ in urban fog. J. Geophys. Res. 88, pp. 6611-6621.

Jiusto J.E. (1981), Clouds: Their Formation, Optical Properties and Effects. (Edited by Hobbs P.V. and Deepak A.) Academic Press, New York.

Jiusto J.E. and Lala G.G. (1983), Radiation fog field programs - recent studies. Report No. 869, ASRC - SUNY, Albany, New York.

Katz U. (1980), A droplet impactor to collect liquid water from laboratory clouds for chemical analysis. VIIIeme Conference Internationale sur la Physique des Nuages, July 15 - 19, Clermont Ferrand, France.

Lazrus A.L., Haagenson P.L., Kok G.L., Huebert B.J., Kreitzberg C.W., Likens G.E., Mohnen V.A., Wilson W.E. and Winchester J.W. (1983), Acidity in air and water in a case of warm frontal precipitation. Atmos. Environ. 17, pp. 581-591.

Levine S.Z. and Schwartz S.E. (1982), In-cloud and below-cloud scavenging of nitric acid vapor. Atmos. Environ. 16, pp. 1725-1734.

Mrose H. (1966), Measurements of pH and chemical analysis of rain-, snow- and fogwater. Tellus 18, pp. 266-270.

Munger J.W., Jacob D.J., Waldman J.M. and Hoffmann M.R. (1983), Fogwater chemistry in an urban atmosphere. J. Geophys. Res. 88, pp. 5109-5121.

New York State Department of Conservation (1981), Report on Air Quality in New York State. Albany, New York.

Okita T. (1968), Concentration of sulfate and other inorganic materials in fog and cloud water and in aerosol. J. Met. Soc. Japan 46, pp. 120-126.

Parungo F., Nagamoto D., Nolt I., Dias M. and Nickerson E. (1982), Chemical analysis of cloud water collected over Hawaii. J. Geophys. Res. 87, pp. 8805-8810.

Regione Emilia Romagna (1974), Studio globale sull'inquinamento nel territorio regionale. Bologna, Italy.

Richards L.W., Anderson J.A., Blumenthal D.L., McDonald J.A., Kok G.L. and Lazrus A.L. (1983), Hydrogen peroxide and sulfur (IV) in Los Angeles cloud water. Atmos. Environ. 17, pp. 911-914.

Stone R.G. (1936), Fog in the United States and adjacent regions. The Geographical Review 26, pp. 111-134.

Tampieri F., Trombetti F. and Scarani C. (1981), Summer daily circulation in the Po Valley, Italy. Geophys. Astrophys. Fluid Dynamics 17, pp. 97-112.

Tomasi C. (1983), The nocturnal surface inversion height in the Po Valley. Atmos. Environ. 17, pp. 1123-1129.

Waldman J.M., Munger J.W., Jacob D.J., Flagan R.C., Morgan J.J. and Hoffmann M.R. (1982), Chemical composition of acid fog. Science 218, pp. 677-680.

Wisniewski J. (1982), The potential acidity associated with dews, frosts and fogs. Water, Air and Soil Pollut. 17, pp. 361-377.

INVESTIGATION OF THE WET AND DRY DEPOSITION OF ACIDIC AND HEAVY-METAL COMPONENTS

H.-W. Georgii, C. Perseke, E. Rohbock
Department of Meteorology and Geophysics
University of Frankfurt/M
F.R.G.

ABSTRACT

The transfer of atmospheric trace-gases and of aerosol-particles to the earth surface is determined mainly by two mechanisms: wet and dry deposition. Wet deposition can be described by the product of the precipitation-rate and the concentration of trace-substance in rain. In the following paper the regional distribution of wet and dry deposition of acidic and heavy-metal components will be discussed on the basis of results gained with a deposition-network in the Federal Republic of Germany and compared with measurements from other parts of Europe. Maxima of wet deposition are found either in polluted areas with high concentrations of trace-substances in rain or in non-polluted areas with elevated precipitation-rate, like mountains. With respect to the ratio of wet and dry deposition of heavy-metal aerosols, the size distribution of aerosols concerned is of major importance. This is clearly shown the different heavy-metal components are compared.

1. INTRODUCTION

The study of the budget of atmospheric trace-substances includes research on sources, transport-mechanisms and sinks. Important sinks for trace-gases and aerosols are dry and wet deposition leading to transfer of substance from the atmosphere to the ground. Both mechanisms may cause an accumulation of adverse substances in lakes, rivers or soil. In spite of the great practical importance which is attributed to deposition-processes, there still exists a considerable lack of knowledge of the mechanism and rate of dry and wet deposition. Major interest in the study of deposition-processes originates from reports on damage to eco-systems by acidic deposition.

NATO ASI Series, Vol. G6
Chemistry of Multiphase Atmospheric Systems
Edited by W. Jaeschke

The research on wet and dry deposition was mainly restricted to sulfur compounds up to now. Earlier studies of the wet and dry sulfur-deposition could only estimate the order of magnitude of the regional deposition owing to lack of sufficient data (Perseke et al., (1)). Wet deposition is always connected with rainfall. An intermediate mechanism between wet and dry deposition is the interception of fog- and cloud water by vegetation or at the ground. Interception cannot be neglected in mountain areas. This is proved by earlier investigations. The rate of fog-water which is collected by this process amounts to 40% of the rainfall-rate on Mt. Hohenpeissenberg in Upper Bavaria, to 35% of the annual rainfall rate in Eifel- and Vogelsbergmountains at about 600 mtrs above sealevel and up to 75% of the annual rainfall rate in the Bavarian Forest in 1300 mtrs. altitude according to results gained by Grunow (2) and Grunow and Tollner (3). It should be mentioned that extremely high rates of interception are known for the Table Mountain near Capetown, where the rate of intercepted cloud-water amounts to 3290 mm, at a rainfall-rate of only 1940 mm. Forest-systems collect the water from drifting fog or clouds and increase the amount of water which is transferred from the atmosphere to the ground. In higher regions of the mountains the additive water-supply to the soil by this process can be considerable. In context with the adverse effects of acidic deposition to eco-systems in mountain-regions the process of interception should therefore not be neglected.

2. DRY DEPOSITION OF GASES AND PARTICULATE MATTER

Dry deposition of gases and aerosols describes the transfer of these components from the atmosphere to the ground. For gas-molecules and particles below a radius of 10 µm in the boundary-layer of the atmosphere this process is determined by diffusion. For larger particles than 10 µm radius sedimentation in the gravity-field of the earth is the dominant sink mechanism. Close to the ground the laminar boundary-layer above the relevant surface determines the diffusion-process. After reaching the surface, gases can become adsorbed or dissolved, particles adhere to the surface by impaction. It should be mentioned that the transfer of particles in the size-range 0.1 µm radius to 1 µm radius, is rather ineffective as has been demonstrated by several authors. The effectiveness of the dry deposition of gases is often described by the "deposition-velocity" according to Chamberlain (4):

$$V_d = \frac{F}{C(z)}$$

V_d = deposition velocity (cm sec^{-1})

F = flux of gas-molecules to the ground

C(z) = concentration at reference level

The deposition-velocity has only been determined for few gases. Investigations by Garland (5) indicate that the deposition-velocity of gases varies within a wide range depending on the structure of the surface, the humidity and pH-value of the surface, type of vegetation, roughness, but it is also dependant on meteorological parameters like the vertical temperature- and wind-profile. Table 1 gives a survey of SO_2-deposition velocities and table 2 some data of deposition velocities of different gases:

Table 1: SO_2-deposition-velocities for different surfaces (Perseke et al., (6))

	Deposition velocity (cm sec^{-1})	
Surface	Range	Average
Short grass (0.1 m high)	0.1-0.8	0.5
Corn (1 m high)	0.2-1,5	0.7
Forest (10 m high)	0.2-2.0	?
Soil (pH 7, dry)	0.3-1.0	0.8
Soil (pH 7, moist)	0.3-1.0	0.8
Acidic Soil (pH 4, dry)	0.1-0.5	0.4
Acidic Soil (pH 4, moist)	0.1-0.8	0.6
Snow	-	0.1
Water surface	0.2-1.5	1.0

Table 2: Typical deposition-velocities of different gases (cm sec^{-1})

Component	Vegetation	Soil	Watersurfaces	Snow
SO_2	0.7	0.8	1.0	0.1
NO_2	0.5	-	-	-
O_3	0.8	-	0.8	1.0
NO	-	-	0.01	-

The deposition-velocity for SO_2 shows also a diurnal tendency with a minimum during the night. Garland and Granson (7) measured 0.6 sec^{-1} during the day and 0.1 cm sec^{-1} during the night. Stratification of the temperature and lack of vertical exchange of air-parcels are responsible for the reduced values during night-time. The dry deposition of aerosol-particles is influenced by the size-distribution of the aerosol. Aitken-nuclei (r < 0.1 um) and giant-nuclei (r > 1 um) are deposited more effectively than large nuclei (0.1 < r < 1 um). For Aitken-nuclei molecular diffusion is the dominant transfer-mechanism. For giant nuclei sedimentation and impaction at the surface are of importance. The theoretical scavenging efficiency of a falling raindrop in relation to the aerosol particle-size is plotted in Fig. 1 adopted from Pruppacher and Klett (8). Sulfate-aerosols are considered as a component of acidic deposition, they are predominantly in the size-range between 0.1 µm and 1 µm and a deposition-velocity of 0.1 cm sec^{-1} is suggested. Measurements by Hoefken et al. (9) show that the deposition-velocity in forest-systems is increased. For sulfate-aerosols a deposition-velocity of 1.1 cm sec^{-1} was measured in a spruce-forest.

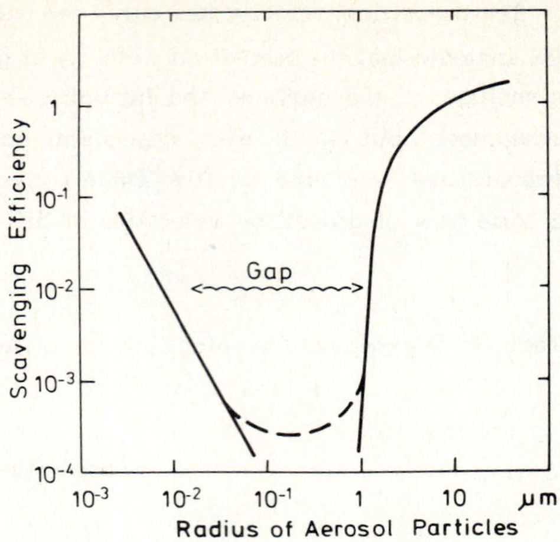

Fig. 1: Scavenging efficiency of falling raindrop as a function of aerosol particle size according to Pruppacher and Klett (1978).

In a recent study the deposition-velocity of heavy metal aerosols was determined from the measured deposition-rate and the measured concentration. Parallel, the size-distribution of the heavy metal aerosols was determined using a cascade-impactor. Table 3 shows the relation between the mass-median diameter of the analyzed components and the deposition-velocity.

Table 3: Characteristics deposition-velocity V_d and average mass-mean diameter of aerosols

Component	V_d (cm sec^1)	MMD (µm)
SO_4^{2-}	0.3	0.4
NO_3^-	0.3	0.7
Pb	0.07	0.4
Cd	0.23	0.85
Mn	0.95	2
Fe	1.5	3

It is realized that the mass-median diameter is only a very rough parameter to be related to the deposition-velocity.

Table 3 indicates that some of the heavy metal aerosols are found in the coarse particle mode. This has influence on the ratio of dry to wet deposition of these components.

3. WET DEPOSITION OF GASES AND AEROSOLS

Investigation of the rate and composition of wet deposition has found increased interest during recent years in context with the occurrence of "acidic precipitation" in Europe and North America. It is emphasized that we understand under the term "wet deposition" the product of precipitation rate and trace-substance concentration in rainwater. This means that the rate of wet deposition increases with rising rainfall-rate and also with rising concentration of trace-material in rainwater. In this context we have to bear in mind that in our regions the annual rainfall-rate is influenced by the orographic structure of the countryside.

4. REGIONAL DISTRIBUTION OF ACIDIC AND HEAVY METAL DEPOSITION

During the period 1979-1981 dry and wet deposition of acidic and heavy metal-components were measured in a network of 12 stations in the Federal Republic of Germany. The selected locations characterize industrial, urban and rural areas as well as mountain stations. A detailed description of the stations has been given elsewhere (Georgii et al., (10)). In the frame of that investigation only wet- and dry-only samplers were used in order to seperate the dry and wet fractions of deposited material. Dry deposition was collected according to the Bergerhoff-method (11). In addition to the deposition collection, airborne aerosol was sampled on filters with a low-volume sampler. Wet deposition was taken on a daily basis, dry deposition over a period of 2 weeks.

4.1 Acidity of precipitation

In many investigations the pH-value of rainwater is taken as an indicator for the acidity of rain. The trend of the pH-value for the period 1979 to 1981 for Schleswig, Braunschweig,

Essen and Deuselbach is shown in Fig. 2 together with the precipitation rate. All stations show a similar tendency of the pH-value between 4 and 5 (average pH was calculated on the basis of H^+-ion concentration). A comparison with monthly rainfall-data does not show a clear relationship. The cumulative frequency-distribution for the stations Frankfurt/M, Deuselbach and Hof for the period 1979-1980 shows a 50%-value of 4.2 for Frankfurt/M. At the more rural stations Deuselbach and Hof the 50%-value is shifted to 4.4 at Deuselbach and 4.6 at Hof. It is interesting to note that the difference of the H^+-concentration by a factor of 2.5 between Frankfurt and Hof is similar to the regional difference of the sulfate-concentration in rain between polluted and unpolluted areas. The pH-value of rainwater should not be overemphasized in its importance. In unpolluted areas in Europe

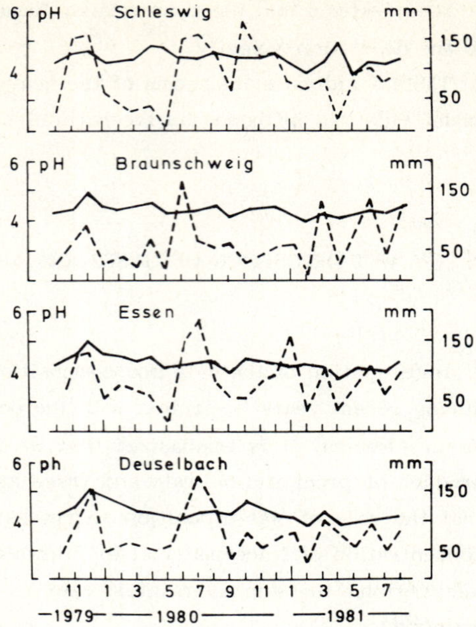

Fig. 2: pH-trend of four stations of the German deposition-network.

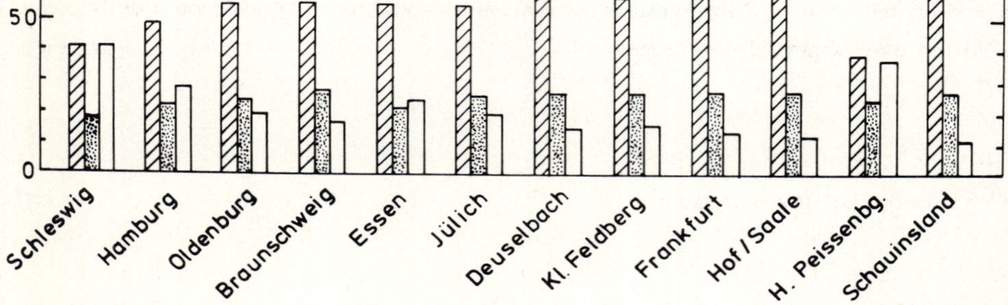

Fig. 3: Ionic composition of the rain at the stations of the German deposition network.

and America as well as on Samoa and Hawaii pH-values between 4.6 and 5.3 are found. This range seems more realistic than the assumed pH-value of 5.6 of water in equilibrium with the atmospheric CO_2.

The determination of acidity in rain has to include the measurement of the concentration of sulfate, nitrate and chloride. Fig. 3 shows that the ionic concentration of these compounds in the Federal Republic of Germany is relatively uniform except at coastal stations. We have assumed for each mole of sulfate an equivalent of two moles of H^+ and for each mole of nitrate and chloride one mole of H^+. On the annual average, the sulfate fraction amounts to 50 to 60%, the nitrate fraction to 25 to 30% of the anion-concentration in rain.

Precipitation analysis on Whiteface Mountain (USA) show a good correlation (r=0.95) for the ions:

$$[H^+] + [NH_4^+] = 2\ [SO_4^{2-}] + [NO_3^-]$$

This correlation indicates that a significant fraction of sulfate and nitrate in rain is present in the form of sulfuric acid and nitric acid (Wilson et al., (12)). In the context of our present study the complete ion-balance could not be assessed owing to the lack of NH_4^+-data. NH_4-analysis was excluded in the present study since the necessary storage and transport of the rain-samples lead to changes of the ammonia content.

5. WET DEPOSITION OF ACIDIFYING IONS AND HEAVY METALS SULFATE

5.1 Sulfate

The wet sulfate-deposition was assessed on the basis of measured concentrations and rates of precipitation. As mentioned above the wet deposition of sulfur is determined by the rainfall-rate and by the concentration of sulfate in rainwater. The regional distribution of the wet sulfate deposition shows the highest values in the Ruhr area with 3.3 mg $Sm^{-2}d^{-1}$ during fall and 9 mg $Sm^{-2}d^{-1}$ during summer. In mountain areas higher values of the sulfate deposition originate from the rise of the precipitation-rate with altitude. So, the sulfate deposition on Mt. Kleiner Feldberg is higher than in Frankfurt/M. When we compare the cumulative rate of the sulfate deposition with the cumulative precipitation-rate for stations with a different grade of pollution we can take from Fig. 4 that in spite of equal precipitation rate on Mt. Kleiner Feldberg and on Mt. Hohenpeissenberg during the measuring period 1979-1981, the rate of sulfate deposition on Mt. Kleiner Feldberg is 1.5 times the sulfate deposition on Mt. Hohenpeissenberg. This reflects the higher grade of pollution in the

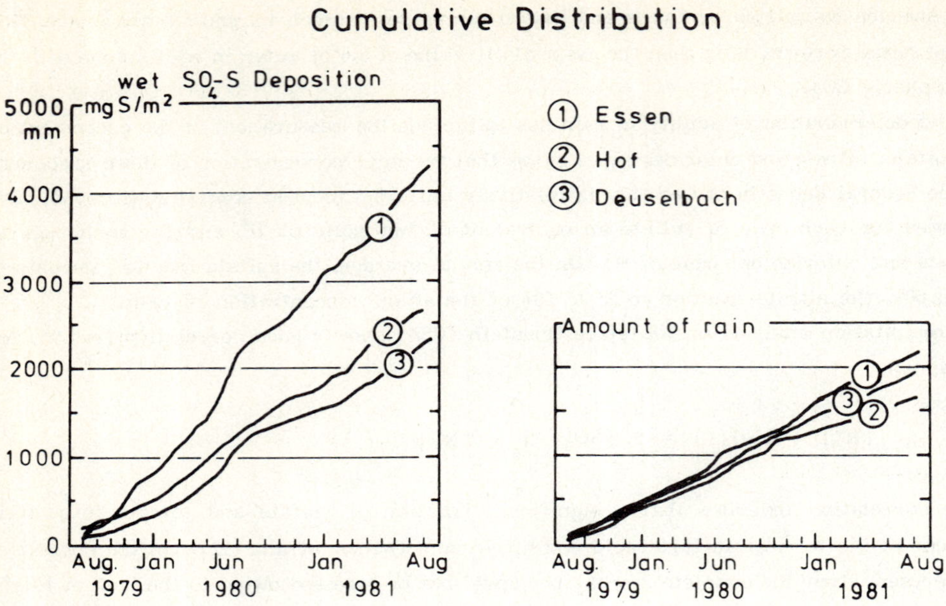

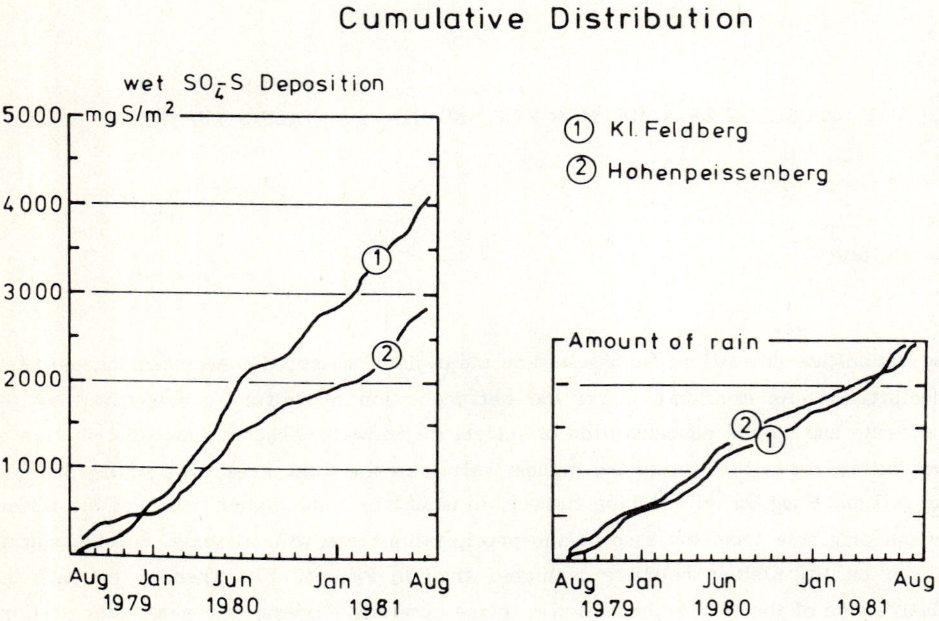

Fig. 4: Cumulative wet deposition of sulfur.

Rhein-Main area. The same explanation can also be given for the difference between the wet sulfate deposition in Essen and Deuselbach.

It should be mentioned that wet deposition comprises the incorporation of SO_2 and sulfate aerosols into the rain-elements.

The annual average of wet sulfate deposition in Germany varies between 1 and 2 g Sm^{-2} yr^{-1}. Meszaros et al. (13) calculated a sulfur budget for Europe and suggested for the territory of the Federal Republic of Germany wet sulfur deposition-rates between 1.6 and 2.7 g $Sm^{-2}yr^{-1}$, which is in good agreement with our results.

Calculations by Perseke et al. (1) of the distribution of the wet sulfate-deposition in the Federal Republic of Germany show annual values of 3 g $Sm^{-2}yr^{-1}$ or more only in the Ruhr area. This refers to the precipitation-distribution of 1974.

Besides the analysis of bulk-samples we have also studied the wet deposition during individual events of rainfall by sequential sampling. For this purpose a sequential deposition-sampler was constructed which operates on the basis of equal samples with a

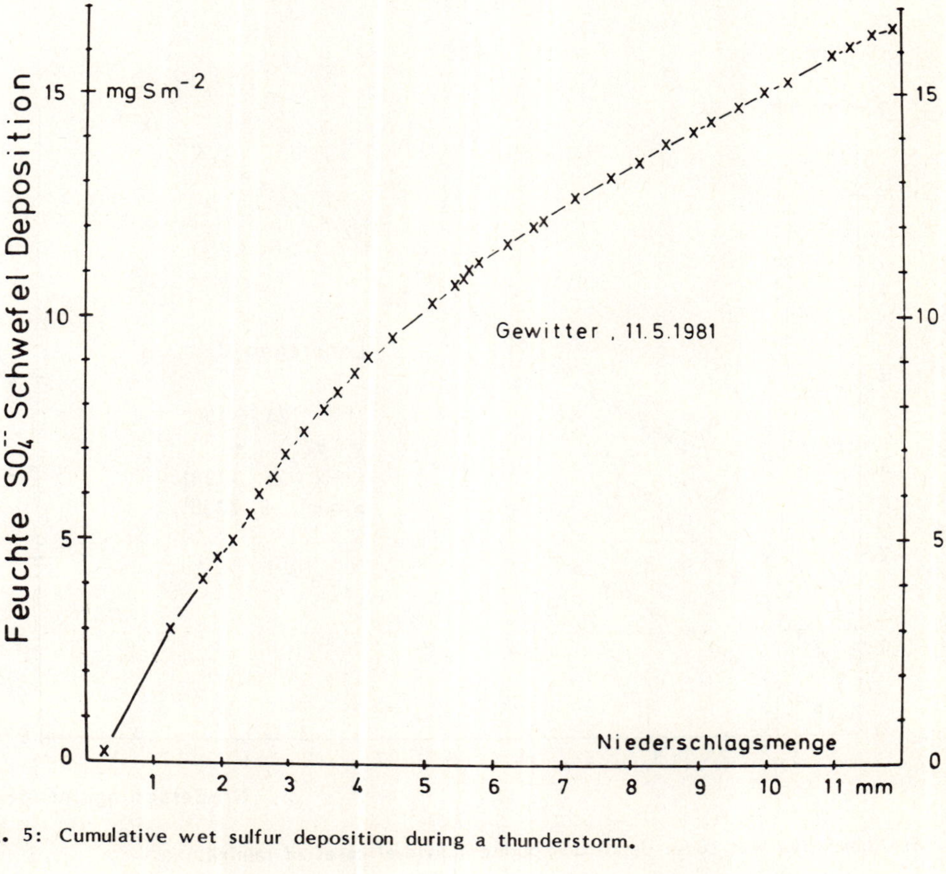

Fig. 5: Cumulative wet sulfur deposition during a thunderstorm.

minimum volume necessary for chemical analysis. Details of the construction are given by L. Kins (14). For the purpose of analysis a volume of 15 ml is necessary, corresponding to a rainfall-rate of 0.36 mm. The sampling dish contains 48 bottles, thus allowing to evaluate a total rainfall of 17.4 mm. It is well known that in convective showers an inverse relationship exists between the trace-substance concentration and the rainfall rate in such a way that a maximum of rain-intensity is accompanied by a minimum of trace-substance concentration. When we translate this result into values of cumulative wet deposition of sulfur during individual cases of rainfall we can see in Fig. 5 that the amount of sulfate deposited on the ground follows during the first 3 mm of rainfall a quasi-linear function. With further continuation of the rainfall the slope of this function becomes more flat. During the thunderstorm on 11 May 1981 6-7 mg Sm^{-2} were deposited during the first 3 mms of rainfall, while the total rainfall of 11 mms was accompanied by a deposition rate of only 16 mg Sm^{-2}. Fig. 6 shows that individual cases of rainfall from layer-clouds are connected with varying rates of deposition. An example shows that a continuous rainfall of also 11 mms is accompanied by a deposition rate of only 7 mg Sm^{-2}. In general we can state for the sulfate-deposition during individual cases of rainfall, that the sulfate-deposition shows an individual character of each case of rainfall depending on the direction of advection on the

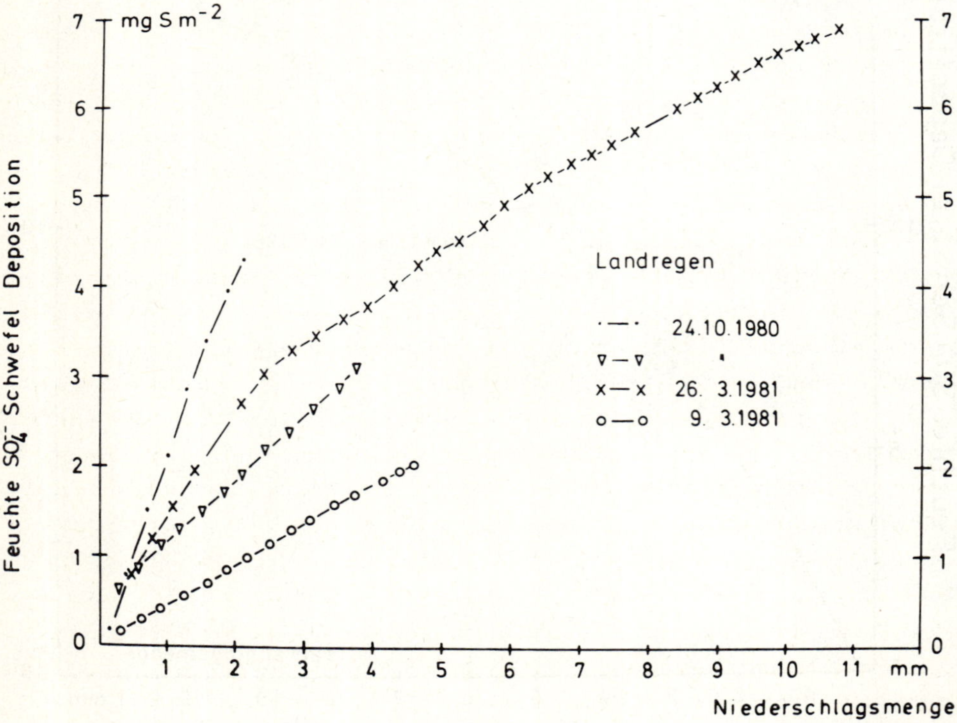

Fig. 6: Cumulative wet sulfur deposition during individual cases of rainfall.

air-parcels, the accumulation of trace-substances therein and the dry period before the onset of rain.

Comparison of the values of wet sulfate-deposition in our country with relevant data from Sweden shows a maximum of sulfate-deposition in the South-West of the Sweden with 1.2 g $Sm^{-2}yr^{-1}$ and a minimum towards the North with less than 0.3 g $Sm^{-2}yr^{-1}$. It has been argued by Swedish scientists that the deposition of sulfur should not exceed 0.5 g $Sm^{-2}yr^{-1}$ in order to substancially protect the vegetation in particular the forests.

5.2 Nitrate

The regional distribution of the wet nitrate-deposition presents a similar picture as the sulfate-deposition. Increased deposition rates are observed at mountain stations. The cumulative distribution of wet nitrate deposition shows a quasi-linear increase with the precipitation-rate. Elevated nitrate depositions at the station Essen in comparison to Deuselbach are produced by higher nitrate-concentration in the Ruhr-area (annual average Essen 0.7 mg N/l, annual average Deuselbach 0.48 mg N/l). When we take Essen as an example for a polluted area we can assume an average wet deposition of 1.9 mg N $m^{-2}d^{-1}$ as representative, while for Deuselbach (Hunsrueck) as a rather unpolluted area an average wet nitrate-deposition of 1.2 mg N $m^{-2}d^{-1}$ is representative. At the mountain station Hohenpeissenberg and Kl. Feldberg/Taunus the nitrate-deposition is higher in accordance with the higher precipitation-rate. However, in contrast to the sulfate deposition, the nitrate deposition on Mt. Kleiner Feldberg does not exceed that on Hohenpeissenberg. The average annual wet nitrate deposition in the Federal Republic of Germany amounts to 0.4 to 0.8 g N $m^{-2}yr^{-1}$. This average is in agreement with estimates by Bonis et al. (15) who note a maximum nitrate deposition of 0.7 g N $m^{-2}yr^{-1}$ for Middle Europe. A comparison between wet sulfate and nitrate deposition shows that the wet sulfate deposition dominates by a factor 2-3.

In this connection the long-term trend of nitrate deposition is of interest which was evaluated by Brimblecombe et al. (16) for the station Rothamsted in the South of England. It shows that during the last 100 years the nitrate deposition rose from 0.1 to 0.3 g N $m^{-2}yr^{-1}$ with particularly high increase during the summer months. The authors discuss the increased use of fertilizer during that period and its contribution to the nitrate deposition. The nitrate deposition in Sweden is considerably lower than the average for Germany. For the South of Sweden values of 0.35 g N $m^{-2}yr^{-1}$ are given and only 0.07 g N $m^{-2}yr^{-1}$ for North Sweden. In a very early paper on precipitation chemistry Mordy (17) investigated the chemical composition of rainwater in Hawaii. He observed that the accumulated amount of nitrogen from the beginning of rainfall over a period of two hours increases rather steady while the rate of rainfall accumulation shows significant variations.

Following the ideas of Mordy, A.N. Hogan (18) recently investigated the wet deposition during individual cases of rainfall. He analyzed the precipitated amount of ions with respect to the precipitated amount of rainwater. While the individual fluctuations of the trace-substances in rainwater are considerable during the storms, the total amount of ions precipitated was found to be in proportion to the total water precipitated. The author concludes that in two cases he evaluated the chemical precipitation rate was greater in the convective rain following frontal passage than in prefrontal rain.

5.3 Chloride

Due to the influence of sea-salt in the coastal region of the Federal Republic of Germany, the regional distribution of wet chloride deposition shows a remarkable difference to the sulfate and nitrate deposition. It should be noted that the chloride-deposition in the Ruhr area is nearly as high as the coastal values. It is emphasized that only a fraction of the chloride in rain originates from HCl, while a varying quantity of Cl^- in rain originates from sea-salt particles. We can assume that in industrial areas, anthropogeneous chloride-sources are predominant, sea-salt there being of minor importance.

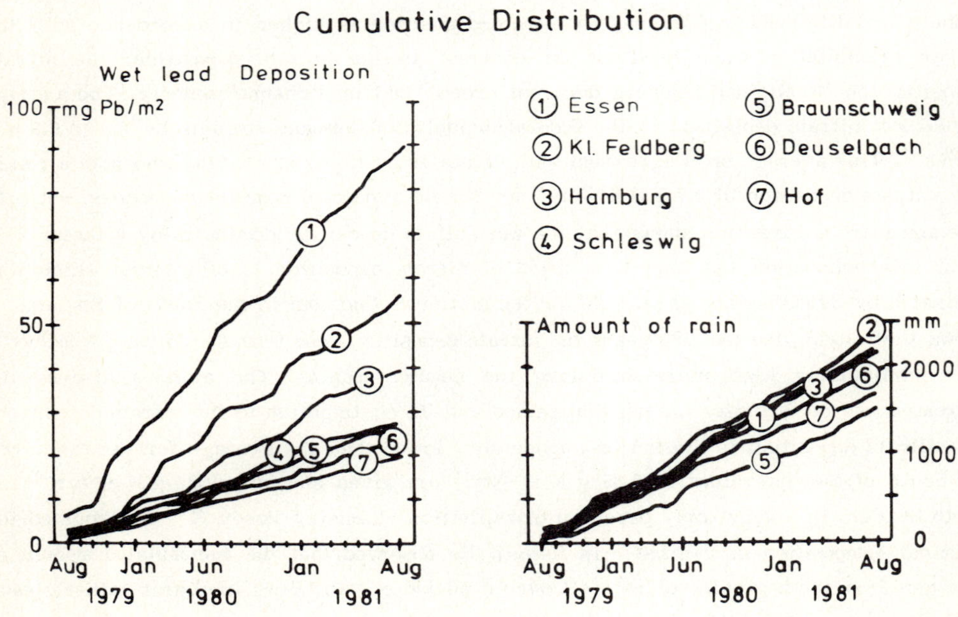

Fig. 7: Cumulative wet deposition of lead.

5.4 Heavy metals

In contrast to sulfate, nitrate and chloride, heavy metals are existent predominantly as aerosol particles. The gas phase is of minor importance and also reactions with other components leading to a transformation of the heavy metal aerosols need not to be considered. Therefore, the removal of the heavy metal aerosols from the atmosphere to the ground is restricted to wet removal by precipitation and dry removal by direct deposition of lead with strong regional differences which are not related to the precipitation-rate but rather to differences in the ambient concentrations. It can be seen that the rainfall rate at the seven stations entered into the figure differs between 1500 mm and 2500 mm for the two-year period, the Pb-deposition, however, differs between 20 and 90 mg Pb m^{-2}, an indication of the rather inhomogeneous distribution of lead in the atmosphere. Lead and cadmium are removed from the atmosphere preferentially by wet deposition, while for manganese and iron different deposition pattern is observed. These metals, mostly bound on coarse particles, are preferentially transferred to the ground by dry deposition. In context with ecological

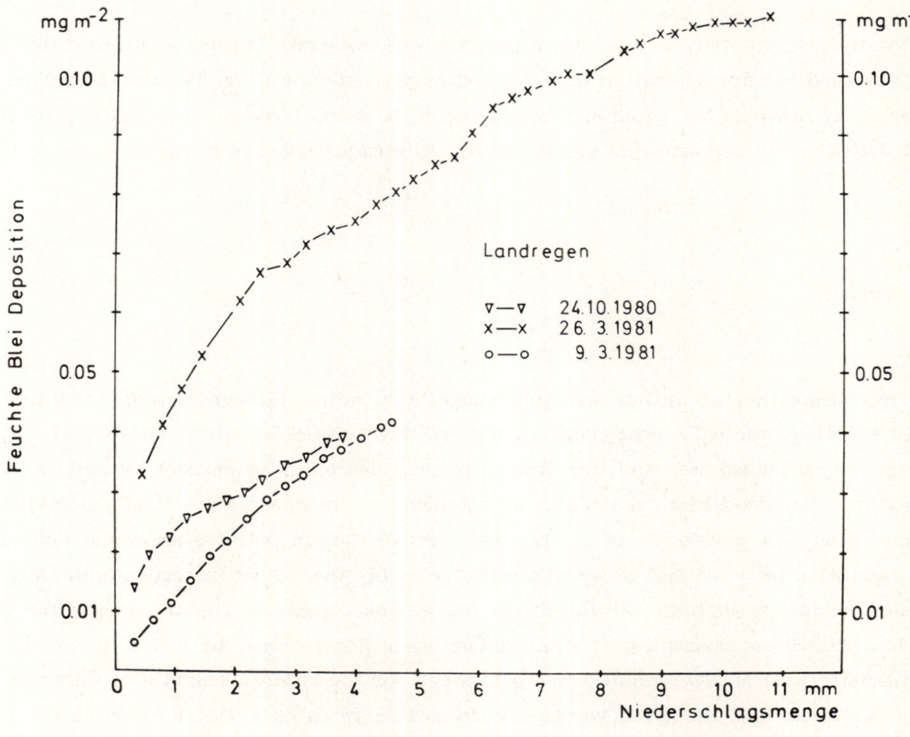

Fig. 8: Cumulative lead deposition during individual cases of rainfall.

hazards the solubility of lead and cadmium should be regarded. It appears from our investigations that the deposition pattern of lead and cadmium follows generally the precipitation pattern while the deposition of manganese and iron is more restricted to the vicinity of sources. It is interesting to note that the lead deposition on Mt. Kleiner Feldberg exceeds the deposition in Frankfurt as a result of the high precipitation rate on the Taunus mountain.

The cumulative deposition of lead during individual cases of rainfall was measured with a sequential deposition sampler. Again, as was shown for sulfate, the deposition rate increases strongly during the first 3 mms of rainfall (for instance on 26 March 1981, Fig. 8) reaching half of the total deposition rate, leaving an equal rate of Pb-deposition for the remaining 8 mm of rainfall. It is emphasized that with the onset of a rainfall a high amount of trace-substances is transferred to the ground and that the first phase of rain transports the dominant fraction of pollutants to the ground.

6. DRY DEPOSITION OF ACIDIFYING IONS AND HEAVY METALS

During the present study wet and dry deposition was measured. It must be stressed that the sampling method for dry deposition is restricted to particulate matter. The dry deposition of SO_2, NO_x and other gases could not be assessed. It must, however, be realized in this context, that the dry deposition of gases constitutes a major fraction of the total deposition.

6.1 Sulfate

The dry deposition of sulfate-aerosols suggests a rather homogeneous distribution of airborne sulfate particles. A comparison of wet and dry sulfate-deposition shows that only 9 to 29% of the combined wet and dry deposition are deposited as aerosol by dry fallout. Estimated dry SO_2-deposition, calculated on the basis of measured ambient SO_2-concentrations and a deposition velocity of 0.8 cm sec^{-1} shows that in polluted areas the total dry sulfur deposition (SO_2 + SO_4^{2-} aerosol) dominates by far the wet sulfur deposition. Only in unpolluted regions with high rainfall-rates the wet deposition of sulfur exceeds the dry deposition. With these assumptions a total sulfur deposition between 2.5 and 10 g S $m^{-2}yr^{-1}$ seems realistic with an average of 5 to 6 g S $m^{-2}yr$ for the Federal Republic of Germany. It must be mentioned that the given values are deposition-rates on a flat horizontal surface. The EMEP-model calculation for the period October 1978 to September 1980 revealed that 48% of the sulfur deposition results from sources within the Federal Republic of Germany, while

45% of the sulfur deposition come from sources in neighbouring countries and 7% are assumed to be background deposition from natural sulfur sources.

6.2 Nitrate

The dry nitrate deposition shows a variation between 0.15 mg N $m^{-2}d^{-1}$ in Schleswig and 0.5 mg N $m^{-2}d^{-1}$ in the Rhein-Main area.

The addition of wet and dry nitrate deposition amounts to values between 1.3 and 2.9 mg N $m^{-2}d^{-1}$. The fraction of dry nitrate-aerosol deposition amounts to 8 to 23% with a minimum in Schleswig (see Fig. 9). Bonis et al. included also the dry NO_2-deposition in their evaluations. On the basis of a deposition-velocity of 0.2 to 0.6 cm sec^{-1} the rate of the dry

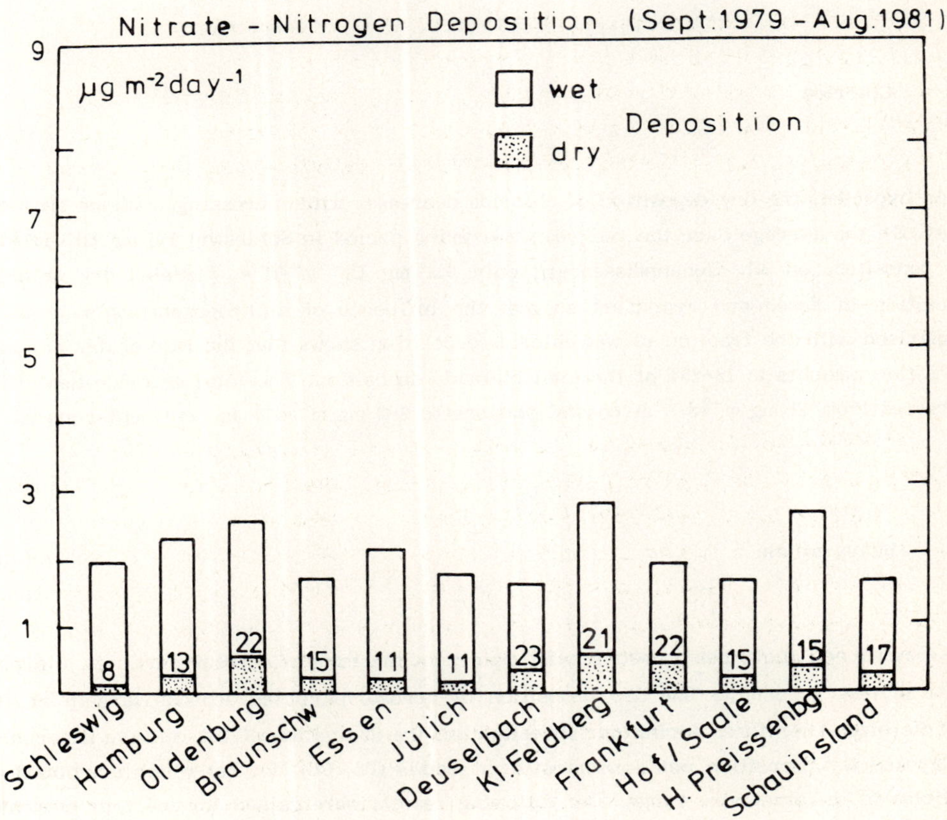

Fig. 9: Combined wet and dry nitrate-nitrogen deposition.

NO_2 deposition had the same order of magnitude as the wet deposition. Perseke (19) calculated the dry NO_2-deposition on the basis of an average NO_2-concentration of 50 ug m^{-3} in polluted areas and 10 ug m^{-3} in rural regions and reached a dry NO_2-deposition of 2.6 to 7.9 mg N $m^{-2}d^{-1}$ resp. 0.5 to 1.5 mg N $m^{-2}d^{-1}$. A comparison with the measured dry deposition of nitrate-aerosols shows clearly the dominance of dry NO_2-deposition in polluted regions. In background regions wet and dry deposition are of the same order of magnitude. With respect to the dry deposition of nitrogen-compounds there exist practically no measurements in other regions. However, it is assumed that dry nitrate deposition in rural areas comes partially from agricultural activities and the use of nitrate-containing fertilizer.

Boettger et al. (20) integrated the total global nitrate deposition over latitudinal belts using precipitation-chemistry measurements. They gained a semi-quantitative overview with maxima of deposition in tropical regions and in continental temperate latitudes. Obviously, the nitrate-deposition was less pronounced over the oceans.

On the whole, the nitrate deposition is of reduced significance compared to the sulfur-deposition. But both contribute to the acidity of precipitation.

6.3 Chloride

As expected, the dry deposition of chloride decreases with increasing distance from the coast. On the average over the two-year measuring period in Schleswig 1.6 mg Cl^- $m^{-2}d^{-1}$ are deposited, on Mt. Hohenpeissenberg only 0.3 mg Cl^- $m^{-2}d^{-1}$. Elevated dry chloride deposition in Essen and Frankfurt suggest the influence of anthropogeneous sources. A comparison with the fraction of wet chloride deposition shows that the rate of dry chloride deposition amounts to 15-25% of the total chloride deposition. The total chloride deposition decreases from 12 mg $m^{-2}d^{-1}$ at coastal stations to 2-3 mg $m^{-2}d^{-1}$ in continental areas.

6.4 Heavy metals

As mentioned above heavy metals exist mainly in the form of aerosol-particles, the gas-phase is less important. The dry deposition of aerosol-particles depends largely on the particle-size. Therefore, during our investigation the evaluation of the relative importance of dry and wet deposition was concentrated on metals (Pb, Cd, Mn, Fe) which are bound on particles of different size range. The following results were gained for the four elements: lead is mainly observed in the submicron particle range. It is preferentially removed by wet deposition. Fig. 10 shows that the regional distribution is rather inhomogeneous with maxi-

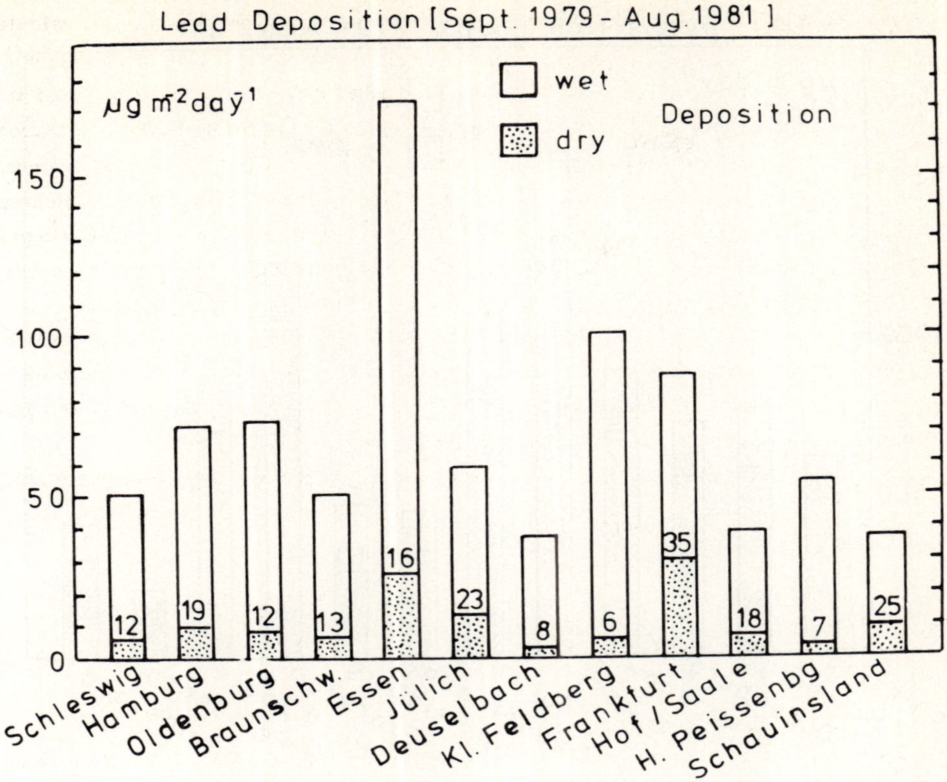

Fig. 10: Combined wet and dry lead deposition.

mum values in areas with high automobile traffic. In unpolluted areas dry deposition of lead amounts to 6-8% of the total deposition. In areas with high traffic like Frankfurt/Main the fraction of dry deposition is higher and reaches 35% of the total deposition. For the Federal Republic of Germany average total lead deposition rates between 40 μg Pb m^{-2} d^{-1} in unpolluted regions and up to 100 μg Pb m^{-2} d^{-1} in the Rhein-Main area were found. These values are below the threshold-limits suggested by the draft of TA Luft with 500 μg Pb m^{-2} d^{-1}. For cadmium, again the wet deposition is the dominant sink, dry deposition accounts to 20% or less of the total deposition. Only in urban areas, like Frankfurt, the dry cadmium deposition is increased to 30%. Cadmium is the most volatile of the heavy metals analyzed and exists to a certain degree also in the gas-phase under atmospheric conditions. Total cadmium depositions were measured in the range 1-4 μg Cd m^{-2} d^{-1}. It appears that the suggested deposition-threshold for cadmium with 7.5 μg m^{-2} d^{-1} is not reached at the stations in our network. Manganese and iron show a different deposition pattern. Manganese and iron-aerosols occur mainly in the coarse particle mode. They are therefore more effectively removed from the atmosphere by dry deposition. For manganese both removal processes are

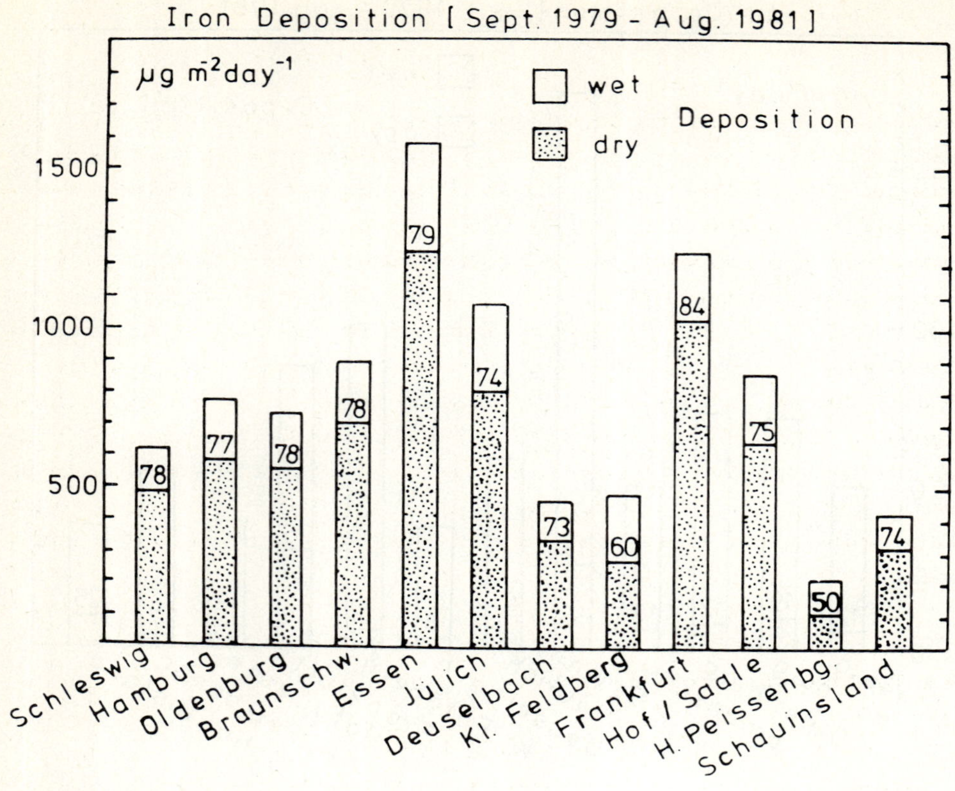

Fig. 11: Combined wet and dry iron deposition.

of equal importance. Only at mountain stations with high precipitation rate, wet manganese deposition dominates. Iron containing particles are coarse. The majority (>70%) is removed by dry deposition (Fig. 11). the total iron deposition varies between 200 and 500 μg Fe $m^{-2}d^{-1}$ in unpolluted areas and 80 to 1500 μg Fe $m^{-2}d^{-1}$ in urban and industrialized area. The investigations in the Federal Republic of Germany show that wet and dry deposition are of different importance for atmospheric metal aerosols. Metals bound to submicron particles are primarily removed by precipitation while coarse metal aerosols are mainly removed by dry deposition. From this follows that manganese and iron particles are primarily deposite in the vicinity of their sources. In contrast to that, the deposition pattern of the toxic lead- and cadmium particles follows the precipitation-pattern. Also the solubility of lead and cadmium has to be taken into consideration. These examples show clearly that in the case of particle deposition the size-range of particles becomes a third important factor besides precipitation rate and distribution of sources.

7. DEPOSITION ON VEGETATION

Measurements of wet and dry deposition have been carried out preferentially under standardized conditions. In reality the structure of the surface influences the mode and rate of deposition. Our knowledge of the mechanism of deposition and on the transport of trace substances to natural surfaces like vegetation is insufficient.

Measurements of aerosol- and precipitation chemistry were made in the Solling-forest by Hoefken et al. (9). Chemical analysis was made of samples collected above and below the canopy. The results show that the concentration of ions and elements in the throughfall was significantly higher (1.5 to 41.2 times) than in the precipitation above the forest. This was particularly the case for sulfate and manganese. The enrichment of the precipitation during throughfall is due to aerosol that was deposited on the branches of the trees during the dry period. Only in the case of manganese a considerable leaching was observed, so that the enrichment of Mn in rain is due to both dry deposition and leaching. Table 4 shows the total deposition above (a) and below (b) the canopy of spruce and beech-forest in the Solling-mountains.

Table 4: Total deposition above (a) and below (b) canopy

Kg ha^{-1} yr^{-1}

		Na	K	Ca	Mg	Cl	S
Solling spruce	a)	7.7	3.3	10.0	1.9	16.9	23.8
	b)	17.6	30.1	34.6	5.0	39.7	88.6
Solling beech	a)	7.7	3.3	10.0	1.9	16.9	23.8
	b)	14.6	29.0	26.9	27.0	33.3	53.4

The soil of the spruce forest received a free acidity input via throughfall about 5 times as high as was brought down in rainwater above the canopy (Gravenhorst et al., (21)). These measurements in the Solling-forest showed that the free proton concentration under the spruce canopy is by a factor of 4.8 ± 1.0 higher than above the canopy, indicating an enormous additional acidification when the rainwater passes through the canopy. In general, the dry deposition of gaseous and particulate matter is much more effective to a spruce than to a beech tree, demonstrating that a spruce forest is an effective filter for the air passing it. One major component of the dry deposition is gaseous and particulate sulfur. We can therefore assume that the dry deposition of sulfur-components is predominantly responsible for the additional acidification of wet throughfall in a spruce-canopy. It should be stressed that it is essential to analyze the throughfall besides the rainfall in an open area in order to assess the input of acid components in a spruce forest. The investigation in the Solling shows that the forest is an effective sink for atmospheric aerosols. The measurements in the Solling-forest showed that a significant fraction of the trace-substances transferred from the atmosphere to the ground were picked up by the branches and needles of the trees through the mechanism of interception. Interception includes the filtering of fog and haze droplets. The effect of interception of fog-, cloud- and haze droplets on e co- systems has not yet been investigated. This is of particular interest since damage to forest-ecosystems has been reported from mountain-regions with frequent occurence of fog and low clouds.

Continuing our previous investigations on wet and dry deposition of acidic components and heavy metal-aerosols we have installed several measuring-sites in mountain areas in Hessen (Taunus, Vogelsberg, Kaufunger Wald) where we have started to take fog-water samples besides wet and dry deposition. For this purpose a special designed fog-water collector was developed on the basis of an all-plastic omnidirectional cloud-water collector designed by Mohnen (22). Preliminary results of a test-series during spring 1983 revealed sulfate-nitrate and chloride-concentrations in fogwater several times higher than in the average rainwater in the Taunus region. These investigations will be intensified since it is assumed that the mechanism of interception is of importance in connection with the observed forest-damages and that acidic fog is an important part of the acidic deposition phenomenon.

ACKNOWLEDGEMENT

This study has been supported by the Umweltbundesamt which is gratefully acknowledged.

REFERENCES

(1) Perseke, C., S. Beilke, H.-W. Georgii, Die Gesamtschwefeldeposition in der Bundesrepublik Deutschland aufgrund von Messdaten des Jahres 1974., Berichte, Inst. f. Met. u. Geophys. Univ. Frankfurt/M, 40, 1980.

2) Grunow, J., Der Niederschlag im Bergwald-Niederschlagzurueckhaltung und Nebelzuschlag, Forstwiss. Cbl. 74, 1955, pp 21-36

3) Grunow, J. und H. Tollner, Nebelniederschlag im Hochgebirge, Arch. Met. Geophys. Biokl. B 17, 1969, pp 201-228

4) Chamberlain, A.C., Aspects of the deposition of radioactive and other gases and particles, Int. J. Air Poll. 3, 1960, pp 63-88

5) Garland, J.A., Field measurements of the dry deposition of small particles to grass, In: H.-W. Georgii and J. Pankrath (eds.) Deposition of atmospheric pollutants, D. Reidel Publ. Comp., Dordrecht, 1982, pp 9-17

6) Perseke, C., Composition of acid rain in the F.R. Germany - Spatial and temporal variations, In: H.-W. Georgii and J. Pankrath (eds.) Deposition of Atmospheric Pollutants, D. Reidel Publ. Comp., Dordrecht, 1982, pp 77-87

7) Garland, J.A. and Branson, J.R., The deposition of sulfurdioxide to pine forest assessed by radio active tracer-method, Tellus 29, 1977, pp 445-454

8) Pruppacher, H.R. and J.D. Klett, Microphysics of clouds and precipitation, D. Reidel Publ. Comp., Dordrecht, 1978

9) Hoefken, K.D., H.-W. Georgii, G. Gravenhorst, Untersuchung ueber die Deposition atmosphaerischer Spurenstoffe an Buchen- und Fichtenwald, Berichte Inst. f. Met. u. Geophys. Univ. Frankfurt/M, 46, 1981

10) Georgii, H.-W., C. Perseke, E. Rohbock, Feststellung der Deposition von sauren und langzeitwirksamen Luftverunreinigungen aus Belastungsgebieten, Abschlussbericht zum Forschungsprojekt 104 02 600 im Auftrag des Umweltbundesamtes, 1982

11) VDI, Richtlinie zur Messung partikelfoermiger Niederschlaege, VDI 2119, (1971)

12) Wilson, J., V. Mohnen, J. Kadlecek, Wet deposition in the northeastern United States, ASRC-Publication 796, Albany, N.Y., 1980

13) Meszaros, E., G. Varhelyi, L. Haspra, On the atmospheric sulfur budget over Europe, Atm. Environment 12, 1978, pp 2273-2277

14) Kins, L., Temporal variation of chemical composition of rainwater during individual precipitation events, in: H.-W. Georgii and J. Pankrath (eds.) Deposition of atmospheric pollutants, D. Reidel Publ. Comp., Dordrecht, 1982, pp 87-96

15) Bonis, K., E. Meszaros, M. Putsay, On the atmospheric budget of nitrogen compounds over Europe, Idoejaras 84, 1980, pp 57-67

16) Brimblecombe, P. and J. Pitman, Long term deposit at Rothamsted, Southern England, Tellus 32, 1980, pp 261-267

17) Mordy, W.A., A note on the chemical composition of rainwater, Tellus 5, 1953, pp 470-474

18) Hogan, A.W., Some characteristics of chemical precipitation, Tellus 35B, 1983, pp 121-130

19) Perseke, C., Die trockene und feuchte Deposition saeurebildender atmosphaerischer Spurenstoffe, Berichte Inst. f. Met. u. Geophys. Univ. Frankfurt 48, 1982

20) Boettger, A., D. Ehhalt, G. Gravenhorst, Atmosphaerische Kreislaeufe von Stickoxiden und Ammoniak, Berichte der KFA Juelich 1558, 1978

21) Gravenhorst, G., K.D. Hoefken, H.-W. Georgii, Acidic Input to a beech and spruce forest, in: S. Beilke and A.J. Elshout (eds.) Acid deposition, D. Reidel Publ. Comp., Dordrecht, 1983, pp 155-172

22) Mohnen, V., Cloudwater collection from aircraft, Atmospheric Technology, NCAR 12, 1980, pp 20-25

ANTARCTIC PRECIPITATION CHEMISTRY

Robert J. Delmas
Laboratoire de Glaciologie et Géophysique de l'Environnement
B.P. 96 38402 St Martin d'Hères Cedex
France

ABSTRACT

Several thousand snow samples collected at various Antarctic locations have been analyzed after melting by ion chromatography and acid base titrimetry in order to obtain the chemical composition of past and present Antarctic precipitation. In particular the ionic balance of major impurities has been obtained for many samples. The concept of ionic budget is presented along with the way to calculate it and the various parameters which are involved, with special attention paid to acidity. The calculations are illustrated by results obtained at five selected Antarctic sites (South Pole, Dome C, Vostok, D 57 and James Ross Island), and for various time periods (from the present back to the last glaciation).

The predominant role played by gas-derived atmospheric acids (H_2SO_4, HNO_3 and HCl) is emphasized.

INTRODUCTION

Antarctica (Figure 1) presents certain special characteristics valuable for environmental studies. Located at high Southern latitudes and isolated far from other continents (the nearest South America foreland lies some 1000 km away from the Antarctic Peninsula), this vast continent (14 x 10^6 km^2) can be considered as an island almost entirely covered by ice. The thickness of the ice sheet reaches more than 4000 m in the central areas. The rare ice-free land surface support virtually no vegetation. Finally, human presence is very recent and still extremely limited. Moreover, more than 90% of the world's industries are located in the Northern Hemisphere and Antarctica may therefore be considered as a virgin continent with respect to the ever growing level of interactions between Man and his Environment.

In a general and simplified manner, atmospheric aerosol is generated by four main sources:

NATO ASI Series, Vol. G6
Chemistry of Multiphase Atmospheric Systems
Edited by W. Jaeschke

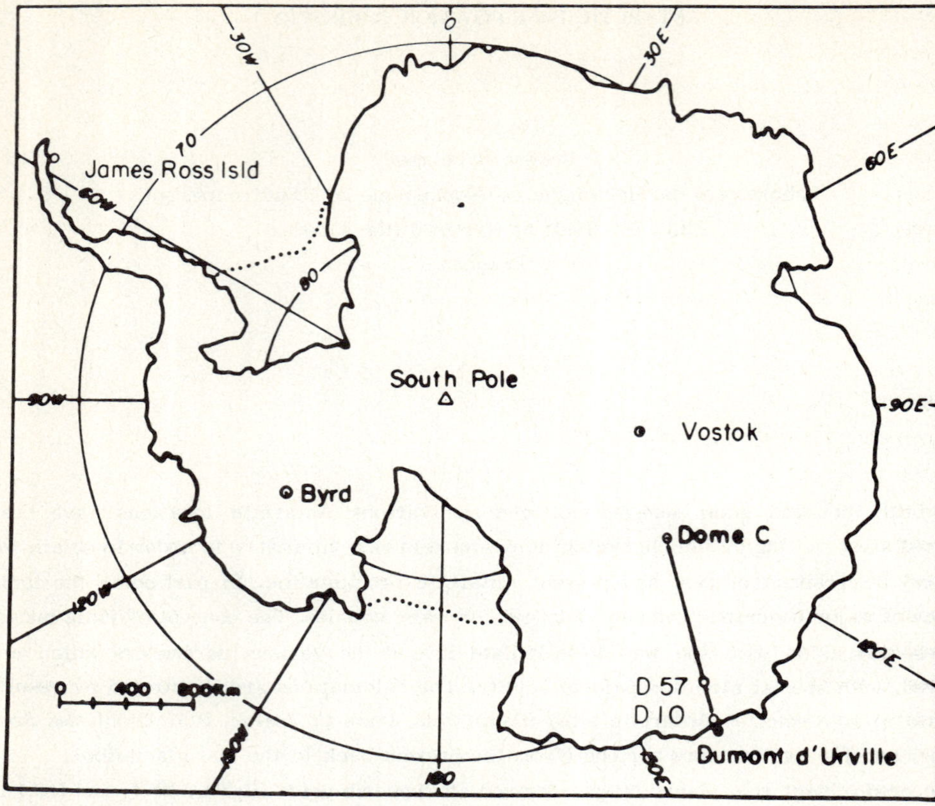

Fig. 1: Map of the Antarctic showing the location of the Antarctic stations studied: South Pole, Dome C, Vostok, James Ross Island and D 57.

First of all, there are the two sources of primary aerosols: continental surfaces (in particular desert areas) and marine surfaces.

Then there is the source of secondary aerosol linked to the in situ transformation of certain atmospheric trace gases such as SO_2 and NO_x (which lead to the two mineral acids H_2SO_4 and HNO_3, partially neutralized by ammonia). Finally, there is the increasingly important source (especially in the Northern Hemisphere) of aerosols created by human activities (dust and gases).

The geographical location of the Antarctic continent and the present day features of atmospheric circulation are such that the only noticeable contributions to atmospheric aerosol loading in these regions are due to gas to particle conversion and sea spray (Delmas et al., 1982a). Moreover when moving inland from coastal areas, the marine component declines considerably. The central Antarctic Plateau is therefore a well suited location for studying the baseline aerosol. In these regions "the background of the backgrounds" of the global atmospheric aerosol may be obtained (Cunningham and Zoller (1981), Delmas and Gravenhorst (1982)).

Glaciochemistry is the study of the chemical information recorded in polar snow and ice in relation to atmospheric composition. The snow data deposited regularly on the surface of the ice cap over many thousands of years represent archives of the chemical composition of the local atmosphere. Because of the low temperatures over the ice sheet, the chemical records have remained unchanged since the time of snow deposition. If we consider that the Antarctic atmosphere is permanently linked to the global environment, the chemical analysis of samples of Antarctic fossil ice should in principle provide information on all events that have affected the terrestrial atmosphere in the past (in particular climatic changes with their consequences on atmospheric circulation and terrestrial or marine biologic acitivity, as well as great natural cataclysms such as volcanic eruptions or falls of giant meteorites). The transfer function between air and snow is a matter of great discussion and is surely an important parameter to take into account in glaciochemical studies. A priori, this function may vary from one site to another, from one compound to another, or according to climatic or meteorological conditions, etc... Unfortunately related knowledge is still very incomplete and glaciochemists presently tend to neglect this parameter and to consider that the chemical composition they obtain in snow layers satisfactorily reflects the composition of the atmospheric aerosol at the place and time of the formation of the precipitation. Recent studies (Lambert et al., 1983; Pourchet et al., 1983) seem to confirm the theoretical consideration made by Junge (1977) regarding the good representativeness of snow with respect to the atmosphere, but further measurements are needed.

Knowledge of the Antarctic aerosol itself is very inadequate. The most recent measurements made on Antarctic aerosol include those of Cunningham and Zoller (1981) based on air filtration at South Pole and those of Shaw (1979) made by optical methods. However certain acid compounds of importance in the ionic budget of snow (such as HNO_3 and HCl) are not taken into account in these types of measurements.

Sample collection is carried out during field work programmes either in snow pits (firn samples down to about 10 m depth) or with the aid of glaciological drillings. A sophisticated methodology has been established for sample collection and several thousands of Antarctic snow and ice samples have been analyzed over recent years for major ion chemistry at our Laboratory in Grenoble. The results reported in this paper were obtained together with M. Legrand, F. Zanolini, A. Aristarain, M. de Angelis and J.M. Barnola.

EXPERIMENTAL METHODS

In a somewhat arbitrary manner, we can distinguish between major and minor impurities. Major impurities ($SO_4^{=}$, NO_3^-, Cl^-, Na^+, NH_4^+, Mg^{++} and H^+) are those with concentrations higher than 0.1 µEq. l^{-1}. Minor impurities have concentrations lower than this figure and have no significant effect on the ionic budget of ice (generally in the range 5-10 µEq. l^{-1}). The case of insoluble microparticles will not be discussed here.

The analysis of minor components requires many precautions and particularly sophisticated methods. The determination of major components also presents certain difficulties and requires the use of special sampling and measurement techniques (Legrand et al., 1984).

Recent snow samples (covering the last few decades) are collected in pits with the aid of precleaned 30 ml plastic vials ("Accuvettes" sold by Coulter) which are pushed directly into the pit wall when the snow is sufficiently soft or used to scrape the snow surface when firn becomes too hard. All samples are placed in double sealed plastic bags and kept frozen until their analysis in Grenoble. Deep firn and ice cores may be collected by electromechanical or thermal augers. The risk of contaminating the cores is particularly high for firn because easily dust penetrates into the open pores of this material. This problem does not exist for ice, where contamination generally remains superficial and can be removed by rinsing with pure water or by a specially designed thermal probe (Legrand et al., 1984) which melts only the central part of the ice core. For contaminated snow and firn cores, a recoring device (a small drill with an internal diameter of 2 cm) is used in a clean air bench located in a cold room (- 20°C).

Major ion concentrations (except the proton concentration which is titrated by a special procedure, Legrand et al., (1982) and Mg^{++} which is calculated from Na^{+} data) are measured by ion chromatography (Dionex 10) the best analytical method for ionic concentrations in the

Site	Coordinates	Elevation (m. a.s.l.)	Mean annual temperature (°C)	Snow accumulation rate g cm^{-2} a^{-1}
South Pole	90° S	2800	- 51	8.5
Dome C	74°42'S, 124°04'E	3240	- 54	3.6
Vostok	78°30'S, 106°54'E	3420	- 56	2.3
D 57	68°11'S, 137°33'E	2050	- 32	20
James Ross Island (Dome Dalinger)	64°13'S, 57°38'W	1640	- 14.5	60

Table 1: Geophysical and physical data of the Antarctic stations studied.

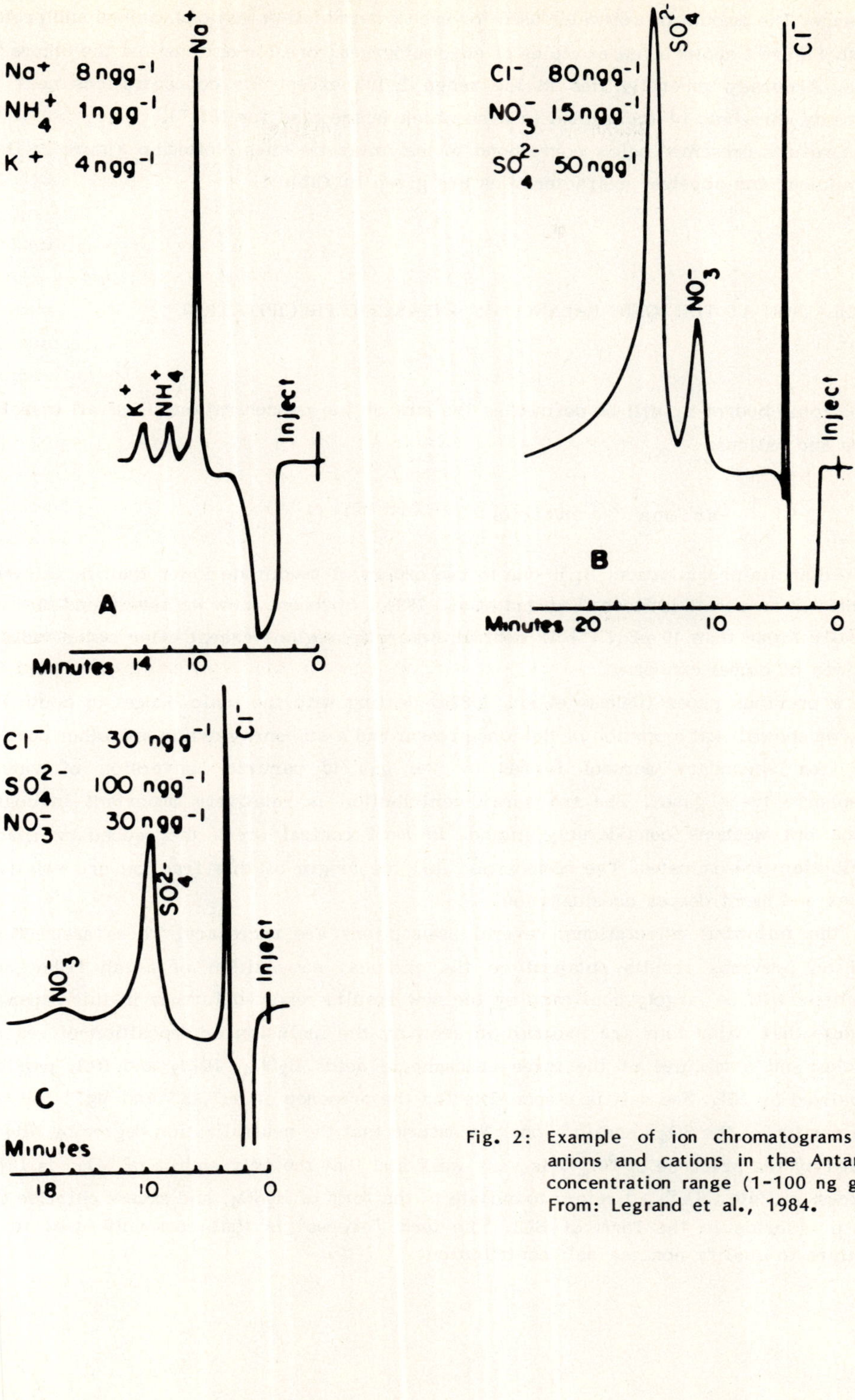

Fig. 2: Example of ion chromatograms for anions and cations in the Antarctic concentration range (1-100 ng g^{-1}) From: Legrand et al., 1984.

ppb range. The amount of meltwater used for each determination (cations, anions and proton) is 5 ml. Figure 2 shows a few examples of chromatograms: one for cations and the others for anions. Accuracy generally lies in the range $\pm$ 10% except for concentrations near the sensitivity threshold of the method (a case which is frequent for NH_4^+).

The results presented below correspond to the Antarctic sites plotted in Figure 1. Their geographical and physical characteristics are given in Table 1.

CALCULATION AT THE IONIC BALANCE OF ANTARCTIC PRECIPITATION

The ionic budget B_i will be defined as the sum of the concentrations C of all dissolved anions and cations:

$$B_i = C_{anions} + C_{cations} \qquad (\text{in Eq. } l^{-1})$$

In Antarctic precipitation, B_i is one to two orders of magnitude lower than in rainwater collected at temperate latitude (Legrand et al., 1984). In recent snow we have found that it is generally lower than 10 μEq. l^{-1} in central Antarctic regions except after major volcanic eruptions of global concern.

In a previous paper (Delmas et al., 1982a) dealing with the ionic budget in South Pole snow, we showed that a portion of the ions present had a sea-spray origin and other portions were from secondary aerosol formed by the gas to particle conversion of various atmospheric trace gases. The sea spray contribution is relatively important in coastal regions but weakens considerably inland. In most central areas the secondary aerosol contribution predominates. The composition and the origin of this fraction are relatively complex and merit deeper consideration.

In the following calculations, several assumptions are necessary. These assumptions, based on previous results obtained on the chemical composition of South Polar snow impurities, will be largely confirmed by the new results reported further in this paper. We postulate that major ions are included in snow by the inclusion or deposition of sea salt particles and a mixture of the three atmospheric acids H_2SO_4, HNO_3 and HCl, partially neutralized by NH_3. Sea salt is responsible for the presence of Na^+, K^+ and Mg^{++} ions and for a portion of the $SO_4^=$ and Cl^- ions. We assume that the neutralization degree by NH_3 (as attested by the presence of NH_4^+) is very weak and that the only source of NO_3^- is HNO_3.

Excess sulfate ($SO_4{}^={}_{ex.}$) refers to sulfate in the form of H_2SO_4, and excess chloride ($Cl^-{}_{ex.}$) to chloride in the form of HCl. The term "excess" is that commonly used in the literature to qualify non sea-salt contribution.

We can write (concentrations in $\mu Eq\ l^{-1}$):

$$[Na^+]+[K^+]+[Mg^{++}]+[NH_4^+]+[H^+] = [SO_4^=]+[NO_3^-]+[Cl^-] \quad (1)$$

(ionic balance)

$$[H^+] = [SO_4^+{}_{ex.}]+[NO_3^-]+[Cl^-{}_{ex.}] \quad (2)$$

(acidity composition)

$$[SO_4^={}_{ex.}] = [SO_4^=]-0.12[Na^+] \quad (3)$$

$$[Cl^-{}_{ex.}] = [Cl^-]-1.17[Na^+] \quad (4a)$$

$$[Na^+{}_{ex.}] = [Na^+]-0.85[Cl^-] \quad (4b)$$

The values of 0.12 and 1.17 appearing in equations (3) and (4a) are those of the ratios $[SO_4^=]$ / $[Na^+]$ and $[Cl^-]$ / $[Na^+]$ in sea water, respectively (Stumm and Morgan, 1970). 0.85 is the reciprocal of 1.17.

Equations 4a and 4b correspond to two consequences of the reactions of sea salt particles with sulfuric acid:

$$H_2SO_4 + 2NaCl \longrightarrow \overset{\nearrow}{2HCl} + Na_2SO_4 \quad (5)$$

It is assumed and was observed that this reaction may occur in the atmosphere with the formation of gaseous HCl. We recently showed that this process seems to occur in Antarctica (Legrand and Delmas, 1983). In certain areas there seems to be a preferential deposition of HCl (excess-Cl^-, equation 4a), at others an excess of sodium by the deposition of Na_2SO_4 (equation 4b). These two cases can be observed in Antarctic precipitation: the former at Dome C and South Pole, the latter a Vostok (see below).

We shall now illustrate these general statements by examples at different Antarctic locations.

IONIC BALANCE AND ACIDITY OF SNOW AT SOUTH POLE

Equations (1), (2), (3) and (4) are used to interpret the results of the chemical analysis of 100 samples collected in 1977 in a snow pit dug in the clean sector of the Amundsen Scott Station (South Pole). The continuous sequence studied (between 2 and 4 m depth) spans the years 1959 to 1969. This time period includes the large volcanic eruption of Mt. Agung (March 1963 in Bali) which significantly disturbed stratospheric chemistry for two years.

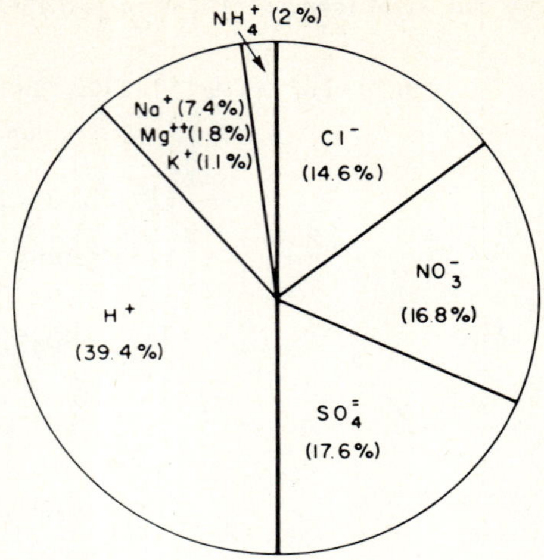

Fig. 3: Mean ionic distribution (% of ionic budget in µEq. l^{-1}) at South Pole From: Legrand et al. (1984)

The dating of snow layers is known accurately (Jouzel et al., 1983). Figure 3 gives the mean ionic distribution obtained for this set of data (Legrand and Delmas, 1984). This distribution is somewhat different from the one reported by Delmas et al. (1982a). The difference is due to errors in the NO_3^- and NH_4^+ data previously used in calculations.

The imbalance between anions and cations now lies within the sum of experimental uncertainties (~2% of the ionic budget), which is very satisfactory. Nearly 80% (39.4 x 2, see Fig. 3) of the ionic budget is linked to the atmospheric acids H_2SO_4, HNO_3 and HCl. This figure confirms our assumption that secondary aerosol contribution is dominant in central Antarctica. We can therefore simplify the checking of the ionic balance by considering only equation (2). Figure 4 shows the result of this calculation on each of the hundred samples analyzed. The calculated acidity (using equation (2)) fits the experimental results very well confirming the assumption relative to the composition of snow acidity.

During 1964 and 1965 (post-Agung years) the influence of the related H_2SO_4 fallout on the chemistry of snow is clearly visible. The concentration of total sulfate increased by a factor of three during these years with respect to the pre-Agung years. This demonstrates that strong volcanic eruptions occurring even at low Southern latitude may have a significant impact on Antarctic atmospheric chemistry (Delmas and Boutron, 1980).

IONIC BUDGET AT DOME C : AN EXAMPLE OF LONG TERM VARIATIONS

A detailed study of the chemical composition of Antarctic precipitation over the last century was made on a continuous profile samples in a snow pit dug at Dome C, a site located on the central Antarctic Plateau (Fig. 1). We calculated a mean ionic budget for two equal time periods of 25 years: 1955-1980 and 1870-1895 in order to see whether or not there had been a change in the local precipitation chemistry (Legrand et al., 1984). Each of these two periods include an explosive volcanic eruption, Agung (1963) and Krakatoa (1883) of the same order of magnitude. The two ionic balances given in Figure 5 lead to the following remarks:

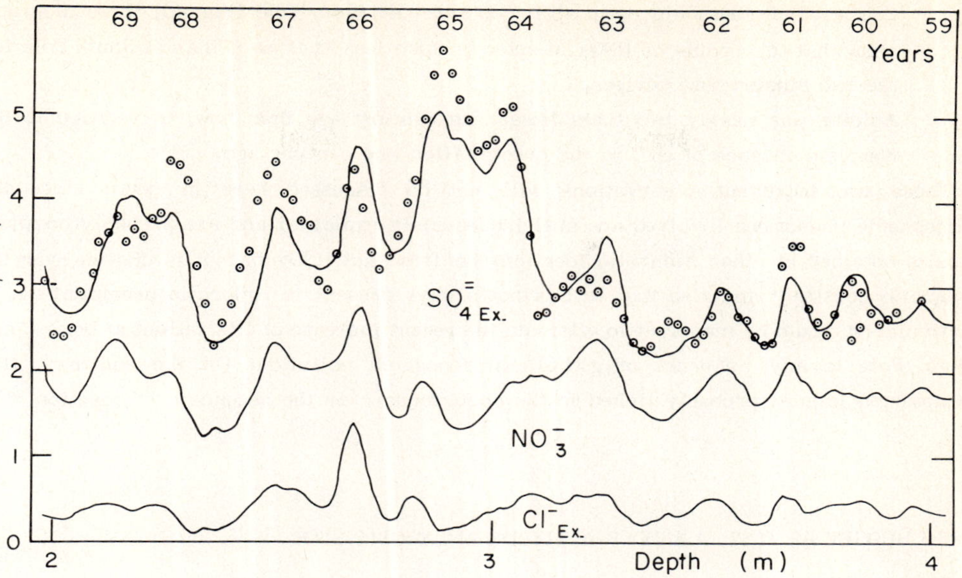

Fig. 4: Reconstitution of snow acidity at South Pole for the time period 1959-69. The curves were obtained by a 3-point running mean calculation. The bottom curve represents $[Cl^-_{ex.}]$, the middle one $[Cl^-_{ex.}] + [NO_3^-]$ and the top one $[Cl^-] + [NO_3^-] + [SO_4^{=}{}_{ex.}]$. Circular points are measured acidities. From Legrand and Delmas, 1984.

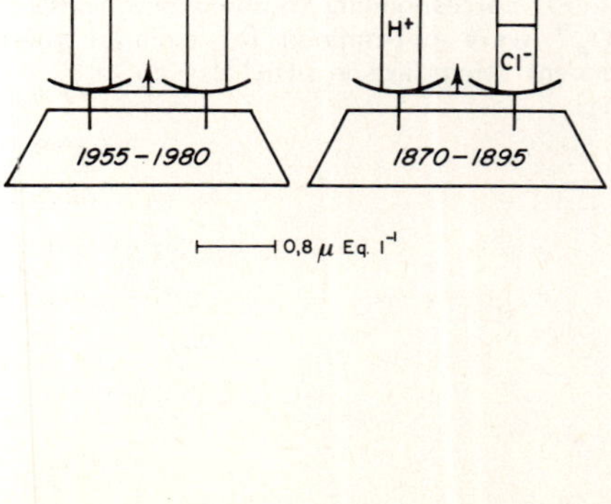

Fig. 5: Evolution of the mean ionic distribution for Dome C snow over the last 100 years for two 25 yr time periods (1955-1980 and 1870-1895). Note that the observed difference is mainly due to HCl content variations. From Legrand et al. (1984).

- Compared to the ionic composition obtained at South Pole (Fig. 4), these diagrams show that the amount of HNO_3 in snow is much less at Dome C than at South Pole for the two studied time periods,
- Acidity was nearly two times lower one century ago than now, in relation to the complete absence of HCl in the precipitation one century ago.

These two important observations will not be discussed here in detail since the atmospheric phenomena involved are still inadequately explained and are probably complex. Results obtained at other Antarctic locations confirm that HNO_3 fallout is uneven over the Antarctic continent and also that HCl is not always present in Antarctic precipitation. In particular it would be incorrect to attribute the recent increase of HCl fallout at Dome C and South Pole to any influence of global anthropogenic pollution. On the contrary, this phenomenon is most probably linked to the occurrence - or the absence - of reaction (5).

IONIC BUDGET AT D 57, A MIDDLE ALTITUDE ANTARCTIC SITE

The electroconductimetric method first developed by Hammer (1977) has been used to study a 200 m deep ice core recovered in Adelie Land at a site located 200 km from the coast (Zanolini, 1983). This method, performed directly on the ice core, is well suited to the rapid detection of exceptional acid levels without melting the ice and thereby destroying the ice core.

One of the goals of this study was to identify the acids responsible for the conductivity peaks. The pieces of ice which contained high conductivity levels were recored using a thermal probe and the meltwater obtained was analyzed for H^+, $SO_4^=$ and NO_3^-. Each individual sample corresponded to a fraction of a year so that each peak was studied in great detail. The acid is identified by correlating, for each of these peaks, the acidity ($[H^+]$) with

Fig. 6: Correlations obtained at D 57 (Terre Adelie) between acidity and nitrate (A) or sulfate (B) content for various acidity peaks detected at several depths along the ice core. Each event, covering about one year, was cut into "slices" corresponding to about one month of precipitation. The values of H^+, NO_3^- and $SO_4^=$ were determined for each of these "slices" after melting (from Zanolini, 1983). Concentrations are in $\mu Eq.l^{-1}$.
The equations of the straight lines are following:

A1: $[H^+] = 0,94\ [NO_3^-] + 1,9$
A2: $[H^+] = 1,02\ [NO_3^-] + 0,5$
B1: $[H^+] = 0,92\ [SO_4^=] + 2,67$
B2: $[H^+] = 0,75\ [SO_4^=] + 3,87$
B3: $[H^+] = 1,02\ [SO_4^=] + 1,67$

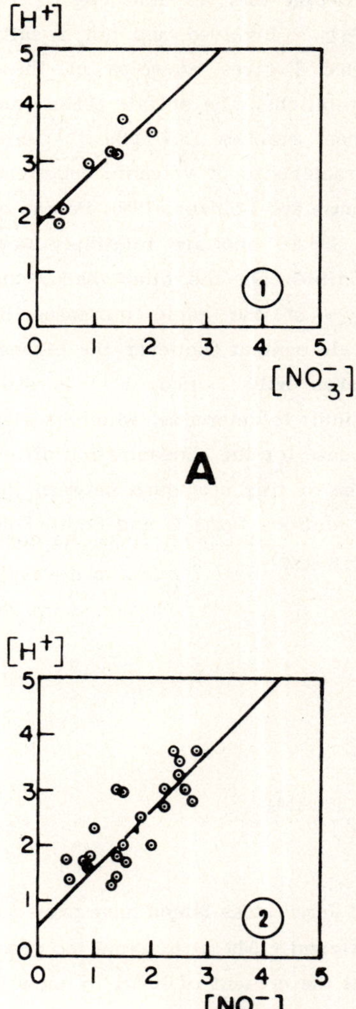
[H+]
[NO3-]
1
2
A

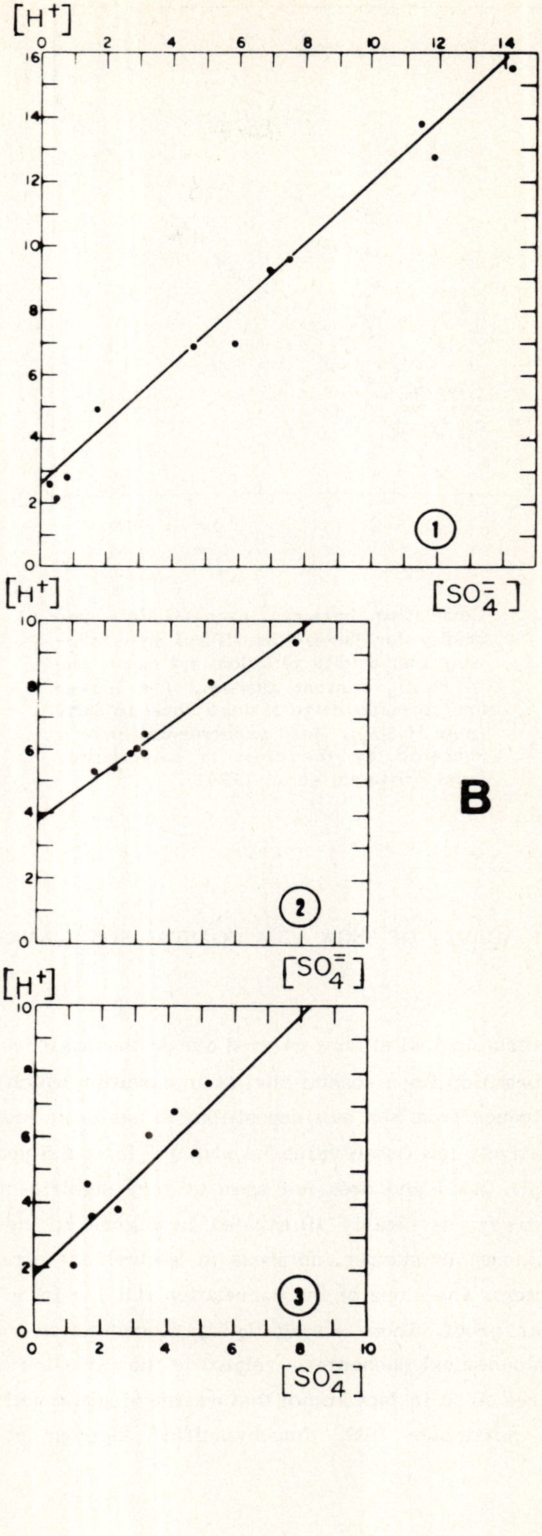
[H+]
[SO4=]
1
2
3
B

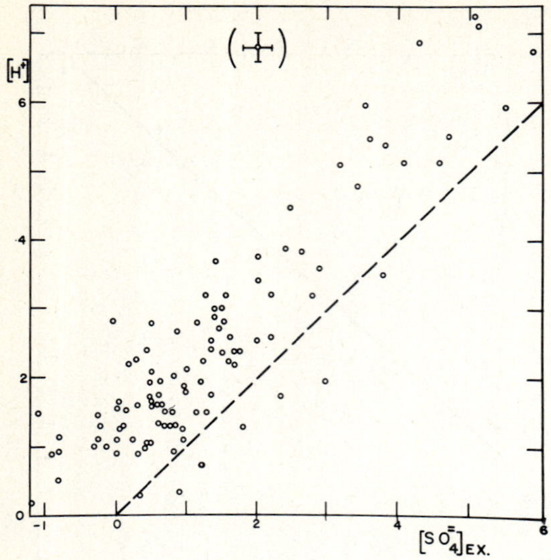

Fig. 7: Correlation between excess-sulfate and acidity for James Ross Island snow showing that acidity variations are mainly due to H_2SO_4 content changes. The broken line corresponds to a slope equal to unity (pure H_2SO_4). Mean experimental error is indicated by the cross in parentheses. From Aristarain et al. (1982).

the sulfate content ($[SO_4^=]$) or nitrate ($[NO_3^-]$) (in μEq. l^{-1}). A slope near or equal to unity is an indication that an acid (H_2SO_4 or HNO_3) is involved and not a salt. Figure 6 gives examples of these correlations. The sulfate peaks (the highest reaches 13.7 μEq l^{-1}) are characteristic of volcanic eruptions (Delmas and Boutron, 1980; Delmas et al., 1982b) and are relatively well explained. On the other hand, the origin of the rapid increases in nitrate content (however the highest reaches only 2 μEq. l^{-1}) is still difficult to determine, which is also the case for the concentration difference of this compound between the two stations Dome C and South Pole (see above).

THE ACIDITY OF SNOW AT A COASTAL SITE (JAMES ROSS ISLAND)

Glaciological studies carried out on the small ice cap of James Ross Island have provided information for a coastal site, at an elevation which is sufficient (1640 m) to avoid too great influence from sea salt deposition. It has been shown that the content of HNO_3 in snow is relatively low (mean value 0.4 μEq. l^{-1} for a 3.5 year sequence studied in detail, Aristarain et al., 1982) and does not seem to vary significantly from one season to another. On the contrary, as clearly illustrated in Figure 7, the strong seasonal variations of acidity (maximum in summer, minimum in winter) are directly linked to the variations of H_2SO_4 content. The slope of the correlation ($[H^+]$ - $[NO_3^-]$) VS $[SO_4{}^={}_{ex.}]$, equal to 0.89 provides clear proof. These strong H_2SO_4 concentration variations in snow were attributed to a photochemical phenomenon related to the oxidation of the biogenic sulfur compounds emitted by sea. It is in fact known that marine biogenic activity generates gaseous sulfur compounds (in particular DMS, dimethylsulfide, Nguyen et al., 1978). The day-night contrast,

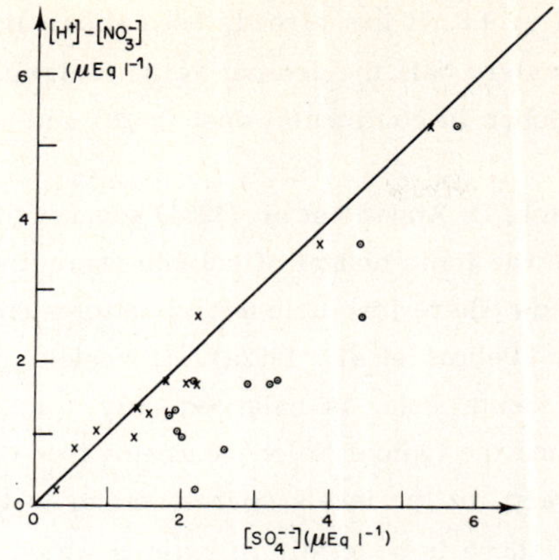

Fig. 8: Correlation between excess-sulfate and acidity (corrected for nitrate content) for the Vostok ice core. Circular points correspond to excess-sulfate calculated by using equation 3 and crosses to excess-sulfate corrected for excess-sodium (equation 3', see text). Straight line would represent the ideal case of pure H_2SO_4 (slope = 1). From De Angelis et al., 1984.

particularly long at polar latitudes (polar days and nights last severals months) may create an important seasonal effect on the generation of OH radicals involved in the production of atmospheric H_2SO_4 over subantarctic oceanic areas. The seasonal modulation of the H_2SO_4 fallout in Antarctic snow seems to be better marked at James Ross Island (coastal region) than inland (for example at the South Pole, Legrand and Delmas, 1984), perhaps in relation to a damping out of seasonal oscillations during air mass transport from the coast to the interior.

ATMOSPHERIC CHEMISTRY AND CLIMATE CHANGE: THE VOSTOK ICE CORE

Vostok Station is located on the Antarctic Plateau 400 km from Dome C. Therefore the chemistry of snow at both these stations should be very similar. That is indeed what is generally observed except however for the particular case of chloride. During the Holocene time period an excess of Na^+ with respect to the marine ratio is found at Vostok as opposed to the deficit found at Dome C. Therefore equation 4b and not 4a must be used in calculating the ionic balance. In this case equation (2) does not include the $[Cl^-_{ex.}]$ term and moreover equation (3) must take into account a correction for the presence of Na_2SO_4 formed in reaction 5:

$$[SO_4{}^{=}{}_{ex.}] = [SO_4{}^{=}] - 0.12\ [Na^+] - [Na^+{}_{ex.}] \qquad (3')$$

This assumption is discussed in detail by De Angelis et al. (1984). In the graph of Figure 8, it appears that the slope of the correlation ($[H^+] - [NO_3{}^-]$) versus $[SO_4{}^{=}{}_{ex.}]$ is near unity (slope corresponding to the compound H_2SO_4) only when the correction for excess-sodium is used. It may therefore be concluded that at this station, during the Holocene, there was no HCl in the precipitation and that the sulfate contribution encompassed a neutral component (Na_2SO_4) which was not directly marine, but rather related to the reaction of H_2SO_4 on sea-salt particles according to (5).

The calculation of the ionic budget during the last ice age is more complicated particularly at the moment of its climax some 20,000 yr BP. It has already been shown that the atmospheric CO_2 content at this time was approximately half the present value (Delmas et al., 1980) and that the Antarctic aerosol was much richer in continental dust (x 20) and sea salt (x 5.7) (Petit et al., 1981).

By analyzing ice cores of the deep drilling at Vostok, De Angelis et al. (1984) suggest that continental impurities are capable of interfering with the ionic budget of soluble impurities, which is not the case in recent central Antarctic snow where insoluble microparticles only represent less than 10% of the total deposited matter (Delmas et al., 1982a). At Vostok, the ionic budget, calculated as already explained for South Pole, is balanced only for the samples which contain aluminium concentrations around the typical Holocene mean value (2.4 ± 1.3 ng g^{-1}, De Angelis et al., 1984). On the contrary, in the levels corresponding to the richest levels in aluminium (the crustal indicator), there is a deficit in cations which is roughly proportional to the Al concentrations. This anomaly could be explained by a partial dissolution of aluminosilicates in meltwater. Metallic ions such as Ca^{++}, Mg^{++} and Fe^{++} could then be freed and thus involved in the ionic budget. Other major changes could also occur during this time period such as those discovered by Danes in Greenland deep ice cores. Very exciting results can still be expected from the study of the chemistry of the glacial age atmosphere.

ACID DEPOSITS FROM CATACLYSMIC VOLCANIC ERUPTIONS: EXAMPLES IN THE DOME C CORE

We have discussed the case of unusually high H_2SO_4 levels in the D 57 ice core, which we attributed to volcanic eruptions of global concern. However Zanolini (1983) has concluded that D 57 is not the best suited site to study paleovolcanic deposits in Antarctic ice cores. Sites with low snow accumulation rates, such as Dome C or Vostok (see Table 1), are probably much better for this purpose.

Large volcanic eruptions generally disturb the stratospheric sulfate budget over a few years and are therefore recorded in relatively thin ice layers in the Dome C ice cores (less than 10 cm in general, one year being represented by 4 to 5 cm of ice thickness). The eruption that was found at 713 m depth (Figure 9) was therefore most probably extremely violent since the sulfate increase peaks at ~ 12 μEq. l^{-1} of sulfate (the increase due to the Agung eruption in 1964-65 was only ~ 2.4 μEq. l^{-1}). The atmospheric effect of this eruption lasted approximately 4 years. A rough comparison with the strength of Agung (source function 20 MT of H_2SO_4) suggests that this unknown eruption could have emitted more than 100 MT of H_2SO_4. Its age is estimated at 23,000 yr BP. At 431 m depth (11,000 yr BP), another very powerful eruption was detected with a maximum sulfate value of 9 μEq. l^{-1}. Most of these volcanic events may have occurred in New Zealand where several thick tephra layers corresponding to cataclysmic eruptions are known and well dated. Further studies are

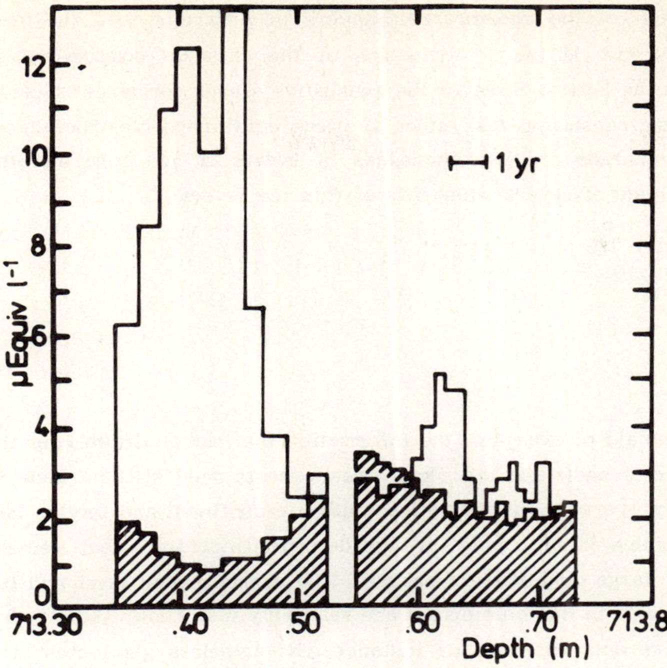

Fig. 9: Two cataclysmic volcanic events recorded in the Dome C ice core by their high sulfate contents (upper and lower curves, 713 m and 431 m depth, respectively). Nitrate profile is not disturbed by volcanic events. (From Delmas et al.,1982 b).

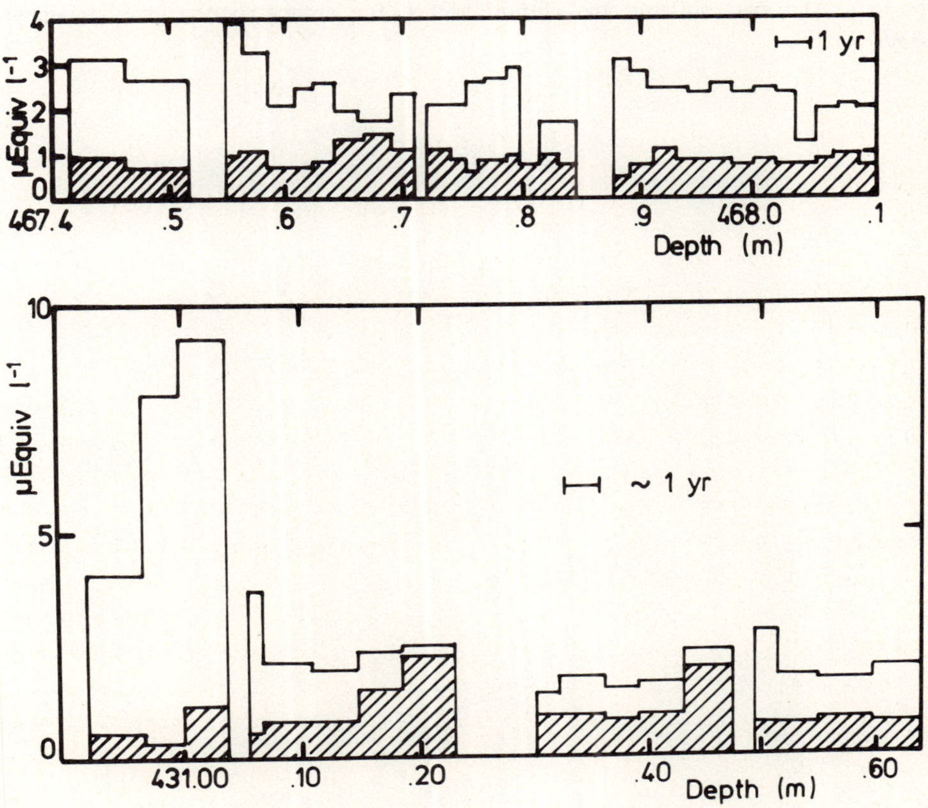

now underway to identify each of the volcanic acid layers in Antarctic ice. The most promising method for this purpose is the determination of the chemical composition of insoluble particles detected in the layers. However the exhaustive search for large volcanic eruptions in the past is a time consuming task since it needs continuous electrochemical measurements over several hundreds or even thousands of meters of ice core and the preparation and chemical analysis of several hundreds of thin ice layers.

CONCLUSIONS

We have illustrated, with the aid of examples, the information that can be drawn from the major ion chemistry of Antarctic snow and ice. Many measurements must still be made in order to provide a satisfactory overview for the entire Antarctic continent and for the last 100,000 years, a time period which is accessible through deep drilling. Analytical methods now exist for the analysis of large quantities of samples. The concentration level and the origin of some of the ions involved in the ionic budget are relatively well known (sulfate for instance). Others require more data (nitrate for instance). Nevertheless glaciochemistry appears to be the most suitable and elegant method for reconstructing past atmospheric chemistry.

REFERENCES

Aristarain A.J., Delmas R.J., and Briat M., (1982). Snow Chemistry on James Ross Island (Antarctic Peninsula). J. Geophys. Res. 87, 11004-11012.

Cunningham W.C. and Zoller W.H. (1981). The chemical composition of remote area aerosols. J. Aerosol Sci. 12, 367-384.

De Angelis M., Legrand M., Petit J.R., Barkov N.I., Korotkevich Y.S., and Kotlyakov V.M. (1984). Soluble and insoluble impurities along the 950 m deep Vostok ice core (Antarctica). Climatic implications. Journal of Atmospheric Chemistry, 1, 215-239.

Delmas R.J. and Boutron C., (1980). Are the past variations of the stratospheric sulfate burden recorded in central Antarctic snow and ice layers? J. Geophys. Res. 85, 5645-5649.

Delmas R.J., Ascencio J.M. and Legrand M. (1980). Polar ice evidence that atmospheric CO_2 20000 yr PB was 50% of present. Nature 274, 155-157.

Delmas R.J., Briat M., and Legrand M., (1982 a). Chemistry of South Polar snow. J. Geophys. Res. 87, 4314-4318.

Delmas R.J., Barnola J.M. and Legrand M., (1982 b). Gas derived aerosol in central Antarctic snow and ice : the case of sulfuric and nitric acids. Annals of Glaciology 3, 71-76.

Delmas R.J., and Gravenhorst G., (1982). Background precipitation acidity. Proc. C.E.C. Workshop "Acid Deposition", Berlin 9-10 Sept. 1982, 82-107.

Hammer C.U. (1977). Past volcanism revealed in Greenland ice sheet impurities, Nature 270, 482-486.

Jouzel J., Merlivat L., Petit J.R., and Lorius ., (1983). Climatic information over the last century deduced from a detailed isotopic record in the South Pole snow. J. Geophys. Res. 8, 2693-2703.

Junge C.E. (1977). Processes responsible for the trace content in precipitation. In: Isotopes et impuretes dans les neige et glace, Compte-rendus du Symposium UGGI Grenoble (1975). Publ. AIHS No 118, 63-77.

Lambert G., Ardouin B., and Mesbah-Bendezu A., (1983). Atmosphere to snow transfers in Antarctica. In: Precipitation scavenging, dry deposition and resuspension, Pruppacher et al. Ed., Elsevier Sc. Publ. Co.

Legrand M., Aristarain A.J., and Delmas J.R., (1982). Acid titration of polar snow. Analyt. Chem. 54, 1336-1339.

Legrand M., and Delmas R.J., (1983); Spatio krustal variations of the Cl/Na ratio in Antarctic snow. Paper presented at the 5th Intern Conf. of CACGP, Oxford, 1983.

Legrand M., Delmas R., and Zanolini F., (1984). Bilan ionique et acidite de la precipitation Antarctique. Proceedings of the third EC Symposium "Physico-chemical behavior of atmospheric pollutants" held in Varese, Italy, April 1984, 441-450.

Legrand M., and Delmas R.J., (1984). The ionic balance of Antarctic snow: a 10 yr detailed record. Atmospheric Environment, 18, 1867-1874.

Legrand M., De Angelis M., and Delmas R.J., (1984). Ion chromatographic determination of major ions at ultratrace levels in Antarctic snow and ice. Analytica Chimica Acta, 156, 181-192.

Nguyen B.C., Gaudry A., Bonsang B., and Lambert G., (1978). Reevaluation of the role of dimethyl sulphide in the sulphur budget, Nature 275, 637-639.

Petit J.R., Briat M., and Royer A. (1981). Ice age aerosol content from East Antarctic ice core samples and past wind strength. Nature 293, 391-394.

Pourchet M., Pinglot F., and Lorius C. (1983). Some meteorological applications of radioactive fallout measurements in Antarctic snows. J. Geophys. Res. 88, 6013-6020.

Shaw G.E. (1979). Considerations on the origin and properties of the Antarctic aerosol. Rev. Geophys. Space Phys. 17, 1983-1998.

Stumm W., and Morgan J.J. (1970). Aquatic Chemistry. Wiley and Sons Ed. 583 p.

Zanolini F. (1983). Conductimetrie et chimie de la glace a D 57 (Terre Adelie). Application a la recherche du paleovolcanisme. Bulletin PIRPSEV No 76, CNRS Paris, 84 p.

Part III

ORIGIN, DISTRIBUTION AND REMOVAL OF ATMOSPHERIC TRACE COMPOUNDS IN THE PRESENCE OF AIRBORNE PARTICLES AND LIQUID WATER

RELATIONS BETWEEN AEROSOL ACIDITY AND ION BALANCE

P. Winkler
Deutscher Wetterdienst
Meteorologisches Observatorium Hamburg
Frahmredder 95
2000 Hamburg 65
F.R.G.

ABSTRACT

Aerosol particles absorb increasing amounts of water with rising relative humidity. Consequently soluble material is in a crystalline state at low humidities and becomes more and more dissolved as more water becomes available with rising humidity. Ion balancing principally could give information of the aerosol acidity, however, this method suffers from various shortcommings: (a) the acid fraction of aerosol particles is rather small so that large errors are obtained from ion balancing, unless a complete analysis of organic and inorganic constituents is available, (b) buffering of H^+-ions occurs, (c) acidity varies with relative humidity due to changing gas-liquid equilibrium conditions for $[NH_3]_{gas}$, and (d) the chemical composition of the dissolved fraction changes with relative humidity e.g. gradual deliquescence of NH_4HSO_4.- From measurements of the pH-value and electrical conductivity of aqueous aerosol solutions the acid fraction of aerosol particles can be derived which is the relative contribution of acid to the total amount of inorganic soluble material present in the particles. This acid fraction provides information on aerosol acidity without detailed knowledge of the chemical composition. Acid fractions are of the order of 5% for Aitken-particles and 1-2% for large and giant particles. pH-values of airborne aerosols are expected to range between pH 1 and 3 depending on the amount of absorbed water. At low relative humidities the pH value can drop below 1. This is also the reason that HNO_3 plays a minor role in aerosol acidity, because at low pH-values the dissociated HNO_3 recombines and may escape if the surrounding HNO_3-vapour pressure is too low. Disregarding of gas-liquid phase equilibria can cause changes in the chemical composition of particles e.g. if they are collected at high humidity and brought into a warm and dry laboratory with low humidity where particles loose their absorbed water. From this reasons we must conclude that our understanding of aerosol acidity is rather incomplete. Inspite of these restrictions some general results are obtained: Increasing humidity and increasing total aerosol concentrations (due to air pollution) correlate positively with the acid fraction of aerosol particles.

NATO ASI Series, Vol. G6
Chemistry of Multiphase Atmospheric Systems
Edited by W. Jaeschke

Aerosol acidity contributes substancially to fog acidity. The contribution of aerosol acidity to precipitation acidity is small in polluted regions and higher but not exclusive in remote regions. Aerosol acidity shows a yearly cycle and even a daily cycle under appropriate meteorological conditions.

1. INTRUDUCTION

During their residence time in the atmosphere, aerosol particles have often been regarded a nearly closed system, not suject to rapid chemical transition. Physical or physicochemical processes like coagulation and changing water absorption at various relative humidities were the only ones from which some change in the chemical composition of an aerosol particle was expected. It will be shown, that with respect to aerosol acidity this picture is too simple and has to be improved for a correct understanding of aerosol acidity.

The knowledge of the acid reaction of aerosol particles is not new. In investigating hygienic aspects of natural and industrial aerosol particles Cauer (1956) developed a chromium covered condensation sphere filled with ice. The intention was, that the movement of the diffusing water vapor molecules would force the aerosol particles to the surface where they become dissolved in the condensing dew. The dropping dew water was acid in almost all cases. Because the aerosol mass which was deposited at the sphere was ill defined and because gases also dissolved in the condensed water, the conclusion on the real acidity of the aerosol was more qualitative than quantitative.

The acidity is caused by the H^+-ions in a solution. The pH-value of a solution depends not only on the activity of H^+-ions, but also on a number of equilibria, which are established when soluble gases such as CO_2, SO_2 or NH_3 dissolve in water and react with the water molecules. Some details will be explained in chapter 5. It could therefore be helpful to discuss aerosol acidity from a standpoint of ion-balance i.e. the ionic composition of the aerosol particles together with the equilibria constants of the relevant gases. The ion balance is based on the rules of electro-neutrality: the sum of the positive charges must be equal to the sum of the negative charges. In the case of precipitation, it has sometimes been tried to derive long term trends of the pH-values from knowledge of the major ion-concentrations, except for the H^+-concentration, by ion-balancing (e.g. Granat, 1972; Liljestrand and Morgan, 1980).

Aerosol particles absorb some water as a function of the relative humidity of the air due to the deliquescent behaviour of the substances which determine their chemical composition. We therefore could be attempted to derive the aerosol acidity from our knowledge of the chemical composition and the absorbed water amount by making use of the method of ion balancing. As we will see in the course of our considerations, ion balancing is extremely difficult to apply to such a complex system as the atmospheric aerosol. Some approximative ways can be used, however, in order to understand some basic features of aerosol acidity.

2. ION-BALANCE

For a sufficiently diluted solution the ion-balance can be described by the following equation which is written as condition for electro-neutrality or charge balance:

$$[H^+] + [NH_4^+] + [Na^+] + [K^+] + 2[Ca^{++}] + 2[Mg^{++}] + \sum_i z_i [Me^{z_i+}] = \\ = [OH^-] + [HCO_3^-] + 2[CO_3^=] + [Cl^-] + [NO_3^-] + 2[SO_4^=] + \sum_j z_j [A_j^{z_j-}] \tag{1}$$

where Me^{z+} denotes the molar concentration of the ith trace cationic species of charge $+z_i$ and A_j^{z-} denotes the molar concentration of the jth trace anionic species of charge $-z_j$. Note that equation (1) is not completely correct, although it is written in its generally used notation. In a strict notation each term has to be multiplied by a factor z·N; z is the number of charges per ion and N is Loschmidt's (or Avogadro's) number. With this factor the right dimension: number of charges per liter (or volume) is obtained.

In the case of aerosol particles where ionic composition except H^+ has been determined, equation (1) could be used to solve for H^+. The carbonate ion and bicarbonate ion concentrations are calculated from equilibrium with the known partial pressure of CO_2.

$$K_1 = \frac{[H^+][HCO_3^-]}{[H_2CO_3]} \tag{2}$$

$$K_2 = \frac{[H^+][CO_3^=]}{[HCO_3^-]} \tag{3}$$

$$K_H = [H\,CO_3] / p_{CO_2} \tag{4}$$

$$K_W = [H^+][OH^-] \tag{5}$$

Usually the assumption

$$\sum_i z_i [Me^{z_i+}] = \sum_j z_j [A^{z_j+}] \tag{6}$$

is made so that

$$[H^+] - \frac{K_w + K_1 K_H p_{CO_2}}{[H^+]} - \frac{2K_1 K_2 K_H p_{CO_2}}{[H^+]^2} = [Cl^-] + [NO_3^-] + 2[SO_4^=] - [NH_4^+] - [Na^+] - [K^+] - 2[Ca^{++}] - 2[Mg^{++}] \tag{7}$$

If other soluble gases which react with water such as SO_2 or NH_3 are present and if their partial pressures are known, they can in principle be treated similarly to CO_2. For reasons of simplicity, this has been omitted, however. The error in H^+ calculated by equation (7) is determined by the uncertainty in the known major ion concentrations and the validity of the assumption (6). From equation (7) the H^+ concentration and thus the pH or the acidity, can in principle be derived from the knowledge of the major ion-concentration.

While in the case of precipitation this method may give reasonable results in many cases, it does not work in the case of aerosol particles for the following reason: Airborne aerosol particles absorb water, depending on the relative humidity. In this absorbed water, soluble material dissolves and except for very high r.h. the concentration of the solution becomes so high that activities instead of concentrations have to be used. In the case of atmospheric aerosol particles, which consist of mixtures of different substances, the activity coefficients are not exactly known, so that ion-balancing in praxis is nearly impossible. In the following two sections, the composition and water absorption of aerosols is explained in some detail, for that reason.

Additionally, buffer effects of still unknown nature are influencing aerosol acidity besides the effects originating from weak acids such as NH_4^+ and H_2SO_3. This means that equation (6) is not justified in this form in the case of aerosol particles but further, however unknown, equilibria had to be taken into consideration in order to treat aerosol acidity in a pure mathematical way. In this context the question of external/internal mixture of substances becomes important. Internal mixture of substances means that the various substances are mixed in each single particle. External mixture means that each particle consists only of one pure substance. Between these two extremes all transitions are imaginable and we could define a degree of internal mixing which can vary between 1 and 0. In the following chapter this question is discussed further.

3. AEROSOL COMPOSITION

In general only single substances are determined during chemical analysis of aerosol particles. So it is well known, for example, that $(NH_4)_2\ SO_4$ and some other inorganic salts make up a major fraction of the total mass. For our present purpose, the consideration of acidity, it is less important to obtain a detailed picture of the inorganic ionic fraction of the aerosol, as we have seen in the previous chapter that ion balancing is not the appropriate tool for understanding aerosol acidity. Instead of this we will consider the aerosol composition from a more qualitative standpoint in order to obtain a complete picture of the relative composition.

The composition is derived from the following experiment (Winkler, 1974): six size segregated impactor samples collected during the same time have been dissolved subsequently in various solvents: water, acetone, methanol, cyclohexane, ether, and have been heated up to 150°C. After each solution process, the sample has been weighted with a microbalance. So the amounts of material dissolvable in a certain solvent could be determined. The special

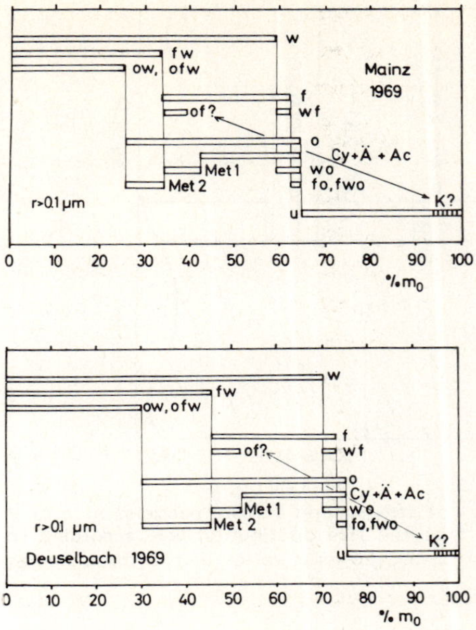

Fig. 1: Average relative composition of atmospheric aerosol particles at an urban (above) and a rural (below) site, fractions:

w	= water soluble
o	= organic soluble
f	= volatile (150°C)
u	= insoluble
ow	= organic insoluble, water soluble
of	= organic insoluble, volatile
fw	= non volatile, water soluble
fo	= non volatile, organic soluble
wo	= water insoluble, organic soluble
wf	= water insoluble
Met_1	= ammonia salts and organics
Met_2	= NaCl, KCl, $MgCl_2$, $CaCl_2$, $NaHO_3$, KNO_3
Cy+Ä+Ac	= Cyclohexane, ether, acetone soluble
K	= crystal water

(Winkler, 1974)

trick of the method now was that each of the six impactor samples was dissolved in a different manner by varying the sequence of the solvents. For example, the first sample was dissolved in the order water, methanol, ether ....., the second in the order ether, acetone, water .... and so on. By combining the solvent properties and the sequence of dissolution the following picture could be derived (fig. 1). On the average the aerosol is composed of a large amount of water soluble matter (w-fraction), of organic soluble material (o-fraction) which turns out to be dissolvable mostly also in water and to a minor extend not in water; the remainder is of insoluble nature (u-fraction) comprising material like minerals, soot and so on. Overlapping of bars in fig.1 means similar or equal properties of different fraction e.g. the overlapping part of the fractions w and f means that this material is water soluble as well as volatile, the overlapping part of the fractions w and o means water soluble as well as soluble in organics. Another example to illustrate how this method works are the methanol soluble fractions Met 1 and 2. One part of the methanol soluble substances (Met 1) is volatile (overlapping with f) as well as water soluble which can be NH_4Cl and NH_4NO_3 and/or organics. The other part of the methanol soluble matter (Met 2) is water soluble but not volatile. $(NH_4)_2SO_4$ may only be contained in the ow fraction.

It could be derived from this figure that about 45% (urban) to 55% (rural) of the total aerosol mass consist of inorganic salts and about 18% consist of organic, water soluble and volatile (150°C) substances. For the acidity considerations the overlapping part of the Cy + A + Ac with the w fraction is of importance. As will be shown later (section 5) this material is most probably causing buffer effects in aerosol particles, influencing their acidity.

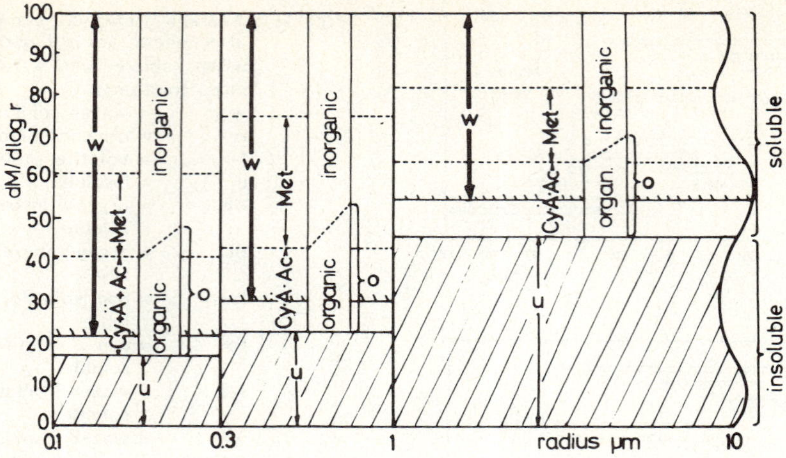

Fig. 2: Radius dependence of the relative composition of atmospheric aerosol particles at a rural site (Deuselbach). The assumption was made that the size distribution was according to Junge's power law ($\nu^* = 3$). (In the actual case, the mass ratios in the three ranges were 2:1:1) (after Winkler, 1974).

We also recognize from fig. 1 that the relative composition of a city and a rural aerosol is rather similar the insoluble fraction being a little bit smaller at the rural station.

Fig. 2 shows the radius dependency of the various fractions and we see that the relative amount of insoluble material increases with increasing radius and that the relative composition of the soluble fraction also varies with particle radius.

With respect to acidity of aerosol particles, we have to conclude that filter samples, where particles of all sizes are collected at the same surface, are not a good primary material for acidity studies.

Consequently the next question is, whether any aerosol sample where many different particles are closely packed together, gives a sample which is representative for the airborne state. In other words there is a question concerning the degree of internal mixture: Are the different materials as seen in fig. 1 mixed together in each single particle or not? Although a clear answer cannot be given, there are some reasons to expect high degree of an internal mixture of substances: particles originating in the fine Aitken mode coagulate rapidly with larger particles. During their residence time, aerosol particles become activated several times as cloud droplets. So gas phase reactions with droplets homogenize the composition after reevaporation of the cloud drops. At least the investigation of single particles (LAMMA), e.g. van Grieken (1984) shows, that single particles contain large amounts of various substances. The maximum of the aerosol surface which is found in the radius range $0.1 \leqq r \leqq 1$ µm is also good evidence that in this range the degree of internal mixture is high. As a consequence, this range is often called the accumulation mode. In the range $r > 1$ µm the degree of internal mixture may be lower as compared to the smaller particles because 1) the residence time of these particles is low, 2) coagulation is slow, 3) many particles are produced due to dispersion of material. In order to illustrate these thoughts we can imagine a sea salt aerosol, where the various substances are mixed within the particles. If this aerosol is mixed with a Sahara dust cloud drifting over the ocean, some external mixture arises, mineral and salt particles being separated. Turnover to mixed particles due to coagulation proceeds slowly as the air mass ages.

While any radius dependency of a substance indicates the existence of some degree of external mixture, we can expect that the degree of internal mixture increases with decreasing radius. An internal mixture, however, increases the probability that buffer mechanisms may be active within the aerosol. It may, however, be more or less pronounced for aerosol particles in different climatic regimes. For our further considerations, we assume the degree of internal mixing to be complete within each of the three radius ranges: Aitken particles, large particles ($0.1 \leqq r \leq 1$ μm), and giant particles ($r > 1$ μm) keeping in mind that this assumption is weakest in the last case.

4. WATER CONTENT OF AEROSOL PARTICLES

An important question to be discussed in connection with aerosol acidity is the water content of the aerosol particles, which depends on the relative humidity. We imagine an internally mixed particle composed of an insoluble nucleus and soluble material and we expose such a particle to rising relative humidity. As shown schematically in fig. 3, we recognize that at low relative humidity no or only a small amount of water is adsorbed onto or absorbed into the particles, either due to physical adsorption or due to dilution of small amounts of acid. Most of the soluble material is crystallized, however. Between roughly 40 and 80% r.h. we observe a transition range, where the dissolvable material gradually becomes dissolved. The particle is composed of an insoluble nucleus surrounded by a solution shell in which the crystals become more and more dissolved as the humidity rises. This picture is highly idealized, because, for example, microscopic photographs of sea salt particles coagulated with Sahara dust particles have shown that the insoluble matter is sitting at the surface of the salt droplet. Whether the insoluble part is inside or at the surface depends on surface tension, wetting angle and so on. Above about 75 or 80% all dissolvable material has dissolved and we now have only the insoluble nucleus surrounded by the solution shell, the concentration of which becomes more and more dilute, according to Raoult's law, as the r.h. increases.

In the right hand part in fig. 3 we have depicted the deliquescence function of a continental and a seasalt particle. This deliquescence function varies between 0 and 1 and describes which amount of the dissolvable material has really dissolved at a certain r.h.. With this deliquescence function we can, for a first approach, describe the water absorption of aerosol particles in a similar manner as the well known Raoult's law:

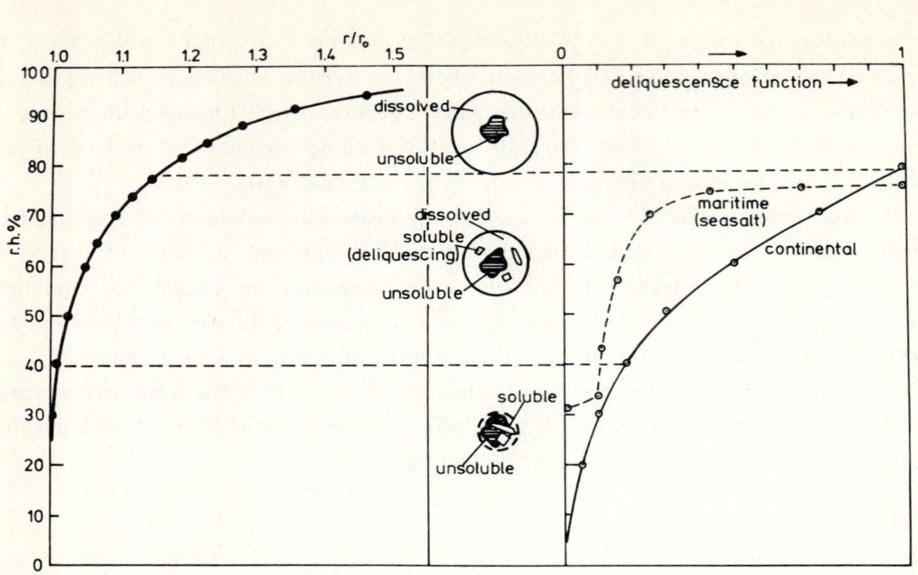

Fig. 3: Left part: Schematic description of the deliquescence of a mixed aerosol particle. Right part: Deliquescence function of the soluble material of continental aerosol and sea salt. This function describes the gradual dissolving of the soluble matter with rising r.h..

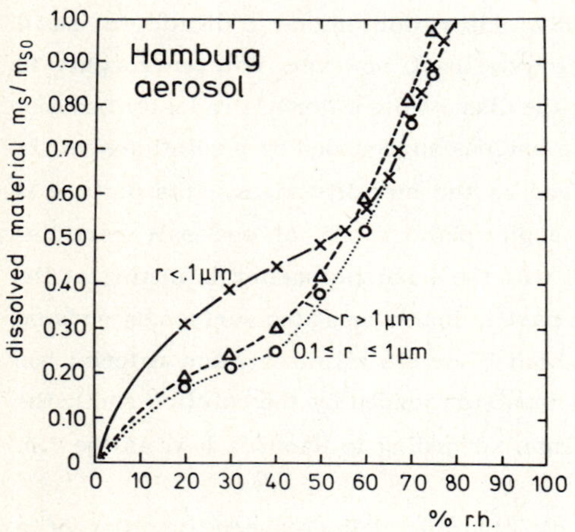

Fig. 4: Average deliquescence function of the soluble matter of aerosol particles for 3 radius ranges. m_s = mass material really dissolved at r.h.; m_{so} = total mass of water soluble material.

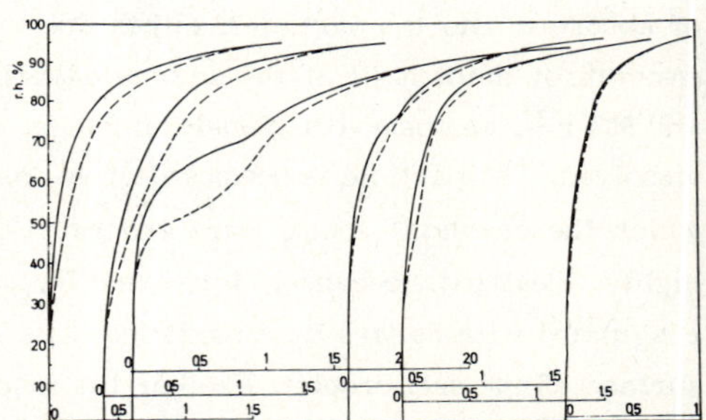

Fig. 5: Some typical growth curves of atmospheric aerosol samples (r > 0.1 μm) collected at Mainz; solid lines refer to increasing, dashed lines to decreasing r.h.. Curves 1, 2, 5, 6 continental air masses, curves 3, 4: maritime air masses (here the jump at 75% r.h. points to a remarcable sea salt fraction, deliquescing at this r.h.).

$$f = \frac{e}{E} = 1 - \frac{C^{*} \; F(f)}{r^3 - r_o^3} \qquad (8)$$

with f = vapor pressure ratio, e = water vapor pressure, E = saturation water vapor pressure, C* = constant depending on the amount of total soluble to insoluble matter, F = deliquescence function, r = particle radius at vapor pressure ratio f ($f \cdot 100$ = r.h.), r_o = dry particle radius. For details of the derivation of eq. (8) see Winkler (1973), where also a procedure for an approximative derivation of F is presented. It would go far too much into detail to derive eq. (8) here. It should be mentioned that a curvature term has to be included if the particle radius becomes smaller than about 0.1 µm.

Average deliquescence functions for 3 size ranges of aerosol particles collected in the outskirts of Hamburg are depicted in fig. 4. We see that the main mass of the soluble material deliquesces between 40 and 80% r.h.. The differences between large ($0.1 \leqq r \leqq 1$ µm) and giant particles are small while the soluble material of the Aitken particles deliquesces more easily i.e. at lower r.h. with a higher degree.

The deliquescence behaviour is likely to have an influence on the acidity of aerosol particles for two reasons: (a) The chemical composition of the remaining solution changes most drastically during the crystallization of dissolvable material. (b) The concentrations of ions found in the remaining solution is to be calculated from the measured amount of absorbed water and the amount of the definitely dissolved material, i.e. the knowledge of not only the deliquescence function but also the knowledge of the changing chemical composition becomes necessary.

Fig. 5 gives a typical example of water absorption by aerosol particles in terms of mass of water related to the dry aerosol mass for two aerosol size ranges collected at a rural station. We see that with rising humidity the amount of absorbed water is slightly smaller than with decreasing humidity indicating some supersaturation of the solution in the latter case.

5. ACIDITY

Acidity in a strict chemical sense is defined as the amount of H^+-ions as determined by a titration procedure (total acidity). The pH is defined as pH = $-\log a_{H^+}$ where a_{H^+} is the activity of the hydronium ion which for dilute solutions as rainwater is nearly identical with the free H^+-concentration (free acidity). Depending on the endpoint of the titration, weak acids as NH_4^+ or H_2CO_3 liberate a portion of the associated H^+-ions, according to their equilibrium constants. Thus the titration "sees" strong acids and some of the weak acids. Since the contribution of organic acids is relatively low (Galloway et al., 1976; Ketseridis et al., 1975, 1976) as can be expected from direct analysis of organic acids or bases, the pH measurement in a fairly acid regime gives in a first approximation the concentration of strong acids. For some difficulties arising with these methods, the reader is referred to Tyree (1981). For the method to determine weak acids we refer to Gran (1952) and Brosset (1976), Brosset and Ferm (1978).

While any difference between total and free acidity can hardly be explained without detailed analysis, some basic general conclusions can be drawn from the following considerations on the relative contribution of acid to the soluble material in aerosols or precipitation.

From the previous section it becomes clear that the water amounts naturally absorbed at aerosol particles are too small for either direct pH-measurements or a titration procedure in an aerosol sample. In chapter 2 it was shown that from ion balancing no realistic results can be expected. The following consideration presents another approach to characterize aerosol acidity.

From simple measurements of the pH-value and the electrical conductivity of an aerosol sample dissolved in some water, the acid fraction can be determined in the following manner:
1) determination of the mass of soluble ions (m_I) from conductivity

a) no H^+-ions, no CO_2 present

$$\Lambda = \sum_i \lambda_i [N_i] \tag{9}$$

Λ = total conductivity, λ_i = ionic conductance (µS/cm or µmho/cm) for i-th species, $[N_i]$ = normal concentrations.

$$\Lambda = \bar{\lambda} [\bar{N}] = \bar{\lambda} \frac{m}{\bar{M}_e V} \tag{10}$$

the bar means average values, $\bar{\lambda}$ = average equivalent conductance, $\overline{Me}$ = mean equivalent weight, m = mass (of ions), V = volume of water in which aerosol sample has been dissolved.

$$m_I = \frac{\Lambda}{\lambda} \overline{Me} \cdot V \tag{11}$$

is the mass of the dissolved ions. It is important to emphasize, that $\overline{Me}$ and $\bar{\lambda}$ do not change very much due to variations of the chemical composition of a solution. In practical calculations $\overline{Me}$ = 35 g and $\bar{\lambda}$ = 60 mho/cm.

b) if H^+-ions and CO_2 are present:

$$\Lambda = \sum_j \lambda_j [N_j] + \lambda_H [H^+] + \lambda_{HCO_3^-} [HCO_3^-] + \lambda_{CO_3^=} [CO_3^=] \tag{12}$$

where the constants can be taken from eqs. (2) to (5).

This equation principally can be treated similar as in case a. In praxis we can make use of some simplifications because the mass of the H^+-ions is negligible and the conductance arising from dissolved CO_2 can be treated as a mere correction term. We can therefore transform (12) without large error to:

$$m_I = \frac{\Lambda - \lambda_{corr}}{\bar{\lambda}} \overline{Me} \cdot V \tag{13}$$

with λ_{corr} = conductivity arising from H^+ and dissolved CO_2. This equation has the advantage that inevitable impurities causing a blind conductivity of the distilled water used to dissolve the aerosol sample can be treated as additive term in λ_{corr}.

The amount of ions derived from conductivity measurements has been compared with a gravimetric determination of the soluble material. It could be shown that the conductivity method on the average yields 80% of the soluble matter as gravimetrically determined (Winkler, 1980), the remainding 20% most probably due to organic water soluble material, which does not dissociate and therefore cannot contribute to the electrical conductivity. In fig. 1 this fraction was characterized as a part of the w-fraction which overlaps with the Cy + Ä + Ac fraction.

2) From the pH-measurement, which is interpreted as primarily caused by strong acids we assess the mass of the acid

$$m_{HA} = \sum_i M_{HA_i} [H_i^+] V_i = \overline{M_{HA}} [H^+] V \tag{14}$$

with M_{HA} = mean equivalent weight of the strong acids (for practical use a composition of 70% H_2SO_4 and 30% HNO_3 is assumed).

If the H^+-concentration is not high enough (< 10^{-5} M), we must correct for the influence of H^+ originating from H_2CO_3. In this case $[H^+]$ in equation (14) must be replaced by $[H^+]_{HA}$

$$[H^+]_{HA} = [H^+] - \frac{10^{-11.3}}{[H^+]}$$

with $[H^+]$ = measured concentration (= 10^{-pH}), pH = measured value. For a detailed derivation see Stumm and Morgan (1970).

By combining eqs. (13) and (14) we can define the acid fraction AF of aerosol particles by

$$AF = \frac{\overline{M_{HA}}\,[H^+]}{\frac{(\Lambda - \lambda cor)}{\overline{\lambda}}\,\overline{Me}} = \frac{C_1[H^+]}{(\Lambda - \lambda cor)\,\overline{Me}} \tag{15}$$

Eq. (15) combines the acid fraction with the measured values of pH and electrical conductivity and in a pH-conductivity diagram we can construct from this equation a set of inclined lines describing the percentage of acid contributing to the soluble matter at a given λ - pH pair (e.g, fig. 6).

The diagram implies some important properties:

1) The range above the 100% line is forbidden, because the conductivity at a fixed pH-value cannot fall below a minimum conductivity caused by the H^+-ions alone. Adding of neutral salts to a pure diluted acid droplet of say pH = 4 increases only the conductivity, the pH remains unchanged.

2) Dilution or concentration of a solution means a movement parallel to the inclined percentage lines as indicated by the double headed arrow in fig. 6. A motion perpendicular to the inclined lines can only be caused by chemical reactions. As we will see later, this is of importance when comparing acidity in aerosol and in precipitation.

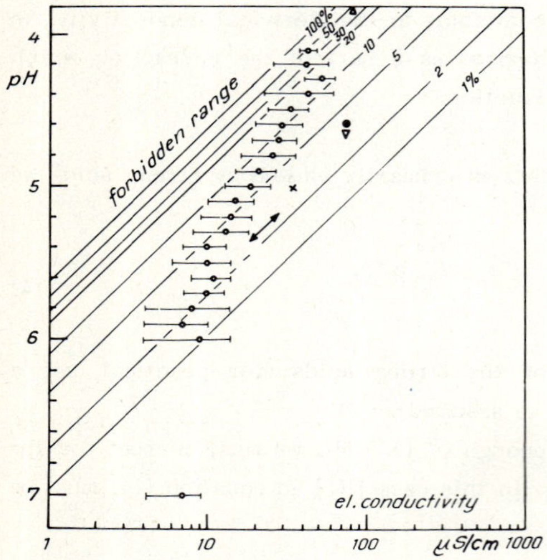

Fig. 6: Simultaneous measurements of pH and electrical conductivity of leaching solutions of Aitken-particles (Winkler, 1980). Open circles with indicated standard deviation: denote Hamburg; after Junge and Scheich (1969) ▲ London, ▼ Mainz, ⊙ Gelsenkirchen, × Goeteborg. Indicates dilution or concentration of the leaching solution.

3) The combined consideration of pH and electrical conductivity makes some use of the concept of ion-balance without need of going to much into detail. The method is limited, however, to acid regimes. The main reason for the reasonable working is the large difference between the equivalent conductances of H^+ and the other ions. In alkaline regimes the equations have to be modified.

This way of describing aerosol acidity differs strongly from the usual methods. The main advantage is, that this "relative free acidity" characterizes a more conservative property of the aerosol. A description in absolute units (e.g. µg/m^3) is less useful because it changes as the aerosol moves away from a source and dilutes due to removal processes. The relative composition is changed to a much smaller extend during removal processes, as can be seen e.g. from fig. 1.

6. OBSERVATIONAL RESULTS

The values depicted in fig. 6 are average results for Aitken particles (Winkler, 1980). We see that the acid fraction varies between 1 and 40%, as read from the inclined lines. During normal meteorological conditions values usually between 0.5 and 5% are observed. The higher percentages connected with conductivities higher than 10 µs/cm belong to pollution episodes. The samples were collected on a filter behind a cascade impactor. The filters were leached with water in such a way that the particles from 1 m^3 of air were dissolved in approximately 0.4 ml of water. This ratio was chosen to compare these results with those of precipitation. Since the liquid water content of precipitating clouds is of the same order of magnitude a direct comparison is reasonable and possible. Furthermore the solutions are so diluted that concentrations rather than activities can be used.

In fig. 7 results of relative acid fractions of particles > 0.1 µm are presented. The acid fractions vary between 1 and 10%. Values between 1 and 2% can be considered as average values, higher figures are observed during pollution episodes. It seems to be important to mention that the points of measurements in figs. 6 and 7 refer to a sample taken from 15 m^3 air volume. The symbols in fig. 7 are results from other authors (Junge and Scheich, 1969; Brosset, 1975, 1976) which have been transformed with some assumptions into this diagram (details of the transformation are given in Winkler, 1980). The good agreement of the measurements from different locations shows that with respect to the relative acid fraction, aerosols are similar in composition and that the measurements of Hamburg are of a more regional and not only local importance. The residence time of aerosol particles being of the order of some days

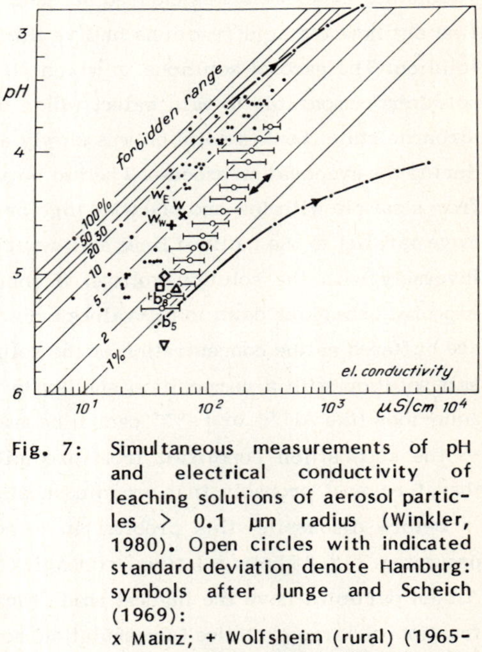

Fig. 7: Simultaneous measurements of pH and electrical conductivity of leaching solutions of aerosol particles > 0.1 µm radius (Winkler, 1980). Open circles with indicated standard deviation denote Hamburg: symbols after Junge and Scheich (1969):
x Mainz; + Wolfsheim (rural) (1965-66); ▵ Mainz (1967-68), ◇ Goeteborg; ○ Gelsenkirchen; □ London; after Brosset (1976, 1978): W_e, W, W_W "white" and b_5, b_8 "black" episodes. Small dots: Hamburg during episodes with high acidity. Heavy lines see text.

(Jaenicke, 1980) supports this conclusion. The dots with percentages above the average line belong to episodes known as acid smog but there are some occasions with high acid fraction occuring during summer. Thus it can be concluded that high acid fractions are reached not only during situations where the pollution is high, i.e. at high conductivities but also during clean situations where the conductivity of the aerosol leaching solution is low. The reasons for the high acid fraction situations are not clear. Low concentrations of ammonia have been found to be able to cause such episodes. After Brosset (1976) such occasions occur at the Swedish west coast during situations with relative pure maritime air reaching the sampling site. His points were marked by w-symbols ("white episodes") in fig. 7. Again it should be noted that the swedish acid episodes are in line with the data obtained during acid episodes in Hamburg. This also means that this behaviour is of more general importance and not only valid for Hamburg.

Near St. Louis, Missouri, Charlson et al. (1978) measured aerosol particles showing humidity growth similar to that of H_2SO_4. By adding small amounts of ammonia to their device, the humidity growth changed drastically and resembled more to $(NH_4)_2/SO_4$. Therefore lack of ammonia is of essential importance for elevated acidity of aerosol particles.

Another experiment was carried out with some of our Hamburg aerosol samples in order to find out how the acid fractions behave during rising concentration of the aerosol leaching solution. The aerosol solutions in which pH and conductance are measured are rather dilute solutions comparable to rain water, which is not very representative for the aerosol in the airborne state. Such a solution was slowly evaporated and pH and conductivity was measured during the evaporation process. The two heavy lines in fig. 7 show two examples, one obtained from a sample with low and one with high original acid fractions. We see that the line does not move parallel to the inclined lines as it would be expected if the H^+-concentration would vary inversely with the solution volume. It should be mentioned that rain water does show the expected behaviour down to pH-values below 2. This means that H^+-ions in aerosol particles are buffered as the concentration of the solution increases. Experiments with artificial mixed salt solution with a composition similar to that of the aerosol allowed the conclusion that aquo-ions like Al^{+++} or Fe^{+++} cannot be made responsible for the buffer behaviour contrary to the supposition concluded from precipitation analysis by Klockow et al., 1978. It is therefore most probably that organic substances cuase the buffer effect.

Before discussing this problem in more detail some other observational results are presented. Fig. 8 shows relative frequencies of the acid fractions of 3 aerosol size ranges. The Aitken particles have the highest acid fractions with a large standard deviation. The acid fraction of large particles (accumulation mode) is distributed more homogeneously. The giant particles (coarse mode r > 1 μm) also show acid reaction in general; however, these particles often react neutral or even alkaline, depicted as negative free acid fractions in fig. 8.

We can draw the conclusion that the acid in aerosol particles has its origin to a large extend in the nucleation mode (Aitken particles). With increasing particle radius the acid

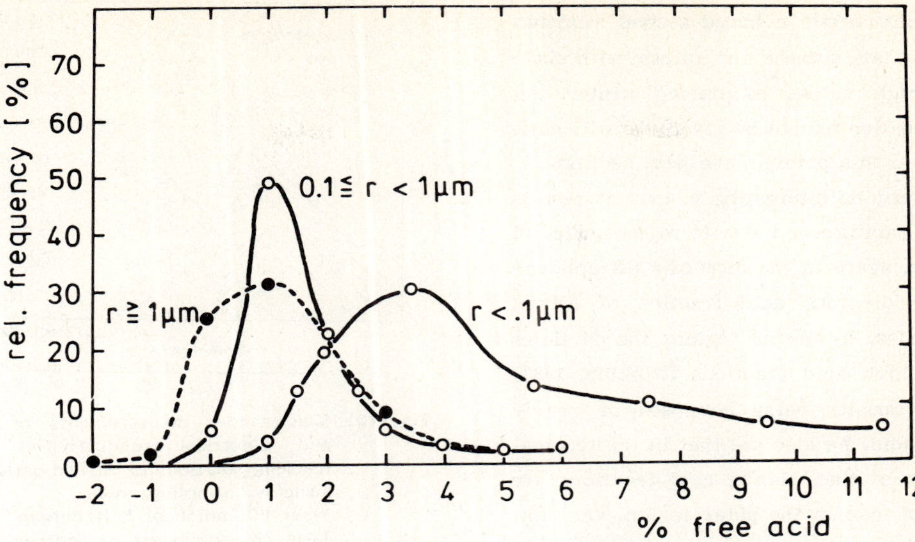

Fig. 8: Relative frequency distribution of free acid (eq. 15) in aerosol particles for three size ranges.

fraction decreases and occasionally may be overcome in the giant particle range by alkaline material. We will come back to this result in section 9.

The yearly cycle (7 years average 1976-1982) of the acid fraction is depicted in fig. 9 for Aitken particles and the particles > 0.1 µm radius. We see that the highest acid fractions are observed during February where the highest total pollution is observed. Stagnant weather conditions are frequent at this time with low level inversions, so that gases like SO_2 and NO_x reach their yearly maximum. Episodes with acid smog are frequent and give rise to the pronounced february maximum of aerosol acidity. The particles > 0.1 µm radius show a flat secondary maximum during July which is most likely due to a type of summer monsoon in Hamburg where frequently fresh maritime air with low NH_3 concentration reaches the measuring site. Similar results were found by Elshout et al. (1978) in the Netherlands. A correlation between NH_3 and acidity has not yet been evaluated up till now. The acid fraction of the Aitken

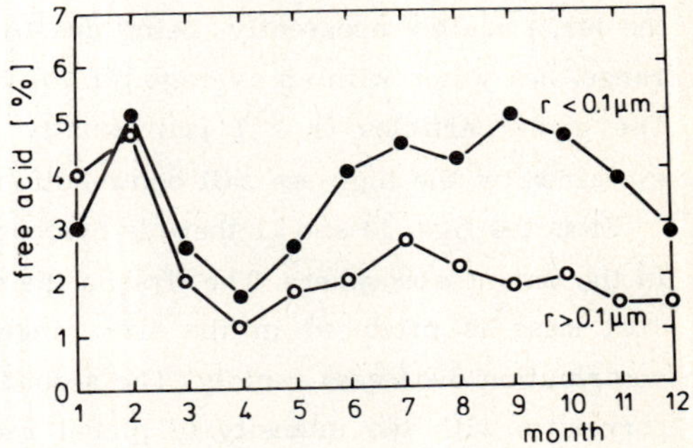

Fig. 9: Yearly cycle of free acid (eq. 15) of atmospheric aerosol particles for two size ranges. Average values over 7 years (1976-1982) of particles collected in Hamburg.

particles shows a broad second maximum during late summer and autumn with similar high values as during winter. An explanation cannot be given; we will come back to this point in the next section.

It may be interesting to look at results from maritime and remote regions. Fig. 10 shows, again in the form of a pH-conductivity diagram, acid fraction of Aitken particles. In marine regions the particles have high acid fractions (Winkler, 1983) some samples being composed of nearly pure acid. We also see that in the tropical region of the Atlantic acid fractions are higher than in the polar region. Very low acid fractions or even alkaline fractions were found over the Sahara. However, these low acid fractions are still in line with the observations made in Hamburg. Since by elemental analysis with an X-ray fluorescence technique it has been found that sulfur is the most abundant element, most probably in the form of $SO_4^{=}$, we can conclude that sulfuric acid is a major component of marine Aitken particles. In accordance to this finding Ono (1978) found acid particles (H_2SO_4) in subsiding unpolluted background air at Fuji in Japan.

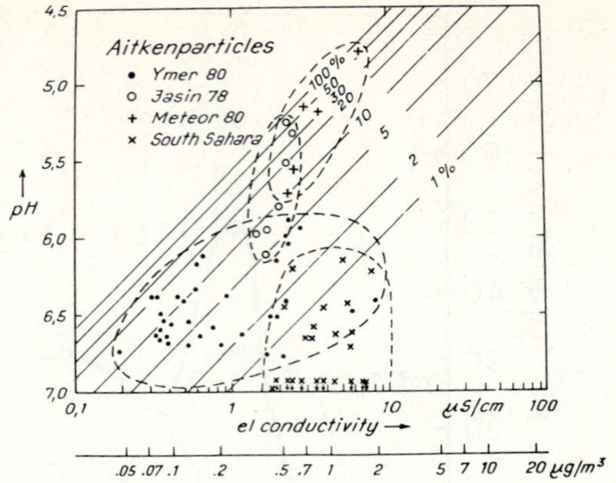

Fig. 10: Simultaneous measurements of pH and electrical conductivities of leaching solutions of Aitken particle samples. Sampling areas:
Ymer 80: north of Spitzbergen,
Jasin 78: northwest of Scotland,
Meteor 80: cruise between Hamburg and Montevideo, samples taken between 30°N and 30°S.
South Sahara: Tombuctu, Senegal and Matam, Mali.
(Winkler, 1983)

In fig. 11 results of the large and giant particles, which again were collected over the Atlantic ocean, are depicted. While the aerosol radius fraction $0.1 \leqq r \leqq 0.3$ µm usually shows acid fractions between 5 and 50%, the fraction $0.3 \leqq r < 1$ µm reacts acid to neutral, the large scatter apparently being due to changes in the sea salt fraction in this radius range. Sea water with an average pH value slightly above 8 produces an alkaline aerosol. The giant particles ($r > 1$ µm) usually react neutral or alkaline which can be easily explained by the high sea salt contribution to this particle fraction.

From the figs. 10 and 11 there is evidence for the existence of two major aerosol sources in the marine atmosphere. The first is the sea spray due to bubble bursting from which the most mass is produced in the size range $r \sim 1$ µm. With decreasing radius the sea salt contribution decreases rapidly. The second source is the gas to particle conversion which correlates with the intensity of global radiation (Haaf and Jaenicke, 1980). Due to rapid coagulation, these particles, being most probably H_2SO_4-particles, accumulate in the size range around 0.1 µm radius. Depending on the availability of ammonia these particles become partly neutralized. Thus, the marine aerosol can be understood as a well aged background aerosol, superimposed by sea salt at larger radii and by the more or less neutralized H_2SO_4

at the smaller radii with some overlap in the range $0.3 \leqq r \leqq 1$ µm. Depending on the weather conditions either the sea salt or the acid component dominates in the overlapping range.

We now consider daily variations of acidity observed in Hamburg. With a high volume Aitken particle collector, short time Aitken nuclei samples have been collected during an episode with heavy pollution in January 1982. It has been found (fig. 12) that the free acid fractions showed a daily variation with a maximum around noon (dashed line). While none of the other trace constituents like Aitken particle-number concentration, SO_2, NO, NO_2, O_3 or meteorological parameters like relative humidity showed a regularly daily cycle during that episode, only sun radiation and temperature showed a similar behaviour. The acid fraction of the large particles showed a variation (not depicted) similar to the Aitken particles probably due to rapid coagulation of Aitken particles with the large particles. The reason for the daily variation of the acid fraction is not clear. The meteorological conditions during the episode were rather

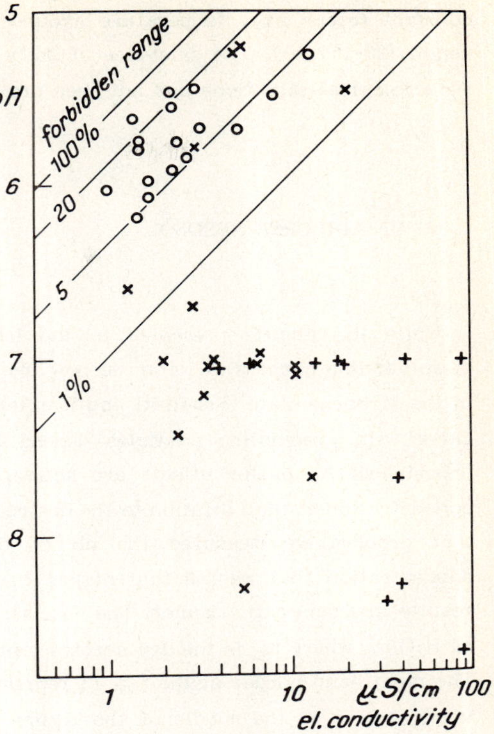

Fig. 11: Simultaneous measurements of pH and electrical conductivities of leaching solutions of maritime aerosol particles of different size ranges.
o particles with $0.1 \leqq r < 0.3$ µm
x particles with $0.3 \leqq r \leqq 1$ µm
+ particles with $r > 1$ µm.
Sampling area: Atlantic ocean between 65°N and 65°S.
(Winkler, 1980)

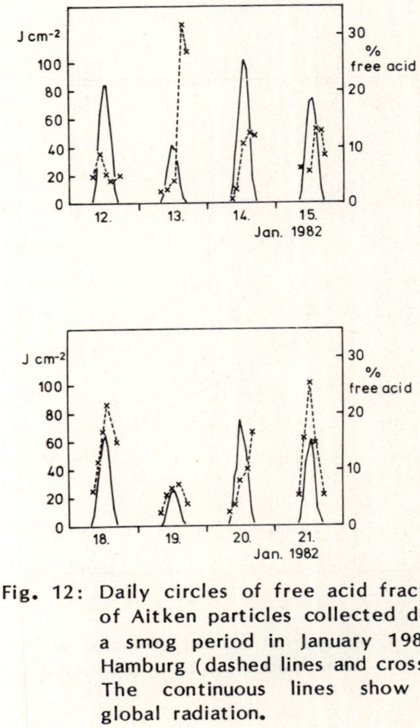

Fig. 12: Daily circles of free acid fractions of Aitken particles collected during a smog period in January 1982 in Hamburg (dashed lines and crosses). The continuous lines show the global radiation.

constant (clear sky, temperature around -10°C, constant wind direction, low inversion height (80-100 m)). The existence of daily cycles during other times of the year or other meteorological situations has not been proved.

7. pH OF AIRBORNE AEROSOL

While the results presented in the last chapter were derived from aerosol samples dissolved in a surplus of water we now try to find out figures of the pH-value for particles in the airborne state (Kaminski and Winkler, 1983). Because many heterogeneous reactions of gases with preexisting particles depend on the pH-value, this question is of general importance. As buffer effects are active, a simple extrapolation from the measured H^+-concentration at high dilution to the desired high concentration is not possible. To approach that problem we measured the pH of the aerosol solution with a micro-probe at a concentration that was 60-fold higher than at the routine procedure. Fig. 13 shows the results in a schematic manner. The abscissa is divided in units of a reciprocal concentration ml H_2O/m_o where m_o is the dry aerosol mass of 1 m^3 of air; here a value of 40 µg was used. The right hand border of the fig. 13 represents the pH-values measured at high dilution, the vertical line in the middle of the figure represents the micro-electrode measurements at high concentrations. Cases "B" and "C" represent typical and extreme values respectively.

In order to find pH-values of the airborne aerosol particles we extrapolate from the two measured concentration values to solution concentrations to be expected in aerosol particles. The left hand border of the figure therefore represents the concentrations to be expected in the solution shell of a particle in equilibrium with a surrounding humidity of 90%. At this humidity the mass of absorbed water is comparable to the dry mass. Two types of buffer behaviour are considered. In case "A" a continuously increasing buffer effect is assumed, similar as was described in fig. 7. We arrive at point 1 at a pH slightly above 3.

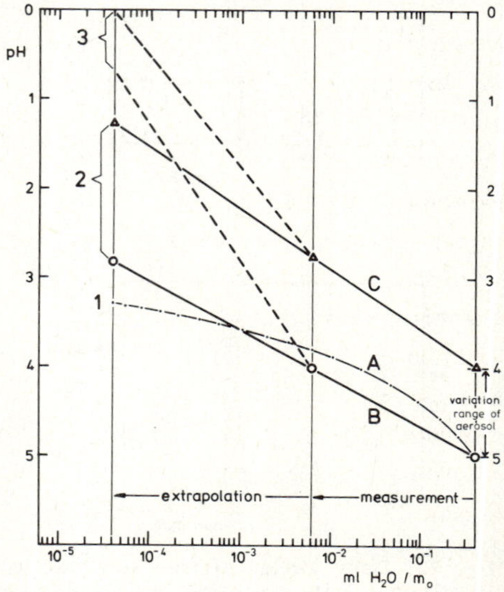

Fig. 13: Estimation of the pH-value of airborne particles. For explanation see text.

In the cases "B" and "C" we use the measured buffer effect and extrapolate linearly. We arrive at pH-values between 1.2 and 2.8, designated as range "2". None of the extrapolation procedures (case A or B or C) are realistic, moreover we would expect a combined behaviour which resembles case A at the more dilute concentrations and case B at the higher concentrations. Nevertheless "A" and "B" do not result in very different value so that range "2" is rather likely. If we would extrapolate without taking any buffering into account we would follow the dashed lines and arrive at range "3" with pH-values between 0 and 1 which seems to be very unrealistic.

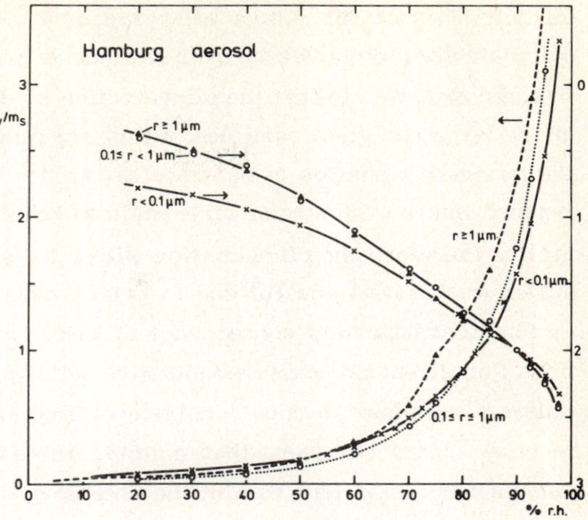

Fig. 14: Average water absorption of aerosol particles as a function of the relative humidity for particles of three radius ranges (arrow pointing to left hand ordinate). Variation of the pH-value with relative humidity if no buffer mechanisms were active (arrows pointing to the right hand ordinate).
m_w = amount of observed water,
m_s = amount of dry water soluble material.

It becomes clear that airborne particles are quite acid, especially the particles in a radius range where the maximum of the aerosol surface usually is found. This is true for urban, rural and marine environments. Nevertheless neutral or basic reacting particles are found sometimes in the size range r > 1 µm. It is obvious that the occurance of basic particles can be subject to local conditions. For heterogeneous reactions these particles may become important in spite of their small contribution to the aerosol surface, if a high pH-value is a prerequisite for reactions to take place.

In fig. 14 we present the pH-variation of aerosol particles at ambient relative humidities. The continuous, dashed and dotted lines represent average water absorption curves as function of the r.h. for the three particles ranges: Aitken, large and giant particles. Note that the amount of absorbed water refers to the mass of water soluble material m_s. These "reduced" curves reflect better the influence of the chemical composition on the water absorption than a relation to the total mass (m_w/m_o) which would additionally depend on the amount of unsoluble matter contributing nearly nothing to the water absorption. The Aitken particles absorb the lowest, the giant particles the highest amounts of water per unit mass of soluble material at relative humidities above 80%. The little jump in the curve for the giant (r > 1 µm) particles at 75% r.h. shows that small amounts of sea salt were present due to some marine influences.

To find pH-variations with r.h. we assume from the results of fig. 12 that an aerosol particles has at 90% r.h. a pH-value of 2. From the measured water absorption we now calculate the pH-variation over the humidity range without taking any buffer mechanisms into account. We see that the pH-variation of the Aitken particles is slightly lower than that of the large and giant particles, which are relatively similar. The pH-variation calculated is the maximum variation to be expected. We see that at low relative humidities pH-values are reached where even strong acids begin to recombine (HSO_4^-: pK = 1.92, HNO_3: pK = -1.32). In reality therefore the pH-variation with r.h. is much more smaller than indicated (a) due to buffer mechanisms and (b) due to partly recombination of strong acids.

The most important consequence of these findings is that neither ion-balance considerations nor direct pH-measurements give satisfying results for a description of aerosol acidity unless a complete chemical analysis of inorganic and organic constituents is available. It becomes clear, however, that acidity, relative humidity and pH-variation with humidity plays a very important role in the chemistry of atmospheric aerosols. It also becomes clear that dissolved SO_2 respectively HSO_3^- plays only a minor role unless the r.h. is above 90% or if single particles react alcaline when absorbing water.

8. THE SYSTEM NH_3 - $SO_4^=$ - NO_3^- -H_2O

The basic considerations over this system have been made by Tang (1976, 1980), Lee and Brosset (1979) and Brosset (1983). We start with the system NH_3 - $SO_4^=$ - H_2O, referring for the detailed mathematics to the mentioned authors and restricting our considerations to air-chemical consequences in relation with aerosol acidity.

The system can be studied in a phase diagram (fig. 15, after Brosset, 1983). This diagram is based on the rather likely assumption that a particle forms as H_2SO_4 solution droplet which equilibrates with respect to water vapor and ammonia. The H_2SO_4 in part becomes neutralized to $(NH_4)_2SO_4$ or other solid phases. The significance of the points and lines in the diagram is explained as follows:

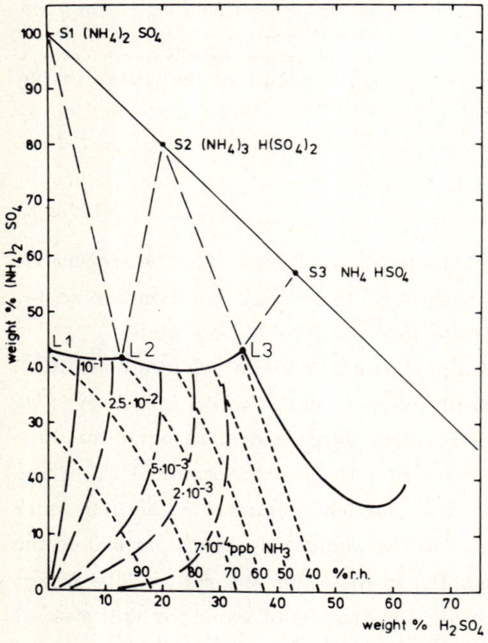

Fig. 15: Phase diagram of the system $(NH_4)_2SO_4$ - H_2SO_4 - H_2O at 25°C (after Brosset 1983). For explanation see text.

a) The line S1-S2-S3

The points S1, S2, S3 correspond to the solid phases as indicated and the lineparts between the points correspond to mixtures of the solid phases, respectively. No water vapor is absorbed at a particle having a composition corresponding to a point at this line.

b) The curve L1-L2-L3

This line describes the solubility curve, where the solid phase is in equilibrium with the saturated solution (25°C). At point L2 equilibrium exists with the solid phases S1 ($(NH_4)_2SO_4$) and S2 ($(NH_4)_3H(SO_4)_2$) and at point L3 equilibrium exists with the points S2 ($(NH_4)H(SO_4)_2$) and S3 (NH_4HSO_4). At points L2 or L3 there are four phases in a three component system which means according to Gibbs phase rule 1 degree of freedom. At a constant temperature both r.h. and the partial pressure of $[NH_3]_g$ is given. At the tie-lines L1-L3 there are 3 phases and therefore 2 degrees of freedom i.e. at a given temperature one of the parameters r.h. or $[NH_3]_g$ can be chosen independently.

c) The area below the solubility curve

In this area no solid phases are present. The number of phases is two and the number of degrees of freedom is three, which means that at a given temperature r.h. and $[NH_3]_g$ can vary independently. However, any chosen values of r.h. and $[NH_3]_g$ define the composition of the solution.

We now perform some thought experiments in the area below the solubility curve:

(1) If we vary the relative humidity the movement of our point is along one of the lines of constant NH_3-partial pressure. This means that at a fixed NH_3-partial pressure the chemical composition of the solution must change as the r.h. varies.

(2) If we vary the NH_3-partial pressure at constant relative humidity we follow along a line of constant r.h. and again the chemical composition, especially the acidity of a particle changes.

As we know from a $[NH_3]_g$ source inventory (e.g. Russel et al. 1983) the NH_3-partial-pressure of an air parcel can vary as it moves over different source regions. Additionally we have the daily variation of the relative humidity. Thus, the chemical compositions and especially the acidity of aerosol particles must change in accordance with the phase diagram.

Measured NH_3-concentrations are in most cases above 10^{-1} ppm over land areas. From the phase diagram we can thus obtain an explanation for the finding, that the acid fraction of aerosol particles is rather low. At this high NH_3 partial-pressure only small amounts of H_2SO_4 are in equilibrium with $(NH_4)_2\ SO_4$. There is, however, still some descrepancy in NH_3 vapor pressure measurements (Lau and Charlson, 1977). Over the oceans the NH_3 partial pressure is much lower than over the continents thus explaining the high acid fractions in aerosol particles found by Brosset (1976) during the so-called white episodes.

If we add to the system $SO_4^{=}$ - NH_3 - H_2O also the component HNO_3, the composition of the solution of aerosol particles will also be determined by NO_3^{-}-ions, the amount of which depends on the HNO_3-partial-pressure. Tang (1980) has shown, that the molar ratio $NO_3^-/SO_4^=$ in the solution depends on $[HNO_3]_g$ and the relative humidity. The ratio remains rather small, however, and seems to be of minor importance. On the other side Russel et al. (1983) in investigating formation and transport of NH_4/NO_3 aerosols found qualitative agreement between model calculations and measured values for a site in California.

9. DISCUSSION

We have seen that acidity of atmospheric aerosol particles is influenced by many parameters:

a) chemical composition, causing buffering effects, most probably due to organic substances.

b) relative humidity.

c) partial pressure of $[NH_3]_g$ and $[HNO_3]_g$.

We have further seen that the relative acid fraction as derived from pH and conductivity measurements (eq. 15) gives additional information beyond that obtained from investigations of the chemical composition. Here it may be informative to discuss differences between the relative acid fractions found in atmospheric aerosols, in fog and precipitation water.

In fig. 16 we compare the acid fractions of aerosol leaching solutions and precipitation samples. We see that the precipitation has higher acid fractions. Imagine that an aerosol particle becomes activated as a cloud droplet and that water condenses onto it, the concentrations of the diluting solution would change along one of the inclined lines. Incorporation of additional particles into cloud drops can also not increase the acid fraction, because the ion-balance cannot be altered due to the condensation process.

The additional acidity of rain drops can therefore only be explained by liquid phase reactions within the droplet. This conclusion is confirmed by comparing $SO_4^=/Pb$ or $SO_4^=/Fe$ ratios. The Pb and Fe content of rainwater almost exclusively originates from incorporation of aerosol particles while $SO_4^=$ comes from both, aerosols and liquid phase oxidation. $SO_4^=$

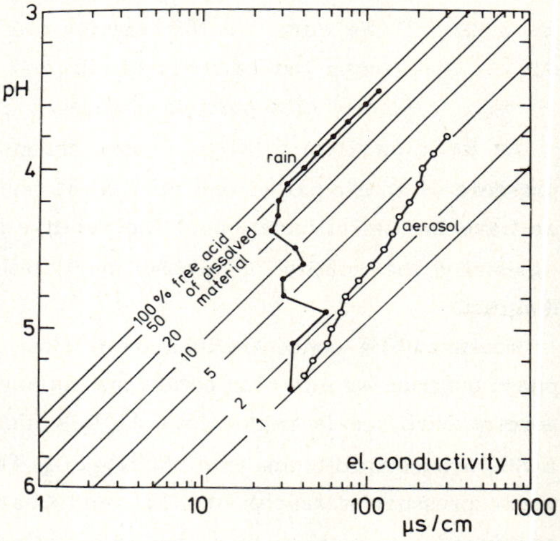

Fig. 16: Comparison of the average acid fractions (inclined lines) of aerosol leaching solutions (open circles) and of precipitation (full dots).

and Pb are associated with the fine, Fe with coarse aerosol mode (data after Georgii et al., 1982; Robock, 1982).

ratio	aerosol	precipitation
$SO_4^=/Pb$	25 (fine mode)	100
$SO_4^=/Fe$	13 (coarse mode)	50

Since aerosol particles are the only source for Fe and Pb in rainwater the ratios to sulfate did not change if $SO_4^=$ in rainwater would come only from the aerosol.

It may be important to mention that similar characteristic curves as presented in fig. 16 have been derived for precipitation of Scandinavian samples (Winkler, 1982) with the only difference that the leg of curve where the acid fraction lowers rapidly, is shifted to somewhat smaller average conductance. Such a characteristic curve seems to reflect an equilibrium between the cloud physical processes at the one side (drop size distribution, lifetime of cloud drops, temperature and so on) and the more chemical processes or physicochemical processes on the other side (vertical distribution of trace gases, reaction rates, availability of oxidants and so on).

In remote regions the contribution of aerosol acidity to precipitation acidity is larger as became evident from figs. 10 and 11. However, also in remote regions a remarkable contribution of precipitation acidity originates from liquid phase processes.

In fig. 17 the comparison is made for fog samples and aerosol acidity (Kaminski and Winkler, 1983). The acid fractions of fog samples are higher as compared with the aerosol samples but are lower than those of precipitation. Although there is large scatter for single cases a grouping mainly around the 10% line is recognized. (The extreme points measured by Houghton were observed in Brooklyn and may be attributed to local pollution.) In the case of fog we recognize a much closer connection between aerosol and fog acidity. The main reason is that fog drops are much

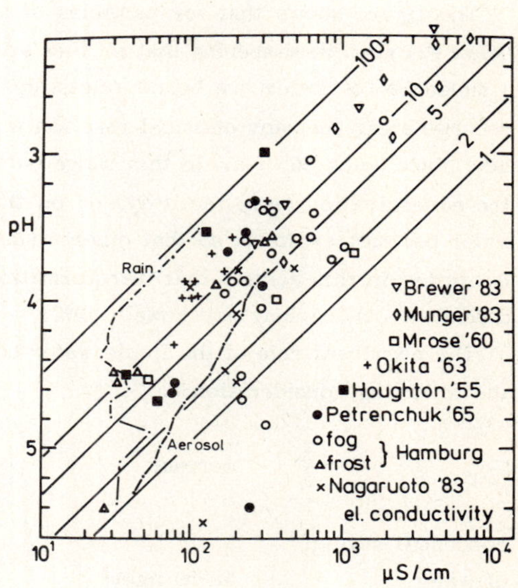

Fig. 17: Comparison of the acid fractions of fog samples with those of aerosol leaching solutions (Kaminski and Winkler, 1983).

much more persistent than cloud drops and that aerosol particles are scavenged very effectively by fog drops. Similar conclusions can be drawn from the studies of Fuzzi et al. (1983) and Fuzzi et al. (1984), who found that trace constituents primarily vary inversely to the liquid water content, but that there was evidence for an increase of the "acid" ions $SO_4^=$ and NO_3^- relative to the liquid water content obviously due to liquid phase oxidation.

What is the reason that we find so large differences in the relative acidity of particles, fogs and precipitation. One answer can be deduced from fig. 18 and the following consideration:
The highest solution concentrations are found in aerosol solutions. If we calculate the mole ratio of water to dissolved ions we obtain a curve as depicted in Fig. 18. The only assumption is the average ionic weight which was calculated for 44 g/mol.

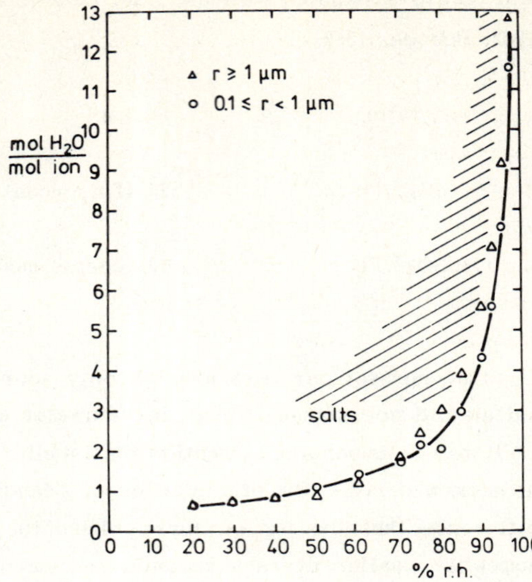

Fig. 18: Number of water molecules per ion in deliquescing atmospheric aerosol particles, assuming an average weight of the ions of 44 g/mol.

The figure shows that for particles of any size the ratio remains below 4 at humidities below 90% r.h.. Remembering that all ions are hydrated in aqueous solutions and usually 6 to 8 molecules of water are bound relatively firm to one ion in the first hydration layer it becomes clear that any chemical reaction which needs water as reaction medium can hardly take place below 90% r.h.. In this range only shifting of equilibria either being dependent on the concentration (resp. activity) or on the pH-value will play a role. Above 90% enough water becomes available so that other mechanisms like oxidation can play a role. It may be mentioned in this context that transformation rates of SO_2 have been proven to be humidity dependent (Alkezweeny and Powell, 1977).

The prominent role of the liquid water content on acidity formation becomes obvious from the following consideration:

	aerosol	fog	raining cloud
condensed amount of water	10-50 µg/m³	10-500 mg/m³	1 g/m³
acid fraction (equ. 15)	2%	10-20%	50%

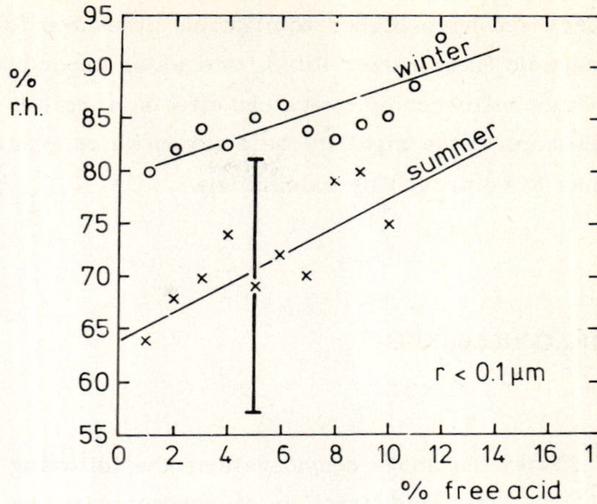

Fig. 19: Increase of the acid fraction of Aitken particles with relative humidity during sampling. Every point is an average value of 10 to 50 single values.

We can conclude from this table and the previous discussion that aerosol acidity is also influenced by some competition between more physicochemical principles (like water content) and more chemical principles (like oxidation potential, buffer capacity). If more water becomes available dissolving of gases like SO_2 can take place with subsequent oxidation. As a consequence the pH rises due to dilution (fig. 14) and HNO_3 can also dissolve more easily. This may explain the higher acid fractions in fog as compared with the aerosol. In the precipitation the still higher acid fractions (at least in polluted areas) may additionally be caused due to the influence of vertical distributions of trace gases which may be different for different gases.

The influence of r.h. on the acid fraction of aerosol particles can be demonstrated in figs. 18 and 19. For the measured acid fractions in Aitkęn particles (fig. 19) and in the large and giant particles (fig. 20), the mean relative humidities measured during the collection period have been averaged as a function of the amount of the acid fraction analyzed.

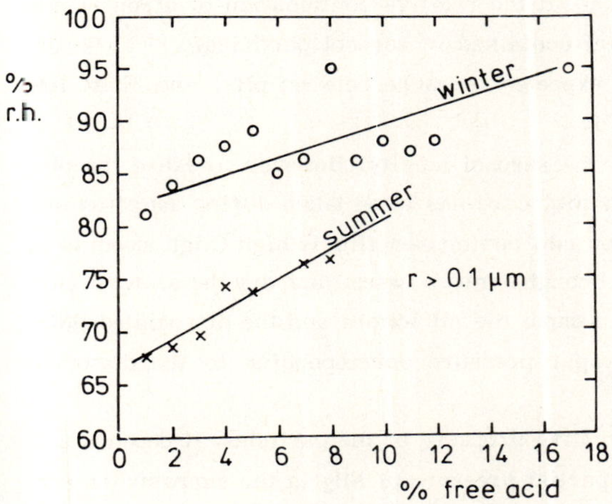

Fig. 20: Same as fig. 19 but for particles with r > 0.1 µm.

We see that the higher acid fractions belong to higher r.h. independent of particle size and time of the year. Although the scatter is so large that one single curve cannot be regarded as statistically conclusive, the similar behaviour of the 4 different populations allows the conclusion, that the acid fraction of aerosol particles increases with increasing r.h..

One final remark concerns the finding that the acid fraction of Aitken particles is higher than that of the larger particles. Along with the phase diagram (fig. 15) we try the following qualitative explanations: We divide hypothetically an aerosol particle of 0.1 µm radius in two parts. Due to increased curvature effects the smaller particles have higher equilibrium pressures for H_2O and $[NH_3]_g$. This means the smaller particle "sees" lower $[NH_3]_g$ and water vapor pressure. In comparison with a particle which is not influence by curvature effects, a smaller particle adopts its equilibrium point in fig. 15 more to the right in the diagram which results in a higher acid fraction. Of course this has to be proven by calculation.

10. CONCLUSIONS

From the above considerations the following conclusions can be drawn:

1) The acid fraction of aerosol particles is radius dependent. Any investigation of acidity has to take this into consideration. During filter sampling without size segregation chemical reactions within the sample may occur.

2) Ion balancing, unless a complete analysis of inorganic and organic constituents is available, is not an appropriate tool for determining the aerosol acidity because
   a) the acid fraction of the aerosol is rather small, so that large errors are to be expected from ion balance considerations,
   b) buffering of H^+-ions by (most probably) organic material occurs,
   c) under appropriate atmospheric conditions the acidity varies with humidity and $[NH_3]_g$ partial pressure in such a way that the acid component varies with humidity when the $[NH_3]_g$ partial pressure remains constant.

3) Besides the known parameters total acidity and free acidity, the introduction of the relative acidity or acid fraction is useful. The acid fraction can be determined from simple measurements of the pH-value and electrical conductivity of aerosol leaching solutions. The acid fraction informs about the relative contribution of strong acids to the total soluble inorganic material contained in aerosol particles.

4) pH-values of airborne aerosols are expected to range between pH 1 and 3. At low humidity the pH can decrease below 1.

5) HNO_3 seems to play a minor role in the aerosol acidity. However, because the pH-values in aerosol particles are rather low, care has to be taken during sampling and analytical determination in cases when r.h. during sampling is high (high amounts of HNO_3 in sample) and the sample is brought into a warm and dry laboratory (low r.h.). Due to loss of water from the sample the pH lowers and the dissociated HNO_3 recombines and may escape if the vapor pressure corresponding to the dissolved amount is too low.

6) Aerosol acidity is not a very conservative property of the particles. It depends e.g. on water content of particles and on partial pressure of NH_3 in the surrounding air. Increasing humidity and increasing total aerosol concentration correlate positively with the acid fraction of aerosol particles.

7) Aerosol acidity contributes substantially to fog acidity. The contribution of aerosol acidity to precipitation acidity is small in polluted regions and higher but not exclusive in remote regions.

Finally some recommendations for future investigations are given. In order to better understand aerosol acidity more complete data sets of the aerosol composition are necessary. For this purpose not only the major but also the minor inorganic ions have to be determined. For example the solubility of metal oxides depends on the pH of the solution and complex formation of these metals (e.g. aquo-ions) may occur, influencing the pH-value again. Additionally, information is necessary on organic substances which are responsible for buffer mechanisms. Since these substances are not equally distributed, it is necessary to determine their radius dependency. At least some micro analysis of single particles is important, because this is the only direct way from which the existence of mixed particles can definitely be confirmed.

Long term observations of aerosol acidity are necessary in order to obtain a data set large enough, to find out relations between aerosol acidity and meteorological conditions, maximum values, frequency and duration of episodes with acid aerosol particles and so on.

REFERENCES

Alkezweeny, A.J. and Powell, D.C. (1978): Estimation of transformation rate of SO_2 to SO_4 from atmospheric concentration data. Atm. Environm. 11, 179-182.

Brewer, R.L., Gordon, R.J., Shepard, L.S. and Ellis, E.C. (1983): Chemistry of mist and fog from the Los Angeles urban area. Atm. Environm. 17, 2267-2270.

Brosset, C. (1976): Black and white episodes. Ambio 5, 157-163.

Brosset, C. (1978): Water soluble sulfur compounds in the aerosols. Atm. Environm. 12, 25-38.

Brosset, C. and Ferm, M. (1978): Man made airborne acidity and its determination. Atm. Environm. 12, 909-916.

Brosset, C. (1983): Characterization of acidity in natural waters. In: Acid Deposition (Edited by S. Beilke and A.J. Elshout) pp. 44-55, Reidel Publ. Comp., Dordrecht.

Cauer, H. (1956):: Die pH-Werte von Aerosolen und Niederschlaegen und ihre lufthygienische und bioklimatische Indikatorbedeutung. Z. f. Aerosolforsch. u. Therapie 5, 459-509.

Charlson, R.J., Covert, D.S., Larson, T.V. and Waggoner, A.P. (1978): Chemical properties of tropospheric sulfur aerosols. Atm. Environm. 12, 39-53.

Elshout, A.J., Viljeer, J.W. and van Duuren, H. (1978): Sulfates and sulfuric acid in the atmosphere in the years 1971-1976 in the Netherlands. Atm. Environm. 12, 785-790.

Fuzzi, S., Orsi, G. and Mariotti, M. (1982): Radiation fog liquid water acidity at a field station in the Po-valley. J. Aerosol Sci. 14, 135-139.

Fuzzi, S., Castillo, R.A., Jiusto, J.E., Lala, G.G. (1984): Chemical composition of radiation fog water at Albany, New York and its relationship to microphysics. J. Geophys. Res. (in press).

Galloway, J.N., Likens, G.E. and Edgerton, E.S. (1976): Acid precipitation in the northeastern United States: pH and acidity. Science 194, 722-724.

Georgii, H.W., Perseke, C. and Robock, E. (1982): Feststellung der Deposition von sauren und langzeitwirksamen Luftverunreinigungen aus Balastgebieten. Final Report; Project No. 10402600 of Environmental Protection Agency, Berlin.

Gran. G (1952): Determination of the equivalence point in potentiometric titrations. Analyst 77, 661-671.

Granat, L. (1972): On the relation between pH and the chemical composition in atmospheric precipitation. Tellus 24, 550-560.

Haaf, W. and Jaenicke, R. (1980): Results of improved size distributions measurements in the Aitken range of atmospheric aerosols. J. Aerosol Sci. 11, 321-330.

Houghton, H.G. (1955): On the chemical composition of fog and cloud water. J. Meteor. 12, 355-357.

Jaenicke, R. (1980): Atmospheric aerosols and global climate. J. Aerosol Sci. 11, 577-588.

Junge, C. and Scheich, G. (1969): Studien zur Bestimmung des Saeuregehaltes von Aerosolteilchen. Atm. Environm. 3, 423-442.

Kaminski, U. and Winkler. P. (1983): Saure Aerosolpartikel und Nebel und ihre Wirkung auf die Biosphaere. Ann. Meteor. (N.F.), Nr. 20, 149-150.

Ketseridis, G. and Hahn, J. (1975): Bestimmung organischer Bestandteile von Aerosolpartikeln in Reinluft. Z. Anal. Chem. 273, 257-261.

Ketseridis, G., Hahn, J., Jaenicke, R. and Junge, Chr. (1976): The organic constituents of atmospheric particulate matter. Atm. Environm. 10, 603-610.

Klockow, D., Denzinger, H. and Roenicke, G. (1978): Zum Zusammenhang zwischen pH-Wert und Elektrolytzusammensetzung von Niederschlaegen. VDI-Berichte Nr. 134.

Lau, N. and Charlson, R.J. (1977): On the discrepancy between background ammonia gas measurements and the existence of acid sulfates as a dominant atmospheric aerosol. Atm. Environm. 11, 475-478.

Lee, Y.H. and Brosset, C. (1979): Interaction of gases with sulfuric acid aerosol in the atmosphere. WMO-Symposium Sofia, WMO-Publ. Nr. 358,181-192.

Liljestrand, H.M. and Morgan, J.J. (1980): Modelling the chemical composition of acid rain in southern California. In: Polluted Rain (Edited by E.Y. Toribara, M.W. Miller, P.E. Morrow), pp. 109-122, Plenum Press, New York.

Mrose, H. (1966): Measurements of pH and chemical analysis of rain, snow and fogwater. Tellus 18, 266-270.

Munger, J.W., Jacob, D.J., Waldman, J.M. and Hoffman, M.R. (1983): Fogwater chemistry in an urban atmosphere. J. Geophys. Res. 88, 5109-5121.

Nagamoto, C.T., Parungo, F., Reinking, R., Pueschel, R. and Gerich, T. (1983): Acid clouds and precipitation in eastern Colorado. Atm. Environm. 17, 1073-1082.

Okita, T. (1968): Concentration of sulfate and other inorganic materials in fog and cloud water and in aerosol. J. Met. Soc. Japan 46, 120-127.

Ono, A. (1978): Sulfuric acid particles in subsiding air over Japan. Atm. Environm. 12, 753-757.

Petrenschuk, O.P. and Drodzdova, V.M. (1966): On the chemical composition of cloud water. Tellus 18, 280-286.

Robock, E. (1982): Atmospheric removal of airborne metals by wet and dry deposition. In: Deposition of atmospheric pollutants (Edited by H.W. Georgii and J. Pankrath) pp 159-172 Reidel Publ. Company, Dordrecht.

Russel, A.G., McRae, G.J. and Cass, G.R. (1983): Mathematical modelling of the formation and transport of ammonium nitrate aerosol. Atm. Environm. 17, 949-964.

Stumm, W., and Morgan, J.J. (1970): Aquatic chemistry. Wiley and Sons, Chichester, New York.

Tang, I.N. (1976): Phase transformation and growth of aerosol particles composed of mixed salts. J. Aerosol Sci. 7, 361-371.

Tang, I.N. (1980): On the equilibrium partial pressure of nitric acid and ammonia in the atmosphere. Atm. Environm. 14, 819-828.

Tyree, S.Y. (1981): Rainwater acidity measurement problems. Atm. Environm. 15, 57-60.

van Grieken, R. and Adams, F. (1984): Characterization of particulate matter by X-ray, Raman and mass spectrometric techniques. (This issue).

Winkler, P. (1973): The growth of atmospheric aerosol particles as a function of the relative humidity. J. Aerosol Sci. 4, 373-387.

Winkler, P. (1974): Die relative Zusammensetzung des Aerosols in Stoffgruppen. Meteor. Rdsch. 27, 129-136.

Winkler, P. (1975): Chemical analysis of Aitken particles (> 0.2 µm Radius) over the Atlantic ocean. Geophys. Res. Letters 2, 45-48.

Winkler, P. (1980): Observations on acidity in continental and in marine atmospheric aerosols and in precipitation. J. Geophys. Res. 85, 4481-4486.

Winkler, P. (1982): Deposition of acid in precipitation. In: Deposition of Atmospheric Pollutants. (Edited by H. W. Georgii and J. Pankrath) pp. 67-76, Reidel Publ. Company, Dordrecht.

Winkler, P. (1983): Acidity of aerosol particles and precipitation in the north polar region over the Atlantic. Tellus 35, 25-30.

THE ORIGIN OF NITROUS AND NITRIC ACID IN THE ATMOSPHERE

Ulrich Platt
KFA Juelich
Institut fuer Chemie 3: Atmosphaerische Chemie
5170 Juelich, Postfach 1913, FRG

ABSTRACT

The various homogeneous and heterogeneous reaction pathways leading to the formation of nitrous acid and nitric acid in the atmosphere as well as possible sources of HNO_2 emission are discussed. While the origin of nitrous acid can still not be explained satisfactorily, recent experimental results show that besides the formation of HNO_3 via OH reactions, nitrate radicals (NO_3) can initiate a reaction sequence of comparable efficiency.

INTRODUCTION

While the oxides of nitrogen (NO and NO_2) in the atmosphere have been extensively studied for more than a decade, the oxacides of nitrogen (HNO_2, HNO_3, HO_2NO_2) are just beginning to receive attention. Of these HNO_2 [Perner 1979, Platt 1980a,b] and HNO_3 [Huebert 1978, Tuazon 1981, Spicer 1982, Cadle 1982] to date have actually been detected in the troposphere. Their significance in the chemistry of the atmosphere is quite different, however. The latter is omnipresent even in unpolluted air and its concentration usually shows no pronounced daily variations. Since HNO_3 (or the NO_3^- ion) is the thermodynamically most stable oxidized nitrogen compound it is the "final fate" of most of the atmospheric NOx. Thus, once formed, HNO_3 and nitrates will be removed from the atmosphere by gas-phase or particle dry deposition and wet removal processes. Thereby HNO_3 deposition will also contribute to the acid rain problem.

The framework of chemical reactions linking the various oxidized nitrogen species occuring in the atmosphere is outlined in fig. 1. Recent research [Platt 1981, 1984a, Noxon 1983] has shown, that besides the "classic" formation mechanism for HNO_3 - the recombination of NO_2 with OH radicals - equally important non-photochemical pathways exist.

NATO ASI Series, Vol. G6
Chemistry of Multiphase Atmospheric Systems
Edited by W. Jaeschke

Nitrous acid on the other hand is a rather transient species during daytime due to its rapid photolysis. To date it has only been detected in polluted air masses, where it probably plays a major role in the photochemical generation and the storage of OH radicals. In addition HNO_2 as a nitrating agent may lead to the formation of nitrosamines.

The origin of atmospheric HNO_2 is less clear: The recombination of NO and OH (see fig. 1), while probably important near strong sources of NO (traffic), cannot explain the observed formation of the species at night. Although direct emission has been shown to play a certain role, and homogeneous formation mechanisms are being speculated about, heterogeneous pathways seem to be more likely.

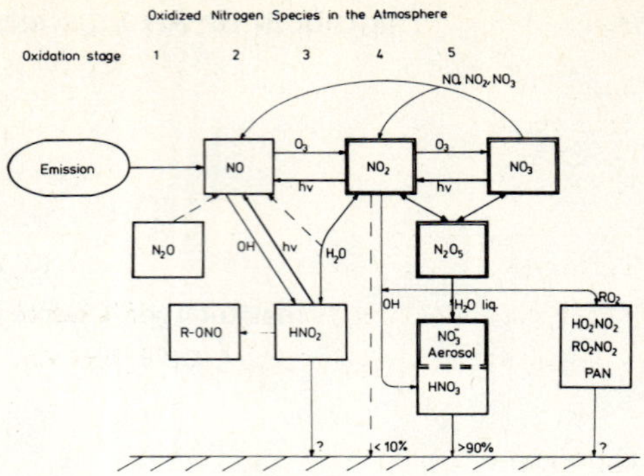

Fig. 1 Reaction scheme of the oxidized nitrogen compounds in the atmosphere.

Table 1: Measurement Techniques for HNO_2

Technique	Interference	Det. Limit ppb typ.	Author
Differential wet chemical analysis	NO_x, PAN	0.1	Nash, 1974
Denuder (Na_2CO_3)/ wet chemical analysis	NO_x, PAN	0.02	Ferm, 1983
Condensation Sampling/ ion chromatography	NO_x, PAN	0.1	Farmer, 1982
Long path UV spectroscopy	NO_2	0.04	Perner, 1979 Harris, 1982, 1983 Platt, 1983
Long path IR spectroscopy	H_2O	~2.5	Hanst, 1982

Table 2: Measurement Techniques for HNO_3

Technique	Interference	Det. Limit ppb typ.	Author
Teflon/Nylon filter +			
- wet chemical analysis	NO_x, PAN	0.01-0.1	Spicer, 1979, 1982
- Ion chromatographic analysis	N_2O_5		Appel, 1980
Impregnated filter/	NO_x, PAN	0.01-0.1	Huebert, 1978
colorimetric analysis	NO_2O_5		Forrest, 1980
1981			Appel, 1980,
			Spicer, 1982
Denuder/chemiluminescence	NO_x, PAN	0.01	Braman, 1982
	N_2O_5		Niessner, 1980
			Klockow, 1983
Denuder/ ion chromatography	NO_x, PAN		Slanina, 1980, 1983
	N_2O_5		
Denuder/ colorimetric analysis	NO_x, PAN		Forrest, 1982
	N_2O_5		
Denuder 'difference'	NO_x, PAN		Shaw, 1982
	N_2O_5		
Condensation Sampling/	NO_x, PAN	0.2	Farmer, 1982
ion chromatography			
Mo-converter/ chemolum.	NO_x, PAN	0.2	Kelly, 1979
differential detection	H_2O	<4	Joseph, 1978
Fourier transform IR	H_2O	6	Tuazon, 1980, 1982
spectroscopy		4	Hanst, 1982
IR diode laser	wall effects	0.4	Brockman, 1978
			Streit, 1979
			Schiff, 1983

MEASUREMENT TECHNIQUES

An overview of the various methods used to measure atmospheric HNO_2 and HNO_3 is given in tables 1 and 2.

For the measurement of gaseous HNO_2 only a small number of techniques have been reported: Wet chemical methods employing bubblers [Nash 1974], denuders [Ferm 1983], or condensation sampling [Farmer 1982] and two spectroscopic methods utilizing the molecule's IR [Hanst 1982] or UV absorption structure [Perner 1979, Harris 1982, Platt 1983]. While the wet chemical methods generally show good sensitivity (<0.1 ppb), their main disadvantage lies in the fact that heterogeneous formation of HNO_2 from other atmospheric oxidized nitrogen components (see for instance reactions 5 and 6 below) at inner surfaces of the sampling device cannot be excluded. From the spectroscopic methods only long path UV - spectroscopy seems to provide sufficient sensitivity to detect ambient levels of HNO_2 (table 1). Here some limitations are due to NO_2 absorption bands overlapping the HNO_2 absorption structures. However, since overlapping bands of known species can be substracted, HNO_2 can still be detected at NO_2/HNO_2 ratios in excess of 400 [Platt 1983]. While the spectroscopic methods are in principle free from wall and chemical interference effects, particulate nitrite cannot be analyzed by these techniques.

Atmospheric HNO_3 concentrations have been measured by a wide variety of methods: The apparently most popular technique employs filters (either pure nylon or impregnated) to scavenge gaseous HNO_3 from the air [Huebert 1978, Spicer 1979, Appel 1980, Grennfeld 1980, Forrest 1981, Harrison 1983]. In order to distinguish between gaseous HNO_3 and particulate nitrate an additional inert (i.e. Teflon) filter is usually placed in front of the HNO_3 scavenging filter. For analysis both filters are leached with water and the nitrate concentration is determined either by wet chemical (colorimetric) methods or by ion chromatography. With this arrangement problems arise from three sources: HNO_3 might evaporate from particles already sampled on the first filter and can be trapped on the second filter [Appel 1981, Shaw 1982], on the other hand, particulate matter collected on the first filter might scavenge part of the gaseous HNO_3 [Appel 1980]. In both cases the apparent NO_3^-/HNO_3 - ratio is changed (while total nitrate is still determined correctly). In addition to those effects - as in the case of HNO_2 - on either of the filters additional nitrate may form from atmospheric NOx thereby leading to an overestimation of the amount of total nitrate. The first two problems can be overcome by the use of denuders or condensation sampling to separate gaseous HNO_3 from particles. The first technique makes use of the fact that aerosol particles diffuse much slower than gases, thus in a gas stream flowing laminarly through a tube, most of the gaseous nitric acid is trapped at the suitable coated (tungstic acid of NaF) walls, while the aerosol particles pass essentially unaffected. A similar effect is used in condensation sampling of atmospheric water vapour at a plate cooled slightly below the dew point of the ambient air, atmospheric concentrations of HNO_3 can then be infered from analyzing the condensate [Farmer 1982]. Later analysis can be done by ion chromatography

[Slanina 1980, 1983] or - in case of the denuder - by heating the tube to drive out the nitrate as NO, which then is analyzed by a chemiluminescence detector [Niessner 1980, Klockow 1983, Braman 1982].

Continuous determination of HNO_3 is possible by "differential" chemiluminescent analysis, where a Molybdenum catalyst is used to convert all oxidized nitrogen species to NO. The fraction being HNO_3 is then determined by forming the difference of instrument readings with and without passing the inlet gas stream through a nylon filter [Kelly 1979, Joseph 1978].

While all the above methods suffer from possible interferences by N_2O_5 and PAN due to HNO_3 production by the hydrolysis of those compounds in the instrument, spectroscopic methods are inherently free from this problem. Since HNO_3 has no structured absorption in the UV and visible spectral regions, only IR spectroscopy has been applied for its detection. Long path Fourier Transform-IR spectroscopy at atmospheric pressure yields relatively high detection limits of a few ppb [Tuazon 1980, 1982, Hanst 1982], partly due to an overlap of HNO_3 lines with H_2O lines. Newer developments use diode laser spectroscopy at reduced pressure, which leads to much less pressure broadening of the absorption lines, thereby eliminating the problem of overlapping bands. Consequently the sensitivity can be improved by about one order of magnitude [Brockman 1978, Streit 1979, Schiff 1983]. Remaining problems are here the possible losses of HNO_3 in the inlet nozzle or at the walls of the chamber housing the optical system.

HNO_2 IN THE ATMOSPHERE

Diurnal concentration profiles of HNO_2 observed in various urban areas (Riverside, fig. 2, Juelich, fig. 3, Cologne, fig. 4) follow similar patterns: After a gradual increase during the night the HNO_2 concentration decreases rapidly after sunrise. Peak levels of up to 8 ppb have been observed in Los Angeles smog [Harris 1982]. A different behaviour of HNO_2 concentration was only observed in the vicinity of freeways carrying heavy traffic, here considerable concentrations (~1 ppb) were found even during daytime [Platt 1984b].

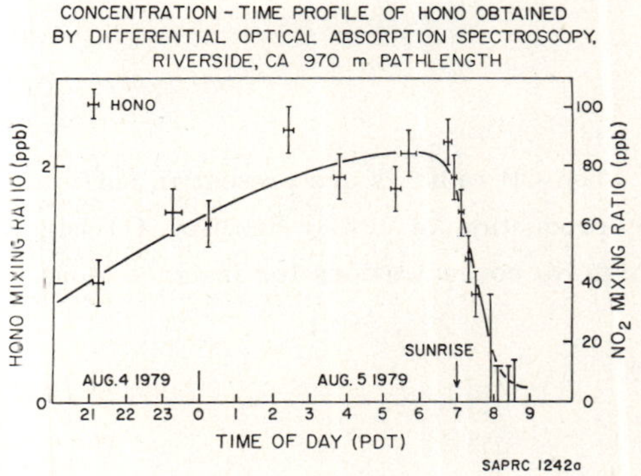

Fig. 2 Nighttime profile of HNO_2 at Riverside/California.

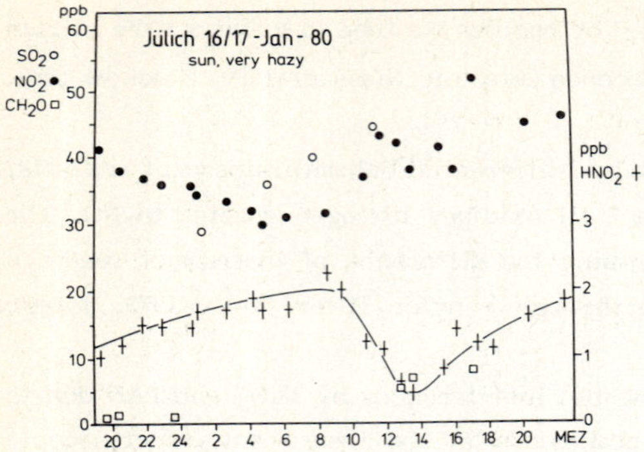

Fig. 3 Nighttime profile of HNO_2 at Juelich/Germany, slow decay of HNO_2 after sunrise due to low light intensity (overcast weather).

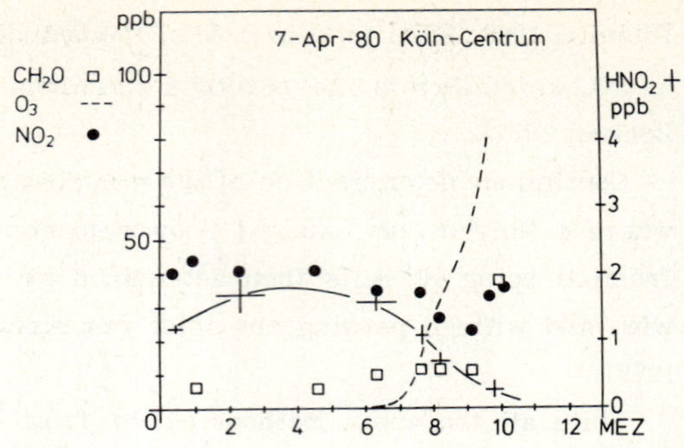

Fig. 4 Nighttime profile of HNO_2 at Cologne/Germany.

Sources of HNO_2

Three groups of possible source mechanisms have been suggested:

1.) HNO_2 may be generated by homogeneous gas phase reactions [Harris 1982, Stockwell 1983, Calvert 1983].

2.) By heterogeneous reactions at aerosol or ground surfaces [Perner 1979, Platt 1980b].

3.) It may be directly emitted as a "by product" from certain sources of NOx [Platt 1984b, Kessler 1984a,b].

The homogeneous recombination of OH radicals with NO, analogous to the reaction of OH with NO_2 will produce HNO_2:

$$OH + NO \xrightarrow{(M)} HNO_2 \qquad (1)$$

This reaction can proceed only during daytime, when OH radicals are present in sufficient quantity, thus it cannot explain the nighttime production of HNO_2. Reaction (1) might, however, be of importance under conditions of high NO concentrations for instance close to routes of automobile traffic.

The reaction involving HO_2 radicals:

$$HO_2 + NO_2 \longrightarrow HNO_2 + O_2 \qquad (2a)$$

has been suggested [Stockwell 1983], however the main product (>99%) of this reaction has been found to be pernitric acid [Howard 1977, Graham 1978b]:

(-2b) $$HO_2 + NO_2 \xleftarrow{(M)}\!\!\rightarrow HO_2NO_2 \qquad (2b)$$

Since HO_2NO_2 is very unstable under tropospheric conditions, undergoing thermal decomposition (reaction -2b) within seconds, there will be essentially no loss of HO_2 radicals through reaction (2b) and even a slow reaction (2a) could be of importance as a source of HNO_2, if there is a non-photochemical formation mechanism for HO_2.

Possibly reactions with alkoxy radicals lead to the formation of HNO_2 [Batt 1979]:

$$R\text{-}CH_2\text{-}O + NO_2 \longrightarrow R\text{-}CHO + HNO_2 \qquad (3)$$

Reactions 1 to 3 are limited by the availability of radicals, especially during periods of low photochemical acitivity, and thus are unlikely candidates for the nighttime HNO_2 formation. Although a radical source mechanism involving the reaction of nitrate radicals with formaldehyde has been postulated by Stockwell and Calvert [1983]:

$$NO_3 + CH_2O \longrightarrow HO_2 + \text{products } (HNO_3) \qquad (4)$$

Further laboratory studies will be required to asses the importance of this reaction sequence (4 followed by 2a).

In any case heterogeneous reactions involving the disproportionation of NOx can proceed without light. These reactions are, however, second order in NOx and thus slow at the low atmospheric NOx levels (table 6) [Chan 1976a,b, Kaiser 1977]:

(-5) $$NO + NO_2 + H_2O \longleftrightarrow 2\ HNO_2 \qquad (5)$$

(-6) $$NO_2 + NO_2 + H_2O \longleftrightarrow HNO_3 + HNO_2 \qquad (6)$$

This also is true for the formation of nitrosylsulfuric acid [Schroeder 1978], which could proceed at sulfuric acid aerosol particles:

$$NO + NO_2 + 2\ H_2SO_4 \longrightarrow 2\ HSO_4NO + H_2O \qquad (7a)$$

$$NO_2 + NO_2 + H_2SO_4 \longrightarrow HSO_4NO + HNO_3 \qquad (7b)$$

Nitrosylsulfuric acid can hydrolyze to form HNO_2:

$$HSO_4NO + H_2O \longrightarrow H_2SO_4 + HNO_2 \qquad (7c)$$

The second order reaction constant for reaction (6) in the liquid phase has been determined by several groups, which essentially agree [Ottolenghi 1968, Sada 1979, Komiyama 1980, Lee

Table 6: Reaction Constants

No.	Reaction	Constant (25C, 1atm) (cm^3/molec. sec^{-1})	Author
1	$OH + NO \xrightarrow{(M)} HNO_2$	$4.9x10^{-12}$	*
2a	$HO_2 + NO_2 \xrightarrow{(M)} HNO_2 + O_2$	$<3.0x10^{-15}$	*
2b	$HO_2 + NO_2 \xrightarrow{(M)} HO_2NO_2$	$1.5x10^{-12}$	*
4	$NO_3 + CH_2O \longrightarrow HO_2 + prod.$	$\sim 1.4x10^{-15}$	Stockwell 1983
5	$NO + NO_2 + H_2O \longrightarrow 2\ HNO_2$	<4.4x10-40**	Kaiser 1977
8	$HNO_3 + NO \longrightarrow HNO_2 + NO_2$	$5.0x10^{-19}$	McKinnon 1979
12	$NO_2 + OH \xrightarrow{(M)} HNO_3$	$1.2x10^{-11}$	*
13a	$N_2O_5 + H_2O \longrightarrow 2\ HNO_3$	$<1.5x10^{-21}$	Platt 1980c
14	$NO_2 + NO_3 \xrightarrow{(M)} N_2O_5$	$8.7x10^{-13}$	*
15	$NO_2 + O_3 \longrightarrow NO_3 + O_2$	$3.0x10^{-17}$	*
17	$NO_3 + NO \longrightarrow 2\ NO_2$	$2.0x10^{-11}$	*

* NASA evaluation [DeMore 1982]

** $cm^6\ molec^{-2}\ sec^{-1}$

1981]. These values lead, even under assumption of favourable conditions (10^{-2} ml/m^3 liquid water in the atmospheric aerosol [Ho 1974] and equilibrium between phases), to a HNO_2 rate of formation about four orders of magnitude lower than observed. It is possible, however, that reaction (5 - 7) may be surface catalyzed as indicated by Chan et al. [1976] and Kaiser and Wu [1977]. Also the reactions might be accelerated catalytically or could proceed in larger amounts of liquid water present at the ground.

Finally HNO_2 may be formed through reduction of nitric acid [McKinnon 1979] in the gas or liquid phase:

$$HNO_3 + NO \longrightarrow HNO_2 + NO_2 \tag{8}$$

In conclusion it can be said, that there is as yet no satisfactory explanation for the origin of nitrous acid in the atmosphere, however, combining the results of atmospheric observations and smog chamber experiments [Kessler 1984b] a first order dependence of the rate of HNO_2 formation from the NO_2 (not NO) mixing ratio is suggested.

In principle HNO_2 can be formed during all NOx producing combustion processes, thus in polluted areas HNO_2 could be emitted from home heating, car exhaust, and power plants.

Because of the low furnace temperatures and associated low NOx levels, home heating is probably not a significant source of HNO_2. Also, measurements at a central heating plant showed no measurable HNO_2 concentration in the stack gases (HNO_2/NOx < 2* 10^{-4}, Kessler

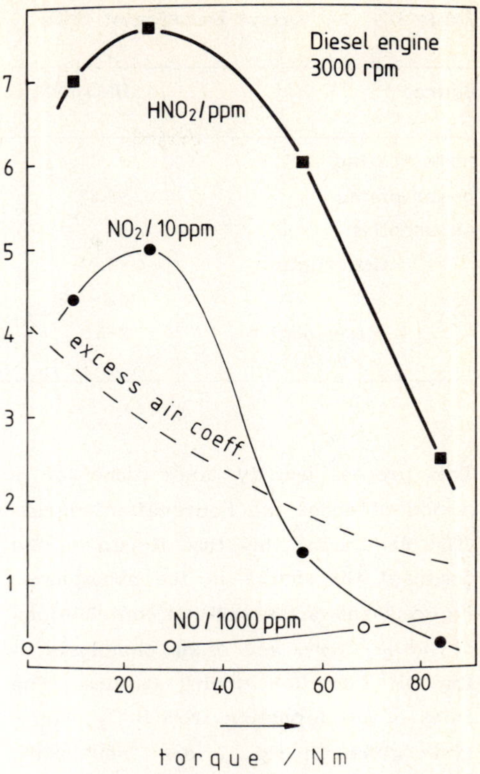

Fig. 5 Dependence of HNO_2, NO_2, and NO mixing ratios in the exhaust gas of an automobile Diesel engine from torque (load).

1984b). The contribution from automobile exhaust is probably small: Nitrous acid can be detected in the exhaust gas of both spark ignition and Diesel type internal combustion engines [Hilliard 1979, Pitts 1982, 1984, Kessler 1984a,b]. While the spark ignition engine normally shows very low HNO_2/NOx ratios (below $2 * 10^{-4}$, only by tuning the carburator to produce lean fuel/air mixtures, HNO_2/NOx ratios up to $2 * 10^{-3}$ could be obtained) with Diesel engines HNO_2/NOx ratios of 0.02 to 0.05 were observed (fig. 5). While the fraction of Diesel-fuel from total fuel used by automobiles is in the range of 10 to 30%, due to the lower NOx emission factor of Diesel engines, the fraction of NOx emitted by Diesel cars is only in the range of 2 to 6%. Consequently HNO_2/NOx ratios not exceeding $3 * 10^{-3}$ are to be expected in the mixed exhaust of normal traffic, which is about one order of magnitude below those frequently found in the atmosphere (0.05).

Table 3 summarizes the various possible sources of direct HNO_2 emission. It can be concluded, that direct emission from automobiles contributes to the HNO_2 budget to some extent, at least near traffic ways, on the other hand, direct emission cannot be the only source of HNO_2.

A possible consequence of the presence of HNO_2 in the atmosphere is the formation of nitrosamines from secondary amines:

$$HNO_2 + RR'NH \longrightarrow RR'NNO + H_2O \qquad (9)$$

[Famy 1976, Hanst 1977, Pitts 1978]. Reaction (9) might proceed either in the atmosphere or after inhalation of HNO_2 with the air.

An important reaction for the chemistry of the polluted atmosphere is the generation of OH radicals by the photolysis of HNO_2:

$$HNO_2 + h\nu \longrightarrow OH + NO \qquad (10)$$

Table 3: Direct Emission of HNO_2

Source	HNO_2/NO_x(%)	Remark
home heating	-	probably low due to low NO_x
power plants	<0.02	observed
automobiles		
Otto engine	<0.01	normal operation
	0.1-0.2	lean operation
Diesel engine	2-5	

This process usually takes place for a period of about two hours after sunrise (fig. 6). During this time it can be the dominant OH source in the atmosphere. Figure 7 shows the relative contributions of HNO_2, CH_2O, and ozone photolysis to the OH formation during daytime. The rates of OH formation from HNO_2 photolysis given in fig. 7 are "net" rates calculated from the observed decay of HNO_2, the actual rate of HNO_2 photolysis (10) can be about a factor of two larger, but a fraction of the OH radicals generated will immediately reform HNO_2 (reaction 1). Due to this effect as well as due to the lower UV intensity the apparent rate of HNO_2 photolysis in the morning is slower than expected from the noontime photolysis frequency of $1.7 * 10^{-3}$ sec^{-1} [Stockwell 1978]. Model calculations confirm, that the formation of photochemical smog can be accelerated by this "morning pulse" of OH [Harris 1982, Pitts 1983], for instance, depending on NOx and hydrocarbon concentrations there can be up to 50% more ozone production during the course of the day (fig. 8), if 4% of the initial NOx are present as HNO_2.

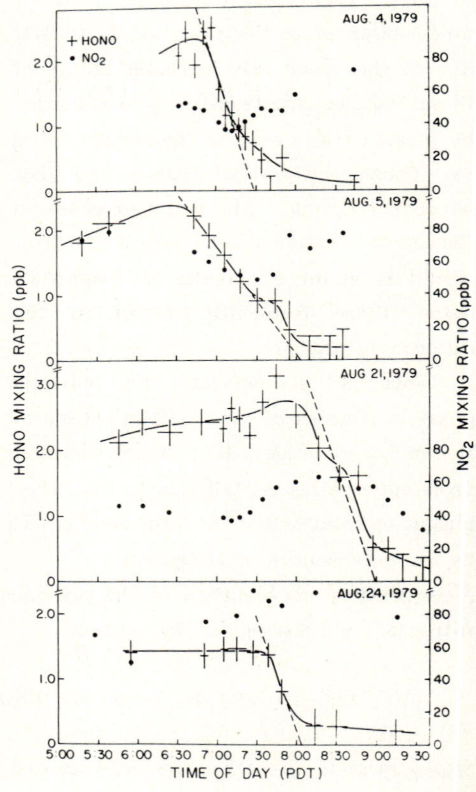

Fig. 6 The photolysis of HNO_2 in the morning (Riverside/USA).

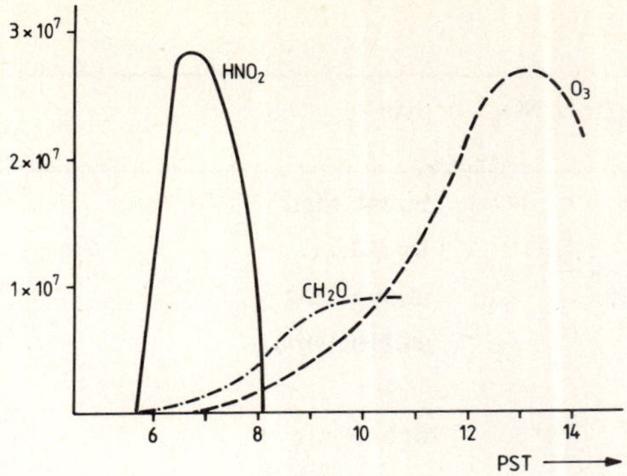

Fig. 7 Comparison of the diurnal variation of the OH formation rates from HNO_2-, CH_2O-, and O_3-photolysis (Los Angeles). The rates are calculated from the observed slope of HNO_2 decrease and measured CH_2O and O_3 concentrations, respectively.

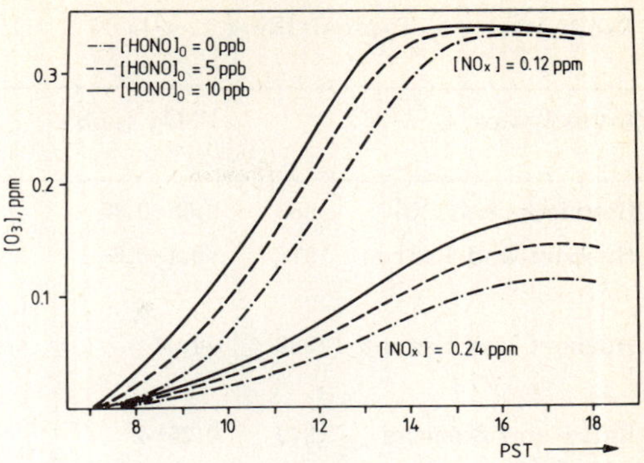

Fig. 8 O_3 formation modelled for 0, 5, and 10 ppb of the NOx initially present as HNO_2 (after Harris et al. (1982)).

While photolysis clearly is the dominating sink for HNO_2 during daytime, reactions of higher order in HNO_2 could constitute significant sink mechanisms [Andrew 1961] at night:

$$HNO_2 + HNO_2 \longrightarrow NO + NO_2 + H_2O \qquad (-5)$$

$$3\ HNO_2 \longrightarrow HNO_3 + 2\ NO + H_2O \qquad (11)$$

HNO_3 IN THE ATMOSPHERE

Unlike HNO_2, nitric acid has not only been found in polluted areas [see for instance Miller 1975, Tuazon 1981, Spicer 1982, Harrison 1983] but also in relatively clean air masses [Huebert 1978, 1980a,b, Kelly 1979], although at very low concentrations (table 5). Due to its stability against chemical degradation and its slow photolysis HNO_3 will have a relatively long atmospheric lifetime, essentially only being removed by dry or wet deposition and incorporation into aerosol particles, which then are deposited.

Table 4: Typical HNO_3 Levels

Investigator		HNO_3 (ppb)	HNO_3/tot. NO_x	Site
Harrison and Pio	1983	0.05-0.36	~0.12	rural atm.
Huebert and Lazrus	1978	<0.1-0.9	< 1	in B.L.
			0.5-10	above B.L.
Huebert and Lazrus	1980	~0.04	0.3	marine atm.
Kelly and Stedman	1979	0.25-4		urban air
Tuazon et al.	1980	up to 20	~0.1	urban air

Sources of HNO_3

Again three mechanisms of HNO_3 formation are of importance:
Homogeneous reaction of OH radicals, hydrolysis of N_2O_5, and heterogeneous reactions of NOx (see reactions 6, 7b).

At noontime and clear sky conditions, the lifetime of NO_2 against conversion by reaction with OH

$$NO_2 + OH \xrightarrow{(M)} HNO_3 \qquad (12)$$

is only a few hours (below 3h at sea level, 10^7 OH cm^{-3}). However, at night, during overcast weather, or at winter time at higher latitudes, OH levels become very low and consequently the effectiveness of reaction (12) decreases considerably [Logan 1981].

During these periods of time, or in air masses with high liquid water contents (i.e. in clouds or fog) the heterogeneous hydrolysis of N_2O_5 or a heterogeneous reaction of NO_3 with liquid water may become an important source of HNO_3:

$$N_2O_5 + H_2O \longrightarrow 2\ HNO_3 \qquad (13a)$$

$$NO_3 + H_2O \longrightarrow HNO_3 + \text{other products} \qquad (13b)$$

Where N_2O_5 is formed in the equilibrium reaction [Graham 1978a]:

$$(-14) \qquad NO_2 + NO_3 \xleftarrow{(M)}\!\!\rightarrow N_2O_5 \qquad (14)$$

from nitrate radicals generated by the reaction of NO_2 with ozone

$$NO_2 + O_3 \longrightarrow NO_3 + O_2 \tag{15}$$

Since the equilibrium (-14, 14) is reached within minutes at conditions of the lower troposphere [Malko and Troe 1982], the rate limiting steps for the conversion of NOx to HNO_3 can either be reaction (13) or reaction (15) depending on atmospheric conditions, which will be discussed in the following:

While reaction (13) is known to proceed rapidly with liquid water, the reaction of N_2O_5 with water vapour seems to be slow. Morris and Niki [1973], Platt et al. [1980c], and Tuazon et al. [1982] estimated upper limits for k_{13a} of 1.3×10^{-20}, 1.5×10^{-21}, and 1.3×10^{-21} cm^3 $molec^{-1}$ sec^{-1}, respectively, from laboratory and field investigations. Also, from the nighttime NO_3 observations made at Deuselbach [Platt 1981], NO_3 lifetimes of the order of 20 to 60 minutes are derived, applying the method used by Platt et al. [1980c]. This corresponds to upper limits for k_{13a} of 3×10^{-22} cm^3 $molec^{-1}$ sec^{-1}. Thus reaction (13) proceeding with atmospheric water vapour will be of minor importance in most cases.

On the other hand recent observations have shown, that under conditions of high relative humidity (R.H. > 50%) the lifetime of atmospheric NO_3 becomes very short (of the order of minutes) [Platt 1981, 1982, 1984a], which has been attributed to the rapid heterogeneous reaction (13) removing N_2O_5 and thus also NO_3 [Ehhalt 1982, Richards 1982, Heikes 1983, Platt 1981, 1984a]. Figures 9 and 10 show two nighttime profiles of NO_3, where its concentration sharply drops after fog (fig. 9) or rain (fig. 10) comes up. While during conditions of fog or rain clearly a large surface area of liquid water is present, the observation of strongly reduced NO_3 lifetime at R.H. above 50% shows, that even deliquescent aerosol [Winkler 1973] (or perhaps moist ground surface) is efficient in reacting with N_2O_5. In fact from simultaneously measured aerosol light scattering data [Platt 1984a] reaction probabilities between 10^{-3} and 10^{-2} per molecule - particle collision can be estimated. Those reaction probabilities are quite reasonable in view of values between 0.2 and 0.4, which have been observed in the reaction of gaseous NH_3 with H_2SO_4 aerosol [McMurry 1983]. Under conditions of high R.H., which are normal in moderate climates at night, the conversion of NOx to HNO_3 is therefore probably only limited by reaction (15). This reaction has been studied by a large number of investigators (see for instance the compilation by Baulch et al. [1980]) with agreeing results, thus for those con-

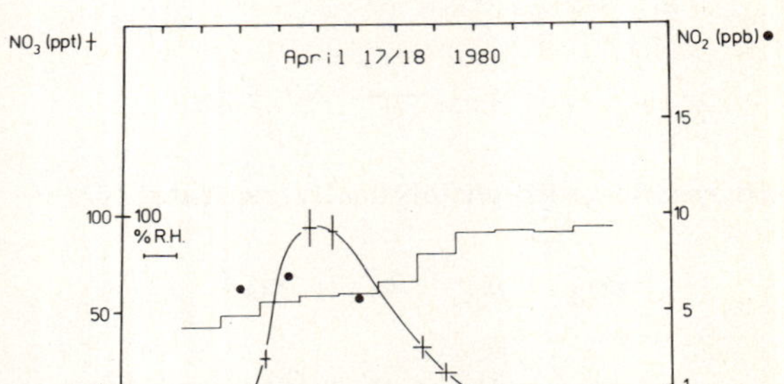

Fig. 9 NO_3 night profile, Deuselbach, 1981, haze after midnight.

Table 5: Different reaction pathways converting NO_x to HNO_3

Mechanism	OH conc. molec/cm^3	O_3 conc. ppb	NO_x conversion rate %/h	remark
$OH + NO_2$	5×10^5 a	-	2	daytime
	2×10^6 b	-	8	daytime
$O_3 + NO_2$ followed by:				
$NO_3 + H_2O_{liq.}$	-	30	8	predominantly
	-	50	13	at night
$NO_3 + NO_2 \longrightarrow N_2O_5$				
and	-	30	15	predominantly
$N_2O_5 + H_2O_{liq.}$	-	50	26	at night
$NO_2 + NO_2 + H_2O$	-	-	0.25	day/night

a typical average OH level [Volz 1981].

b typical surface daytime OH level [Huebler 1984].

ditions the NO_2 to HNO_3 conversion rate can be estimated from temperature and O_3 concentration alone.

It should be noted, that HNO_3 is also a product of NO_3 reacting with aldehydes [Morris 1974] like reaction (3) and possibly with olefines [Hoshino 1978].

During daytime photolysis [Magnotta 1980]

$$NO_3 + h\nu \longrightarrow NO_2 + O \qquad (16a)$$

$$\longrightarrow NO + O_2 \qquad (16b)$$

and reaction with photolytically generated NO

$$NO_3 + NO \longrightarrow 2\ NO_2 \qquad (17)$$

compete with reaction (14). The lifetime of NO_3 then is only of the order of seconds (0.2 sec^{-1} photolysis frequency), so that the reaction sequence described above becomes less effective, except, perhaps, inside clouds, were collision frequencies for molecule - droplet collisions are of the order of 10 sec^{-1}.

Heterogeneous formation of HNO_3 can occur through the reactions (6) and (7b). If those reactions take place, they will produce HNO_3 in amounts at least equal to those of HNO_2. Thus from observations of HNO_2 a lower limit of this reaction pathway can be estimated. Since HNO_2 might autodecompose immediately after formation in the liquid phase (reactions -5 and 11), while HNO_3 does not, there can be much more HNO_3 produced than HNO_2 is observed in the gas phase.

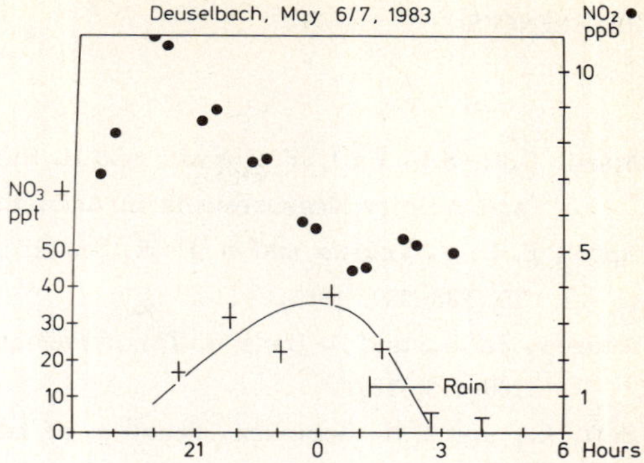

Fig. 10 NO_3 night profile, Deuselbach, 1983, rain after 3:00 in the morning.

As can be seen from table 5 comparing the various mechanisms converting NOx to HNO_3, the OH reaction will be the dominating mechanism under conditions of high photochemical activity, while dark reactions initiated by reaction (15) can be equally important during nighttime.

CONCLUSION

The overall mechanism of the HNO_3 formation and its removal from the atmosphere seems to be reasonable well understood. The relative importance of photochemical and recently recognized non-photochemical formation pathways is roughly equal, although large variations with UV illumination and humidity conditions can be expected. The role of clouds and fog in the HNO_3 formation needs further clarification.

Nitrous acid has been shown to be an important initiator of photochemical smog, however the known source mechanisms of atmospheric HNO_2 (direct emission, homogeneous and heterogeneous reactions) cannot account for the amounts observed in polluted areas.

REFERENCES

Appel, B.R., S.M. Wall, Y. Tokiwa, and M. Haik, Simultaneous Nitric Acid, Particulate Nitrate and Acidity Measurements in Ambient Air. Atmosph. Environ. 14, 549-554, 1980.

Appel, B.R., Y. Tokiwa and M. Haik, Sampling of Nitrates in Ambient Air. Atmosph. Environ. 15, 283-289, 1981.

Andrew, S.P.S., and D. Hanson, The dynamics of nitrous gas absorption, Chem. Eng. Sci. 14, 105-113, 1961.

Batt, L., and G.N. Robinson, Reaction of methoxy radicals with oxygen. I. Using dimethyl peroxyde as a thermal source of methoxy radicals. Int. J. Chem. Kin. 11, 1045-1053, 1979.

Baulch, D.L., R.A. Cox, R.F. Hampson, J.A. Kerr, J. Troe, and R.T. Watson, Evaluated kinetic and photochemical data for atmospheric chemistry. J. Phys. Chem. Ref. Data. 9, 295-471, 1980.

Braman, R.S., and T.J. Shelley, W.A. McClenny, Tungstic Acid for Preconcentration and Determination of Gaseous and Particulate Ammonia and Nitric Acid in Ambient Air. Anal. Chem. 54, 358-364, 1982.

Brockman, P., C.H. Bair, and F. Allario, High Resolution Spectral Measurement of the HNO_3 11.3µm Band Using Tunable Diode Lasers, Appl. Optics 17, 91-100, 1978.

Cadle, S.H., R.J. Countess, and N.A. Kelly, Nitric acid and ammonia in urban and rural locations. Atm. Environ. 16, 2501-2506, 1982.

Calvert, J.G. and W.R. Stockwell, Acid Generation in the Troposphere by Gas-Phase Chemistry. Environ. Sci. Technol. 17, 428A-443A, 1983.

Chan, W.H., R.J. Nordstrom, J.G. Calvert, and J.H. Shaw, Kinetic study of HONO formation and decay reactions in gaseous mixtures of HONO, NO, NO_2, H_2O, and N_2. Environ. Sci. Techn. 10, 674-682, 1976a.

Chan, W.H., R.J. Nordstrom, J.G. Calvert, and J.H. Shaw, An IRFTS spectroscopic study of the kinetics and the mechanism of the reactions in the gaseous system, HONO, NO, NO_2, H_2O. Chem. Phys. Lett. 37, 441-446, 1976b.

DeMore, W.B., R. Watson, D. Golden, R. Hampson, M. Kurylo, C. Howard, M. Molina, Chemical kinetics and photochemical data for use in stratospheric modeling Evaluation Number 5. JPL Publication 82-57, NASA, 1982.

Ehhalt, D.H., and J.W. Drummond, The tropospheric cycle of NOx: Chem. Unpoll. Poll. Troposphere, Proc. NATO Adv. Study Inst., Corfu, Greece, 28 Sept.-10 Oct., 1981, H.W. Georgii, W. Jaeschke (Ed.), Reidel, Dordrecht, Boston, London, 219-251, 1982.

Famy, O.G., and M.J. Famy, Mutagenicity of N-a-Acetoxyethyl-N-ethylnitrosamine and N, N-Diethylnitrosamine in Relation to the mechanism of metabolic activation of dialkyl-nitrosamines. Cancer Res. 36, 4504-4512, 1976.

Farmer, J.C., and G.A. Dawson, Condensation sampling of soluble atmospheric trace gases. J. Geophys. Res. 87, 8931-8942, 1982.

Ferm, M., and A. Sjoedin, A simple method for determination of gaseous nitrous acid in the atmosphere. Report IVL Publ 718, Swed. Environ. Res. Inst., Gothenburg, Swe., 1983.

Forrest, J., R.L. Tanner, D. Spandau, T. D'Ottavio, and L. Newman, Determination of Total Inorganic Nitrate Utilizing Collection of Nitric Acid on NaCl - Impregnated filters. Atmosph. Environ. 14, 137-144, 1980.

Forrest, J., D.J. Spandau, R.L. Tanner, and L. Newman, Determination of Atmospheric Nitrate and Nitric Acid Employing a Diffusion Denuder with a Filter Pack. Atmosph. Environ. 16, 1473-1485, 1982.

Graham, R.A., and H.S. Johnston, The photochemistry of NO_3 and the kinetics of the N_2O_5 - O_3 system. J. Phys. Chem. 82, 254-268, 1978a.

Graham, R.A., A.M. Winer, and J.N. Pitts, Pressure and temperature dependence of the unimolecular decomposition of HO_2NO_2. J. Chem. Phys. 68, 4505-4510, 1978b.

Grennfelt, P., Investigation of gaseous nitrates in an urban and rural area. Atmosph. Environ. 14, 311-316, 1980.

Hanst, P.L., N.W. Wong, and J. Bragin, A long-path infra-red study of Los Angeles smog. Atm. Env. 16, 969-981, 1982.

Hanst, P.L., J.W. Spence, and M. Miller, Atmospheric chemistry of N-nitroso dimethylamine. Env. Sci. Technol. 11, 403-405, 1977.

Harris, G.W., W.P.L. Carter, A.M. Winer, J.N. Pitts, U. Platt, and D. Perner, Observations of nitrous acid in the Los Angeles atmosphere and implications for the predictions of ozone-precursor relationships. Environ. Sci. Techn. 16, 414-419, 1982.

Harris, G.W., A.M. Winer, J.N. Pitts, U. Platt, and D. Perner, Measurement of HONO, NO_3 and NO_2 by long-path differential optical absorption spectroscopy in the Los Angeles basin. Optical and Laser Remote Sensing, D.K. Killinger and A. Mooradian (Eds.) Springer Ser. Optical Sciences 39, 106-113, 1983.

Harrison, R.M., C.A. Pio, An Investigation of Atmospheric HNO_3-NH_3-NH_4NO_3 Equilibrium Relationship in a Cool Humid Climate, Tellus 35B, 155-159, 1983.

Heikes, B.G. and A.M. Thompson, Effects of heterogeneous processes on NO_3, HONO, and HNO_3 chemistry in the troposphere. J. Geophys. Res. 88, 10883-10895, 1983.

Hilliard, J.C., and R.W. Wheeler, Nitrogen dioxide in engine exhaust. Paper 790691, SAE, Warrendale, Pennsylvania, 1979.

Ho, W., G.M. Hidy, and R.M. Gowan, Microwave measurements of the liquid water content of atmospheric aerosols. J. Appl. Meteorol. 13, 871-879, 1974.

Hoshino, M., T. Ogata, H. Akimoto, G. Inoue, F. Sakamak, and M. Okuda, Gas Phase Reactions of N_2O_5 with Propylene. Chemistry Letters, Chem. Soc. Japan, 1367-1370, 1978.

Howard, C.J. and K.M. Evenson, Kinetics of the reaction of HO_2 with NO. Geophys. Res. Lett. 4, 437-440, 1977.

Huebert, B.J. and A.L. Lazrus, Global Tropospheric Measurements of Nitric Acid Vapor and Particulate Nitrate. Geophys. Res. Lett. 5, 577-580, 1978.

Huebert, B.J., Nitric Acid and Aerosol Nitrate Measurements in the Equatorial Pacific Region. Geophys. Res. Lett. 7, 325-328, 1980a.

Huebert, B.J., and A.L. Lazrus, Tropospheric gas-phase and particulate nitrate measurement. J. Geophys. Res. 85, 7322-7328, 1980b.

Huebler G., D. Perner, U. Platt, A. Toennissen, and D.H. Ehhalt, Groundlevel OH radical concentration: New measurements by optical absorption. J. Geophys. Res. 89, 1309, 1984.

Joseph, D.W., and C.W. Spicer, Chemiluminescence Method for Atmospheric Monitoring of Nitric Acid and Nitrogen Oxides. Analyt. Chem. 50, 1400-1403, 1978.

Kaiser, E.W., C.H. Wu, A kinetic study of the gas phase formation and decomposition reactions of nitrous acid. J. Phys. Chem. 81, 1701-1706, 1977.

Kelly, T.J., D.H. Stedman, and G.L. Kok, Measurements of H_2O_2 and HNO_3 in rural air. Geophys. Res. Lett. 6, 375-378, 1979.

Kelly, T.J., D.H. Stedman, J.A. Ritter, and R.B. Harvey, Measurements of oxides of nitrogen and nitric acid in clean air. J. Geophys. Res. 85, 7417-7425, 1980.

Kessler, C., D. Perner, and U. Platt, Spectroscopic measurements of nitrous acid and formaldehyde - Implications for urban photochemistry. Proc. 2nd Europ. Symp. on Physico-Chemical Behaviour of Atm. Pollut., Varese, Italy. 29. Sept - 1. Oct, CEC, B. Versino and H. Ott (ed.), 393-400, 1981.

Kessler, C., D. Perner, U. Platt, and K.-H. Ertl, The emission of nitrous acid and nitrogen oxides by automobile engines - A spectroscopic study. In preparation 1984a.

Kessler, C., Gasfoermige salpetrige Saeure (HNO_2) in der belasteten Atmosphaere. Ph. D. Thesis, Univ. Koeln, 1984b.

Kley, D., J.W. Drummond, M. McFarland, and S.C. Liu, Tropospheric profiles of NOx. J. Geophys. Res. 86, 3153-3161, 1981.

Klockow, D., R. Niessner, M. Richter, G. Schmidt, Grundlagen der diffusionskontrollierten Probennahme von Luft und ihre Anwendung zur selektiven Erfassung atmosphaerischer Spurenstoffe. Symp. Anorg. Analytik in Umweltforschung und Umweltschutz, Juelich 13.-16. Juni, 1983.

Komiyama, H., and H. Inohue, Absorption of nitrogen oxides into water. Chem. Eng. Sci. 35, 154-161, 1980.

Lee, Y.N., and S.E. Schwartz, Reaction Kinetics of Nitrogen Dioxide with liquid water at low partial pressure. J. Phys. Chem. 85, 840-848, 1981.

Logan, J.A., M.J. Prather, S.C. Wofsy, and M.B. McElroy, Tropospheric chemistry: A global perspective. J. Geophys. Res. 86, 7210-7254, 1981.

Magnotta, R., and H.S. Johnston, Photodissociation quantum yields for the NO_3 free radical. Geophys. Res. Lett. 7, 769-772, 1980.

Malko, M.W., and J. Troe, Analysis of the unimolecular reaction $N_2O_5 + M \rightarrow NO_2 + NO_3 + M$. Int. J. Chem. Kin. 14, 399-416, 1982.

McKinnon, J.G. Mathieson, and I.R. Wilson, Gas phase reaction of nitric oxide with nitric acid. J. Phys. Chem. 83, 779-780, 1979.

McMurry, P.H., H. Takano, and G.R. Anderson, Study of the Ammonia (Gas)-Sulfuric Acid (Aerosol) Reaction Rate. Environ. Sci. Technol. 17, 347-352, 1983.

Miller, D.F., and C.W. Spicer, Measurement of nitric acid in smog. J. Air Poll. Ass., 940-942, 1975.

Morris, E.D., and H. Niki, Reaction of Dinitrogen Pentoxide with Water. J. Phys. Chem. 77, 1929-1932, 1973.

Morris, E.D., and H. Niki, Reaction of the nitrate radical with acetaldehyde and propylene. J. Phys. Chem. 78, 1337-1338, 1974.

Nash, T., Nitrous acid in the atmosphere and laboratory experiments on its photolysis. Tellus 26, 175-179, 1974.

Niessner, R., and D. Klockow, A thermoanalytical approach to speciation of atmospheric strong acid. Intern. J. Environ. Anal. Chem. 8, 163-175, 1980.

Noxon, J.F., NO_3 and NO_2 in the Mid-Pacific Troposphere. J. Geophys. Res. 88, 11017-11021, 1983

Ottolenghi, M. and J. Rabani, Photochemical Generation of Nitrogen Dioxide in Aqueous Solutions. J. Phys. Chem. 72, 593-598, 1968.

Perner, D., and U. Platt, Detection of nitrous acid in the atmosphere by differential optical absorption. Geophys. Res. Lett. 6, 917-920, 1979.

Pitts, J.N., D. Grosjean, K. van Cauwenberghe, J.P. Schmid, and D.R. Fritz, Photooxidation of aliphatic amines under simulated atmospheric conditions: Formation of nitrosamines, nitramines, amides, and photochemical oxidants. Environm. Sci. Technol. 12, 946-953, 1978.

Pitts, J.N., H.W. Biermann, E.C. Tuazon, U.F. Platt, D.B. Duignan, Direct measurements of nitrous acid, nitrogen dioxide, and formaldehyde in auto exhaust by differential optical absorption spectroscopy. Progress Report for ARB, 1982.

Pitts, J.N., A.M. Winer, R. Atkinson, and W.P.L. Carter, Comment on "Effect of nitrogen oxide emissions on ozone levels in metropolitan regions", "Effect of NOx emission rates on smog formation in the California South Coast Air Basin", and "EFfect of hydrocarbon and NOx on photochemical smog formation under simulated transport conditions". Environm. Sci. Technol. 17, 54-57, 1983.

Pitts, J.N., H.W. Bierman, A.M. Winer, and E.C. Tuazon, Spectroscopic identification and measurement of gaseous nitrous acid in dilute auto exhaust. Atmosph. Environm. in press, 1984.

Platt, U., and D. Perner, Direct measurements of atmospheric CH_2O, HNO_2, O_3, NO_2, and SO_2 by differential optical absorption in the near UV. J. Geophys. Res. 85, 7453-7458, 1980a.

Platt, U., D. Perner, G.W. Harris, A.M. Winer, and J.N. Pitts, Observation of HONO in an urban atmosphere by differential optical absorption. Nature 285, 312-314, 1980b.

Platt, U., D. Perner, G.W. Harris, A.M. Winer, and J.N. Pitts, Detection of NO_3 in the polluted troposphere by differential optical absorption. Geophys. Res. Lett. 7, 89-92, 1980c.

Platt, U., D. Perner, J. Schroeder, C. Kessler, and A. Toenissen, The diurnal variation of NO_3. J. Geophys. Res. 86, 11965-11970, 1981.

Platt, U., D. Perner, and C. Kessler, The importance of NO_3 for the atmospheric NOx cycle from experimental observations. Preprints 2nd symp. Compos. Nonurban Trop., May, 25-28, Williamsburg, Va., AMS, AGU, NASA, 21-24, 1982.

Platt, U., and D. Perner, Measurements of atmospheric trace gases by long path differential uv/visible absorption spectroscopy: Optical and Laser Remote Sensing, D.K. Killinger and A. Mooradian (Eds.) Springer Ser. Optical Sci. 39, 95-105, 1983.

Platt, U., A.M. Winer, H.W. Bierman, R. Atinson, and J.N. Pitts, Measurement of Nitrate Radical Concentrations in Continental Air. Environm. Sci. Techn., submitted 1984a.

Platt, U., J.N. Pitts,, and A.M. Winer, Spectroscopic Measurements of Diurnal Profiles of HONO, HCHO, and NO_2 along a California Freeway. Environm. Sci. Techn., to be submitted 1984b.

Richards, L.W., Comments on the oxidation of NO_2 to nitrate - day and night. Atmosph. Environm. 17, 397-402, 1983.

Sada, E., H. Kumazawa and M.A. Butt, Single and Simultaneous Absorptions of Lean SO_2 and NO_2 into Aqueous Slurries of Ca(OH)2 or Mg(OH)2 Particles. J. Chem. Ing. Japan 12, 111-117, 1979.

Schiff, H.I., D.R. Hastie, G. I. Mackay, T. Iguchi, and B.A. Ridley, Tunable Diode Laser Systems for Measuring Trace Gases in Tropospheric Air. Environm. Sci. Technol. 17, 352A-364A, 1983.

Schroeder, W.H., and P. Urone, Isolation and identification of nitrosonium hydrogen sulfate ($NOHSO_4$) as a photochemical reaction product in air containing sulfur dioxide and nitrogen oxide. Environm. Sci. Techn. 12, 545-550, 1978.

Schwartz, S.E., and W.H. White, Solubility equilibria of the nitrogen oxides and oxyacids in dilute aqueous solution. Adv. Environ. Sci. Eng. 4, in press. J.R. Pfafflin and E.N. Ziegler, Eds., Gordon and Breach, New York, 1980.

Shaw, R.W. Jr., R.K. Stevens, J. Bowermaster, J.W. Tesch, and E. Tew, Measurements of Atmospheric Nitrate and Nitric Acid: The Denuder Difference Experiment. Atmosph. Environm. 16, 845-853, 1982.

Slanina, J., L. V. Lamoen-Doornbal, W.A. Lingerak and W. Meilof, Application of a thermo-denuder analyser to the determination of H_2SO_4, HNO_3 and NH_3 in air. Stichting Energieonderzoek Centr. Nederl. 1-21, 1980.

Spicer, C.W., J.E. Howes, T.A. Bishop and L.H. Arnold, Nitric acid measurement methods: An intercomparison. Atmosph. Environ. 16, 1487-1500, 1982.

Stockwell, W.R., J.G. Calvert, The near ultraviolet absorption spectrum of gaseous HONO and N_2O_3. J. Photochem. 8, 193-203, 1978.

Stockwell, W.R., and J. Calvert, The mechanism of NO_3 and HNO_2 formation in the nighttime chemistry of the urban atmosphere. J. Geophys. Res. C88, 6673-6682, 1983.

Streit, G.E. and J.S. Wells, A tuneable diode laser study of the reaction of nitric and nitrous acids: HNO_3 + NO and HNO_2 + O_3, J. Chem. Phys. 70, 3439-3443, 1979.

Tang, I.N., Phase transformation and growth of aerosol particles composed of mixed salts. J. Aerosol Sci. 7, 361-371, 1976.

Tang, I.N., H.R. Munkelwitz, and J.G. Davis, Aerosol Growth Studies - II. Preparation and growth measurements of monodisperse salt aerosols, J. Aerosol Sci. 8, 149-159, 1977.

Tuazon, E.C., A.M. Winer, R.A. Graham, and J.N. Pitts, Jr., Atmospheric measurements of trace pollutants by kilometer-pathlength FT-IR spectroscopy. Environ. Sci. Technol. 10, 259-299, 1980.

Tuazon, E.C., A.M. Winer, and J.N. Pitts, Trace Pollutant Concentrations in a Multiday Smog Episode in the California South Coast Air Basin by Long Path Length Fourier Transform IT Spectroscopy. Environm. Sci. Technol. 15, 1232-1237, 1981.

E.C. Tuazon, R. Atinson, C.N. Plum, A.M. Winer, and J.N. Pitts, The Reaction of gas phase N_2O_5 with water vapor. Geophys. Res. Lett. 10, 953-956, 1983.

Volz, A., D.H. Ehhalt, and R.G. Derwent, Seosonal and latitudinal variation of ^{14}CO and the tropospheric concentration of OH radicals. J. Geophys. Res. 86, 5163-5171, 1981.

Winkler, P., The growth of atmospheric particles as a function of the relative humidity. II. Improved concept of mixed nuclei. Aerosol. Sci. 4, 373-387, 1973.

ON THE IMPORTANCE OF LIGHT HYDROCARBONS IN MULTIPHASE ATMOSPHERIC SYSTEMS

D.H. Ehhalt, J. Rudolph, and U. Schmidt
Institut fuer Chemie 3: Atmosphaerische Chemie
der Kernforschungsanlage Juelich GmbH
D-5170 Juelich
F.R.G.

ABSTRACT

The role of organic molecules in multiphase atmospheric systems is illustrated by two examples. The first draws attention to the possible role of an organic surface film on aerosol particles; the second discusses the role of "in cloud" chemistry in the formation of HCOOH and CH_3COOH from light alkanes. Vertical and meridional distributions of C_2 to C_5 hydrocarbons are presented which underline their importance at the midlatitudes of the northern hemisphere. An attempt is made to quantify the global emissions of individual light nonmethane hydrocarbons, and the various sources of C_2 - C_5 hydrocarbons are briefly discussed.

INTRODUCTION

Organic compounds are present in all phases of the atmospheric chemical system. They constitute a major class of gaseous trace species, they occur in precipitation, and they form a significant fraction of the atmospheric aerosol.

Obviously, there must exist heterogeneous chemical pathways which interconnect the organic molecules in those three "phases". In fact there is a number of them and it is not possible to cover them all. Instead we will concentrate on two examples, where multiphase chemical reactions of organic molecules seem to be of more general importance. In the second part of the paper we will investigate the question of mass balance: Are there enough organic molecules around to make them globally important? Here also it is not possible to cover all classes of organic molecules, and we will restrict the discussion to the light hydrocarbons, the class of organic molecules which is most abundant in the atmosphere.

NATO ASI Series, Vol. G6
Chemistry of Multiphase Atmospheric Systems
Edited by W. Jaeschke

MULTIPHASE CHEMISTRY OF ORGANIC MOLECULES

One of the most interesting interactions of organic molecules is based on the fact that higher organic molecules form monolayers which have lower free surface energies than most other substances. Thus they tend to remain and accumulate on surfaces (Somorjai, 1982). Aerosol particles from many locations have been reported to contain organic matter (Ketseridis and Eichmann, 1978; Winkler, 1983) on the order of 10 %. A fair fraction of that organic matter is insoluble in water (Winkler, 1983), and we may expect that much of that fraction is concentrated on the surface of the particles. Indeed, such an organic surface layer was already considered by Junge and Scheich (1969), and such considerations led Graedel and Weschler (1981) to publish the aerosol model which is shown in Figure 1.

The model particle consists of an insoluble core. This core is surrounded by a layer of water which may account for 30 % or more of the particles weight, depending on the relative humidity. This layer, actually an aqueous solution, contains significant amounts of dissolved inorganic and organic compounds. It is covered by a surface film of organic material consisting of polar molecules. These can be visualized as fatty acids, alcohols etc. with their polar functional groups (indicated by the circles) oriented towards the liquid core. The non polar part made up by an aliphatic chain of C- and H-atoms is oriented towards the surface. Clearly such a veneer of oriented organic molecules - if it could be maintained for an appreciable fraction of the lifetime of the aerosol - would determine the surface properties of the aerosol particles. It would influence the sticking probability of gaseous molecules to the surface of the aerosol. The hydrophobic ends would influence the wettability of the aerosol surface. The film would act as a barrier for certain molecules, e.g. H_2O. And finally the C-H chains would react selectively with other molecules. For example, attack by hydroxyl radicals and subsequent oxygen attachment could lead to highly reactive surface sites of peroxy groups.

Despite its potential importance the existence of that organic surface layer, although plausible on general grounds, is by no means established. The techniques for its identifi-

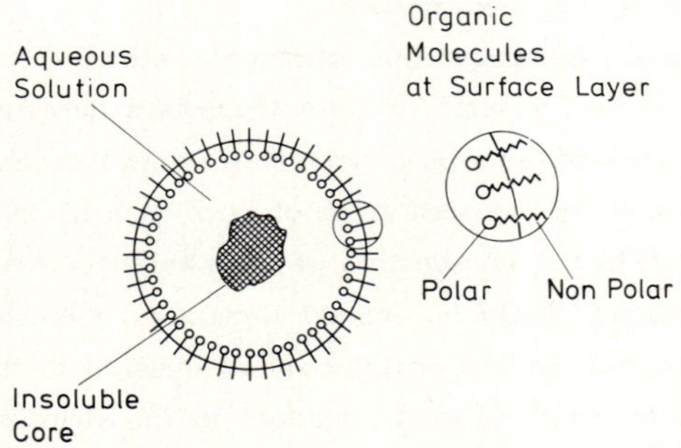

Figure 1: Schematic model of atmospheric aerosol particles (after Graedel and Weschler, 1981).

cation had not been available. It is therefore all the more important to know that with the "Lamma Sonde" a technique has become available which allows just that (van Grieken and Adams, 1983). There is hope now, to determine what kind of aerosol particles, if any, carry such an organic surface film and to obtain a clue of what its nature may be.

Our next example is from the field of cloud and precipitation chemistry. Precipitation has been found to contain formaldehyde (CH_2O) and acetaldehyde (CH_3CHO) - CH_2O at a concentration of 0.05 to 0.15 mg (Klippel and Warneck, 1978; Thompson, 1980; Lowe et al., 1981; Lowe and Schmidt, 1983); CH_3CHO about a factor of five less.

Formic acid, HCOOH, and acetic acid, CH_3COOH, have been reported at even higher concentrations and that at background stations. Galloway et al. (1982) published precipitation measurements which give an average concentration of 0.7 mg/l for HCOOH and 0.27 mg/l for CH_3COOH for four background stations. Moreover Keene et al. (1983) point out that in one of the stations, Katherine, Australia, these two acids contribute 63 % of the total acidity. The authors conclude that organic acids may play an important part in regulating the acidity in the precipitation of remote continental areas.

The available sets of measurements of CH_2O, CH_3CHO, HCOOH and CH_3COOH are rather limited and not sufficient to draw general conclusions. Nevertheless they indicate one feature which is rather interesting. Whereas the amounts of CH_2O and CH_3CHO found in precipitation can easily be explained by the gas phase production of these gases from light hydrocarbons, this becomes difficult when we turn to the corresponding acids. The quoted organic acid concentrations in rain are quite high. To illustrate that, we assume for the moment that they are globally representative. Assuming further a global precipitation rate of 100 cm/yr these concentrations would correspond to a global wet deposition of HCOOH of about

$$300 \times 10^{12} \text{ g/yr},$$

and a wet deposition of CH_3COOH of about

$$100 \times 10^{12} \text{ g/yr}.$$

These are quite large figures. And though as global average, they may be quite uncertain, they indicate that the deposition of formic and acetic acids is much higher than the presently known surface emissions. Unless the precipitation measurements are wrong, both substances must be generated within the atmosphere and - judging from their large amounts - they must be generated from molecules with a large global source strength.

It also appears that there are just not enough olefins around, which through reaction with O_3 and via the Criegee intermediate are known to yield HCOOH and CH_3COOH. Thus one must look for additional precursors for the production of the organic acids. And in the following we will investigate, whether or not the alkanes might serve that purpose.

For clarity we will do it for the simplest molecule that would allow the production of acetic acid, namely ethane, C_2H_6. The reaction sequence is described in the following (cf. Aikin et al., 1980). The oxidation of C_2H_6 begins with a reaction with the OH-radical which abstracts an H-atom and leaves and ethyl radical.

$$\begin{matrix} H & & H \\ HC & - & CH \\ H & & H \end{matrix} + \cdot OH \longrightarrow \begin{matrix} H & & H \\ HC & - & C\cdot \\ H & & H \end{matrix} + H_2O \qquad (1)$$

$$\begin{matrix} H & & H \\ HC & - & C\cdot \\ H & & H \end{matrix} + O_2 \xrightarrow{M} \begin{matrix} H & & H & & \\ HC & - & C & - & OO\cdot \\ \underline{H} & & \underline{H} & & \end{matrix} \qquad (2)$$

Because of the large abundance of O_2, the next step, the addition of O_2, to form the ethylperoxy radical is very fast. There is no chance for competing reactions. The ethyl peroxy radical is polar and should be water soluble, a fact which we indicate by underlining. It is also relatively longlived, since the subsequent reaction with HO_2 giving ethyl hydroperoxide or, with NO giving ethoxy radicals are both relatively slow.

$$\underline{CH_3CH_2OO\cdot} \begin{cases} + \; HO_2\cdot \longrightarrow \underline{CH_3CH_2OOH} + O_2 & (3a) \\ + \; \cdot NO \longrightarrow \underline{CH_3CH_2O\cdot} + \dot{N}O_2 & (3b) \\ + \; O_3 \longrightarrow \underline{CH_3CH_2O\cdot} + 2O_2 & (3c) \end{cases}$$

$$\underline{CH_3CH_2OOH} + h\nu \longrightarrow \underline{CH_3CH_2O\cdot} + \underline{\cdot OH} \qquad (4)$$

Depending on the HO_2 and NO concentrations CH_3CH_2OO could live between a few hundred to a few thousand seconds, enough time to allow collisions with aerosol particle surfaces or with cloud droplets in a cloud. Thus heterogeneous reactions are possible, and in a cloud even probable.

The products, CH_3CH_2OOH and CH_3CH_2O, of the gas phase reaction are also water soluble, and CH_3CH_2OOH is longlived enough to make heterogeneous removal possible, especially in a cloud. The ethoxy radical is shortlived, with a lifetime of a few milli-seconds only; it will react within the gas phase.

$$\underline{CH_3CH_2O\cdot} \longrightarrow \begin{matrix} H \diagdown \\ \underline{C{=}O} \\ H \diagup \end{matrix} + CH_{\underline{3}}\underline{\cdot} \qquad (5a)$$

$$\underline{CH_3CH_2O\cdot} + O_2 \longrightarrow \begin{matrix} H_3C \diagdown \\ \underline{C{=}O} \\ H \diagup \end{matrix} + \underline{HO_2\cdot} \qquad (5b)$$

Unimolecular decay of CH_3CH_2O gives formaldehyde, its reaction with O_2 yields acetaldehyde. At groundlevel the rate of the latter reaction is about 6 times faster. We also note that in this case another peroxy radical is formed: HO_2. The resulting CH_2O and CH_3CHO are the longest lived intermediate products of the ethane oxidation. They live about 10 hours against gas phase reaction with OH or against photolysis and are water soluble; heterogeneous reactions are possible and very likely in a cloud. The methyl radical formed in reaction (5a) reacts further to form also CH_2O.

$$\underline{CH_3\cdot} + O_2 \xrightarrow{M} \underline{CH_3O_2\cdot} \quad (9)$$

$$\underline{CH_3O_2\cdot} + \cdot NO \longrightarrow \underline{CH_3O\cdot} + \dot{N}O_2 \quad (10)$$

$$\underline{CH_3O\cdot} + O_2 \longrightarrow H_2\underline{CO} + \cdot HO_2 \quad (11)$$

The gas phase reactions will eventually degrade H_2CO and CH_3CHO to CO_2 and H_2O. Acetaldehyde can further react to eventually form peroxyacetylnitrate, PAN. The sequence contains 3 steps.

$$\underline{CH_3CHO} + \cdot OH \longrightarrow \underline{CH_3\dot{C}O} + H_2O \quad (6)$$

$$\underline{CH_3\dot{C}O} + O_2 \xrightarrow{M} \underline{CH_3C(O)OO\cdot} \quad (7)$$

$$\underline{CH_3C(O)OO\cdot} + \dot{N}O_2 \xrightarrow{M} \underline{CH_3C(O)OONO_2} \quad (8)$$

An abstraction of the H-atom by OH to form an acetyl radical is followed by the fast addition of O_2 to form the peroxyacetyl radical which reacts with NO_2 to form PAN. The dissociation to NO_2 and the peroxyacetyl radical depends strongly on temperature. At 20°C the lifetime of PAN against unimolecular decay is a few hours.

Up to this point in the reaction sequence the various reaction products have been: Peroxy radicals, HO_2, CH_3O_2, $CH_3CH_2O_2$, and CH_3COO_2, the peroxides H_2O_2 (not explicitly mentioned) and CH_3O_2H, the aldehydes, H_2CO and CH_3CHO, and finally PAN. All these substances are water soluble and long lived enough to be incorporated into cloud droplets when produced within the cloud. Us the presence of ethane (and other light hydrocarbons) will add substantially to the balance of HO_2 and H_2O_2 in a cloud. Moreover it adds a whole host of higher radicals and peroxides to the wet chemistry which had so far not been considered (cf. Chameides, 1983).

But up to this point neither formic nor acetic acid have been formed in the reaction sequence. There is a good reason for this: the gas phase production of formic and acetic acid from C_2H_6 is not very likely. One can construct gas phase pathways leading to those acids, but they do not seem to proceed with high efficiencies. There is, however, a good chance that aqueous phase reactions do convert CH_2O to HCOOH and CH_3CHO to CH_3COOH. Thus both acids can be formed from C_2H_6 oxidation inside a cloud.

In the following we will outline the process for H_2CO. H_2CO is highly soluble in water. Upon dissolution it readily forms a hydrate, methylene glycol $H_2C(OH)_2$ (at room temperature 99.9 % of the dissolved H_2CO are in the form of $H_2C(OH)_2$). There are several pathways how methylene glycol can convert to formic acid. One has been suggested by Graedel and Weschler (1981). It relies on the reaction with dissolved OH, which during daytime is likely to be present at high concentrations in cloud water (Chameides and Davis, 1982; see also Chameides and Davis, 1983; Chameides, 1983). The OH-radical abstracts a hydrogen atom, the resulting radical $CH(OH)_2$ reacts with dissolved O_2 to form HCOOH and a HO_2 radical (Chameides and Davis, 1983).

Another possibility is the reaction of $H_2C(OH)_2$ with H_2O_2, also known to be present in cloud and precipitation water, both, from experimental and theory (cf. Chameides and Davis, 1982). The direct reaction appears to be slow. In any case there is at least one efficient reaction in aqueous phase to convert CH_2O via methylene glycol to formic acid. With a simple estimate we can convince ourselves that this path may be of global importance. Assuming OH concentration in the range of 6×10^7 to 6×10^8 cm^{-3} in the liquid gas phase as estimated by Chameides and Davis (1982) and the reaction rate constant of 2×10^9 M^{-1} sec^{-1} quoted by Chameides and Davis (1983) for OH + $CH_2(OH)_2$, one obtains a lifetime of 500 to 5000 sec for CH_2O in cloud droplets. With an effective Bunsen solubility of 4×10^5 for CH_2O with respect to dissolved $H_2C(OH)_2$ and a liquid water content of 1 g/m^3 in the cloud one obtains a lifetime for CH_2O in the cloud between 1.5×10^3 and 1.5×10^4 sec, about a factor of 3 to 20 shorter than in outside air. Since clouds make up about 10 % of the tropospheric volume, it is quite conceivable that a significant fraction, on the order of 10 %, of the tropospheric CH_2O is converted to HCOOH via cloud chemistry. Some of the HCOOH will be rained out with the forming precipitation. Another fraction of the HCOOH will be oxidized to CO_2 and H_2O by OH in the cloud droplets. However, since most of the clouds evaporate, much of the thus formed HCOOH will end up in aerosol particles or evaporate also to form gaseous HCOOH, whose final fate is gas phase oxidation to H_2O and CO_2.

A quite analogous process exists for the conversion of CH_3CHO to CH_3COOH. It should be slower because CH_3HCO is much less soluble than H_2CO. Thus gas phase production of CH_3COOH may still be required to explain the high concentration in rain observed by Galloway et al. (1982).

In the case of acetic acid, there is of course also the possibility of the uptake of PAN by cloud droplets and its subsequent hydrolysis to acetate and nitrate. But PAN is even less soluble in water than CH_3CHO and its hydrolysis is slow. The time constant for hydrolysis is about 1 h (at 22°C and a pH of 5.6 (Lee et al. 1983)). Though possible, this process appears too inefficient to produce the large concentration of acetic acid observed in background precipitation.

As far as CH_3COOH is concerned we are left with a twofold problem. Firstly, the presently known conversion processes of hydrocarbons to CH_3COOH are too inefficient to produce the large CH_3COOH concentrations reported in rain, even including multiphase reactions. Secondly, the global emissions of hydrocarbons appear to be insufficient to balance the deposition of CH_3COOH (see Table I below).

Our second example showed how gas phase chemistry and wet phase chemistry interplay to oxidize an alkane to formic and acetic acid. The gas phase oxidation of C_2H_6 worked to produce soluble organic compounds: the aldehydes. The aqueous phase chemistry completed the oxidation to acids. We also saw that this chemistry which occurs in clouds might be of global importance.

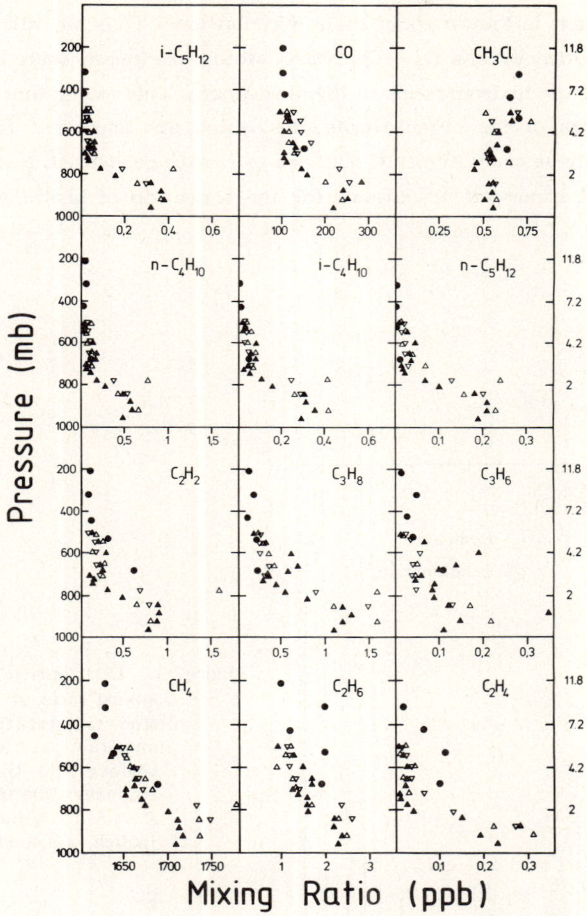

Figure 2: Vertical profiles of hydrocarbons over Europe (cf. Rudolph et al., 1982a). The triangles represent samples collected during October and November 1979 over the Eifel mountains, Germany, 52°N latitude; the dots represent samples collected in June 1979 over Southern France, 43°N.

What we have shown so far is that organic molecules are potentially important for atmospheric multiphase chemistry. The actual importance of organic molecules depends of course on the concentrations of the gaseous precursors in the atmosphere. And this point we would like to address next. There is a great variety of organic molecules in the atmosphere and not all that much is known about their distributions. Thus we will concentrate on the light nonmethane hydrocarbons C_2 - C_5, NMHC, which are known best. These molecules also make up the bulk of the hydrocarbons in the atmosphere. This means that we learn very little about the precursors of the organic molecules which are important for aerosol particles because these must have much higher C numbers to remain condensed. But we will have a fair picture of the most important precursors for the formation of acetic and formic acid and peroxy radicals.

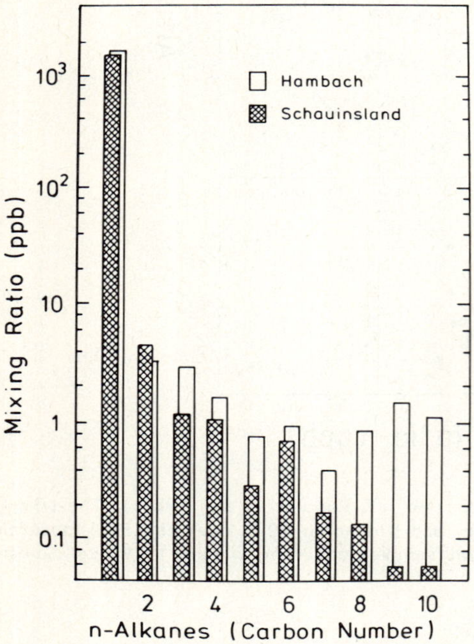

Figure 3: Distribution of the atmospheric mixing ratio of straight chain alkanes from two locations. Schauinsland is a mountain station (1200 m above sealevel) in the Black Forest area, Germany; the other station refers to a small wooded area close to Juelich, Germany.

We will first take a look at the vertical distribution of light hydrocarbons. Vertical profiles of C_1 - C_5 hydrocarbons are summarized in Figure 2 (Rudolph et al., 1982a). They were obtained using grab sampling aboard an aircraft and subsequent analysis by gas chromatography in the laboratory (Rudolph et al., 1982a). The triangular symbols represent samples collected over Northern Germany during October and November 1979; the dots represent samples collected over Southern France in June 1979. Of course, the data points do show some scatter. This is to be expected for trace gases whose atmospheric lifetime ranges from a few months, C_2H_2, C_2H_6, to a week or so for the alkanes C_3H_8, C_4H_{10}, C_5H_{12}, and a day for the olefins C_2H_4, C_3H_6, and whose sources are highly local. Nevertheless, all these gases clearly present a common picture: Higher mixing ratios, nearly constant with altitude, in the boundary layer, a sharp dropoff at the boundary layer which was at 800 mb pressure or 2 km altitude, a slower decrease above that altitude to a background value which seems to be reached by about 400 mb or 7.2 km altitude. This points to strong sources for these gases at the continental surface which in this case must be clearly industrial. The source strengths for the various gases as expressed by the mixing ratio enhancement in the boundary layer are quite different. For CH_4 the enhancement amounts to 100 ppb, for CO to 150 ppb, for C_2H_6 only 1 ppb, for C_2H_2 0.7 ppb and for the C_2H_4 only 0.3 ppb. The only exception is CH_3Cl whose gradient is inverted indicating the nonexistence of a surface

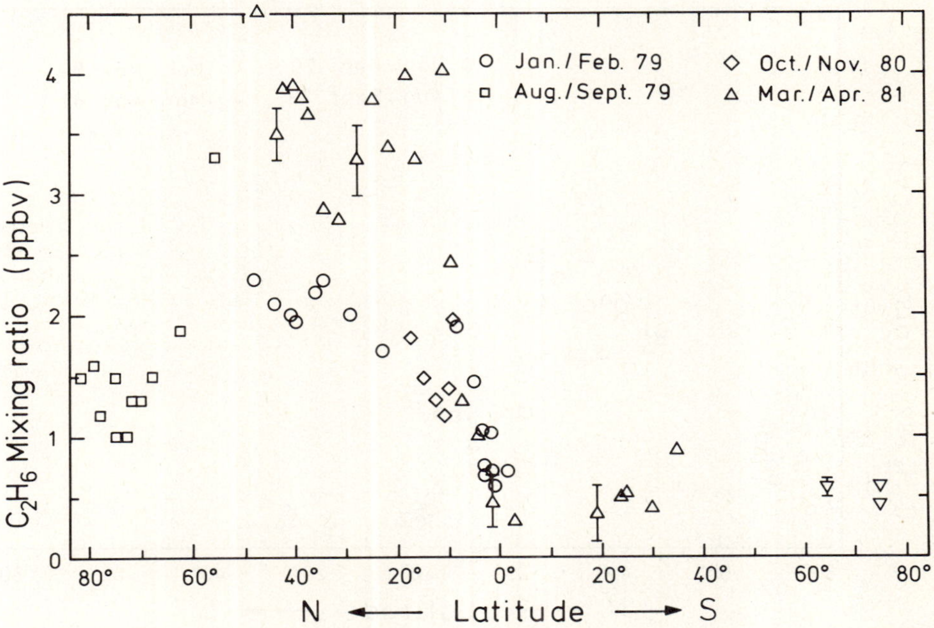

Figure 4: Latitudinal distribution of C_2H_6 in surface air over the Atlantic. The dates of the ship cruises are indicated by different symbols. The inverted triangles represent samples from the Ross and Faraday base, Antarctica, collected during December 1980 and January 1981 (after Rudolph et al., 1982b).

source in this region. We also note that the absolute mixing ratios are quite different. They range from 1600 ppb for CH_4 to 0.2 ppb for n-C_5H_{12}. The mixing ratios of hydrocarbons show a rather regular pattern: a decrease from low C numbers to high C numbers. For straight chain alkanes this is illustrated in Figure 3. This bar graph shows the mixing ratio of alkanes beginning from CH_4 (C=1), C_2H_6 (C=2) to n-decane (C=10) for two stations. One, indicated by crosshatched bars, is a remote mountain station in Germany, which at the time of sampling was above boundary layer and should be as close to background conditions as you can get in middle Europe. The big jump in concentration between CH_4 and C_2H_6 is followed by a more or less regular decrease of a factor 1.5 or so with increasing carbon number.

Next we investigate the latitudinal distribution of light hydrocarbons. This is done with the help of samples taken during a number of cruises of the research vessel F.S. Meteor in the Atlantic. The results for C_2H_6 are shown in Figure 4 (Rudolph and Ehhalt, 1981, Rudolph et al., 1982b). The cruise data are supplemented by a few data from Antarctica indicated by the inverted triangles. The scatter within each cruise is relatively small. But the data for the March cruise are significantly higher than the others. In principle this could be assigned to a systematic seasonal variation of the hydrocarbon mixing ratio. Such a seasonal

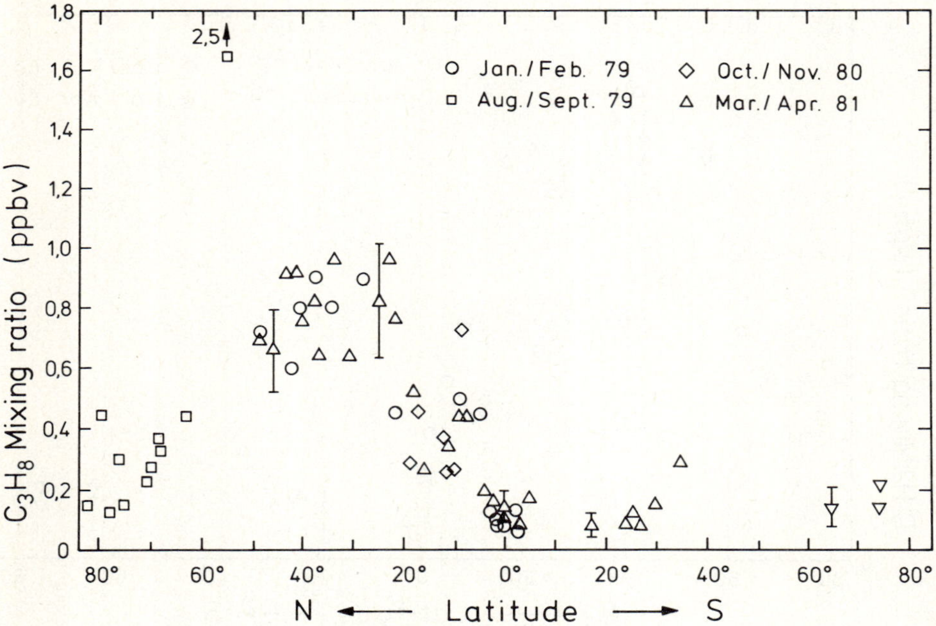

Figure 5: Latitudinal distribution of C_3H_8 in surface air over the Atlantic. The dates of the ship cruises are indicated by different symbols. The inverted triangles represent samples from the Ross and Faraday base, Antarctica, collected during December 1980 and January 1981 (after Rudolph et al., 1982b).

variation would not be unexpected and has been observed for C_3H_8, for example (Singh and Salas, 1982). However, as we shall see, the other hydrocarbons do not show this difference which seems to exclude that explanation.

Despite this scatter the data show a clear and systematic latitudinal variation: There are lower mixing ratios of about 1 ppb at high latitudes, the values increase towards midlatitudes. A broad maximum is reached between 50° and 10° N with mixing ratios between 2 and 4 ppb. South of 10° N the mixing ratios show a very rapid decrease towards the equator reaching values below 0.5 ppb. The southern hemisphere exhibits uniformly low C_2H_6 mixing rations between 0.5 and 1 ppb.

This behaviour clearly indicates strong C_2H_6 sources at northern midlatitudes. It also confirms the expected strong sinks in the tropics, which are due to the elevated OH concentrations there. We also note that the mixing ratio maximum is shifted to the lower latitudes, i.e. the areas where most of the crude oil and natural gas is produced rather than to the latitudes where most of the industries and heaviest car traffic are located. This becomes even more evident when we consider the latitudinal distribution of C_3H_8 (Figure 5) which exhibits less scatter and therefore a much better defined maximum of nearly 1 ppb between 40° and 20° N. The rest of the latitudinal distribution of C_3H_8 is also quite similar to that of C_2H_6, with low values in all of the southern hemisphere and low values at polar

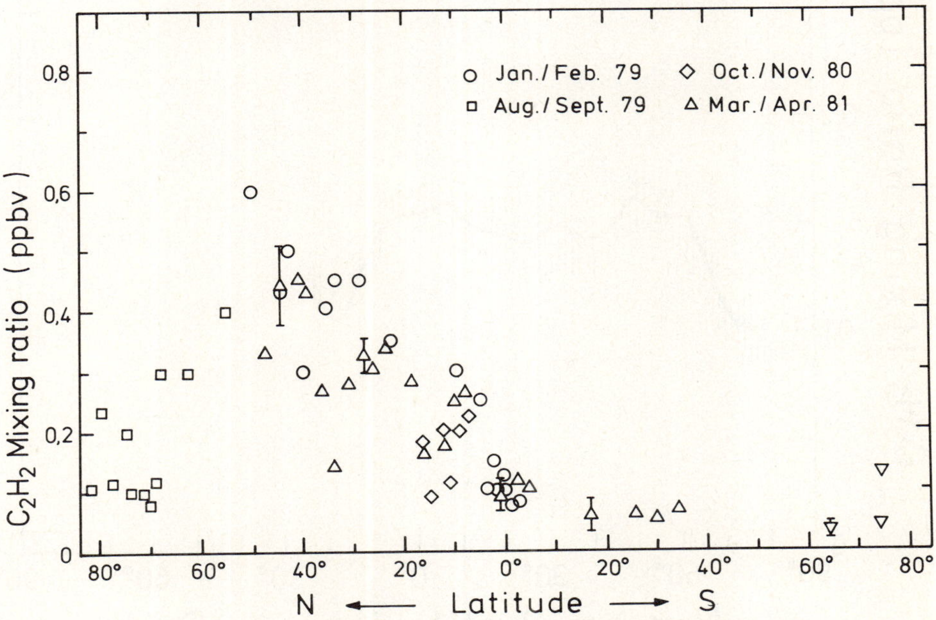

Figure 6: Latitudinal distribution of C_2H_2 in surface air over the Atlantic. The dates of the ship cruises are indicated by different symbols. The inverted triangles represent samples from the Ross and Faraday base, Antarctica, collected during December 1980 and January 1981 (after Rudolph et al., 1982b).

latitudes. A quite similar behaviour was also found for butane.

Our next example is the distribution of C_2H_2, a gas which is mostly produced by incomplete combustion. On the first glance it looks quite similar to those of the light alkanes: low values in the southern hemisphere, a strong, well defined maximum at northern midlatitudes. On closer inspection, however, we detect a few differences. The increase in mixing ratio from the equator to midlatitudes is less abrupt, starting at the equator and reaching the maximum at 40° N instead of 10° of 20° N as for C_2H_6. In addition the maximum of the C_2H_2 mixing ratio straddles 40° N instead of 30° in the case of the alkanes. That behaviour is quite plausible. As mentioned C_2H_2 is emitted during incomplete combustion of fossil fuel - mostly from automobiles. Its latitudinal distribution, therefore, reflects the consumption of fossil fuel rather than the production of fossil fuel as is the case for C_2H_6. And the consumption of fossil fuel takes place mostly in the industrial countries around 40°N.

So we see that the latitudinal distribution affords a first glimpse on the nature of the

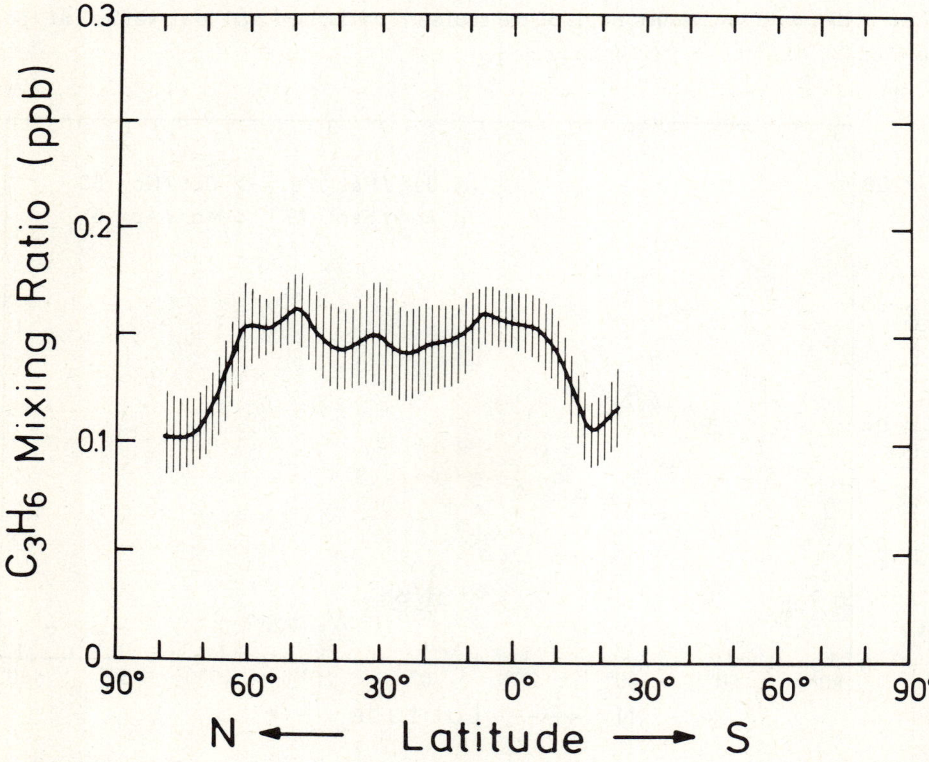

Figure 7: Running average of the latitudinal distribution of C_3H_6 in surface air over the Atlantic using a 15° averaging interval. The curve is based on data from the same cruises as listed in Figure 4. The vertical bars indicated the mean standard deviation.

sources of the light hydrocarbons. This becomes even more apparent when we consider the latitudinal distribution of propylene, C_3H_6 (Figure 7). The data for propylene scatter somewhat more strongly and show less systematic variation than the alkanes do, so for clarity we show a running average of the data as indicated by the curve. The vertical bars indicate the mean standard deviation. Clearly C_3H_6 shows a different latitudinal distribution than C_2H_2 or the light alkanes: Between 60° N and nearly 10° S the C_3H_6 mixing ratios over the Atlantic Ocean are more or less constant with latitude. What is even more surprising, these mixing ratios are relatively high, 0.15 ppb, very close to what we found over continental Europe. On the other hand, C_3H_6 is highly reactive and thus shortlived. Its lifetime is about 0.6 days in midlatitudes and 0.3 days in the tropics. Therefore transport from continental sources cannot explain the C_3H_6 mixing ratios found over the ocean, and a local i.e. oceanic source of C_3H_6 seems required. This is indeed the case. Measurements by Swinnerton and Lamontagne (1974) and Lamontagne et al. (1974) have shown that the ocean is highly supersaturated in C_3H_6-by about a factor of 100. In fact the observed mixing ratios of C_3H_6 can be fully explained by oceanic emissions (Rudolph and Ehhalt, 1981). The same is true for ethylene.

As interesting and informative as the NMHC measurements over the ocean are, they only provide one dimension in the required global distribution. What one really would like is a picture of the NMHC in at least 2 dimensions, namely latitude and altitude. In April and May, 1980, we had the opportunity to participate in the "Stratoz II" flight of the French Caravelle research aircraft. The flight track is shown in Figure 8. It took about 3 weeks to complete it. The plane landed about every 3000 km which allowed to take vertical profiles during the approach to the airports, whose positions are indicated by the dots. Because of the expected pollution close to the airport no samples were taken below 800 mb. The meridional distribution shown in the following are divided into a southbound leg leading from Greenland to Puntas Arenas, and a northbound flight segment from Puntas Arenas to Paris.

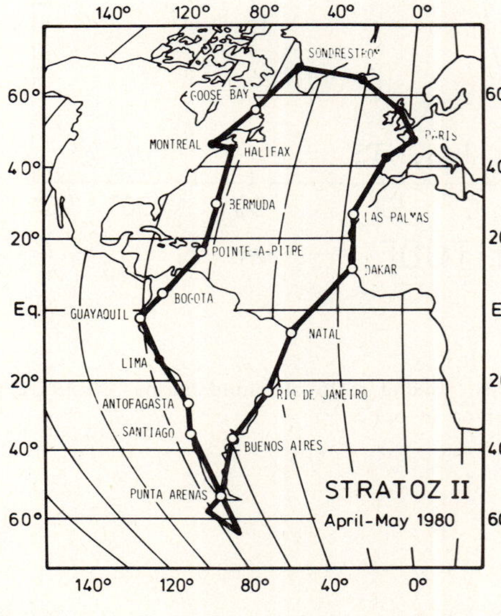

Figure 8: Flight track of the STRATOZ II experiment.

Since it is not possible to present all of the data, only a few selected examples are shown. The other NMHC cross sections will be published by Ehhalt et al. (1985). Figure 9 gives the meridional distribution of ethane during this flight. For clarity the figure does not contain the individual data points, but rather the isolines of constant mixing ratio interpolated bet-

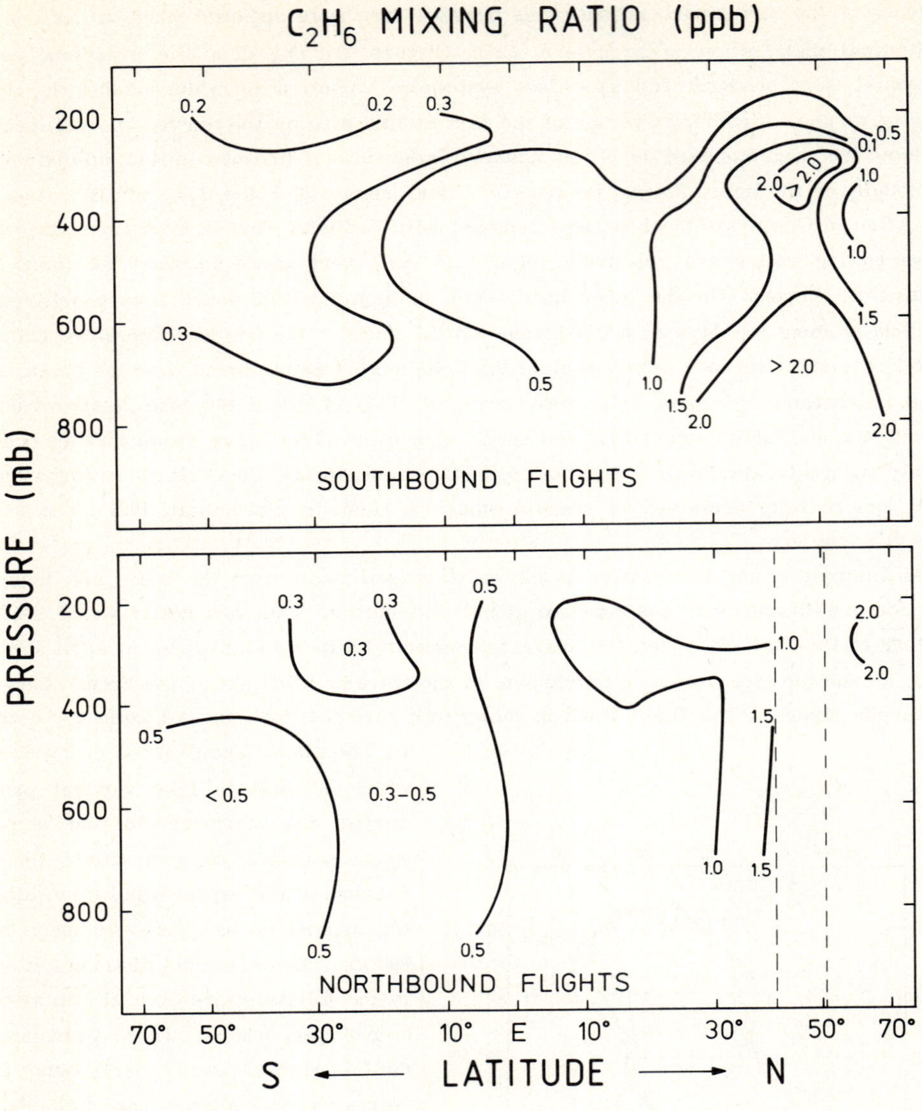

Figure 9: Latitude-altitude cross section of the C_2H_6 mixing ratio obtained during STRATOZ II. The contours of constant mixing ratio are given in ppb.

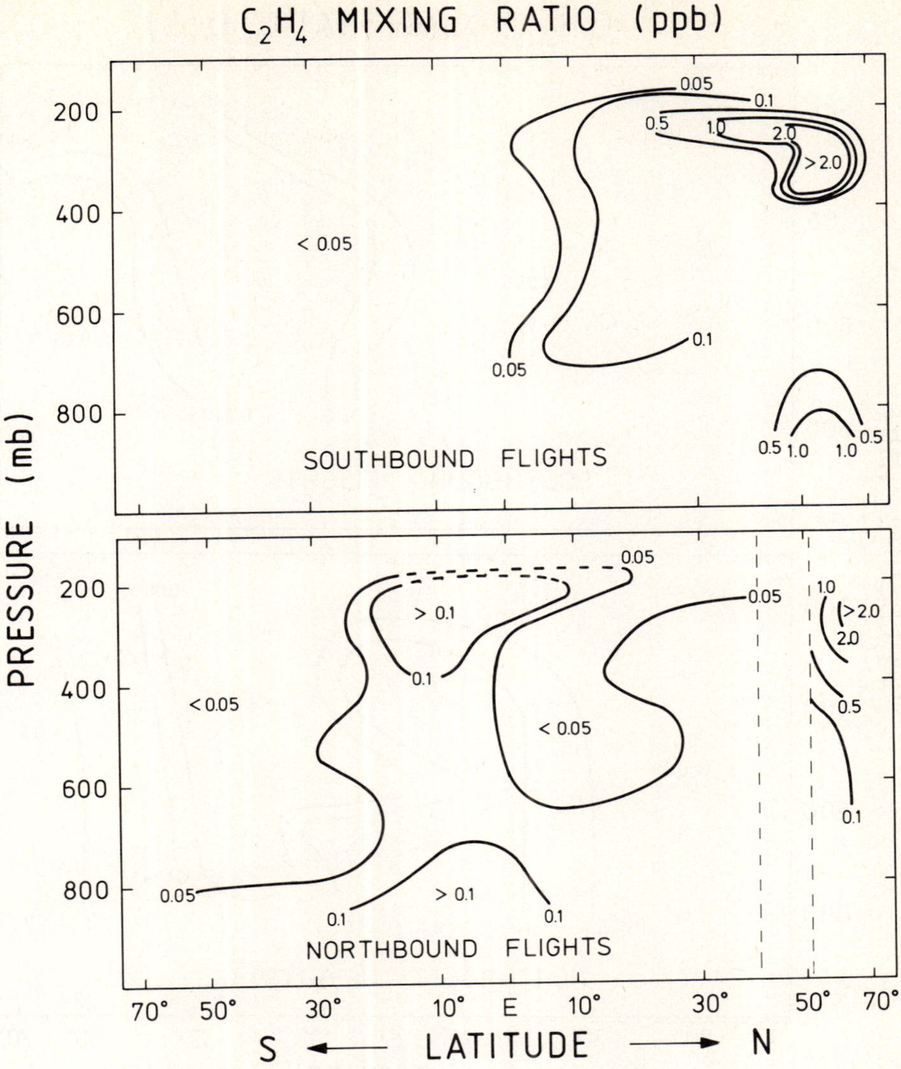

Figure 10: Latitude-altitude cross section of the C_2H_4 mixing ratio obtained during STRATOZ II. The contours of constant mixing ratio are given in ppb.

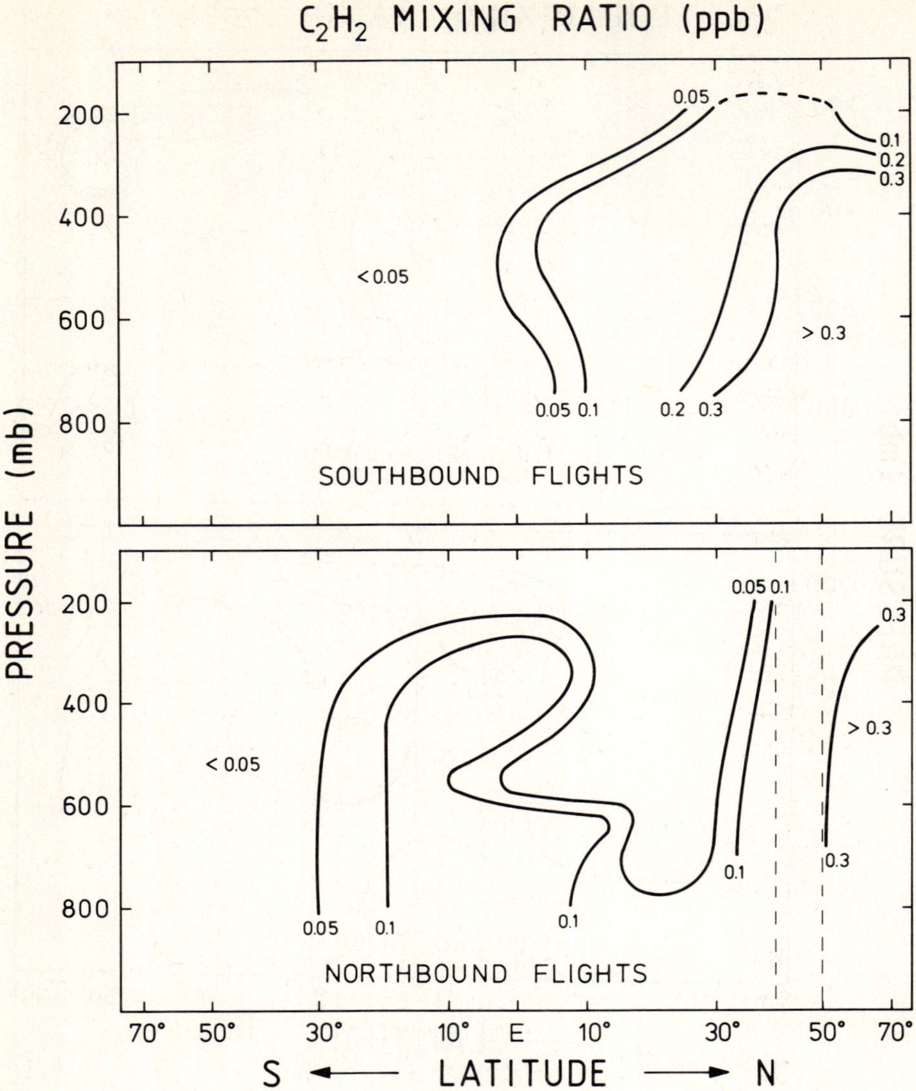

Figure 11: Latitude-altitude cross section of the C_2H_2 mixing ratio obtained during STRATOZ II. The contours of constant mixing ratio are given in ppb.

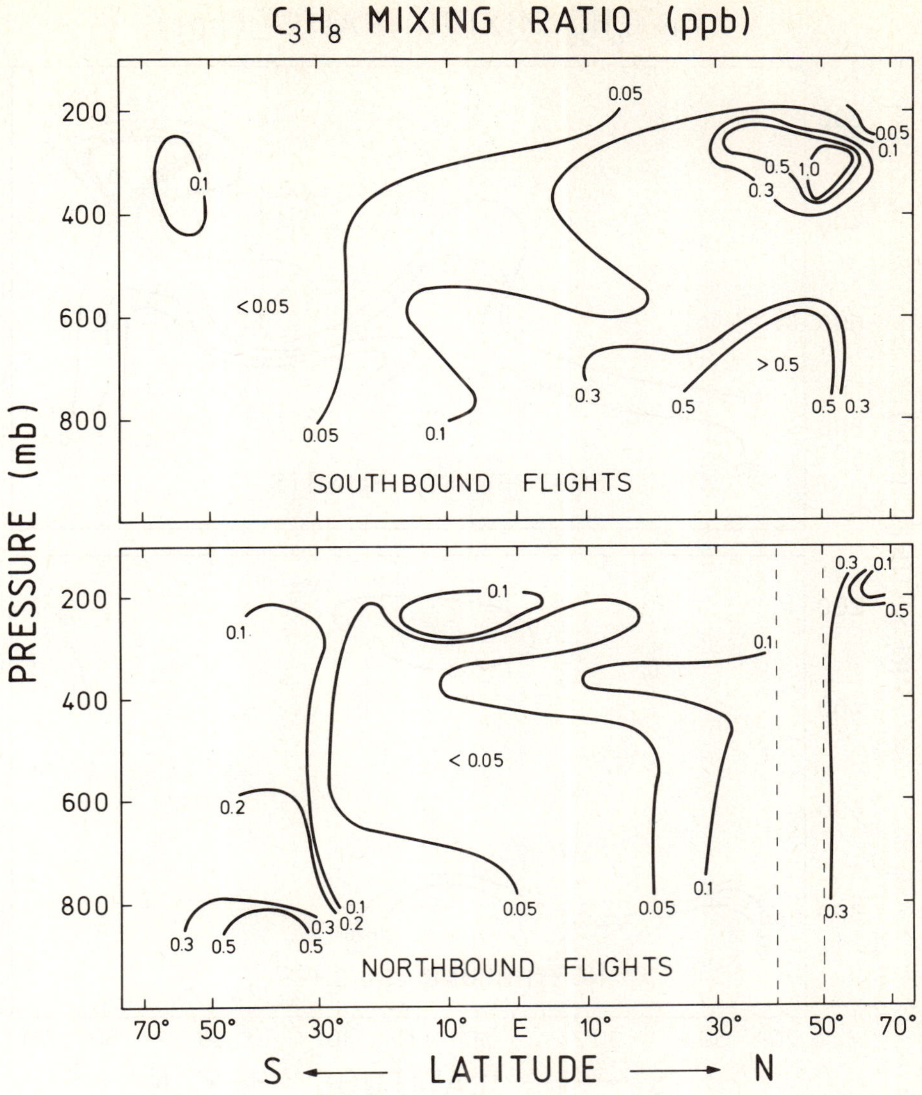

Figure 12: Latitude-altitude cross section of C_3H_8 mixing ratio obtained during STRATOZ II. The contours of constant mixing ratio are given in ppb.

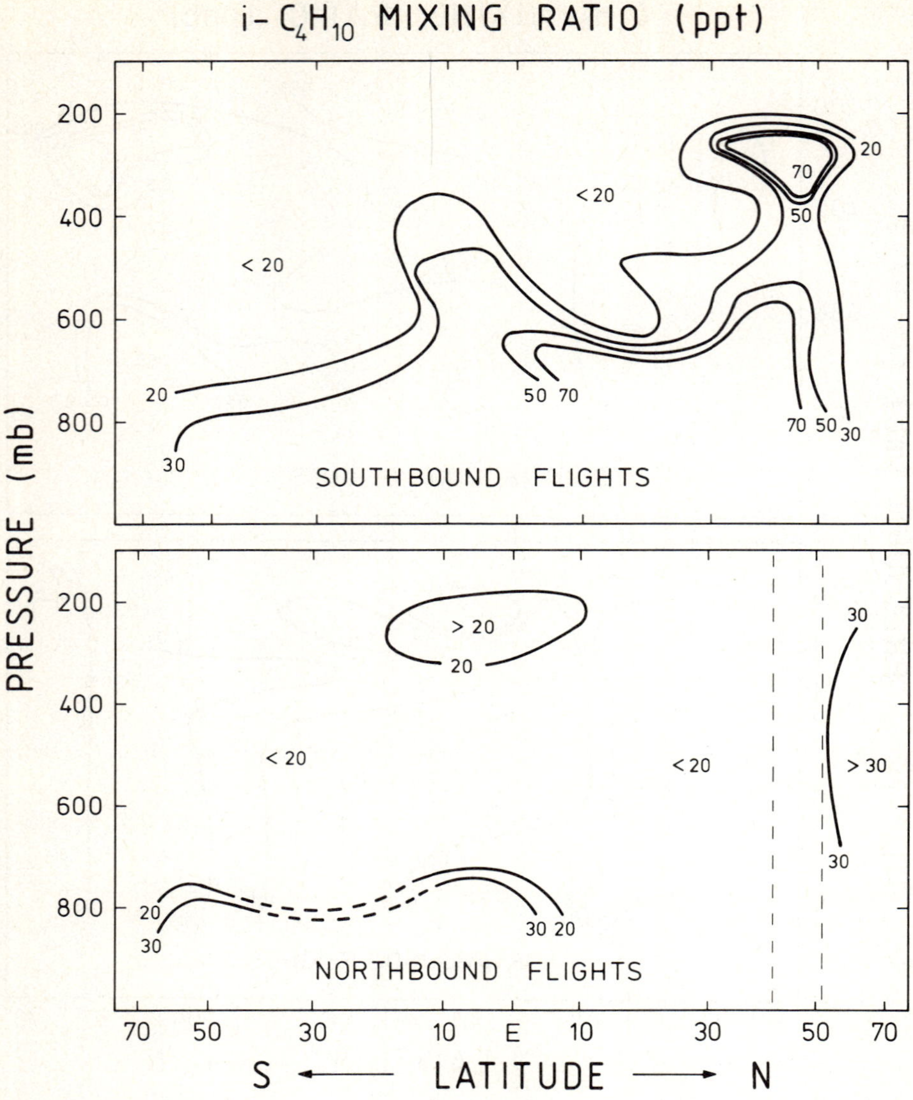

Figure 13: Latitude-altitude cross section of $i\text{-}C_4H_{10}$ mixing ratio obtained during STRATOZ II. The contours of constant mixing ratio are give in ppt.

ween them. The meridional distribution of C_2H_6 shows a number of characteristic features which are repeated in the distributions of the other trace gases as well. As in the measurements over the ocean surface we note a mean latitudinal gradient now at all altitudes, the mixing ratio decreasing as we go from mid-north latitudes to the equator and into the southern hemisphere. We also have generally a clear vertical decrease of mixing ratio with altitude. There is however one conspicuous exception. There is a definite cell of maximum mixing ratios around 50° N latitude and about 10 km altitude in the southbound flight. Such a maximum - although much less pronounced - is also present over the tropics during the northbound flight. The maximum at 50° N causes a very steep vertical gradient towards the tropopause. It also influences the isolines further south in the sense that they run more or less trapezoidal with the exception of some wiggles probably superimposed by air mass movements. These are the regular distribution features which we can extract from Figure 9. At the same time we note a substantial variability which is expressed in the bends and folds of the lines of constant mixing ratios. They are caused by large scale advection. We further note that the south and northbound cross sections differ significantly indicating at least temporary zonal variations. For C_2H_6 this is not very conspicuous, but generally the northbound leg over the African Coast shows lower mixing ratios than the southbound leg.

The same features are visible for ethylene (Figure 10): latitudinal gradient, vertical gradient, a cell of maximum mixing ratios at 10 km altitude at 50° N. Here this maximum is even more surprising, since C_2H_4 is shortlived and not likely to be transported over large distances. We also note how the high mixing ratios in the planetary boundary layer cause a convex bulge pushing the 1 ppb isoline upward to 3 km altitude at midlatitudes. C_2H_2 is the only case where the cell of maximum mixing ratios is so weak that it cannot be traced with certainty. But mixing ratios of 0.3 ppb, i.e. as high as in the boundary layer, are certainly found at 50° N and 10 km altitude (Figure 11). Here, the upwelling in the tropics is also clearly visible; we further note an upward thrust of higher mixing ratios over the equator.

All the above mentioned features are also found for propane (Figure 12) and for isobutane (Figure 13). Indeed they are present for all the NMHC (C_2H_6, C_2H_4, C_2H_2, C_3H_8, C_3H_6, nC_4H_{10}, iC_4H_{10}, nC_5H_{12}, iC_5H_{12}) which we measured, as well as for CO (Schmidt et al., 1982).

Whereas the latitudinal or vertical gradients can easily be explained by surface sources, the cell of maximum mixing ratios at 10 km altitude, 50° N presents somewhat of a puzzle. The common appearance in all NMHC cross sections points to a common mechanism.

It is quite tempting to assume that the maximum of NMHC mixing ratios observed at 10 km altitude and 50° N latitude is caused by exhaust emissions of commercial jet aircraft. The flight path of the Caravelle aircraft followed over large parts the trans-Atlantic flight pattern of that day. It is even conceivable that the Caravelle was following another plane relatively closely and thus encountering relatively high levels of pollution. And this was indeed our first thought when we saw these data. There is, however, a serious inconsistency in that explanation: The ratio of excess CO to excess NMHC (C_2 - C_5) in the maximum, expressed as the ratio of carbon atoms, C_{CO}/C_{NMHC}, amounts to about 5. A jet engine, however, emits very few NMHC in the C_2 - C_5 range (cf. Figure 14). Most of the exhausted hydrocarbons have carbon numbers above 7, and the ratio of CO to NMHC (C_2 - C_5) in jet exhaust is very large namely about 100 which seems to rule out a large contribution by jet exhaust. On the other hand most of the NMHC emitted by a gasoline engine are in the range C_2 - C_9 (cf. Figure 14) and the carbon ratio of CO to the NMHC with C_2 - C_5 is much lower namely about 10. This value is also observed in the boundary layer as a whole. Thus NMHC observed in the maximum show a pattern more consistent with boundary layer air.

If this interpretation is correct there must be rather fast transport processes to carry the observed high concentrations of olefins to these altitudes. For example an extended front with high reaching thunderstorms which could pump boundary layer air quickly to altitudes close to the tropopause. This possibility has been investigated by Ehhalt et al., (1985).

The current explanation of the maximum introduces another point. The two dimensional cross sections given above represent a more or less instantaneous snapshot of the atmospheric distribution of NMHC. Rather shortlived meteorological processes could have influenced them. To extract an average meridional distribution from them can be misleading.

We will need quite a few more of these profiles to obtain a good global average. At the same time, the observations indicate the potential value of measurements of NMHC concentrations as traces of atmospheric transport processes. What are then the conclusions on the NMHC distributions which will stand up? We still can say with certainty that the highest mixing ratios of NMHC are found in the planetary boundary layer, i.e. close to the earth's surface and at the midlatitudes of the northern hemisphere, We can also state that there is a general decrease in mixing ratio with altitude which is the steeper the shorterlived the hydrocarbon compound is. But occasionally relatively high mixing ratios of NMHC in the middle and upper troposphere of the northern midlatitudes must be expected. The mixing ratios in the tropics and in the southern hemisphere are much lower than in the northern hemisphere, except for C_2H_4 and C_3H_6 or others with a large oceanic surface source.

That is to say that much of the multiphase atmospheric chemistry of NMHC must go on in northern midlatitudes, where most of the current investigations take place. It seems important

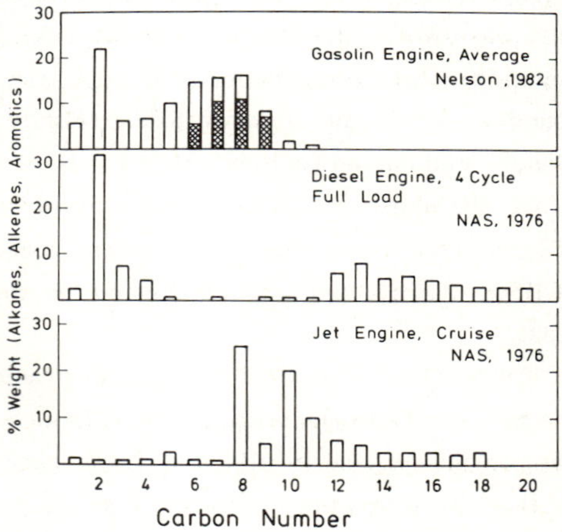

Figure 14: The distribution of hydrocarbons in the exhaust of a gasoline engine, a diesel engine and a jet engine, under specified operating conditions. Hydrocarbons with the same carbon number are lumped together. The full portion of the bars for the gasoline engine indicates the fraction of aromatic compounds.

to keep this bias in mind.

SOURCES OF NMHC

We have repeatedly talked about the sources of NMHC, and in this last chapter we will attempt to quantify these sources. This is difficult. There are regional studies which provide detailed budgets for the individual hydrocarbons, for Los Angeles for example or for Sydney. But there are no detailed global estimates. All published estimates of global HC emissions are bulk numbers giving only the sum of all hydrocarbons (cf. NAS, 1976; Duce et al., 1982). As a consequence we are forced to prepare a detailed budget without precedents. This is not an easy task, and the uncertainties in such an estimate are large. The uncertainties in the CH_4 budget for example are about a factor of 2. This, despite the fact that CH_4 with global emissions totaling about 500 x 10^{12} g/yr is the most important and the best investigated of the HC.

The estimates for the C_2 - C_5 hydrocarbons which we will present must be much more uncertain, and one may question the rationality behind such crude estimates. Our excuse is that we have to start somewhere and that these crude estimates still allow us to determine which sources appear more important globally and should therefore be investigated with highest priority. Table I summarizes the emission estimates for the C_2 - C_5 hydrocarbons. For comparison we have added some estimates of a few aromatic compounds and of the naturally emitted isoprene and terpenes. The various source estimates of the hydrocarbons will be briefly discussed below.

AUTOMOBILE EXHAUST

The most important anthropogenic source of light NMHC are the emissions from automobiles. Most of these emissions result from the incomplete combustion in gasoline engines. To estimate the strength of this source we have to know the global consumption of gasoline by cars and the globally averaged emission factor for NMHC. Both numbers are available from the literature. They are, however, not up to date.

For the global gasoline consumption we used the estimate for the year 1976 by Logan et al. (1981) which amounts to 620 x 10^{12} g gasoline/yr. The emission factor was derived from the estimates of NMHC emissions from gasoline powered cars and the gasoline consumption in West Germany for 1978 (Luftreinhaltung '81). It corresponds to 2.6 % by weight. This value is close to emission factors quoted for American cars (NAS, 1976). It does not take into account any reduction of NMHC emission due to the installation of catalytic converters. The product of these two numbers gives the total NMHC emission of 16 x 10^{12} g/yr shown in the second column of Table I.

To assign individual emission rates to the various NMHC, the average composition of the exhaust has to be known. The composition is variable depending on the type of gasoline utilized and the operating conditions of the engine. We used the exhaust composition measurements given by Nelson et al. (1982) for Sidney. They agree reasonably well with the exhaust composition from US studies (cf. NAS, 1976).

As Table I shows, C_2H_4. C_2H_2, C_3H_6, C_4H_{10}, and n-, i-C_5H_{12} are among the dominant species. The aromatic hydrocarbons, benzene, toluene, and the xylenes, are also abundant. The lighter hydrocarbons are generated in the combustion process, unburned fuel is responsible for the hydrocarbons with carbon numbers $C \geqq 4$.

The emission factor for Diesel engines is lower, only about a factor 0.5 of that of gasoline engines. In addition the global use of Diesel fuel was only 158 x 10^{12} g/yr in 1976 (Logan et al., 1981). Thus the total NMHC emission from Diesel exhaust is only 3 x 10^{12} g/yr. The composition of Diesel exhaust is different from that of gasoline exhaust especially for carbon numbers $C > 5$.

EVAPORATION OF FUEL FROM AUTOMOBILES

Evaporation of fuel in automobiles is another source of NMHC. It occurs in two ways: a) The socalled heat-soak loss from carburettors. This term describes the evaporation from the carburettor as it heats up after engine shutoff. Since the evaporation is essentially complete the resulting emissions have the composition of the fuel itself. b) Losses from fuel tanks. Since that loss is via the vapor phase, the emissions depend also on the vapor pressure of the individual fuel components, i.e. the light, low boiling NMHC escape preferentially. The total evaporative loss is estimated to be about 25 % of the exhaust emissions (Stern, 1968; NAS, 1976; Nelson et al., 1982). It consists mostly of the butanes and i-pentane (NAS, 1976; Nelson et al., 1982).

EVAPORATION LOSSES FROM DISTRIBUTION AND REFINERIES

Emissions also occur during distribution and refining of fuels. The emission factors quoted in the literature vary considerably (Robinson and Robbins, 1969; Stern, 1968). Here we use a total NMHC emission factor of 0.5 % weight NMHC per weight fuel for distribution and 0.1 % for refining, which seem to apply for West Germany (Luftreinhaltung '81). Assuming the same composition as for the evaporation losses from cars, we obtain the numbers shown in Table I.

LOSSES OF NATURAL GAS

Besides Ch4, natural gas also contains higher alkanes notably C_2H_6, C_3H_8, iC_4H_{10}, nC_4H_{10}. Thus losses and seepage of natural gas lead to emissions of NMHC. They are more difficult to estimate. Firstly the composition of natural gas varies widely. Dry natural gas, i.e. natural gas resulting from biogenic decomposition of organic matter contains only traces of NMHC, whereas natural gas associated with crude oil deposits, socalled wet natural gas can have substantial amounts of NMHC, and so has gas from coal deposits. In the following we assume an average methane to NMHC ratio of CH_4/NMHC = 5 as it is found for the average natural gas in the US.

Little is also known about the average loss rate of natural gas. CH_4 budgets have assumed factors of 1 - 5 %. We will assume that 3 % of the total natural gas production is lost to the atmosphere. The resulting emissions of C_2H_6, C_3H_8 and C_4H_{10} are shown in Table I.

CHEMICAL INDUSTRY, SOLVENTS

Somewhat smaller losses occur in the chemical industry, and mostly in form of compounds not of interest here. The emission in form of solvents on the other hand is quite large. But again most of the solvents (used in paints, dry cleaning etc.) are halogenated compounds, alcohols or esthers of lesser interest here and are not detailed in Table I.

COMBUSTION OF OIL AND COAL

Combustion of oil and coal also releases NMHC at an emission rate shown in Table I. Since the hydrocarbon emissions are relatively low and their composition poorly known, no detailed estimates are given in Table I.

BIOMASS BURNING

Under this heading we include the emissions by slash and burn agriculture, deforestation, burning of savanna and bushland, forest wild fires, burning of fuel wood, burning of agricultural wastes, and burning of industrial wood. These add up to the burning of 6800 x 10^{12} g/yr of biomass (Crutzen et al., 1981). There are various estimates for the emission factors (Robinson and Robbins, 1968; Duce et al., 1982; NAS, 1976). We assume a relatively low average emission factor of 0.2 % weight NMHC per weight fuel. For the composition we used the relative values given by Stephens and Burleson (1969), Since this is such a mixture of sources with different emission factors and composition, the estimates given in Table I are very tentative and uncertain. They also involve large extrapolations of local measurements to global areas. But they indicate that despite the low emission factor biomass burning is potentially a big source and merits further investigation.

FOLIAGE

The emissions of hydrocarbons by foliage are apparently enormous (Zimmerman et. al. 1981) But the gases emitted, isoprene and terpenes do not fit into NMHC discussed here. Moreover they are very shortlived and measurable concentrations exist only in the immediate environment of the sources. Their main impact on atmospheric chemistry is local or regional and there it may be overwhelming. Their global impact proceeds through the products formed during their oxidation, CH_2O, H_2, CO, and possibly O_3. They will not be discussed further. However, leaves, especially when damaged also emit minor amounts of C_2H_6 and C_2H_4. An estimate for these emissions is included.

MICROBIAL PRODUCTION OF NMHC

Most of the CH_4 found in the atmosphere has been released after microbial production in anoxic environments (Ehhalt, 1974). This emission rate totals about 500 x 10^{12} g CH_4/yr. Under the same conditions NMHC may also be produced and released. Van Cleemput et al. (1981) report the production of CH_4, C_2H_4, C_2H_6, C_3H_8 after incubation of soil samples. The NMHC production is however small. Relative to CH_4 the production of C_2H_4 was on the order of 5 x 10^{-3}, that of C_2H_6 was 7 x 10^{-4}, and that of C_3H_8 was 4 x 10^{-4} on a volume to volume bases. Extrapolated to the globe with the help of above CH_4 emission one would obtain 4.4 x

10^{12} g C_2H_4/yr, 0.7 x 10^{12} g C_2H_6/yr and 0.5 x 10^{12} g C_3H_8/yr. These estimates are entered also into Table I.

OCEANS

Another important source of NMHC is the ocean. Light NMHC (C_2H_6. C_2H_4, C_3H_8, C_3H_6) are dissolved in surface ocean water at large supersaturations with respect to their atmospheric concentrations (Lamontagne et al., 1974; Swinnerton and Lamontagne, 1977). Thus they effuse into the atmosphere. For certain ocean areas the fluxes can be estimated (Rudolph and Ehhalt, 1981). At 70° N latitude they amount to a few mg/m^2/yr. Extrapolated to the globe we obtain the numbers shown in Table I.

There are indications, however, that surface waters of the tropical oceans are by a factor 3 more supersaturated in C_2H_4 and C_3H_6 (Lamontagne et al., 1974). This means that the tropical ocean could be a stronger source for C_2H_4 and C_3H_6 (but not for C_2H_6 and C_3H_8) than estimated from the data from 70° N. With the atmospheric mixing ratios of C_2H_4 and C_3H_6 given in the previous chapter, and the estimated lifetimes of 0.7 days for C_2H_4 and 0.3 days for C_3H_6 in the tropics, we can also calculate the sink strength over the tropical ocean between 10° N to 10° S, assuming a well mixed boundary layer of 1 km depth. The resulting values are 5 x 10^{12} g C_2H_4/yr and 10 x 10^{12} g C_3H_6/yr. These values are entered separately in Table I. As in most of the previous estimates the extrapolation involved in the estimate of the oceanic source strengths are enormous. The uncertainties are large. Nevertheless the estimates indicate that the oceans may be a large source of NMHC and merit further investigation.

This concludes the list of NMHC sources known to us. There might be others which have not been identified yet, and the ones presented, especially the natural sources have very large uncertainties. It should be reemphasized that the present exercise was a first attempt at a budget of the NMHC and that the estimates presented, sometimes nothing more than guesses, are preliminary, meant more as guidelines for future research than as final answers.

Disregarding isoprene and terpenes, the organic emission from natural and anthropogenic sources appear approximately the same. They total about 100 x 10^{12} g/yr, or about 20 % of the global CH_4 emission rate. However, only 11 x 10^{12} g/yr of the anthropogenic emissions are in the C_2 to C_5, whereas all of the natural emission, 42 x 10^{12} g/yr, fall in that category. They combine to 50 x 10^{12} g/yr or 10 % of the CH_4 emissions. Based on this balance the global impact of C_2 - C_5 hydrocarbons on the tropospheric chemistry would be on the order of 10 % of that of CH_4. However much of these emissions take place in the midlatitudes of the northern hemisphere. The photochemistry in the boundary layer of a large part of that region may indeed be dominated by NMHC rather than by CH_4, and as pointed out above most of the experiments on tropospheric chemistry have been and are carried there, a point which

Table Ia: Global Estimates of anthropogenic hydrocarbon emissions into the atmosphere (10^{12} g/yr)

Source	CH_4	NMHC	C_2H_6	C_2H_4	C_2H_2	C_3H_8	C_3H_6	$n\text{-}C_4H_{10}$	$n\text{-}C_5H_{12}$	$i\text{-}C_5H_{12}$	Benzene	Toluene	p, m, o- Xylene	Isoprene	Terpenes
1. Automobile exhaust a. gasoline engines	.9	16	.2	1.9	1.4	.02	.8	.4	.5	.8	.8	1.7	1.5		
b. diesel engines		3													
2. Evaporation losses from cars	-	4	-	-	-	-	-	.4	.4	.7	.06	.3	.2		
3. Evaporation losses during distribution and refining	-	6	-	-	-	-	-	.6	.5	1.0	.08	.4	.3		
4. Natural gas losses	25	5	1	-	-	1	-	1*							
5. Coal burning	-	3	-												
6. Oil burning	-	1	-												
7. Chemical industry	-	3	.1	.3							.1	.1			
8. Solvents	-	15	-									1	.5		
		56													

*including $i\text{-}C_4H_{10}$

Table Ib: Global Estimates of natural hydrocarbon emissions into the atmosphere (10^{12} g/yr)

Source	CH_4	NMHC	C_2H_6	C_2H_4	C_2H_2	C_3H_8	C_3H_6	$n\text{-}C_4H_{10}$	$n\text{-}C_5H_{12}$	$i\text{-}C_5H_{12}$	Benzene	Toluene	p, m, o- Xylene	Isoprene	Terpenes
9. Biomass burning	60	14	4.1	4	.6	1.7	3.8								
10. Foliage		(1.5)	.5	1										350	480
11. Microbial Production	500	5.7	.7	4.4			.6								
12. Ocean	5														
tropical		15		5			10								
rest		6	1	2		1.3	1.6								
		42													

must be kept in mind when interpreting these data.

CONCLUSION

Our knowledge about the atmospheric chemistry and budget of hydrocarbons is still extremely patchy. It resembles a puzzle where many of the pieces are missing. This makes it difficult to present a systematic picture. At the same time there is sufficient information to establish that NMHC are certainly of high regional importance and very probably of global importance. In both environments multiphase chemistry of organic molecules seems to play an important role. The many gaps in our knowledge of the chemistry and budget of organic compounds in the atmosphere present a challenge. A challenge that we can begin to meet, since with the advance of the various highly sensitive chromatographic techniques and others like the Lamma Sonde the tools for this trade have become available.

EFERENCES

Aikin, A.C., J.R. Herman, and E.J. Maier, 1980, Ethane and ethylene behaviour in the atmosphere, Proc. NATO Advanced Study Institute on Atmospheric Ozone: Its variation and Human Influences, 1.-13. Oct. 1979, Aldeia das Acoteias, Algarve, Portugal, Report No. FAA-EE-80-20, Federal Aviation Administration Washington, D.C., pp 191-208.

Chameides, W.L., and D.D. Davis, 1982, The free radical chemistry of cloud droplets and its impact upon the composition of rain. J. Geophys. Res. 87, 4863 - 4877.

Chameides, W.L., and D.D. Davis, 1983, Aqueous-phase source of formic acid in clouds, Nature, 304, 427 - 429.

Chameides, W.L., 1985, Photochemistry in the atmospheric aqueous phase, this volume.

Crutzen, P.J., L.E. Heidt, J.P. Krasnec, W.H. Pollock, and W. Seiler, 1979, Biomass burning as a source of atmospheric gases CO, H_2, N_2O, NO, CH_3Cl and COS, Nature 282, 253 - 256.

Duce, R.A., V.A. Mohnen, P.R. Zimmerman, D.Grosjean, W. Cantreels, R. Chatfield, R. Jaenicke, J.A. Ogven, E.D. Pellizzari, and G.T. Wallace, 1983, Organic material in the global troposphere, Rev. Geophys. Space Phys. 21, 921 - 952.

Ehhalt, D.H., 1974, The atmospheric cycle of methane, Tellus 26, 58 - 70.

Ehhalt, D.H., J. Rudolph, F. Meixner, and U. Schmidt, 1985, Measurements of selected C_2-C_5 hydrocarbons in the background troposphere: Vertical and latitudinal variations, J. Atmos. Chem. 3, in press.

Galloway, J.N., G.E. Likens, W.C. Kenne, and J.M. Mueller, 1982, The composition of precipitation in remote areas of the world. J. Geophys. Res. 87, 8771 - 8786.

Graedel, T.E., and C.J. Weschler, 1981, Chemistry within aqueous atmospheric aerosols and raindrops, Rev. Geophys. Space Phys. 19, 505 - 539.

Junge, C. and G. Scheich, 1969, Studien zur Bestimmung des Saeuregehaltes von Aerosolteilchen, Atmospheric Environment 3, 432 - 441.

Keene, W.C., J.N. Galloway, and J.D. Holden Jr., 1983, Measurements of weak organic acidity in precipitation from remote areas of the world. J. Geophys. Res. 89, 5122 - 5130.

Ketseridis, G., and R. Eichmann, 1978, Organic compounds in aerosol samples, Pure Appl. Geophys. 116, 274 - 282.

Klippel, W., and P. Warneck, 1978, Formaldehyde in rainwater and on the atmospheric aerosol, Geophys. Res. Lett. 5, 177 - 179.

Lamontagne, R.A., J.W. Swinnerton, and V.J. Linnenbom, 1974, C_1 - C_4 hydrocarbons in the North and South Pacific, Tellus 26, 71 - 77.

Lee, Y.-N., G.I. Senum, and J.S. Gaffney, 1983, Peroxyacetyl-nitrate (PAN) stability, solubility and reactivity - implications for tropospheric nitrogen cycles and precipitation chemistry. CACGP Symposium on Tropospheric Chemistry, 28. August - 3. September 1983, Christchurch, Oxford, England.

Logan, J.A., M.J. Prather, S.C. Wofsy and M.B. McElroy, 1981, Tropospheric chemistry: A global perspective, J. Geophys. Res. 86, 7210-7254.

Lowe, D.C. and U. Schmidt, 1983, Formaldehyde (HCHO) measurements in the non-urban atmosphere, J. Geophys. Res. 88, 10844-10858.

Lowe, D.C., U. Schmidt, and D.H. Ehhalt, 1981, The tropospheric distribution of formaldehyde, Berichte der Kernforschungsanalage Juelich, No. Juel-1756. 5170 Juelich, F.R.G.

Luftreinhaltung, 1981, Entwicklung, Stand, Tendenzen. Umweltbundesamt ed. Erich Schmidt-Verlag, Berlin.

NAS, 1976, Vapor phase organic pollutants, National Academy of Sciences, Washington D.C., 1976.

Nelson, P.F., S.M. Quigley, and M.Y. Smith, 1982, Sources of atmospheric hydrocarbons in Sidney: A quantitative determination using a source reconciliation technique. Atmos. Environ. 17, 439 - 449.

Robinson, E., and R.C. Robbins, 1968, Source, abundances and the fate of atmospheric pollutants, Final report, Stanford Research Institute, Project PR-6755 for the American Petroleum Institute, New York.

Rudolph, J., and D.H. Ehhalt, 1981, Measurements of C_2 - C_5 hydrocarbons over the North Atlantic. J. Geophys. Res. 86, 11959 - 11964.

Rudolph, J., D.H. Ehhalt, U. Schmidt, and A. Khedim, 1982a, Vertical distributions of some C_2 - C_5 hydrocarbons in the nonurban troposphere. Proc. 2nd Symp. on Composition of the Nonurban Troposphere, 25-28 May, 1982, Williamsburg, Va., USA, 56 - 59.

Rudolph, J., D.H. Ehhalt, A. Khedim, and C. Jebsen, 1982b, Latitudinal profiles of some C_2 - C_5 hydrocarbons in the clean troposphere over the Atlantic. Proc. 2nd Symp. on Composition of the Nonurban Troposphere, 25-28 May, 1982, Williamsburg, Va., USA, 284 - 286.

Schmidt, U., A. Khedim, F.J. Johnen, J. Rudolph, and D.H. Ehhalt, 1982, Two dimensional meridional distributions of CO, CH_4, N_2O, $CFCl_3$, and CF_2Cl_2 in the remote troposphere over the Atlantic Ocean. Proc. 2nd Symp. on Composition of the Nonurban Troposphere, 25-28 May, 1982, Williamsburg, Va., USA, 52-55.

Singh, H.B., and L.J. Salas, 1982, Measurements of selected light hydrocarbons over the Pacific Ocean: Latitudinal and seasonal variations. Geophys. Res. Lett. 9, 842 - 845.

Somorjai, G.A., 1982, The surface science of heterogeneous catalysis: Possible applications in atmospheric sciences. in: Heterogeneous Atmospheric Chemistry, Geophys. Monograph 26, 88 - 92, D.R. Schyer ed.

Stephens, E.R., and F.R. Burleson, 1969, Distribution of light hydrocarbons in ambient air. J. Air. Poll. Control. Ass. 19, 929 - 936.

Stern, A.C., editor, 1968, Air Pollution III, Sources of Air Pollution and their control. Academic Press, New York.

Swinnerton, J.W., and R.A. Lamontagne, 1974, Oceanic distribution of low-molecular-weight hydrocarbons, baseline measurements. Environ. Sci. Technol. 8, 657 - 663.

Thompson, A.N., 1980, Wet and dry removal of tropospheric formaldehyde at a coastal side, Tellus 32, 376 - 383.

van Cleemput, O., A.S. El-Sebaay, and L. Baert, 1981, Production of gaseous hydrocarbons in soil. in: Physico-Chemical Behaviour of Atmospheric Pollutants, B. Versino and H. Ott ed., Reidel Publishing Company, Dordrecht, Holland, pp 349-355.

van Grieken, R., and F. Adams, 1985, Characterization of particulate matter by X-ray, Raman and mass spectrometric techniques, this volume.

Winkler, P., 1983, Relations between aerosol acidity and ion balance, this volume.

Zimmerman, P.R., R.B. Chatfield, J. Fishman, P.J. Crutzen, and P.L. Hanst, 1978, Estimates of the production of CO and H_2 from the oxidation of hydrocarbon emissions from vegetation, Geophys. Res. Lett. 5, 679 - 682.

AMBIENT RADICAL CONCENTRATIONS IN THE PRESENCE OF AIRBORNE LIQUID WATER

D.H. Stedman and J.G. Walega
University of Denver,
Denver CO 80208, U.S.A.

C.A. Cantrell
National Center for Atmospheric Research,
Boulder CO 80303, U.S.A.

J.P. Burrows and G. Tyndall
Max-Plank-Institut fuer Chemie
D 6500 Mainz, F.R.G.

ABSTRACT

A new technique for the continuous measurement of peroxy- and oxy-radicals in air is based on the chain reaction in which these molecules participate to oxidize NO and CO to NO_2 and CO_2. The NO_2 produced is measured by luminol chemiluminescence, based on the gas-surface reaction of NO_2 with a pH 12, 3 mM aqueous luminol solution. NO_2 produced from sources other than the chain reaction is measured by substitution of N_2 for CO one-half of the time. Therefore, the radical signal is modulated and determined by difference. Somewhat discrepant calibration is obtained by measuring the time dependence of free radical decay, and by direct interface with a chemical modulation system in which HO_2, CH_3O_2 and $CH_3CO{\cdot}O_2$ were produced. The detection limit of radicals with this system is a strong function of the variability of ambient NO_2 and O_3, but usually less than 1 pptv can be detected. Under conditions when there is significant liquid water content in the air (fog/rain) no free radical signal could be detected.

INTRODUCTION

The chemistry of trace species in polluted and unpolluted atmospheres is fundamentally dependent on the reactions of free radicals. In air with low hydrocarbon content, the most

NATO ASI Series, Vol. G6
Chemistry of Multiphase Atmospheric Systems
Edited by W. Jaeschke

important radicals are predicted to be the odd hydrogen radicals: hydroxyl, HO, and hydroperoxy, HO_2. Organic peroxy radicals become more abundant as the hydrocarbon concentration increases, and possibly at night. Major loss mechanisms for many species are initiated by reaction with HO (e.g. CH_4 and higher hydrocarbons, SO_2 and NO_2). Hydroperoxy oxidizes NO to NO_2 and is the major homogeneous source of H_2O_2 in clean atmospheres. There has been considerable interest in the measurement and modeling of HO and HO_2, since 1969 when the importance of odd hydrogen free radicals to smog chemistry was recognized by Weinstock et al. (1969) and Heicklen et al. (1969). Shortly thereafter the interaction between the odd hydrogen family and other atmospheric cycles was recognized as important in the clean troposphere (Crutzen, 1970; Levy, 1971).

Laser induced fluorescence has been attempted by several groups around the world for atmospheric OH (Heaps et al., 1981; Davis et al., 1976; Wang et al., 1975). Long path absorption spectroscopy (LPAS) is used to measure a wide variety of atmospheric constituents, including HO, as reported by Perner, et al. (1976). A chemical method has been used to measure tropospheric HO by Campbell, et al. (1979). This involves allowing an air sample to react with a small amount of ^{14}CO and measuring the yield of $^{14}CO_2$.

Atmospheric peroxy radicals, including HO_2, have been measured by only one technique other than our method, as reported by Mihelcic, et al. (1978). Their method involves trapping ambient free radicals on a cold finger in a matrix of CO_2 and H_2O (recently D_2O), and returning the sample to the laboratory for radical detection by electron paramagnetic resonance (EPR). Recently, ambient HO has been sampled from an aircraft platform by absorption on a filter paper coated with a spin trap, followed by measurement of the spin-trapped HO complex using EPR (Watanabe et al., 1982).

Our technique of PEroxy Radical measurement by Chemical Amplification (PERCA) uses the oxidizing chain reaction in which HO and HO_2 participate directly (Cantrell and Stedman, 1982; Stedman and Cantrell, 1982; Cantrell et al., 1983). Other RO_2 free radicals are predicted to become HO_2 after a few reaction steps. The yield of NO_2 produced by the ambient radicals is measured, and an ambient total peroxy plus hydroxyl radical concentration is inferred, which, in clean air, is expected to be mostly HO_2. Presented here are the results of recent calibrations using laboratory sources of HO_2, CH_3O_2 and $CH_3CO{\cdot}O_2$, as well as some preliminary results of ambient measurements.

EXPERIMENTAL

Apparatus

The PERCA instrumentation is composed of three components which provide for: 1) reagent addition (CO or N_2 and dilute NO in N_2), 2) a glass reactor in which the chain reaction takes place, and 3) NO_2 detection.

The flow rate of the reagent gases is controlled through flow restrictors (Mott Metallurgical), and delivered through 0.25 inch (6 mm) o.d. Teflon tubing to two three-way solenoid valves (Fluorocarbon). The reagent mixture is delivered to the interior of a glass sphere in the intake region of the Pyrex glass reactor. The incoming sample air passes over the 1 cm diameter glass sphere which is pierced with seventeen 1 mm diameter holes from which the reagent gases emerge. The reactor has a total volume of 150 cm^3, which results in a reaction time of 4.2 seconds using a flow of 2.0 liters min^{-1}. The resultant mix is drawn into the NO_2 detector, which is based on the chemiluminescent reaction between NO_2 and luminol, described recently by Wendel et al. (1983). The photocurrent generated is amplified by an electrometer (Keithley Instruments 410A) and read on a chart recorder (Heath Schlumberger SR-204). Electrometer output is also collected by an IBM Personal computer. The solenoid valve timing sequence is controlled by a Chrontrol CT-4 table top timer, or if the data is computer acquired, by parallel I/O connections through opto-isolator relays.

The reagent gases used in the detection system are carbon monoxide, CP grade (Air Products); nitrogen, prepurified, (AP); 100 ppmv (parts per million by volume) nitric oxide (NO) in nitrogen (purchased from Air Products or synthesized in the laboratory in aluminum cylinders from high purity nitric oxide and oxygen-free nitrogen).

Procedure

The basis of the chemical amplifier technique has been described by Cantrell et al. (1982). Briefly, air containing oxy- (such as HO) or peroxy- (HO_2 and RO_2) radicals is drawn into the glass reactor to which are added the appropriate reagent gases. A partial mechanism for the reactions of the radicals and the reagents is:

$$HO_2 + NO \longrightarrow NO_2 + HO \qquad 1)$$

$$HO + CO \longrightarrow H + CO_2 \qquad 2)$$

$$H + O_2 + M \longrightarrow HO_2 + M \qquad 3)$$

Proper choice of [NO], [CO] ([ ] = concentration), and reaction time, can produce several NO_2 molecules for each radical entering the system. Typical conditions are [NO] = 3 ppmv, [CO] = 10% and t_{rxn} = 5 seconds. The amount of NO_2 produced from these reactions is limited by reactions which terminate the chain reaction. The most important radical sinks for reagent concentration and reaction times in the range described above are:

$$HO + NO + M \longrightarrow HONO + M \qquad 4)$$

$$HO_2 + HO_2 \longrightarrow H_2O_2 \qquad 5)$$

However under normal HO_2 concentrations (< 200 pptv), reaction 5 is very slow and not very important in this system. Chemical kinetic modelling of reactions 1 - 4 using up to date rate constants (DeMore et al., 1982) leads to a predicted increase in nitrogen dioxide concentration ($[NO_2]$) = 1100 x $[HO_x]_o$.

Since there are other sources of NO_2 in the atmosphere and in this system including ambient NO_2, ad NO_2 produced from ambient ozone:

$$NO + O_3 \longrightarrow NO_2 + O_2 \qquad 6)$$

a modulation technique is used. Nitrogen is substituted for carbon monoxide one-half of the time. The radical signal is then determined from the difference in NO_2 signal with and without CO present.

RESULTS AND DISCUSSION

Theory

RO_2 radicals, where R is a hydrocarbon or R'CO group, react rapidly with NO and, apart from a small probability of pernitrite formation for larger hydrocarbons (Atkinson et al., 1983), indulge in a series of reaction steps leading directly to the production of an HO_2 radical which then participates in the chain reaction. An analytical solution for the NO_2 yield was derived previously (Cantrell and Stedman, 1982), and resulted in the following equation for the yield of NO_2 from reactions 1) through 4) as a function of [NO], [CO], reaction time, initial radical concentration, and rate constants:

$$[NO_2] = k_2[CO][HO_2](1-\exp(-q)) \; / \; k_4[NO] \qquad \text{a)}$$

where:

$$q = k_1k_4[NO]^2 t / k_2[CO]$$

The NO_2 yield of this system can be described relative to the initial radical concentration rather than in concentration units. The "chain length" (CL) is defined as the increase in NO_2 concentration divided by the initial radical concentration:

$$(CL) = [NO_2] \; / \; [HO_2]_o \qquad \text{b)}$$

Figure 1 shows the calculated chain length versus [NO] for various reaction times from a), using the most recent updated values for k_1 through k_5. The values of the rate constants for reactions 1) through 5) used are shown in Table 1. This is plotted for a constant [CO] = 10%. Figure 2 shows the chain length dependence on [CO]. Some reduction in chain length is expected because of reactions of radicals with the reactor walls. At a 4.5 second reaction time with [CO] = 10%, the calculated reduction in chain length is about 15% for a reasonable

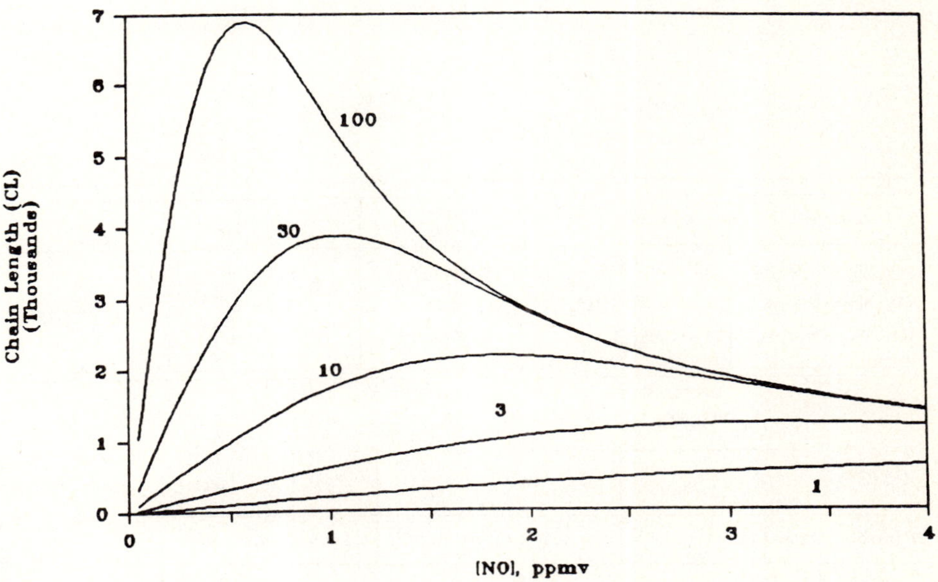

Figure 1. Chain length ($[NO_2]/[HO_2]_o$) versus [NO] for various reaction times from equation (a). Shown are reaction times of 1, 3, 10, 30 and 100 seconds, with [CO] = 10%, and a temperature of 298 K. Wall reactions and other minor reaction paths have been neglected in this model.

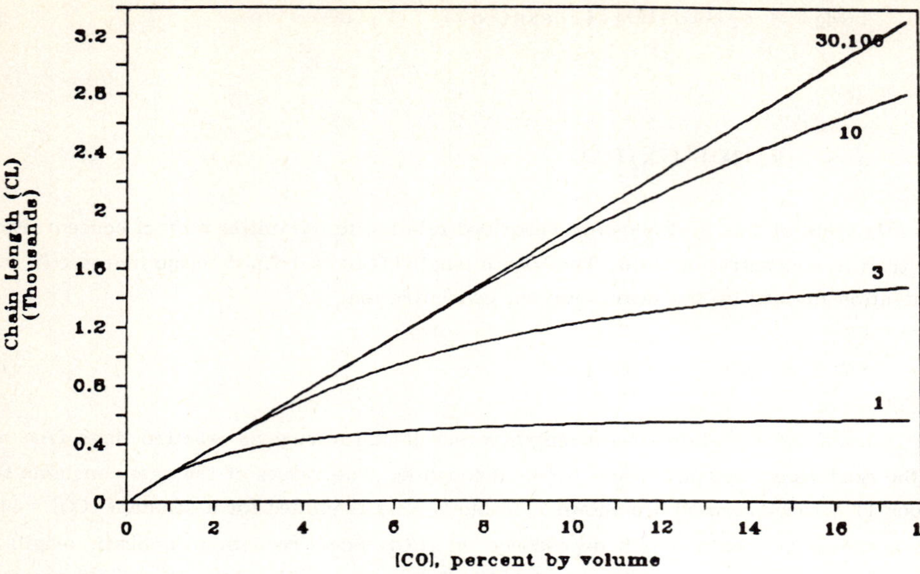

Figure 2. Calculated chain length versus [CO] at reaction times of 1, 3, 10, 30 and 100 seconds, for [NO] = 3 ppmv, from equation (a). A practical maximum [CO] in air is the lower explosion limit of 24% by volume.

REACTION	A-FACTOR	$(E/R) \pm \Delta(E/R)$	k_{1atm}^{298}	UNCERTAINTY
(1) $HO_2+NO \rightarrow NO_2+HO$	3.7×10^{-12}	$-(240 \pm 80)$	8.3×10^{-12}	1.2
(2) $HO+CO \rightarrow CO_2+H$	$1.35 \times 10^{-13}(1+P_{atm})$	(0 ± 200)	2.7×10^{-13}	1.25
(4) $2HO_2 \rightarrow H_2O_2+O_2$	$(3.4+2.5P_{atm}) \times 10^{-14}$	$-(1150 \pm 600)$	2.8×10^{-12}	1.5

REACTION	LOW PRESSURE $k_0^{300}=k^{300}(T/300)^{-n}$		HIGH PRESSURE $k_{oo}^{300}=k^{300}(T/300)^{-m}$		K_{1atm}^{298}
	k_0	n	k_{oo}	m	
(3) $HO+NO(M) \rightarrow HONO$	$(6.7 \pm 1.2) \times 10^{-31}$	2.5 ± 1.0	$(1.5 \pm 1.0) \times 10^{-11}$	0.5 ± 0.5	4.8×10^{-12}

Table I. Summary of pertinent rate constants used for the analytical solution of the chemical amplifier equation from DeMore et al., (1982). The uncertainty values were used to evaluate the confidence in model description.

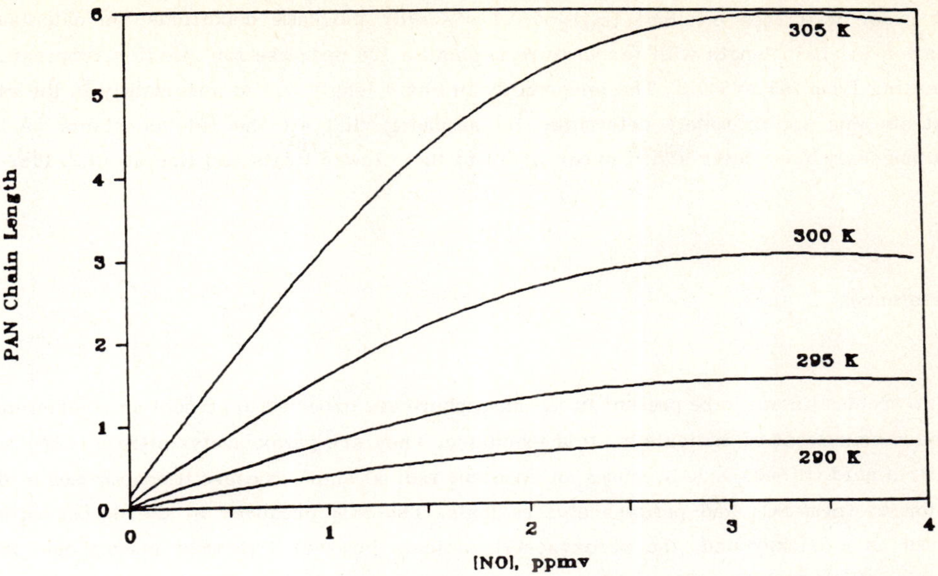

Figure 3. Calculated chain length for PAN versus [NO] for temperatures of 290, 295, 300, and 305 K. The model is for a flow time of 4.2 seconds at [CO] = 10%.

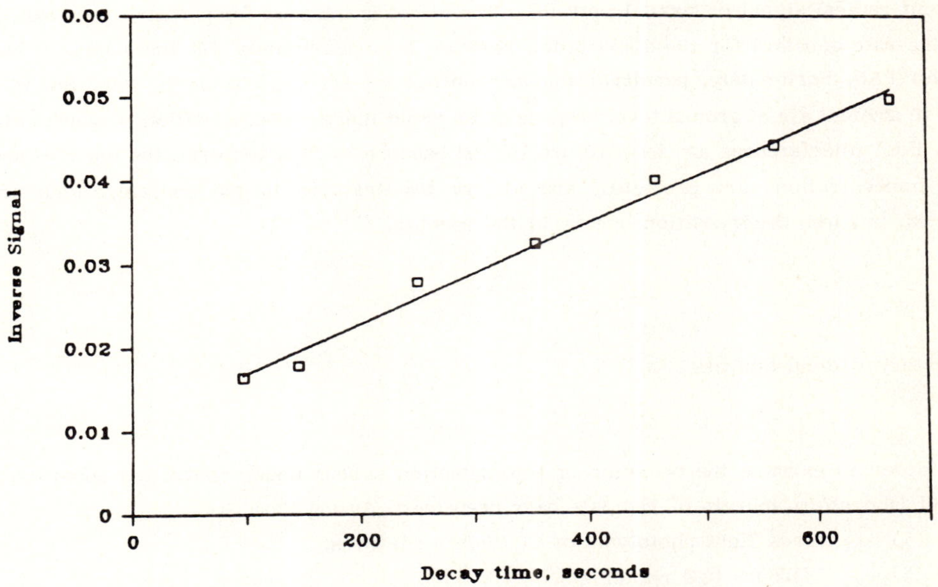

Figure 4. Inverse radical signal versus decay time using a hot nichrome wire as the source of peroxy radicals. This is one method used as a calibration technique. The slope of plots such as these yield a chain length for the chemical amplifier.

first order wall loss constant (k_W) of 0.1 s^{-1}. For the same conditions the calculated variation in chain length with temperature is about a 10% decrease for reaction temperature increasing from 260 to 340 K. The uncertainty in chain length due to uncertainty in the rate constants was approximately determined by assuming that all the rate constants varied simultaneously from their stated upper limits to their lower limits and was about $\pm$ 10%.

Interferences

Two species known to be present in the atmosphere are expected to present an interference to the measurement of radicals by this technique. They are peroxyacetyl nitrate (PAN) and pernitric acid (HO_2NO_2). PAN causes an apparent radical signal because it decomposes in the reactor to form NO_2 and peroxyacetyl radicals. The NO_2 produced in the decomposition appears as a background, the peroxyacetyl radicals however like most atmospheric RO_2 radicals, ultimately form HO_2 under normal measurement conditions. The calculated effect of PAN in the chain reactor is shown in Figure 4 plotted as the "PAN chain length" ($[NO_2]/[PAN]_o$) versus [NO], at various temperatures. The PAN dissociation equilibrium is very temperature dependent, and thus results in a temperature dependent PAN sensitivity. Notice that at 300 K and typical HO_2 sensitivity ((CL) = 1100), 1 ppbv of PAN would give an apparent radical signal of about 1 pptv HO_2. The effect is the same for pernitric acid except that the rate constant for the dissociation reaction is approximately 200 times larger than that for PAN. Fortunately, pernitric acid concentrations are expected to be very low (< 1 pptv) in ambient air at ground level because of its rapid thermal decomposition (Logan et al., 1981). Both interferences are temperature limited because at high temperatures low PAN and HNO_4 concentrations are predicted, and at low temperatures larger concentrations are expected, but less decomposition occurs in the reactor.

Laboratory Radical Sources

In order to examine the behavior of this detection system under controlled conditions, several laboratory sources of radicals have been used. Among them are:

1) black light photolysis of Cl_2/H_2/air mixtures,
2) 184.9 nm H_2O photolysis,
3) methyl nitrite photolysis,
4) passage of dry air over a hot nichrome wire (26),

and 5) ultraviolet photolysis of O_3/H_2O/air mixtures.

Some of these techniques are either H atom or HO sources. Hydrogen atoms rapidly form HO_2 in air, and hydroxyl radicals are measured by this as HO_2. These sources were used to determine experimentally the dependance of the radical signal on reagent gas concentrations and reaction times. They were also used in the radical calibration procedure.

Peroxy and hydroxyl radicals are short lived and thus no method exists for producing a standard amount of these species. However, hydroperoxy radicals decay by reaction with each other at a known rate (DeMore et al., 1982), yielding a loss which is second order in HO_2. This fact can be used to provide a radical calibration for the detection system. If reaction 5) is the only loss route for HO_2 under a given set of conditions, then the time dependent loss can be expressed as a differential equation:

$$-d[HO_2] \;/\; dt = 2k_5[HO_2]^2 \qquad \text{c)}$$

The time dependence of the radical concentration is:

$$[HO_2]_t^{-1} = [HO_2]_o^{-1} + 2k_5 t \qquad \text{d)}$$

If it is assumed that the radical signal of the chemical amplifier is linear in $[HO_2]$ (true when the second order reaction is negligibly slow compared to the reaction time), then $[HO_2]_t = [NO_2]_t/(CL)$. Substituting b) into d) yields an expression for observed radical signal (NO_2) in terms of the decay time, t, and the second order rate constant. If the inverse radical signals for a given decay study are plotted versus decay time, the result is a straight line with:

$$\text{slope} = 2k_5 \;/\; (CL) \quad \text{and} \quad \text{intercept} = 1 \;/\; [NO_2]_o$$

A sample decay is shown in Figure 4, plotted as described above. Experimentally these decays were monitored by producing an arbitrary concentration of radicals in an air flow in a 6 inch (15 cm) diameter Pyrex pipe. The radical signal decay was monitored by sampling the air along the axis at various distances down the tube and converting the distance into time from the volumetric flow rate. The values for (CL) found by this technique were found to agree within about 20% over a period of several weeks for a given set of reagent concentrations and reactor configuration using the hot wire source. The other sources of free radicals were tested briefly, however the most consistent data were obtained with the hot wire. A preliminary study of solar photolysis of water saturated ozone at 100 ppb, using our own measurement technique for $j(O_3\text{-}O(^1D))$ gave data consistent with the previous decay studies, and shows promise of becoming a portable field calibration method.

Instrumental Precision and Accuracy

Figure 5 shows radical signals from laboratory measurements of a synthetic radical source. The one hour data set in Figure 5a from a constant radical source gives a relative standard deviation of 1.1%. This is about 0.5 pptv (1.2×10^7 molecules cm^{-3}) in absolute terms. This long term stability of the peroxy radical detector is evidenced by observations of signals which vary by less than 1% per hour, such as shown in Figure 5a, for periods of 12 hours or more. Ambient air data from Ann Arbor are given in Cantrell et al. (1983). A conservative estimate of the instrument precision in the field is the observed relative standard deviation of an apparently constant signal because included in the standard deviation are fluctuations in the concentration of both the species causing the background signal and any fluctuations in the radical concentration. The measurement site on the roof of the laboratory in an urban area is a site where large fluctuations are both observed and expected. A calculation of the relative standard deviation of very low radical concentrations such as were usually observed at night, results in an absolute standard deviation of 0.24 pptv (6×10^6 molecules cm^{-3}). These results indicate the good instrumental precision of this radical monitoring technique.

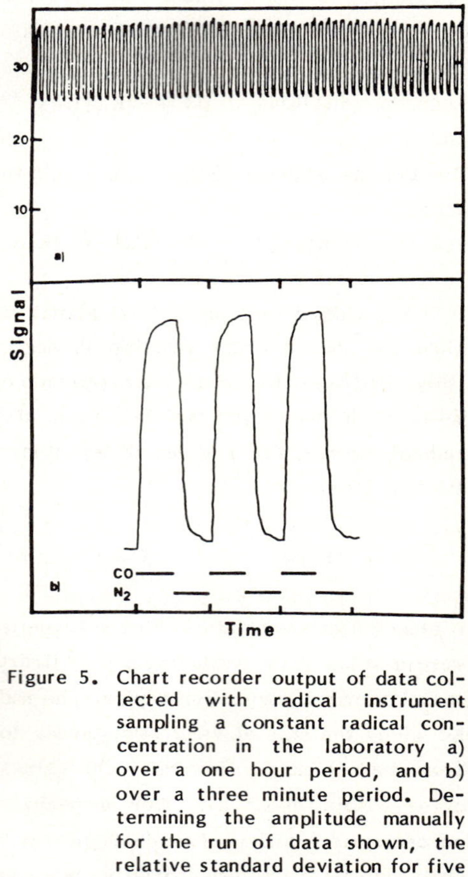

Figure 5. Chart recorder output of data collected with radical instrument sampling a constant radical concentration in the laboratory a) over a one hour period, and b) over a three minute period. Determining the amplitude manually for the run of data shown, the relative standard deviation for five minute averages is 1.1%. Figure 5b demonstrates the short term temporal behavior of an approximately 150 pptv (3.8×10^9 molecules cm^{-3}) radical signal. The tic marks on the time scale are one minute apart, and the bars at the bottom show which reagent gas (N_2 or CO) was flowing at any given time.

The accuracy of radical measurements obtained with this method is highly dependent on the accuracy of the calibration procedure which is performed in the laboratory. The calibration factor (the chain length, $[NO_2]/[HO_2]_o$) is the value of the rate coefficient for the disproportionation of HO_2 divided by the slope of the second order decay plot. The accuracy of each of these terms is estimated in the following example, using the data shown for the sample decay in Figure 4. As shown in Table 1, the uncertainty factor in the slope of the data as determined by the least squares method is 1.14. The relative uncertainty in the chain length is estimated by calculating the contribution to the variance in the chain length by the slope and the rate constant. This results in a relative uncertainty factor of 1.52, most of which is contributed by the uncertainty in the rate constant.

Calibration Using Cl_2 Photolysis

At the Max-Planck-Institut in Mainz (FRG) is a system for free radical generation and studies using chemical modulation based on chlorine photolysis (M.P.I.). The addition of H_2, CH_4 and CH_3CHO allows for production of HO_2, CH_3O_2 and $CH_3CO{\cdot}O_2$ radicals. As discussed elsewhere, (for instance Cox and Tyndall, 1979; Cox and Burrows, 1979) the predicted radical concentration is

$$[RO_2] = (j(Cl_2)[Cl_2] \; / \; k_{(recomb)})^{1/2}$$

where $j(Cl_2)$ is the chlorine photolysis frequency and $k_{(recomb)}$ is the second order recombination rate constant. HO_2 and $CH_3CO{\cdot}O_2$ recombination produce peroxide products such that each absorbed photon produces one product molecule thus

$$[Products] = j(Cl_2)[Cl_2]t$$

where t is the total photolysis time.

The M.P.I. system is designed to generate and study RO_2 radicals in the 10^{11} cm^{-3} concentration range. The PERCA is designed to study the 10^7 to 10^9 cm^{-3} concentration range. To handle this descrepancy a dilution system was installed in which all effluent from the M.P.I. cell (1-4 liter min^{-1}) came out a 6 mm id by 100 mm long exit port where it encountered a large flow (~45 liter min^{-1}) of filtered laboratory air. This large dilution flow was then sampled by the PERCA. Radical signals were observed at high signal to noise and the behavior of the PERCA signal and background, relative to the M.P.I experimental conditions was studied. Typical data are shown in Figure 6. PERCA modulation and background were measured as a function of light intensity (I), M.P.I flow (F) and PERCA NO flow using the Cl_2/H_2 system. The PERCA background was shown to be sensitive to Cl_2 (slightly) and to H_2O_2 produced by the radical recombination. The observed d.c. background was linear in M.P.I. photolysis frequency and inversely proportional to M.P.I. flow, as expected for product detection by the luminol system. The observed modulation increased linearly with the square root of light intensity (again as expected). Based on all of the above and a PERCA NO_2 calibration using the UV absorption system as a standard, a chain length of

60 was determined. This value is very much less than the predicted value (~1100). It was thought that radicals might be lost to the walls of the (Teflon coated) M.P.I. exit tube, thus the flow F was increased. According to the theory if F increases then $[Cl_2]$ decreases as $F^{-1/2}$, $[HO_2]$ as $F^{-1/2}$ and $[H_2O_2]$ as F^{-2} since it is in steady state with HO_2 formation and loss by flow out.

Since the PERCA measures the mixing ratio after dilution in a constant flow of room air then it should see HO_2 increase as $F^{1/2}$ and H_2O_2 decrease as F^{-1}. The latter prediction was confirmed by the observations but the former was not observed. An $F^{1/2}$ increase in modulation would be expected if there were no diffusion controlled radical losses, and an even larger increase if there were, because the time in the narrow exit flow tube is reduced by increasing flow. Hardly any increase is observed. This corresponds to a lowering of the determined chain length to 40-45. Since the status of the PERCA is not effected at all by flow changes in the M.P.I., the chain lengths ought to be constant. The observed low values and variability implies that the interface between the M.P.I. and PERCA was not functioning as designed. Nevertheless a few studies on Cl_2/CH_4 and Cl_2/CH_3CHO were undertaken. In both cases strong modulation was observed (Figure 6). In arbitrary units a signal of 1.35 was observed for Cl_2/H_2, 1.0 for Cl_2/CH_4 and 1.4 for Cl_2/CH_3CHO under otherwise identical conditions. Based upon recombination rate constants of 2.6×10^{-12} for HO_2, 3.4×10^{-13} for CH_3O_2 and 9.8×10^{-13} cm^3 molecule^{-1} s^{-1} for $CH_3CO{\cdot}O_2$ the observed modulations should have been 0.36, 1.0 and 1.9 (normalized to CH_4 = 1.0).

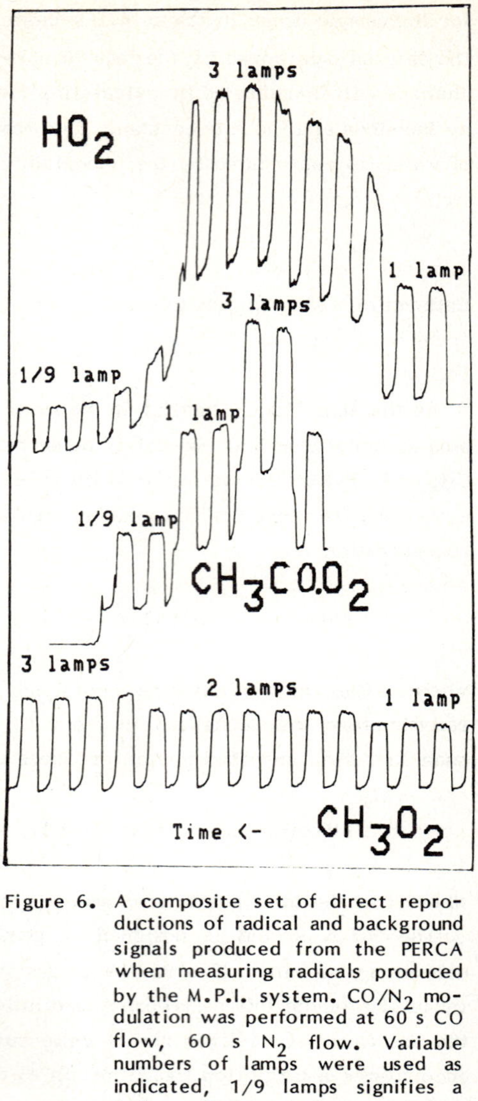

Figure 6. A composite set of direct reproductions of radical and background signals produced from the PERCA when measuring radicals produced by the M.P.I. system. CO/N_2 modulation was performed at 60 s CO flow, 60 s N_2 flow. Variable numbers of lamps were used as indicated, 1/9 lamps signifies a 1:8 on:off duty cycle repeated at 1 Hz. a) HO_2 from photolysis of Cl_2/H_2 mixtures, b) CH_3O_2 from Cl/CH_4 mixtures, c) $CH_3CO{\cdot}O_2$ from Cl_2/CH_3CHO mixtures.

In the case of CH_4 (Figure 6b) no background was observed. Since $2CH_3OO \longrightarrow CH_2O + CH_3OH + O_2$ this again would be expected. The very limited data we have for CH_4 suggests that the radical signal was approximately proportional to $I^{1/2}$ and $F^{1/2}$ as would be expected.

For acetaldehyde the data showed very good signal/noise (Figure 6c) and both background and signal were measured as a function of flow and light intensity. The background may arise through a series of complex steps from a sensitivity to peracetic acid.

Quite contrary to expectation the background showed excellent linearity with $I^{1/2}$ and complete independence on F. The modulated signal showed dependence on I to a power < 1/2 and on F roughly to 1/2. An explanation for this observation is not apparent. It is possible there was enough CH_3CHO in the system to take over chain carrying from the CO when the N_2 was on.

Several studies of the effect of PERCA NO flow were conducted and a very flat maximum was observed at about 3 ppmv NO in the PERCA, only decreasing by 1/2 at about 0.3 ppmv NO.

Ambient Radical Measurements

The instrument as described was placed on the roof of the Space Research Building at the University of Michigan in Ann Arbor on 26-29 October 1982 and 21-27 April 1983, for ambient sampling. The results are given in Cantrell et al. (1983). Noon peaks between 2 and 7 pptv and nocturnal values < 0.3 pptv were typical. The last two days of the October sample however show very low radical signals, while the UV peaked at levels only a factor of two less than the previous days. These days were moderately overcast, foggy and had some drizzle. Since the estimated source strength (ozone photolysis) had not changed greatly, this seems to indicate a change in the atmospheric sinks for these radicals. Chameides and Davis (1982) have presented calculations which indicate that the transport of radicals to water droplets could represent an important factor. Depending on the cloud droplet size and the sticking coefficient, the atmospheric lifetime of HO_2 can be reduced from about 100 seconds, in the absence of heterogenous loss processes, to about 3 seconds, significantly reducing the expected ambient concentration. The October radical observations lend credence to these calculations.

It is difficult to comment on the value of the maximum HO_2 concentrations as compared to models, because the concentrations of other important species which influence the radical concentration were not measured, also the October 1982 data were about a factor of 6 larger than those in April 1983. This can be explained in part by the fact that in the former measurements the glass inlet was exposed to solar radiation, and in the latter the inlet was covered with black velvet. It was found that exposing the inlet to sunlight on a clear day increased the amplitude of the radical signals by about a factor of 3. Since the dark inlet more closely simulates the conditions under which radical calibration was done, more confidence is placed in the more recent data. The remaining differences between these data sets can be potentially explained by real differences in radical concentrations, due to changes in source or sink rates. Models which simulate very clean air (Logan et al., 1981), and very polluted air (Calvert et al., 1981), predict maximum values for HO_2 of 10 and 200 pptv, respectively.

Studies of Surface Loss Rates of HO_2

Laboratory experiments were performed to estimate the loss rate of HO_2 on the walls of the glass reaction vessel. At the same time, loss rates of HO_2 on glass coated with various substances was also determined. The experimental procedure was as follows. A stable radical signal was achieved from a synthetic HO_2 source, and a length of glass tubing coated with the substance of interest was placed in front of the reactor. The reduction in radical signal was noted, and from the diameter of glass inlet and the flow rate of gas through this inlet, a first order loss rate coefficient can be calculated. Sticking coefficients were also calculated using the Kennedy-Gormley equation describing wall collisions in a tubular flow system.

Table II shows the results of measurements of surface loss of HO_2 with several surfaces, indicating fairly large sticking probabilities for all of them. These types of measurements demonstrate the fact that our method can be used to study the heterogeneous behavior of gas phase free radicals.

SURFACE	C/C°	k_w (s^{-1})	ϕ
Glass only	0.40	0.35	0.24
NaOH	0.12	0.82	0.81
Teflon®	0.41	0.34	0.23
H_3PO_4 (85%)	0.54	0.23	0.18
Warm glass	0.50	0.27	0.19

Table II. Summary of studies of HO_2 reactivities on surfaces. NaOH was added by coating the glass with 10% aqueous solution, which subsequently dried out yielding a powdery coating. H_3PO_4 was washed on as concentrated reagent. Teflon was 1 mm thick sheet of FEP lining the glass reactor. "Warm" glass is 70°C. Experiments were performed with a detector flow rate of 2.0 liters min^{-1}, reagent concentrations: [NO] = 1.5 ppmv, [CO] = 10%, reagent time 3.8 seconds, HO_2 was produces from the 184.9 nm photolysis of a water vapor/hydrogen mixture. $[HO_2]$ = 7 pptv.

CONCLUSIONS

A continuous technique for the measurement of peroxy radicals in the atmosphere and in laboratory systems has been presented and shown to detect HO_2, CH_3O_2 and $CH_3O{\cdot}O_2$ radicals. A model of the detection system has been developed which describes the behavior of the instrument as a function of various system parameters. The method has been used to measure peroxy radicals in the atmosphere and to measure surface loss rates of HO_2 on various substances. This method has applicability to the determination of several homogeneous and heterogeneous kinetic parameters of free radicals. This measurement method also exhibits several advantages over current radical measurement techniques for ambient air, not the least of which are sensitivity and continuous detection ability. This technique has a detection limit which is dependent on the variability of the components which contribute to the background in the system being measured. This typically yields a lower limit of 1 pptv in ambient air, and 0.1 pptv with a response of 0.1 Hz in a quiet laboratory system. Accurate absolute calibration for this radical detector remains a serious problem. By contrast the system of Mihelcic et al. (1978) measures total "spins" collected on a cold finger. For a typical volume of air (10 liters) their "sensitivity" of 10^{12} spins yields a lower limit of detection for HO_2 of about 4 pptv for a typical collection time of 30 minutes.

ACKNOWLEDGEMENTS

The author appreciate many helpful discussions with R.E. Shetter, G. Wendel, and B. Walunas, and support from the U.S. EPA under Grant EPA R-808706-01-0.

REFERENCES

Atkinson, R.; Pitts, J.N., Jr.; and Winer, A.M. (Aug.28-Sept.2, 1983), Kinetics and Mechanisms of Selected Atmospheric Reactions Involving Nitrogenous Species, Am. Chem. Soc. National Meeting Washington, D.C.

Calvert, J.G. (1984), The Chemistry of the Polluted Troposphere, Proceedings NATO Adv. Study Inst. on Chem. of the Poll. and Unpoll. Troposphere, Corfu, Greece.

Campbell, M.J.; and Sheppard, J.C.; Au, B.F. (1979), Geophys. Res. Lett., 3, 175.

Cantrell, C.A.; and Stedman, D.H. (1982), Geophys. Res. Lett., 9, 846-849.

Cantrell, C.A.; Stedman, D.H.; and Wendel, G.J. (1984), Anal. Chem., 56, 1494-1502.

Chameides, W.L.; and Davis, D.D. (1982), J. Geophys. Res., 87, 4863-4677.

Cox, R.A. and Burrows, J.P. (1979), J. Phys. Chem., 83, 2560.

Cox, R.A. and Tyndall, G. (1979), Chem. Phys. Lett., 65, 357.

Crutzen, P. (1970), Q.J.R. Met. Soc., 96, 320.

Davis, D.D.; Heaps, W.; and McGee, T. (1976), Geophys. Res. Lett., 3, 331.

DeMore, W.B.; Watson, R.T.; Golden, D.M.; Hampson, R.F.; Kurylo, M.; Howard, C.J.; Molina, M.J.; and Ravishankara, A.R. (1982) Chemical Kinetics and Photochemical Data for Use in Stratospheric Modeling, Evaluation No. 5, IPL Publication.

Gormley, P.G.; and Kennedy, M. (1949), Proc. R. Ir. Acad., Sect. A. 52A, 163-169.

Heaps, W.S.; McGee, T.J.; Hudson, R.D.; and Caudill, L.O. (1981), NASA Report No. X-963-81-27.

Heicklen, J.; Westburg, C.; and Cohen, N. (1969), from the discussion section of Stevens, E.R. in Chemical Reactions in Urban Atmospheres, Symposium at General Motors Laboratories, 55-58, Tuesday, C.S., ed.

Logan, J.A.; Prather, M.J.; Wofsy, S.C.; and McElroy, M.B. (1981), J. Geophys. Res., 886, 7210-7254.

Levy, H., II (1971), Science, 173, 141:143.

Mihelcic, D.; Ehhalt, D.H.; Kulessa, G.F.; Klomfass, J.; Trainer, M.; Schmidt, A.; and Rohrs, H. (1978), Pure and Applied Geophysics, 116, 530.

Perner, D.; Ehhalt, D.H.; Patz, H.W.; Platt, U.; Roth, E.D.; and Volz, A. (1976), Geophys. Res. Lett., 3, 466-468.

Stedman, D.H., and Cantrell, C.A. (May 25-28, 1982), Second Symposium on the Composition of the Nonurban Troposphere, Williamsburg, VA.

Wang, C.C.; Davis, L.I., Jr.; Wu, C.H.; Japar, S.; Niki, H.; and Weinstock, B. (1975), Science, 189, 797-800.

Watanabe, T.; Yoshida, M.; Fujiwara, S.; Abe, K.; Onoe, A.; Hirota, M.; and Igarashi, S. (1982), Anal. Chem., 54, 2470-2474.

Weinstock, B.; Daby, E.; and Niki, H. (1969), from the discussion section of Stevens, E.R. in Chemical Reactions in Urban Atmospheres, Symposium at General Motors Laboratories, 54-55, Tuesday, C.S., ed.

Wendel, G.J.; Stedman, D.H.; Cantrell, C.A.; and Damrauer, L.D. (1983), Anal. Chem. 55, 937-940.

Part IV

CHEMICAL CONVERSIONS IN MULTIPHASE ATMOSPHERIC SYSTEMS AND THEIR PHYSICO CHEMICAL INVESTIGATION

PHOTOCHEMISTRY OF THE ATMOSPHERIC AQUEOUS PHASE

William L. Chameides
School of Geophysical Sciences
Georgia Institute of Technology
Atlanta, Georgia 30332
U.S.A.

ABSTRACT.

The coupled gas- and aqueous-phase photochemistry of a stratiform cloud in a remote region of the marine atmosphere is investigated with a time dependent box model. Both scavenging of ambient acidic aerosols and gases as well as aqueous-phase chemical reactions within droplets are found to be important sources of acidity to cloudwater and can lead to pH levels in cloudwater in the remote marine atmosphere well below 5.6. The major sources of acidity via aqueous-phase chemical reactions are the generation of sulfuric acid from dissolved SO_2 and the generation of formic acid from dissolved formaldehyde. In both cases aqueous-phase free radicals can play a significant role either by directly oxidizing dissolved SO_2 and HCHO or indirectly by producing the aqueous-phase oxidant H_2O_2. While the transfer of HO_2 and OH from the gas to the aqueous phase during the daylight hours can represent a significant source of free radicals to cloudwater, calculations indicate that another, potentially important source of aqueous-phase free radicals may arise from coronal discharges from droplets in electrified clouds. The rate of SO_2 conversion to sulfuric acid is sensitive to a variety of parameters including the accommodation or sticking coefficient for SO_2, H_2O_2, O_2, HO_2, and OH, the droplet growth rate and liquid water content, and the ambient levels of SO_2, HNO_3, and other acidic or basic gases. Because the SO_2-to-sulfuric acid conversion rate is inhibited by low pH, the possibility exists that the pH of precipitation in polluted regions will respond non-linearly to reduced S emissions.

INTRODUCTION

The growing concern that acidic rainfall may be causing significant damage to terrestrial ecosystems has led to an intense scientific interest in the chemical and physical processes which control the acidity of precipitation (c. f. Eriksson, 1952; Barret and Brodin, 1955;

NATO ASI Series, Vol. G6
Chemistry of Multiphase Atmospheric Systems
Edited by W. Jaeschke

Cogbill and Likens, 1974; Oden, 1976; Scott and Hobbs, 1967; Penkett et al., 1979). Much of this interest has focussed on H_2SO_4 and HNO_3; these acids are most often responsible for lowering the pH of rain below 5.6, the value water attains in equilibrium with atmospheric CO_2 (Granat, 1972). Also of interest has been NH_3, often the predominant basic species in rainwater. In addition there has been growing interest in organic acids in light of the recent finding that HCOOH and CH_3COOH can be the major acidic species in rainwater in remote regions of the globe (Galloway et al., 1982).

Field measurements as well as theoretical calculations have indicated that the pH of precipitation is largely determined by the concentrations of the acidic and basic species in the cloud droplets from which the rainwater was initially formed (Petranchuk and Drozdova, 1966; Garland, 1978; Parungo et al., 1982; Levine and Schwartz, 1982; Lazrus et al., 1983). As a result, a significant advance in our understanding of the chemistry of precipitation will likely result from an accurate and comprehensive understanding of the chemistry of clouds. However, from a chemical point of view clouds represent a highly complex environment where both the gas phase and aqueous phase co-exist and as a result homogeneous gas-phase and aqueous-phase, as well as heterogeneous reactions can occur simultaneously. During the daylight hours, the penetration of actinic radiation into clouds can trigger photochemical reactions further complicating the chemistry by leading to the presence of free radical species in both phases (Graedel and Weschler, 1981; Chameides and Davis, 1982). Furthermore, as will be demonstrated in this work, similar effects can apparently result from in-cloud electrical phenomena such as coronal discharges from droplets.

The acidity of cloud droplets can be influenced by a wide variety of chemical and physical processes. For instance, acids can be generated in clouds by the simple physical dissolution of acidic gases of particulates which existed in the air parcel before condensation. Alternatively, homogeneous gas-phase reactions within the cloud can generate an acidic gas which is then scavenged by and incorporated into a cloud droplet. It is also possible for a soluble gas to be dissolved into a droplet and then chemically transformed to an acid by a homogeneous aqueous-phase reaction. These gas-phase and aqueous-phase reactions can often involve free radical species or species originally generated from free radical species. Finally, the acidity of a droplet can be affected by condensation and evaporation processes which tend to dilute and concentrate the solutes in the drop, respectively.

In order to further clarify our understanding of this highly complex chemical system, a series of calculations using a time-dependent, chemical model of a cloud were carried out and are discussed here. The purpose of these model calculations was not exactly simulate the chemistry of a specific cloud; indeed the paucity of detailed data on cloud composition and the large uncertainties in many aspects of cloud chemistry precludes such an exercise. Instead the calculations presented here should be viewed as a series of sensitivity studies aimed at identifying the major chemical pathways by which acids are generated in clouds as well as determining the key uncertainties in this chemical system. In this light, the model calculations should prove useful in the formulation of future field operations and laboratory experiments designed to study various aspects of the chemistry of clouds. The model simulates many of the key features of the coupled gas- and aqueous-phase chemistry of a cloud including the highly complex free radical mechanism described by Chameides and Davis (1982). Rather than treat cloud microphysics explicitly, these processes are crudely parameterized in the model by allowing for temporal variations in droplet radii and liquid water content. The model calculations discussed in this work are specifically designed to simulate the chemistry of a daytime stratiform cloud in the remote marine troposphere. A remote, marine, stratiform cloud was chosen for this study because, compared to other cloud environments, it is relatively simple; the remote, marine atmosphere has low levels of organics, whose gas- and aqueous-phase chemistry is highly uncertain, and stratiform clouds are characterized by fairly constant liquid water contents and temperatures.

NOMENCLATURE

In order to clearly distinguish between gas-phase and aqueous-phase species, concentrations, rate constants, and reactions, a relatively detailed nomenclature has been adopted. As an aid to the reader, this nomenclature is outlined below.

Gas-Phase Parameters:

$(I)_g$	denotes a species I in the gas-phase
$n(I)$	denotes the number density (molecules/cm^3) of gas-phase species I.
$X(I)$	denotes the mixing ratio (v/v) of gas-phase species I.
n_M	denotes the total atmospheric number density (molecules/cm^3).

Aqueous-Phase Parameters:

$(I)_{aq}$	denotes a species I dissolved in the aqueous-phase
[I]	denotes the concentration (M=moles/liter) of aqueous-phase species I.
X_{eq} (I)	denotes the equivalent mixing ratio (v/v) of aqueous-phase species I if the species were completely transferred to the gas-phase. (see discussion below).

Chemical Notation:

(RGn)	denotes the nth gas-phase reaction involving the CH_4-CO-NO_x-H_xO_y system
k_{Gn}	denotes the rate constant for (RGn)
(RGCn)	denotes the nth gas-phase reaction involving non-methane hydrocarbons
k_{GCn}	denotes the rate constant for (RGCn)
(RAQn)	denotes the nth aqueous-phase reaction
k_{AQn}	denotes the rate constant for (RAQn)
(REQn)	denotes the nth aqueous-phase chemical equilibrium
K_{EQn}	denotes the nth equilibria between the gas and aqueous phase
(RSn)	denotes the nth dissolution reaction
K_{Sn}	denotes the solubility or Henry Law constant for (RSn)

Cloud Parameters:

W_L	denotes the liquid water content of cloud (g m^{-3})
a	denotes the droplet radius (μ)
N	denotes the number of droplets (drops/cm^3)

Some further discussion of $X_{eq}(I)$ the equivalent mixing ratio of aqueous-phase species I is in order. As described by Schwartz et al. (1982), $X_{eq}(I)$ is the gas-phase mixing ratio an aqueous-phase species I would attain if the species were completely transferred from the aqueous phase to the gas phase and is defined by

$$X_{eq}(I) = \frac{[I](W_L \times 10^{-9})(6.02 \times 10^{23})}{n_M} \tag{1}$$

For a liquid water content of 0.5 g m^{-3} and an ambient number density of 2.2×10^{19} molecules cm^{-3}, a species with an aqueous-phase concentration of 10^{-6} M will have an equivalent mixing ratio of about 14 pptv. Because $X_{eq}(I)$ has the same units as X(I), the gas-phase mixing ratio of species I, it is an useful indicator of the relative partioning of the species between the aqueous and gas phases.

MODEL DESCRIPTION

We simulate the chemistry of a cloud with a time-dependent box model. The model treats the chemistry of the gas and aqueous-phase, as well as the dynamic exchange of species between the two phases, in a homogeneous parcel of air in a cloud. The cloud is assumed to be saturated with water vapor at a temperature T and to contain a monodisperse, chemically homogeneous distribution of N droplets with a radius a; in the standard model the droplet radii are held fixed at a constant value of 10 μ however sensitivity calculations in which the radii are allowed to vary with a prescribed time derivative are also presented. Transport processes such as entrainment and mixing of outside air into the cloud are neglected as is the effects of temporal variations in temperature.

Chemical Mechanism:

The gas- and aqueous-phase species included in the model are listed in Table 1 and the model's chemical reactions and equilibria, along with their rate and equilibrium constants are listed in Table 2. The homogeneous gas-phase reactions consist of the well-known CH_4-CO-NO_x-H_xO_y chemical system with rate constants adopted from Demore et al. (1982) and a condensed, carbon-bond mechanism similar to that proposed by Whitten et al. (1980) to simulate the chemistry of non-methane hydrocarbons. The CH_4-CO-NO_x-H_xO_y chemical system has been described in detail elsewhere (c.f. Chameides and Tan, 1981; Logan et al., 1981) and is therefore not documented here. The central features of this system is the generation and cycling of gaseous OH, HO_2, and H_2O_2 from reactions such as

(RG1) $$O_3 + h\nu \longrightarrow O(^1D) + O_2$$

(RG2) $$O(^1D) + H_2O \longrightarrow 2OH$$

(RG3) $$OH + CO \xrightarrow{O_2} HO_2 + CO_2$$

(RG4) $$HO_2 + NO \longrightarrow OH + NO_2$$

(RG5) $$HO_2 + HO_2 \longrightarrow H_2O_2 + O_2$$

and

(RG6) $$H_2O_2 + h\nu \longrightarrow 2OH$$

Table 1. Chemical Species, Initial Concentrations, and Parameters Adopted in Standard Model

A. Gas-Phase Species

Species	Initial Concentration (v/v)	footnote #
CH_4	1.65 (-6)*	1
CO	6.5 (-8)	2
NO	4 (-12)	3
NO_2	1.1 (-11)	4
HNO_3	2.5 (-11)	5
NO_3	9 (-16)	4
N_2O_5	4 (-18)	4
HNO_2	1 (-13)	4
OH	3 (-13)	4
HO_2	3 (-11)	4
H_2O_2	1 (-9)	4
CH_3O_2	2.5 (-11)	4
CH_3OOH	2 (-9)	4
HCHO	4.5 (-10)	4
C_2H_4	2 (-10)	6
OLE	1.5 (-10)	6
C_2H_6	7 (-10)	6
PAR	6 (-10)	6
RO_2	7 (-12)	4
ROOH	6 (-10)	4
PAC	1 (-11)	4
PAN	1 (-11)	4
HCOOH	3 (-11)	4
CAR	4 (-10)	4
SO_2	5.5 (-11)	7
NH_3	5 (-11)	8
O_3	3 (-8)	9
H_2	5 (-7)	10

* 1.65 (-6) = 1.65×10^{-6}

B. Aqueous-Phase Species

Species	Initial Concentration (M)	footnote #
Cl^-	1×10^{-4} M	11
Na^+	1×10^{-4} M	11
$SO_4^=$	6×10^{-6} M	11
$NH_4^+ + NH_3$	6×10^{-6} M	11
H^+	6×10^{-6} M	11
NO_3^-	0	12
$CH_2(OH)_2 + CH_2HSO_3^-$	0	12
$HCOOH + HCOO^-$	0	12
O_3	0	12
OH	0	12
$HO_2 + O_2^-$	0	12
H_2O_2	0	12
$S\overline{IV} = SO_2 + HSO_3^- + SO_3^= + CH_2HSO_3^-$	0	12
$CO_2 + HCO_3^- + CO_3^=$	0	12
$Cl + Cl_2^-$	0	12
SO_4^-	0	12
NO_3	0	12

C. Model Parameters

W_L = Liquid Water Content = 0.5 g m^{-3}
N = droplet concentration = 100 cm^{-3}
a = droplet radii = constant = 10 u
O = midday solar zenith angle = 30°
$\mathcal{E}$ = cloud transmissivity factor = 0.5
T = temperature = 289°K
RH = relative humidity = 100%

1 = Rasmussen and Khalil (1981); 2 = Heidt et al. (1980); 3 = McFarland et al. (1979); 4 = Calculated by model for cloud-free conditions, T = 295°K, RH = 80% (see text for details); 5 = Huebert and Lazrus (1980); 6 = Rudolph and Ehhalt (1981); 7 = Maroulis et al. (1980); 8 = Ayers and Gras (1983); 9 = U.S. Stand. Atm.; 10 = Scholz et al. (1970); 11 = Initial levels of Cl^-, Na^+, $SO4^=$, $NH4^+$, and H^+ assumed to arise from nucleation scavenging of ambient NaCl and NH_4HSO_4 aerosols as described in text; 12 = Assumed to be zero.

as well as the production of HCHO from the methane oxidation chain, i.e.

$$\text{(RG7)} \quad CH_4 + OH \xrightarrow{O_2} CH_3O_2 + H_2O$$

$$\text{(RG8)} \quad CH_3O_2 + NO \longrightarrow CH_3O + NO_2$$

and

$$\text{(RG9)} \quad CH_3O + O_2 \longrightarrow HCHO + HO_2$$

As will be described later, each of these gaseous species can be rapidly scavenged by droplets within a cloud and incorporated into the aqueous phase where they can play an important role in the generation of acids in cloudwater.

In the carbon-bond mechanism, selected organic molecules are sub-divided into their respective carbon-bonding groups (i.e. paraffinic, olefinic, etc.). Because of the similarities in the chemistry and reaction rates of different organic molecules with similar bonding, equivalent carbon bonding groups are lumped together and treated as a single reactive species, thereby greatly reducing the total number of reactants and reactions that need to be explicitly included in the model. In the carbon-bond mechanism adopted here, seven types of carbon-bonding groups are treated. These are: PAR for paraffinic groups (i.e. $-CH_2$ or $-CH_3$); OLE for olefinic groups (i.e. $-CHCH_2$); RO_2 for organic peroxy radical groups (i.e. $-CH_2O_2$); ROOH for organic peroxide groups (i.e. $-CH_2OOH$); CAR for carbonyl groups (i.e. $-CHO$); PAC for peroxyakyl radical groups (i. e. $-CH_2OO_2$); and PAN for peroxyakylnitrate groups (i. e. $-CH_2OO_2NO_2$), Thus, 1 ppbv of propylene is treated in this scheme as 1 ppbv of OLE and 1 ppbv of PAR, while 1 ppbv of butane is equivalent to 4 ppbv of PAR. Because of the vastly lower reactivities of ethane (C_2H_8) and ethylene (C_2H_6) as compared to higher molecules weight alkanes and alkenes, respectively, these species are treated explicitly and are not classified as PAR and OLE.

There are two major differences between the mechanism adopted here and that used by Whitten et al. (1980): 1) All species involved in the methane oxidation sequence (i.e. CH_4, CH_3, CH_3O_2, CH_3OOH, and HCHO) are treated explicitly here and are not lumped together with the species arising from the oxidation of non-methane hydrocarbons; and 2) because formic acid (HCOOH) can be a major acidic component of rain in remote locations (Galloway et al., 1982) a scheme to roughly simulate the photochemistry of gaseous formic acid has been incorporated into the carbon-bond mechanism. The production of gaseous HCOOH arises from the reactions of ozone with olefins via

$$\text{(RGC3)} \quad C_2H_4 + O_3 \longrightarrow HCHO + HCO_2H^+$$

$$\text{(RGC4)} \quad OLE + O_3 \longrightarrow CAR + HCO_2H^+ .$$

The excited intermediate, HCO_2H^+, formed by (RGC3) and (RGC4) rapidly decomposes via

$$\text{(RGC5)} \quad HCO_2H^+ \longrightarrow 0.15\ HCOOH + 0.3\ (H\ O + CO) + 0.2(CO_2 + H_2) + 0.1(CO_2 + 2H) + 0.25\ \text{ozonides}$$

to form a variety of products including HCOOH. The assumed HCOOH quantum yield of 0.15 from the above reactions is based upon the product distribution observed by Kan et al. (1981) for the O_3-C_2H_4 system. While it is likely that higher molecular weight organic acids (e.g. CH_3COOH) are also produced from O_3-olefins reactions, HCOOH was assumed to be the only organic acid produced from these reactions in the calculations presented here. The gas-phase, photochemical sink for HCOOH is

$$\text{(RGC22)} \quad HCOOH + OH \xrightarrow{O_2} H_2O + CO_2 + HO_2$$

with a rate constant of 3×10^{-13} cm^3 $molecule^{-1}$ s^{-1} (Zetsch and Stuhl, 1981). Given the present uncertainties in the photochemistry of gaseous formic acid, the above chemical scheme must be viewed as being only a very crude approximation of the actual gas-phase photochemistry of this molecule. Nevertheless, as will become apparent in our later discussion, this chemical mechanism provides a convenient mathematical framework for examining the relative roles of gas-phase and aqueous-phase reactions in controlling the levels of formic acid within a cloud.

The solubility constants and aqueous-phase equilibria listed in Table 2C and 2D, respectively were derived from existing thermodynamic data using the formula

$$K(T) = \exp\left(\frac{-\Delta G_f(298°K)}{KT}\right) \exp\left(\frac{-\Delta H(298°K)}{K}\left(\frac{1}{T} - \frac{1}{298°K}\right)\right) \qquad (2)$$

where K(T) is the equilibrium or solubility constant as a function of temperature T, ΔG_f is the change in Gibbs Free Energy in going from the initial state to the final state, and ΔH is the equivalent change in the enthalpies. In the case of the dissolution of OH (i.e. (RS2)) no data is available to obtain a solubility constant; and a value of 10^5 M atm^{-1} has been assumed. However, because of the extremely short chemical lifetime of this species in the aqueous-phase, an equilibrium between the two phases for OH is never established and as a result the results are independent of the value assumed for K_{S2} (see Chameides and Davis, 1982).

The set of aqueous-phase reactions listed in Table 2E and included in the model is largely based upon previous calculations (i.e. Chameides and Davis, 1982, 1983). (However, the reader will note that several reactions and reactants included in the original model of Chameides and Davis (1982) but found to have a negligible effect upon the overall chemistry of the system have been deleted). With the exception of formaldehyde and formic acid, the aqueous-phase chemistry of organic species are neglected as are the possible effects of organic films on droplets. In the remote marine atmosphere, where organic species are relatively low in abundance, the impact of organics other than formaldehyde and formic acid should be small. For a detailed discussion of the effects of organics and organic films upon the chemistry of atmospheric hydrometeors, the reader is referred to Graedel and Weschler (1981), Graedel and Goldberg (1983), and Gill et al. (1983).

Following Chameides and Davis (1982) photodissociation rates within a cloudy air parcel were determined by first calculating these rates for a cloud-free atmosphere using a two-stream algorithm to simulate the transfer of radiation through an absorbing and multiply-scattering atmosphere (see Dickerson et al., 1979). To take into account the reduction in the solar flux within a cloud, the clear-sky photolysis rates were then reduced by a cloud transmissivity factor of 0.5. As previously argued by Chameides and Davis (1982) this factor seems reasonable for a stratoform cloud in light of both observations and calculations of the transfer and reflection of radiation through clouds (c.f. Conover, 1965; Borkowski et al., 1978; Nack and Green, 1978; Spinhurne and Green, 1978; McCartney, 1976).

Mathematical Formulation:

The concentrations of the species in Table 1 were solved for as a function of time by integrating a coupled set of mass balance equations using a standard Gear (1971) routine while demanding that the aqueous-phase equilibria of Table 2 and charge neutrality were satisfied. The mass balance equations describing the temporal variations in a gas-phase species I with number density n is given by

$$\frac{dn(I)}{dt} = P_g(I) - D_g(I) - N\Phi_{\alpha}(I) \tag{3}$$

where P_g and D_g are the gas-phase photochemical production and destruction rates. The term on the right-hand side of (1) represents the rate of loss of species I via heterogeneous scavenging where Φ_a is the rate of loss to a single cloud droplet of radius a. Following Fuchs and Sutugin (1971)

$$\Phi_a(I) = \frac{4}{3}\pi \ell \left(\frac{8KT}{m(I)\pi}\right)^{\frac{1}{2}} \frac{a}{1+\left(0.7+\frac{4(1-\alpha)}{3\alpha}\right)\frac{\ell}{a}}(n(I)-[I]H_{eff}(I)/R/T)V(\alpha) \tag{4}$$

Table 2. Chemical Mechanism

A. Gas-Phase Reactions for the CH_4-CO-NO_x-H_xO_y Systems

Reaction set taken from Chameides and Tan (1980) with rate constants from Demore et al. (1982)

B. Gas-Phase Reactions for Non-Methane Hydrocarbons - Carbon-Bond Mechanism

	Reaction	Rate Constant*	Reference
RGC1	$C_2H_4 + OH \xrightarrow{O_2} CH_3O_2 + HCHO$	2.1×10^{-12} exp(400/T)	DeMore et al., 1982
RGC2	$OLE + OH \xrightarrow{O_2} RO_2 + CAR$	4.1×10^{-12} exp(544/T)	Atkinson and Pitts, 1975
RGC3	$C_2H_4 + O_3 \longrightarrow HCHO + HCO_2H^{*}$	2.6×10^{-14} exp(-2828/T)	Kan et al., 1981
RGC4	$OLE + O_3 \longrightarrow CAR + HCO_2H^{*}$	6.1×10^{-15} exp(-1900/T)	Herron and Huie, 1974; Kan et al., 1981
RGC5	$HCO_2H^{*} \rightarrow$ Multi Steps $\rightarrow 0.3(H_2O + CO)$		
	$+ 0.4\ HCOOH + 0.2(CO_2 + H_2) + 0.1(CO_2 + 2H)$	Fast	Kan et al., 1981
RGC6	$C_2H_6 + OH \xrightarrow{O_2} RO_2 + H_2O$	2×10^{-11} exp(-1260/T)	DeMore et al., 1982
RGC7	$PAR + OH \xrightarrow{O_2} RO_2 + H_2O$	5×10^{-12} exp(-800/T)	DeMore et al., 1982
RGC8	$RO_2 + NO \xrightarrow{O_2} CAR + NO_2 + HO_2$	4.2×10^{-12} exp(180/T)	assumed
RGC9	$RO_2 + RO_2 \xrightarrow{2O_2} 2CAR + 2HO_2$	1.6×10^{-13} exp(220/T)	assumed
RGC10	$RO_2 + CH_3O_2 \xrightarrow{2O_2} CAR + HCHO + 2HO_2$	1.6×10^{-13} exp(220/T)	assumed
RGC11	$RO_2 + HO_2 \longrightarrow ROOH + O_2$	7.7×10^{-14} exp(1300/T)	assumed
RGC12	$ROOH + OH \longrightarrow RO_2 + H_2O$	2.6×10^{-12} exp(-190/T)	assumed
RGC13	$ROOH + h\nu \xrightarrow{O_2} CAR + OH + HO_2$	3.6×10^{-6}	assumed
RGC14	$CAR + h\nu \xrightarrow{2O_2,\ PAR} HO_2 + CO + CH_3O_2$	4.4×10^{-6}	Horowitz and Calvert, 1982
RGC15	$CAR + OH \xrightarrow{O_2} H_2O + PAC$	1.5×10^{-11}	Atkinson and Pitts, 1975
RGC16	$PAC + NO_2 \longrightarrow PAN$	1.4×10^{-12}	Cox and Roffey, 1977
RGC17	$PAN \longrightarrow PAC + NO_2$	8×10^{14} exp(-12516/T)	Cox and Roffey, 1977
RGC18	$PAC + NO \xrightarrow{O_2,\ PAR} NO_2 + CH_3O_2 + CO_2$	2.5×10^{-12}	Cox and Roffey, 1977
RGC19	$PAC + HO_2 \longrightarrow$ non-volatile alcohol	$(3.4+2.5P_{atm})10^{-14}$exp(1150/T)	assumed
RGC20	$PAC + CH_3O_2 \xrightarrow{PAR,\ 2O_2} HCHO + O_2 + CH_3O_2 + CO_2 + HO_2$	1.6×10^{-12}	Dodge and Arnts, 1979
RGC21	$PAC + RO_2 \xrightarrow{PAR,\ 2O_2} CAR + O_2 + CH_3O_2 + CO_2 + HO_2$	1.6×10^{-12}	Dodge and Arnts, 1979
RGC22	$HCOOH + OH \xrightarrow{O_2} H_2O + CO_2 + HO_2$	5×10^{-12}	estimated

Table 2 (cont.)

C. Solubility Constants

Reaction				Solubility Constant	Ref.
RS1	$(HO_2)_g$	$\rightleftharpoons$	$(HO_2)_{aq}$	10^5 M/atm	Schwartz (1983)
RS2	$(OH)_g$	$\rightleftharpoons$	$(OH)_{aq}$	10^5 M/atm	assumed
RS3	$(H_2O_2)_g$	$\rightleftharpoons$	$(H_2O_2)_{aq}$	$9.7\times10^4\exp[6600(\frac{1}{T} - \frac{1}{298})]$ M/atm	NBS
RS4	$(O_3)_g$	$\rightleftharpoons$	$(O_3)_{aq}$	$1.15\times10^{-2}\exp[2360(\frac{1}{T} - \frac{1}{298})]$ M/atm	NBS
RS5	$(HNO_3)_g$	$\rightleftharpoons$	$H^+ + NO_3^-$	$2.6\times10^6\exp[8700(\frac{1}{T} - \frac{1}{298})]$ M^2/atm	NBS
RS6	$(H_2CO)_g$	$\overset{H_2O}{\rightleftharpoons}$	$CH_2(OH)_2$	$7\times10^3\exp[6425(\frac{1}{T} - \frac{1}{298})]$ M/atm	NBS
RS7	$(NH_3)_g$	$\rightleftharpoons$	$(NH_3)_{aq}$	$5.8\times10^1\exp[4085(\frac{1}{T} - \frac{1}{298})]$ M/atm	NBS
RS8	$(SO_2)_g$	$\rightleftharpoons$	$(SO_2)_{aq}$	$1.23\exp[3120(\frac{1}{T} - \frac{1}{298})]$ M/atm	NBS
RS9	$(CO_2)_g$	$\rightleftharpoons$	$(CO_2)_{aq}$	$3.11\times10^{-2}\exp[2423(\frac{1}{T} - \frac{1}{298})]$ M/atm	NBS
RS10	$(HCOOH)_g$	$\rightleftharpoons$	$(HCOOH)_{aq}$	$3.7\times10^3\exp[5700(\frac{1}{T} - \frac{1}{298})]$ M/atm	CRC
RS11	$(HNO_2)_g$	$\rightleftharpoons$	$(HNO_2)_{aq}$	$4.9\times10^1\exp[4781(\frac{1}{T} - \frac{1}{298})]$ M/atm	NBS
RS12	$(NO_2)_g$	$\rightleftharpoons$	$(NO_2)_{aq}$	$1.2\times10^{-2}\exp[2500(\frac{1}{T} - \frac{1}{298})]$ M/atm	NBS
RS13	$(NO_3)_g$	$\rightleftharpoons$	$(NO_3)_{aq}$	10^5 M/atm	assumed

D. Aqueous-Phase Equilibria

Reaction				Solubility Constant	Ref.
REQ1	H_2O	$\rightleftharpoons$	$H^+ + OH^-$	$1\times10^{-14}\exp[-6716(\frac{1}{T} - \frac{1}{298})]$ M	NBS
REQ2	HO_2	$\rightleftharpoons$	$H^+ + O_2^-$	1.3×10^{-5} M	CD
REQ3	$(SO_2)_{aq}$	$\overset{H_2O}{\rightleftharpoons}$	$HSO_3^- + H^+$	$1.7\times10^{-2}\exp[2090(\frac{1}{T} - \frac{1}{298})]$ M	NBS
REQ4	HSO_3^-	$\rightleftharpoons$	$H^+ + SO_3^=$	$6\times10^{-8}\exp[1120(\frac{1}{T} - \frac{1}{298})]$ M	NBS
REQ5	Cl_2^-	$\rightleftharpoons$	$Cl + Cl^-$	5×10^{-6} M	CD
REQ6	NH_3	$\overset{H_2O}{\rightleftharpoons}$	$NH_4^+ + OH$	$1.7\times10^{-5}\exp[-4325(\frac{1}{T} - \frac{1}{298})]$ M	NBS
REQ7	HCOOH	$\rightleftharpoons$	$HCOO^- + H^+$	1.8×10^{-4} M	NBS
REQ8	$CH_2(OH)_2 + HSO_3^-$	$\rightleftharpoons$	$H_2O + CH_2OHSO_3^-$	10^5	Dasgupta et al. (1980)
REQ9	$(CO_2)_{aq}$	$\overset{H_2O}{\rightleftharpoons}$	$HCO_3^- + H^+$	$4.3\times10^{-7}\exp[-913(\frac{1}{T} - \frac{1}{298})]$ M	NBS
REQ10	$(HNO_2)_{aq}$	$\rightleftharpoons$	$H^+ + NO_2^-$	$5.9\times10^{-4}\exp[-1760(\frac{1}{T} - \frac{1}{298})]$ M	NBS

Table 2. (cont.)

E. Aqueous-Phase Reactions

	Reactions			Rate Constants*	Ref.
RAQ1	$OH+HO_2$	$\longrightarrow$	H_2O+O_2	$7x10^{9}$	CD
RAQ2	$OH+O_2^{-}$	$\longrightarrow$	$OH^{-}+O_2$	$1x10^{10}$	CD
RAQ3	$O_2^{-}+O_3$	$\xrightarrow{H^{+}}$	$2O_2+OH^{**}$	$1x10^{9}$	Bahnemann and Hart, 1982
RAQ4	$O_2^{-}+HO_2$	$\xrightarrow{H^{+}}$	$H_2O_2+O_2^{**}$	$8.5x10^{7}$	CD
RAQ5	HO_2+HO_2	$\longrightarrow$	$H_2O_2+O_2$	$7.5x10^{5}$	CD
RAQ6	$OH+H_2O_2$	$\longrightarrow$	HO_2+H_2O	$4.5x10^{7}$	CD
RAQ7	$HCO_3^{-}+OH$	$\longrightarrow$	$H_2O+CO_3^{-}$	$1.5x10^{7}$	CD
RAQ8	$O_2^{-}+HCO_3^{-}$	$\xrightarrow{H^{+}}$	$H_2O_2+CO_3^{-**}$	$1.5x10^{6}$	CD
RAQ9	$CO_3^{-}+H_2O_2$	$\longrightarrow$	$HCO_3^{-}+HO_2$	$8x10^{5}$	CD
RAQ10	$CO_3^{-}+O_2^{-}$	$\xrightarrow{H^{+}}$	$HCO_3^{-}+O_2^{**}$	$4x10^{8}$	CD
RAQ11	$OH+HSO_3^{-}$	$\xrightarrow{O_2,HSO_3^{-}}$	$OH^{-}+2H^{+}+SO_4^{=}+SO_4^{-**}$	$9.5x10^{9}$	CD
RAQ12	$SO_4^{-}+Cl^{-}$	$\longrightarrow$	$SO_4^{=}+Cl$	$2x10^{8}$	CD
RAQ13	$SO_4^{-}+H_2O_2$	$\longrightarrow$	$H^{+}+SO_4^{=}+HO_2$	$1.2x10^{7}$	CD
RAQ14	$(SO_2)_{aq}+H_2O_2$	$\longrightarrow$	$SO_4^{=}+2H^{+**}$	$\frac{8x10^{4}exp[-3650(1/T-1/298)]}{(0.1+[H^{+}])}$	Martin, 1983
RAQ15	$S\overline{\underline{IV}}+O_3$	$\longrightarrow$	$H^{+}+SO_4^{=}+O_2^{**}$	$\frac{(4.39x10^{11}exp(-4131/T)+2.56x10^{3}exp(-966/T)}{[H^{+}]}$	Maahs, 1983
RAQ16	$SO_4^{-}+HSO_3^{-}$	$\xrightarrow{O_2,HSO_3^{-}}$	$2SO_4^{=}+2H^{+}\cdot\cdot SO_4^{-**}$	$1x10^{9}$	Maruthamuthu and Neta, 1978
RAQ17	$Cl^{-}+OH+H_3O^{+}$	$\longrightarrow$	$Cl+2H_2O^{**}$	$2x10^{10}$	CD
RAQ18	$Cl_2^{-}+HO_2$	$\longrightarrow$	$2Cl^{-}+H^{+}+O_2$	$(4.5+\frac{K_{EQ2}}{[H^{+}]})10^{9}$	CD
RAQ19	$Cl_2^{-}+HSO_3^{-}$	$\xrightarrow{HSO_3^{-},O_2}$	$2Cl^{-}+2H^{+}+SO_4^{-}+SO_4^{=**}$	$7x10^{7}$	see footnote a
RAQ20	$Cl_2^{-}+H_2O_2$	$\longrightarrow$	$2Cl^{-}+H^{+}+HO_2$	$1.4x10^{5}$	Hasegawa and Neta, 1978

Table 2. (cont.)

E. Aqueous-Phase Reactions (cont.)

	Reactions	Rate Constants*	Ref.
RAQ21	$OH+CH_2OH+OH \longrightarrow CH(OH)_2+H_2O$	$2x10^9$	Markovic and Sehested, 1972
RAQ22	$CH(OH)_2+O_2 \longrightarrow HCOOH+HO_2$	$4.5x10^9$	Bothe and Schultz-Frohlinde, 1980
RAQ23A	$HCOOH+OH \xrightarrow{O_2} H_2O+CO_2+H_2O_2$	$1.5x10^8$	Nenadovic et al., 1972
RAQ23B	$HCOO^-+OH \xrightarrow{H^+,O_2} H_2O+CO_2+HO_2$	$3x10^9$	Nenadovic et al., 1972
RAQ24	$H_2O_2+h\nu \longrightarrow 2OH$	$2.5x10^{-6}$	CD
RAQ25	$NO_3+Cl^- \longrightarrow NO_3^-+Cl$	$1x10^8$	CD
RAQ26	$NO_3+HSO_3^- \xrightarrow{HSO_3^-,O_2} NO_3^-+2H^++SO_4^-+SO_4^=$**	$1x10^8$	see footnote a
RAQ27	$NO_3+H_2O_2 \longrightarrow NO_3^-+H^++HO_2$	$1x10^6$	see footnote a
RAQ28	$NO_2+HSO_3^- \xrightarrow{NO_2,H_2O} 2NO_2^-+SO_4^=+3H^+$	$2x10^6$	Lee and Schwartz (1983)
RAQ29	$NO_2+OH \longrightarrow H^++NO_3^-$	$1.3x10^9$	CD
RAQ30	$NO_2^-+OH \longrightarrow OH^-+NO_2$	$1x10^{10}$	CD
RAQ31	$(e^-)_{aq}+O_2 \longrightarrow O_2^-$	$2x10^{10}$	Draganic and Draganic, 1971

* Units for gas-phase reactions are s^{-1} for photolytic processes and $cm^3\ s^{-1}$ for two-body reactions.
Units for aqueous-phase reactions are s^{-1} for photolytic processes, $M^{-1}\ s^{-1}$ for two-body reactions and $M^{-2}\ s^{-1}$ for three body-reactions.
Photolytic rate constants quoted here are for midday equinox at 30° latitude with a cloud transmissivity factor of 0.5.

** These aqueous-phase reactions actually consist of a combination of two or more elementary reactions and/or equilibria; see text and Chameides and Davis (1982) for the complete chemical mechanism.

NBS = National Bureau of Standards Technical Note 270-1, 1965.

CRC = Chemical Rubber Company, Handbook of Chemistry and Physics, 59th Edition, Edited by R. C. Weast, 1978.

CD = see references cited in Chameides and Davis (1982).

(a) the rate constants for the reaction of Cl_2^- with HSO_3^- and for NO_3 with HSO_3^- and H_2O_2 were estimated by assuming these reactions to be about 1/10 as fast as the equivalent SO_4^-, (Maruthamutu and Neta (1978), Dogliotti and Hayon (1967), and Hasegawa and Neta (1978)).

where ℓ is the mean free path, K is the universal gas constant, m(I) is the molecular weight of I, α is the accommodation coefficient, [I] is the aqueous-phase concentration of species I in the cloud droplet in M(moles/liter), $H_{eff}(I)$ is the effective Henry's Law or solubility constant for I in M atm^{-1}, and $R=1.36 \times 10^{-22}$ atm-sm^3/molec/°K. The term V(a) in Eqn. (4) is a small correction to Φ_a due to ventilation effects and is given by (Harding, 1972)

$$V(a) = 1.0 + 36.8(\alpha \times 10^{-4}) + 3012(\alpha \times 10^{-4})^2 \tag{5}$$

where a is in units of μ.

The effective Henry's Law or solubility constant H_{eff} differs from the standard solubility constant K_S in that it takes into account the effects of dissociation and complex formation upon the species' solubility. For instance in the case of HNO_3,

$$H_{eff}\ (HNO_3) = K_{S5}/[H^+] \tag{6}$$

while H_{eff} for S $\overline{\underline{IV}}$ is given by

$$H_{eff}(S\ \overline{\underline{IV}}) = K_{S8}\left(1+\frac{K_{EQ3}}{[H^+]}(1+[CH_2(OH)_2]K_{EQ8})+\frac{K_{EQ}\ K_{EQ4}}{[H^+]^2}\right) \tag{7}$$

One of the key parameters in the formulation for Φ is α, the accommodation or sticking coefficient. Unfortunately, however, while several measurements have been made of the accommodation coefficient for H_2O molecules impinging upon water droplets, virtually no data exists for the α-values of other chemical species impinging upon a water surface. Citing measurements of α for OH and HO_2 impinging upon non-aqueous surface, Chameides and Davis (1982) have argued that the value of α for OH and HO_2 impinging upon water is likely between 1 and 10^{-4} and most probably about 10^{-2}, similar to the α-value for H_2O impinging upon water (Pruppacher and Klett, 1978). Following Chameides and Davis (1982), a value of 0.01 was adopted for α in the Standard Model and sensitivity calculations with α=1.0, 10^{-4}, and 10^{-6} were also carried out and will be presented. Given the dearth of data for α it is not yet possible to reliably estimate how the parameter will vary from one species to another and thus, for simplicity, α was assumed to be the same for all soluble species.

In the aqueous-phase, all drops were assumed to be homogeneously mixed, a reasonable assumption for the drop sizes treated (Schwartz and Freiberg, 1981). The mass balance equation for an aqueous-phase species I is then given by

$$\frac{d\ [I]}{dt} = P_{aq}(I) - D_{aq}(I) + S_a(I) - \frac{3}{a}\ [I]\frac{d^a}{dt} \tag{8}$$

where P and D are the aqueous phase, chemical-production and destruction rates of species I in a droplet of radius a and S is the source of I to the droplet via scavenging from the gas-phase. Following Chameides and Davis (1982), S is given by

$$S_a\ (I) = \frac{I_a\ (I)}{\frac{4}{3}\pi\ a^3\ 6.02\ x\ 10^{23}} \tag{9}$$

The last term on the right-hand side of Eqn. (8) takes into account the dilution or enhancement in I that occurs as the droplet radius changes due to condensation or evaporation at some prescribed rate "da/dt". While "da/dt" was taken to be 0 for the Standard Model, a calculation with a growing droplet distribution has also been carried out and will be discussed here.

Model Conditions:

Table 1 outlines the conditions and initial concentrations assumed for the Standard Model. The initial levels of the gas-phase species roughly correspond to the concentrations that might be typically encountered in the remote marine boundary layer at midday. The initial concentrations of the long-lived species (CH_4, C_2H_6, C_2H_8, PAR, OLE, CO, O_3, HNO_3, SO_2, NH_3, H_2) were chosen to match the levels observed for these species in the remote marine boundary layer, while the initial concentrations of the short-lived species (OH, HO_2, H_2O_2, NO, NO_2, NO_3, N_2O_5, HNO_2, CH_3O_2, CH_3OOH, HCHO, RO_2, ROOH, CAR, PAC, PAN, HCOOH) were calculated by the model under the assumptions of cloud-free conditions and a boundary layer temperature of 295°K and relative humidity of 80%. For this calculation the concentrations of the long-lived species were held fixed at the values specified in Table 1 and the gas-phase mass balance equations for cloud-free conditions (i.e. Eqn. 3 with N=O) for the short-lived species were integrated over several diurnal cycles until convergence was obtained. Convergence was assumed to occur when the calculated noon-time concentrations of all species varied by less than 1% from the levels calculated 24 hours earlier. In addition to the gas-phase photochemical sinks, soluble species (i.e. H_2O_2, CH_3OOH, HCHO, ROOH, CAR, PAN, HCOOH) were allowed to be deposited on the ocean surface; a deposition velocity of 0.5 cm s^{-1} and a mixing height of 1 km were assumed to simulate this process. And in order to obtain realistic NO_x levels, a local source of NO was assumed to exist and its magnitude was adjusted to yield a noon-time NO level of 4 pptv in accordance with the observations of McFarland et al. (1979). For a mixed layer height of 1 km, the NO source thus determined corresponds to a NO flux of $1x10^9$ molecules $cm^{-2}s^{-1}$; a similar flux was inferred by Liu et al. (1983) in their diagnostic study of the gas-phase photochemistry of the remote marine boundary layer.

At time t=0, an air parcel containing the initial gaseous mixture described in Table 1 and a boundary layer temperature of 295°K and relative humidity of 80% was assumed to be suddenly cooled to a temperature of 289°K by an adiabatic/pseudo-adiabatic process; this leads to an adiabatic liquid water content of 0.5 g m^{-3}. In the Standard Model the adiabatic liquid water content was assumed to be attained instantaneously and thus at t=0 a monodisperse distribution of 100 droplets cm^{-3} with radii of 10 u was introduced into the gaseous mixture. As indicated in Table 1 initial levels of Cl^-, Na^+, $SO_4^=$, and H^+ were also assumed to be present in the droplets; these levels equal those that would arise from the complete nucleation scavenging of an aerosol suspension containing 4 ug m^{-3} NaCl and 0.33 μg m^{-3} NH_4HSO_4, which crudely corresponds to the particulate Na^+, Cl^-, excess $SO_4^=$, and NH_4^+ levels typically observed in marine environments (c.f. Bigg, 1980; Bonsang et al., 1980; Andreae, 1982). For sensitivity calculations in which different or variable liquid water content was assumed, the initial values of the droplet radius a and the initial levels for $[Cl^-]$, $[Na^+]$, $[SO_4^=]$, $[NH_4^+]$, and $[H^+]$ were appropiately adjusted.

Following the introduction of the droplets into the gaseous mixture at t=0, the concentrations of all gas-phase and aqueous-phase species were calculated as a function of time, t, by integrating the coupled set of equations of the form of (3) and (8) using a standard Gear (1971) routine while maintaining the equilibria of Table 2c and charge neutrality.

AQUEOUS-PHASE, FREE RADICAL SOURCES

The calculations to be presented here indicate that free radical species dissolved in droplets can play an important role in the chemistry of a cloud and thus before presenting the results of these calculations it is helpful to briefly discuss the processes that control the levels of these species. (For a more detailed discussion the reader is referred to Chameides and Davis, (1982).) The production of aqueous-phase, free radicals in the Standard Model is triggered by the penetration of actinic, solar radiation into the cloud. In situ production of OH radicals within the droplets themselves can occur from the photolysis of dissolved H_2O_2, i.e.

$$\text{(RAQ24)} \quad H_2O_2 + h\nu \longrightarrow 2\ OH$$

as well as from the photolysis of dissolved O_3, producing $O(^1D)$ which then reacts with the solvent cage to produce OH, i.e.

$$\text{(RAQ32)} \quad O_3 + h\nu \xrightarrow{H_2O} 2OH + O_2.$$

These two processes are estimated to produce OH radicals at rates of about 2.5×10^{-10} $M^{-1}s^{-1}$ and 1×10^{-14} $M^{-1}s^{-1}$, respectively. However, as noted by Chameides and Davis (1982) a much larger source of aqueous-phase, free radicals arises from the production of gas-phase OH and HO_2 from reactions

$$\text{(RG1)} \quad O_3 + h\nu \longrightarrow O(^1D) + O_2$$

$$\text{(RG2)} \quad O(^1D) + H_2O \longrightarrow 2OH$$

and

$$\text{(RG3)} \quad OH + CO \xrightarrow{O_2} HO_2 + CO_2$$

and the subsequent transfer of these radicals to the aqueous-phase via droplet scavenging. The magnitude of the source due to scavenging by a droplet of radius a is given by the quantities $S_a(HO_2)$ and $S_a(OH)$, as defined by Eqn. 9. For the Standard Model conditions, the average HO_2 and OH source rates to a 10 u droplet via droplet scavenging is calculated to be about 9×10^{-9} M s^{-1} and 1.4×10^{-9} M s^{-1}, respectively, yielding a total radical source rate of $\sim 10^{-8}$ M s^{-1}. (The rate of scavenging of other radicals such as NO_3 is found to be negligibly small compared to that of HO_2 and OH for the conditions adopted here). That a radical source of 10^{-8} M s^{-1} is potentially quite significant to the chemistry of clouds can be illustrated by noting that if every radical transferred to the aqueous phase were to oxidize one S atom from a $+\overline{\underline{IV}}$ oxidation state to a $+\overline{\underline{VI}}$ state, the corresponding SO_2 oxidation rate would be about 0.5 ppbv per hour.

As noted by Chameides and Davis (1982) however, the magnitude of the free radical source is sensitive to a variety of parameters, including the cloud transmissivity, $\mathcal{E}$, and the accommodation coefficient, α. If, for instance, $\mathcal{E}$ is taken to be 1.0, a value probably appropriate for the upper layers of a cloud, instead of the Standard Model value of 0.5, the above radical source rates increase by a factor of 2. When $\alpha = 10^{-4}$ the source of $(HO_2)_{aq}$ and $(OH)_{aq}$ decreases to a value of about 1.7×10^{-9} M s^{-1}, while for $\alpha = 1$ the source increases slightly over that obtained for the Standard Model case to about 1.3×10^{-8} M s^{-1}. Of course as the free radical source rates change so do the $(HO_2)_{aq}$ and $(OH)_{aq}$ concentrations and thus, to the extend that α and $\mathcal{E}$ are uncertain, the levels of these species calculated in the model are similarly uncertain. The effects of these uncertainties upon the model calculations will be discussed further in subsequent sections of this work.

A potentially important but highly speculative source of free radicals to cloudwater is that arising from electrical phenomena such as coronal discharges from droplets. Suppose a continuous, low-level current of positive-point coronal discharges from droplets existed in a cloud. These discharges would cause 1 electron to be deposited on the droplet and 1 positive ion (most often O_2^+) to be produced in the gas-phase for each ion pair produced (Loeb, 1965). The electrons deposited on the droplet would be incorporated into the droplet and become hydrated electrons (i.e. $(e^-)_{aq}$). In the presence of dissolved O_2, these hydrated electrons would rapidly form O_2^- via

$$\text{(RAQ31)} \quad (e^-)_{aq} + (O_2)_{aq} \longrightarrow O_2^-.$$

The O_2^+ ions produced in the gas-phase would lead to the eventual formation of an OH radical and a hydrated oxonium ion, $H_3O^+ \cdot n\ H_2O$ ion (Good et al., 1978). Heterogeneous scavenging of the $H_3^+ \cdot nH_2O$ ion and its incorporation into the droplets to form H^+ would maintain the nominal charge neutrality of the droplets. A sizable fraction of the gas-phase OH radicals produced by the discharge would also be scavenged, either as $(OH)_g$ or $(HO_2)_g$, and incorporated into the droplet representing an additional radical source to the aqueous-phase. For $\alpha=10^{-2}$, about 1.5 aqueous-phase free radicals would be produced for each ion pair generated. Thus for this process to be competitive with the 10^{-8} M s^{-1} source of aqueous-phase free radicals from droplet scavenging of photochemically produced $(OH)_G$ and $(HO_2)_G$, an ion-pair-production rate from coronal discharges of about $2x10^6$ $cm^{-3}s^{-1}$ would be needed. (While continuous low-level corona of this magnitude has never been observed in a cloud, it has not yet been proven to be not present either. One unique feature of positive-point corona is the production of OH radicals in the gas-phase from the conversion of O_2^+ to $H_3O^+ \cdot n\ H_2O$. This suggests that it would be possible to chemically test for the existence of coronal discharges in clouds by determining if significant levels of gas-phase OH radicals are present in clouds during the night, when photochemical sources of $(OH)_g$ are inactive.) As will be demonstrated in a subsequent section the production of radicals from coronal discharges can have a significant effect upon the chemistry and acidity of a cloud.

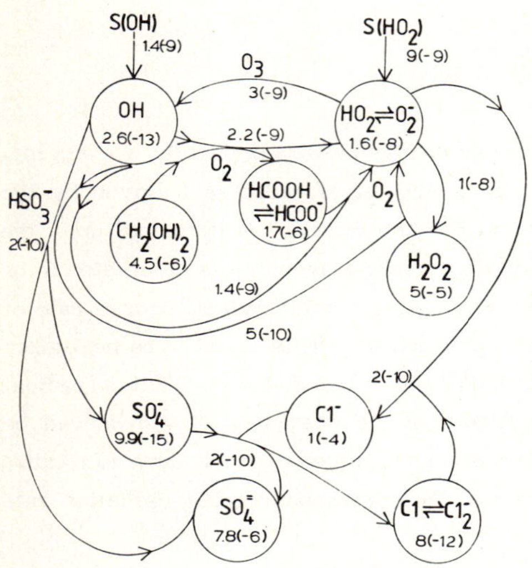

Figure 1. The primary chemical pathways for aqueous-phase, free radicals in a remote marine cloud. SOH and SHO_2 represent sources of OH and HO_2 to cloud droplets as a result of scavenging from the gas phase. The circles represent reservoirs of dissolved species and the arrows reactive pathways with reactants indicated along the arrow. Model calculated concentrations and reaction rates, in units of M and M s^{-1}, respectively, for the Standard Model at t = 10 minutes are indicated.
Note that $1(-10) = 1 \times 10^{-10}$.

Once in the aqueous phase, $(HO_2)_{aq}$ and $(OH)_{aq}$ initiate a highly complex series of chemical reactions. The primary chemical pathways of these species are illustrated in Figure 1. Note that OH and HO_2 may be rapidly cycled without radical termination via the sequence of

$$\text{(REQ2)} \qquad HO_2 \rightleftharpoons H^+ + O_2^-$$

$$\text{(RAQ3)} \qquad O_2^- + O_3 \xrightarrow{H^+} 2O_2 + OH$$

followed by

$$\text{(RAQ21)} \qquad OHCH_2OH + OH \longrightarrow CH(OH)_2 + H_2O$$

and

$$\text{(RAQ22)} \qquad CH(OH)_2 + O_2 \longrightarrow HCOOH + HO_2$$

or

$$\text{(RAQ23B)} \qquad HCOO^- + OH \xrightarrow{H^+,O_2} H_2O + CO_2 + HO_2,$$

the latter three reactions belonging to the aqueous-phase formaldehyde/formic acid oxidation scheme described by Chameides and Davis (1983). A major OH sink is by reaction with HSO_3^- via (RAQ11); with a rate constant of $9.5x10^9$ $M^{-1}s^{-1}$ and a $[HSO_3^-]$ level of $\sim 10^{-7}$ M, this reaction implies a $(OH)_{aq}$ chemical lifetime of about 10^{-3}s. Note in Figure 1, that the OH + HSO_3^- reaction does not terminate the radical chain but rather gives rise to other radical species such as SO_4^-, Cl, and Cl_2^-.

For the conditions adopted here radical termination most frequently occurs from the conversion of $(HO_2)_{aq}$ and O_2^- to $(H_2O_2)_{aq}$ via

$$\text{(RAQ4)} \qquad O_2^- + HO_2 \xrightarrow{H^+} H_2O_2 + O_2$$

or

$$\text{(RAQ8)} \qquad HCO_3^- + O_2^- \xrightarrow{H^+} CO_3^- + H_2O_2$$

followed by

$$\text{(RAQ10)} \qquad CO_3^- + O_2^- \xrightarrow{H^+} HCO_3^- + O_2.$$

These reactions lead to an HO_2/O_2^- aqueous-phase chemical lifetime of about ls. An additional chain termination reaction arises from

$$\text{(RAQ18)} \qquad Cl_2^- + HO_2 \longrightarrow 2Cl^- + H^+ + O_2.$$

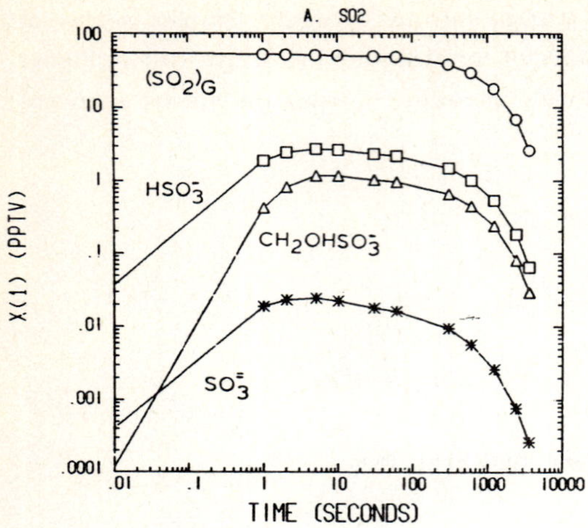

Figure 2.A

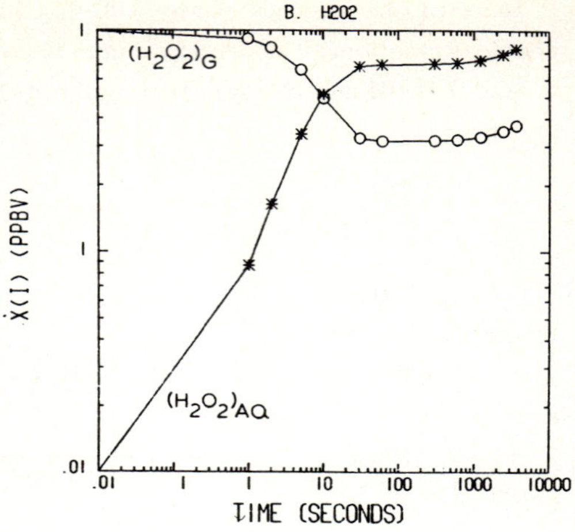

Figure 2.B

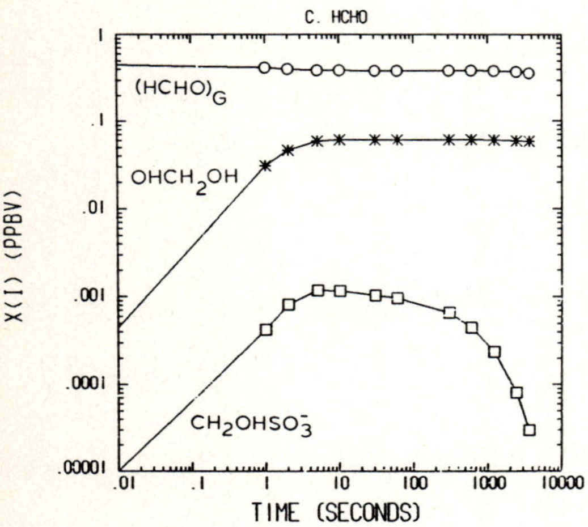

Figure 2.C

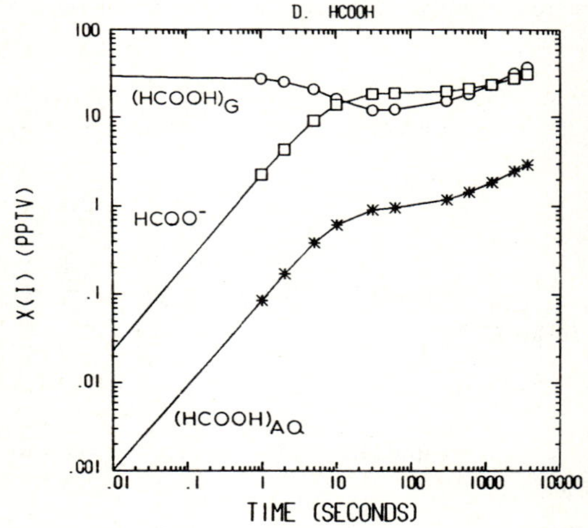

Figure 2.D

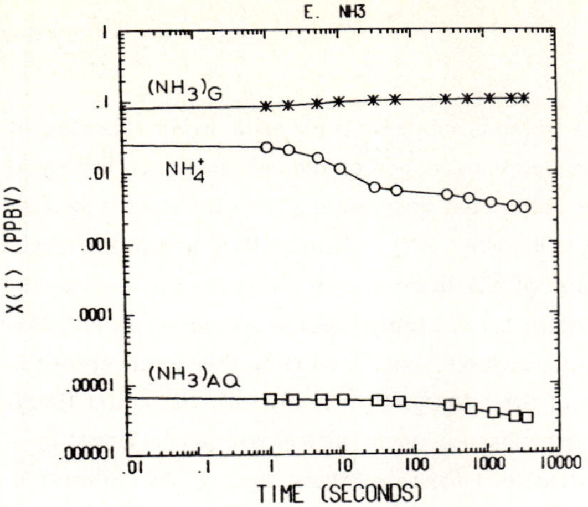

Figure 2.E

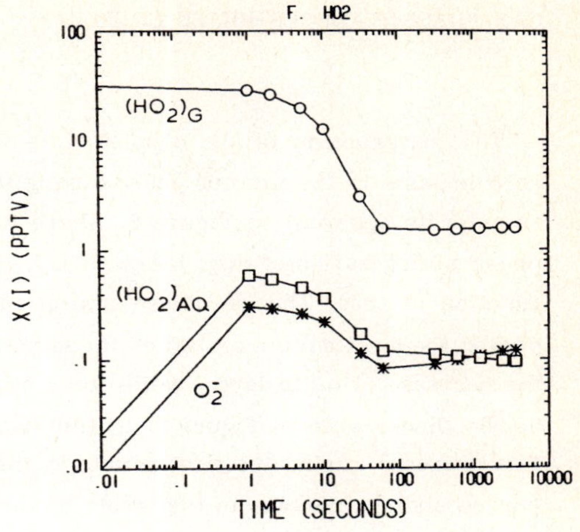

Figure 2.F

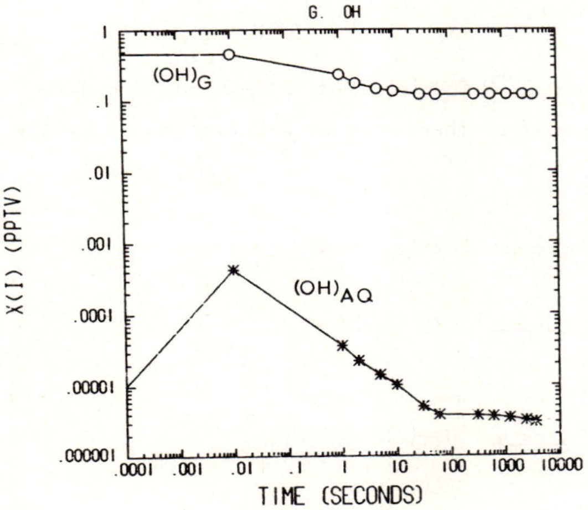

Figure 2.G

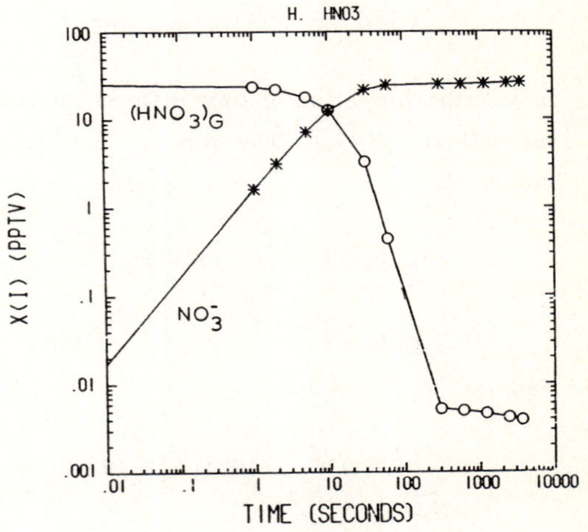

Figure 2.H

Fig. 2.A to 2.H The calculated gas-phase and equivalent aqueous- phase mixing ratios, X(I), as a function of time for: (A) SO_2 and related species; (B) H_2O_2; (C) HCHO and related species; (D) HCOOH and related species; (E) NH_3 and related species; (F) HO_2 and related species; (G) OH; and (H) HNO_3 and related species.

GAS-PHASE / AQEOUS-PHASE COUPLING

The introduction of the droplets into the gaseous mixture leads to a rapid transfer of soluble gases to the aqueous phase during the first 60 seconds of the calculation. This rapid transfer is apparent in Figure 2, where the calculated gas-phase and equivalent aqueous-phase mixing ratios of SO_2, H_2O_2, HCHO, HCOOH, NH_3, HO_2, OH and HNO_3 are plotted as a function of time. The gas-phase mixing ratios of all these species were found to decrease during the first minute or two of the calculation; for a highly soluble species such as HNO_3 the decrease is quite large, while for a slightly soluble species such as SO_2 the decrease is barely discernable in Figure 2. In this case of SO_2, H_2O_2, HCHO, HCOOH, NH3, and HNO_3, this decrease during the first minute in the gas-phase concentrations was accompanied by a corresponding increase in the levels of the dissolved forms of these species. In contrast to the above species, the aqueous-phase levels of HO_2 and OH stopped increasing within the first second of the calculation and then decreased as their corresponding gas-phase levels decreased. As will be discussed later in this section, the unique behaviour of HO_2 and OH arises from their high reactivities in the aqueous phase.

Because of the highly complex equilibria that occur in the aqueous-phase, the dissolution of a gas into a droplet can lead to the formation of a variety of aqueous-phase species. For instance, the dissolution of SO_2 via

$$\text{(RS8)} \qquad (SO_2)_g \rightleftharpoons (SO_2)_{aq}$$

causes the formation of four different dissolved S $\overline{\underline{IV}}$ species: HSO_3^-, $SO_3^=$, and $CH_2OHSO_3^-$ as well as $(SO_2)_{aq}$. The relative concentrations of these species are determined by the equilibria

$$\text{(REQ3)} \qquad (SO_2)_{aq} + H_2O \rightleftharpoons HSO_3^- + H^+$$

$$\text{(REQ4)} \qquad HSO_3^- \rightleftharpoons H^+ + SO_3^=$$

and

$$\text{(REQ8)} \qquad HSO_3^- + OHCH_2OH \rightleftharpoons H_2O + CH_2OHSO_3^-.$$

For the conditions adopted here the major S $\overline{\underline{IV}}$ species are HSO_3^- and $CH_2OHSO_3^-$ (see Figure 2). Similarly the two species produced from the dissolution of HCHO via

$$\text{(RS6)} \qquad (HCHO)_g \xrightleftharpoons{H_2O} OHCH_2OH$$

are $OHCH_2OH$, hydrated formaldehyde, and $CH_2OHSO_3^-$ formed in (REQ8).

While the dissolution of NH_3 forms two species, $(NH_3)_{aq}$ and NH_4^+, NH_4^+ is the predominant form of dissolved N $\overline{III}$ for a pH below 8. Similarly in the cases of HNO_3 and HCOOH the major dissolved forms are NO_3^- and $HCOO^-$, respectively. On the other hand, for the pH range modelled here, the dissolution of HO_2 results in the formation of roughly equivalent amounts of $(HO_2)_{aq}$ and O_2^-; the HO_2/O_2^- ratio, determined by

$$\text{(REQ2)} \quad (HO_2)_{aq} \rightleftharpoons H^+ + O_2^-,$$

is calculated to be 0.77 for a pH of 5.

In addition to establishing the above chemical equilibria within the aqueous phase, the transfer of a species from the gas to aqueous phase tends to establish a thermodynamic equilibrium between the phases as well. When this thermodynamic equilibrium is attained the aqueous-phase concentration [I] is related to the gas-phase concentration n[I] by Henry's Law, i.e.

$$[I] = n(I) \quad H_{eff}(I)RT. \tag{10}$$

Note from Eqn (4) that when Eqn (10) holds, $\Phi_a(I)=0$ and thus at equilibrium the transfer of the species between the two phases ceases. If we define a partitioning function, PF(I), to represent the relative amounts of species (I) in each phase, i.e.

$$PF(I) = X_{eq}(I)/X(I), \tag{11}$$

then from Eqns (1) and (10) it follows that at thermodynamic equilibrium

$$(PF(I))_{equil.} = \left(\frac{X_{eq}(I)}{X(I)}\right)_{equil.} = H_{eff}(I)RT(W_L \times 10^{-9}) \; 6.02 \times 10^{23} \tag{12}$$

The values of PF(I) at thermodynamic equilibrium for the conditions adopted in the Standard Model are listed in Table 3 for several species of interest.

The degree to which a species has reached thermodynamic equilibrium can be inferred from the relative partitioning function, $PF^1(I)$, defined by

$$PF^1(I) = PF(I)/(PF(I))_{equil.} = \frac{[I]}{n(I) \; H_{eff} \; (I) \; RT} \tag{13}$$

When $PF^1 = 1$, the species is at equilibrium, when $PF^1 < 1$ the species is undersaturated in the aqueous-phase and $\Phi_a > 0$, and when $PF^1 > 1$ the species is supersaturated and thus $\Phi_a < 0$. Figure 3 illustrates the calculated values of PF^1 as a function of time for SO_2, NH_3, and H_2O_2; similar results were obtained for HCOOH and HCHO. Note that each of these species

Table 3. Calculated Values of $(PF(I))_{equil}$, the Aqueous/Gas Phase Partitioning Function at Thermodynamic Equilibrium*

Gas-Phase Species	Aqueous-Phase Species	$(P(I))_{equil}$
SO_2	$S\overline{\underline{IV}} = SO_2 + HSO_3^- + SO_3^= + CH_2OHSO_3^-$	0.06
HCHO	$CH_2(OH)_2 + CH_2OHSO_3^-$	0.16
HCOOH	$HCOOH + HCOO^-$	0.17
H_2O_2	H_2O_2	2.3
HO_2	$HO_2 + O_2^-$	2.9
NH_3	$NH_3 + NH_4^+$	19
HNO_3	NO_3^-	9×10^6

*calculated for pH=5.1, T=289°K, W_L = 0.5g m^{-3}

attains thermodynamic equilibrium (i.e. $PF^1 = 1$) within 100s after the introduction of the droplets into the gaseous mixture thereby defining a gas/aqueous-phase equilibration time of 100s for the conditions adopted here. In the absence of any perturbing chemical reactions the gas- and aqueous-phase concentrations of these species would remain constant after the establishment of thermodynamic equilibrium at $t \sim 100s$. This is in fact the case for NH_3 (see Figure 2F) which is assumed to be inert in the aqueous-phase. However note in Figure 2 that the levels for SO_2, H_2O_2, HCHO, and HCOOH all vary with time for $t > 100s$; SO_2 and HCHO decrease, while H_2O_2 and HCOOH increase. These slow variations are caused by chemical reactions within the aqueous-phase, which can have significant effects on time scales of tens of minutes. As these chemical reactions slowly perturb the aqueous-phase levels of these species, the species' gas-phase concentrations adjust in order to maintain thermodynamic equilibrium between the two phases. As will be described in a later section, these chemical reactions play an integral role in the generation of acidity within cloudwater.

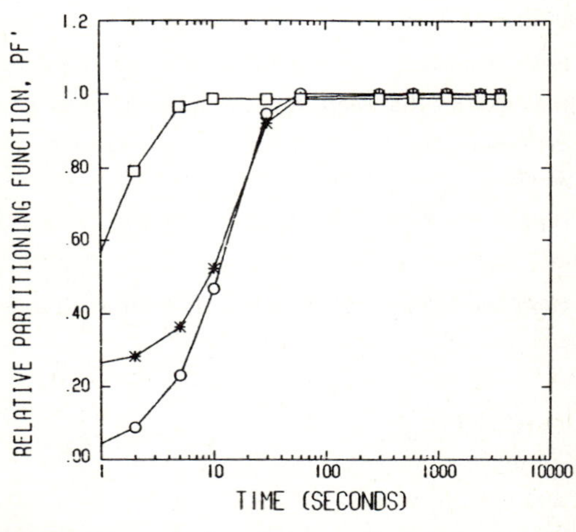

Fig. 3 The calculated temporal variation in the relative partitioning function, PF', for SO_2 (- -), H_2O_2 (O-O-O), and NH_3 (*-*-*). Similar results were obtained for HCHO and HCOOH.

In contrast to SO_2, H_2O_2, HCHO, and H_2O_2, and NH_3, photochemical rather than thermodynamic processes tend to control the relative amounts of HO_2 and OH in the two phases. Unlike the former species which have aqueous-phase, chemical lifetimes of the order of tens of minutes, the chemical lifetimes of $(OH)_{aq}$ and $(HO_2)_{aq}$ are considerably shorter than the 100s gas/aqueous phase equilibration time. As described earlier the chemical lifetime for $(OH)_{aq}$ is about 10^{-3}s, its major sinks being

$$\text{(RAQ11)} \quad OH + HSO_3^- \xrightarrow{O_2,\, HSO_3^-} OH^- + 2H^+ + SO_4^= + SO_4^-$$

$$\text{(RAQ21)} \quad OH + CH_2(OH)_2 \longrightarrow CH(OH)_2 + H_2O$$

and

$$\text{(RAQ23B)} \quad OH + HCOO^- \longrightarrow H_2O + CO_2 + H_2O$$

The chemical lifetime for $(HO_2)_{aq}$ is about 1s, its major sinks being (REQ2) followed by

$$\text{(RAQ4)} \quad HO_2 + O_2^- \xrightarrow{H^+} H_2O_2 + O_2$$

and

$$\text{(RAQ3)} \quad O_2^- + O_3 \xrightarrow{H^+} 2O_2 + OH.$$

Because of their very short chemical lifetimes, these species are chemically destroyed as rapidly as they are transferred to the aqueous-phase by droplet scavenging and as a result they never build up sufficient concentration levels in the aqueous-phase to approach thermodynamic equilibrium. This fact is illustrated in Figure 4 where PF^1 for HO_2 and OH are plotted as a function of time; note that $PF^1 << 1$ for these two species throughout the calculation. Instead of being controlled by thermodynamics, the relative levels of HO_2 and OH in the gas and aqueous phase are controlled by a photochemical steady state in which photochemical production in the gas-phase, loss from the gas-phase and incorporation into the aqueous-phase via heterogeneous scavenging, and chemical destruction in the aqueous-phase are all in balance. (For a more detailed description of this photochemical steady state the reader is referred to Chameides and Davis, 1982). Thus rather than continue to increase in concentration in response to the continuous transfer of material from the gas- phase,

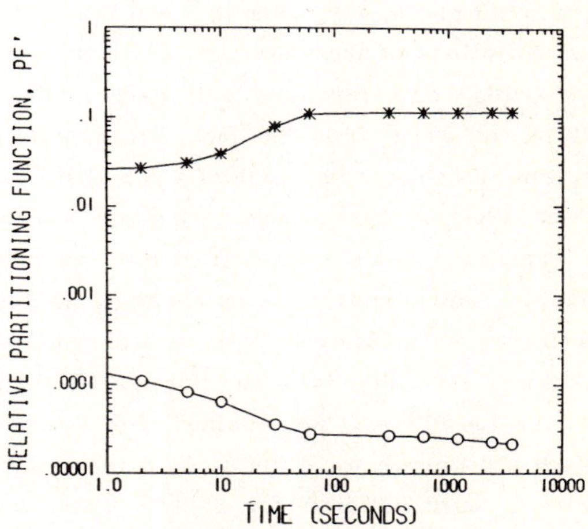

Fig. 4 The calculated temporal variation in the relative partition function, PF', for HO_2 (*-*-*) and OH (O-O-O).

the levels of $(HO_2)_{aq}$ and $(OH)_{aq}$ reach a photostationary state within the first second and then decrease as the concentrations of their gas-phase counterparts decrease (see Figure 2).

Similar to OH and HO_2, HNO_3 never attains thermodynamic equilibrium between the gas and aqueous phases. As illustrated in Figure 2h, X(HNO_3), falls from an initial value of $2.5x10^{-11}$ v/v to $\sim10^{-15}$ v/v after ~1 minute and then remains fairly constant over the rest of the simulation. However, as indicated in Table 3 for HNO_3 to be in equilibrium its mixing ratio would have to fall to a value of $\sim10^{-18}$ v/v since $X_{eq}(HNO_3)$ $\sim10^{-11}$ v/v. This absence of equilibrium between the gas and aqueous phase for HNO_3 is caused by photochemical processes in the gas-phase. The gas-phase HNO_3 mixing ratio of 10^{-13} v/v is maintained by a balance between photochemical production of gaseous HNO_3 from reaction of $(NO_2)_g$ with $(OH)_g$ and loss of $(HNO_3)_g$ by dissolution into cloud droplets. This process leads to a continuous source of NO_3^- to the aqueous-phase. However this source is small and has negligible effect upon $[NO_3^-]$. As result $[NO_3^-]$ remains virtually constant after the initial rapid transfer of ambient HNO_3 to the droplets. However, note that in regions with elevated levels of NO_x this process plus the scavenging of $(NO_3)_g$ and $(N_2O_5)_g$ can represent a significant source of NO_3^- to cloudwater (Heikes and Thompson, 1984).

Sensitivity to Variations in α:

The rapidity with which species are calculated to be transferred from the gas to aqueous phase is highly sensitive to uncertainties in α, the accommodation coefficient. In the case of HO_2 and OH, which never reach equilibrium and are continuously transferred from the gas to aqueous phase, variations in can lead to large changes in the calculated aqueous-phase concentrations of these species, as illustrated in Figure 5. The variations in the $(HO_2)_{aq}$ and $(OH)_{aq}$ concentrations with α are similar to those obtained by Chameides and Davis (1982) and arise from the fact that decreases in α cause decreases in the HO_2 and OH aqueous-phase sources, $S_a(HO_2)$ and $S_a(OH)$, as described in the previous section, and thus lower levels of these species in the aqueous phase.

Variations in α can also affect the concentrations of species such as SO_2, H_2O_2, HCHO, and NH_3, which equilibrate, by changing the gas/aqueous-phase equilibration time. This fact is illustrated in Figure 6 where the effects of varying α upon NH_3 are shown. Note in Figure 6A where PF^1 (NH_3) is plotted for α =1.0 and 10^{-4}, that while equilibrium between the two phases (i.e. PF^1 = 1) was attained after only 10s for α =1, about 50 minutes were needed to reach equilibrium for $\alpha=10^{-4}$. As a result significantly lower NH_3 aqueous-phase concentrations were obtained for $\alpha=10^{-4}$ than for α=1 during the first 50 minutes of the calculation (see Figure 6B). However, since α has no effect upon the equilibrium NH_3 levels, the same concentrations were obtained for the two cases when t > 50 minutes.

These results serve to emphasize the need for accurate determinations of α for non-ra-

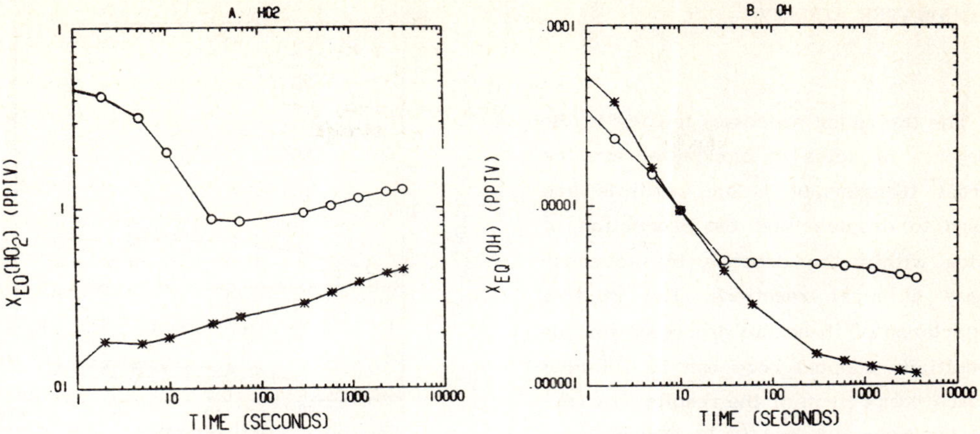

Figure 5. The effect of varying a on (A) the HO_2 equivalent aqueous-phase mixing ratio $X_{eq}(HO_2)$ and (B) the OH equivalent aqueous-phase mixing ratio $X_{eq}(OH)$; $\alpha = 1$ is shown by O-O-O and $\alpha = 10^{-4}$ is shown by *-*-*.

dical species as well as for HO_2 and OH. While the assumption commonly adopted in many cloud and fog models of nearly instantaneous equilibrium between the gas and aqueous phases for species such as SO_2, H_2O_2, and NH_3 may prove to be correct, it has yet to be adequately supported by laboratory data on the appropriate accommodation coefficients and is therefore presently somewhat questionable. On the other hand, should future field measurements in clouds indicate a very short gas/aqueous-phase equilibration time it will imply that the -values for most soluble species and radicals are of the order of 10^{-2} or larger.

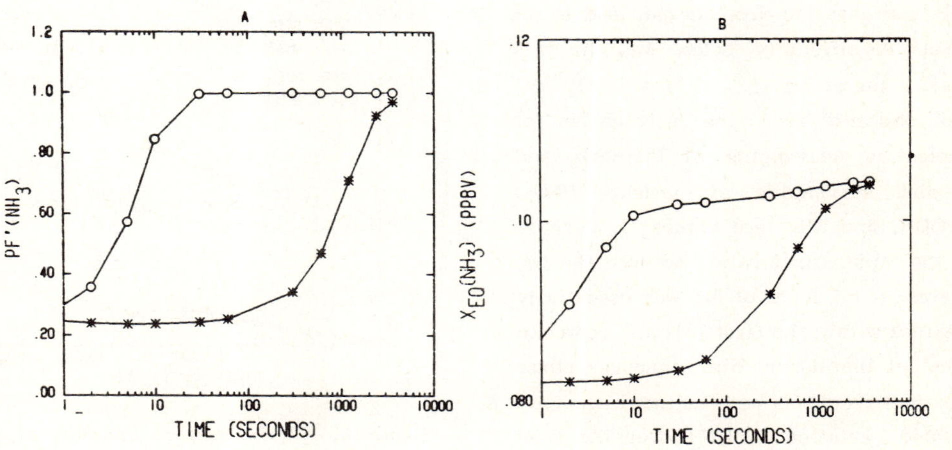

Figure 6. The effect of varying α on (A) the NH_3 relative partitioning function PF', and (B) the NH_3 equivalent aqueous-phase mixing ratio X_{eq}; $\alpha = 1$ is shown by O-O-O and $\alpha = 10^{-4}$ is shown by *-*-*.

The two major processes leading to the presence of acids in cloudwater are the direct transfer of acidic aerosols and gases to droplets and the production of acids within the droplets by aqueous-phase chemical reactions. The relative importance of these two processes for the conditions adopted here can be inferred from a comparison of the results illustrated in Figures 7 and 8. In Figure 7 the concentrations of the major acidic and basic species are plotted for calculation in which the aqueous-phase is assumed to be inert (i.e. all reactions in Table 2E are neglected) and hence all aqueous-phase species arise solely from nucleation and droplet scavenging processes. In Figure 8 the concentrations of the primary acidic and basic species are plotted for a calculation in which the complete aqueous-phase chemical reaction set is included.

As pointed out by Charlson and Rodhe (1982), the simple transfer of acidic aerosols and gases to droplets can lead to pH levels significantly below 5.6. In this case, a pH of 5.1 (i.e. $[H^+] = 8.6 \times 10^{-6}M$) was obtained from the combination of nucleation scavenging of NH_4HSO_4 and droplet scavenging of gaseous HNO_3, HCOOH, and NH_3 (see Figure 7). Because of the rapid equilibration between the two phases, a pH level of 5.1 was essentially attained within the first 100s and remained constant thereafter. When aqueous- phase chemical reactions were included, the calculated variation in $[H^+]$ over the first 100s was virtually the same as that obtained without aqueous-phase chemical re-

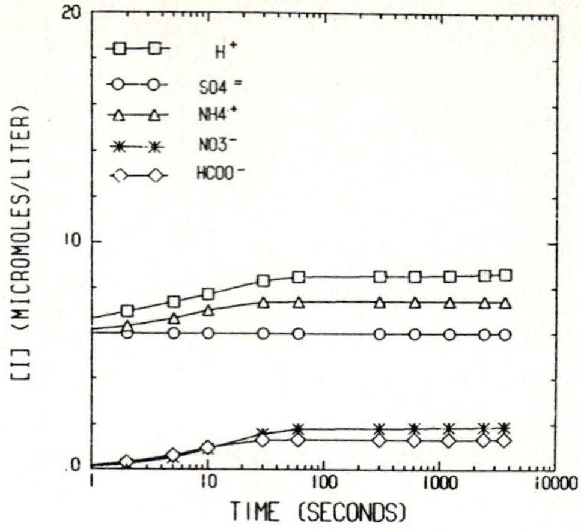

Figure 7. The temporal variation in the concentrations (I) of the major acidic and basic species for a calculation in which all aqueous-phase chemical reactions were neglected.

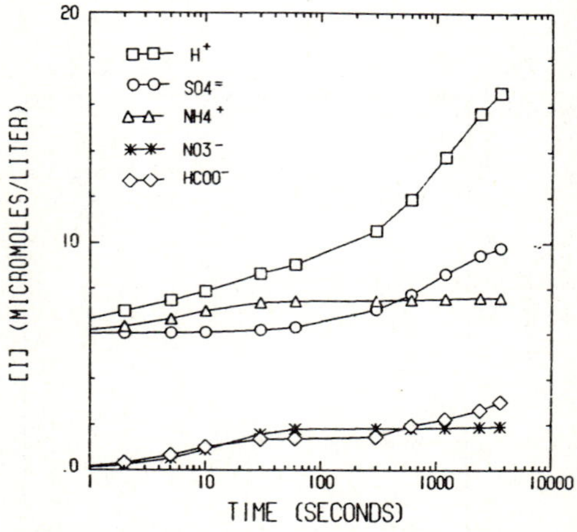

Figure 8. The temporal variation in the concentrations (I) of the major acidic and basic species for a calculation in which all aqueous-phase chemical reactions were included.

actions. However for t > 100s, the [H^+] levels calculated with aqueous-phase chemistry continued to rise as dissolved acids were generated by chemical transformations occuring within the droplets. (see Figure 8). After one hour the pH in this case was about 4.8, which corresponds to a doubling in [H^+] over that obtained without aqueous-phase reactions. This increase in acidity was caused primarily by the aqueous-phase oxidation of dissolved S IV to sulfuric acid (i.e. S VI). Thus in Figure 8, it can be seen that the rise in [H^+] from 9×10^{-6} M at t=100s to 1.65×10^{-5} M at t=3600s is closely correlated with the increase in [$SO_4^=$] calculated to occur during this same time period. (Recall that for the pH range considered here sulfuric acid is completely dissociated into $2H^+$ and $SO_4^=$). The mechanisms responsible for the conversion of S IV to $SO_4^=$ are discussed below.

S IV $\longrightarrow$ S VI Conversion:

The chemical mechanisms by which dissolved S IV is oxidized to S VI in the model presented here include the oxidation of S IV by $(H_2O_2)_{aq}$ and $(O_3)_{aq}$; these reactions both involve a sequence of elementary reactions which can be represented by the overall equations

$$\text{(RAQ17)} \qquad (SO_2)_{aq} + H_2O_2 \longrightarrow SO_4^= + H^+ + HO_2$$

and

$$\text{(RAQ18)} \qquad (SO_2)_{aq} + O_3 \xrightarrow{H_2O} 2H^+ + SO_4^= + O_2 .$$

The conversion of S IV to S VI can also be initiated by OH radicals via the elementary reactions.

$$OH + HSO_3^- \longrightarrow HSO_3 + OH^-$$

followed by (Hayon et al., 1972)

$$HSO_3 + O_2 \longrightarrow HSO_5$$

and

$$HSO_5 + HSO_3^- \longrightarrow 2H^+ + SO_4^= + SO_4^- .$$

These three steps are represented here by a single overall reaction of the form of

$$(\text{RAQ11}) \quad OH + HSO_3^- \xrightarrow{HSO_3^-,\ O_2} SO_4^= + 2H^+ + SO_4^-.$$

The SO_4^- radical produced by (RAQ11) can either propagate the chain by a reaction sequence of

$$SO_4^- + HSO_3^- \longrightarrow SO_4^- + HSO_3$$

$$HSO_3 + O_2 \longrightarrow HSO_5$$

and

$$HSO_5 + HSO_3^- \longrightarrow 2H^+ + SO_4^- + SO_4^=$$

which is represented here by the single reaction

$$(\text{RAQ16}) \quad SO_4^- + HSO_3^- \xrightarrow{HSO_3^-,\ O_2} 2H^+ + 2SO_4^= + SO_4^-$$

or terminate the chain by reactions such as

$$(\text{RAQ12}) \quad SO_4^- + Cl^- \longrightarrow SO_4^= + Cl$$

followed by

$$(\text{REQ5}) \quad Cl^- + Cl \rightleftharpoons Cl_2^-$$

and

$$(\text{RAQ18}) \quad Cl_2^- + HO_2 \longrightarrow 2Cl^- + H^+ + O_2.$$

The oxidation of S $\overline{\text{IV}}$ can also be triggered by nitrogen oxide reactions. NO_3 radicals scavenged from the gas-phase can initiate a chain oxidation scheme similar to that of OH via

$$(\text{RAQ26}) \quad NO_3 + HSO_3^- \xrightarrow{HSO_3^-,\ O_2} NO_3^- + SO_4^= + SO_4^- + 2H^+.$$

In addition dissolved NO_2 can react with HSO_3^- via (Lee and Schwartz, 1983)

$$(\text{RAQ28}) \quad NO_2 + HSO_3^- \xrightarrow{NO_2,\ HSO_3^-} 2NO_2^- + SO_4^= + 3H^+$$

The calculated per cent of the initial gas phase SO_2 that is converted to $SO_4^=$ by the

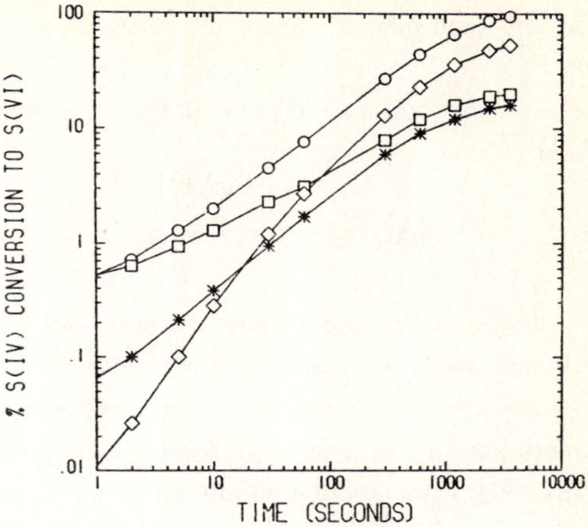

Figure 9. The calculated per cent of the initial S $\overline{IV}$, as $(SO_2)_g$, that was converted to S $\overline{VI}$, as a function of time. The total conversion is shown by O-O-O, the conversion due to the reaction of S $\overline{IV}$ with $(H_2O_2)_{aq}$ is shown by ◇-◇-◇, the conversion due to the reaction of S $\overline{IV}$ with $(OH)_{aq}$ is shown by □-□-□, reaction of S $\overline{IV}$ with $(O_3)_{aq}$ is shown by *-*-*.

aqueous-phase reactions discussed above is plotted as a function of time in Figure 9. A major feature of these results is the rapid rate of SO_2 oxidation; 20 minutes after the introduction of the droplets over 65% of the SO_2 has been converted to $SO_4^=$ and after 1 hour 95% has been converted. Note that rapid S $\overline{IV}$ conversion rates, similar to the rate calculated here, have also been reported by other investigations (c.f. Hegg and Hobbs, 1981; Penkett et al., 1979; Martin, 1983). As indicated in Figure 9, the primary SO_2 oxidant for the conditions adopted here is found to be $(H_2O_2)_{aq}$, with about 60% of the initial SO_2 being oxidized via (RAQ17). Als important are $(OH)_{aq}$, and $(0_3)_{aq}$ which caused about 20% and 15%, respectively, of the initial SO_2 to be oxidized to $SO_4^=$ after one hour. For $\mathcal{E}$=1.0, about 30% of the SO_2 is oxidized by $(OH)_{aq}$. For the conditions adopted here, the reactions of dissolved S $\overline{IV}$ with NO_3 and NO_2 do not significantly contribute to the production of $SO_4^=$; however in regions affected by anthropogenic NO_x emissions these reactions can, under some circumstances, affect the rate of S $\overline{IV}$ oxidation and thus need be considered.

Sources of $(H_2O_2)_{aq}$:

In light of the importance of $(H_2O_2)_{aq}$ in the conversion of SO_2 to $SO_4^=$, It Is relevant to examine the processes which controls the levels of H_2O_2 in cloud droplets. In the Standard Model, $(H_2O_2)_{aq}$ was produced from two processes: 1) the direct dissolution of H_2O_2 from the gas phase; and 2) the aqueous-phase production of $(H_2O_2)_{aq}$ from free radical reactions (Chameides and Davis, 1982). In this later mechanism, aqueous-phase free radicals derived from the scavenging and dissolution of gas-phase HO_2 and OH, react to form $(H_2O_20_{aq}$ through reactions such as (REQ2) followed by

$$\text{(RAQ4)} \qquad HO_2 + O_2^- \xrightarrow{H^+} H_2O_2 + O_2$$

or the sequence of

$$(RAQ8)\quad O_2^- + HCO_3^- \xrightarrow{H^+} H_2O_2 + CO_3^-$$

and

$$(RAQ10)\quad CO_3^- + O_2^- \longrightarrow CO_3^= + O_2.$$

The relative importance of these two sources is illustrated in Figure 10 in which the $[H_2O_2]$ levels are shown for a calculation in which al aqueous-phase free-radical $[H_2O_2]_{aq}$ sources within the droplets were neglected, as well as ones in which these sources were included. In the absence of the aqueous-phase sources, $[H_2O_2]$ reaches a peak level of about $5x10^{-5}$ M after about 1 minute and then slowly decays due to its reaction with S $\underline{IV}$ to a level of about $4.5x10^{-5}$ M after 1 hour. When the free radical aqueous-phase sources were included $[H_2O_2]$ also reaches a level of about $5x10^{-5}$M after 1 minute, but thereafter, rather than decreasing, $[H_2O_2]$ continues to increase and reaches a level of about $6x10^{-5}$M after one hour for the Standard Model and about $7x10^{-5}$M for a calculation with $\mathcal{E}=1.0$. Roughly 25% of the $(H_2O_2)_{aq}$ calculated to be present in the droplet after one hour was produced from free radical reactions for the Standard Model and about 35% for $\mathcal{E}=1.0$. Thus, free radical reactions not only directly contribute t S $\overline{\underline{IV}}$ oxidation by the reaction of OH with HSO_3^-, but by producing significant quantities of $(H_2O_2)_{aq}$ they also indirectly enhance the S $\overline{\underline{IV}}$ to S $\overline{\underline{VI}}$ conversion rate.

While the calculated concentration of H_2O_2 is consistent with measurements of $[H_2O_2]$ in rainwater (Kok, 1980; Zika et al., 1982), the fraction of H_2O_2 generated within the aqueous-phase does not appear to be large enough to completely explain the measurements of Heikes et al. (1982) and Zika et al. (1982). These investigators found evidence for an aqueous-phase H_2O_2 source in clouds that dominated over the source from the dissolution of $(H_2O_2)_g$, while in the model the aqueous-phase source only contributed about 30% of the total H_2O_2. However, the relative amounts of H_2O_2 generated from dissolved H_2O_2 and HO_2 radicals is highly sensitive to the initial level of $(H_2O_2)_g$ assumed to be present in the gas-phase. Thus a definitive test of the importance of the H_2O_2 ge-

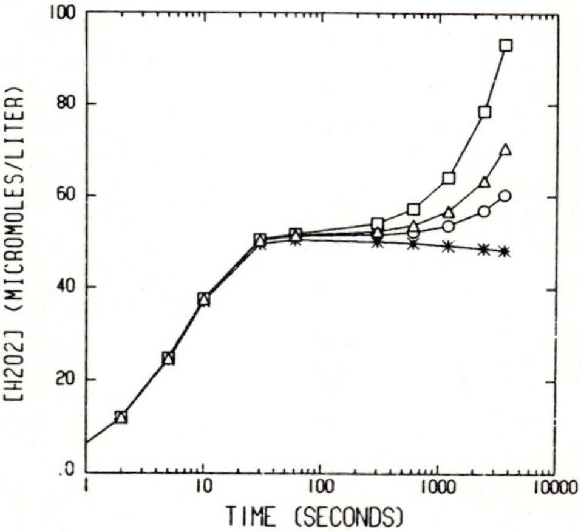

Figure 10. The temporal variation in (H_2O_2), the aqueous- phase concentration of H_2O_2, for a calculation in which all aqueous-phase sources of $(H_2O_2)_{aq}$ were neglected (*-*-*), for a calculation in which the aqueous-phase sources of $(H_2O_2)_{aq}$ were included (O-O-O), for a calculation with all aqueous-phase sources and with $\mathcal{E} = 1.0$ (Δ-Δ-Δ), and for a calculation in which a free radical source due to coronal discharges was assumed (□-□-□).

nerative mechanism from dissolved HO_2 will have to await simultaneous measurements of gaseous and aqueous-phase H_2O_2 in a cloud. Nevertheless, the possibility that other aqueous-phase H_2O_2 generative mechanisms exist should not be discounted.

Kok (1982), citing a correlation between $[H_2O_2]$ levels observed in rain and thunderstorm activity and laboratory experiments showing $[H_2O_2]$ production in water drops subjected to discharges, has proposed that H_2O_2 is produced from lightning strokes. However, it seems unlikely that H_2O_2, which is unstable at high temperatures, would be produced in the hot air in and around the lightning channel. It is possible that the H_2O_2 production observed by Kok arose from a different but related electrical phenomenon; coronal discharges from droplets. Recall that micro-electrical discharges such as positive point coronal discharges from droplets can generate significant quantities of aqueous-phase free radicals. These radicals, in turn, can produce dissolved H_2O_2 via reactions such as (RAQ4). That this or a similar electrical process could represent a significant source of $(H_2O_2)_{aq}$ in clouds is illustrated in Figure 10, where the calculated $[H_2O_2]$ levels are shown for a cloud with an assumed ion-pair-production rate due to coronal discharges of 5×10^6 cm^{-3} s^{-1}. (As described previously, each ion pair is assumed to produce one dissolved O_2^- ion and one gaseous OH radical). Clearly if these discharges do in fact occur they can efficiently produce H_2O_2 in droplets and thus possibly play a significant role in the conversion of S $\underline{\overline{IV}}$ to S $\underline{\overline{VI}}$.

Formic acid:

In addition to the generation of sulfuric acid, the aqueous-phase production of formic acid also contributes to the increase in the droplet acidity over that attained from the direct scavenging of acidic aerosols and gases. Similar to the calculations of Chameides and Davis (1983), the HCOOH levels calculated in the model are controlled by the formaldehyde/formic acid oxidation scheme in which hydrated formaldehyde is oxidized by $(OH)_{aq}$

$$\text{(RAQ21)} \qquad OHCH_2OH + OH \longrightarrow OHCHOH + H_2O$$

and ultimately forms $(HCOOH)_{aq}$ via

$$\text{(RAQ22)} \qquad OHCHOH + O_2 \longrightarrow HCOOH + HO_2.$$

The $(HCOOH)_{aq}$ produced by (RAQ21) + (RAQ22) is then removed from the droplet either through evaporation via (RS10) or through oxidation by $(OH)_{aq}$ via the sequence of

$$\text{(REQ7)} \qquad HCOOH \rightleftharpoons HCOO^- + H^+$$

and

$$(\text{RAQ23B}) \quad HCOO^- + OH \xrightarrow{O_2} HO_2 + OH + CO_2 .$$

After about 30-50 minutes a photostationary state is reached in which the production of $(HCOOH)_{aq}$ from (RAQ21) is balanced by its loss from (RAQ23B), i.e.

$$[OHCH_2OH]\ [OH]\ k_{AQ21} = [HCOO^-]\ [OH]\ k_{AQ23B} \tag{14}$$

Thus the ratio of dissolved formic acid and formaldehyde becomes independent of [OH], such that (Chameides and Davis, 1983)

$$\frac{[HCOO^-]}{[OHCH_2OH]} = \frac{k_{AQ21}}{k_{AQ23B}} \sim 1 . \tag{15}$$

Since the $[OHCH_2OH]$ level is strongly related to the gas-phase concentration of HCHO assumed to be initially present in the air parcel, Eqn. (15) predicts that at photostationary state the $[HCOO^-]$ level well be dependent on the initial gas-phase concentrate of HCHO but independent of the initial level of gaseous HCOOH. Since the gas-phase HCOOH level is related to $[HCOO^-]$ via (REQ7) and (RS10), once thermodynamic equilibration is established X(HCOOH) should have a similar dependence. The extent to which these relationships hold is indicated in Figure 11, where the calculated levels of $[HCOO^-]$ and X(HCOOH) are illustrated

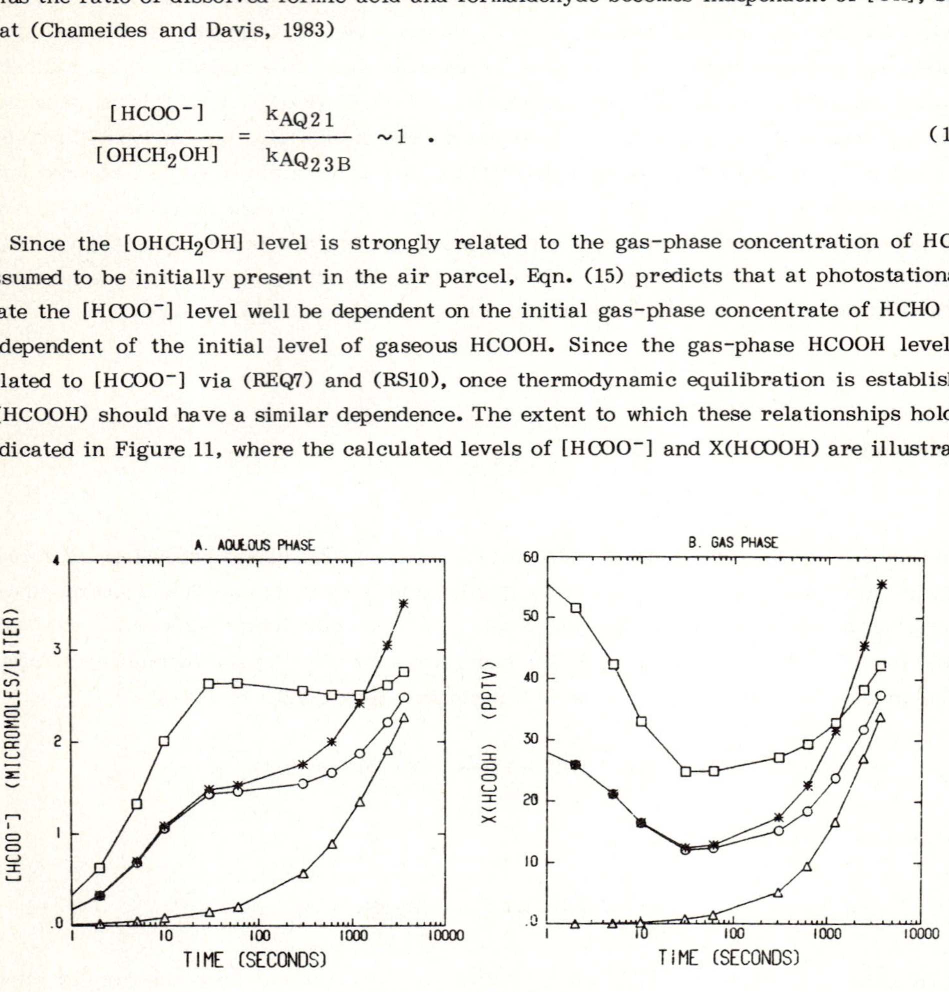

Figure 11. The temporal variation in (A) ($HCOO^-$), the aqueous-phase $HCOO^-$ concentration, and (B) X(HCOOH), the gas-phase HCOOH mixing ratio, for the Standard Model (O-O-O), for a calculation with an assumed doubling in the initial HCHO mixing ratio to 0.9 ppbv (*-*-*), for a calculation with an assumed doubling in the initial HCOOH mixing ratio to 60 pptv (□-□-□), and for a calculation in which no initial HCOOH was assumed to be present (Δ-Δ-Δ).

for: the Standard Model; a calculation in which the initial gas-phase HCHO mixing ratio was doubled; a calculation in which the initial gas-phase HCOOH mixing ratio was doubled; and a calculation in which no initial gaseous HCOOH was assumed to be present. While a doubling in the initial X(HCHO) resulted in a substantial increase in the final calculated values of [$HCOO^-$] and X(HCOOH), essentially the same values were calculated for these two parameters for the cases with different initial $(HCOOH)_g$ mixing ratios. Thus it appears that the aqueous-phase formic acid/formaldehyde oxidation scheme may dominate over the gas-phase photochemical reactions such as ozone-olefine reactions in controlling formic acid levels in clouds in the presence of free radical sources. These results serve to illustrate the possible danger of using data on the chemical composition of cloud and rainwater to determine the abundances of soluble species in the ambient atmosphere. In the case of a species such as HCOOH, which can be produced and destroyed by chemical reactions in the aqueous-phase, the levels observed in clouds and rain may have very little relation to the actual atmospheric abundance of this species.

SENSITIVITY OF S $\overline{\underline{IV}}$ ⟶ S $\overline{\underline{VI}}$ CONVERSION RATE TO MODEL ASSUMPTIONS

Similar to measurements of cloudwater acidity in relatively polluted environments (c.f. Hegg and Hobbs, 1981; Daum et al., 1982; Lazrus et al., 1983), the calculations presented here indicate that one of the major sources of free acidity in a remote marine cloud arises from the conversion of gaseous SO_2 to dissolved sulfuric acid. For the Standard model, this conversion was found to proceed quite rapidly with essentially all of the initial SO_2 being oxidized within one hour after the cloud droplets were introduced into the gaseous mixture. In this section we examine the sensitivity of this result to variations in several model parameters.

Accommodation Coefficient:

As noted earlier the lack of reliable data on α, the accommodation or sticking coefficients for soluble species impinging on water droplets implies a large uncertainty in the rate at which species are transferred from the gas to aqueous phase. Recall that as α was varied from 1 to 10^{-4} the time required for a soluble species to equilibrate varied from about 10 seconds to about 50 minutes. Since the conversion of SO_2 in a cloud depends upon the dissolution of SO_2 as well as S $\overline{\underline{IV}}$ oxidants such as H_2O_2, O_3, and OH, the S $\overline{\underline{IV}}$ conversion rate is also affected by α. This dependence is indicated in Table 4. For $\alpha \geq 10^{-2}$ the transfer of gases to the aqueous phase is sufficiently fast so that the conversion rate is controlled by the S $\overline{IV}$ aqueous-phase chemical lifetime and thus the conversion rate is hardly changed by increasing α from 10^{-2} to 1. However for $\alpha \leq 10^{-2}$. the conversion rate becomes increasingly limited by the rate of mass transfer between the phases and as a result in decreasing below 10^{-2} the S $\overline{IV}$ conversion rate is slowed. While for $\alpha = 10^{-2}$, essentially all the SO_2 has been oxidized after 1 hour, for $\alpha = 10^{-4}$ only 80% has been oxidized after an hour. For $\alpha = 10^{-6}$, only 4 % of the SO_2 was calculated to be oxidized after an hour. While the observations of Hegg and Hobbs (1981) of rapid S $\overline{IV}$ oxidation in clouds would appear to preclude such a slow conversion rate, these calculations clearly indicate the need for laboratory experiments to determine the accommodation coefficients for a large array of soluble gases, and especially for SO_2, H_2O_2, and NH_3 as well as HO_2 and OH.

Liquid Water Content:

The liquid water content of a cloud, W_L, can vary considerably from cloud to cloud or even within a single cloud as droplets alternately condense and evaporate. Typically, observations indicate a range for W_L in stratus clouds of 0.05 - 0.5 g m^{-3} (c.f. Mason, 1957). Ryan et al. (1972), for example, found W_L to range from 0.01 to 0.35 g m^{-3} in shallow, non-precipitating clouds over the Pacific Ocean. Observations in stratus clouds of W_L in excess of 0.5 g m^{-3} have also been reported (c.f. Neiburger, 1949). In light of this variability it is interesting to examine how variations in W_L from the Standard Model case of 0.5 g m^{-3} affect

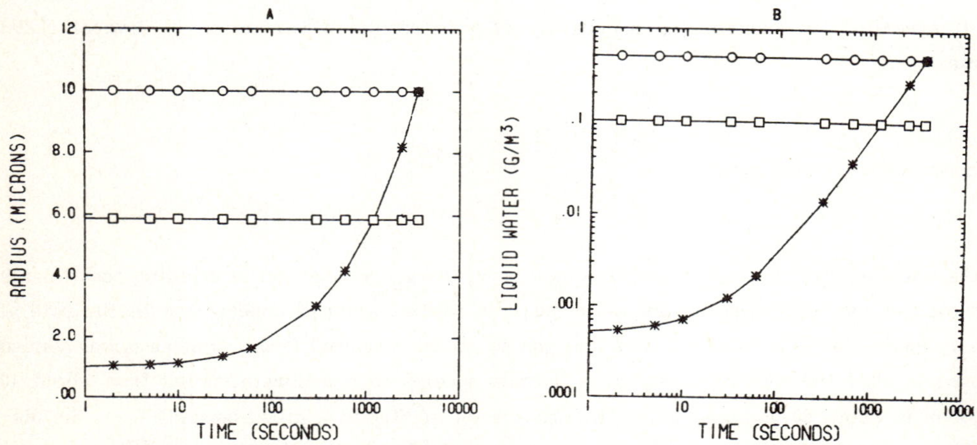

Figure 12. The temporal variations in (A) droplet radius and (B) liquid water content for the Standard Model (O-O-O), for a calculation with reduced liquid water (□-□-□), and for a calculation with a growing droplet distribution (*-*-*).

the SO_2 conversion rate. As illustrated in Figure 12, two other cases were considered: one in which W_L was reduced to a constant value of 0.1 g m^{-3} by assuming a constant droplet radius of 5.8 μ and a droplet concentration of 100 cm^{-3}; and the other in which W_L was allowed to increase with time to a final value of 0.5 g m^{-3} by assuming a monodisperse distribution of growing droplets with a concentration of 100 cm^{-3}. In the later case, a parabolic growth rate, as might be expected for a condensing drop (c.f. Pruppacher and Klett, 1978), was assumed i.e.

$$\frac{da}{dt} = \frac{1.38\times10^{-2}}{a} \qquad (16)$$

where a has units of μ.

The resulting SO_2 conversion rates are listed in Table 4. Clearly a reduction in W_L causes a reduction in the rate of SO_2 oxidation; while 95% of the SO_2 was found to be

Table 4. The Sensitivity of the Calculated % Conversion of S $\underline{IV}$ to S $\underline{VI}$ to Various Model Parameters

Model Description	% of Initial S IV Converted to S VI		
	t=20 minutes	t= 40 minutes	t= 60 minutes
Standard Model: $\alpha = 10^{-2}$, W_L=0.5 gm^{-3}, Initial $X(SO_2)$ = 55 pptv, Initial $X(HNO_3)$ = 25 pptv	66	87	95
α= 1	69	89	96
α= 10^{-4}	41	66	80
α= 10^{-6}	1.5	2.9	4.2
W_L = 0.1 g m^{-3}	28	48	63
W_L = variable with time	14	48	77
Initial $X(SO_2)$ = 5 ppbv, Initial $X(HNO_3)$ = 25 pptv	22	25	27
Initial X (SO_2) = 55 pptv, Initial $X(HNO_3)$ = 1 ppbv	51	77	89
Initial $X(SO_2)$ = 5 ppbv, Initial $X(HNO_3)$ = 1 ppbv	21	24	26

oxidized after 1 hour for W_L = 0.5g m^{-3} (the Standard Model case), only 63% was oxidized for W_L = 0.1g m^{-3} and 77% for the variable W_L calculation. This occurs because as W_L decreases, the total amount of SO_2 dissolved in the droplets decreases (see Eqn. 12), and thus the S $\overline{\text{IV}}$ oxidation rate slows. The fact that the total amount of SO_2 oxidized after one hour for the variable W_L case was significantly less than that of the Standard Model case, even though both had the same liquid water contents at t=1 hour, indicates the need to properly treat the relevant microphysical processes when attempting to simulate the oxidation of S $\overline{\text{IV}}$ in a specific cloud.

Initial SO_2 and HNO_3 Levels:

Another important variable affecting the S $\overline{\text{IV}}$ ⟶ S $\overline{\text{VI}}$ conversion rate is the initial levels of gaseous or particulate acids and acid precursors present in the air parcel. Calculations indicate that as the levels of these species are increased, the relative rate of SO_2 oxidation is slowed. This trend is apparent in Table 4 and in Figure 13, where the calculated percentage of SO_2 converted to $SO_4^=$ is listed and illustrated for initial $(SO_2)_g$ concentrations ranging from 50 pptv to 5 ppbv and for initial $(HNO_3)_g$ levels of 25 pptv and 1 ppbv. For initial $(SO_2)_g$ levels less than 0.5 ppbv, the conversion rate is quite rapid (in excess of 200% per hour during the early stages of the integration) and essentially goes to completion within 1 hour. However, for higher $(SO_2)_g$ levels, the percentage conversion rate slows; for an initial $(SO_2)_g$ level of 5 ppbv less than 30% of the $(SO_2)_g$ is converted to $SO_4^=$ after 60 minutes for the conditions adopted here. Similarly, increasing the initial $(HNO_3)_g$ concentration also results in a decrease in the percentage of SO_2 oxidized.

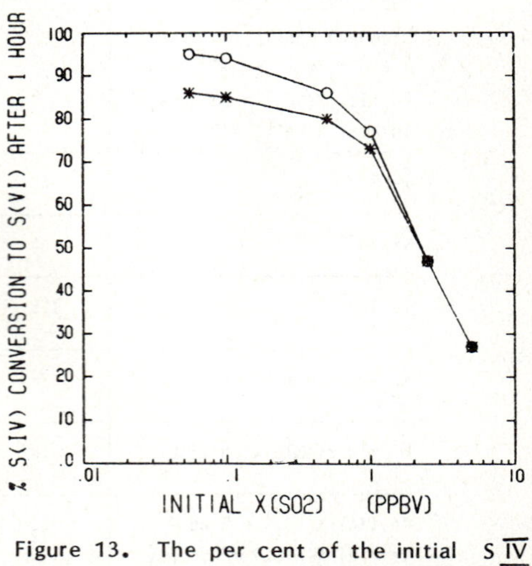

Figure 13. The per cent of the initial S $\overline{\text{IV}}$ as $(SO_2)_{aq}$ that was calculated to be converted to S $\overline{\text{VI}}$ as dissolved $SO_4^=$ as a function of the initial gas-phase SO_2 mixing ratio, $X(SO_2)$, when the initial gas-phase HNO_3 was assumed to be 25 pptv (o-o-o) and 1 ppbv (*-*-*).

The tendency for the relative SO_2 conversion rate to slow as the initial levels of $(SO_2)_g$ and $(HNO_3)_g$ increase arises from the inhibiting effect of H^+ ions. As $(SO_2)_g$ and $(HNO_3)_g$ increase, $[H^+]$ increases. As $[H^+]$ increases in a droplet: 1) the effective Henry's Law constant for SO_2 decreases (see Eqn 7) and thus the total amount of S $\overline{\text{IV}}$ in solution decreases; 2) the O_3-S $\overline{\underline{IV}}$ rate constant decreases; 3) the $(OH)_{aq}$ concentration is depleted due to the enhanced rate of

$$\text{(RAQ17)} \quad Cl^- + H^+ + OH \longrightarrow Cl + H_2O;$$

and 4) the H_2O_2 - S $\overline{\underline{IV}}$ rate constant increases. While effects 1,2, and 3 tend to slow the SO_2 oxidation rate, effect 4 tends to increase the rate as H^+ increases. However, the inhibiting effects dominate, and thus the relative rate of SO_2 oxidation tends to slow as $[H^+]$ increases. This tendency for SO_2 conversion to slow as the pH decreases has been observed in clouds by Hegg and Hobbs (1981) and Lazrus et al. (1983); the calculations of Martin (1983) also indicate a similar tendency.

The inhibiting effect of $[H^+]$ upon the SO_2 conversion rate could have significant implications for pollution control strategies. Note in Figure 12 that as the initial $(SO_2)_g$ level decreases the net percentage of this SO_2 that is converted to sulfuric acid increases. For instance when the initial $(SO_2)_g$ is decreased by a factor of five from 5 ppbv to 1 ppbv the SO_2-to-$SO_4^=$ conversion increases from 27% to 75% and thus the $[SO_4^=]$ and $[H^+]$ levels after 1 hour decrease by less than a factor of 2 from $1x10^{-4}$M to $6x10^{-5}$M and $2x10^{-4}$M to $1.2x10^{-4}$M, respectively. This result implies that the application of emission controls to decrease ambient levels of SO_2 would result in far less of a reduction in $[H^+]$ in a given cloud than one might predict on the basis of simple stoichiometric considerations.

CONCLUSION

Even in remote locations of the marine atmosphere, pH levels significantly below 5.6, the value nominally assigned to "neutral" rain, can be generated in cloudwater. While a major source of this acidity can arise from the scavenging of pre-existing acidic aerosols and gases, another major source is the generation of acids (e.g. sulfuric and formic acid) in cloudwater from aqueous-phase reactions. Dissolved H_2O_2 and, to a lesser extend, O_3 play an important role in the system by oxidizing dissolved SO_2 and producing sulfuric acid. However, calculations which simulate the coupled gas/aqueous-phase free radical chemistry of a cloud indicate that HO_2 and OH can also be important by triggering the conversion of S $\overline{\underline{IV}}$ to S $\overline{\underline{VI}}$ and of HCHO to HCOOH as well as by generating H_2O_2 within droplets. The possibility that the generation of acids in clouds is affected by electrical/chemical processes, such as the generation of aqueous-phase free radicals by coronal discharges, also needs to be considered.

The aqueous-phase generation of acids in stratoform cloudwater can be affected by a variety of parameters. In the case of the conversion of S $\overline{\underline{IV}}$ to S $\overline{\underline{VI}}$, the conversion rate was found to be dependent upon the assumed value of the accommodation coefficients for SO_2,

H_2O_2, O_3, OH, and HO_2, the assumed droplet growth rate and liquid water content, and the initial levels of SO_2 and HNO_3, as well as other acidic or basic species. The calculated levels of formic acid were found to be sensitive to the ambient HCHO level, but only weakly dependent upon the assumed ambient HCOOH level.

In the past it has often been tacitly assumed that the atmospheric aqueous-phase is a relatively inert chemical medium, where soluble atmospheric species are dissolved and eventually transported to the earth's surface in precipitation without significant chemical changes. However, as we study the chemistry of this system in more detail, a quite different picture is beginning to emerge. This and other studies (c.f Penkett et al., 1979; Graedel and Weschler, 1981; Heikes et al., 1982; Zika et al., 1982; Chameides and Davis, 1982) have shown cloudwater to be a very active chemical medium where significant levels of free radicals can exist, where many soluble species can be rapidly oxidized, and where other species can be rapidly generated. As a result some caution should be exercised when attempting to use data on the chemical composition of precipitation to infer information on the composition of the ambient atmosphere. Similarly, some considerations should be given to the possibility that aqueous-phase chemical transformations also act to obscure the results of wet chemical techniques in which the atmospheric abundance of a soluble gas is determined by first passing air through an aqueous trap and then chemically analyzing the solution. The tacit assumption that the aqueous trap is chemically inert may prove to be as erroneous as earlier assumptions of the inert properties of cloudwater.

ACKNOWLEDGEMENT:

The author would like express his gratitude to Dr. D. D. Davis for his many helpful comments and suggestions. This work was supported in part by funds from the National Science Foundation under Grant ATM 8208828 and the National Aeronautics and Space Administration under Grant NAG-1-385.

REFERENCES

Andreae, M.O., Marine aerosol chemistry at Cape Grim, Tasmania, and Townsville, Queensland, J. Geophys. Res., 87, 8875-8885, 1982.

Atkinson R., and J.N. Pitts, Rate constants for the reaction of OH radicals with propylene and butenes over the temperature range of 297-425 K, J. Chem. Phys., 63, 3591-3595, 1975.

Ayers, G.P., and J.L. Gras, The concentrations of ammonia in southern ocean air, J. Geophys. Res., in press, 1983.

Barret, E., and G. Brodin, The Acidity of Scandinavian Precipitation, Tellus, 7, 251-257, 1955.

Bigg, E.K., Comparison of aerosol at four baseline atmospheric monitoring stations, J. Appl. Met., 19, 521-533, 1980.

Bonsang, B., B.C. Nguyen, A. Gaudry, and G. Lambert, Sulfate enrichment in marine aerosols owing to biogenic gaseous sulfur compounds, J. Geophys. Res., 85, 7410-7416, 1980.

Borkowski, J., A.T. Chai, T. Mo, and A.E.S. Green, Cloud Effects on Middle Ultraviolet Global Radiation, Acta Geophysica Polonica, 25, 287-301, 1977.

Bothe, E. and D. Schulte-Frohlinde, Reactions of dihydroxylmethyl radical with molecular oxygen, Z. Naturforsch, 35b, 1035-1039, 1980.

Chameides, W.L. and D.D. Davis, The free radical chemistry of cloud droplets and its impact upon te composition of rain, J. Geophys. Res., 87, 4863-4877, 1982.

Chameides, W.L. and D.D. Davis, Cloud chemistry: An aqueous-phase source of formic acid, Nature, 304, 427-429, 1983.

Chameides W.L. and A. Tan, The two-dimensional diagnostic model for tropospheric OH: An uncertainty analysis, J. Geophys. Res., 86, 5209-5223, 1981.

Charlson, R.J., and H. Rodhe, Factors controlling the acidity of natural rainwater, Nature, 295, 683-685, 1982.

Cogbill, C.V., and G.E. Likens, Acid precipitation in the Northeastern United States, Water Resour. Res., 10, 1133-1137, 1974.

Conover, J., Cloud and Terrestrial Albedo Determinations from TIROS Satellite Pictures, J. Applied Met., 4, 378-396, 1965.

Cox, R.A., and M.J. Roffey, Thermal decomposition of peroxyacetyl nitrate in the presence of nitric oxide, Environ. Sci. Technol., 9, 900-906, 1977.

Dasgupta, P.L., K. DeCesare, and J.C. Ullrey, Determination of atmospheric sulfur dioxide without tetrachloromercurate (II) and the mechanism of the Schiff reaction, Anal. Chem., 52, 1912-1922, 1980.

Daum, P.H., S.E. Schwartz, and L. Newman, Studies of the gas and aqueous phase composition of stratoform clouds, Proceedings of the 4th International Conference on Precipitation Scavenging, Dry Deposition, and Resuspension, Elsevier Press, in press, 1983.

Dawson, G.A., J.C. Farmer, and J.L. Moyers, Formic and acetic acids in the atmosphere of the southwest U.S.A., Geophys. Res. Lett., 7, 725-728, 1982.

DeMore, W.B., R.T. Watson, C.J. Howard, D.M. Golden. M.J. Molina, R.F. Hampson, M. Kurylo, A.R. Ravishankara, Chemical Kinetics and Photochemical Data for Use in Stratospheric Modeling, Jet Propulsion Laboratory, California Institute of Technology, Pasadena, CA., JPL Publication 82-57, 186 pp, 1982.

Dickerson, R.R., D.H. Stedman, W.L. Chameides, P.J. Crutzen. and J. Fishman, Actinometric measurements and theoretical calculations of $J(O_3)$, the rate of photolysis of ozone to O('O), Geophys. Res. Lett, 6, 833-836, 1979.

Dodge, M.C., and R.R. Arnts, A new mechanism for the reaction of ozone with olefins, Int. J. Chem. Kinetics, 11, 399-410, 1979.

Dogliotti, L. and E. Hayon, Transient species produced in the photochemical decomposition of ceric salts in aqueous solution. Reactivity of NO_3 and HSO_4 free radicals, J. Phys. Chem., 71, 3802-3808, 1967.

Draganic, I.G. and Z.D. Draganic, The Radiation Chemistry of Water, Academic Press, New York, 242 pp., 1971.

Eriksson, E., Composition of atmospheric precipitation I. nitrogen compounds, Tellus, 4, 215-232, 1952.

Fuchs, N.A. and A.G. Sutugin, High-dispersed aerosols, Intern Reviews of Aerosol Physics and Chemistry, 2, edited by G.M. Hidy and J.R. Brock, Pergamon Press, 1-60, 1971.

Galloway, J.N., G.E. Likens, W.C. Keene, and J.M. Miller, The composition of precipitation in remote areas of the world, J. Geophys. Res., 87, 8771-8786, 1982.

Garland, J.A., Dry and wet removal of sulfur from the atmosphere, Atmos. Env. 12, 349-362, 1978.

Gear, C.W., Numerical Initial Value Problems in Ordinary Differential Equations, Englewood Cliffs, New Jersey, Prentice Hall, Inc., 1971.

Gill, P.S., T.E. Graedel, and C.J. Weschler, Organic films on atmospheric aerosol particles, fog droplets, cloud droplets, raindrops and snowflakes, Rev. Geophys. Space Phys, 21, in press, 1983.

Good, A., A. Durden, and P. Kebarle, Mechanism and rate constants of ion-molecule reactions leading to formation of $H^+(H_2O)_n$ in moist oxygen and air, J. Chem. Phys., 52, 222-229, 1970.

Graedel, T.E. and C.J. Weschler, Chemistry within aqueous atmospheric aerosols and raindrops, Rev. Geophys. Space Phys., 19, 505-539, 1981.

Graedel, T.E., and K.I. Goldberg, Kinetic studies of raindrop chemistry I. Inorganic and organic processes, J. Geophys. Res., in press, 1983.

Granat L., On the relation between pH and the chemical composition in atmospheric precipitation, Tellus, 24, 550-560, 1972.

Harding, D.D., Microphysical response of cloud droplets in a fluctuating updraft, Ph.D. thesis, University of Michigan, 1977.

Hasegawa, K., and P. Neta, Rate constants and mechanisms of reaction of Cl_2^- radicals, J. Phys. Chem, 82, 854-857, 1978.

Hayon, E., A. Treinin, and J. Wilf, Electronic spectra, photochemistry, and autoxidation mechanism of the sulfite-bisulfite-pyrosulfite systems. The SO_2^-, SO_3^-, SO_4^-, and SO_5^- radicals, J. Am. Chem. Soc., 94, 47-57, 1972.

Hegg, D.A., and P.A. Hobbs, Cloudwater chemistry and the production of sulfates in clouds, J. Atmos. Sci., 15, 1597-1604, 1981.

Heidt, L.E., J.P. Krasnec, R.A. Loeb, W.H. Pollock, B.E. Henry, and P.J. Crutzen, Latitudinal distributions of CO and CH_4 over the Pacific, J. Geophys. Res., 85, 7285-7572, 1980.

Heikes, B., and A.M. Thompson, Effects of heterogeneous processes on NO_3, HONO, and HNO_3 chemistry in the troposphere, J. Geophys. Res., in press, 1984.

Heikes, B.G., A.L. Lazrus, G.L. Kok, S.M. Kunen, B.W. Gandrud, S.N. Gitlin, P.D. Sperry, Evidence for aqueous phase hydrogen peroxide synthesis in the troposphere, J. Geophys. Res., 87, 3045-3051, 1982.

Herron, J.T., and R.E. Huie, Rate constants for the reaction of ozone with ethene and propene from 235.0 to 362.0 K, J. Phys. Chem., 78, 2085-2088, 1974.

Horowitz, A., and J.G. Calvert, Wavelength dependence of the primary processes in acetaldehyde photolysis, J. Phys. Chem., 86, 3105-3114, 1982.

B.J. Huebert and A.L. Lazrus, Tropospheric gas-phase and particulate nitrate measurements, J. Geophys. Res., 85, 7322-7328, 1980.

Kan, C.S., F. Su, J.G. Calvert, and J.H. Shaw, Mechanism of the ozone-ethene reaction in dilute N_2/O_2 mixtures near 1-atm., J. Phys. Chem., 85, 2359-2363, 1981.

Kok, G.L., Measurements of hydrogen peroxide in rainwater, Atmos. Env. 14, 653-656, 1980.

Kok, G.L., The production of hydrogen peroxide by lightning, EOS, Transaction of the Amer. Geophys. Union, 63, 891, 182.

Lazrus, A.L., P.L. Haagenson, G.L. Kok, B.J. Huebert, C.W. Kreitzberg, G.E. Likens, V.A. Mohnen, W.E. Wilson, and J.W. Winchester, Atmos. Env. 17, 581-591, 1983.

Lee, Y.N., and S.E. Schwartz, Kinetics of oxidation of aqueous sulfur (IV) by nitrogen dioxide, Proceedings of the 4th International Conference on Precipitation Scavenging, Dry Deposition, and Resuspension, Elsevier Press, in press, 1983.

Levine, S.Z., and S.E. Schwartz, In-cloud and below-cloud scavenging of nitric acid vapor, Atmos. Env., 16, 1725-1734, 1982.

S.C. Liu, M. McFarland, D. Kley, O. Zafiriou, and B. Huebert, Tropospheric NO_x and O_3 budgets in the equatorial Pacific, J. Geophys. Res., 88, 1360-1368, 1983.

Loeb, L.B., Electrical Coronas. Their Basic Physical Mechanics, Univ. of California Press, Berkeley, Ca., 694 pp, 1965.

Logan, J.A., M.J. Prather, S.C. Wofsy, and M.B. McElroy, Tropospheric chemistry: A global perspective, J. Geophys. Res. 86, 7210-7254, 1981.

Maahs, H.G., The importance of ozone in the oxidation of sulfur dioxide in nonurban tropospheric clouds, J. Geophys. Res, in press, 1983.

Markovic, V. and K. Sehested, Radiolysis of aqueous solutions of some simple compounds containing aldehyde groups. Part I: Formaldehyde, in Proceedings of the Third Tihany Symposium on radiation chemistry, edited by J. Dobos and P. Hedvig, Akademai Kiado, Budapest, 1243-1253, 1972.

Maroulis, P.J., A.L. Torres, A.B. Goldberg, and A.R. Bandy, Atmospheric SO_2 Measurements on Project GAMETAG, J. Geophys. Res., 85, 7345-7349, 1980.

Martin, L.R., Kinetic studies of sulfite oxidation in aqueous solution, in Acid Precipitation, ed. by J.G. Calvert, Ann Arbor Science, Ann Arbor, Michigan, 1983.

Maruthamuthu, P. and P. Neta, Phosphate radicals, spectra, acid-base equilibria and reactions with inorganic compounds, J. Phys. Chem., 82, 711-713, 1978.

Mason, B.J., The Physics of Clouds, Oxford University Press, London, 481 pp, 1957.

McCartney, E.J., Optics of the Atmosphere, Scattering by Molecules and Particles, John Wiley, New York, 1976.

McFarland, M., D. Kley, J.W. Drummond, A.L. Schmeltekopf, and R.H. Winkler, Nitric oxide measurements in the equatorial pacific region, Geophys. Res. Letters, 6, 605-608, 1979.

Nack, M.L. and A.E.S. Green, Influence of Clouds, Haze, and Smog on the Middle Ultraviolet Reaching the Ground, Applied Optics, 13, 2405-2415, 1974.

Nenadovic, M.T., Z.D. Draganic, I.G. Draganic and B. Kidric, Radiolysis of formic acid - oxygen solutions at pH 1.3-13 and the yields of primary products in -radiolysis of heavy water, in Proceedings of the Third Tihany Symposium on Radiation Chemistry, edited by J. Dobos and P. Hedvig, Akademai Kiado, Budapest, 1269-1280, 1972.

Oden, S., The acidity problem - An outline of concepts, Water, Air, Soil Pollution, 6, 137-166, 1976.

Parungo, F., C. Nagamoto, I. Nolt, M. Dias, and E. Nickerson, Chemical analysis of cloud water collected over Hawaii, J. Geophys. Res., 87, 8805-8810, 1982.

Penkett, S.A., B.M.R. Jones, and A.E.J. Eggleton, A study of SO_2 oxidation in stored rainwater samples, Atmospheric Environment, 13, 139-147, 1979.

Petranchuk, O.P., and V.M. Drozdova, On the chemical composition of cloud water, Tellus, 28, 280-285, 1966.

Pruppacher, H.R. and J.D. Klett, Microphysics of Clouds and Precipitation, D. Reidel Publishing Co., Boston, MA. 714 pp., 1978.

Rasmussen, R.A., and M.A.K. Khalil, Atmospheric methane (CH_4): trends and seasonal cycles, J. Geophys. Res., 86, 9826-9832, 1981.

Rudolph, J., and D.H. Ehhalt, Measurements of C_2-C_5 hydrocarbons over the North Atlantic, J. Geophys. Res., 86, 959-964, 1981.

Ryan, R.T., H.H. Blau, P.C. von Thuna, and M.L. Cohen, Cloud microstructure as determined by an optical cloud particle spectrometer, J. Appl. Meteor., 11, 149-156, 1972.

Scholz, T.G., D.H. Ehhalt, L.E. Heidt, and E.A. Martell, Water vapor, molecular hydrogen, methane, and tritium concentrations near the stratopause, J. Geophys. Res., 75, 3049-3054, 1970.

Schwartz, S.E., and J.F. Freiberg, Mass-transport limitation to the role of reaction of gases in liquid droplets: Application to oxidation of SO_2 in aqueous solutions, Atmos. Env., 15, 1129-1144, 1981.

Schwartz, S.E., and P.H. Daum, M.R. Hjelmfelt, and L. Newman, Cloudwater acidity measurements and formation mechanisms - Experimental design, Proceedings of the 4th International Conference on Precipitation Scavenging, Dry Deposition, and Resuspension, Elsevier Press, 1983.

Schwartz, S.E., Gas- and aqueous-phase chemistry of HO_2 in liquid-water clouds, Presented at the meeting of the American Chemical Society, Division of Environmental Chemistry, Washington, D.C., September, 1983.

Scott, W.D. and P.V. Hobbs, The formation of sulphate in water droplets, J. Atmospheric Sci., 24, 54-57, 1967.

Spinhurne, J.D., and A.E.S. Green, Calculations of the relative influence of cloud layers on received ultraviolet and integrated solar radiation, Atmospheric Environment, 12, 2449-2454, 1978.

Whitten, G.Z., H. Hogo, and J.P. Killus, The carbon-bond mechanism: A condensed kinetic mechanism, Env. Sci. and Techn., 14, 690-700, 1980.

Zika, R., E. Saltzman, W.L. Chameides, and D.D. Davis, H_2O_2 Levels in Rainwater Collected in South Florida and the Bahama Islands, J. Geophys. Res., 87.

MASS-TRANSPORT CONSIDERATIONS PERTINENT TO AQUEOUS PHASE REACTIONS OF GASES IN LIQUID-WATER CLOUDS

Stephen E. Schwartz
Environmental Chemistry Division
Department of Applied Science
Brookhaven National Laboratory
Upton, New York 11973
U.S.A.

ABSTRACT.

Reactions of gases in liquid-water clouds are potentially important in the transformation of atmospheric pollutants affecting their transport in the atmosphere and subsequent removal and deposition to the surface. Such processes consist of the following sequence of steps: Mass-transport of the reagent gas or gases to the air-water interface; transfer across the interface and establishment of solubility equilibria locally at the interface; mass-transport of the dissolved gas or gases within the aqueous phase; aqueous-phase chemical reaction(s); mass-transport of reaction product(s) and possible subsequent evolution into the gas-phase. Description of the rate of the overall process requires identification of the rate-limiting step (or steps) and evaluation of the rate of such step(s). Identification of the rate-limiting step may be achieved by evaluation and comparison of the characteristic times pertinent to the several processes and may be readily carried out by methods outlined herein, for known or assumed reagent concentrations, drop size, and fundamental constants as follows: gas- and aqueous-phase diffusion coefficients; Henry's law coefficient and other pertinent equilibrium constants; interfacial mass-transfer accommodation coefficient; aqueous-phase reaction rate constants(s). A graphical method is described whereby it may be ascertained whether a given reaction is controlled solely by reagent solubility and intrinsic chemical kinetic or is mass-transport limited by one or another of the above processes. In the absence of mass-transport limitation, reaction rates may be evaluated uniformly for the entire liquid-water content of the cloud using equilibrium reagent concentrations. In contrast, where appreciable mass-transport limitation is indicated, evaluation of the overall rate requires knowledge of and integration over the drop-size distribution characterizing the cloud.

NATO ASI Series, Vol. G6
Chemistry of Multiphase Atmospheric Systems
Edited by W. Jaeschke

1. INTRODUCTION

A major emphasis of current atmospheric research is directed to acquiring an understanding of the processes giving rise to the composition of liquid-water clouds. Such research is motivated by the desire to understand the role of clouds in the cycling of atmospheric constituents and more specifically the role played by clouds in the delivery of soluble substances to the surface in precipitation.

The present paper focuses on the reactive uptake of gaseous substances by liquid-water clouds. Such reactive uptake has been implicated in field studies of the chemical composition of clouds (Lazrus et al., 1983; Daum et al., 1984) that have shown higher concentrations of aqueous sulfuric and nitric acids than can be accounted for by dissolution of soluble particulate matter present in clear air. The importance of in-cloud reaction of sulfur and nitrogen oxides is suggested as well on thermochemical and chemical-kinetic grounds (e.g., Schwartz, 1984a; Martin, 1984). Moreover, clouds, by their very nature, i.e., consisting of highly dispersed small droplets suspended in air, would appear to be an ideal reaction medium for promoting interphase mixing necessary for gas-aqueous reactions. These observations suggest that rates of uptake of gases by cloud droplets and of aqueous-phase reaction can be evaluated on the basis of thermochemical and chemical-kinetic considerations, provided appropriate account is taken of the rates of mass-transport processes within each of the two phases and at the interface. The objective of this paper is to outline methods for carrying out such evaluations, with emphasis on the mass-transport processes. These methods facilitate evaluation of the rate of aqueous-phase reactions in clouds and in turn of the coupling of the chemical kinetics in the two phases, suitable for incorporation in numerical models.

In addition to development of such methodology, considerable insight may be gained regarding rate-controlling processes by defining and comparing the characteristic times of the several processes. As it turns out, frequently one or another process is rate controlling so that the rate of the overall process may, to good approximation, be treated as the rate of the controlling process, thereby simplifying the analysis. In particular, if mass-transport processes are sufficiently rapid compared to aqueous-phase reactions that mass-transport limitation is negligible, then the rate of the overall process is governed entirely by chemical kinetics. In this limit the rate of reactive uptake of the gas becomes independent of the drop-size distribution characterizing the cloud, leading to substantial simplification of the description. At the other extreme, for very rapid reactions, the rate of reactive uptake is controlled entirely by the rate of gas-phase mass transport, which is strongly dependent on cloud microphysical properties but only weakly dependent on the identity of the transported species. In order to facilitate identification of such situations, we develop readily applicable criteria to ascertain whether these situations obtain or conversely whether the more complete treatment is required.

In addition to the application of the methods presented here to description of atmospheric systems, it should be noted that these methods are applicable as well to interpretation of laboratory studies of reactions of gases with spherical drops. In particular, considerations such as the dependence of the observed reaction rate on drop size may indicate the presence and nature of mass-transport limitation in the system under investigation.

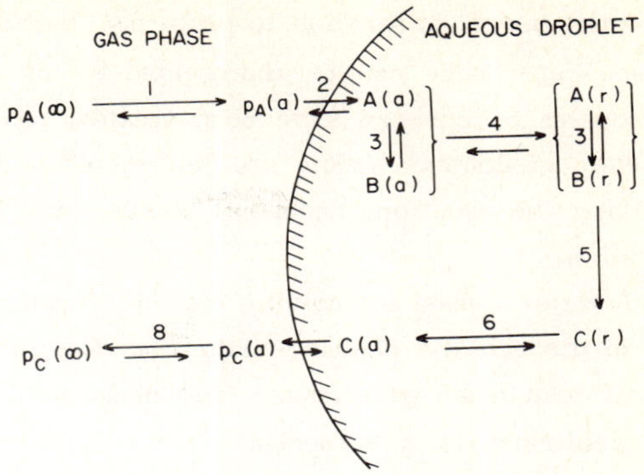

Figure 1. Schematic illustration of the sub-processes comprising the overall gas-aqueous reaction process under examination in this paper. Numbers labelling the several processes correspond to the listing presented in the text. A represents aqueous phase reagent species transferred from the gas phase; B, species in rapid equilibrium with A; and C, product species, at the surface of the drop (a) or in the interior (r); p_A and p_C represent gas-phase partial pressures of A and C at the surface of the drop (a) and at large distances from the drop (∞).

2. STATEMENT OF THE PROBLEM

In order to describe the overall process of a gas-aqueous reaction in a cloud, it is useful to consider this process as consisting of a sequence of steps or sub-processes. By achieving a description of the several sub-processes it becomes possible to describe the overall process. For this purpose we describe the overall process as consisting of the following sub-processes:

1. Gas-phase mass transport of the reagent gas (or gases) to the air-water interface.
2. Mass transport across the interface and possible establishment of solubility equilibria locally at the interface.
3. Establishment of rapid aqueous-phase equilibria involving the dissolved gas(es), e.g., hydration and acid-dissociation equilibria.
4. Aqueous-phase mass transport of the dissolved species within the droplet.
5. Aqueous-phase chemical reaction(s).
6. Mass transport of reaction product(s) in the aqueous phase.
7. Possible evolution of volatile product species into the gas phase.
8. Subsequent gas-phase mass transport of the evolved gaseous product.

These processes are illustrated schematically in Figure 1. The objective of this paper will be to describe these several processes and the coupling of these processes to constitute the overall process. The approach will be by means of description of the space and time dependence of species concentrations in terms of partial differential equations. We further restrict our consideration to problems characterized by spherical symmetry, in which the space dependence can be represented by the single radius variable r. Since most of the processes of concern have been treated, separately, by a variety of investigators in a variety of contexts (many are "classical" textbook examples), we make no apology for not solving the equations here, but merely state results (often by a figure) with appropriate citation.

A major reason for solving the time-dependent (i.e., partial) differential equation system is to discern the characteristic time for relaxation of transients. These "characteristic times" can often be surmised from dimensional considerations, although exact specification is arbitrary to a numerical factor of order unity. At times long compared to the characteristic time for a given process, the system reaches a "steady state" in which species concentrations, fluxes, reaction rates, etc., no longer vary as a function of time. In this steady-state limit, species concentrations are represented by an ordinary differential equation in the radius variable r. Treatment in the steady-state limit greatly simplifies the description of the system and, additionally, provides much insight.

For the system of interest here, which consists of coupled sub-processes, approach to the steady state will be governed by the slowest of the several processes. However, for sub-processes having greatly differing characteristic times, the simplification afforded by the steady-state treatment nonetheless obtains, since the "faster" system(s) will rapidly adjust to changes imposed by the time variation of the "slower" system, i.e., they will "follow" the slower system. In practice, for cloud systems (in contrast to some laboratory studies) such quasi-steady-state situations will almost invariably be the rule, as the cloud itself relaxes slowly to externally imposed conditions, e.g., adiabatic cooling imposed by lifting that leads to condensational growth of cloud droplets.

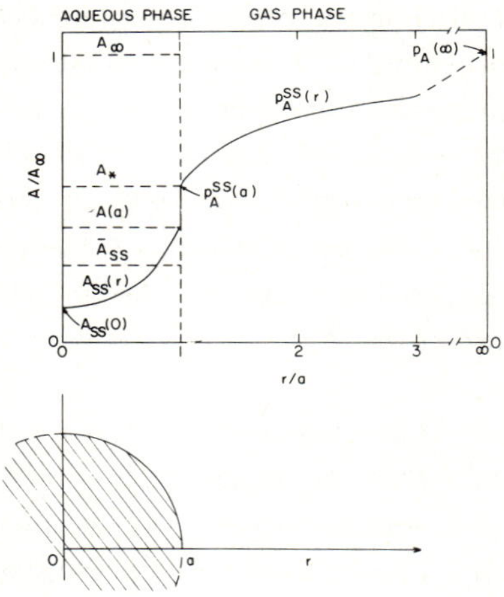

Figure 2. Hypothetical concentration profiles in gas and aqueous phases indicating gradient in reagent concentration due to flux of material into and within drop. Concentration scales of aqueous phase (r < $\underline{a}$) left ordinate and gas-phase (r > $\underline{a}$) right ordinate are chosen so that the same coordinate on each scale represents the condition of phase equilibrium. Departure from the uniform profile at the "bulk" (r = ∞) value represents the inability of mass transport to maintain the reagent concentration as the reagent is consumed by aqueous-phase reaction. After Schwartz and Freiberg (1981).

A schematic illustration of the spatial dependence within and in the vicinity of a single drop of the steady-state concentration of a gas undergoing aqueous-phase reaction is shown in Figure 2. Here the region r > $\underline{a}$ represents the gas phase surrounding the drop and the region r < $\underline{a}$

represents the interior of the drop. The scales representing aqueous concentration (left ordinate) and gas-phase partial pressure (right ordinate) are chosen such that equilibrium at the interface is represented by the same coordinate on each scale. For reactive uptake of the gaseous reagent negative departure from the uniform profile (at a value corresponding to the large-radius or "bulk" concentration) may occur in either or both of the two phases and/or at the interface because of the inability of mass-transport to maintain the uniform profile in the face of the aqueous reaction sink. For reaction higher than zeroth order in the dissolved reagent (i.e., essentially all cases of interest), this negative departure from a uniform profile results in a decrease in the reaction rate from that characteristic of a uniform profile, i.e., there is mass-transport limitation to the rate of reactive uptake of gas.

A second motivation for introducing the characteristic times of the several processes is that by appropriately comparing these quantities it is possible to discern whether the system, at steady state, is mass-transport limited and to identify the limiting process, or, alternatively, to establish that the rate of the overall process is governed by aqueous-phase chemical kinetics. This approach was introduced in earlier papers from the author's laboratory (Schwartz and Freiberg, 1981; Freiberg and Schwartz, 1981; Schwartz, 1984a) that serve as the basis for much of the present paper.

The foregoing discussion has outlined what may be referred to as the microscale problem, determination of the time and space dependence of concentrations, fluxes, and reaction rates within and in the vicinity of a single drop. In order to describe the rates of these processes in actual or model clouds (the "macro-scale problem") it is necessary as well to take into account pertinent physical properties of the cloud, most importantly the liquid water content and the drop-size distribution. By considering the droplets that comprise the cloud to be non-interacting it is possible to adduce a description of the coupling of the chemical kinetics in the two phases. The importance of this coupling has been pointed out in recent work of Chameides and Davis (1982, 1983); see also the review by Chameides (1985) in this volume. Description of chemical kinetics in the coupled two-phase system is achieved by means of a set of ordinary differential equations in species concentrations in the two phases. The transfer of material between phases is treated by a formalism involving phenomenological mass-transfer rate coefficients whose definition takes into account the pertinent cloud properties. The resulting system of ordinary differential equations may, under certain conditions, be treated entirely like systems of coupled differential equations for a single-phase situation that are conventionally employed in the description of gas-phase or aqueous-phase chemical kinetics.

The next several sections treat the individual sub-processes that comprise the overall gas-aqueous reaction pertinent to liquid-water clouds. First, we set the scene with a brief description of the properties of liquid-water clouds. The next section treats gas-aqueous equilibria, in particular the Henry's law equilibrium between a gaseous species and its dissolved counterpart. In Section 5 we define several expressions for chemical kinetic rates in mixed phase systems generally as well as in clouds specifically and define the relevant characteristic times. In Section 6 we examine the mass-transport processes pertinent to gas-aqueous reactions in clouds, treating gas-phase and aqueous-phase mass-transport mechanisms as well as the mechanisms of transport across the interface. Section 7 examines the kinetics of these mass-transport processes and develops expressions for evaluation of mass-transport limitation. In Section 8 these results are applied to develop a formalism for describing coupled gas-aqueous reaction systems in clouds.

3. CLOUD MICROPHYSICAL PROPERTIES

We confine this discussion to a bare-bones description of physical properties of liquid-water clouds pertinent to description of gas-aqueous reactions. For a more complete description the reader is referred to appropriate monographs (Pruppacher and Klett, 1978, pp. 9-21; Rogers, 1979, pp. 87-29; Mason, 1971, pp. 92-121).

Liquid Water Content.

The liquid water content is a key property of a cloud, governing the partition of a gas between gaseous and aqueous phases. The liquid water content of clouds ranges from one-tenth to several cm^3/m^3, typically 0.3 to 1 cm^3/m^3, or 0.3 to 1 parts per million by volume (Clouds are mostly air!). In description of partition of material and of gas-aqueous kinetics, it is convenient to express the liquid water content as a dimensionless volume fraction, L, typically (0.3 to 1) x 10^{-6}.

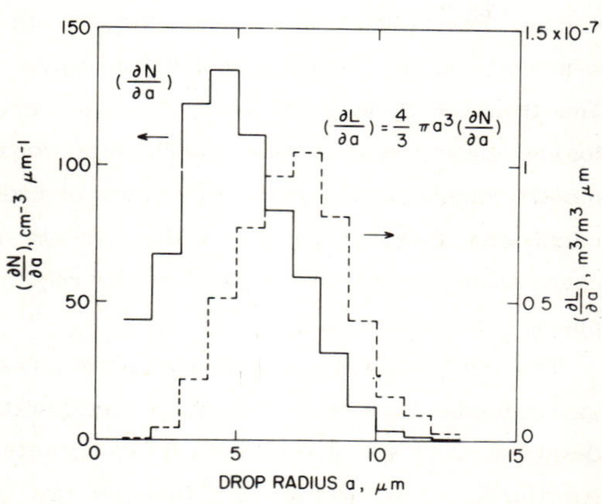

Figure 3. Spectra of droplet number density and volume fraction for continental cumulus. N = 700 cm^{-3}; L = 0.51 x 10^{-6}. Data of Knollenberg (1981).

Drop-Size Distribution.

The distribution of drop sizes within a cloud exerts a major influence on the kinetics of mass-transport processes within clouds. Observed cloud-droplet size distributions dN/da are typically rather narrow, peaking at some low value (below 8 μm diameter) and decreasing rather strongly with increasing size. (Drop sizes are most commonly referred to by diameter, but algebraic expressions, spectra, etc., are frequently given as a function of radius a, a convention we retain here.) Examples of measured cloud droplet size distributions are presented in Figures 3 and 4. For other examples see, e.g., Pruppacher and Klett (1978, pp. 10-21) and Mason (1971, pp. 98-113). The total number of drops per unit volume is given as the integral over the number distribution,

$$N = \int (\frac{dN}{da})\, da .$$

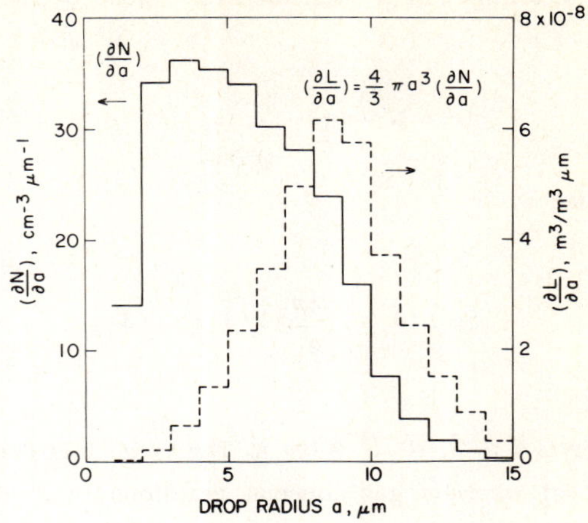

Figure 4. Spectra of droplet number density and volume fraction for aged stratus. $N = 300\ cm^{-3}$; $L = 0.34 \times 10^{-6}$. Data of Knollenberg (1981).

In addition to the number density distribution it is also useful, particularly in the context of aqueous-phase reactions, to examine the distribution of liquid water content as a function of drop size,

$$\frac{dL}{da} = \frac{4}{3}\pi a^3 \frac{dN}{da} ,$$

also shown in Figures 3 and 4. Because of the greater volume of the larger droplets, the distribution of liquid-water content peaks at diameter values substantially greater than the number distribution. The liquid-water content is given as the integral over the partial volume distribution,

$$L = \int (\frac{dL}{da})\, da = \frac{4}{3}\pi \int a^3 (\frac{dN}{da})\, da .$$

Examination of representative cloud drop spectra permits us to bound the size range of drops concern in examination of gas-aqueous reactions in clouds. We take this range to be roughly 1-30 μm radius or 2-60 μm diameter.

It is of interest to consider the mean separation of cloud droplets $N^{-1/3}$, relative to the dimension of these droplets, to examine the adequacy of treating cloud droplets as non-interacting. For a "monodisperse" cloud of radius a

$$L = N \frac{4}{3}\pi a^3$$

whence

$$N^{-1/3} = \left(\frac{4\pi}{3}\right)^{1/3} L^{-1/3} a = 1.61 \; L^{-1/3} a.$$

For $L \approx 10^{-6}$, $N^{-1/3} \approx 10^2$ a. The large interdrop spacing relative to drop size supports the model whereby gas-aqueous reactions in a cloud are treated as the summation of the processes taking place in the vicinity of and within non-interacting drops.

It is often convenient to represent the cloud drop-size distribution by an algebraic expression. One convenient such representation is that of Khrigian and Mazin (see Pruppacher and Klett, 1978, pp. 11),

$$\frac{dN}{da} = A \; a^2 \; e^{-Ba}$$

where A and B are parameters that may be adjusted to match the mean drop radius $\bar{a}$ and liquid water volume fraction L, viz.,

$$B = 3/\bar{a}, \qquad \text{and}$$

$$A = \left(\frac{3}{4\pi}\right) \frac{3^6}{5!} \; L \; \bar{a}^{-6} \approx 1.45 \; L \; \bar{a}^{-6} \; .$$

For this distribution

$$N = \left(\frac{3}{4\pi}\right) \frac{2 \cdot 3^3}{5!} \; L \; \bar{a}^{-3} \approx 0.107 \; L \; \bar{a}^{-3} \; .$$

For $L = 0.5 \times 10^{-6}$ and $\bar{a} = 5 \times 10^{-4}$ cm, $N = 430$ cm^{-3}. Model distributions such as this are useful for the insight that may be gained in modeling gas-aqueous reactions in clouds.

Sedimentation Velocities.

The gravitational sedimentation of cloud droplets is important in the present context because of the convective flows induced thereby and the possible resultant enhanced mixing rates in either the gas or aqueous phase. This sedimentation, of course, is ultimately responsible for collisional growth of droplets and the delivery of precipitation to the surface. The focus here, however, is on the implications for mass-transport.

A first approximation to the terminal sedimentation velocity of cloud droplets is given by Stokes law for flow of a sphere in a viscous medium (e.g., Pruppacher and Klett, 1978, p. 322),

$$U_\infty = \frac{2\ a^2\ g\ (\rho_w - \rho_a)}{9\ \eta_a} \quad \frac{2\ a^2\ g\ \rho_w}{9\ \eta_a} \ . \tag{3.1}$$

Here g is the gravitational constant, ρ_w and ρ_a are the densities of water and air, respectively, and η_a is the viscosity of air. Departures from the velocity are found at both ends of the drop-size range of concern here (Pruppacher and Klett, loc. cit.). At the low end an increase in velocity is caused by the departure from pure continuum flow; under atmospheric conditions this departure amounts to 17% for $\underline{a}$ = 1 um and 1% for $\underline{a}$ = 10 um. At the high end of the drop-size range the Stokes law overestimates actual sedimentation velocities as drag forces on the drop are increasingly influenced by the drop velocity; this departure amounts to -6% for $\underline{a}$ = 30 μm and -13% for $\underline{a}$ = 42 μm. Nonetheless, at least for first approximation the Stokes formula may be employed for evaluation of sedimentation velocities of cloud drops.

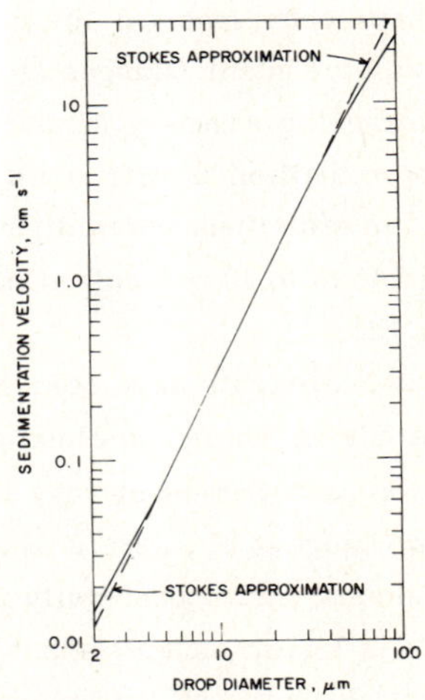

Figure 5. Droplet gravitational sedimentation velocity as a function of diameter. Departure from Stokes formula is indicated at both ends of the diameter range. Evaluated for 10 °C.

In evaluation of sedimentation velocities it is necessary to employ the viscosity of air. A convenient approximation is given by Pruppacher and Klett (1978, p. 323).

$$\eta_a(\text{c.g.s.}) = (1.72 + 0.0049\ T) \times 10^{-4},\ T\ (^\circ\text{C})\ .$$

The sedimentation velocity as a function of drop radius is shown in Figure 5.

Cloud Composition.

Measurements of the concentrations of dissolved constituents in cloudwater indicate, not surprisingly, a similarity in composition to that of rain (Daum et al., 1984, and references to earlier work therein). Principal cations are H^+ and NH_4^+; principal anions are $SO_4^=$ and NO_3^-, the distributions varying with location and altitude, as well as with time at any given location. Concentrations range from high values $\sim 10^{-3}$ equivalents/liter to substantially lower concentrations in remote locations. There is some indication of an inverse variation of concentration with liquid water content suggestive of dilution or concentration of non-volatile solutes associated with condensation or evaporation of water, respectively. Considerably higher concentrations of dissolved constituents (10^{-2} equivalents/liter or more) have been reported for ground fogs that are evidently characterized by rather low liquid water content (Munger et al., 1983).

The foregoing summary supports the approach adopted here of treating cloud water in the first approximition as sufficiently dilute to be treated as ideal solutions. However, it should be noted in situations where high accuracy is necessary that departures from ideality become appreciable (5 to 10 percent) at concentrations greater than 10^{-3} equivalents/liter (Adamson, 1973, p. 521).

Few measurements have been reported of concentrations of trace gases in interstitial air, i.e., the air in which the cloud drops are suspended. Evidently highly soluble gases (e.g., NH_3, HNO_3) are present at very low concentration consistent with rapid scavenging by cloud droplets (Daum et al., 1984). Gases of low solubility (e.g., O_3, NO_2) are commonly present within clouds, and occasionally appreceable concentrations of SO_2 are encountered.

Little is known as well about the concentration and composition of interstitial aerosol, i.e., particles present within clouds that have not been activated to become cloud drops. Several recent studies suggest that most but not all (by mass) of soluble aerosols present prior to cloud formation is rapidly (seconds to minutes) incorporated into cloud droplets by nucleation (Leaitch et al., 1983; Radke, 1983; Hegg and Hobbs, 1983). Measurements of the composition of the interstitial aerosol and comparison with corresponding cloudwater composition support this supposition (Daum et al., 1984). Presumably the non-nucleated interstitial aerosol is concentrated in the smaller size range, for which supersaturation sufficiently great to achieve nucleation has not been attained.

4. GAS-AQUEOUS EQUILIBRIA

A major consideration in describing mass transport and chemical kinetics of gas-liquid reactions is the equilibrium distribution of the reagent gas or gases between the two phases. For a detailed discussion of such equilibria pertinent to atmospheric gases, see the review by Warneck (1985) in this volume. For dilute solutions the distribution of a gas between the two phases is given by Henry's law, which states that the equilibrium concentration of a gas dissolved in a liquid is proportional to the partial pressure of the gas,

$$[X(aq)] = H_X p_X, \tag{4.1}$$

where H_X is the so-called Henry's law coefficient. A variety of units have been and continue to be employed for these several quantities. A particularly convenient set of units is p_X in atmospheres and [X(aq)] in molar, consistent with the thermodynamic standard states chosen as 1 atm and 1 M, respectively, and H_X, therefore, in units of M atm^{-1}. H_X is thus seen to be the equilibrium constant of the reaction

$$X(g) = X(aq)$$

and is related to the difference in standard free energies of formation of the gaseous and aquated species as

$$H_X = \exp(-\Delta G^\circ/\underset{\sim}{R}T)$$

where

$$\Delta G^\circ = \Delta_f G^\circ[X(aq)] - \Delta_f G^\circ[X(g)] \ .$$

In the absence of specific chemical forces, Henry's law coefficients are fairly well correlated with the physical properties of solute and solvent. However, for polar, protic solvents (e.g., water) chemical forces frequently dominate, and Henry's law coefficients can differ greately for "similar" solutes. Examples of Henry's law coefficients of several atmospheric gases are given in Table 1. Henry's law coefficients generally increase with decreasing temperature, reflecting a negative enthalpy of solution.

In the case of gases which physically dissolve in water without further reaction the Henry's law coefficient may be measured directly, by allowing the system to reach equilibrium and measuring p_X and [X(aq)]. Similarly, in the case of a gas that establishes a solution-phase chemical equilibrium, e.g.,

$$SO_2(aq) = H^+ + HSO_3^-$$

$$HSO_3^- = H^+ + SO_3^=$$

the Henry's law coefficient may be deter mined from consideration of the several equilibria (e.g., Johnstone and Leppla, 1934). However, in the case of a gas that reacts rapidly and irreversibly to form much more highly soluble products, e.g., NO_2 to form NO_2^- and NO_3^-, the Henry's law coefficient can generally be determined only by indirect methods. Such methods include thermochemical cycles involving the aqueous-phase species (e.g., Schwartz

Table 1

Henry's Law Coefficients of Some Atmospheric Gases Dissolving in Liquid Water[a]

Gas	H, $M\ atm^{-1}$	Reference
O_2	1.3(-3)[b]	1
NO	1.9(-3)	1
C_2H_4	4.9(-3)	1
NO_2[c]	1 (-2)	2
O_3	1.3(-2)	3
N_2O	2.5(-2)	1
CO_2[d]	3.4(-2)	1
SO_2[d]	1.3	4
CH_3ONO_2[e]	2.6	5
PAN[c,e]	3.6	5
HNO_2	4.9(1)	6
NH_3[d]	6.2(1)	7
H_2CO	6.3(3)	8
H_2O_2	1 (5)	9
HNO_3[d]	2.1(5)	6

a. T = 25°C except as noted.
b. The notation 1.3(-3) represents 1.3×10^{-3}.
c. Physical solubility; reacts with liquid water.
d. Physical solubility, i.e., exclusive of acid-base equilibria.
e. T = 22°C.

References: 1, Loomis (1928); 2, Schwartz and White (1983); 3, Briner and Perrottet (1939); 4, Johnstone and Leppla (1934); 5, Lee et al. (1983); 6, Schwartz and White (1981); 7, van Krevelen et al. (1949); 8, Blair and Ledbury (1925); 9, Martin and Damschen (1981).

and White, 1981; Schwartz, 1984b) and kinetic studies of the reactive uptake (e.g., Schwartz and White, 1983).

In considering the solubility of gases that undergo rapid reversible aqueous-phase reactions such as acid-base ionization equilibria, it is frequently convenient to extend the definition of Henry's law by defining a pseudo Henry's law coefficient that encompasses the totality of the dissolved species. For example, in the case of SO_2 we define

$$H^*_{S(IV)} \equiv [S(IV)]/p_{SO_2}$$

$$= ([SO_2(aq)] + [HSO_3^-] + [SO_3^=])/p_{SO_2}$$

$$= H_{SO_2}\left(1 + \frac{K_1}{[H^+]} + \frac{K_1K_2}{[H^+]^2}\right) ,$$

where K_1 and K_2 are the first and second ionization constants of sulfurous acid. It is seen that $H^*_{S(IV)}$ so defined depends on the hydrogen ion concentration, and thus hat a linear relation between [S(IV)] and p_{SO_2} will obtain only for a solution that is adequately buffered compared to the incremental hydrogen ion contributed by the acid dissociation. Examples of pseudo Henry's law coefficient for some acidic and basic gases of atmospheric interest are shown as a function of solution pH in Figure 6. It is seen that the total amount of dissolved material may greatly exceed that dissolved by Henry's law alone, i.e., that H^* may greatly exceed H.

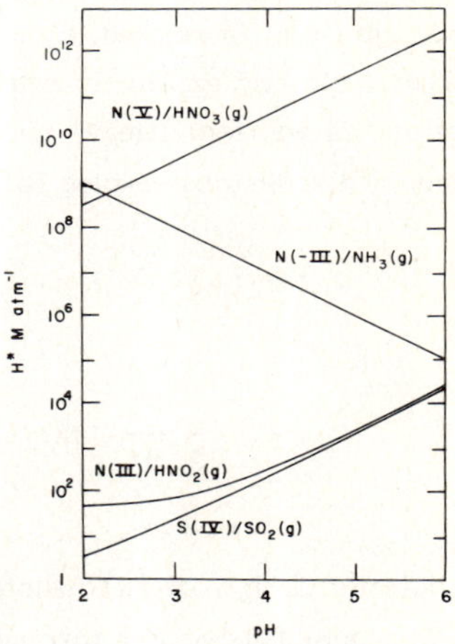

Figure 6. Effective Henry's law coefficients for gases which undergo rapid acid-base dissociation reactions in aqueous solution, as a function of solution pH. Buffer capacity of solution is assumed to greatly exceed incremental concentration from uptake of indicated gas. From Schwartz (1984a).

The equilibrium ratio of concentration of a species in the gaseous and aqueous phases is of particular importance in mass-transport considerations. This ratio is given according to Henry's law as

$$\frac{[X(aq)]}{[X(g)]} = \frac{H_X p_X}{p_X/\underset{\sim}{R}T} = H_X\underset{\sim}{R}T , \qquad (4.2)$$

where we have made use of the ideal gas law. For H_X in M atm^{-1} the universal gas constant R is taken as 0.082 atm M^{-1} K^{-1}, and $\underset{\sim}{R}T \approx 25$ atm M^{-1}. For a gas undergoing rapid aqueous-phase equilibrium, H_X must be replaced by H_X^*.

5. GAS-AQUEOUS CHEMICAL KINETICS

In view of the multi-phase nature of the system and of potential spatial nonuniformities, it is necessary to specify expressions for rates of chemical reactions with some precision. We restrict this discussion to reactions which occur in the aqueous solution phase. However, it is useful to express the reaction rate either in aqueous-phase units (M s^{-1}) or, for comparison to gas-phase rates, in equivalent gas-phase units (atm s^{-1}).

The local instantaneous aqueous-phase reaction rate R(a,r,t) is, in principle, a function of drop radius a, radius within the drop r, and for non-steady-state problems, of time t. It is this rate that is related, by a chemical rate law, to the local concentrations of reagents, catalysts, etc. we explicitly employ the chemical kineticist's convention of rate--moles (of reaction, as written) liter^{-1} second^{-1}, as distinct from that frequently employed in mass-transport discussions--moles (of reagent of interest) liter^{-1} second^{-1}. Thus for the reaction

$$2\ NO_2(aq) + H_2O \longrightarrow 2\ H^+ + NO_2^- + NO_3^- , \qquad [1]$$

$$R_1 = -\frac{1}{2}\frac{d_1}{dt}[NO_2(aq)] = \frac{d_1}{dt}[NO_3^-] .$$

The notation $\pm\, d_1 / dt\, [X]$ denotes the rate of increase/decrease of [X] due to reaction [1], i.e., the local flux of [X] through [1]. This would differ from the net rate of change in [X] if other processes were occurring simultaneously, e.g., other reactions or mass-transport processes. Lee and Schwartz (1981a) report a rate law for [1],

$$R_1 = k_1\ [NO_2(aq)]^2$$

by which they mean that the flux of $NO_2(aq)$ through [1] is

$$\frac{d_1}{dt}[NO_2] = -2\ R_1 = -2\ k_1\ [NO_2(aq)]^2 .$$

For a reaction at steady state we may drop the explicit dependence on t and write R(a,r) or occasionally $R_{SS}(a,r)$.

The average rate over the droplet volume is given as

$$\bar{R}(a) = (\frac{4}{3}\pi a^3)^{-1} \int_0^a 4\pi r^2 \, R(a,r) \, dr \; . \tag{5.1}$$

For a cloud characterized by a distribution of drop radii, the average rate, on a liquid water basis, is

$$\langle R\rangle = L^{-1} \int (\frac{4}{3}\pi a^3) \, (\frac{dN}{da}) \, \bar{R}(a) \, da \; . \tag{5.2}$$

For a "mono-disperse" cloud $\langle R\rangle = \bar{R}(a)$.

It is often useful to express the reaction rate in partial pressure units, i.e., atmospheres (of reaction, as written) second^{-1}, which is evaluated as

$$(R) = L\underset{\sim}{R}T \, \langle R\rangle . \tag{5.3}$$

Finally we introduce the fractional rate of removal of a gas-phase reagent X, having units s^{-1} (or commonly % h^{-1})

$$\rho_X = \nu_X(R)/p_X \; ; \tag{5.4}$$

here ν_X is the stoichiometric coefficient of X in the reaction as written.

Phase-mixed System.

It is instructive to consider these kinetic expressions in the limit of Henry's law equilibrium being satisfied. Reaction rates in this "phase-mixed limit", which are readily evaluated in terms of gas-phase reagent concentrations by means of the Henry's law solubility equilibria and aqueous-phase kinetic expressions, represent an upper-limit reaction rate, i.e., no mass-transport limitation. We consider initially a first-order reaction

$$R(a,r) = k^{(1)} \, [X(aq)](a,r) \; . \tag{5.5}$$

Here we explicitly note that [X(aq)] in principle depends on a and r; however, by assumption of Henry's law equilibrium,

$$[X(aq)](a,r) = H_X p_X \,, \tag{5.6}$$

the aqueous-phase reagent concentration is independent of a and r, and

$$\langle R \rangle = k^{(1)} H_X p_X \,.$$

In turn,

$$(R) = k^{(1)} L\underset{\sim}{R} T\, H_X\, p_X \,.$$

Finally, the fractional rate of reaction is

$$\rho_X = k^{(1)} L\underset{\sim}{R} T\, H_X \,.$$

Similar expressions obtain for situations other than simple first-order reaction of a single reagent. For example, the reaction of dissolved sulfur(IV) with H_2O_2,

$$S(IV) + H_2O_2 \longrightarrow S(VI) + H_2O$$

is characterized by a pH-dependent effective second-order rate constant:

$$R(a,r) = k^{(2)} [H_2O_2]\, [S(IV)] \,.$$

By assumption of Henry's law equilibrium for both species,

$$\langle R \rangle = k^{(2)} H_{H_2O_2} H^*_{S(IV)}\, p_{H_2O_2}\, p_{SO_2} \,.$$

The fractional reaction rates for the two reagents are

$$\rho_{SO_2} = k^{(2)} H_{H_2O_2} H^*_{S(IV)}\, L\underset{\sim}{R} T\, p_{H_2O_2} \,.$$

and

$$\rho_{H_2O_2} = k^{(2)} H_{H_2O_2} H^*_{S(IV)}\, L\underset{\sim}{R} T\, p_{SO_2} \,.$$

Similarly, in the case of second-order reaction of NO_2,

$$R(a,r) = k^{(2)} [NO_2(aq)]^2, \tag{5.7}$$

$$\langle R \rangle = k^{(2)} H_{NO_2}{}^2 \, p_{NO_2}{}^2 \; , \tag{5.8}$$

and

$$\rho_{NO_2} = 2 \; k^{(2)} \; H_{NO_2}{}^2 \; p_{NO_2} \; ,$$

where the factor of 2 represents the stoichiometric coefficient of NO_2 in [1].

Characteristic Time.

It is useful to define a characteristic time of aqueous-phase reaction that permits examination of mass-transport limitation by comparison to characteristic times of mass-transport processes. This can be defined on dimensional grounds as the ratio of the aqueous-phase concentration of a species X to the steady-state flux of X through reaction,

$$\tau_r = - [X(aq)] / \left(\frac{d_R[X]}{dt}\right)_{ss} = [X(aq)] / \nu_X R_{ss}(a,r) \; . \tag{5.9}$$

For the H_2O_2-S(IV) reaction the effective first-order rate coefficients for reaction of S(IV) and H_2O_2 are

$$k^{(1)}_{S(IV)} = k^{(2)} \; [H_2O_2] \; . \tag{5.10}$$

and

$$k^{(1)}_{H_2O_2} = k^{(2)} \; [SO_2] \; .$$

Similarly, for the NO_2-NO_2 reaction

$$k^{(1)}_{NO_2} = 2 \; k^{(2)} \; [NO_2 \; (aq)] \; . \tag{5.11}$$

These quantities may be expressed in terms of gas-phase partial pressures by means of Henry's law equilibria.

For reaction by simple first-order kinetics,

$$\frac{d_R[X]}{dt} = - k^{(1)}[X]$$

such that

$$\tau_r = [k^{(1)}]^{-1} \quad . \tag{5.12}$$

i.e., the inverse of the first-order rate coefficient. For more complex rate expressions it is useful to define an effective first-order rate coefficient,

$$k^{(1)} = \frac{d_R\,[X]}{dt} \;/\; [X(aq)] \tag{5.13}$$

such that τ_r continues to be represented by eq. (5.12).

6. MASS TRANSPORT PROCESSES

We outline here the pertinent processes responsible for mass transport separately in the gas and aqueous phases and across the gas-aqueous interface.

Gas Phase.

For systems with dimension much greater than the molecular mean free path ($a \gg \lambda$, or Knudsen number $Kn \equiv \lambda/a \ll 1$) mass-transport is considered to be adequately described as that of a continuum fluid, i.e., without explicit recognition of the motion of individual molecules (e.g., Hidy and Brock, 1970, p. 15). In air under standard conditions the mean free path is approximately 6 x 10^{-6} cm (0.06 μm) well less than the radius of cloud particles ($\gtrsim$ 1 μm). Hence the mass transport of gas molecules to (or from) cloud droplets may be treated as a continuum process.

Transport of a reagent gas to a cloud droplet will occur at least at the rate governed by molecular diffusion, as augmented by any organized flows, e.g., turbulent motions of convection induced by the sedimentation velocity of the cloud droplet. Evaluation of the mass-transport rate as governed by molecular diffusion alone, therefore, yields a lower bound to this rate, and if no gas-phase mass-transport limitation is indicated by this rate, that conclusion holds a fortiori in the presence of organized flows. In fact, however, the influence of such flows relative to diffusion would appear to be minimal for the range of drop sizes characteristic of non-precipitating clouds.

To examine the importance of turbulence-enhanced mass transport, Brock and Durham (1984) note that the characteristic scale length of atmospheric turbulence is

$$r_t = (D_g{}^3 / \mathcal{E})^{1/4}$$

where D_g is the gas-phase diffusion coefficient and $\mathcal{E}$ is the turbulent energy dissipation rate. The criterion for the unimportance of turbulent mass transfer relative to diffusive is

$$a \ll r_t.$$

For $D_g \approx 0.1$ cm^2 s^{-1} and $\mathcal{E} = 10$ cm^{-2} s^{-3} (low atmospheric turbulence) or 10^4 cm^2 s^{-3} (intense turbulence), $r_t = 0.18$ and 0.018 cm, respectively, indicating appreciable turbulence-induced mass transport only for the larger drops at intense levels of turbulence. By this criterion we may neglect turbulence-induced mass transport in evaluation of gas-phase mass transport rates.

Convective mass-transport induced by droplet sedimentation is treated by Pruppacher and Klett (1978. p. 440 ff.) in terms of measured evaporation rates of falling droplets. For low Reynold number (Re < 2.5) the ratio of convective to diffusive mass-transport in air is approximated as 0.086 Re. Thus the criterion that convective mass transport not exceed 10% diffusive mass transport is satisfied for Re $\leq$ 1.16. This value of Reynold number correspond to a = 42 um (Pruppacher and Klett, 1978, p.326). Thus again molecular diffusion is seen to represent the major gas-phase mass-transport mechanism for droplets in the size range pertinent to clouds.

The gas-phase mass-transport of SO_2 to water droplets falling freely in air was modeled numerically by Baboolal et al. (1981) taking into account the flow field induced in the gas by the falling drop. Drop radii treated were 10, 71, 142, and 306 um. Significant enhancement of the steady-state uptake rate beyond that due to diffusion in stagnant air was found for all drop radii except 10 um. This enhancement amounted to approximately 50% at a = 71 um. For a = 40 um the enhancement would appear to be about 20% (Baboolal et al., 1981, Figure 2). It would thus appear that the treatment of gas-phase mass transport by diffusion alone yields a close approximation to the uptake rate throughout the size range of interest for liquid-water clouds.

Gas-Phase Diffusion Coefficients.

Measurements are available for the diffusion coefficients of a number of species of atmospheric interest in air, some which are listed in Table 2. (Since diffusion coefficients vary inversely with total pressure it is commonly the product P D_g that is tabulated.) It is seen that values of D_g (at 1 atm) are of the order of 0.1 $cm^2\ s^{-1}$, a value that we shall employ in examples treated below.

For gases whose diffusion coefficient in air has not been measured there are a number of closely related semi-empirical equations by which binary diffusion coefficients (including diffusion coefficients in air) may be estimated from the properties of the individual molecules. (For review see Sherwood et al., 1975, pp. 10-24; Marrero and Mason, 1982.) Lugg (1968) has compared the performance of 9 such equations for up to 147 compounds and found that the formula of Wilke and Lee (1955) yields a value within 5% of the observed value in 70% of the cases. A similar, somewhat easier to apply formula given by Fuller et al. (1966) also appears to yield good predictions for D_g.

The temperature dependence of the diffusion coefficient at 1 atm, at least for relatively narrow ranges of temperature, is given as

$$D_g \propto T^n$$

with values of n ranging from 1.5 to 2 (Chapman and Cowling, 1970, p. 264).

Aqueous Phase.

Mass transport within aqueous cloud droplets will, analogously to gas-phase mass transport, take place by molecular diffusion as enhanced by any organized flow. Such an organized flow may be induced by coupling of momentum from the relative motion of the falling drop within the air medium, and thus is increasingly important for larger drops. Pruppacher and colleagues (cf. Pruppacher and Klett, 1978, pp. 305-310) have examined internal circulation induced by droplet sedimentation and conclude, at least for smaller drops (< 100 um radius), that the Stokes-flow approximation accurately approximates the actual flow field. This approximation yields for the radial velocity within the drop

$$u_{ir} \approx \frac{\eta_a}{2\eta_w} U_\infty \left(\frac{r^2}{a^2} - 1\right) \cos\Theta \qquad (6.1)$$

where U is the terminal fall velocity of the drop, and n_a and n_w are the viscosities of air and water, respectively. Θ is the angle from the flow direction. In turn, from the expression (3.1) for U_∞

$$u_{ir} \approx \frac{g\ a^2\ \rho_w}{9\ \eta_w}\left(\frac{r^2}{a^2} - 1\right) \cos \Theta \ . \tag{6.2}$$

Equation (6.2) may be used to estimate the Peclet number characteristic of the system, which is a dimensionless index of the relative importance of convective to diffusive motion--Pe $\gtrsim$ 1 is indicative of significant convective mass transport (Brock and Durham, 1984)--

Table 2

Diffusion Coefficients of Some Representative Gases in Air

Gas	$P\ D_g$ atm cm^2 s^{-1}	Reference
H_2	0.712	1
He	0.687	2
H_2O	0.264	2
NH_3	0.234	3
CH_4	0.228	1
O_2	0.207	1
CO_2	0.159-0.170	1,2,3
HCOOH	0.153	4
CH_3OH	0.152	4
CH_3CH_2OH	0.118-0.133	2,4
Cl_2	0.128	3
SO_2	0.126	3
CH_3COOH	0.124	4
Br_2	0.094-0.106	3,4
CS_2	0.105	4

References: 1, Chapman and Cowling (1939). Data for 273K scaled to 298K as $T^{1.75}$; 2, Sherwood et al. (1975). Data for 313-317K scaled to 298K as $T^{1.75}$; 3, Andrew (1955). Data for 293K scaled to 298K as $T^{1.75}$; 4, Lugg (1968).

$$Pe = \frac{a\ u_{ir}(max)}{D_a}$$

$$= \frac{g\ \rho_w\ a^3}{9\ \eta_w\ D_a}\ .$$

For a = 1, 10, 40 μm, Pe evaluated in this way has values 10^{-3}, 1, and 30, respectively, suggesting significant convective mass transport for a $\gtrsim$ 10 μm.

One additional point to be noted about equations (6.1) and (6.2) is the dependence of internal velocity on the viscosities of the two media, which reflects the mechanism of coupling of momentum between the two phases. As pointed out by Brock and Durham (1984) use of the bulk viscosity of water in these expressions assumes applicability of this quantity to the interface. This condition might not hold in the presence of a surfactant film (Gill et al., 1983), which would reduce the surface viscosity and thereby decrease the internal mixing from that evaluated using the bulk viscosity.

The mass transport of S(IV) within droplets having internal flow fields characteristic of drops falling freely in the atmosphere has been modeled by Baboolal et al. (1981). Drop radii treated were 10, 71, 142, and 306 μm. Significant enhancement of the aqueous-phase mass-transport rate over that due to diffusion alone as calculated for a stagnant drop was found for all drop radii except 10 μm. This enhancement amounted to approximately 50% at a = 71 um. It is not possible based on the data of the paper to estimate the enhancement for a = 40 um, but it would appear to be substantially less. This treatment would thus suggest that for the present it is appropriate to proceed with analysis of mass-transport rates based on molecular diffusion only, but with recognition that convective enhancement may be appreciable for a > 10 μm.

Aqueous-phase Diffusion Coefficients.

Diffusion coefficients of gases and other solutes in dilute solutions have been the subject of much study (Himmelblau, 1964; Wilke and Chang, 1955). Examples of some measured diffusion coefficients of dissolved gases are given in Table 3, and are seen to be ~2 x 10^{-5} $cm^2\ s^{-1}$ for many gases at 25 C. For simple solutes, i.e., solutes which do not undergo chemical reactions in solution, the semiempirical expression of Wilke and Chang (1955) gives a good estimate of the diffusion coefficient in terms of solute and solvent properties. In particular we note that there are two contributions to the temperature dependence,

Table 3

Diffusion Coefficients of Some Representative Gases in Water

Gas	D_a 10^{-5} cm^2 s^{-1}
He	5.8
H_2	3.4 - 4.1
N_2O	2.5
O_2	2.1 - 2.4
N_2	1.8 - 2.2
C_2H_2	2.0
SO_2	1.8 - 2.0
Ar	1.5 - 2.0
CO_2	1.8 - 1.9
Cl_2	1.4 - 1.7
CH_3Cl	1.5
H_2S	1.4

Source: Himmelblau (1964); T ≈ 25°C.

$$D \propto T\,\eta^{-1}$$

since the viscosity of solvents is generally temperature dependent. For water as solvent this viscosity has a temperature variation

$$\frac{\eta(T)}{\eta(25°)} = \exp\left[2140\left(\frac{1}{T} - \frac{1}{298}\right)\right]$$

that substantially exceeds the fractional variation in T.

In the case of solutes that dissociate into ionic constituents, e. g., $SO_2 = H^+ + HSO_3^-$, the diffusion coefficients of the ionic species may be calculated from the equivalent ionic conductivities $_i$ by the Nernst-Einstein relation (e.g., Adamson, 1973, pp. 546-547) as

$$D_i = \frac{\lambda_i RT}{z_i F^2}$$

where z_i is the ionic charge, and F is the Faraday constant; here $\underset{\sim}{R}$ is in joule mol^{-1} K^{-1}.

The ambipolar diffusion coefficient of the two ions (which must diffuse together) is

$$D = \frac{2\ D_1\ D_2}{D_1 + D_2}\ .$$

Diffusion coefficients of 1-1 electrolytes evaluated in this way are also of the order 2 x 10^{-5} $cm^2\ s^{-1}$.

Interfacial Mass Transport.

Despite the fact that for the total pressure and droplet size range characteristic of cloud gas-phase mass transport to (and from) cloud droplets is squarely in the continuum regime, nonetheless mass transport across the air- water interface, as across any interface is, ultimately, a process that involves individual molecules. Recognition of this principle, in conjunction with the kinetic theory of gases, permits the number of collisions of a gas-phase species X per unit interface area per unit time (the flux to the surface) to be evaluated as 1/4 $n_X\ \bar{v}_X$, where n_X is the number density of gas molecules and $\bar{v}_X$ is the average molecular speed (Hertz, 1882; Knudsen, 1913). In turn on a molar basis the collision flux is

$$\sigma_X^{coll} = \frac{1}{4}\ [X(g)]\ \bar{v}_X = \frac{1}{4}\ p_X\ \bar{v}_X/\underset{\sim}{R}T\ , \qquad (6.3)$$

where

$$\bar{v}_X = (8\ \underset{\sim}{R}T/\pi M_X)^{1/2}\ . \qquad (6.4)$$

Here M_X is the molar mass ("molecular weight") of species X; for M_X in g mol^{-1} and $\bar{v}_X$ in cm s^{-1}, $\underset{\sim}{R}$ has units erg mol^{-1} K^{-1}. This expression was initially applied to a discussion of condensation and evaporation rates of pure substances by introduction of a mass accommodation coefficient α representing the fraction of collisions resulting in condensation, yielding for the condensation flux

$$\sigma_X^{cond} = \frac{1}{4}\ \alpha\, p_X\ \bar{v}_X/\underset{\sim}{R}T \qquad (6.5)$$

As noted by Knudsen (1913) it is not possible a priori to specify the value of the accommodation coefficient. This point was underscored six decades later by Sherwood et al. (1975, p. 183), who observe that "Not only is there no useful theory to employ in predicting α, there is no easy way to experimentally measure it."

In the present context we are interested not so much in the self accommodation coefficient, as defined above, but in the accommodation coefficient of a solute gas impinging on a solvent, specifically water. However we note that equations (6.3) and (6.5) are applicable as well to partial pressure (specified for the interface) so that the formalism initially developed for condensation is applicable also to dissolution.

With respect to measurement we note that accommodation coefficients may be determined from measurement either of the uptake rate or of the evaporation or desorption rate, the latter by means of detailed balance arguments--at equilibrium the rate of evaporation is equal to the rate of condensation.

The values of reported accommodation coefficients vary considerably. In a number of studies with pure substances including metals (Knudsen, 1913; Langmuir, 1913), solid organic fatty acids and alcohols (Littlewood and Rideal, 1955) and liquid organics (Maa, 1967, 1970; Chang and Davis, 1976) the accommodation coefficient was determined to be unity. Other investigators, however, report values of α as low as 10^{-9} (see the discussion and references in Sherwood et al., 1975, pp. 182-184; also, Paul, 1962). Experimental difficulties associated with the measurements include both accurate knowledge of the surface temperature and maintenance of a surface that is free from surfactant films. It is suggested (Sherwood et al., 1975; Chang and Davis, 1976) that if both of these problems are resolved, the available data suggest that the evaporation coefficient or self-accommodation coefficient for liquids is essentially unity.

With respect to the self-accommodation coefficient of water there is a disquieting lack of agreement between investigators. Pruppacher and Klett (1978, pp. 133-135) summarize a number of these studies, reporting values of α ranging from ~ 0.01 to 1. In addition to the studies cited there one would call attention also to the work of Maa (1967, 1970), which reports $\alpha = 1$ for a study with a water jet, and of Wagner (1982) monitoring droplet growth in an expansion cloud chamber, also reporting $\alpha = 1$. Pruppacher and Klett observe that the data fall into two sets, noting that studies performed with rapidly renewed surfaces exhibit higher values of α (0.2 to 1). This distinction would be consistent with inhibition of mass transport at the interface in static experiments by unavoidable surfactant films. It should be noted that even monomolecular films can inhibit the evaporation rate from water surfaces (La Mer, 1962). It should also be noted that surfactant materials might be commonly present on atmospheric aerosols (Graedel et al., 1983) and thereby serve to inhibit mass transfer to cloud droplets formed from such aerosols. Pruppacher and Klett suggest that the condition of a rapidly renewed surface would be unlikely to be realized in clouds and propose that the lower values of α (0.01 to 0.07) be employed in cloud physics computations.

Relatively few studies have been performed examining the accommodation coefficient for uptake of solute gases by water. Danckwerts (1970, p. 69) suggest that there is no appreciable surface resistance for uptake of gases such as CO_2 by water, except in the presence of surfactants. However examinations of the papers cited by Danckwerts in support of this statement (Raimondi and Toor, 1959; Harvey and Smith, 1959) suggests that these studies were insensitive to values of α greater than perhaps 10^{-3}, although indeed a severe decrease in α (to 6×10^{-6}) was indicated in the presence of surfactants (Harvey and Smith, 1959).

In summary, there is a growing body of information suggestive that the self accommodation coefficient of liquids approaches unity for clean surfaces. However relatively little information is available regarding the magnitude of the accommodation coefficient of potential reagent gases either on clean liquid-water surfaces or on surfaces of natural cloud droplets. For this reason we must treat α somewhat as an adjustable parameter and examine the sensitivity of mass-transport rates to the value of this quantity. As it turns out, for representative atmospheric conditions (as tested in Section 7), overall mass-transport rates are relatively insensitive to values $\alpha \geq 0.01$, but become increasingly sensitive to lower values of α. These observations suggest the importance of laboratory study of for various solute gases both on clean surfaces and on surfaces representative of natural clouds.

7. MASS-TRANSPORT KINETICS

In this section we examine the kinetics of mass-transport processes in the gas-phase, at the gas-water interface, and in the aqueous phase from the perspective of ascertaining the influence of mass-transport kinetics upon the overall rate of the gas-aqueous reaction. Specifically, it is the objective of this section to develop approaches to evaluating the departure of the overall rate from that which would obtain in the absence of mass-transport limitation as a function of the pertinent conditions--solubilities, rate constants, drop size, accommodation coefficient, reagent concentrations. This discussion thus focuses upon concentration profiles about and within a single drop of radius $\underline{a}$ (cf. Figure 2) leading to an evaluation of $\bar{R}(a)$ (as defined in equation 5.1). A secondary objective of the discussion is to define the conditions where mass-transport limitation is sufficiently small, say 10%, that the kinetics of the system may be approximated as phase-mixed. In this limit description of the kinetics is considerably simplified.

In this discussion we consider first gas-phase mass transport, then interfacial, and finally aqueous-phase. Mass-transport in both phases is assumed to occur only by molecular diffusion, as discussed in Section 6. Mass transport at the interface occurs by the mechanism of gas-kinetic collisions with arbitrary accommodation coefficient α.

The coupling of gas-phase and interfacial mass-transport kinetics to the rate of aqueous-phase reaction is entirely general, depending only on the reaction rate averaged over the drop, $\bar{R}$, i.e., independent of the details of the reaction mechanism. In contrast, the aqueous-phase mass-transport and kinetics are intimately coupled, and a different treatment is required for each mechanism for quantitative description of mass-transport limitation.

Happily, the case of a first-order reaction (or pseudo-first-order in diffusing species) is analytically soluble. We thus focus on the example of a first-order mechanism both for the quantitative results afforded for this mechanism and for the qualitative insights provided for other kinetic rate expressions.

Gas Phase.

We first consider gas-phase diffusion of a species to the surface of a drop, a problem initially treated by Maxwell (1877). In the absence of gas-phase sources or sinks, the differential equation for gas-phase concentration G in spherical symmetry is (e.g., Crank, 1975, p. 89)

$$\frac{\delta G}{\delta t} - D_g \frac{1}{r^2} \frac{\delta}{\delta r} (r^2 \frac{\delta G}{\delta r}) = 0 \; . \tag{7.1}$$

We first consider a situation for which the concentration is initially uniform ($t \leq 0$) in which diffusion to a perfectly absorbing sphere of radius a is initiated at $t = 0$. The long-time (steady-state) solution to (7.1) is obtained by setting $dG/dt = 0$, which yields the ordinary differential equation

$$D_g \frac{1}{r^2} \frac{d}{dr} (r^2 \frac{dG}{dr}) = 0 \tag{7.2}$$

having solution (still for $G(a) = 0$)

$$G_{ss}(r) = G_\infty (1 - \frac{a}{r}) \; . \tag{7.3}$$

The flux corresponding to this profile is

$$F_{ss}(r) = - D_g \frac{dG_{ss}(r)}{dr} = - D_g \; G_\infty \frac{a}{r^2} \; . \tag{7.4}$$

At steady state the flux of reagent into the drop is equivalent to a reaction rate averaged over the drop as

$$\bar{R}_{ss} = -\left(\frac{4}{3}\pi a^3\right)^{-1} (4\pi r^2)\ F_{ss}(r), \tag{7.5}$$

whence, for G in mol/liter, we may equate the flux (7.4) to an average molar uptake rate

$$R^{(g)}_{max} = \frac{3\ D_g\ G_\infty}{a^2}\ . \tag{7.6}$$

Here the superscript (g) denotes gas-phase mass transport and the subscript max denotes that the flux (7.4) for G(a) = 0 represents the maximum steady-state flux into the drop that can be achieved by gas-phase diffusion. In terms of gas-phase partial pressure

$$R^{(g)}_{max} = \frac{3\ D_g\ p_A(\infty)}{\tilde{R}T\ a^2}\ . \tag{7.7}$$

This maximum uptake rate is shown as a function of drop diameter in Figure 7. It is seen that the maximum uptake rate, expressed per ppb of gas-phase reagent permits quite substantial reaction rates in the context of concentrations of cloudwater-dissolved materials, e.g., 5 x 10^{-5} M s^{-1} or 0.18 M h^{-1} at 10 μm diameter.

The recognition that there is a maximum rate at which material can be diffused into a

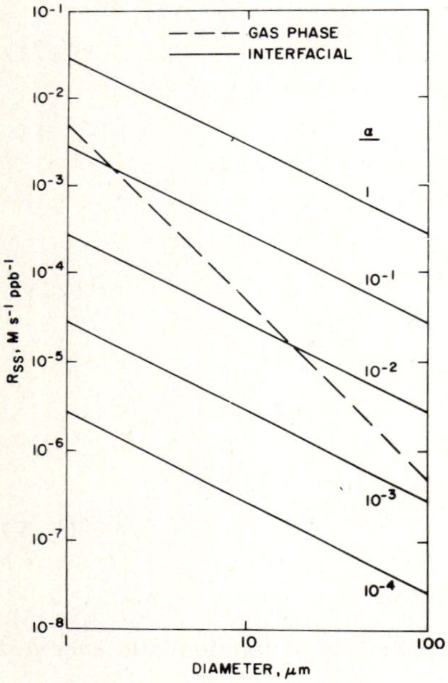

Figure 7. Maximum molar uptake rate per ppb of gas-phase reagent as a function of cloud drop diameter, as controlled by gas-phase diffusion (- - -) or interfacial collision rate (——) for indicated values of the accommodation coefficient α. Conditions: T = 298K, D_g = 0.1 $cm^2\ s^{-1}$, M = 30 g mol^{-1}.

droplet leads to identification of an important characteristic time associated with gas-aqueous reaction, viz., the time constant associated with saturating the drop in the reagent gas in the absence of any aqueous-phase sink,

$$\tau_{sat}^{(g)} = \frac{[A]_{eq}}{R_{max}^{(g)}} = \frac{H_A \, \underset{\sim}{R}T \, a^2}{3 \, D_g} \, . \tag{7.8}$$

(For a gas such as SO_2 that undergoes rapid aqueous-phase equilibrium, H in (7.8) would be replaced by H*.) As shown by Freiberg and Schwartz (1981) this characteristic time was sufficiently great under the conditions of certain laboratory kinetic studies of the oxidation of S(IV) to have appreciably decreased the S(IV) concentration and in turn the oxidation rate from the steady-state values that had been assumed in the analysis of the kinetic data.

Identification of the characteristic time associated with saturating the drop by gas-phase diffusion permits us to anticipate a further condition that must be satisfied in order that the rate of reaction at steady state not be restricted by the rate of diffusive mass transport, viz., that the time required to saturate the drop be short compared to the characteristic time of reaction of the dissolved material, i.e.,

$$\tau_{sat}^{(g)} << \tau_r \, . \tag{7.9}$$

This supposition is verified shortly.

We now return to the time-dependent equation. Based upon dimensional considerations, we anticipate the characteristic time for approach to the steady-state solution to be of order (a^2/D_g); we arbitrarily define the characteristic time associated with gas-phase diffusion as

$$\tau_{d.g.} = \frac{a^2}{3 \, D_g} \, ; \tag{7.10}$$

this definition differs slightly from that introduced by Schwartz and Freiberg (1981). The characteristic time for gas-phase diffusion (as well as those for aqueous phase diffusion and for inter-

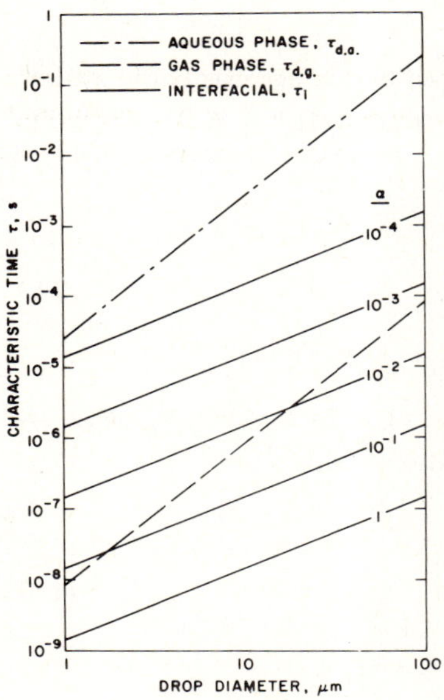

Figure 8. Characteristic times associated with diffusion in aqueous-phase ($\tau_{d.a.}$) and gas-phase ($\tau_{d.g.}$) and with interfacial mass transport (τ_i), as a function of diameter for a spherical droplet. Diffusion coefficients are 1×10^{-5} and $0.1 \ cm^2 \ s^{-1}$ for aqueous and gas phase, respectively; τ_i, evaluated for molecular weight 30, is shown for indicated values of the accommodation coefficient α.

facial mass transport, to be introduced below) is shown as a function of drop diameter in Figure 8. The solution to the time-dependent problem has been given by various authors (e.g., Beilke and Gravenhorst, 1978; Seinfeld, 1980, pp. 76-79). One convenient way to depict this solution is as the flux into the drop as a function of time, Figure 9. This flux is initially quite high but asymptotically approaches the steady-state flux given by equation (7.4). In fact, the asymptotic approach to the steady-state flux is quite slow; nonetheless, as may be seen by comparison of the characteristic times of gas and aqueous diffusion in Figure 8, the relaxation of the gas-phase system is still sufficiently rapid that the gas-phase system may be considered to rapidly adjust to changes in conditions imposed by the aqueous phase system.

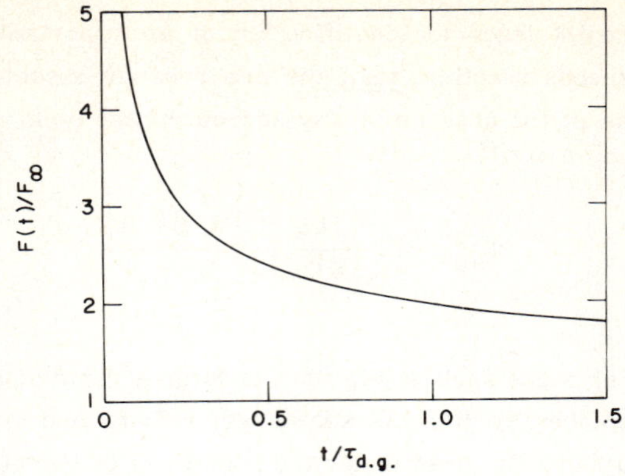

Figure 9. Time dependence of flux of gas into irreversible absorbing solution. Abscissa is expressed as dimensionless ratio of time to $\tau_{d.g.} = a^2/3D_g$; ordinate is expressed as ratio of flux to infinite-time value, $F_\infty = D_g G(\infty)/a$.

We now consider the steady-state equation (7.2) for arbitrary G(a), $0 \leq G(a) \leq G(\infty)$. The resulting concentration profile is (cf. Figure 2)

$$G(r) = G(\infty) - \frac{a}{r}\left[G(\infty) - G(a)\right] \quad . \tag{7.11}$$

Equating the resultant flux with a rate of reactive uptake of A in the drop ν_A R(a), we obtain for the departure from uniform gas-phase concentration

$$G(\infty) - G(a) = \frac{a^2}{3D_g} \nu_A \bar{R}(a) \tag{7.12}$$

or, in terms of partial pressure

$$p_A(\infty) - p_A(a) = \frac{a^2 \underset{\sim}{R}T}{3\ D_g} \nu_A \bar{R}(a) \quad . \tag{7.13}$$

Equation (7.13) permits us to formulate a criterion for absence of gas-phase mass-transport limitation, viz.,

$$\frac{p_A(\infty) - p_A(a)}{p_A(\infty)} \leq \mathcal{E}_g << 1 \ . \qquad (7.14)$$

This criterion sets an upper bound on the aqueous-phase reaction rate

$$\nu_A \ \bar{R}(a) \leq \mathcal{E}_g \frac{3D_g}{a^2 \ \underset{\sim}{R}T} p_A(\infty) \ . \qquad (7.15)$$

It is readily established that the bound (7.15) is equivalent to the intuitive supposition (7.9) for the absence of gas-phase mass-transport limitation.

The bound (7.15) may expressed in terms of solubility and rate coefficients for various mechanisms. For first-order kinetics (eq. 5.5, $\nu = 1$). We obtain the bound

$$k^{(1)} \ H_A \leq \mathcal{E}_g \frac{3D_g}{\underset{\sim}{R}T \ a^2} \ . \qquad (7.16)$$

This inequality is illustrated in Figure 10 for various values of drop diameter for $\mathcal{E}_g$ arbitrarily taken as 0.1. In preparing this figure we have employed $D_g = 0.1$ cm^2 s^{-1}, taking advantage of the relative insensitivity of D_g to the identity of the diffusing gas, as illustrated in Table 2. For the coordinates of the figure taken as log $k^{(1)}$ and log H_A, the inequality (7.16) is represented by a region in the $k^{(1)}$--H_A plane below and to the left of the line representing the condition of equality. To employ Figure 10 the point ($k^{(1)}$, H_A) corresponding to a situation of interest is located on the figure. For a reagent gas undergoing rapid aqueous-phase equilibrium H* must be employed. Similarly, for a second-order rate coefficient (5.10) must be employed. The point ($k^{(1)}$, H_A) is then compared to the location of the bound (7.16) for various drop sizes; for the point below and to the left of the line gas-phase mass-transport limitation to the rate of reaction does not exceed 10%.

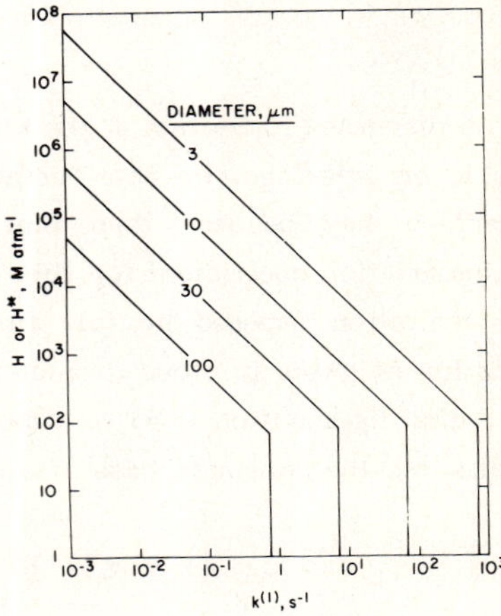

Figure 10. Gas-phase and aqueous-phase mass-transport limitation. The several lines represent the onset (10%) of mass-transport limitation for the indicated values of drop diameter. Diagonal sections represent gas-phase limitation; vertical sections represent aqueous-phase limitation. For situations such that ($k^{(1)}$, H) is below and to the left of a given line mass-transport limitation is less than 10% for the indicated drop diameter. Evaluated for $D_g = 0.1$ cm^2 s^{-1}, $D_a = 1 \times 10^{-5}$ cm^2 s^{-1}.

Figures such as this are useful (Schwartz, 1984a) as they permit a readily applicable means by which it may be ascertained that a situation of interest is free of mass-transport limitation or, conversely, that an evaluation of mass-transport limitation is required in an examination of the overall gas-aqueous reaction kinetics.

A similar treatment can be given for second-order kinetics (eq. 5.8, = 2), yielding

$$k^{(2)} H_A{}^2 \leq \mathcal{E}_g \frac{3\ D_g}{2\ \underset{\sim}{R}T\ a^2\ p_A} \ . \tag{7.17}$$

It is seen that for other conditions held constant there may or may not be gas-phase mass-transport limitation depending on the reagent partial pressure. Example of second-order reactions of atmospheric interest include $2NO_2$ [+ H_2O (l)] $\longrightarrow 2H^+ + NO_2^- + NO_3^-$ (Lee and Schwartz, 1981b) and $2HO_2 \longrightarrow H_2O_2(aq) + O_2$ (Schwartz, 1984b).

Interfacial.

As discussed in Section 6, the kinetic theory of gases sets an upper limit to the flux of a gas to an interface. We note further that the effective flux of a reactive gas across this interface may be less than that given by the gas- kinetic collision rate, i.e., the accommodation coefficient for this mass transport may be less than unity. We examine here the limitation imposed by this finite rate of mass transport on the rate of uptake and reaction of gases in cloud droplets.

Following equation (6.5) we denote the effective flux of reagent species A from the gas-phase into the aqueous-phase of a droplet of radius a as

$$\sigma_+ = p_A(a)\ \bar{v}\ \alpha / 4\ \underset{\sim}{R}T \ . \tag{7.18}$$

Here the subscript + explicitly denotes that this flux represents the gross (or one-way) flux into the drop. Also we explicitly employ $p_A(a)$, the partial pressure at the interface, which may not be approximately equal to the bulk partial pressure because of the gas-phase concentration gradient.

The flux (7.18) sets an upper limit to the rate of uptake of a gas by a droplet of radius a, exactly analogously to the maximum diffusive flux, equation (7.4), yielding for the maximum molar uptake rate

$$R^{(i)}_{max} = \frac{3\ p_A(a)\ \bar{v}\alpha}{4\ \underset{\sim}{R}T\ a} \tag{7.19}$$

This maximum uptake rate is shown as a function of drop diameter in Figure 7 for various values of the accommodation coefficient α. Comparison of $R_{max}^{(i)}$ with $R_{max}^{(g)}$, the maximum uptake rate permitted by gas-phase diffusion, allows identification of which of these two mass-transport processes is more restrictive and therefore controlling. It is seen that for $\alpha \gtrsim 0.1$ gas-phase diffusion is more restrictive over the drop-size range of interest, whereas for $\alpha \lesssim 10^{-3}$ interfacial mass transport is more restrictive.

The recognition that there is a maximum rate of interfacial mass transport into a drop leads to identification of a characteristic time associated with saturating the drop in the reagent gas as controlled by interfacial mass transport (cf., $\tau_{sat}^{(g)}$, eq. 7.8) in the absence of any aqueous-phase sink,

$$\tau_{sat}^{(i)} = \frac{[A]eq}{R_{max}^{(i)}} = \frac{4\ H_A\ \underset{\sim}{R}T\ a}{3\ \bar{v}\,\alpha} . \qquad (7.20)$$

Depending on whether $R_{max}^{(i)}$ or $R_{max}^{(g)}$ is controlling, the time to saturate the drop in the reagent gas will be controlled by $\tau_{sat}^{(i)}$ or $\tau_{sat}^{(g)}$, respectively.

Analogous to the argument above for gas-phase mass transport, identification of the characteristic time associated with saturating the drop by interfacial mass transport permits us to anticipate the condition that must be satisfied in order that the rate of reaction at steady state not be restricted by the rate of interfacial mass transport, viz., that the time required to saturate the drop be short compared to the characteristic time of reaction of the dissolved material, i.e.,

$$\tau_{sat}^{(i)} << \tau_r \qquad (7.21)$$

This supposition is verified shortly.

We now wish to address the net flux across the interface, which will differ appreciably from the gross flux when the aqueous-phase concentration is appreciable. We observe that we may express the gross flux (7.18) equally well in terms of $[A]^*$, the aqueous-phase concentration of A in equilibrium with $p_A(a)$ as

$$\sigma_+ = [A]^*\ v\,\alpha/4\ H_A\ \underset{\sim}{R}T . \qquad (7.22)$$

The actual gross flux of material leaving the aqueous phase into the gas phase is given by the same expression but employing the actual aqueous-phase concentration at the surface instead of the equilibrium concentration,

$$\sigma_- = [A]_a\ \bar{v}\,\alpha/4\ H_A\ \underset{\sim}{R}T . \qquad (7.23)$$

This is so because, by microscopic reversibility, σ_+ must equal σ_- at equilibrium and because the flux is linear in concentration. The net flux of material transfer into solution is

$$\sigma = \sigma_+ - \sigma_- = \frac{\bar{v}\,\alpha}{4\ H_A\ \underset{\sim}{R}T}\,(H_A\ p_A(a) - [A]_a\ . \tag{7.24}$$

This net flux may be equated to the rate of reactive uptake of A in the drop $_A\bar{R}(a)$ to optain the departure from Henry's law equilibrium at the surface that results from this uptake rate:

$$H_A p_A(a) - [A]_a = \frac{4a\ H_A\ \underset{\sim}{R}T}{3\ \bar{v}\ \alpha}\ \nu_A\ \bar{R}(a)\ . \tag{7.25}$$

The difference between $H_A p_A$ and $(A)_a$ is represented by the step in concentration profile at the interface indicated in Figure 2.

Equation (7.25) permits us to define a criterion for the absence of interfacial mass-transport limitation, viz.,

$$\frac{H_A\ p_A(a) - [A]_a}{H_A\ p_A(a)} \leq \varepsilon_i << 1 \tag{7.26}$$

This criterion also sets an upper bound on the aqueous-phase reaction rate, in this case

$$\nu_A\ \bar{R}(a) \leq \varepsilon_i\ \frac{3\ \bar{v}\ \alpha}{4\ a\ \underset{\sim}{R}T} p_A(a)\ . \tag{7.27}$$

It is readily established that the bound (7.27) is equivalent to the intuitive supposition (7.21) for the absence of interfacial mass-transport limitation.

The bound (7.27) may be expressed in terms of solubility and rate coefficients for various mechanisms just as was done in the examination of gas-phase mass-transport limitation. For first-order kinetics (eq. 5.5, ν = 1) we obtain the

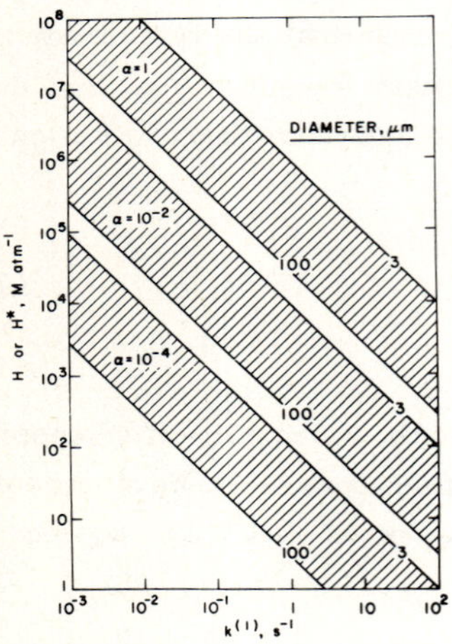

Figure 11. Interfacial mass-transport limitation. The several bands represent the onset (10%) of mass-transport limitation for the indicated values of accommodation coefficient α; the width of each band corresponds to the drop diameter range 3 to 100 μm. Mass-transport limitation is less than 10% if the point $(k^{(1)}, H)$ is below and to the left of the bound representing the applicable values of α and drop diameter. Evaluated for gas molecular weight = 30 g mol^{-1}.

bound

$$k^{(1)} H_A \leq \mathcal{E}_i \frac{3 \bar{v} \alpha}{4 a \underset{\sim}{R}T} \tag{7.28}$$

This inequality is illustrated in Figure 11, for $\mathcal{E}_i$ taken as 0.1. Here a range of bounds must be indicated reflecting both the pertinent range of drop size and the uncertainty in knowledge of α. It is seen by comparison to Figure 10 that interfacial mass-transport may be more or less controlling than gas-phase, depending largely upon the value of α. A similar treatment can be given for second-order kinetics yielding

$$k^{(2)} H_A{}^2 \leq \mathcal{E}_i \frac{3 \bar{v} \alpha}{8 a \underset{\sim}{R}T p_A} , \tag{7.29}$$

also exhibiting an inverse dependence on the partial pressure of the reacting gas. Such a situation has been treated by Schwartz (1984b).

Gas-Phase plus Interfacial.

We have observed that the description of both gas-phase and interfacial mass-transport is independent of the details of the aqueous-phase reaction mechanism. This observation suggests the utility of developing a single expression that combines both processes. Moreover, for the purpose of developing a formalism for treating mixed-phase kinetics at steady state, we shall find it useful to express the rate of these mass-transport processes by a single phenomenological "rate coefficient." We develop the expression for the mass-transport rate coefficient by analogy to the following model of a two-step reaction mechanism

$$X \underset{-1}{\overset{1}{\rightleftharpoons}} Y \qquad [1]$$

$$Y \xrightarrow{R_2} Z \qquad [2]$$

Reaction [1] is reversible with forward and reverse rate coefficient k_1 and k_{-1}; we note that the equilibrium constant for reaction [1] is $K_1 = k_1/k_{-1}$. Reaction (2) is irreversible, proceeding at steady state with rate R_2. The usual steady-state treatment for the concentration [Y] yields

$$\frac{[Y]}{K_1} = [X] - \frac{R_2}{k_1} \ . \tag{7.30}$$

We observe that an equation formally identical with (7.30) is obtained by combining equations (7.13) and (7.25) for gas-phase and interfacial mass transport, respectively, viz.,

$$\frac{[A(aq)]_a}{H_a} = p_A(\infty) - \underset{\sim}{R}T \left(\frac{a^2}{3\ D_g} + \frac{4a}{3\ \bar{v}\alpha}\right) \nu_A\ \bar{R}(a) \ . \tag{7.31}$$

We may thus make the following identification between elements of the reaction mechanism [1] - [2] and the mass-transport system:

X : $A(g)_{r=\infty}$

Y : $A(aq)_{r=a}$

[1] : $A(g) \rightleftharpoons A(aq)_{r=a}$

R_1 : $\dfrac{d_1[A(aq)]_a}{dt} = k_1\ p_A(\infty)$

R_{-1} : $\dfrac{d_{-1}[A(aq)]_a}{dt} = k_{-1}[A(aq)]_a$

R_2 : $\nu_A\ \bar{R}(a)$

k_1 : $\dfrac{1}{\underset{\sim}{R}T}\left(\dfrac{a^2}{3\ D_g} + \dfrac{4a}{3\ \bar{v}\alpha}\right)^{-1}$

K_1 : H_A

k_{-1} : $\dfrac{1}{H\underset{\sim}{R}T}\left(\dfrac{a^2}{3\ D_g} + \dfrac{4a}{3\ \bar{v}\alpha}\right)^{-1}$

This identification is of considerable utility and exhibits as well a number of interesting features that we briefly discuss here. Before that, however, we introduce for convenience an additional definition, viz.,

$$k_{mt} = \left(\frac{a^2}{3\,D_g} + \frac{4a}{3\,\bar{v}\alpha}\right)^{-1} ; \tag{7.32}$$

k_{mt} represents the rate coefficient for gas-phase plus interfacial mass transport. With this definition equation (7.31) becomes

$$\frac{[A(aq)]_a}{H_A} = p_A(\infty) - \frac{\nu_A\,\bar{R}(a)}{(\underset{\sim}{R}T)^{-1}\,k_{mt}} . \tag{7.33}$$

A first observation concerns units, although to be sure the units of the several quantities are somewhat arbitrary, reflecting convention and definition. We note that we have chosen to express rates in aqueous-phase units, i.e., M(aq) s^{-1}. Thus for the phenomenological mass-transport reaction

$$[1] \qquad A(g)_{r=\infty} \rightleftharpoons A(aq)_{r=a}$$

the rate R_1 represents moles of reagent A entering the drop per liter of drop volume per second. Since, in keeping with the standard state convention we have adopted for gases, we wish to express the reagent gas activity as a partial pressure in units of atmospheres, the "rate coefficient" k_1 must have units M atm^{-1} s^{-1}. This falls out naturally in the identification

$$k_1 = \frac{1}{\underset{\sim}{R}T}\left(\frac{a^2}{3\,D_g} + \frac{4a}{3\,\bar{v}\alpha}\right)^{-1} . \tag{7.34}$$

In defining k_{mt} according to (7.32) the choice was made to reflect the kinetics by k_{mt} (units: s^{-1}) and to effect the conversion of the concentration units by the factor $(RT)^{-1}$ (units: M atm^{-1}). It may be observed that

$$k_{mt}^{-1} = \frac{a^2}{3\,D_g} + \frac{4a}{3\,\bar{v}\alpha} , \tag{7.35}$$

where the first term on the RHS is the characteristic time associated with gas-phase diffusion, $\tau_{d.g.}$, equation (7.10). We thus identify the second term as a characteristic time associated with interfacial mass transport,

$$\tau_i = \frac{4a}{3\ \bar{v}\alpha} \ . \tag{7.36}$$

We therefore define a characteristic time associated with gaseous plus interfacial mass transport,

$$\tau_{mt} = k_{mt}^{-1} = \tau_{d.g.} + \tau_i \tag{7.37}$$

i.e., the sum of the characteristic times of the two processes. Equations (7.35) and (7.37) reflect the usual combinatorial algebra associated with sequential mass-transport processes, i.e., of rate quantities "adding" like electrical conductances in series, or time quantities adding like resistances in series.

The dependence of the mass-transfer rate coefficient k_{mt} on accommodation coefficient and drop size is indicated in Figure 12. It is seen that for high values of , k_{mt} approaches a constant value that represents diffusion-controlled uptake, whereas at small k_{mt} decreases linearly with decreasing α, indicative of collision-controlled uptake. The a dependence at high α is quadratic, whereas at low α it is linear. We return to further consideration of k_{mt} in Section 8.

An important point to emphasize about the treatment that has been given thus far is that it is completely general with respect to the mechanism of the aqueous-phase reaction and with regard to whether or not the concentration profile of the reagent in the drop is spatially uniform. All that is assumed is that the aqueous-phase reaction is at steady state, proceeding with a spatially averaged rate $\bar{R}$. We would note, however, that in many cases of interest the aqueous-phase concentration profile is uniform, for which situation additional simplification obtains. Considerable simplification obtains as well in the case of a first-order aqueous-phase reaction (whether or not the aqueous concentration profile is uniform). For reaction [2] first-order in [Y], i.e.,

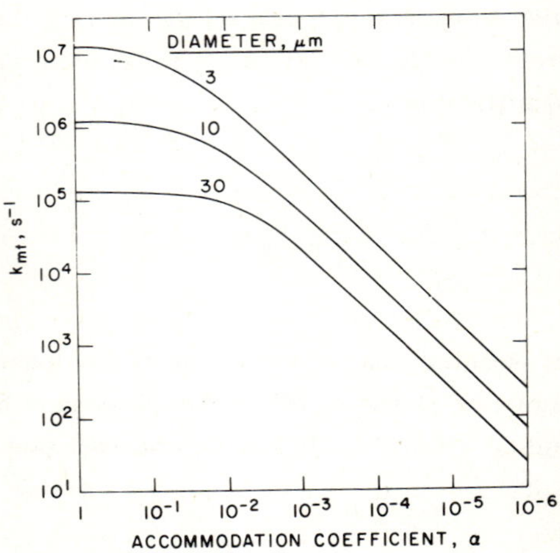

Figure 12. Dependence of the mass-transfer rate coefficient on accommodation coefficient and drop diameter. Note sense of log scale for α.

$R_2 = k_2[Y]$, we obtain from (7.30) the customary steady-state concentration

$$[Y] = \frac{k_1[X]}{k_{-1} + k_2} . \qquad (7.38)$$

We make the further identification with the mass-transport system for first-order reaction with $\nu_A = 1$,

$$k_2 : k_e^{(1)}$$

$$R_2 : k_e^{(1)}[A(aq)]_{r=a} ;$$

here $k_e^{(1)}$ is an effective first-order rate coefficient, which would be less than the intrinsic $k^{(1)}$ in the case of a non-uniform aqueous-phase concentration profile. We thus obtain

$$[A(aq)]_{r=a} = \frac{(\underset{\sim}{R}T)^{-1} k_{mt} \, p_A(\infty)}{(H_A \underset{\sim}{R}T)^{-1} k_{mt} + k_e^{(1)}} , \qquad (7.39)$$

and in turn for the overall rate of reaction

$$R = \frac{k_e^{(1)} (\underset{\sim}{R}T)^{-1} k_{mt} \, p_A(\infty)}{(H_A \underset{\sim}{R}T)^{-1} k_{mt} + k_e^{(1)}} . \qquad (7.40)$$

Equations (7.39) and (7.40) have two interesting limits. For $(H_A RT)^{-1} k_{mt} >> k_e^{(1)}$, $[A(aq)] \approx H_A p_A(\infty)$, and $R \approx k_e^{(1)} H_A p_A(\infty)$, i.e., the phase-mixed limit, in which the aqueous-phase (at least at the interface, for non-uniform aqueous-phase profile) is in equilibrium with the reagent gas at the bulk partial pressure. At the other extreme, $(H_A \underset{\sim}{R}T)^{-1} k_{mt} << k_e^{(1)}$, $R \approx k_{mt} p_A/\underset{\sim}{R}T$, i.e., the reactive uptake is controlled completely by mass transport.

Some further comments are in order regarding equation (7.31) and in particular the term for mass-transport. We note that quite a few expressions of form similar to (7.32) have been advanced by various authors to describe mass-transport in the "Knudsen" regime, i.e., the regime intermediate between the continuum regime treated here and the free-molecular regime, in which the radius of the particle is much smaller than the gas mean free path (Wagner, 1982; Davis, 1983; Dahneke, 1983, and references therein). Of these expressions that due to Fuchs and Sutugin (1971), which was based on a numerical solution to a neutron scattering problem, is perhaps the most widely employed in consideration of mass-transport of minor solute species (e.g., Chameides and Davis, 1982). However, as emphasized by Davis (1983) all such expressions for mass-transport kinetics resulting from various simplified treatments are of necessity somewhat artificial because of inherent approximations. Undoubtedly the expression (7.32) would be subject to the same sort of criticism. However in support of the present treatment and the resulting expression for k_{mt} we would offer the following arguments:

1. Simplicity. Only continuum diffusion theory and the gas kinetic surface collision rate have been employed. No ad hoc assumptions have been made.

2. Comprehensiveness. The present model incorporates both reversible dissolution (according to Henry's law) and variable mass-accommodation coefficient.

3. Generality. The model is independent of the mechanism of aqueous-phase reaction and of any assumption regarding the aqueous-phase concentration profile.

4. Accuracy in the limits. The expression for the uptake rate approaches in the appropriate limit the correct value for the uptake rate in the continuum regime, in the free-molecular regime, and in the regime corresponding to Henry's law equilibrium between the concentration at surface of the drop and the bulk partial pressure.

To the above list one might also add the observation that comparison with the formula of Fuchs and Sutugin (1971) shows quite a close agreement over the entire range of mutual applicability, the maximum departure between the two expressions being 15%. These arguments certainly lend support to the present approach to evaluation of mass-transport kinetics for uptake and reaction of gases in aqueous droplets, and in particular to the mass-transport rate coefficient defined by equation (7.32).

Aqueous-Phase Diffusion and Reaction.

We now consider the problem of diffusion of a dissolved reagent from the surface of a drop toward the interior coupled with aqueous-phase reaction in the drop. We focus principally upon the steady-state treatment of this problem but briefly consider the time-dependent problem for the insights afforded. The objective of the analysis is to evaluate $\bar{R}_{ss}(a)$, the steady-state rate of reaction averaged over the volume of the drop, for the assumed chemical-kinetic rate law, as a function of the steady-state reagent concentration at the surface of the drop, $[A]_a$ (or $A_{ss}(a)$; Figure 2). This average rate may be compared with the rate evaluated under the assumption of a uniform reagent profile in the drop (at concentration $[A]_a$); the quotient of these two rates is a measure of the limitation imposed on the rate of reaction by the finite rate of aqueous-phase diffusion. We note further that evaluation of $\bar{R}_{ss}(a)$ as a function of $[A]_{r=a}$ permits evaluation of $[A]_{r=a}$ by (7.31) in terms of of the bulk partial pressure $p_A(\infty)$, thereby solving the entire problem of mass-transport (gas, aqueous, interfacial) coupled with aqueous-phase chemical reaction. The quotient of the average rate $\bar{R}_{ss}(a)$ to the rate evaluated for a uniform aqueous-phase concentration in Henry's law equilibrium with the bulk gas-phase concentration (A_∞ in Figure 2) is a measure of the overall mass-transport limitation.

We address the problem of coupled mass-transport and reaction by means of the partial differential equation in the time- and space-dependent aqueous-phase concentration A(r,t), analogous to (7.1) but inclusive of a term representing a reactive sink,

$$\frac{\delta A}{\delta t} - D_a \frac{1}{r^2} \frac{\delta}{\delta r} \left(r^2 \frac{\delta A}{\delta r}\right) + R(r,t) = 0 \; . \tag{7.41}$$

Initially the concentration A is uniformly 0. At the time t = 0 the drop is exposed to a gas-phase concentration that establishes and maintains a constant surface concentration A(a). (We thereby assume that transients due to gas-phase and interfacial mass transport relax rapidly.) In the absence of the chemical sink term the concentration approaches a constant value equal to the surface concentration, with a time constant characteristic of aqueous-phase diffusion

$$\tau_{d.a.} = \frac{a^2}{\pi^2 D_a} \tag{7.42}$$

(Crank, 1975, p. 92). For the reactive problem, at long time the system achieves a steady-state that is inherently characterized by a non-uniform concentration profile as the reagent is consumed by reaction in the interior of the drop. Introduction of the characteristic time associated with aqueous-phase diffusion allows us to anticipate the condition that must be satisfied in order that reaction not greatly deplete the aqueous-phase concentration from its surface value, viz.,

$$\tau_{d.a.} << \tau_r \tag{7.43}$$

We return now to the reactive problem. At steady state $\delta A/\delta t = 0$ and the concentration profile is governed by the ordinary differential equation

$$D_a \frac{1}{r^2} \frac{d}{dr} (r^2 \frac{d A_{ss}(r)}{dt}) = R_{ss}(r) \tag{7.44}$$

Equations (7.41) and (7.44) are not in general analytically soluble. However in the case of a first-order rate law

$$R(r,t) = k^{(1)} A(r,t) \tag{7.45}$$

an analytical solution exists for both the time-independent and time-dependent problems. We therefore examine this system in some detail, not only to obtain a quantitative measure of the effect of mass-transport limitation but also for qualitative insight into other rate laws whose kinetics can be expressed in terms of an effective first order rate coefficient, as introduced in Section 5.

We first consider the steady-state solution to (7.44) with rate law (7.45), and return presently to the time-dependent solution. This solution is given (Schwartz and Freiberg, 1981) as

$$A_{ss}(r)=A(a) \frac{a}{r} \frac{\sinh (qr/a)}{\sinh q} \tag{7.45}$$

where we introduce the dimensionless diffuso-reactive parameter

$$q \equiv a(k^{(1)}/D_a)^{1/2} . \tag{7.46}$$

As we shall see, knowledge of the value of q characterizing a given physical system is fundamental to the understanding and description of the system.

In order to enable appreciation of the role of the parameter q upon the steady state concentration profiles within the drop, we show these profiles in Figure 13 for several values of q. It is seen that at low value of q, the steady state-reaction profile is nearly uniform. This reflects the rapidity of diffusion relative to reaction. However, for values of q of the order of unity and greater there is an appreciable depletion of A_{ss} towards the center of the drop, reflecting the inability of diffusion to restore the decrease in A_{ss} caused by chemical reaction. In other words, with increasing q there is decreasing penetration by A_{ss} from the surface of the drop into the interior. This effect is displayed also in Figure 14, in which the concentration at the center of the drop,

$$A_{ss}(0) = A(a) \frac{q}{\sinh q} \tag{7.47}$$

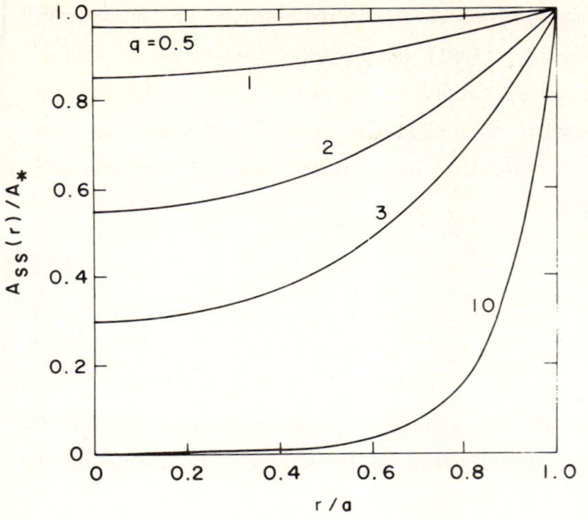

Figure 13. Steady-state radial concentration profiles of concentration of a reactant species A(r), relative to the concentration at the surface of the drop A_*, for a drop of radius a, for indicated values of the dimensionless parameter $q = a(k^{(1)}/D_a)^{1/2}$. In the notation of the present paper $A_* = A(a)$. From Schwartz and Freiberg (1981).

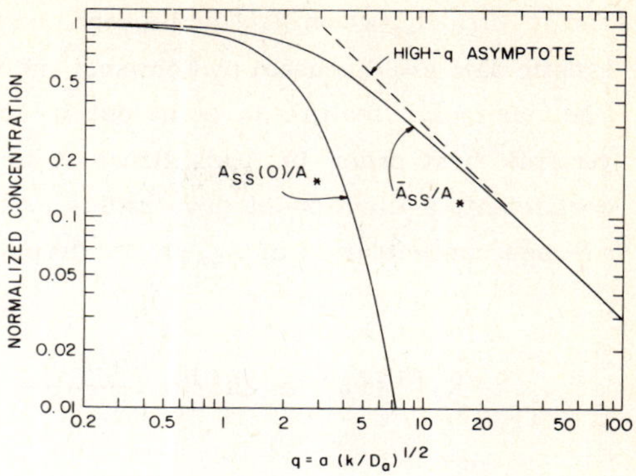

Figure 14. Average concentration of reactant species $\bar{A}$ and the concentration at the center of the drop A(0), relative to the concentration at the surface of the drop A_*, at steady state, as a function of q. From Schwartz and Freiberg (1981).

is given as a function of q.

More important than the dependence on q of the concentration at the center of the drop is the dependence on q of the average concentration $\bar{A}_{ss}$, since the rate of reaction averaged over the drop, $\bar{R} = k\bar{A}$, will at steady state be proportional to this average. $\bar{A}_{ss}$ is evaluated as

$$\bar{A}_{ss} = (4\pi a^3/3)^{-1} \int_0^a 4\pi r^2 \, A_{ss}(r)dr \quad . \tag{7.48}$$

Integration of $A_{ss}(r)$ given in Eq. (7.45) yields the result

$$\bar{A}_{ss}/A(a) = 3 \left(\frac{\coth q}{q} - \frac{1}{q^2}\right) \quad . \tag{7.49}$$

This steady-state average concentration is also shown in Figure 14 as a function of the parameter q. For $q \gtrsim 1$ the steady-state average concentration $\bar{A}_{ss}$ is seen to be significantly less than the equilibrium concentration A(a). The fall-off of $\bar{A}_{ss}$ with increasing q is slower than that of the concentration at the center, $A_{ss}(0)$. This slower fall-off reflects the large fraction of the volume of the drop that is "near" the surface. The dependence of $\bar{A}_{ss}$

and in turn $\bar{R}_{ss}(a)$ on $k^{(1)}$ and a, and the implications on the interpretation of laboratory kinetic data are discussed by Schwartz and Freiberg (1981) and Freiberg and Schwartz (1981). One particular feature to point out in the present context, however, is that the uptake remains first order in A(a). Hence it is useful to define an effective first-order rate coefficient, a function of drop radius a, that takes into account the non-uniform spatial-average concentration of A_{ss} in the drop, viz.,

$$k_e^{(1)}(a) \equiv 3k^{(1)} \left(\frac{\coth q}{q} - \frac{1}{q^2}\right) . \qquad (7.50)$$

This effective first-order rate coefficient, which was suggested in the discussion above, is useful in consideration of reaction rates for distribution of drop sizes, as discussed in Section 8.

An important result that may be obtained from equation (7.49) is the value of q for which $\bar{A}_{ss}/A(a)$ commences to depart appreciably from unity. We observe that for low q (7.49) may be approximated as

$$\frac{\bar{A}_{ss}}{A(a)} \approx 1 - \frac{q^2}{15} + \frac{2q^4}{315} \cdots , \quad q \lesssim 1.5 . \qquad (7.51)$$

Equation 7.51 permits a bound to be set on q, and in turn on k and that the fractional departure of the average concentration from that at the surface not exceed a specified value, viz., for

$$1 - A_{ss}/A(a) \leq \varepsilon_a << 1 ,$$

$$k^{(1)} \leq 15\, \varepsilon_a \frac{Da}{a^2} , \quad \varepsilon_a \leq 0.1 . \qquad (7.52)$$

It is readily established that this bound is equivalent to the intuitive supposition (7.43) for the absence of aqueous-phase mass-transport limitation. The bound (7.52) is illustrated in Figure 10 for various values of drop diameter, for ε_a arbitrarily taken as 0.1. For values of $k^{(1)}$ less than the bound indicated for the drop diameter aqueous-phase mass-transport limitation is minimized. The combined bounds represented by gas- and aqueous-phase mass-transport considerations restrict the values of ($k^{(1)}$, H_A) for situations free of mass-transport. Additionally, interfacial mass-transport limitation would also have to be addressed, Figure 11.

We note that the value of the Henry's law coefficient at the "knee" of the bound representing the onset of gas- and aqueous-phase mass-transport limitation is independent of drop radius, a. This value of H is given by

$$H_{knee} = \frac{1}{5\,\tilde{R}T}\,\frac{D_g}{D_a} ,$$

for $\mathcal{E}_a = \mathcal{E}_g$. For $D_g = 10^4\ D_a$, $H_{knee} \approx 80\ \mathrm{M\ atm^{-1}}$. For values of $H < H_{knee}$, aqueous-phase mass transport is more controlling than gas-phase, and for $H > H_{knee}$, gas-phase mass transport is more controlling. Examples of the application of these relationships to measurement of mass-transport limitation for situations of interest in the ambient atmosphere have been given by Schwartz (1984a).

We now return briefly to the time-dependent problem, which has been treated in detail by Schwartz and Freiberg (1981) as well as other authors cited therein. An example of this time dependence is shown in Figure 15, which shows the concentration A(r,t) for the value q = 1.5, representative of a slight (~15%) steady-state mass-transport limitation. The several curves indicate the penetration of A into the interior of the drop and the eventual establishment of the steady-

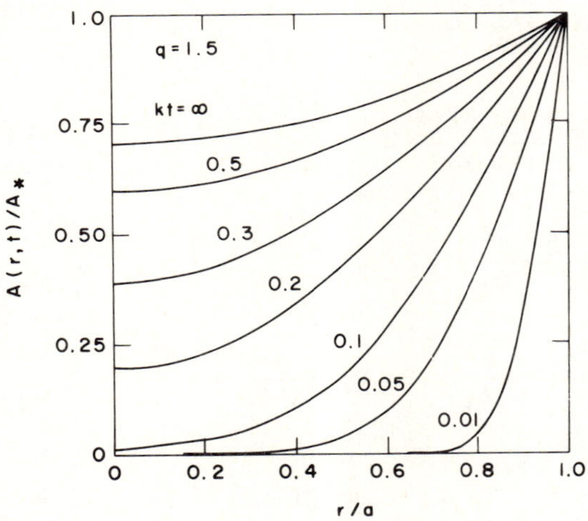

Figure 15. Values of the time- and space-dependent reagent concentration A(r,t) as a function of r for indicated values of $k^{(1)}t$, and for the diffuso - reactive parameter $q=a(k^{(1)}/D_a)^{1/2} = 1.5$. With increasing time the radial concentration function approaches the non-uniform steady-state profile; contrast Fig. 6.1 of Crank (1975) displaying profiles for diffusion in the absence of reaction, i.e., q = 0 in the present notation. From Schwartz and Freiberg (1981).

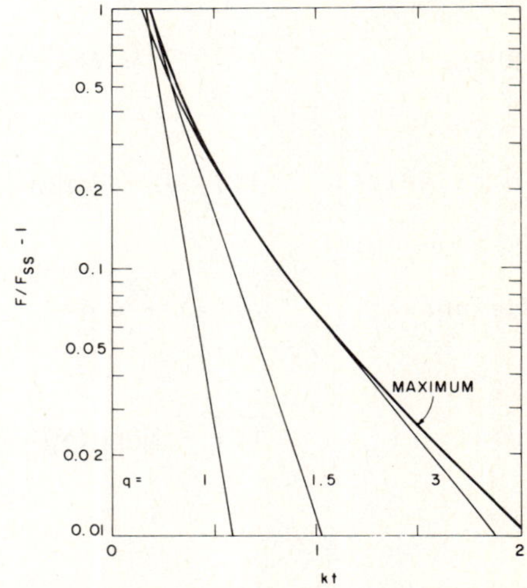

Figure 16. Time dependence of the relative difference between the flux into the drop at time t and the steady-state flux into the drop, for indicated values of q. Dark curve at right of figure represents the envelope of the several curves and constitutes an upper bound on the relative difference for all values of q. From Schwartz and Freiberg (1981).

state relationship. Perhaps more important is the present context in Figure 16, which shows the approach of the flux into the drop to the steady-state flux as a function of time for various values of q. This approach is seen to be rapid (relative to $\tau_r = k^{(1)-1}$) for all values of q, supporting the steady-state approach.

Aqueous-Phase Source.

For a reaction forming a product species C that diffuses within the aqueous phase and evolves into the gas phase the opposite situation obtains from that thus far treated, i.e., the normalized concentration profile on a plot such as Figure 2 decreases with increasing r. We briefly consider this problem to assess the magnitude of steady-state concentrations build-up as a function of reaction rate. As before gas-phase and interfacial results are independent of the details of R, but the aqueous-phase result depends on the radial distribution of R(r). For this discussion we take R to be constant.

Solution of the appropriate differential equations yields the following results:

Aqueous: $$\overline{[C]} - [C]_{r=a} = \frac{1}{15}\frac{a^2}{D_a} R$$

Interfacial: $$[C]_{r=a} - H_C p_C(a) = \frac{4\ H_C\ \underset{\sim}{R}T\ a}{3\ \bar{v}\alpha} R$$

Gas phase: $$p_C(a) - p_C(\infty) = \frac{\underset{\sim}{R}T\ a^2}{3\ D_g} R$$

Total: $$\overline{[C]} - H_C p_C(\infty) = [\frac{1}{15}\frac{a^2}{D_a} + H_C\ \underset{\sim}{R}T\ (\frac{4a}{3\ \bar{v}\alpha} + \frac{a^2}{3\ D_g})]\ R$$

$$= [\frac{\pi^2}{15}\tau_{d.a.} + H_C\ \underset{\sim}{R}T\ (\tau_i + \tau_{d.g.})]\ R$$

These expressions permit evaluation of the magnitude of product build-up as a function of formation rate and pertinent mass-transport properties. Expression of each of the ΔC terms as a product of rate x characteristic time is particularly heuristic and therefore useful for acquiring insight into the system.

Summary.

In this section we have outlined a procedure for evaluation of mass-transport limitation to the reactive uptake of gases by aqueous droplets, treating gas-phase, interfacial, and aqueous-phase mass transport. The treatment was both heuristic, in terms of characteristic times (as summarized in Table 4), and quantitative, in terms of steady-state solutions to the pertinent differential equations. The treatment of gas-phase and interstitial mass-transport was general with respect to the aqueous-phase rate law, whereas the treatment of coupled aqueous-phase mass transport and reaction was restricted to a first-order rate law. Expressions were presented for the magnitude of departure from uniform concentration profile attributable to the several mass-transport processes; these are summarized in Table 5. Bounds representing the onset of mass-transport limitation were presented, as summarized also in Table 5. Finally the concentration profiles associated with aqueous-phase formation of product species were examined.

Table 4

Characteristic Times Pertinent to Reactions of Gases in Aqueous Droplets

Process	Characteristic Time	Definition	Equivalence
Reaction	τ_r	$[A]/\nu_A R$	
Aqueous-phase Diffusion	$\tau_{d.a.}$	$a^2/\pi^2 D_a$	
Gas-phase Diffusion	$\tau_{d.g.}$	$a^2/3D_g$	
Saturation by Gas-phase Diffusion	$\tau_{sat}^{(g)}$	$H\underset{\sim}{R}T \dfrac{a^2}{3\ D_g}$	$H\underset{\sim}{R}T\ \tau_{d.g.}$
Interfacial Mass Transport	τ_i	$\dfrac{4a}{3\ \bar{v}\alpha}$	
Saturation by Interfacial Mass Transport	$\tau_{sat}^{(i)}$	$H\underset{\sim}{R}T \dfrac{4a}{3\ \bar{v}\alpha}$	$H\underset{\sim}{R}T\ \tau_i$
Gas-phase + Interfacial Mass Transport	τ_{mt}	$\tau_{d.g.} + \tau_i$	

Table 5

Departure from Uniform Concentrations in Reactions of Gases in Aqueous Droplets

Phase	Magnitude of Departure	Uniformity Condition: Heuristic	Uniformity Condition: Quantitative
Gas	$p(\infty) - p(a) = \frac{a^2 \underset{\sim}{R}T}{3\,D_g}\,\nu R$	$\tau_{sat}^{(g)} \ll \tau_r$	$Hk^{(1)} \leq \varepsilon \frac{3\,D_g}{\underset{\sim}{R}T\,a^2}$
Interfacial	$H_p - [A]_{r=a} = \frac{4a}{3\,\bar{v}\alpha} H\underset{\sim}{R}T\,\nu R$	$\tau_{sat}^{(i)} \ll \tau_r$	$Hk^{(1)} \leq \varepsilon \frac{3\,\bar{v}\alpha}{4a\,\underset{\sim}{R}T}$
Aqueous	$\overline{[A]}/[A]_{r=a} = 3\left(\frac{\coth q}{q} - \frac{1}{q^2}\right)$	$\tau_{d.a.} \ll \tau_r$	$k^{(1)} \leq 15\,\varepsilon \frac{D_a}{a^2}$

Aqueous phase departure and quantitative uniformity criteria are expressed for first-order rate law. Diffuso-reactive parameter $q = a(k^{(1)}/D_a)^{1/2}$.

$\varepsilon \leq 0.1$ represents maximum allowable fractional departure from uniformity condition.

8. GAS-AQUEOUS KINETICS IN CLOUDS

In this section we develop an approach to the description of reaction kinetics of gases in cloud droplets and the coupling of aqueous- and gas-phase kinetics. Such an approach in principle permits treatment of reactions in the two phases by means of a set of coupled ordinary differential equations.

Phase-mixed Limit.

Provided it can be established, by the methods outlined in Section 7, that the gas-aqueous reaction under examination is phase-mixed, then evaluation of aqueous-phase rates may proceed by the method outlined in Section 5, i.e, evaluation of aqueous-phase rates in terms of gas-phase partial pressures by Henry's law. This approach has been employed in our previous evaluations of the rates of aqueous-phase in-cloud reactions of nitrogen and sulfur oxides (Lee and Schwartz, 1981b; Schwartz, 1984a).

Monodisperse Cloud.

For non-phase mixed systems explicit consideration must be given to mass-transport kinetics. This treatment is heuristic for a monodisperse cloud, so we begin with this model.

Consider a monodisperse cloud, characterized by drop radius a and liquid-water volume fraction L. Consider a reaction mechanism

$$Q \longrightarrow A(g) \qquad [Q]$$

$$A(g) \longrightarrow \text{gaseous products} \qquad [S]$$

$$A(g) \rightleftharpoons A(aq) \qquad [1]$$

$$A(aq) \longrightarrow \text{products} \qquad [2]$$

Here [Q] and [S] represent gas-phase source and sink reactions of the reagent of concern A(g), having rates of q and s, respectively (atm s^{-1}). The following rate expression obtains for $[A]_a$ the aqueous-phase concentration of A, at the surface of the drop,

$$\frac{d[A]_a}{dt} = (\underset{\sim}{R}T)^{-1} k_{mt} p_A - (H_A\underset{\sim}{R}T)^{-1} k_{mt}[A]_a - \bar{R}_2, \qquad (8.1)$$

where we have used equation (7.34) for k_1. $\bar{R}_2$ represents the rate of reaction [2] averaged over the volume of the drop. The corresponding rate expression for p_A is

$$\frac{d\, p_A}{dt} = q - s - L\, k_{mt}\, p_A + \frac{L}{H_A} k_{mt}[A]_a\ , \qquad (8.2)$$

where we have used equation (5.3). Equations (8.1) and (8.2) form the basis for coupling of gas- and aqueous-phase chemical kinetics by the kinetics of inter-phase mass transport.

Before proceeding further it is useful to comment on the magnitude of the gas-aqueous exchange rate as represented by the term $L\, k_{mt}\, p_A$ in equation (8.2). We have noted that the cloud liquid-water volume fraction L is of order 10^{-6}, and we observe from Figure 12 that k_{mt} is of order 10^{-3} s^{-1} ($\alpha \approx 10^{-5}$) to 10^{6} s^{-1} ($\alpha \geq 10^{-2}$). Thus the one-way fractional removal rate of a gas into the aqueous phase is of order 10^{-3} to 1 s^{-1}. This order-of-magnitude argument establishes potentially very high removal rates of reactive gases by aqueous-phase reactions in clouds. In this context we would note that Levine and Schwartz

(1982) have calculated aqueous-phase uptake rates of HNO_3 for representative cloud drop size spectra of 0.2 to 1 s^{-1}, under assumption of unit accommodation coefficient and irreversible dissolution. Such potentially high removal rates also suggest that aqueous-phase reactions would be competitive with gas-phase processes, including photolysis, provided aqueous-phase reactions are sufficiently fast. A possible example might be aqueous-phase reaction of N_2O_5 competing with photolysis of NO_3, and has recently been suggested by Heikes and Thompson (1983).

Polydisperse Cloud.

We now explicitly recognize that k_{mt} is a function of drop radius $\underline{a}$ and as a consequence $[A]_a$ and $\bar{R}_2$ are functions of $\underline{a}$ as well. Thus for each drop radius cut equation (8.1) becomes

$$\frac{d[A]_a(a)}{dt} = (\underset{\sim}{R}T)^{-1} k_{mt}(a)\, p_A - (H_A \underset{\sim}{R}T)^{-1} k_{mt}(a)[A]_a(a) - \bar{R}_2(a) \, . \qquad (8.3)$$

In turn for the contribution to the mass-transport kinetics of A(g) due to cloud drops of size $\underline{a}$ we obtain

$$\frac{d_{mt}\, p_A}{dt} = - p_A\, k_{mt}(a)\, \delta L(a) + \frac{1}{H_A} k_{mt}(a)\; [A]_a(a)\, \delta L(a) \;, \qquad (8.4)$$

where $\delta L(a)$ is the partial liquid-water volume fraction in the size cut corresponding to $\underline{a}$,

$$\delta L = (\frac{4}{3}\pi \;\; a^3)\; (\frac{dN}{da})\; \delta a,$$

see Figures 3 and 4. The total mass transport rate expression is obtained by integration over the partial volume spectrum of the cloud,

$$\frac{d_{mt} p_A}{dt} = - p_A \int k_{mt}(a)\; dL + \frac{1}{H_A} \int k_{mt}(a)\; [A]_a\; dL \;, \qquad (8.5)$$

Equations 8.3 and 8.5 provide the basis for a set of coupled differential equations in the gas-phase and aqueous species. In practice, application of these equations to evaluation of rates for real or model systems will require either use of a model drop-size distribution of summation over size cuts as indicated in Section 3.

For aqueous-phase reaction by first-order reaction it is useful to treat the aqueous-phase concentration by the steady-state approximation. We note that for a first-order reaction $R_2(a) = k_e^{(1)}(a)\ (A)_a$, where $k_e^{(1)}$ is the effective first-order rate coefficient defined relative to the surface concentration (eq. 7.50).

Setting $\dfrac{d[A]_a(a)}{dt} = 0$ in equation (8.3) we obtain

$$[A]_a(a) = (\underset{\sim}{R}T)^{-1}\, p_A \frac{k_{mt}(a)}{(H_A\underset{\sim}{R}T)^{-1}\, k_{mt}(a) + k_e^{(1)}(a)}\,, \tag{8.6}$$

and in turn

$$\bar{R}_2(a) = (\underset{\sim}{R}T)^{-1}\, p_A \frac{k_e^{(1)}(a)\, k_{mt}(a)}{(H_A\underset{\sim}{R}T)^{-1}\, k_{mt}(a) + k_e^{(1)}(a)}\,. \tag{8.7}$$

Equation (8.7) embodies the dependence of the rate of aqueous-phase reaction on the size of the drop and provides a basis for anticipating that gas-aqueous reactions will proceed at a faster rate in smaller drops, expressed on a molar basis. In turn one would anticipate that the concentrations of non-volatile product species will increase in smaller drops at faster rates than in larger drops, and that this higher concentration might be measurable by means of a fractionating cloudwater collector. On the other hand, the increased solute concentration of the smaller drops might lead to an enhanced accretion rate of water vapor, at the expense of the supersaturation or even of the larger drops. This process would result in a leveling of the solute concentration distribution and a narrowing of the drop-size distribution.

9. SUMMARY

A formalism exists that permits description of the kinetics of mass-transport processes in the uptake of gases by aqueous droplets and aqueous-phase chemical reaction for droplets in the size range pertinent to liquid-water clouds. These mass-transport processes are gas-phase diffusion, gas-kinetic collision at the interface, and aqueous-phase diffusion coupled with chemical reaction. Criteria are presented whereby it may be readily discerned, as a function of drop size, reaction rate coefficient, and mass-accommodation coefficient whether a system under examination may be treated as phase mixed or whether more detailed treatment of mass-transport limitation is required, and expressions are presented by which mass-transport limitation can be evaluated.

One finding that emerges from the treatment of mass-transport kinetics is the dominance of interfacial mass-transport limitations for values of the mass-accommodation coefficient less than 10^{-2}. Unfortunately, little information is available pertinent to this quantity for solute gases with liquid water. The potential importance of interfacial mass-transport limitation to the rate of gas-aqueous reactions in clouds indicates the need for laboratory studies directed to determination of this quantity.

ACKNOWLEDGMENT

This work was supported in part by the Electric Power Research Institute (Project RP2023-1) and was performed under the auspices of the Department of Energy under Contract No. DE-AC02-76CH00016.

PRINCIPAL NOMENCLATURE

a	Drop radius, cm.
A,[A]	Aqueous-phase concentration of species A, M.
D	Diffusion coefficient, $cm^2\ s^{-1}$.
F	Flux, mol $cm^{-2}\ s^{-1}$.
g	Gravitational constant, $cm^2\ s^{-1}$.
G	Gas-phase concentration, M.
H_X	Henry's law coefficient of species X, M atm^{-1}.
$k^{(m)}$	Reaction rate coefficient for m^{th} order reaction, $M^{-(m-1)}\ s^{-1}$.
k_n	Reaction rate coefficient of reaction [n].
k_{mt}	Rate coefficient for gas-phase plus interfacial mass transport, s^{-1}.
L	Liquid water volume fraction (of cloud).
N	Drop number density (of cloud), cm^{-3}.
p_X	Partial pressure of species X, atm.
q	Diffuso-reactive parameter, $a(k^{(1)}/D_a)^{1/2}$
r	Radius variable, cm.
R_n	Rate of reaction [n], M s^{-1}.
$\underset{\sim}{R}$	Universal gas constant, $atm^{-1}\ M^{-1}\ K^{-1}$ except where indicated.
T	Absolute temperature, K.
$\bar{v}$	Mean molecular speed, cm s^{-1}.
α	Mass accommodation coefficient.
$\mathcal{E}$	Maximum allowable departure from uniformity condition.
η	Dynamic viscosity, g $cm^{-1}\ s^{-1}$.
νX	Stoichiometric coefficient of species X in reaction.
ρX	Fractional rate of reaction of gaseous species X, s^{-1}.
σ	Interfacial collision flux, mol $cm^{-2}\ s^{-1}$.
τ	Characteristic time, s (see Table 4).

REFERENCES

Adamson, A.W. (1973), A Textbook of Physical Chemistry. Academic Press, New York.

Andrew, S.P.S. (1955), A Simple Method of Measuring Gaseous Diffusion Coefficients. Chem. Eng. Sci. 4, pp. 269-272.

Baboolal, L.B., H.R. Pruppacher, J.H. Topalian (1981), A Sensitivity Study of a Theoretical Model of SO_2 Scavenging by Water Drops in Air. J. Atmos. Sci. 38, pp. 856-870.

Beilke, S., G. Gravenhorst (1978), Heterogeneous SO_2 Oxidation in the Droplet Phase. Atmos. Environ. 12, pp. 231 - 239.

Blair, E.W., W. Ledbury (1925), The Partial Formaldehyde Vapor Pressures of Aqueous Solutions of Formaldehyde, Part I. J. Chem. Soc. 1925, pp. 26-40.

Briner, E., E. Perrottet (1939), Détermination des Solubilités de l'Ozone dans l'Eau et dans une Solution Aqueuse de Chlorure de Sodium; Calcul des Solubilités de l'Ozone Atmospherique dans les Eaux. Helv. Chim. Acta 22, pp. 397-404.

Brock, J.R., J.L. Durham (1984), Aqueous Aerosol as a Reactant in the Atmosphere, In: SO_2, NO, and NO_2 Oxidation Mechanisms: Atmospheric Considerations, J.G. Calvert, ed., pp. 209-250, Butterworth, Boston.

Chameides, W.L. (1985), Photochemistry in the Atmospheric Aqueous Phase. This volume.

Chameides, W.L., D.D. Davis (1982), The Free Radical Chemistry of Cloud droplets and its Impact Upon the Composition of Rain. J. Geophys. Res. 87, pp. 4863-4977.

Chameides, W.L., D.D. Davis (1983), Aqueous-Phase Source of Formic Acid in Clouds. Nature 304, pp. 427-429.

Chang, R., E.J. Davis (1976), Knudsen Aerosol Evaporation. J. Colloid Interface Sci. 54, pp. 352-363.

Chapman, S., T.G. Cowling (1939), The Mathematical Theory of Non-Uniform Gases. Cambridge University Press, Cambridge.

Chapman, S., T.G. Cowling (1970), The Mathematical Theory of Non-Uniform Gases, 3rd ed. Cambridge University Press, Cambridge.

Crank, J. (1975), The Mathematics of Diffusion, 2nd ed. Clarendon Press, Oxford.

Dahneke, B. (1983), Simple Kinetic Theory of Brownian Diffusion in Vapors and Aerosols. In: Theory of Dispersed Multiphase Flow. R.E. Meyer, ed., Academic Press, Inc., New York, pp. 97-133.

Danckwerts, P.V. (1970), Gas-Liquid Reactions. McGraw-Hill, New York.

Daum, P.H., S.E. Schwartz, L. Newman (1984), Acidic and Related Constituents in Liquid Water Clouds. J. Geophys. Res., 89, pp. 1447-1485

Davis, E.J. (1983), Transport Phenomena with Single Aerosol Particles. Aerosol Sci. Technol. 2, pp. 121-144.

Freiberg, J.E., S.E. Schwartz (1980), Oxidation of SO_2 in Aqueous Droplets: Mass Transport Limitation in Laboratory Studies and the Ambient Atmosphere. Atmos.Environ. 15, pp. 1145-1154.

Fuchs, N.A., A.G. Sutugin (1971), High-Dispersed Aerosols. In: Topics in Current Aerosol Research, G.M. Hidy, J.R. Brock, eds., pp. 1-60, Pergamon, Oxford.

Fuller, E.N., D. Schettler, J.C. Giddings (1966), A New Method for Predictation of Binary Gas-Phase Diffusion Coefficient. Ind. Eng. Chem. 58 (5), pp. 19-27.

Gill, P.S., T.E. Graedel, C.J. Weschler (1983), Organic Films on Atmospheric Aerosol Particles, Fog Droplets, Cloud Droplets, Raindrops, and Snowflakes. Rev. Geophys. Space. Phys. 21, pp. 903 - 920

Graedel, T.E., P.S. Gill, C.J. Weschler (1983), Effects of Organic Surface Films on the Scavenging of Atmospheric Gases by Raindrops and Aerosol Particles. In: Precipitation Scavenging, Dry Deposition and Resuspension, Vol. 1, H.R. Pruppacher, R.G. Semonin, W.G.N. Slinn, eds., pp. 417-430, Elsevier, New York.

Harvey, E.A., W. Smith (1959), The Absorption of Carbon Dioxide by a Quiescent Liquid. Chem. Eng. Sci. 10, pp. 274-280.

Hegg, D.A., P.V. Hobbs (1983), Preliminary Measurements on the Scavenging of Sulfate and Nitrate by Clouds. In: Precipitation Scavenging, Dry Deposition and Resuspension, Vol. 1, H.R. Pruppacher, R.G. Semonin, W.G.N. Slinn, eds., pp. 79-89, Elsevier, New York.

Heikes, B.G. and A.M. Thompson (1983), Effects of Heterogeneous Processes on NO_3, HONO and HNO_3 Chemistry in the Troposphere. J. Geophys. Res., 88, pp 10883 - 10895

Hertz, H. (1882), Ueber die Verdunstung der Fluessigkeiten, insbesondere des Quecksilbers, im luftleeren Raume. Ann. Phys. 17, pp. 177-193.

Hidy, G.M., J.R. Brock (1970), The Dynamics of Aerocolloidal Systems. Pergamon, Oxford.

Himmelblau, P.M. (1964), Diffusion of Dissolved Gases in Liquids. Chem. Rev. 64, pp. 527-550.

Johnstone, H.F., P.W. Leppla (1934), The Solubility of Sulfur Dioxide. J. Am. Chem. Soc. 56, pp. 2233-2238.

Knollenberg, R.G. (1981), Techniques for probing cloud microstructure. In: Clouds -- Their Formation, Optical Properties, and Effects, P.V. Hobbs, A. Deepak, eds., pp. 15-91, Academic Press, New York.

Knudsen, M. (1915), Die maximale Verdampfungsgeschwindigkeit des Quecksilbers. Ann. Phys. 47, pp. 697-708.

La Mer, V.K. (162), Retardation of Evaporation by Monolayers: Transport Processes. Academic Press, New York.

Langmuir, I. (1913), The Vapor Pressure of Metallic Tungsten. Phys. Rev. 2, pp. 329-342.

Lazrus, A.L., Haagenson, P.L., Kok, G.L., Huebert, B.J., Kreitzberg, C.W., Likens, G.E., Mohnen, V.A., Wilson, W.E., Winchester, J.W. (1983), Acidity in air and water in a case of warm frontal precipitation. Atmos. Environ. 17, pp. 581-591.

Leaitch, W.R., J.W. Strapp, H.A. Wiebe, G.A. Isaac (1983), Measurements of Scavenging and Transformation of Aerosol Inside Cumulus, In: Precipitation Scavenging, Dry Deposition and Resuspension, Vol. 1, H.R. Pruppacher, R.G. Semonin, W.C.N. Slinn, eds., pp. 53-69, Elsevier, New York.

Lee, Y.-N., S.E. Schwartz (1981a), Reaction Kinetics of Nitrogen Dioxide with Liquid Water at Low Partial Pressure. J. Phys. Chem. 85, pp. 840-848.

Lee, Y.-N. and S.E. Schwartz, (1981b), Evaluation of the Rate of Uptake of Nitrogen Dioxide by Atmospheric and Surface Liquid Water, J. Geophys.Res. 86, pp. 11971-11983.

Lee, Y.-N., G.I. Senum, J.S. Gaffney (1983), Peroxyacetyl Nitrate (PAN) Stability, Solubility, and Reactivity - Implications for Tropospheric Nitrogen Cycles and Precipitation Chemistry. Fifth International Conference of the Commission on Atmospheric Chemistry and Global Pollution; Symposium on Tropospheric Chemistry, Oxford, England, August 28 - September 2, 1983.

Littlewood, R., E. Rideal (1956), On the Evaporation Coefficient. Trans. Faraday Soc. 52, pp. 1598-1608.

Loomis, A.G. (1928), Solubilities of Gases in Water. In: International Critical Tables, Vol. III. pp. 255-261, McGraw-Hill, New York.

Lugg, G.A. (1968), Diffusion Coefficients of Some Organic and Other Vapors in Air. Anal. Chem. 40, pp. 1072-1077.

Maa, J.R. (1967), Evaporation Coefficient of Liquids. Ind. Eng. Chem. Fundam. 6, pp. 504-518.

Maa, J.R. (1970), Rates of Evaporation and Condensation between Pure Liquids and Their Own Vapors. Ind. Eng. Chem. Fundam. 9, pp. 283-287.

Marrero, T.R., E.A. Mason (1972), Gaseous Diffusion Coefficients. J. Phys. Chem. Ref. Data 1, pp. 3-118.

Martin, L.R. (1984), Kinetic Studies of Sulfite Oxidation in Aqueous Solution. In: SO_2, NO, and NO_2 Oxidation Mechanisms: Atmospheric Considerations, J.G. Calvert, ed., 63-100, Butterworth, Boston.

Martin L.R., D.E. Damschen (1981), Aqueous Oxidation of Sulfur Dioxide by Hydrogen Peroxide at Low pH. Atmos. Environ. 15, pp. 1615-1621.

Mason, B.J. (1971), The Physics of Clouds, 2nd ed. Clarendon Press, Oxford.

Maxwell, J.C. (1877), Diffusion, In: Encyclopedia Britannica, Vol. 2. p. 82. Reprinted in: The Scientific Papers of James Clerk Maxwell, W.D. Niven, ed., Dover, New York, Vol. 2, pp. 625-646.

Mohnen, V. (1969), Die Radioaktiv-Markierung von Aerosolen. Z. Phys. 229, pp. 109-122.

Munger, J.W., D.J. Jacob, J.M. Waldman, M.R. Hoffmann (1983), Fogwater Chemistry in an Urban Atmosphere. J. Geophys. Res. 88, pp. 5109-5121.

Paul, B. (1962), Compilation of Evaporation Coefficients. Amer. Rocket. Soc. J. 32, pp. 1321-1328.

Pruppacher, H.R., J.D. Klett (1978), Microphysics of Clouds and Precipitation. D. Reidel Publishing Co., Boston, MA.

Radke, L.F. (1983), Preliminary Measurements of the Size Distribution of the Cloud Interstitial Aerosol. In: Precipitation Scavenging, Dry Deposition and Resuspension, Vol. 1, H.R. Pruppacher, R.G. Semonin, W.G.N. Slinn, eds., pp. 71-78, Elsevier, New York.

Raimondi, P., H.L. Toor (1959), Interfacial Resistance in Gas Absorption. Am. Inst. Chem. Eng. J. 5, pp. 86-92.

Rogers, R.R. (1979), A Short Course in Cloud Physics, 2nd ed. Pergamon Press, Oxford.

Schwartz, S.E. (1984b), Gas- and Aqueous-Phase Chemistry of HO_2 in Liquid- Water Clouds. J Geophys. Res. 89, pp/ 11589 - 11598

Schwartz, S.E., (1984), Gas-Aqueous Reactions of Sulfur and Nitrogen Oxides in Liquid-Water Clouds. In: SO_2, NO, and NO_2 Oxidation Mechanisms: Atmospheric Considerations, J.G. Calvert, ed., pp. 173-208, Butterworth, Boston.

Schwartz, S.E., J.E. Freiberg (1981), Mass-Transport Limitation to the Rate of Reaction in Liquid Droplets: Application to Oxidation of SO_2 in Aqueous Solutions. Atmos. Environ. 15, pp. 1129-1144.

Schwartz, S.E., W.H. White (1981), Solubility Equilibria of the Nitrogen Oxides and Oxyacids in Dilute Aqueous Solution, In: Advances in Environmental Science and Engineering, Vol. 4, J.R. Pfafflin and E.N. Ziegler, eds., pp. 1-45, Gordon and Breach, New York.

Schwartz, S.E., W.H. White (1983), Kinetics of Reactive Dissolution of Nitrogen Oxides into Aqueous Solution. In: Advances in Environmental Science and Technology, Vol. 12, S.E. Schwartz, ed., pp. 1-116, John Wiley and Sons, Inc., New York.

Seinfeld, J.H. (1980), Lectures in Atmospheric Chemistry, American Institute of Chemical Engineers, Monograph Series, No. 12, Vol. 76, New York.

Sherwood, T.K., R.L. Pigford, C.R. Wilke (1975), Mass Transfer. McGraw-Hill, New York.

van Krevelen, D.W., P.J. Hoftijzer, F.J. Huntens (1949), Composition and Vapour Pressures of Aqueous Solutions of Ammonia, Carbon Dioxide and Hydrogen Sulphide. Recueil Trav. Chim. Pays Bas 68, pp. 191-216.

Wagner, P.E. (1982), Aerosol Growth by Condensation. In: Aerosol Microphysics II: Chemical Physics of Microparticles, W.H. Marlow, ed., Springer- Verlag, Berlin, pp. 129-178.

Warneck, P. (1985), The Equilibrium Distribution of Atmospheric Gases between the Two Phases of Liquid Water Clouds. This volume.

Wilke, C.R., P. Chang (1955), Correlation of Diffusion Coefficients in Dilute Solutions, Am. Inst. Chem. Eng. J. 1, pp. 264-270.

Wilke, C.R., C.Y. Lee (1955), Estimation of Diffusion Coefficients for Gases and Vapors, Ind. Eng. Chem. 47, pp. 1253-1257.

THE EQUILIBRIUM DISTRIBUTION OF ATMOSPHERIC GASES BETWEEN THE TWO PHASES OF LIQUID WATER CLOUDS

Peter Warneck
Max Planck Institut fuer Chemie
Saarstr. 23
D-6500 Mainz 31
F.R.G.

ABSTRACT

The partitioning of water-soluble gases between the aqueous and gaseous phases of clouds is treated by assuming that the system is near thermodynamic equilibrium. Henry's law can then be applied to determine the degree of in-cloud scavenging of atmospheric volatiles, the resulting concentrations of the substances in the aqueous phase, and the associated residence times in the atmosphere due to rain-out. Gases reacting with water to form ions require the definition of a modified Henry coefficient which depends on the pH of the solution. It is shown that, potentially, the pH of individual cloud drops can vary over a wide range. The true distribution of pH among cloud drops is presently not known, so that the application of the formalism developed to determine the scavenging efficiency for reactive gases must await further data. Adduct formation between formaldehyde and bisulfite is discussed as an example fo solute-solute interaction which are not accounted for by Henry's law. Such interactions can be treated provided equilibria are established and the equilibrium constants are known.

1. INTRODUCTION

Clouds represent multi-phase systems consisting of liquid water drops and/or ice particles dispersed in air. The dissolution in cloud water may cause a significant fraction of an atmospheric trace gas to move from the gas to the condensed phase, a process usually referred to as in-cloud scavenging (Slinn, 1974). This article deals with the partitioning of gases between the two phases of liquid water clouds on the premise that the system is kept close to thermodynamic equilibrium during the life time of the cloud. The fraction of a gas

NATO ASI Series, Vol. G6
Chemistry of Multiphase Atmospheric Systems
Edited by W. Jaeschke

or vapor residing in aqueous solution is then determined by Henry's law.

It will be useful to begin with a brief description of the mechanism of cloud formation before entering into discussions of Henry's law and its applications. While many atmospheric gases undergo merely a physical dissolution in cloud water, there are a number of important atmospheric constituents which interact with water to form ions. Their treatment is more involved and leads to a modified form of Henry's law with a partitioning coefficient that depends on the pH. Although this aspect is, in principle, well understood, we are presently lacking information on the frequency distribution of pH values among cloud drops of varying sizes that would allow us to calculate the partitioning of reactive gases between the gas and the liquid phases of clouds with some confidence. Nevertheless, factors determining the pH in individual cloud drops can and will be discussed. Finally, in the last section, adduct formation from formaldehyde and bisulfite is considered as an example for solute interactions in the aqueous phase causing deviations from Henry's law behavior. If such reactions are reversible and equilibrium is established, one can take their effects into account provided the equilibrium constants are known.

2. MECHANISM OF CLOUD FORMATION

Cloud drops are generated by the condensation of water onto aerosol particles when in a rising air parcel the relative humidity transcends the saturation level due to adiabatic cooling. Aerosol particles are important in two ways. On one hand, they serve as condensation nuclei enabling clouds to form at fairly low degrees of H_2O supersaturation. On the other hand, they supply water-soluble material such as inorganic salts which contribute significantly to the ionic composition of cloud water. Although with rising relative humidity all particles acquire a certain share of liquid water, it is only the larger particles with radii greater than 0.1 µm which become activated and grow into cloud drops once the relative humidity passes beyond the 100% mark (Junge, 1963, Pruppacher and Klett, 1978). During the condensation stage, droplet growth is mediated by the diffusive transport of water vapor. The growth rate is inversely proportional to the droplet's radius, however, so that the rate slows down as the drop size increases. For this reason the size distribution of cloud drops develops initially in a narrow range of radii in the vicinity of 5-10 µm. Larger drops with sizes up to 100 µm are produced later by the coalescence of water drops as the cloud ages and increases in depth.

The dissolution of gases in cloud water likewise is a kinetic process involving molecular diffusion of a substance from the gaseous to the liquid phase. The diffusion coefficients of common gases have magnitudes similar to that of water vapor, so that the gases are incorporated into cloud drops with nearly the same rate as water. Accordingly, one expects the dissolution of gases in cloud water to take place concurrently with cloud drop formation.

In fact, trace gases may temporarily reach a certain supersaturation in the liquid phase during the condensation stage owing to the comparatively massive flux of water vapor which obstructs the re-evaporation of gases from the liquid. This concept suggests that most gases equilibrate quickly with the liquid phase during the initial stage when the cloud forms. Alternatively, one may consider a pure water drop of given size and examine the time constant for the approach to solution equilibrium by diffusional transport of material from the gas to the liquid phase. The kinetic processes involved are discussed by Schwartz (1985) elsewhere in this volume. The time constants for mass transfer increases with the size of the cloud drop and the amount of the substance that is to be incorporated into the drop. Usually, the time constant for approach to equilibrium is less than 100 s, whereas most clouds live at least ten times longer before they dissolve again or rain out. The assumption of solution equilibrium thus seems to provide a useful approximation to the real situation.

3. HENRY'S LAW FOR UNREACTIVE GASES AND VAPORS

For ideal conditions which are realized for sufficiently small concentrations, thermodynamics predicts a linear relationship between the mole fraction x_s of a substance dissolved in water and its partial pressure P_s in the gas phase above the solution

$$P_s = H x_s = H \nu_s/(\nu_w + \sum \nu_i) \approx H \nu_s/\nu_w \qquad (1)$$

where H is a coefficient that is independent of concentration but varies with temperature (approximately $\sim \exp(-A/T)$). The relation has been verified experimentally for slightly soluble gases and is known as Henry's law. ν_s denotes the number of moles of the solute of interest and $\nu_w + \sum \nu_i$ is the total number of moles of liquid water and all solutes present. Textbooks usually treat only binary systems. Henry's law remains valid for multi-component systems, as indicated above, provided there is no interaction between the solute molecules. This condition exists in dilute solutions and fortunately for our further considerations, cloud water represents a solution sufficiently dilute to be treated in this manner. The mole fraction then approximates to ν_s/ν_w and is proportional to concentration. Accordingly, Henry's law often is written in the forms

$$P_s = H_c c_s \qquad \text{or} \qquad [s] = K_H P_s \qquad (1a)$$

Here, c_s denotes the concentration in $\mathrm{Mol\,kg^{-1}}$ (molality scale), and [s] is the concentration in Mol liter $^{-1}$ (molarity scale) here denoted by M. Both units are related in that $[s] = \rho_w c_s$ where $\rho_w \approx 1$ kg dm^{-3} is the density of water. Its variation with temperature causes the molarity scale to be temperature dependent whereas the molality scale is not. In the temperature range 0-25°C, the density of water differs from unity by less than 0.3%, however, and since most Henry constants are less well known, it is permissible to set [s] = c_s with reasonable accuracy. The number of moles of water contained in 1 kg H_2O is ν_w = $10^3/M_w$ where M_w is the molecular weight of water. Therefore, one has

$$c_s = (10^3/M_w)x_s$$

$$H = (10^3/M_w)H_c = (10^3/M_w)(\rho_w/K_H) \quad \text{with } 10^3/M_w = 55.51 \qquad (2)$$

as convenient conversion formulas for Henry coefficients.

Wilhelm et al.,(1977) have tabulated Henry coefficients for many gases, expressing their temperature dependence in a suitable parametric form. Data for water-soluble vapors are scattered throughout the literature. At high aqueous concentrations, their vapor pressures often deviate considerably from Henry's law. Due to inadequacies of the employed analytical techniques or due to different research goals, many literature reports of vapor pressures above aqueous solutions refer to concentrations outside the Henry law regime. An example is given in Figure 1 for acetonitrile. The curvature exhibited in the plot of P_s versus x_s makes it difficult to extrapolate the data toward lower concentrations where the relation is expected to be linear. As a consequence, Henry law coefficients for many substances of interest to atmospheric chemistry are not really well known.

Another problem encountered in the application of these data is the phenomenon of supercooling in deep clouds. The spontaneous glaciation of cloud drops at temperatures below the freezing point of water generally is a rare event unless the temperature approaches about -20°C. This value corresponds to an altitude of about 6 km. In the middle troposphere, therefore, supercooled clouds are the rule rather than an exception. Yet we have no measurements of Henry coefficients for supercooled solutions. Numerical values can be deduced only by extrapolation of the data available for temperatures above 273 K.

A few selected Henry coefficients are collected in Table 1. The examples given refer to the temperature at the cloud condensation level, i.e. the altitude where a rising air parcel with an initial relative humidity of 60% passes the saturation point. For illustration, a temperature of 283 K is here adopted. The corresponding

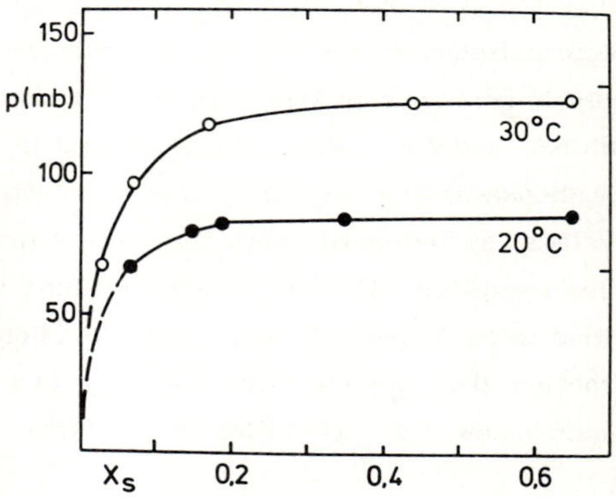

Figure 1. Vapor pressure of acetonitrile according to Vierk (1950). The non-linearity observed indicates that the data fall outside the Henry law regime. It is difficult to extrapolate the data toward smaller mole fractions where linearity is expected to exist.

Table 1 Henry law coefficients for selected atmospheric gases, in-cloud scavenging efficiencies ϵ, concentrations c_s of substances in cloud water for initial gas phase mixing ratios m_o, and residence times τ for rainout. The latter data for a liquid water content L=2 g m^{-3}. Powers of ten are given in parentheses.

Constituent	H(bar)[a]	ϵ L=0.1	ϵ L=1.0	m_o	c_s(Mol kg^{-1})	τ L=2.0		References
N_2	6.66(4)	2.0(-9)	2.0(-8)	0.78	5.6(-4)	4.0(5)	yrs	Wilhelm et al(1977)
O_2	3.89(4)	4.0(-9)	4.0(-8)	0.21	3.0(-4)	2.0(5)	"	"
O_3	3.09(3)	4.3(-8)	4.3(-7)	4.0(-8)	6.2(-10)	1.9(4)	"	"
N_2O	1.43(3)	9.2(-8)	9.2(-7)	3.0(-7)	1.0(-8)	8.6(3)	"	"
CH_3Cl	3.19(2)	4.1(-7)	4.1(-6)	5.0(-10)	7.8(-11)	1.8(3)	"	"
NO	2.21(4)	5.9(-9)	5.9(-8)	5.0(-10)	1.1(-12)	1.3(5)	"	Winkler(1901)
NO_2	7.93(3)[b]	1.6(-8)	1.6(-7)	1.0(-9)	6.3(-12)	4.5(4)	"	Lee & Schwartz(1981)
HCHO	3.85(-3)	3.3(-2)	2.5(-1)	5.0(-10)	3.7(-6)	14.2	days	Walker(1975)
H_2O_2	2.29(-4)	5.7(-1)	9.3(-1)	1.0(-9)	1.7(-5)	6.2	"	Martin & Damschen(1981)

a T=283 K except b: 295 K

elevation is roughly 1 km and the associated pressure is 900 mb.

4. IN-CLOUD SCAVENGING EFFICIENCY AND TIME CONSTANT FOR RAINOUT.

In order to determine the equilibrium partitioning between the two phases of a cloud for a water-soluble atmospheric component, we consider a fixed volume V_c of cloud-filled air containing ν_o number of moles of the substance in question. Let ν_s moles of it reside in solution and ν_g moles in the gas phase. We make use of the ideal gas law in form

$$\nu_g = P_s V_c / R_g T \tag{3}$$

where $R_g = 8.3 \times 10^{-5} m^3$ bar $deg^{-1} Mol^{-1}$ is the gas constant and T the absolute temperature. Further, we set

$$L = \nu_w M_w / V_c \quad (g\ m^{-3}) \tag{4}$$

to represent the liquid water content of the cloud. The following relation then holds:

$$\nu_o = \nu_s + \nu_g = \nu_s + P_s V_c / R_g T = \nu_s + \frac{V_c H \nu_s}{R_g T \nu_w} = \nu_s (1 + HM_w / R_g TL) \tag{5}$$

The fraction of the substance incorporated into solution is given by

$$\varepsilon = \nu_s/\nu_o = 1/(1 + HM_w/R_gTL) \quad (6)$$

and the corresponding concentration in the aqueous phase is

$$c_s = \frac{\nu_s 10^3}{\nu_w M_w} = \frac{10^3 \varepsilon \nu_o}{LV_c} = \frac{10^3 \varepsilon m_o P_{air}}{R_gTL} \quad (7)$$

Here, P_{air} is the atmospheric pressure at the condensation level, and m_o is the mixing ratio of the substance in the gas phase before condensation took place.

The liquid water content of clouds ranges from 0.1 g m^{-3} for continental fair weather cumuli to 1-3 g m^{-3} for precipitating clouds. We select 0.1 and 1 g m^{-3} for the purpose of calculation and list in Table 1 the corresponding values of ε and of c_s for the gases shown. The fraction is small for most gases, but becomes significant for some important atmospheric constituents such as hydrogen peroxide and formaldehyde. The concentrations depend on the product εm_o, so that nitrogen and oxygen occur with fairly high concentrations even though ε is small. Note that oxygen is somewhat more soluble than nitrogen, whereby the O_2/N_2 ratio in solution increases to 0.54 compared with 0.27 in the gas phase.

The above results may be utilized to derive the residence time of a water-soluble gaseous constituent in the troposphere due to rainout. We must assume that the cloud water concentration c_s is approximately preserved during precipitation, i.e. we neglect processes such as below-cloud scavenging and evaporation of water from falling rain drops. The average mass flux of the substance being removed by rainout is

$$F_s = 10^{-3} c_s M_s \overline{R} \quad (8)$$

where M_s is the molecular weight of the substance and $\overline{R}$ is the average annual precipitation rate within the region considered. The average residence time is given by the ratio of the column density of the substance in the troposphere to the mass flux:

$$\tau_{rainout} = \frac{G_s}{F_s} = \frac{(M_s/M_{air})\overline{m}G_T}{10^{-3}c_s M_s \overline{R}} = \frac{\overline{m}G_T R_g TL}{m_o \varepsilon M_{air} P_{air} \overline{R}} = 1.5\times10^{-2}\frac{m}{m_o}\frac{1}{\varepsilon}(yrs) \quad (9)$$

G_T is the column density of air and $\overline{m}$ is the height-averaged mixing ratio of the substance in the troposphere. Averaged over a hemisphere we have approximately: $G_T = 8.3\times10^3$kg m^{-2}, $\overline{R} = 10^3$kg $m^{-2}yr^{-1}$, M_{air}=29 $gMol^{-1}$, and we adopt L=2 g m^{-3} to derive the numerical factor shown. For constituents which are fairly evenly distributed in the troposphere one has $\overline{m}$ = m_o so that that the residence time becomes independent of the ratio $\overline{m}/m_o$. In this case it depends only on the fraction of the substance that is dissolved in cloud water.

Values for $\tau_{rainout}$ calculated with this provision are included in Table 1 to indicate the orders of magnitude obtained. For slightly soluble gases the time scale is too long to have any significance. Permanent gases, moreover, are not retained by ground surface materials and are soon released again to the atmosphere. To reach residence times of one year or less requires an in-cloud scavenging such that the fraction ε becomes 2% or greater. This condition is met for only a few highly soluble vapors. However, ε cannot exceed unity so that the lower limit residence time according to the preceding equation is 1.5×10^{-2} years or 5.5 days. This value should not be taken as an absolute lower limit because $\bar{m}$ may be and frequently is smaller than m_o. In addition, we have ignored below-cloud scavenging in our considerations.

In the considerations leading to equ (9) it was convenient to use an hemispherically averaged precipitation rate. On the local scale, rainfall is a statistical event admitting dry periods longer than a week. This is longer than the global residence time calculated for those trace gases whose in-cloud scavenging efficiency is near unity. Clearly in such cases, considerable local fluctuations of residence time and mixing ratio must occur. Rhode and Grandell (1972) have dealt with this aspect.

5. MODIFICATION OF HENRY'S LAW FOR REACTIVE GASES

A number of gases and vapors interact with water to form ions upon dissolution. To this category belong several important atmospheric constituents, among them CO_2, SO_2, NH_3, HCl and HNO_3 which we now consider. Their reactions with water are reversible and we shall again assume that equilibrium conditions are set up in clouds. Table 2 summarizes the reactions involved and the associated equilibrium constants. The values shown refer to temperatures of 283 K and 298 K. Values for other temperatures can be calculated from the thermochemical data tabulated by the National Bureau of Standards. Most of the equilibria describe dissociation reactions in which a proton is released to the aqueous medium. Rate coefficients for many of the individual forward and reverse reactions are known. Their values are included in Table 2 to demonstrate that these reactions are fast. The associated time constants are of the order of microseconds or less so that such equilibria are rapidly established.

As Table 2 shows, the dissolution of reactive gases proceeds by at least two consecutive steps. The first involves the hydration of the substance, the second the formation of the ionic species from the hydrate. A third step, the donation of another proton to the aqueous medium, exists for diprotic acids. To illustrate these effects, we select CO_2 as an example. The equilibrium relations then are

Table 2 Equilibrium constants for reactive gases and the ions involved; calculated mostly from thermochemical data(Wagman et al,1968). Data for HCl are from Fritz and Fuget(1956), data for HNO_3 from Schwartz and White(1981). Rate coefficients were taken from the compilation of Graedel and Weschler(1981). Equilibrium constant for reaction 9 see text.Powers of ten are shown in parentheses.

No	Reaction		K_{298}	K_{283}	units[a]	H[b]	k_f[c]	k_r[c]
w	$H_2O \rightleftharpoons OH^- + H^+$	K_w	1.0(-14)	3.0(-15)	M^2		1.3(-3)	1.3(11)
H1	$CO_{2g} + H_2O \rightleftharpoons CO_2 \cdot H_2O$	K_{H1}	3.4(-2)	5.2(-2)	$Mbar^{-1}$	1.07(3)		
1	$CO_2 \cdot H_2O \rightleftharpoons H^+ + HCO_3^-$	K_1	4.3(-7)	3.6(-7)	M		0.04	5.6(4)
2	$HCO_3^- \rightleftharpoons H^+ + CO_3^=$	K_2	4.5(-11)	3.3(-11)	M		~2.5	~5 (10)
H2	$SO_{2g} + H_2O \rightleftharpoons SO_2 \cdot H_2O$	K_{H2}	1.3	2.2	$Mbar^{-1}$	2.52(1)		
3	$SO_2 \cdot H_2O \rightleftharpoons H^+ + HSO_3^-$	K_3	1.3(-2)	2.4(-2)	M		3.4(6)	2.0(8)
4	$HSO_3^= \rightleftharpoons H^+ + SO_3^=$	K_4	6.4(-8)	7.8(-8)	M		3.0(3)	5.0(10)
H3	$NH_{3g} + H_2O \rightleftharpoons NH_3 \cdot H_2O$	K_{H3}	5.8(1)	1.2(2)	$Mbar^{-1}$	4.77(-1)		
5	$NH_3 \cdot H_2O \rightleftharpoons NH_4^+ + OH^-$	K_5	1.8(-5)	1.7(-5)	M		6.0(5)	3.4(10)
H4	$HNO_{2g} \rightleftharpoons HNO_{2aq}$	K_{H4}	4.9(1)	1.1(2)	$Mbar^{-1}$	4.93(-1)		
6	$HNO_{2aq} \rightleftharpoons H^+ + NO_2^-$	K_6	6.0(-4)	4.4(-4)	M		3 (5)	5(10)
H5	$HCOOH_g \rightleftharpoons HCOOH_{aq}$	K_{H5}	3.8(3)	1.0(4)	$Mbar^{-1}$	5.50(-3)		
7	$HCOOH_{aq} \rightleftharpoons H^+ + HCO_2^-$	K_7	1.8(-4)	1.8(-4)	M		8.6(6)	~5(10)
H6	$HCl_g + H_2O \rightleftharpoons H^+ + Cl^-$	K_{H6}	2. (6)	1.4(7)	M^2bar^{-1}			
H7	$HNO_{3g} + H_2O \rightleftharpoons H^+ + NO_3^-$	K_{H7}	3.2(6)	1.5(7)	M^2bar^{-1}			
8	$HSO_4^- \rightleftharpoons H^+ + SO_4^=$	K_8	~1 (-2)	-	M			
9	$HCHO_{aq} + HSO_3^- \rightleftharpoons CH_2(OH)SO_3^-$	K_9	~5(4)	-	M^{-1}			

a M = mol liter^{-1}; b for T=283 K; c in units of Mol liter^{-1} and second.

$$K_H P_{CO_2} = [CO_2 \cdot H_2O]$$

$$K_1 [CO_2 \cdot H_2O] = [H^+][HCO_3^-]$$

$$K_2 [HCO_3^-] = [H^+][CO_3^=]$$

Here, the concentration units are Mol liter^{-1} (indicated by brackets) and we assume as previously the existence of an ideal solution. The first product is written in the form of an adduct, because it is difficult to distinguish analytically between a CO_2 molecule enclosed in a solvent cage and carbonic acid, H_2CO_3. Both are present in solution and may undergo ion pair formation. Accordingly, K_1 is a composite equilibrium constant which refers to the sum of CO_{2aq} and H_2CO_3. The concentration of the former is dominant, but the latter is the stronger acid (Stumm and Morgan, 1981). A similar situation exists for SO_2. The equilibrium constants for the individual species are not sufficiently well known, however, to make their use profitable. Note also that hydrogen ions in aqueous solution always occur in the form of H_3O^+ (hydronium) and that the notation H^+ is merely a convention.

The total solute concentration of carbonate species following from the dissolution of CO_2 in water is

$$[s] = \rho_w c_s = [CO2 \cdot H2O] + [HCO_3^-] + [CO_3^=] \tag{10}$$

$$= [CO_2 \cdot H_2O] \left\{ 1 + \frac{K_1}{[H^+]} + \frac{K_1 K_2}{[H^+]^2} \right\}$$

$$= [CO_2 \cdot H_2O] \frac{1}{f_{CO_2}} = K_H P_{CO_2} \frac{1}{f_{CO_2}}$$

where we have assumed that the hydrate obeys Henry's law, and f_{CO_2} is a function solely of the hydrogen ion concentration, identical to the reciprocal of the expression in the curved brackets. Combining eqs (1), (2) and (11) yields.

$$P_{CO_2} = f_{CO_2} H_{CO_2} x_s \tag{11}$$

Equivalent results are obtained for other gases whose reactions with water produce weak acids. The relations obtained are similar to Henry's law except for a factor f_a which contains the dependence on hydrogen ion concentration. Formally, a new generalized Henry coefficient H^* may be defined by setting

$$P_s = f_a H \, x_s = H^* x_s \tag{12}$$

Here, the asterisk is to serve as a reminder that the new coefficient varies with the pH of the solution.

A similar procedure is applicable to ammonia. In this case, the pertinent equilibrium relations are

$$K_H \, p_{NH_3} = [NH_3 \cdot H_2O]$$
$$K_5 [NH_3 \cdot H_2O] = [NH_4^+][OH^-]$$
$$K_w = [H^+][OH^-]$$

The ensuing solute concentration is obtained as

$$[s] = [NH_3 \cdot H_2O] + [NH_4^+] = [NH_3 \cdot H_2O] \left(1 + \frac{K_5}{K_w} [H^+]\right) \tag{13}$$

and the factor f_a with which the true Henry constant is to be weighted becomes

$$f_a = 1/\left(1 + \frac{K_5}{K_w} [H^+]\right) \tag{14}$$

The procedure fails when one tries to apply it to the strong acids HCl and HNO_3. Their dissociation in aqueous solution is essentially complete in the accessible pH range, and it is difficult to determine K_H and the dissociation constant separately in laboratory studies. Particularly for dilute solutions we have an adequate knowledge only of the product of the two constants so that, as a consequence, we must write

$$K_{HX} = K_{diss} K_H = [H^+][X^-]/p_{HX}$$

where X^- stands for either Cl^- or NO_3^- and HX refers to HCl and HNO_3, respectively. The concentration of X^- in solution is

$$[s] = [X^-] = K_{HX} \, p_{HX}/[H^+]$$

and by comparison with equs (2) and (3) we can define the modified Henry coefficient as

$$H_{HX}^* = \frac{10^3 \rho_w [H^+]}{K_{HX} M_w} \tag{15}$$

Table 3 Definitions of modified Henry coefficients for reactive gases.

CO_2	$\mathbb{H}^* = (10^3 \rho_w/K_{H1}M_w)/(1+\frac{K_1}{[H^+]}+\frac{K_1K_2}{[H^+]^2})$
SO_2	$\mathbb{H}^* = (10^3 \rho_w/K_{H2}M_w)/(1+\frac{K_3}{[H^+]}+\frac{K_3K_4}{[H^+]^2})$
NH_3	$\mathbb{H}^* = (10^3 \rho_w/K_{H3}M_w)/(1+(\frac{K_5}{K_w})[H^+])$
HCl	$\mathbb{H}^* = (10^3 \rho_w/K_{HCl}M_w)[H^+]$
HNO_3	$\mathbb{H}^* = (10^3 \rho_w/K_{HNO_3}M_w)[H^+]$

The individual expressions obtained for the generalized Henry coefficients are summarized in Table 3 for several gases of interest. The fraction of any of these substances that is incorporated into cloud water can be calculated from equ (6) provided the true Henry coefficient $\mathbb{H}$ is replaced by $\mathbb{H}^*$. Equation (6) thus modifies to

$$\varepsilon = 1/(1 + H^*M_w/R_gTL) \qquad (16)$$

It should be re-emphasized that the degree of in-cloud scavenging now depends on the pH of the solution as an additional parameter.

The behaviour of ε as a function of the pH is shown in Figure 2 a-d for CO_2, SO_2, HCOOH and NH_3. The acid-forming gases are poorly absorbed by cloud water when the pH values are low. CO_2 at pH < 5 and HCOOH at pH < 2 are essentially undissociated in solution so that ε is independent of the pH. Raising the pH increases the degree of dissociation and causes more and more of the substance to be incorporated into cloud water. In the pH range shown, the fraction of CO_2 entering the aqueous phase remains small, whereas SO_2 and HCOOH eventually become fully dissolved when the pH reaches values around pH = 7. Ammonia, whose dissolution in water forms a base shows the opposite behavior. NH_3 is almost completely scavenged at pH values less than six but the fraction decreases substantially with increasing alkalinity. Table 4 is added to provide in a similar fashion as in Table 1 modified Henry coefficients for several reactive gases of interest (selecting a pH value of five), the resulting total concentration of the substances in the aqueous phase of precipitating clouds, and the associated residence time in the atmosphere. Most importantly, it is found that NH_3,

HCl and HNO_3 are essentially completely scavenged from the gas phase. Concerning HCl and HNO_3, the liquid water content of the system would have to decrease appreciably before the vapors are revolatilized from the liquid. In fact, it is necessary to proceed all the way back to the aerosol stage ($L \approx 10^{-5}$) if the acids are to be returned to the vapor phase. Although in principle, the present formalism may be extended to aerosol particles, the solutions then become too concentrated to allow the use of ion concentrations and in their place the appropriate activities must be inserted. Due to the complexity of the multi-component system represented by aerosol particles we have no knowledge of the required activity coefficients

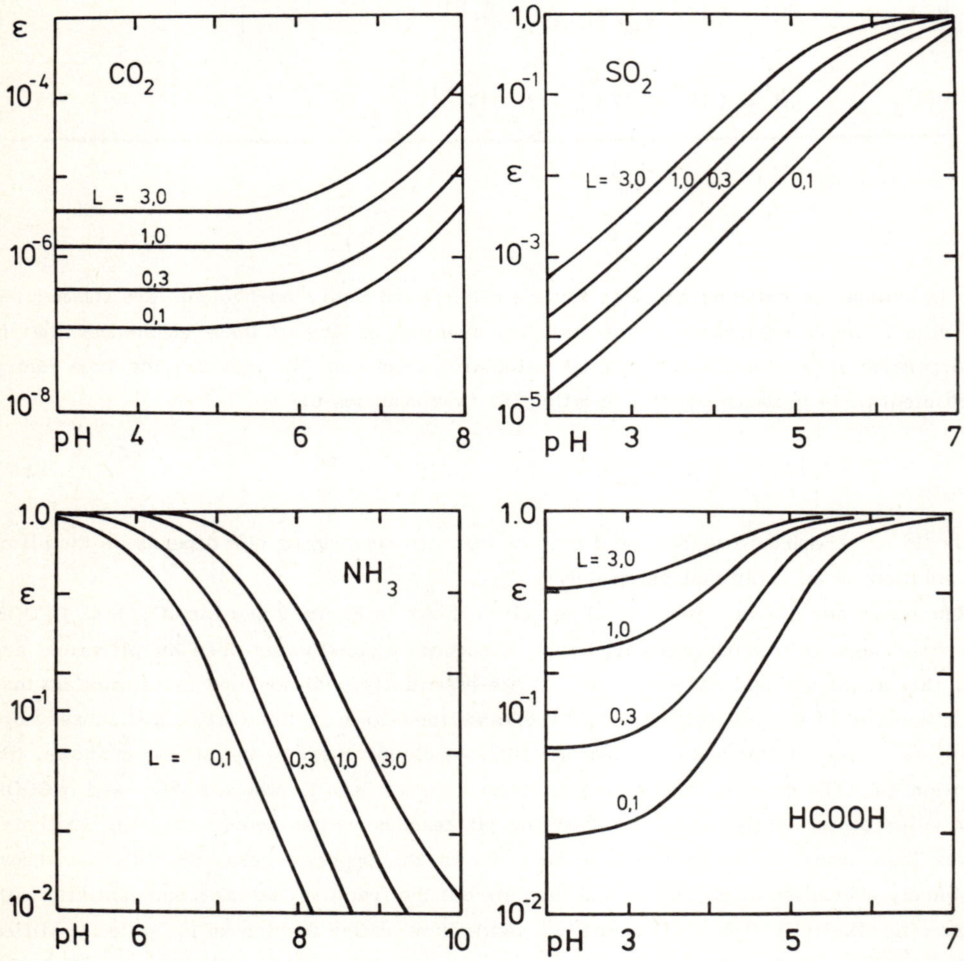

Figure 2. In-cloud scavenging efficiencies for CO_2, SO_2, NH_3, and HCOOH as a function of the pH of cloud water and the liquid water content L of the cloud.

Table 4 Modified Henry law coefficients for reactive gases, fractions ϵ incorporated into cloud water, concentrations c_s of substances in the aqueous phase for initial gas phase mixing ratios m_o, residence times τ for rainout. The latter data for a liquid water content of L=2 gm^{-3}. Powers of ten are shown in parentheses.

Constituent	pH	H(bar)	ϵ L=0.1	ϵ L=1.0	m_o	c_s(Mol kg^{-1}) L=2.0	τ	
CO_2	≤5	1.07(3)	1.2(-7)	1.2(-6)	3.3(-4)	1.5(-5)	6.0(3)	yrs
SO_2	5	1.06(-2)	1.2(-2)	1.1(-1)	1.0(-9)	3.6(-6)	29	days
HNO_2	5	1.09(-2)	1.2(-2)	1.0(-1)	5.0(-10)	1.8(-6)	28	"
HCOOH	5	2.83(-4)	3.1(-1)	8.2(-1)	5.0(-10)	8.6(-6)	6	"
NH_3	5	8.4(-6)	9.4(-1)	9.9(-1)	1.0(-9)	1.9(-5)	5.5	"
HCl	≥1	≤4(-7)	≥9.9(-1)	~1.0	5.0(-10)	9.6(-6)	5.5	"
HNO_3	≥1	≤4(-7)	≥9.9(-1)	~1.0	5.0(-10)	9.6(-6)	5.5	"

and cannot pursue the subject here.

6. WHAT DETERMINES THE pH OF CLOUD WATER?

In keeping with our concept of dilute solutions we define the pH = -log $[H^+]$ in terms of proton concentration $[H^+]$ rather than activity a_{H^+} as required for concentrated solutions. A number of studies have explored pH and ionic composition of bulk samples of cloud water. Falconer and Falconer (1980) provided a tabular summary of the investigations, the sampling methods used, and the pH values obtained. To give a few examples, Oddie (1962) found a range of 4.4-7.2 with a frequency maximum in the region 5.5-6.5 for clouds in Southern England; Petrenchuk and Drozdova (1966) reported values between 3.4 and 5.4 over the European Soviet Union; and Okita (1968) measured pH values of 3.4-6.5 over Japan. Falconer and Falconer (1980) described the frequency distribution of pH values for clouds appearing at the summit of Whiteface Mountain in New York State. The observed range was from about 2.5 to 6.0 with a maximum in the region 3.6-4.2. The occurrence of precipitation generally increased the pH by about half an unit. Similar frequency distributions of pH values have been observed for bulk samples of rain water (e.g. Mrose, 1966), although in this case the maximum usually appears in the pH range 4 - 5. Esmen and Fergus (1976) investigated the frequency spectrum of pH values in individual rain drops and found a bimodal distribution with maxima in the ranges 2.5-3.0 and 4.0-5.5. Similar observations on clouds are not available.

This brief review suggests that cloud and rain water represent acidic solutions, that the pH values are quite variable, and that the variability is inherent already in the individual cloud and precipitation elements. The basic condition which governs the hydrogen ion concentration in cloud water is the charge balance between positive and negative ions. In addition, one must take into account the most important reaction equilibria existing between the various ions. These aspects will now be considered.

The major ions present in cloud water are $SO_4^{=}$, Cl, NO_3^-, H^+, NH_4^+, Na^+, K^+, Ca^{++}, and Mg^{++}. Typical concentrations are shown in Table 5. Other ions that are known to be present include F^-, Br^-, I^- and $HCOO^-$, but these generally are minor species and we shall ignore them here for simplicity. HSO_3^- and $SO_3^{=}$ have been determined only recently by Richards et al., (1983). On this basis, the charge balance equations reads:

$$[OH^-] + 2[CO_3^{=}] + [HCO_3^-] + 2[SO_3^{=}] + [HSO_3^-] + 2[SO_4^{=}] + [HSO_4^-] + [Cl^-] + [NO_3^-] \quad (17)$$

$$= [H^+] + [NH_4^+] + [Na^+] + 2[Ca^{++}] + 2[Mg^{++}]$$

Table 5 Concentrations of major ions in cloud and fog waters (in units of μMol liter^{-1})

Ion	European USSR[a] frontal precipitating	European USSR[a] non-precipitating	S. England[b] non-precipitating	Germany[c] mountain fog	Pasadena, USA[d] urban fog
$SO_4^{=}$	29	117	40	387	240-472
Cl^-	22.2	76.8	94	205	480-730
NO_3^-	3.2	16.	18.6	450	1220-3250
HCO_3^-	11.5	16.4	-	-	-
NH_4^+	28.8	11.1	22.1	710	1290-2380
Na^+	17.0	29.8	95.2	295	320-500
K^+	5.1	20.4	12.5	85	33-53
Ca^{++}	10.0	29.3	33.2	110	70-265
Mg^{++}	12.3	40.0	12.3	-	45-160
H^+	5.0	94.4	0.06-40	12.6	14-1200
pH	5.3	4.0	4.4-7.2	5.1	4.85-2.92
anions[e]	95	327	193	1429	2180-4924
cations[e]	100	294	220	1383	1887-4983
number of samples	125	194	23	19	4

a Petrenchuk & Selezneva(1970); b Oddie(1962); c Mrose(1966)
d Munger et al(1983); e sum of charges in ueq liter^{-1}.

By means of the ion and gas-liquid equilibria listed in Table 2 one obtains the equation in the modified form

$$\frac{K_W}{[H^+]} + \left(\frac{2K_2}{[H^+]} + 1\right) \frac{K_1K_{H1}}{[H^+]} P_{CO_2} + \left(\frac{2K_4}{[H^+]} + 1\right) \frac{K_3K_{H2}}{[H^+]} P_{SO_2} \quad (17a)$$

$$+ \left(\frac{[H^+]}{K_8} + 2\right) [SO_4{}^=] + [Cl^-] + [NO_3{}^-]$$

$$= [H^+] + [NH_4{}^+]_s + \frac{K_5K_{H3}}{K_W}[H^+]P_{NH_3} + [Na^+] + [K^+] + 2[Ca^{++}] + 2[Mg^{++}]$$

Here, it is assumed that HCl and HNO_3 originally present in the gas phase are fully dissolved in cloud water, and $[NH_4]_s$ represents that portion of ammonium incorporated by cloud condensation nuclei. The difficulty in applying the equation lies in the fact that ions are provided both by the dissolution of gases and by solutes derived from cloud condensation nuclei, and that the contribution to H^+, $NH_4{}^+$, Cl^- and $NO_3{}^-$ from both sources are not easily distinguished.

All of the ions shown in Table 5 (with the possible exception of $HCO_3{}^-$) occur in aerosol particles in the form of inorganic salts, usually in relative proportions very similar, on average, to those evident from Table 5 for clouds. A few observations on fogs are included in the table to show that electrolyte concentrations are higher in fogs, presumably due to the higher concentrations of particulate matter and scavengable trace gases at the ground level.

The chemical composition of the aerosol varies with the size of the particles. For the present discussion it is convenient to subdivide particulate matter into coarse and fine particles with radii in the size ranges above and below 1 µm, respectively. Alkali and alkaline earth metals are associated largely with coarse particles which derive from the wind erosion of soils (on the continents) and from sea spray (over the oceans). The dominant salts in sea spray are NaCl and $MgSO_4$, so that in coastal areas the coarse particle contain much chloride and some sulfate. Soils contain little chloride and sulfate but provide Na, K, Ca and Mg in comparable amounts. Most of the sulfate in continental aerosols (and to some extent also in marine aerosols) occurs in the fine particle fraction. This portion of $SO_4{}^=$ is formed by the gas phase oxidation of SO_2 to sulfuric acid and its partial neutralization by ammonia. Indeed, one finds ammonium mainly associated with sulfate in the fine particle size fraction. The larger population of fine particles compared with coarse particles provides an optimum surface area for the interception of gas phase condensation products in the 0.1-1.0 µm size range. It is interesting to note that nitrate is attached mainly to coarse particles although nitric acid resulting from the oxidation of NO_2 is an important gaseous constituent of air which one would expect to condense into the fine particle size range.

These considerations suggest that the ionic composition of the solution contained in an individual cloud drop depends greatly on the type of the particle that had served as the condensation nucleus. The pH of the solution will vary accordingly. To illustrate the range of possibilities we consider two extreme cases representing the fine and the coarse fraction of the continental aerosol: (i) a particle consisting predominantly of ammonium bisulfate (NH_4HSO_4), and (ii) a particle composed solely of calcium carbonate. The latter substance is poorly soluble in pure water, but it dissolves readily in the presence of H_2CO_3. We exclude initially the interaction with all gases except CO_2. The charge balance equations can then be cast in the forms:

$$\text{(i)}\quad [H^+]^2 - \frac{c_o}{C_o(1+[H^+]/K_8)[H^+]}[H^+] = K_w + \left(\frac{2K_2}{[H^+]} + 1\right) K_1 K_{H1} P_{CO_2} \qquad (18)$$

$$\text{(ii)}\quad [H^+]^2 - 2c_o[H^+] = K_w + \left(\frac{2K_2}{[H^+]} + 1\right) K_1 K_{H1} P_{CO_2} \qquad (18a)$$

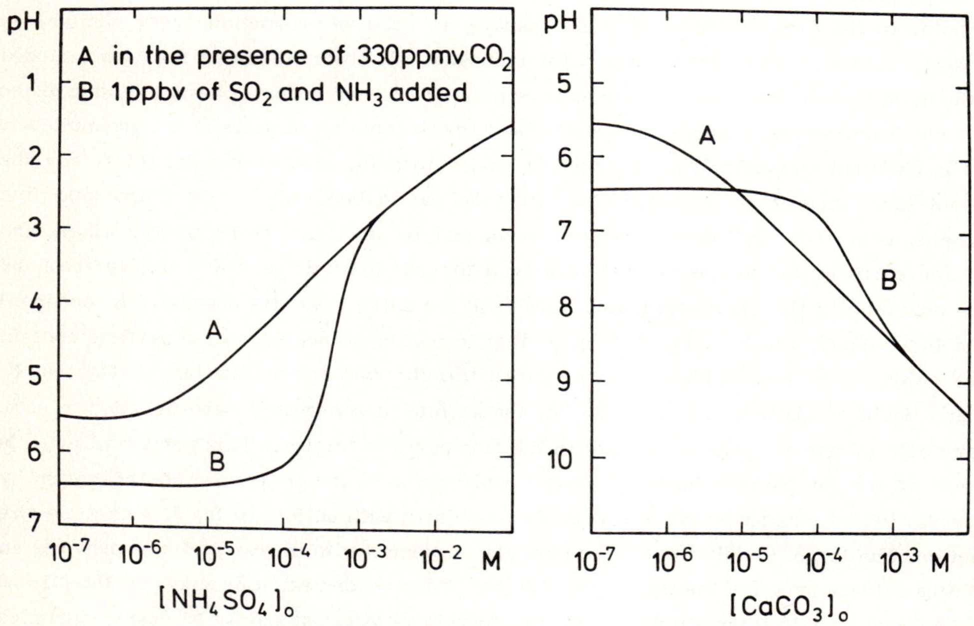

Figure 3. The pH of a cloud drop having formed on ammonium bisulfate (left) and calcium carbonate (right) particles serving as condensation nuclei. A in the presence of 330 ppmv CO_2 alone, B after the addition of SO_2 and NH_3 with initial mixing ratios of 1 ppbv each.

where c_o is the concentration of the dissolved salt. The equations are quadratic in form except for small correction terms (small in the pH range 2-10), and they are most conveniently solved by iteration. Results for a range of c_o values are shown in Figure 3. NH_4HSO_4 causes cloud drops to become more acidic, whereas $CaCO_3$ in sufficient concentrations produces alkaline solutions. The limiting value of pH=5.62 for low concentrations results from the interaction of CO_2 with water in the absence of other solutes. Since during the process of cloud formation coarse particles are preferentially activated, they are expected to acquire more liquid water than fine particles and the resulting cloud drops should become somewhat more dilute. Accordingly, we estimate the pH of individual cloud drops to range from about 2 to 8 depending on the liquid water content of the cloud. Note, however, that smaller aerosol particles are more abundant than larger ones, so that, on average, the pH will tend to lie far in the acidic region. The pH values derived from averaging the hydrogen ion concentrations in individual cloud drops may be quite different from that observed in bulk samples of cloud water, as ionic interactions undoubtedly occur upon mixing. These problems remain to be explored.

It should be re-emphasized that the foregoing samples are extreme cases. The aqueous solutions resulting from both types of particles undergo modification by the dissolution of reactive gases. Consider, for the purpose of illustration, the presence of SO_2 and NH_3 with initial mixing ratios of 1 ppbv each. This requires the addition of the terms

$$\left(\frac{2K_4}{H^+} + 1\right) K_3 K_{H2} p_{SO_2} \quad \text{and} \quad \frac{K_5 K_{H3} [H^+]^2}{K_w} p_{NH_3}$$

on the right and left hand sides of equs (18). Unlike CO_2, the gases SO_2 and NH_3 undergo changes in pressure in clouds, and these must be taken into account by setting

$$p_{SO_2} = p^o_{SO_2} (1 - \varepsilon_{SO_2}) \quad , \quad p_{NH_3} = p^o_{NH_3} (1 - \varepsilon_{NH_3})$$

The scavenging efficiencies must be expressed by equ (16). With these modifications, the charge balance equation can again be solved by iterative procedures. Results for the indicated conditions and assuming L-0.1 g m^{-3} are included in Figure 3. The pH value of 6.43 that adjusts for low concentrations of salt is due to the interaction of CO_2, SO_2 and NH_3 with cloud water alone. If SO_2 and NH_3 were fully dissolved, their concentrations in the aqueous phase would be 383 µM. The actual concentrations at pH = 6.43 are 109 and 142 µM, respectively. The values vary little so long as c_o stays small indicating that the system is buffered, but a drastic titration-like change occurs when c_o reaches and then exceeds the concentrations of dissolved SO_2 and NH_3. At high concentrations of salt the influence of the dissolved gases becomes unimportant.

In the presence of neutral salts such as NaCl the gas-solution equilibria are reversible and SO_2 and NH_3 are recovered when the cloud drops dry up. The situation is different for

Table 6 pH values in cloud water and in-cloud scavenging efficiencies for SO_2 calculated from the dissolution of SO_2 and HNO_3, respectively, in the aqueous phase for different conditions. ($m(CO_2)$=330 ppmv). Powers of ten are shown in parentheses.

Mixing ratio m_o			10^{-11}	10^{-10}	10^{-9}	10^{-8}	10^{-7}
SO_2	L=0.1	pH	5.61	5.49	5.14	4.66	4.16
	$m_o(NH_3)=0$	∈	4.9(-2)	3.8(-2)	1.7(-2)	5.9(-3)	1.8(-3)
	L=1.0	pH	5.61	5.52	5.16	4.67	4.16
	$m_o(NH_3)=0$	∈	3.4(-1)	2.9(-1)	1.5(-1)	5.4(-2)	1.8(-2)
	L=0.1	pH	6.97	6.85	6.43	5.77	4.90
	$m_o(NH_3)=10^{-9}$	∈	6.7(-1)	5.8(-1)	2.9(-1)	7.1(-2)	9.9(-3)
	L=1.0	pH	6.71	6.66	6.16	5.04	4.29
	$m_o(NH_3)=10^{-9}$	∈	8.9(-1)	8.8(-1)	6.7(-1)	1.2(-1)	2.3(-2)
HNO_3	L=0.1	pH	5.30	4.42	3.42	2.42	1.42
	$m_o(SO_2)=m_o(NH_3)=0$						
	L=1.0	pH	5.59	5.30	4.42	3.42	2.42
	L=0.1	pH	6.42	6.35	4.84	2.46	1.42
	$m_o(SO_2)=m_o(NH_3)=10^{-9}$	∈	2.8(-1)	2.5(-1)	5.8(-3)	4.1(-5)	8.5(-6)
	L=1.0	pH	6.15	6.09	5.17	3.46	2.42
	$m_o(SO_2)=m_o(NH_3)=10^{-9}$	∈	6.6(-1)	6.3(-1)	1.6(-1)	3.7(-3)	3.2(-4)

non-neutral salts to the extent that they become neutralized. In the case of NH_4HSO_4 the amount of ammonia needed to form $(NH_4)_2SO_4$ becomes permanently fixed; and in the case of $CaCO_3$, the absorption of SO_2 displaces CO_2 and forms $CaSO_3$ (or $Ca(HSO_3)_2$ in the place of $CaCO_3$ when the cloud evaporates again). It is conceivable that the association of nitrate with coarse aerosol particles results from the interaction of nitric acid from the gas phase with calcium and magnesium-containing particles when these undergo a cloud (or fog) condensation-evaporation cycle. However, this matter remains to be clarified.

In contrast to the preceding two cases in which the pH of a cloud drop is significantly influenced by the substances supplied with the condensation nucleus, one may also start with a neutral salt

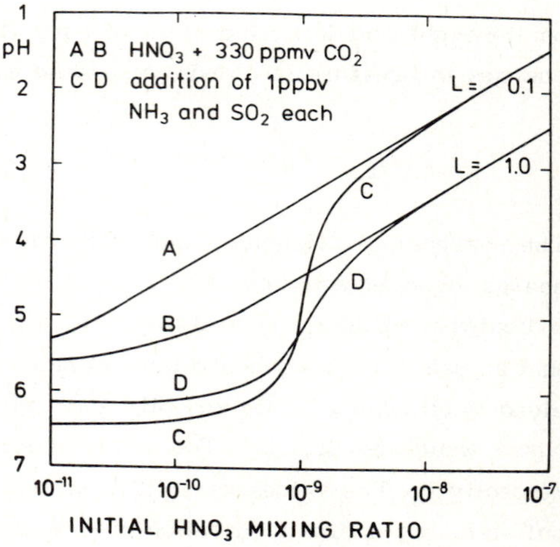

Figure 4. The pH of cloud water resulting from the dissolution of HNO_3 as a function of its initial mixing ratio. Substances incorporated into cloud water by condensation nuclei are assumed to have a negligible effect (in contrast to Figure 3).

such as seasalt and inquire about the pH resulting from the dissolved gases alone. In this case, the positive and negative ions provided by the salt can be omitted from the charge balance equation because they do not contribute to the H^+ concentration and their charges cancel. pH values derived from the dissolution of SO_2 and HNO_3 in the presence of CO_2 and NH_3 are shown in Table 6 and Figure 4 for a range of mixing ratios (prior to condensation). The table includes the associated $\mathcal{E}$ values for the scavenging of SO_2. For nitric acid one may assume that $\mathcal{E}=1$. At the same mixing ratio, the dissolution of HNO_3 causes the pH to decrease more strongly than the dissolution of SO_2. The reason lies in the poorer solubility of SO_2 which decreases as the pH is lowered. The addition of NH_3 neutralize the ions derived from SO_2 or HNO_3 to the extent that NH_3 is incorporated into cloud water and the pH is correspondingly raised. The neutralizing capacity is exhausted when NH_3 becomes fully dissolved (at pH = 5) and the mixing ratios of SO_2 and HNO_3 are sufficiently high for the dissolved acids to exceed NH_4^+ in concentration. At this point a titration-like behavior is again observed. Increasing the liquid water content of the cloud from L=0.1 to L=1.0 g m^{-3} causes more SO_2 to become dissolved, but the pH of the solution is not too much affected apparently because the concentration of sulfite and bisulfite ions stay about the same. In the case of HNO_3, however, which is fully scavenged by cloud water, the concentration decreases when L rises. The hydrogen ion concentration correspondingly decreases and the pH is raised.

The preceding discussion can be summarized as follows:

1. The pH of cloud water is determined both by the dissolution of gases (SO_2, NH_3, HNO_3, etc.) and the dissolution of electrolytes supplied with the cloud condensation nuclei.
2. The pH can be calculated from the charge balance of all ions in the solution provided the initial concentrations of the principal substances involved are known.
3. The variability in the chemical composition of cloud condensation nuclei causes the pH of individual cloud drops to vary over a wide range.
4. Increasing the liquid water content of the cloud dilutes the concentration of the substances incorporated with the cloud condensation nuclei, but at the same time it increases the scavenging efficiency for moderately soluble gases such as SO_2.
5. Most effective in decreasing the pH below that due to dissolved CO_2 are the gaseous acids HNO_3 and HCl, and acidic salts such as ammonium bisulfate. The resulting acidification of cloud water is in part compensated by the dissolution of ammonia.

7. MODIFICATION OF IN-CLOUD SCAVENGING FOR NON-UNIFORM pH

The recognition that clouds feature a broad distribution of the pH among individual cloud elements requires a revision of equ (18) which describes the scavenging of reactive gases based on the assumption that conditions are uniform. Therefore, let us subdivide the liquid water content of the cloud into a suitable number of fractions L_i, each falling into a well-defined interval of pH values such that the entire pH range is covered. We then have

$$L = \sum_i L_i \tag{19}$$

and the procedure described in section 4 leads to an expression for the scavenging efficiency in the form

$$\mathcal{E} = 1/(1 + M_w/R_gT \sum_i \frac{L_i}{H_i^*}) \tag{20}$$

In a similar manner, one may express the distribution of pH values in terms of a normalized distribution function

$$f(pH) = dL/dpH \tag{21}$$

such that

$$L = L \int f(pH)dpH \tag{21a}$$

where the integral is to cover the entire pH range. In this case, the scavenging efficiency becomes

$$\mathcal{E} = 1/(1 + M_w/R_gTL \int (f(pH)/H^*(pH))dpH) \tag{22}$$

At present, we have no opportunity to apply these equations in as much as the distribution of pH values among the individual cloud elements is unknown.

8. THE REACTION OF SULFITE AND BISULFITE WITH FORMALDEHYDE

As explained in Section 3, the validity of Henry's law relies on the assumption that the various solute molecules do not interact with each other. Such interactions do occur, however, and this final section is to show that one can take solute interactions into account provided they involve reversible reactions. The specific example chosen for illustration is the reaction of sulfite and bisulfite ions with dissolved formaldehyde. In aqueous solutions, formaldehyde hydrates to form methyleneglycol

$$CH_2O + H_2O \rightleftharpoons CH_2(OH)_2$$

with the equilibrium completely shifted toward the right hand side at temperatures below 50 C. The reactions with sulfite and bisulfite

$$HSO_3^- + CH_2(OH)_2 \overset{K_9}{\rightleftharpoons} CH_2(OH)SO_3^- + H_2O \qquad \text{R 9}$$

$$SO_3^= + CH_2(OH)_2 \overset{K_{9a}}{\rightleftharpoons} CH_2(OH)SO_3^- + OH^- \qquad \text{R 9a}$$

lead to an adduct which in its protonated form is known as methanehydroxylsulfonic acid. By adding the reactions

$$SO_3^= + CH_2(OH)_2 \rightleftharpoons CH_2(OH)SO_3^- + OH^-$$

$$OH^- + H^+ \rightleftharpoons H_2O$$

$$HSO_3^- \rightleftharpoons SO_3^= + H^+$$

$$HSO_3^- + CH_2(OH)_2 \rightleftharpoons CH_2(OH)SO_3^- + H_2O$$

it can be seen that the equilibria (9) and (9a) are equivalent, and only one of them is needed to describe the system. We select reaction (9). The two equilibrium constants are related by $K_{9a} = K_9K_w/K_4$. In the following, the concentrations of formaldehyde and the adduct are denoted by [FA] and [Add], respectively.

A value for the equilibrium constant of reaction (9) was first reported by Kerp (1904). He determined the concentration of bisulfite and sulfite by iodometric titration, assuming that the rate of dissociation of the adduct is sufficiently slow. The value given ($K_g \sim 8 \times 10^6$ M^{-1}) is too high, however. A redetermination has been performed only recently by Dasgupta et al.,

(1980), who prepared millimolar solutions of the sodium salt of the adduct and measured by means of ultraviolet spectrophotometry the concentrations of S_{IV} = H_2SO_3+HSO_3^- + $SO_3^=$ which are present in equilibrium with the adduct. A variety of buffers were used to adjust the pH to integral values between 2 and 9 (with exception of pH=3). The equilibrium constant was reported in the pH-dependent composite form K_9 = [Add]/[FA][S_{IV}]. The results are displayed in Figure 5. Klippel and Warneck (unpublished) also used ultraviolet spectrophotometry (at 205 nm) to determine concentration of bisulfite near pH = 4. They prepared unbuffered, 0.5-1 millimolar solutions of HSO_3^- from sodium pyrosulfite ($Na_2S_2O_5$) and measured the decrease in the HSO_3^- concentration upon the addition of formaldehyde. Their value for the equilibrium constant is K_9 = [Add]/[FA][HSO_3^-] = 2.5 x $10^4 M^{-1}$. The result is included in Figure 5.

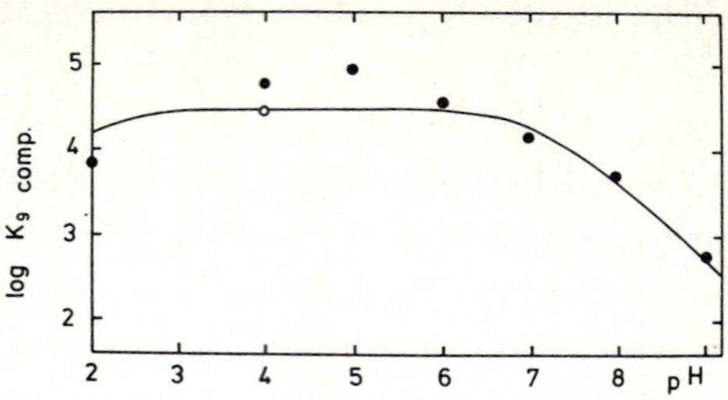

Figure 5. Composite equilibrium constant for reaction 9 as a function of pH. Solid points are measurements of Dasgupta et al. (1980). The open point is a determination of K_9 by Klippel and Warneck (unpublished). The line is calculated from equ. (24) and K_9=3 x $10^4 M^{-1}$.

We must now discuss the pH dependence of the composite equilibrium constant. The total concentration of S_{IV} is given by an expression similar to equ (11) and the fraction present as HSO_3^- is

$$f = \frac{[HSO_3^-]}{[S_{IV}]} = \frac{[HSO_3^-]}{[H_2SO_3]} \left\{1 + \frac{K_3}{[H^+]} (1 + \frac{K_4}{[H^+]}) \right\}^{-1} \qquad (23)$$

$$= \frac{K_3}{[H^+]} \left\{1 + \frac{K_3}{[H^+]} (1 + \frac{K}{[H^+]}) \right\}^{-1}$$

The equilibrium constant accordingly may be written

$$K_{9\,comp} = \frac{[Add]}{[FA][S_{IV}]} = \frac{[Add]\ f}{[FA][HSO_3^-]} = K_9\ f \qquad (24)$$

The solid line in Figure 4 was calculated from this expression with K_9=3 x $10^4 M^{-1}$. It shows that the suggested representation of the composite equilibrium constant is in reasonable agreement with the experimental data. HSO_3^- is the dominant species in the pH range 3.6-6.0.

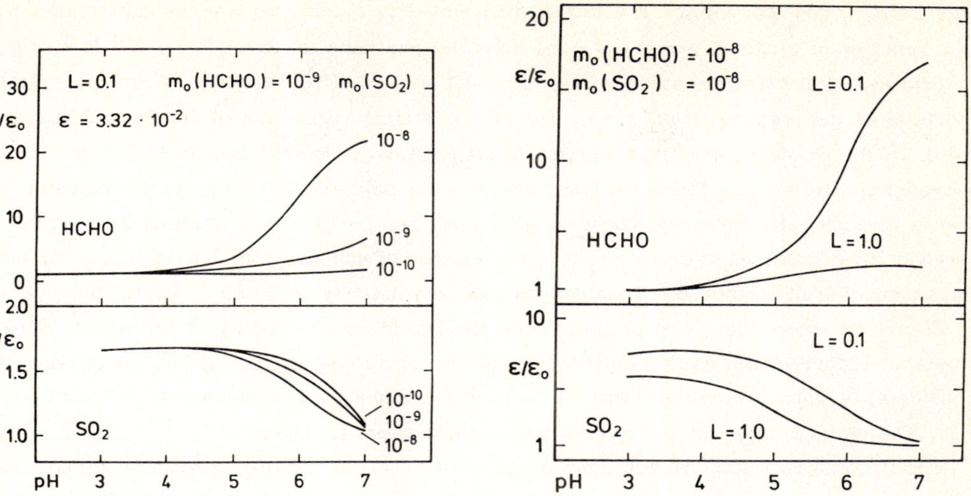

Figure 6. Enhancement of in-cloud scavenging efficiency for SO_2 and HCHO by adduct formation in the aqueous phase. (a) variation of the ratio $\mathcal{E}/\mathcal{E}_0$ as a function of pH and $m_0(SO_2)$ for L=0.1 and m_0(HCHO)= 1 ppbv, where $\mathcal{E}_0$ denotes the scavenging efficiency without adduct formation. (b) influence of the liquid water content L on the ratio $\mathcal{E}/\mathcal{E}_0$. The initial mixing ratios for SO_2 and HCHO are set to 10 ppbv each.

In this region, the composite equilibrium constant varies little and essentially equals K_9. The experimental data nevertheless show appreciable scatter indicating that the measurements are still inadequate. Moreover, the temperature dependence of K_9 is not well known.

It will be obvious that the interaction of SO_2 and formaldehyde in cloud water and the resulting adduct increases the in-cloud scavenging efficiency for both gases. The total concentration of formaldehyde in cloud water and that of sulfur (IV) species in the presence of the adduct are:

$$[HCHO]_{tot} = [FA]\left\{1 + K_9 \frac{K_3}{[H^+]}[SO_2 \cdot H_2O]\right\} = [FA]/\ f_{HCHO} \qquad (25)$$

$$[S_{IV}]_{tot} = [SO_2 \cdot H_2O]\left\{1+\frac{K_3}{[H^+]}(1+\frac{K_4}{[H^+]})+K_9 \frac{K_3}{[H^+]}[FA]\right\} = [SO_2 \cdot H_2O]/f_{SO_2}$$

They define the factors f_{HCHO} and f_{SO_2} with which the Henry coefficients for formaldehyde and SO_2 are to be multiplied in order to calculate from equ. (1) the fractions of these gases residing in cloud water. Again, it is convenient to use an iterative procedure in solving the equations. We are mainly interested in the increase of $\mathcal{E}$ values over those obtained in the absence of adduct formation. The latter are denoted by $\mathcal{E}_o$. Figure 6 shows their ratios $\mathcal{E}/\mathcal{E}_o$ as a function of pH for a variety of conditions. K_9 was here set equal to 5 x 10^4 M^{-1}. Figure 6a depicts results for constant values of L=0.1 and m_o (HCHO)= 10^{-9}. Near pH = 7, the increase of the scavenging efficiency for SO_2 is slight, regardless of its gas phase mixing ratio. In the pH range 4-5, it maximizes to $\mathcal{E}/\mathcal{E}_o$ = 1.66. A more pronounced increase in the scavenging efficiency is found for formaldehyde, but only at pH > 5. Figure 2b indicates that SO_2 is already efficiently scavenged at pH values greater than six, whereas the fraction of formaldehyde dissolved in cloud water in the absence of adduct formation is only 3.3%. Thus, the scavenging efficiency of formaldehyde can be raised by a factor of up to 30.

Figure 6b shows the effect of increasing the liquid water content of the cloud. Here, a somewhat higher initial mixing ratio was assumed for formaldehyde. Its $\mathcal{E}/\mathcal{E}_o$ value decreases considerably when L is raised from 0.1 to 1g m^{-3} because $\mathcal{E}_o$ is then increased from 0.033 to 0.25. The general shape of the curves remains the same, however.

Finally, one may compare the concentration of S(IV), calculated to exist in cloud water, with the measurement of Richards et al. (1983). The data are shown in Table 7. Here, we have based the calculation on the actually observed values for the gas phase mixing ratio of SO_2, the aqueous concentration of formaldehyde and the pH of cloud water. A gas phase mixing ratio for HCHO of about 6 ppbv is inferred. Two values of K_9 were used, K_9=5 x 10^4 and K_9=1 x 10^5. The former should be considered a lower limit to the true value at the observed temperature of 286 K. The latter is based on K_9=5 x 10^4 at room temperature and a

Table 7 Effect of formaldehyde-bisulfite adduct formation on the scavenging of SO_2 in clouds. Comparison of calculated aqueous concentrations of S(IV) with observational data of Richards et al.(1983).

	L ($g\ m^{-3}$)	pH	$m(SO_2)$ (ppbv)	m(HCHO) (ppbv)	$[S_{IV}]$ (uM)	[HCHO] (uM)
observed[a]	-	2.82	18	-	105	87
calculated[b]	0.1		20	6.5	3.4	85
$K_9=5\cdot10^4$	0.3	"	20	6.5	3.5	87
	1.0	"	20	6.5	3.5	88
calculated[b]	0.1	"	20	5.8	6.6	90
$K_9=1\cdot10^5$	0.3	"	20	5.7	6.5	88
	1.0	"	20	5.7	6.5	89

a average of nine observations in clouds; b values for L. pH and $m(SO_2)$ are adopted, m(HCHO) was adjusted to match observed [HCHO].

temperature dependence in the form exp ($\Delta H/RT$) with $\Delta H = 51.5$ kJ Mol^{-1} equivalent to that observed by Stewart and Donnally (1932) for the benzaldehyde-bisulfite addition complex. The liquid water content of the cloud was varied, but it has only a minor influence on the total S(IV) concentration. The calculated concentration ranges from 3.5 to 6.3 µM, a factor of at least sixteen less than is actually found. It is apparent that adduct formation with formaldehyde alone is insufficient to explain the observed enhancement of S(IV) in cloud water and other processes must be explored to account for it. The importance of adduct formation increases with decreasing temperature, however, so that in deep clouds it should not be neglected.

ACKNOWLEDGEMENT

This work was performed within the program of the Sonderforschungsbereich 73 (Atmospheric Trace Components).

REFERENCES

Dasgupta P.K., DeCesare K. and Ullrey J.C. (1980) Determination of atmospheric sulfur dioxide without tetrachloromercurate (II) and the mechanism of the Schiff reaction. Anal Chem. 52 1912-1922.

Esmen N.A. and Fergus R.B. (1976) Rain water acidity: pH spectrum of individual drops. Sci. Total Environment 6 223-226.

Falconer R.E. and Falconer P.D. (1980) Determination of cloud water acidity at a mountain observatory in the Adirondack Mountains of New York State. J. Geophys. Res. 85 7465-7470.

Fritz J.J and Fuget C.R. (1956) Vapor pressure of aqueous hydrogen chloride solutions, 0-50 C. Ind. Eng. Chem. 1 10-12

Graedel T.E. and Weschler C.J. (181) Chemistry within aqueous atmospheric aerosols and raindrops. Rev. Geophys. Space Phys. 19 505-539.

Junge C. (1963) Air Chemistry and Radioactivity, Academic Press.

Kerp W. (1904) Zur Kenntnis der gebundenen Schwefligen Saeure. Arbber. Kais. Gesundheitsamt A21 180-225, 372-376.

Lee Y-N. and Schwartz S.E. (1981) Reaction kinetics of nitrogen dioxide with liquid water at low partial pressure. J. Phys. Chem. 85 840-848.

Martin L.R. and Damschen D.E. (1981) Aqueous oxidation of sulfur dioxide by hydrogen peroxide at low pH. Atmospheric Environment 15 1615-1621.

Mrose H. (1966) Measurements of pH, and chemical analyses of rain- snow- and fogwater. Tellus 18 266-270.

Munger J.W., Jacob D.J., Waldmann J.M. and Hoffmann M.R. (1983) Fogwater chemistry in an urban atmosphere. J. Geophys. Res. 88 5109-5121.

Oddie B.C.V. (1962) The chemical composition of precipitation at cloud levels. Quart. J. Roy. Met. Soc. 88 535-538.

Okita T. (1968) Concentrations of sulfate and of the inorganic materials in fog and cloud water and in aerosol. J. Meteorol. Soc. Japan 46 120-126.

Petrenchuk O.P. and Drozdova V.M. (1966) On the chemical composition of cloud water. Tellus 18 280-286.

Petrenchuk O.P. and Selezneva E.S. (1970) Chemical composition of precipitation in regions of the Soviet Union. J. Geophys. Res. 75 3629-3634.

Pruppacher H.R. and Klett J.D. (1978) Microphysics of Clouds and Precipitation. Reidel Publ. Comp.

Rhode H. and Grandell J. (1972) On the removal of aerosol particles from the atmosphere by precipitation scavenging. Tellus 24 442-454.

Richards L.W., Anderson J.A., Blumenthal D.L., McDonald J.A., Kok G.L. and Lazrus A.L. (1983) Hydrogen peroxide and sulfur (IV) in Los Angeles cloud water. Atmospheric Environment 17 911-914.

Schwartz S.E. (1985) Mass Transport considerations pertinent to aqueous phase reactions of gases in liquid-water clouds. This Volume.

Schwartz S.E. and White W.H. (1981) Solubility equilibria of the nitrogen oxides and oxyacids in dilute aqueous solution. Adv. Environm. Sci. Eng. 4 1-45.

Slinn W.G.N. (1974) Proposed terminology for precipitation scavenging. in R.G. Semonin and R.W. Beadle (coordinators) "Precipitation Scavenging" ERDA Symposion Series CONF-741003, US Dept. of Commerce, Springfield, Virginia, pp 813-818.

Stewart T.D. and Donnally L.H. (1932) The aldehyde-bisulfite compounds II. The effect of varying the hydrogen ion concentration and temperature upon the equilibrium between benzaldehyde and bisulfite ion. J. Am. Chem. Soc. 54 3555-3569.

Stumm W. and Morgan J.J. (1981) Aquatic Chemistry, 2nd ed. Wiley-Interscience, New York.

Vierk A-L. (1950) Experimentelle Untersuchungen an den Zweistoffsystemen: Wasser-Acetonitril, Wasser-Dioxan, Aethanol-Acetonitril und Cyclohexan-Dioxan. Z. anorg. Chem. 261 283-309.

Wagman D.D., Evans W.H., Parker V.B., Halow I., Bailey S.M. and Schumm R.H. (1968) Selected values of chemical thermodynamic properties NBS Technical Note 270-3, US Dept. of Commerce, National Bureau of Standards.

Walker J.F. (1975) Formaldehyde, Krieger Publ. Co., Huntington, N.Y. 3rd edition. p.113.

Wilhelm E., Battino R. and Wilcock R.J. (1977) Low pressure solubility of gases in liquid water. Chem. Rev. 77 219-262.

Winkler L.W. (1901) Die Loeslichkeit der Gase im Wasser. Ber. 34 1408-22.

THE INTERACTION OF SULFUR TRIOXIDE WITH WATER CLUSTERS

R. G. Keesee, R. Sievert, and A. W. Castleman, Jr.

Department of Chemistry
The Pennsylvania State University
University Park, PA 16802
U.S.A.

ABSTRACT

Sulfur trioxide may play an important role in the production of atmospheric aerosols, particularly when gas phase sulfur chemistry can lead to supersaturation in sulfuric acid and eventually nucleation. To improve our understanding of some of the mechanisms involved, the products of the reaction of sulfur trioxide with water clusters have been studied using molecular beam techniques. Sulfur trioxide was introduced through a coaxial tube or through an annular ring around an inner nozzle from which the water clusters emanated. The beam of clusters was passed through an electric quadrupole focusing region (to analyze the polar nature of the clusters) and then was ionized using electron impact. The resulting distribution of ions was analyzed with a quadrupole mass spectrometer. Beside protonated water clusters, a less abundant series of ions of the form $SO_3(H_2O)_nH^+$ was observed. Protonated water clusters produced from ionization of neutral $SO_3(H_2O)_n$ clusters could not be discerned from the ionized products of unreacted water clusters. Electrostatic quadrupole focusing experiments of the neutral beam indicated that any $SO_3 \cdot H_2O$ adduct rapidly isomerized to sulfuric acid.

INTRODUCTION

The gas phase reaction of SO_3 with water was reported [1] as early as 1934, but the molecular details are still not well known. Interest stems in part due to the belief that SO_3 plays a role in the production of H_2SO_4 [2], which is an important pollutant as well as a

NATO ASI Series, Vol. G6
Chemistry of Multiphase Atmospheric Systems
Edited by W. Jaeschke

nucleation agent in the atmosphere [3,4]. The clustering of water with H_2SO_4 is considered in various nucleation models [5,6], but there is still a lack of knowledge of the details of the mechanism of H_2SO_4-aerosol production.

One pathway for the production of atmospheric sulfate is believed to involve homogeneous gas-phase reactions initiated by the reaction

$$SO_2 + OH \xrightarrow{M} HSO_3$$

which has been well studied [7]. This reaction is particularly important in the lower stratosphere [2]. The fate of the product HSO_3 is less certain but reactions such as

$$HSO_3 + OH \longrightarrow SO_3 + H_2O$$

or

$$HSO_3 + O_2 \longrightarrow HSO_5{}^* \longrightarrow SO_3 + HO_2$$

may produce gaseous SO_3.

Some evidence [8] has supported the suggestion [1] that the gas-phase reaction of SO_3 and water is fast. Since direct reaction to form H_2SO_4 would have to proceed via a four-center mechanism, the findings have been rationalized [8-10] by speculation that an adduct is first formed. Thereafter, the adduct may be expected to rearrange to form the H_2SO_4 molecule [10] which is subsequently incorporated into aerosol particles.

Aqueous-phase and heterogeneous processes are believed to contribute significantly to sulfate production in the troposphere [11]. An investigation of the interaction of sulfur trioxide with small water clusters provides an example to aid in understanding heterogeneous processes on a molecular scale where the water clusters may be thought of as microdrops whose properties approach fog or aerosol particles as the clusters increase in size. The aim of the present study is to investigate the mechanism of the reaction of SO_3 with water and water-clusters and to determine the molecular nature of the products.

EXPERIMENTAL

A beam of the clusters of interest was produced, passed through an electrostatic quadrupole focusing field, and then ionized using electron impact. The resulting distribution of ions was analyzed with a quadrupole mass spectrometer. The experimental apparatus is described in more detail elsewhere [12,13]. The water clusters were produced by coexpanding the vapor with argon through a sonic nozzle (100 μm diameter) to form a free jet; the stagnation chamber was generally set at one atmosphere pressure. The stagnation chamber as well as the nozzle were heated to prevent condensation. The amount of water vapour was controlled by the temperature of the liquid reservoir through which the argon was bubbled.

Sulfur trioxide was introduced in two ways: one, where SO_3 was admitted through a metal capillary bent as a ring containing a series of small holes coaxial with and a few millimeters from the nozzle from which the water clusters were formed; the other one, where the SO_3 was introduced through a coaxial glass tube that was fitted around the inner nozzle. In the latter case, the area of the annular opening was about 5-6 times larger than the area of the inner capillary hole. Most experiments were performed with the coaxial glass system since it was more resistive to the extremely corrosive SO_3. The sulfur trioxide flow was adjusted by controlling the temperature of the stabilized liquid SO_3 source. No evidence of polymers or sulfuric acid was found when the SO_3 was introduced in the absence of water.

Klemperer and coworkers [14] have given the theoretical background and experimental procedure for the electric deflection studies of the molecular beams. Briefly, species with permanent dipole moments can be focused around an obstacle by an electrostatic quadrupole field. Consequently, if the precursor neutral is polar, an increase in ion intensity is observed when sufficient voltage is applied to the quadrupole rods. In the case where no focusing is observed, the obstacle is removed. The beam then becomes defocused if the species are polarizable. This occurs since these species are deflected toward the rods when a voltage is applied, leading to a corresponding decrease in the observed ion intensity.

RESULTS

The ion clusters observed in these experiments were a series of protonated water clusters $H^+(H_2O)_n$ with n up to 14 and another less abundant series of the form $SO_3(H_2O)_nH^+$ with n up to 9. The $H^+(H_2O)_n$ series could result from the ionization of unreacted water clusters and also from the ionization of $SO_3(H_2O)_m$ clusters. The relative distribution of the $H^+(H_2O)_n$ series, however, was not appreciably affected by the introduction of SO_3. The $SO_3(H_2O)_nH_+$ distribution was very similar in form to the $H^+(HNO_3)(H_2O)_n$ distribution observed from the expansion of aqueous nitric acid vapour [15] in that a local intensity minimum (at n=4 for the $SO_3(H_2O)_n$ distribution) and a local maximum (at n=6) were observed. No masses corresponding to more than one SO_3 molecule in a cluster were detected. It should be noted that the size of the clusters may be decreased at two different stages. The first fragmentation can occur upon the reaction with SO_3 where the exothermicity of the reaction could lead to boiling off of water molecules from the clusters and the second fragmentation occurs during the electron impact ionization.

A primary aim of studying the SO_3-H_2O system was to determine whether the adduct $SO_3 \cdot H_2O$ due to the reaction of SO_3 with H_2O is long-lived or whether isomerization to sulfuric acid is rapid. A calculation by Holland and Castleman [10] indicated a small barrier to isomerization of about 3.3 kcal/mole. The dipole moment of $SO_3 \cdot H_2O$ was calculated to be 7.4 Debye compared to an experimental value [16] of 2.7 Debye for sulfuric acid. The corresponding calculated value was 2.9 Debye. So the focusing characteristics of the adduct and acid are expected to be different. Focusing experiments were performed from 0 to 5 kV monitoring the HSO_3^+ and $H_2SO_4^+$ ions in the SO_3+H_2O case and monitoring the $H_2SO_4^+$ ion for the vapor of concentrated sulfuric acid. The behaviour, based on all three monitored ions, was identical within experimental scatter. Refocusing was about 10% at 1 kV, increasing to 20% at 5 kV. Therefore, the $SO_3 \cdot H_2O$ adduct did indeed appear to isomerize to H_2SO_4 during the transit times (on the order of a few hundred microseconds) to the focusing region. An estimate based on an RRKM calculation suggested that an isomerization barrier of about 13 kcal/mole would be necessary to observe any adducts under the experimental conditions. The details of the RRKM calculation are given elsewhere [17]. Both the experiment and calculations suggest that the observed clusters are actually sulfuric acid-water clusters.

In the case where $SO_3(H_2O)_nH^+$ clusters were produced, the precursor neutrals (for n=1, 2, and 3) defocused. A similar result was found earlier [18] for $HNO_3(H_2O)_nH^+$ clusters.

ACKNOWLEDGEMENTS

Support of the Department of Energy under Grant No. DE-AC02-82-ER60055, The Department of the Army under Grant No. DAAG29-82-K-0160, and the National Science Foundation under Grant No. ATM-82-4010 is gratefully acknowledged.

REFERENCES

1. C. F. Goodeve, A. S. Eastman, and A. Dooley, Trans. Faraday Soc. 30, 1126 (1937).
2. R. P. Turco, P. Hamill, O. B. Toon, and R. C. Whitten, J. Atmosph. Sciences 36, 699 (1979).
3. R. H. Heist and H. Reiss, J. Chem. Phys. 61, 573 (1974).
4.(a) C. S. Kiang and P. Middleton, Geophys. Res. Lett. 4, 17 (1977).
4.(b) C. S. Kiang and R. D. Cadle, Atmosph. Sciences 34, 150 (1977).
5. W. Roedel, J. Aerosol Science 10, 375 (1979).
6. K. Suzuki and V. Mohnen, J. Aerosol Sci. 12, 61 (1981).
7. A. W. Castleman, Jr., and I. N. Tang, J. Photochem. 6, 349 (1976/77); G. K. Moortgat and C. E. Junge, Pure Appl. Geophys. 115, 759 (1977); G. Paraskevopoulas, D. L. Singleton, and R. S. Irwin, Chem. Phys. Lett. 100, 83 (1983).
8. A. W. Castleman, Jr., R. E. Davis, H. R. Munkelwitz, I. N. Tang, and W. P. Wood, Int. J. Chem. Kinet. Symp. 1, 629 (1975).
9. J. G. Calvert, Private Communication.
10. P. M. Holland and A. W. Castleman, Jr., Chem. Phys. Lett. 56, 511 (1978).
11. See for example S. Beilke and G. Gravenhorst, Atmos. Environ. 12, 231 (1978); T. Novakov, S. G. Chang, and A. B. Harker, Science 186, 259 (1974); S. A. Penkett, Nature 240, 105 (1972).
12. R. Sievert, I. Cadez, J. Van Doren, and A. W. Castleman, Jr., J. Phys. Chem. 88, 4502 (1984).
13. B. D. Kay, Ph.D. Thesis, University of Colorado (1982).
14. W. E. Falconer, A. Baechler, J. L. Stauffer, and W. Klemperer, J. Chem. Phys. 48, 312 (1968).
15. B. D. Kay, V. Hermann, and A. W. Castleman, Jr., Chem. Phys. Lett. 80, 469 (1981).
16. R. L. Kuczkowski, R. D. Suenram, and F. J. Lovas, J. Am. Chem. Soc. 103, 2561 (1981).
17. R. Hofmann-Sievert and A. W. Castleman, Jr., Reaction of SO_3 with Water Clusters and the Formation of H_2SO_4, J. Phys. Chem. 88, 3329 (1984).
18. A. W. Castleman, Jr., B. D. Kay, F. J. Schelling, and R. Sievert, Farad. Disc. 73, 124 (1982).

LABORATORY STUDIES OF THE MULTIPHASE $S^{IV} \longrightarrow S^{VI}$ CONVERSION RATE

S. A. Penkett
Environmental and Medical Sciences Division
Atomic Energy Research Establishment
Harwell, Oxfordshire OX11 ORA
England

ABSTRACT

Oxidation reactions of dissolved sulphur dioxide in an atmospheric context are considered. Expressions for the rate of SO_2 oxidation by oxygen in the presence of metal catalysts by hydrogen peroxide, by ozone and by the hydroxyl radical, are presented. Experiments which are designed to collect kinetic information in carefully controlled conditions i.e. classical rate phase studies, are to be preferred in most cases to those which attempt to simulate oxidation in cloud droplets.

1 INTRODUCTION

The most common form of S^{IV} in the atmosphere is gaseous sulphur dioxide. This is a very soluble gas and two new species, bisulphite and sulphite ions, are formed by hydrolysis. This allows multiphase oxidation, which takes place in fog, cloud or rain droplets, to predominate over homogeneous gas phase oxidation in certain circumstances.

Atmospheric pollution associated with emissions of sulphur gases in humid atmospheres has been recognised for centuries and was particularly bad in London in the first part of the twentieth century. Similarly a study of the oxidation of sulphur dioxide in aqueous solution has featured in the chemical literature since the early nineteenth century when the nature of the products formed was investigated by such notables as Gay-Lussac and Berzelius. Modern understanding of the chemistry of S^{IV} oxidation in solution really starts with the work of Backstrom (1927) and Frank and Haber (1931) who clearly recognised that aerobic oxidation was a chain reaction and who proposed a mechanism involving free radical species. Many studies have been made of this reaction since, both in the presence and absence of catalysts, which are known to be present in particles in polluted atmospheres. These included a number of examples of experiments designed to study the complete process of transfer of reactant

NATO ASI Series, Vol. G6
Chemistry of Multiphase Atmospheric Systems
Edited by W. Jaeschke

from gas phase to droplets and subsequent oxidation, which may genuinely be described as multiphase experiments.

Another major form of air pollution was discovered in Los Angeles about thirty years ago (Haagen-Smit (1952). This results in the photochemical formation of oxidants, such as ozone and hydrogen peroxide in significant concentrations in the atmosphere (Demerjian et al. (1974)). The underlying chemistry is controlled by gaseous free radicals which cause many species to be oxidised, including sulphur dioxide which forms a sulphate aerosol. This is now recognised as a widespread phenomenon capable of generating strong oxidant species and free radicals throughout the sunlit atmosphere.

Several of these oxidants are extremely soluble in water and conversion of S^{IV} to S^{VI} by oxidants of this type has therefore been studied using apparatus capable of measuring rates of fast reactions (Penkett et al. (1979)). Most of these studies have been made using bulk solutions and few attempts have been made to study the multiphase oxidation process. In the case of hydrogen peroxide, this was done inadvertantly to demonstrate the efficiency of droplet reactions, (Beilke and Georgii (1968)), before it was realised that H_2O_2 was a common trace species in the atmosphere.

There have been a number of reviews recently concerned with droplet chemistry in the atmosphere (Beilke and Gravenhorst (1978), Hegg and Hobbs (1978), National Academy Press, Washington DC (1983)). These tend to have a standard form dealing with transport of gaseous reactants into droplets, diffusion and equilibria in the droplet, and followed by a review of reaction rate data. This article will emphasise the chemical aspects and will deal almost entirely with chemically ideal systems where the concentrations do not exceed 10^{-3} M. These are by far the most common type of situation which occur in the atmosphere. Clouds and fogs are examples of dilute solutions of ions, only in deliquescent aerosols does the ionic concentration rise to high values. The extent of water phase associated with these aerosols is many orders of magnitude lower than that present of cloud water.

2 S^{IV} EQUILIBRIUM AND THERMOCHEMISTRY

When sulphur dioxide dissolves in water it hydrolyses to form bisulphite and sulphite ions. The distribution of the three main species (SO_2, H_2O), HSO_3^- and $SO_3^=$ is very sensitive to the hydrogen ion concentration via the following equilibria (Sillen (1964)).

$$SO_{2(g)} + H_2O \rightleftharpoons SO_2 \cdot H_2O \quad : K_{H(288°K)} = 1.83 \text{ M atm}^{-1}$$

$$SO_2 \cdot H_2O \rightleftharpoons H^+ + HSO_3^- \quad : K_{1(288°K)} = 2.19 \times 10^{-2} \text{ M}$$

$$HSO_3^- \rightleftharpoons H^+ + SO_3^= \quad : K_{2(288°K)} = 7.9 \times 10^{-8} \text{ M}$$

The main form of S^{IV} in solution in the atmospheric situation is the bisulphite ion, HSO_3^-, which predominates between pH 2 and pH 6. Above pH 7 the sulphite ion is the main form of S^{IV} in solution, but pH's of cloud droplets rarely exceed 6.

The heats and free energy change associated with oxidation of both ions are large and negative (Latimer (1952)).

$$HSO_3^- + \tfrac{1}{2}O_2 \longrightarrow HSO_4^- \quad \Delta H = -59.8 \text{ kcals/mole}$$
$$\Delta G = -53.94 \text{ kcals/mole}$$

$$SO_3^- + \tfrac{1}{2}O_2 \longrightarrow SO_4^= \quad \Delta H = -65 \text{ kcals/mole}$$
$$\Delta G = -61.26 \text{ kcals/mole}$$

Both processes are therefore favourable from a thermodynamic standpoint with sulphite ion

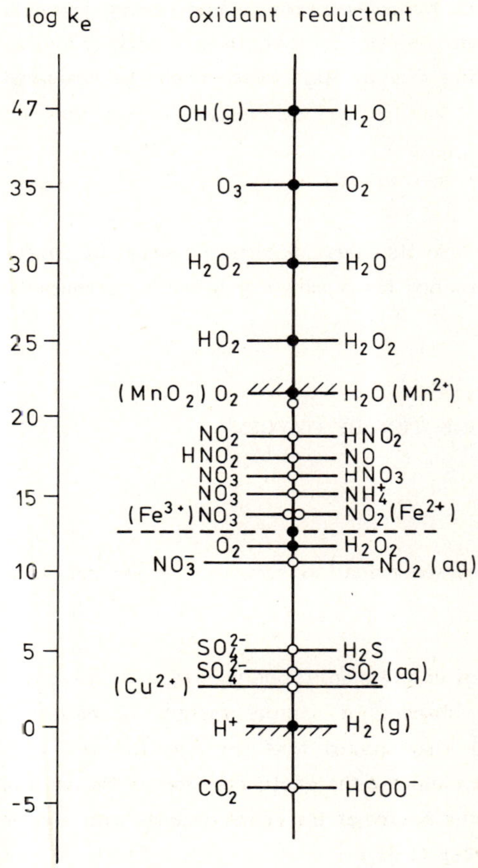

Figure 1: Energetic levels for electron transfer reactions (oxidant + e^- $\rightleftharpoons$ reductant (Ke))

oxidation being slightly more favoured.

In the gas phase the oxidation process

$$SO_2 + \tfrac{1}{2}O_2 \longrightarrow SO_3 \quad \Delta H = -23.69 \text{ kcal/mole}$$
$$\Delta G = -16.73 \text{ kcal/mole}$$

has a lower value for ΔG suggesting a less favoured process. In practice it has been observed that the rate of gas phase oxidation of SO_2 by oxygen is extremely slow at normal atmospheric temperatures and can effectively be ignored.

Before considering in detail the large amount of experimental work that has been performed on the conversion of S^{IV} to S^{VI} in solution, it is useful to consider the energetics involved for the various potential oxidant species. This has been done rather well by Morgan (1982) and a diagram in the form of energy levels for various oxidant reductant couples is shown in Figure 1 where K_e is the standard equilibrium constant for unit electron acceptance. The most powerful oxidants are clearly hydroxyl radicals but their oxidative capacity is closely rivalled by ozone and possibly hydrogen peroxide. Hypochlorite is also a strong oxidant with an oxidative capacity between hydrogen peroxide and peroxy radicals according to Latimer (1952). Oxygen itself is much less potent than these strong oxidants, also nitrogen oxides are less capable of oxidising SO_2 to $SO_4^{=}$ based upon the combined energy level difference for the reaction

$$\text{oxidant} + SO_{2(aq)} \longrightarrow SO_4^{=} + \text{reductant}$$

The rate of oxidation is not necessarily related to its thermodynamic potential of course and individual rates of reaction have to be determined for a range of possible processes.

3 OXIDATION OF S^{IV} TO S^{VI} IN AQUEOUS SOLUTION BY OXYGEN

3(i) General Background: Catalysed vs uncatalysed oxidation

Early studies of the rate of aerobic oxidation of sulphite solutions showed that this could vary markedly depending on the presence or absence of minute amounts of catalytic impurities (Backstrom (1934)). Backstrom (1927) also showed that the reaction was also photosensitive and he estimated that the quantum yield for the chain reaction at 254 nm was in the region of 5 x 10^4. A mechanism describing the nature of the chain process with copper ions as catalysts was proposed by Frank and Haber (1931).

$$Cu^{2+} + SO_3^{=} \longrightarrow Cu^{+} + SO_3^{\overline{\cdot}}$$

$$SO_3^{\overline{\cdot}} + H^{+} \longrightarrow HSO_3\cdot$$

$$HSO_3\cdot + O_2 \longrightarrow HSO_5\cdot$$

$$HSO_5\cdot + SO_3^{=} \xrightarrow{H_2O} 2HSO_4^{-} + OH\cdot$$

$$SO_3^{-} + OH\cdot \longrightarrow SO_3^{\overline{\cdot}} + OH^{-}$$

This was one of the first mechanistic interpretations of a chain reaction in solution and contains many ideas which are in vogue today.

Backstrom modified the mechanism in his 1934 publication preferring a propagation step which included sulphur species entirely and excluded the hydroxyl radical

$$SO_3^{=} + M^{+} \xrightarrow{1} SO_3^{\overline{\cdot}} + M$$

$$SO_3^{\overline{\cdot}} + O_2 \xrightarrow{2} SO_5^{\overline{\cdot}}$$

$$SO_5^{\overline{\cdot}} + SO_3^{=} \xrightarrow{3} SO_5^{=} + SO_3^{\overline{\cdot}}$$

$$SO_5^{=} + SO_3^{=} \xrightarrow{4} 2SO_4^{=}$$

$$SO_5^{\overline{\cdot}} + SO_5^{\overline{\cdot}} \xrightarrow{5} S_2O_6^{=} + 2O_2$$

$$SO_5^{\overline{\cdot}} + SO_3^{=} \xrightarrow{6} S_2O_6^{=} + O_2^{\overline{\cdot}}$$

The rate law derived from this mechanism depends upon the nature of the rate determining step and termination process. If reaction 3 is slow and reaction 5 terminates the chain reaction

$$\frac{d}{dt}(SO_4^{=}) = k_3 \frac{(k_1)^{\frac{1}{2}}}{(k_5)} [M^{+}]^{\frac{1}{2}} [SO_3^{=}]^{\frac{3}{2}} \qquad (1)$$

If reaction 3 is slow and reaction 6 is the termination process

$$\frac{d}{dt}(SO_4^{=}) = k_3 \frac{k_1}{k_6} [M^{+}] [SO_3^{=}] \qquad (2)$$

In addition to the propagation step shown (reaction 3) it is also likely that the equivalent step with bisulphite ion will occur.

$$SO_5^{\cdot-} + HSO_3^- \longrightarrow SO_5^= + HSO_3\cdot$$

The reaction can also be initiated by photons which are absorbed into a charge transfer to solvent band of the sulphite ion.

$$SO_3^= \xrightarrow[\lambda < 260\ nm]{h\nu} SO_3^{\cdot-} + e^-_{aq}$$

In spite of the necessity of an external initiation process demanded by the above mechanism, various authors have maintained that the oxidation proceeds spontaneously. Fuller and Crist (1941) rigorously distilled the water which was used as a reaction medium until a relatively rapid reaction was observed. They claimed that this was due to the removal of inhibitors. They measured the rate of disappearance of oxygen into sulphite solutions using a gas burette and derived the following rate expression

$$\frac{d}{dt}(SO_4^=) = [k_1 + k_2\ (H^+)^{\frac{1}{2}}](SO_3^=) \qquad (3)$$

The values of k_1 and k_2 at 25°C were $1.3 \times 10^{-2}\ s^{-1}$ and $6.6\ M\ \ s^{-1}$ respectively.

The reaction was also studied by Schroeter (1963) who used air rather than oxygen as the oxidant. He obtained a value for k_1 of $1.86\ x^{-3}\ s^{-1}$ at 25°C and concluded that this agreed well with Fuller and Christ's estimate if the reaction was first order in oxygen. Schroeter disagreed with Fuller and Christ as to whether both sulphite and bisulphite ions would react. Schroeter maintained that both ions were equally oxidisable in spite of experimental evidence that oxidation ceases when all sulphite has been removed and bisulphite remains. Other investigators of the reaction, particularly Brimblecombe and Spedding (1974), who used a radioactive tracer technique in dilute solution, and Penkett et al. (1979a) who performed a spectrophotometric study, observed much lower rates. Brimblecombe and Spedding's value for k_1 at pH 8.5 and temperature 20°C was $0.8 \times 10^{-5}\ s^{-1}$ and Penkett et al's value was very similar ($1.55 \times 10^{-5}\ s^{-1}$ at pH 8.5 and 25°C).

A recent study by Clark and Radojevic (1983) also give rise to a low value for k_1 ($1.3 \times 10^{-5}\ s^{-1}$ at 25°C). They took note of comments by Dasgupta et al. (1980) that inhibitors present in distilled water held in plastic containers were responsible for the low rates observed. Taking various samples of freshly distilled water they observed little change in rate. They did observe a considerable increase in rate in the presence of high concentrations of chloride salts. No firm conclusions were drawn as to the detailed effect of chloride ions on the mechanism.

A general conclusion of these studies is that the rates are extremely variable. On balance it seems extremely unlikely that spontaneous oxidation occurs and in all cases the rate is determined by the presence of some external catalyst such as transition metal ions or light.

3(ii) Photo-oxidation

Photo-oxidation by sulphite solutions was noted by Backstrom (1927). He recorded large quantum yields but reported no kinetic detail. There is little information in the literature available on this since most studies have been concerned with thermal processes.

In some previously unreported work Penkett and Jones studied the photo-induced decay of sulphite solutions in a stopped-flow spectrophotometer at 240 nm. The study was rather fortuitous as a thermal study was originally intended and no quantitative assessment was made of the light output of the lamp. This was a low intensity deuterium arc lamp which was used as a monitoring light beam in a stopped-flow machine. The general kinetics of the reaction were determined though. The solutions were made up between 10^{-2}M and 4 x 10^{-2}M sodium sulphite and they were mixed with distilled water that was either saturated with air or oxygen. The sulphite was always greatly in excess therefore. It was observed that the sulphite concentration (observed spectrophotometrically by the monitoring light) declined in a linear manner with time until most of the oxygen had been consumed (Figure 2). The reaction-time plot then became curved. Since sulphite was greatly in excess, the plot reflected the oxygen concentration in solution. The main reaction was shown to be first order with respect to sulphite ions and zero order with respect to oxygen. The specific rate constant $k_{(O_2)}$ defined as $-d/dt(SO_3^{=}) = k_{(O_2)} (SO_3^{=})$ was 6.2 x 10^{-3} sec^{-1} at 25°C which lies between the value obtained by Fuller and Crist (1941) and by Schroeter (1963) for $k_{(O_2)}$. It was found that the reaction stopped completely when the monitoring light was shielded from the reaction cell of the stopped flow machine.

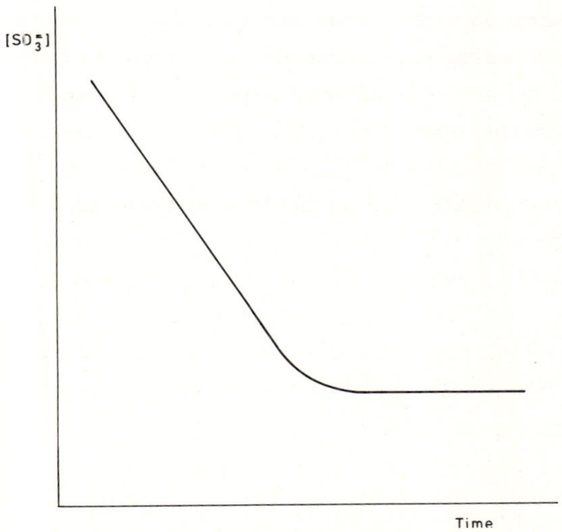

Figure 2: Decay of sulphite absorption at 240 nM during photooxidation. (The decay terminates when the oxygen concentration is reduced to zero.)

Rates recorded in the absence of light were identical to those reported by Penkett et al.[5] from a spectrophotometric study which took place in a standard laboratory spectrophotometer. The significance of the close approximation of $k_{(O_2)}$ to the faster rates determined by other authors is not known.

The kinetic observations can be accounted for by the following mechanism

$$h\nu + SO_3^{=} \xrightarrow{k_0} SO_3^{\cdot -} + e^-_{(aq)}$$

$$SO_3^{\cdot -} + O_2 \xrightarrow{k_2} SO_5^{\cdot -}$$

$$SO_5^{\cdot -} + SO_3^{=} \xrightarrow{k_3} SO_5^{=} + SO_3^{\cdot -}$$

$$SO_5^{=} + SO_3^{=} \xrightarrow{k_4} 2SO_4^{=}$$

$$SO_5^{\cdot -} + SO_3^{=} \xrightarrow{k_6} S_2O_6^{=} + O_2^{\cdot -}$$

and the following rate expression can be derived assuming that reaction 3 is rate determining.

$$\frac{d}{dt}(SO_4^{=}) = \frac{k_0 k_3}{k_6}[SO_3^{=}] \qquad (4)$$

(k_0 includes both the quantum efficiency, absorption cross section and lamp photon output.) This expression is rather similar to equation 2 which was derived for the metal catalysed process. After initiation, the propagation, oxidation and termination steps would be expected to be identical in the two cases. The kinetics changes when the oxygen concentration becomes very small. In this case reaction 2 would perhaps be rate determining and $SO_3^{\cdot -}$ would probably be present in larger concentrations than $SO_5^{\cdot -}$. A probable termination step is then:

$$SO_3^{\cdot -} + SO_3^{\cdot -} \xrightarrow{k_7} S_2O_6^{=}$$

and the rate law for sulphate formation or oxygen decay is

$$\frac{d}{dt}(SO_4^{=}) = k_2\left[\frac{k_0}{k_7}\right]^{\frac{1}{2}}(SO_3^{=})^{\frac{1}{2}}(O_2) \qquad (5)$$

It is not possible from the data available to determine whether an overall 3/2 order decay was being observed when the oxygen in solution was nearly exhausted but it is not unlikely.

The close similarity between the reaction time curves observed in these experiments and those of Fuller and Crist (1941) makes it interesting to speculate on the nature of the process being studied in the two cases. It is highly unlikely that Fuller and Crist's observations were caused by photo-oxidation since light with a maximum wavelength of about 260 nm is required for absorption by the sulphite ion chromaphore. Catalysis by minute concentration: of true metals remains a distinct possibility however in spite of their attempts to remove all trace of catalytic metals. No analytical data were presented on the metal content which therefore remains an unknown quantity.

3(iii) Catalysed Oxidation of S^{IV} by Oxygen

3(iii)a Distilled water studies

This is a very open-ended subject with many publications and much confusion. It was effectively reviewed by Hegg and Hobbs (1978) and they concluded that the kinetic mechanisms involved were complex and depended upon the concentrations and natures of both reactants and catalysts. They thought it was doubtful whether any numerical rate expression could be extracted and utilised for calculating atmospheric rates. This has been one of the basic problems in the past in assessing the relative importance of the multiphase and homogeneous conversion of S^{IV} to S^{VI}. There are some simplifying features however and it is possible to set limits on the efficiency of the catalytic oxidation process.

Some of the better work in this field was performed by a relatively small number of investigators. Junge and Ryan (1958) and Bracewell and Gall (1967) showed that the most effective catalysts for SO_2 oxidation in solution were iron and manganese. Hoather and Goodeve (1934) investigated the manganese assisted rate in acid solutions and found the following rate expression:

$$-\frac{d}{dt}(O_2) = \frac{d}{dt}(SO_4^{=}) = k_e\,[Mn]^{1.7} \qquad (6)$$

The reaction, which was followed by studying the rate of uptake of oxygen, was zero order in oxygen and bisulphite ions. Similar kinetics were found by Johnstone and Coughanour in 1958 and also by Levvy and Penkett in 1972. Levvy and Penkett studied the rate of oxidation of dilute solutions of sodium bisulphite (5×10^{-5}M) in the presence of micromolar concentrations of manganese chloride using a radioactive tracer technique described by Penkett et al. in (1979). The kinetics of the reaction showed a zero order dependence on $[HSO_3^-]$ and $[O_2]$ and a second order dependence on manganese ions.

$$\frac{d}{dt}(SO_4^=) = k_f [Mn]^2 \qquad (6a)$$

A plot of the specific rate constant (k_f) against the pH of the solution is shown in Figure 3 for both the Johnstone and Coughanour and Levvy and Penkett data. The value of k_f increases steadily up to pH 4.5 and then declines.

The overlap of the two independent sets of data suggests that the manganese catalysed oxidation is not difficult to reproduce under laboratory conditions. Bracewell and Gall's specific rate constant at 25°C ($1.84 \times 10^5/M^{-1}$ min^{-1}) is also in close agreement if the reaction took place close to pH 4, and Coughanour and Krause's (1965) rate constant of $2.9 \times 10^4/M^{-1}$ min^{-1} at pH 2.8 agrees well.

The kinetics do not follow the course predicted by the Backstrom mechanism so it is quite likely that the reaction intermediates are quite different to the ones proposed earlier. Bassett and Parker (1951) considered that complexes of the type $[Mn[HSO_3O_2]^=]$ were involved and oxygen transfer took place within the coordination shell of the transition metal ion. Since the rate is proportional to the second power of the manganese concentration it is possible that bridging complexes may be involved.

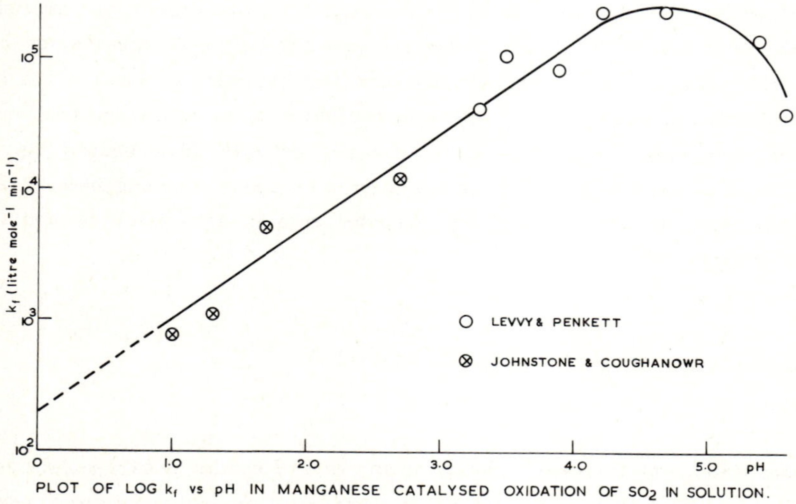

Figure 3: Variation of rate constant with pH for manganese catalysed oxidation of bisulphite ions by oxygen

Matteson et al. (1969) proposed a mechanism of the following type for the oxidation of SO_2 observed in manganese sulphate aerosol droplets.

$$Mn^{2+} + SO_2 \underset{}{\overset{k_1}{\rightleftharpoons}} MnSO_2{}^{2+}$$

$$2MnSO_2{}^{2+} + O_2 \rightleftharpoons [(MnSO_2{}^{2+})_2 \cdot O_2]$$

$$[(MnSO_2{}^{2+})_2 \cdot O_2] \rightleftharpoons 2MnSO_3{}^{2+}$$

$$MnSO_3{}^{2+} + H_2O \rightleftharpoons Mn^{2+} + HSO_4{}^{-} + H^{+}$$

They showed that in conditions where the catalyst concentration was much lower than the SO_2 concentration the rate expression would become

$$\frac{d}{dt}(S^{IV}) = k_1 \; [Mn]^2 \qquad (6b)$$

which accounts for the kinetic observations in dilute solutions. There is no direct evidence of course for the occurrence of the manganese complexes proposed but it seems very likely that oxidation of S^{IV} proceeds in this case in the coordination sphere of the manganese atom rather than by simple electron removal from a sulphite or bisulphite ion.

Perhaps the main problem of extrapolation of laboratory studies into the atmospheric situation is concerned with the efficiency of iron as a catalyst. It is the most abundant trace catalytic metal in the atmosphere and several laboratory studies have shown that ferrous iron (Fe^{II}) has a catalytic efficiency of the same order as manganese (Junge and Ryan (1958), Bracewell and Gall (1967) and Barrie and Georgii (1976)).

In the atmosphere the most abundant form of iron will be Fe^{III} and this creates difficulties in applying laboratory rates to the atmospheric situation since Fe^{III} is distributed amongst many chemical forms over the pH range 2 - 7 (see Hegg and Hobbs (1978)).

Whereas catalytic oxidation of sulphite solutions by manganese has been studied in some detail and the data appears self-consistent, this is not the case for iron. Reinders and Vles (1925) merely reported quantitative results and recent studies have reported conflicting kinetics in the order with respect to dissolved SO_2. Karrahar (1963) and Brimblecombe and Spedding (1974) found a first order variation with respect to dissolved SO_2 and Neytzell de Wilde and Taverner (1958) found a second order dependence. Brimblecombe and Spedding derived a rate law from their kinetic studies using Fe^{III} in dilute solutions of the form

$$\frac{d}{dt}[S^{IV}] = k''[Fe^{III}][S^{IV}] \qquad (7)$$

where $k'' = 100\ M^{-1}\ s^{-1}$ at 25°C.

These studies were made at a pH close to 4.9 when iron is mostly in the form FeOOH according to the authors. The form of iron in solution is one of the most difficult quantities to specify and data obtained from freshly made-up solutions under laboratory conditions are most unlikely to be directly applicable to the atmosphere. The work of Brimblecombe and Spedding showed that extremely small traces of iron, such as will be omnipresent in the atmosphere, had a marked effect on the rate of SO_2 oxidation in solution. Freiberg (1974) considered it unlikely therefore that the uncatalysed mechanism is of any consequence under atmospheric conditions.

3(iii)b Rainwater Studies

A study which set out to determine the rate of catalytic oxidation of SO_2 in samples of atmospheric water was undertaken by Penkett et al. (1979b) at Harwell. Radioactive SO_2 was added to rain samples that had been collected and stored for about a year, and the rate of sulphate formation was correlated with the concentration of a number of trace elements in the rainwater. These included manganese, vanadium, chromium, cobalt, nickel, iron and copper. The correlation coefficients (r^2) for each metal are shown in Table 1. It is very clear that manganese is the controlling element, that vanadium may play some role and that iron is unimportant as a catalyst.

The kinetics were quite different from those observed previously for solutions of manganese in distilled water. This is shown clearly in Figure 4. The rate in rain is proportional to the first power of the manganese concentration whereas in distilled water it is proportional to the second power of the manganese concentrations. Also for any given manganese concentration in the range likely to be found in samples of atmospheric water, the rate of sulphate formation is much higher than would be predicted from the distilled water studies.

No attempt was made to alter the chemical composition of the rain as collected and stored. The pH range in the study was therefore quite narrow (3.7 to 4.4), but this is the most common pH encountered in rain in the United Kingdom and elsewhere. Even so some interesting facts were obtained simply by varying the amount of SO_2 that was added to rainwater to follow its course of oxidation. Figure 5 is a logarithmic plot of the rate of SO_2 oxidation observed in one rainwater sample against the initial SO_2 concentration $[SO_2]_i$. Up

TABLE 1

Regression analysis if trace element content VS rate of SO_2 oxidation

Rate = Slope x Concentration + Intercept

Element	Slope	Intercept	Correlation (r^2)
Mn	8.5	-0.61	0.94
V	62.58	-1.53	0.66
Cr	57.1	-2.35	0.37
Co	-3.24	4.62	0.16
Ni	-2.77	4.19	0.05
Fe	0.26	2.35	0.02
Cu	2.15	2.31	0.01

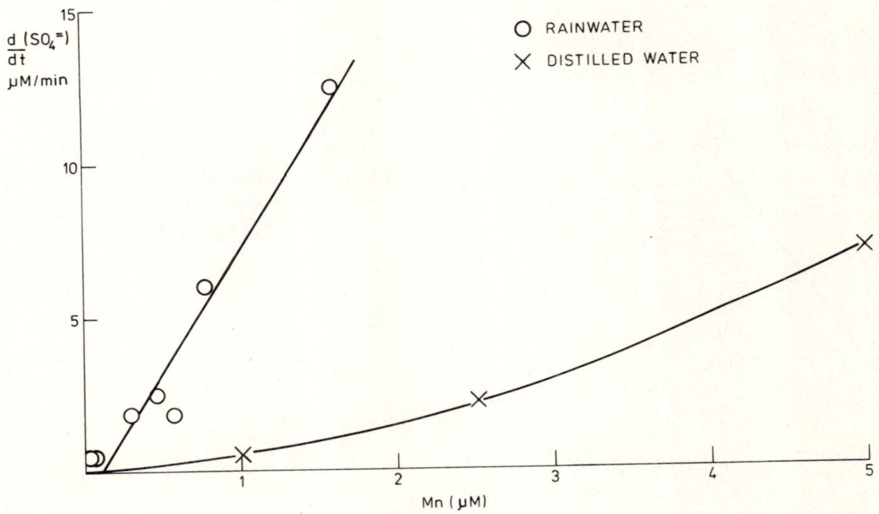

Figure 4: Rate of sulphate formation (d/dt ($SO_4^{=}$)) vs manganese concentration in rainwater and distilled water

to a concentration of about 40 μM the rate is proportional to the three halves power of $(SO_2)_i$. Beyond this the rate is zero order in $(SO_2)_i$.

Expressions for the rate of sulphate formation in these experiments at 25°C are as follows

$$\frac{d}{dt}(SO_4^{=}) = 0.142\ [Mn]\ M\ s^{-1} \tag{8}$$

for the case when $[SO_2]_i$ exceeds 4×10^{-5} M and

$$\frac{d}{dt}(SO_4^{=}) = 1.42 \times 10^5\ [Mn]\ [SO_2]_i^{3/2}\ M.s^{-1} \tag{9}$$

for the case when $[SO_2]_i$ is less than 4×10^{-5} M. The activation energy for the above reactions is 88.2 kJ mole^{-1}. This type of behaviour is best explained by a free radical mechanism of the Backstrom type which was referred to earlier. In the low concentration range

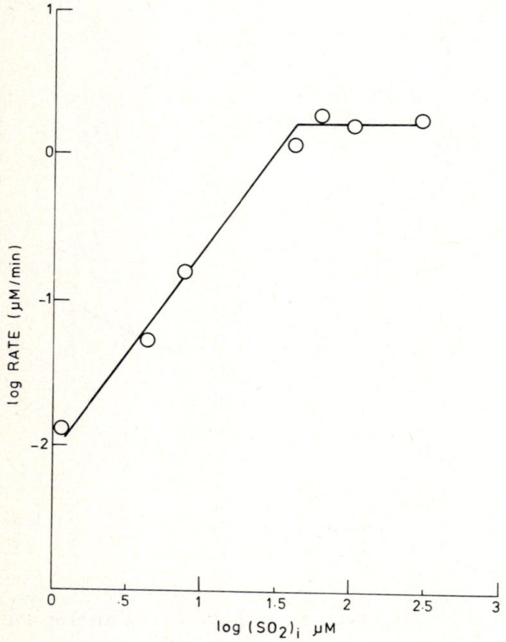

Figure 5: Variation of the rate of sulphate formation with initial SO_2 concentration (SO_2) in a sample of rainwater

$$HSO_3^- + Mn^{n+} \xrightarrow{a} HSO_3\cdot + Mn^{(n-1)+}$$

$$HSO_3\cdot + O_2 \xrightarrow{b} HSO_5\cdot$$

$$HSO_5\cdot + HSO_3^- \xrightarrow{c} HSO_5^- + HSO_3\cdot$$

$$HSO_5^- + HSO_3^- \xrightarrow{d} 2HSO_4^-$$

$$HSO_5\cdot + HSO_5\cdot \xrightarrow{e} S_2O_6^= + 2H^+ + 2O_2$$

The termination products are probably dithionates. If reaction (c) is rate controlling then

$$\frac{d}{dt}(SO_4^=) = k_c \frac{(k_a)^{\frac{1}{2}}}{(k_e)} [HSO_3^=]^{\frac{3}{2}} [Mn]^{\frac{1}{2}} \qquad (10)$$

A fractional order for both [HSO_3] and [Mn] is predicted, and thus equation (10) differs from equation (9).

At increased concentrations of HSO_3^- the kinetics can change.

$$HSO_3^- + Mn^{n+} \xrightarrow{a} HSO_3\cdot + Mn^{(n-1)+}$$

$$HSO_3\cdot + O_2 \xrightarrow{b} HSO_5\cdot$$

$$HSO_5\cdot + HSO_3^- \xrightarrow{c} HSO_5^- + HSO_3\cdot$$

$$HSO_5^- + HSO_3^- \xrightarrow{d} 2HSO_4^-$$

$$HSO_3\cdot + HSO_3^- \xrightarrow{f} H_2S_2O_6 + e^-$$

The termination step involves HSO_3^- and the rate determining step is now reaction (b). In this case

$$\frac{d}{dt}(SO_4^=) = k_b \frac{k_a}{k_f} [Mn]\ [O_2] \qquad (11)$$

The reaction is first order in manganese and oxygen and zero order in bisulphite ion. This type of kinetic scheme could well account for the limited amount of data that was collected in the original study. The overall kinetics of the two regimes are expressed by the two equations 8 and 9 depending on whether the amount of dissolved SO_2 exceeded 40 μM or not. A follow-up study is required to test various points of the mechanism including the fractional order for manganese at lower $[SO_2]_i$ concentrations and the first order prediction of $[O_2]$ at higher $[SO_2]_i$.

The reason for the difference in kinetic behaviour between distilled water and rainwater needs to be investigated. It was suggested originally (Penkett et al. (1976b)) that the effect could be due to promotion, whereby the presence of one catalyst stimulates the effect of another. Solutions containing iron and manganese were shown to be much more effective than manganese or iron alone. The rate in this case was proportional to the product of iron and manganese, which was not observed with rainwater. It is probably more likely that manganese is complexed with a complexing agent in the rain which makes electron transfer an easier process kinetically. The manganese could well be in the trivalent form, making electron transfer a much more likely process.

4 OXIDATION OF S^{IV} TO S^{VI} BY PEROXIDES AND FREE RADICALS

As indicated in Figure 1, oxidation by these species is thermodynamically favoured over reaction by oxygen. It will be shown in due course that oxidation rates are indeed much faster when these species are present. The realisation that mechanisms of this sort could be important under atmospheric conditions was only made in the light of increased understanding of gas phase atmospheric chemistry (Demerjian et al. (1974)).

Ozone is a trace gas which is present throughout the lower troposphere. It is formed in the stratosphere by photolysis of oxygen and can be photolysed in the troposphere with light of wavelength less than 310 nm.

$$O_3 \xrightarrow[\lambda < 310\ nm]{h\nu} O_2 + O(^1D)$$

The reactive oxygen atom produced reacts with water to give rise to hydroxyl radicals

$$O(^1D) + H_2O \longrightarrow 2OH$$

These are very active radicals which react with many atmospheric trace gases in a complex chain reaction. In the course of reaction with gases such as carbon monoxide and hydrocarbons, hydroperoxyl radicals and hydrogen peroxide are produced.

$$OH + CO \longrightarrow CO_2 + H$$

$$M + H + O_2 \longrightarrow HO_2 + M$$

$$HO_2 + NO \longrightarrow OH + NO_2$$

$$HO_2 + O_3 \longrightarrow OH + 2O_2$$

$$HO_2 + HO_2 \longrightarrow H_2O_2$$

The atmosphere thus has means of generating significant quantities of these strong oxidant species. The average concentration of ozone is in the region of 50 parts in 10^9 by volume, which is much higher than the SO_2 concentration found throughout most of the atmosphere. The concentrations of the other species are less well known. A few measurements have been made of the OH radical concentration in polluted air (Perner et al. (1976)) and of the hydrogen peroxide concentration in rain (Zika et al. (1982)). Calculations suggest daytime maximum concentrations in the region of parts in 10^{13} for OH, parts in 10^{11} for HO_2 radicals and parts in 10^9 for hydrogen peroxide (Logan et al. (1981)). These are all significant and can provide effective rates for the oxidation of SO_2 in atmospheric droplets.

The chemistry of peroxides, and free radicals in aqueous solution has been a subject of intense studies for many years. The decomposition of hydrogen peroxide assisted by ferrous iron was interpreted by Haber and Weiss (1934) in terms of a chain mechanism involving interchange of hydroxyl and hydroperoxyl radicals.

$$H_2O_2 + Fe^{2+} \longrightarrow Fe^{3+} + OH^- + OH$$

$$H_2O_2 + OH \longrightarrow HO_2 + H_2O$$

$$H_2O_2 + HO_2 \longrightarrow OH + H_2O + O_2$$

This was the first occasion when this particular type of chain was proposed. It is now a common feature of many reactions in solution and in the gas phase, including those leading to the formation of ozone in photochemical smog. Much of the basic knowledge on free radicals in solution has been gained in the course of studies of the radiolysis of water. There are many good reviews on this subject including those by Uri (1952), and by Bielski and Gebicki (1970). The pH equilibria for all species are clearly defined and rate constants for most species with many substrates have been published in compendiums of tables (see F. and A.B. Ross (1977)).

4(i) Studies of SO_2 Oxidation by Hydrogen Peroxide in Aqueous Solution

The significance of this reaction in the context of formation of acidity in rainwater was first pointed out in 1977 (Penkett et al. (1977) but work on its likely mechanism had been performed earlier. A good review of the oxidation reactions of hydrogen peroxide was written by John O. Edwards in 1961 (Edwards (1961)). This considered that many of the reactions involving hydrogen peroxide were polar and did not involve free radicals. He advanced a general mechanism in terms of nucleophilic reaction on oxygen, with many similarities to those involving a displacement on carbon. This explained many of the known features of hydrogen peroxide reactions including first order dependencies of reactants, a linear free energy correlation with substrate anion basicity and positive catalysis by protons. Edwards showed that for many substances the rate of oxidation was roughly correlated with a reactivity constant calculated from electrode potentials. This is quite useful since it allows estimates to be made of the likely rate of reaction from published thermodynamic quantities.

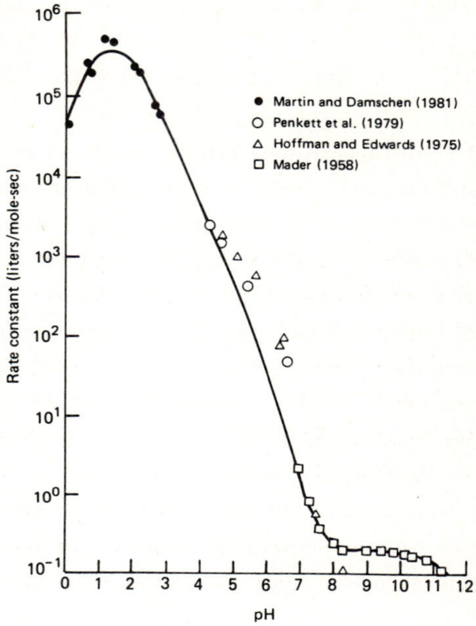

Figure 6: A plot showing the variation of rate constant with pH determined by various authors for the reaction of bisulphite ions with hydrogen peroxide. (Taken from Martin and Damschen, Atmospheric Environment 15, 1615-1621 (1981))

For the specific case of the reaction of hydrogen peroxide and solutions containing bisulphite and sulphite ions, the evidence is strongly in favour of a polar reaction. Halperin and Taube (1952) proposed that the reaction proceeded through a persulphurous acid intermediate, which would explain their observations made with O^{18} enriched peroxide.

Rate studies of this reaction were first performed by Mader (1958) at a pH in the region of 8 where the reaction is slow enough to follow by conventional techniques. At lower pH's the reaction is much faster and could only be followed using the stopped-flow technique.

In their rate studies in acid media Hoffman and Edwards (1975) proposed that the reaction occurred by a nucleophilic displacement by H_2O_2 on HSO_3^- to form a peroxomonosulphurous acid intermediate. This then underwent a rate determining rearrangement which was subject to general acid catalysis

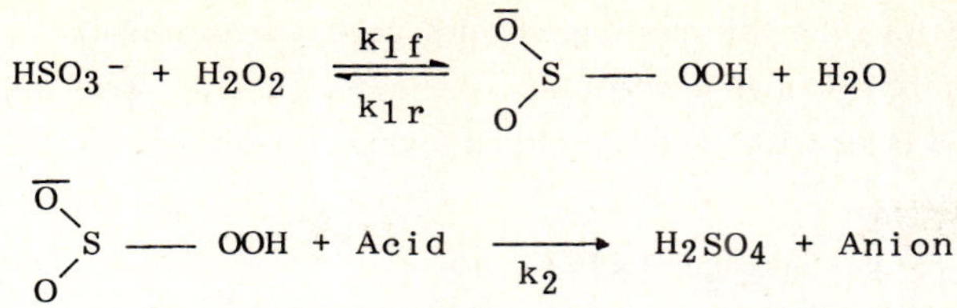

The acid could be H_3O^+ or another species such as H_2PO_4 which was used as a buffer.

The kinetics of this reaction were also studied by Penkett et al. (1979a) over the pH range 4.3 to 8.2 and very similar behaviour was observed. A rate law of the type:

$$-\frac{d}{dt}(HSO_3^-) = k[H_2O_2][HSO_3^-][H^+] + k_a[H_2O_2][HSO_3^-][HA] \qquad (12)$$

was proposed where HA represents any acid other than protons. The reaction order with respect to $[H^+]$ was found to be 0.7 instead of 1, leaving some doubt as to whether the reaction proceeded entirely through a polar mechanism.

A more recent study by Martin and Damschen (1981) provided data at lower pH's (0 - 3) and confirmed that the reaction order with respect to H^+ was 1. From a steady state analysis of the peroxymonosulphurous acid intermediate they derived a general rate expression for catalysis by protons

$$\frac{d}{dt}(SO_4^=) = \frac{k_2\ k_{1f}\ [HSO_3^-][H_2O_2][H^+]}{k_{1r} + k_2[H^+]} \qquad (13)$$

Values were obtained from their data for k_{1r}/k_2, (0.10 M) and $k_{1f} = 5.2 \times 10^6\ M^{-1}\ s^{-1}$ at 25°C. They compared a calculated rate constant k_o with observed values quoted by Hoffman and Edwards, by Penkett et al. and by Mader.

$$k_0 = \frac{k_2\ k_{1f}\ [H^+]}{(k_{1r} + k_2[H^+])(1 + [H^+]/K_1 + K_2/[H^+])}$$

where K_1 and K_2 are the hydrolysis constants referred to in Section 2, corrected to 25°C. The resulting comparison is shown in Figure 6.

McArdle and Hoffman (1984) have also studied the reaction over the pH range 0 to 3 and concluded with Martin and Damschen that the order of reaction with respect to $[H^+]$ was unity. A fractional order for $[H^+]$ could be interpreted in terms of a free radical mechanism perhaps occuring in parallel to the polar mechanism. This was ruled out by McArdle and Hoffman who observed the same rate in the presence of mannitol, which was a very effective free radical scavenger.

These authors concluded that the intermediate species peroxymonosulphurous acid (A^-) could only be formed from HSO_3^- and accordingly the reaction only went through this particular species. They also dealt with general acid catalysis and derived a rate expression

$$\frac{d}{dt}(SO_4^{=}) = \frac{k_{1f}\ K_1\ [H_2O_2]\ [S^{IV}]}{(k_{1r} + k_2[H^+] + k_3[HA])(K_1 + [H^+])} \cdot (k_2[H^+] + k_3[HA]) \qquad (14)$$

where al rate constants refer to the earlier scheme with the addition of k_3 which refers to the process

$$A^- + HA \xrightarrow{k_3} H_2SO_4 + \text{Anion}$$

McArdle and Hoffman's rate expression reduces to a more simple expression which can be used to calculate the rate of SO_2 oxidation by H_2O_2 in atmospheric droplets

$$\frac{d}{dt}(SO_4^{=})(25°C) = \frac{8.8 \times 10^5\ M^{-1}\ s^{-1}\ [H_2O_2]\ [SO_2\ H_2O]}{(1.0 + 20\ M^{-1}\ [H^+])} \qquad (15)$$

This is virtually equivalent to the expression derived by Martin and Damschen for the same process.

$$\frac{d}{dt}(SO_4^{=})(25°C) = 8.3 \times 10^5\ M^{-1}\ s^{-1}\ [H_2O_2][SO_2 \cdot H_2O] \qquad (16)$$

The activation energy to be used for the above reaction is - 28kJ mole^{-1}.

The rate parameters derived from all three studies concerned with peroxide oxidation of SO_2 in aqueous solution give very similar values (see Figure 6). This reaction is therefore well understood kinetically and its application to atmospheric situations is relatively straightforward. The major problem however is associated with the establishment of the concentration of hydrogen peroxide in the atmosphere.

Little is known about the rates of oxidation of SO_2 by organic peroxides in the aqueous phase. It can confidently be assumed that reaction does occur though. Also peroxy acetyl nitrate (PAN) has been proposed as a potential oxidant in the solution phase. Its concentration in the atmosphere is probably higher than hydrogen peroxide on many occasions but it is less soluble and its hydrolysis rate is such that the overall rate of reaction with SO_2 will be slow (Lee et al. (1984)). On balance, therefore, PAN probably does not take part in droplet phase reactions to any great extent.

4(ii) Oxidation by Ozone in Aqueous Solution

Ozone has been known as a trace gas in the atmosphere since the last century. Its potential as an oxidant in atmospheric aqueous chemistry was only established in 1972 when Penkett published the first estimate of its rate of reaction with bisulphite ions (Penkett (1982)).

The chemistry of ozone in water is complex and highly dependent upon pH. In 1935 Weiss proposed that its decomposition was a chain reaction involving free radicals (Weiss (1935)).

$$O_3 + OH^- \longrightarrow HO_2 + O_2^{\bar{\cdot}}$$

$$O_3 + HO_2 \longrightarrow OH + 2O_2$$

overall $$2O_3 + OH^- \longrightarrow OH + O_2^{\bar{\cdot}} + 2O_2$$

The OH yield from decomposed ozone molecules was established to be close to 0.5 by Hoigné

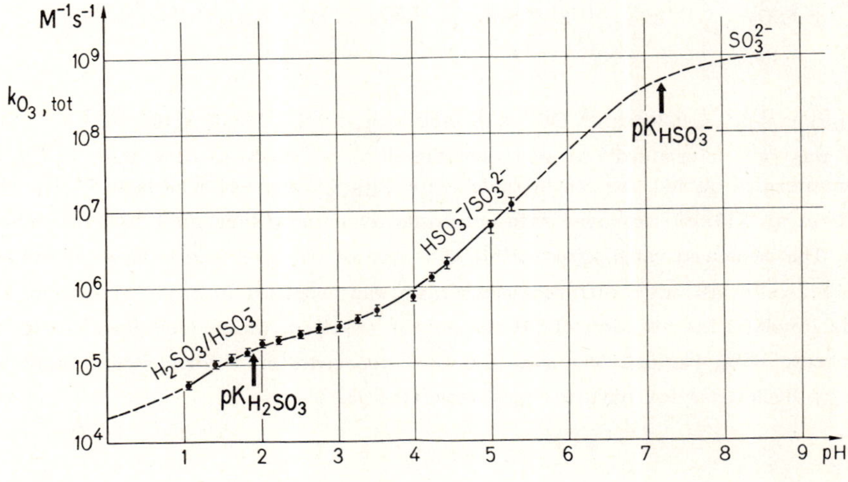

Figure 7: Hoigné et al's experimental value for the apparent rate constants for reactions of ozone with sulphurous acid species. (To be published in Water Research in 1984)

and Bader who also discussed alternative chemical pathways for ozone reactions in aqueous solution (Hoigné and Bader (1976)). These were polar reactions which tended to be slow and highly selective (Figure 7). For many species such as benzene, toluene etc., the free radical pathway in aqueous solution predominated from pH 4 and above. The polar pathway only predominated at pH 2.

Edwards (1961) considered that ozone could act in a similar way to hydrogen peroxide in some reactions. Oxidation of chloride ions for instance is catalysed by protons. Exchange studies of the ozone reaction with dissolved sulphur dioxide species in acidic and basic solutions indicate different mechanisms under different conditions. There are close similarities between the ozone and peroxide reactions in strongly acid conditions (0.1 M $HClO_4$), but in basic solutions the ozone mechanism may involves free radical intermediates (Halperin and Taube (1952) and Espenson and Taube (1965)).

The kinetics of the reaction between ozone and hydrolysis products of SO_2 was first studied in 1972 by Penkett at a fixed pH (4.65). The reaction was shown to be fast and studies made with a stopped-flow machine monitoring ozone decay spectrophotometrically, showed a first order dependence of both reactants (Penkett (1972)). Subsequently studies of this reaction have been made by Erikson et al. in 1977, by Larson et al. in 1978, by Penkett et al. in 1979 and recently by Maahs in 1983 and by Hoigné et al. in 1984. Martin (1983) compared much of this data with his own and concluded that the work of Larson was in error.

Erikson et al. (1977) studied the reaction over the pH range - 0.3 to 4.02 and observed a non-linear increase in rate with increasing pH. This was interpreted in terms of completely different reaction rates of molecular ozone with the two major ions, HSO_3^- and $SO_3^=$. A rate expression was derived of the kind

$$\frac{d}{dt}(SO_4^=) = k_{SO_2(aq)}[SO_{2(aq)}][O_3] + k_{HSO_3^-}[HSO_3^-][O_3] + k_{SO_3^=}[SO_3^=][O_3] \qquad (17)$$

where $k_{HSO_3^-} = 3.1 \pm 0.24 \times 10^5$ ($M^{-1}\ s^{-1}$) and $k_{SO_3^=} = 2.2 \pm 0.35 \times 10^9$ ($M^{-1}\ s^{-1}$) at 25°C. $k_{SO_2(aq)}$ was not determined but was considered to be much smaller than $k_{HSO_3^-}$. The activation energies quoted are -46 kJ mole^{-1} for $k_{HSO_3^-}$ and -44.5 kJ mole $^{-1}$ for $k_{SO_3^=}$.

Penkett et al. (1979a) measured rate constants over the pH range 1 to 5 in unbuffered solutions. The measured rates agree within a factor of two over the common pH range with those of Erikson et al. but a different mechanism was proposed in order to account for the data. This involved the creation of OH radicals in solution, which then reacted with HSO_3^- ions to create HSO_3 radical. The reaction then proceeded in a very similar way to that proposed by Backstrom for oxygen (Backstrom (1943)).

$$2O_3 + OH^- \xrightarrow{a} OH\cdot + O_2^{\overline{\cdot}} + 2O_2$$

$$OH^\cdot + HSO_3^- \xrightarrow{b} HSO_3\cdot + OH^\cdot$$

$$HSO_3\cdot + O_2 \xrightarrow{c} HSO_5^{\overline{\cdot}}$$

$$HSO_5\cdot + HSO_3^- \xrightarrow{d} HSO_5^- + HSO_3\cdot$$

$$HSO_5^- + HSO_3^- \xrightarrow{e} 2HSO_4^-$$

$$HSO_5\cdot + HSO_5\cdot \longrightarrow H_2S_2O_6 + 2O_2$$

If reaction (d) is rate determining then

$$-\frac{d}{dt}(O_3) = K_W k_d \sqrt{\frac{k_a}{k_f}}\,\frac{[O_3][HSO_3^-]}{[H^+]^{\frac{1}{2}}} \qquad (18)$$

where K_W is the hydrolysis constant for water.

A general expression for the rate of sulphate formation by the ozone reaction is

$$\frac{d}{dt}(SO_4^=) = k_{obs}\,[O_3][S^{IV}] \qquad (19)$$

The values of k_{obs} obtained by Penkett et al. (1979a) at 10 C for pH 4 was $7.13 \times 10^5\ M^{-1}\ s^{-1}$, for pH 5 was $1.79 \times 10^6\ M^{-1}\ s^{-1}$ and for pH 6 was $4.5 \times 10^6\ M^{-1}\ s^{-1}$.

The two recent studies of Maahs (1983) and of Hoigne and his coworkers (1984) both used buffers to control the pH of the reactant medium and both report higher rates than Penkett et al. above pH 4. Maahs concluded that it was not possible to distinguish between the free radical and molecular mechanisms but Hoigne et al. favour the molecular mechanism originally proposed by Erikson et al. Hoigne's data is shown graphically in Figure 7. He concludes that the inflection point near pH 4 is caused by a change in control from reaction with HSO_3^- ions to $SO_3^=$ ions. The rate law is identical to reaction 17 i.e.

$$\frac{d}{dt}(SO_4^=) = k_{SO_2(aq)}[SO_{2(aq)}][O_3] + k_{HSO_3^-}[HSO_3^-][O_3] + k_{SO_3^=}[SO_3^=][O_3] \qquad (20)$$

with $K_{SO_2(aq)} = 2 \pm 2 \times 10^4\ M^{-1}\ s^{-1}$, $k_{HSO_3^-} = 3.2 \pm 0.2 \times 10^5\ M^{-1}\ s^{-1}$ and $k_{SO_3^=} = 1.0 \pm 0.2 \times 10^9\ M^{-1}\ s^{-1}$ at 22°C.

4(iii) Oxidation by Chlorine in Solution

Hydrolysis products of chlorine can also act as strong oxidant species. The oxidation process is probably of little relevance in the atmosphere but a kinetic study has shown that the reaction is extremely rapid and a multiphase study can be of some value in understanding the nature of droplet reactions.

The hydrolysis reaction of chlorine can be expressed as follows:

$$Cl_2 + H_2O \overset{K_H}{\rightleftharpoons} H^+ + Cl^- + HOCl$$

where

$$K_H = \frac{[H^+][Cl^-][HOCl]}{[Cl_2]} = 3.94 \times 10^{-4}\ M^2 \text{ at } 298°K.$$

HOCl is a weak acid which dissociates into hypochlorite ions and protons

$$HOCl \overset{K_H}{\rightleftharpoons} OCl^- + H$$

where

$$K_{(HOCl)} = 3.2 \times 10^{-8}\ M \text{ at } 298°K.$$

The reaction between dissolved chlorine and sulphur dioxide species has been studied at pH 13 by a rapid mixing technique. It was assumed that the reaction proceeded in a simple manner (Lister and Rosenblum (1961)).

$$OCl^- + SO_3^= \xrightarrow{k_{(OCl^-)}} SO_4^= + Cl^-$$

The reaction has recently been studied over the pH range 1 to 13 using the stopped-flow technique (Penkett et al. (1983)). It was found that the value of $k_{(OCl^-)}$ at 280°K increased from $1.20 \times 10^4\ M^{-1}\ s^{-1}$ at pH 13 to $7.1 \times 10^4\ M^{-1}\ s^{-1}$ at pH 8. Below pH 8 the reaction became too fast to measure with a second order rate constant in excess of $10^6\ M^{-1}\ s^{-1}$. In this pH range the dominant chlorine species is HOCl and the reaction probably has the form

$$HOCl + HSO_3^- \xrightarrow{k_{(HOCl)}} HSO_4^- + H^+ + Cl^-$$

A comparison of $k_{(HOCl)}$ with $k_{(OCl^-)}$ for the reaction of chlorine with a number of nitrogeneous compounds suggests that HOCl is about 10,000 times more reactive than OCl^-

Morris (1978)). If this were also true for the reaction with hydrolysis products of SO_2 it would suggest that $k_{(HOCl)}$ could be as high as 10^9 M^{-1} s^{-1}.

4(iv) Reactions of Dissolved SO_2 with Free Radicals

The importance of the process of capture by cloud droplets of free radicals generated by gas phase chemistry has been indicated recently (Chameides and Davis (1982)). Radicals such as HO_2 and OH can be captured at a rate which provides a significant source of these species in solution. The HO_2 radicals do not react rapidly with many other substrates but combine with each other to form hydrogen peroxide. The chemistry of this process has been discussed in detail by Chameides and Davis (1982). The OH radicals are very reactive and do react rapidly with both HSO_3^- and $SO_3^=$ ions.

$$\frac{d}{dt} (SO_4^=) = k_{HSO_3^-} [OH] [HSO_3^-] \tag{21}$$

$$\frac{d}{dt} (SO_4^=) = k_{SO_3^=} [OH] [SO_3m^=] \tag{22}$$

A number of techniques have been developed to study the rates of hydroxyl radical reactions in water. Pulse radiolysis is probably the most commonly used for producing the radicals, which are followed directly or indirectly by spectroscopic techniques. The extinction coefficient for the maximum OH absorption in neutral solution is small (370 M^{-1} cm^{-1} at 200 nm). It is therefore usual either to follow the course of product formation or alternatively to make use of competition reactions. In the case of the HSO_3^- and $SO_3^=$ ions the rates were determined in competition with the reactions of OH with $CO_3^=$. This reaction forms a carbonate radical ion

$$OH + CO_3^= \longrightarrow CO_3^{\overline{\cdot}} + OH^-$$

which absorbs strongly at 600 nm. The rate constants derived from this work at 25°C and quoted in F. and A.B. Ross (1977) are 9.5 x 10^9 and 5.5 x 10^9 M^{-1} s^{-1} respectively for $k_{HSO_3^-}$ and $k_{SO_3^=}$ shown in equations (21) and (22). These are extremely high and are very close to the diffusion limit.

5 MULTIPHASE SIMULATION EXPERIMENTS

In the atmosphere reactions take place in the liquid phase after absorption of reactants from the gas phase. This has led many investigators to attempt to simulate the atmospheric process in multiphase experiments. The results of these experiments have often been confusing and rarely lead to a clear understanding of the applicability of individual processes under atmospheric conditions.

The oxidation of SO_2 in droplets of distilled water in the presence of ammonia gas was studied by van den Heuvel and Mason (1963). Their conclusions were confusing in that they claimed that the rate was chemically controlled and then proceeded to show that the overall rate was controlled by diffusion of the reacting gases within the droplets. A rate constant for SO_2 oxidation was derived from this work by Scott and Hobbs (1967), ($k_1 = 1.7x10^{-3}\ s^{-1}$, where k_1 is equivalent to that defined in equation 3), who then proceeded to calculate a potential rate of sulphate formation under atmospheric conditions. Beilke et al. (1975) also derived a value for k_1 from some experiments on the uptake of SO_2 into distilled water in a small chamber. Their rate expression took the form

$$\frac{d}{dt}\,(SO_4^=) = k_o\ [SO_3^=] \qquad (23)$$

where $k_o = 1.2 \times 10^{-4}\ [H^+]^{-0.16}$. At pH 5, $k_o = 7.6 \times 10^{-4}\ s^{-1}$ which is slightly smaller than the value used by Scott and Hobbs but is within the limits for k_1 in equation 3 which has been derived from more classical laboratory studies. Other authors, notably Larson et al. (1968) and Miller and de Pena (1972) have also attempted to derive rate constants from multiphase experiments. The rate constants vary appreciably no doubt reflecting the nature of the chemistry which is involved. As concluded earlier, (Section 3i) the presence of small traces of catalytic substances almost certainly controls the rate, making this type of simulation experiments of very limited value to the real atmospheric situation.

Experiments have been performed in which catalysts have been deliberately added. Experiments in relatively dilute solutions such as those performed by Johnstone and Coughanour (1958) and by Barrie and Georgii (1976) suffer from mass transport limitations. The droplet size in these experiments was of the order of 1 mm and diffusion considerations within the drop are rate limited according to Freiberg and Schwartz (1980). The catalytic efficiency of iron, which was tested in both these experiments is far from clear, with Johnstone and Coughanour suggesting very limited oxidation of SO_2 and Barrie and Georgii suggesting substantial activity. This activity was enhanced in the presence of mixed catalysts such as iron and manganese (Barrie and Georgii (1976)). Again there is a problem of the applicability of the findings of these experiments to the atmospheric situation. Catalysed oxidation of SO_2 in rainwater follows quite different kinetics to the behaviour in distilled water. Also the kinetics can change with dissolved S^{IV} concentration decline (Penkett et al. (1979b)).

Experiments have also been carried out in droplets formed by deliquescent electrolytes, and empirical expressions have been derived for the oxidation rate of SO_2. Extension of these results to the atmospheric situation is complicated by the choice of reactant aerosol. Most studies have used pure manganese or iron salts which are not present as such in the atmosphere.

The advantage of many of these experiments is that the electrolyte droplets are small (1 to 10 µ) and the reaction rate is not limited by diffusion of reactants within the droplet. This was clearly shown in some experiments performed by Penkett, where a fine aerosol of deliquescent sodium chloride droplets containing manganese chloride was exposed to a gas mixture of radioactive S^{35} sulphur dioxide and ammonia (reported by Eggleton and Atkins (1972)). The rate of sulphate formation, which could be studied at relative humidity of 78% was shown to be proportional to the first power of the SO_2 concentration from 0.1 to 10 ppm. The rate was also proportional to the ammonia concentration and to the second power of the manganese concentration in the droplet aerosol.

These results clearly show that the overall rate is chemically controlled in droplets which have a diameter of 3 $\pm$ 2 microns. This is a very important finding since it allows kinetic results obtained in bulk studies to be applied with confidence to cloud situations, provided of course the correct rates are used.

A physical parameter which has a large effect on the rate of uptake of gases into droplets is the accommodation coefficient . This is defined as the probability of a colliding gas molecule being captured by the droplet. If this is low (10^{-4}) equilibrium times can be of the order of minutes. If it is above 10^{-2} the equilibrium time is a second or less. Some data which confirms this second case was obtained recently in a study of the oxidation of SO_2 by chlorine in small ammonium sulphate droplets (Penkett et al. (1983)). 0.1 ppm of radioactively labelled SO_2 was injected into the reaction chamber containing 1 ppm of chlorine and a droplet aerosol of ammonium sulphate at 90% relative humidity. All the SO_2 was oxidised within seconds indicating that both chemical and physical processes must be extremely rapid (see section 4iii). Samples of the initial aerosol, and that present after reaction, were taken and the size of individual droplets had increased from between 2 and 5 µ to between 5 and 15 µ.

An expression for the rate of uptake $\Phi_{a(x)}$ by a single droplet radius (a) was used by Chameides and Davis (1982).

$$\Phi_{a(x)} = 4\pi D \left[\frac{a}{1 + \left(0.7 + \frac{4(1-\alpha)}{3\alpha}\right)\frac{l}{a}}\right] N_{(x)}$$

where D is the gas phase diffusion coefficient, $N_{(x)}$ is the gas phase concentration, l is the mean free path and α is the accommodation coefficient. Using this expression for Φ_a

indicates that α would have to exceed 10^{-1} in order to account for the observations made above. This suggests that physical processes will not normally limit chemical reactions in small droplets.

There are very few other observations of multiphase oxidation of SO_2 by oxidants other than oxygen. Droplets containing hydrogen peroxide solution have also been used in at least two investigations to confirm rapid uptake of SO_2 into the aqueous phase (Johnstone and Coughanour (1958), Beilke and Georgii (1967)). Perhaps the only experiment, which has attempted to confirm a prediction from laboratory kinetic studies, was made by Penkett and Garland in 1974. The rate of SO_2 oxidation by ozone was studied in a fog chamber. The rate was shown to be proportional to the first power of the ozone concentration as predicted by the kinetics. Estimates were made of the liquid water content of the fog using a nephelometer and this was used to compare the multiphase observations with predictions made from the kinetic studies.

6 CONCLUSION

Atmospheric oxidation of S^{IV} in solution can occur by a number of processes. Of the possible oxidants, oxygen almost certainly requires catalysis by traces of transition metals. Manganese is probably the key catalyst and quantitative rate expressions are shown in equations (8) and (9). The mechanism of oxidation under atmospheric conditions probably involves free radical intermediates.

The reaction with hydrogen peroxide has received considerable attention and is understood in the most detail. The mechanism does not involve free radicals and is not subject to trace metal catalysis. Oxidation by ozone probably occurs by a molecular mechanisms although a radical process can occur at higher pH's. Quantitative rate expressions for the peroxide and ozone reactions are shown in equations (15), (16) and (20) respectively.

Other reactions involve both sulphite and bisulphite ions with hydroxyl radicals (equations (21) and (22)). These rates have not been determined directly as yet and rely upon data obtained in competition reactions. It is highly desirable to have direct measurements at a variety of pH's. The reaction of chlorine with S^{IV} has been studied recently over the pH range 0 to 13. The reaction of hypochlorous acid with bisulphite ions could be almost as fast as the reaction of bisulphite ions with OH radicals.

Simulation experiments designed to investigate chemical and physical aspects of S^{IV} oxidation have been examined. Studies where oxygen is the oxidant are probably of less value than studies designed to collect rate data in bulk phase chemical experiments. This is primarily because the chemistry is not sufficiently well defined.

There are very few multiphase studies with oxidants other than oxygen, even though these may dominate the oxidation rate in most atmospheric situations. Of the ones which have been performed, the oxidation of SO_2 by chlorine in electrolyte droplets can be particularly useful for studying the physical aspects of multiphase oxidation.

ACKNOWLEDGEMENT

This work was sponsored by the United Kingdom Department of the Environment.

REFERENCES

Acid Deposition: Atmospheric processes in Eastern North America (1983) National Academy Press, Washington D.C.

Backstrom, H. (1927) Chain Theory of negative catalysis. Am. Chem. J. 49, 1460-1472.

Backstrom, H. (1934) The chain mechanism in the auto oxidation of sodium sulphite solutions. Z. Physik Chem. 1325, 99-121.

Barrie, L.A. and Georgii, H.W. (1976) An experimental investigation of the absorption of sulphur dioxide by water drops containing heavy metal ions. Atmos. Env. 10, 743-749.

Bassett, H. and Parker, W. (1951) The oxidation of sulphurous acid. J. Chem. Soc. 47, 1540-1560.

Beilke, S. and Georgii, H.W. (1968) Investigations on the incorporation of sulphur dioxide into fog and rain droplets, Tellus, 20, 435-441.

Beilke S., Lamb, D. and Muller, J. (1975) On the uncatalysed oxidation of atmospheric SO_2 by oxygen in aqueous systems. Atmos. Env. 9, 1083-1090.

Beilke, S. and Gravenhorst, G. (1978) Heterogeneous SO_2 oxidation in the droplet phase. Atmos. Env. 12, 231-239.

Bielski, B.H.J and Gebicki, J.M. (1970) Species in irradiated oxygenated water-in Adv., in Rad. Chem. Vol. 2. Eds. Burton, M. and Magee, J.L., Wiley Interscience, New York.

Bracewell, J.M. and Gall, D. (1967) The catalytic oxidation of sulphur dioxide in solution at concentrations occuring in fog droplets. Proceedings of the symposium on the Physicochemical Transformations of sulphur compounds in the atmosphere and the formation of acid smogs. Mainz, Germany.

Brimblecombe, P. and Spedding, D.J. (1974) The catalytic oxidation of micromolar aqueous sulphur dioxide I. Oxidation of dilute solutions of iron (III). Atmos. Env. 8, 937-945.

Chameides, W.L. and Davis, D.D. (1983) The free radical chemistry of cloud droplets and its impact upon the composition of rain. J. Geophys. Res. 87, 4863-4875.

Clarke, A.G. and Radojevic, M. (1983) Chloride ion effects on the aqueous oxidation of SO_2. Atmos. Env. 17, 617-624.

Coughanour, D.R. and Krause, F.E. (1965) The reaction of SO_2 and O_2 in aqueous solutions of $MnSO_4$. I. and E.C. Fundamentals 4, 61-68.

Dasgupta, P.K. (1980) Discussion: The importance of atmospheric ozone and hydrogen peroxide in oxidising sulphur dioxide in cloud and rainwater. Atmos. Env. 14, 272-275.

Demerjian, K.L., Kerr, J.A. and Calvert, J.A. (1974) in Advances in Environmental Science and Technology. Vol. 4. Pitts, J.N. and Metcalfe R.C. (Eds.), Wiley Interscience New York.

Edwards, J.D. (1961) Nucleophilic displacement on oxygen in peroxides in Peroxide Reaction mechanisms. Wiley Interscience publishers, New York.

Eggleton, A.E.J. and Atkins, D.H.F. (1972) Results of the Teeside Investigation. Harwell Report AERE R 6983, HM Stationary Office.

Erikson, R.E., Yates, L.M., Clark, R.L. and McEwen, D. (1977) The reaction of sulphur dioxide with ozone in water and its possible atmospheric significance. Atmos. Env. 11, 813-817.

Espenson, J.H. and Taube, H. (1965) Trace experiments with ozone as oxidising agent in aqueous solution. Inorg. Chem. 4, 704-709.

Frank and Haber (1931) Iber Preuss. Adad. Wiss. 250-262.

Freiberg, J.E. and Schwartz, S.E. (1980) Oxidation of SO_2 in aqueous droplets: mass transport limitation in laboratory studies and the ambient atmosphere. Atmos. Env. 15, 1145-1159.

Freiberg, J. (1974) The mechanism of ion catalysed oxidation of SO_2 in oxygenated solutions. Atmos. Env. 9, 661-672.

Fuller, E.C. and Crist, R.H. (1941) The rate of oxidation of sulphite ions by oxygen. Am. Chem. J. 63, 1644-1650.

Haagen-Smit (1952) Chemistry and Physiology of Los Angeles Smog. Ind. Eng. Chem. 44, 1342-1351.

Haber, F. and Weiss, J. (1943) The catalytic decomposition of hydrogen peroxide by ion salts. Proc. Roy. Soc. (London)) A147, 332-351/

Halperin, J. and Taube, H. (1952) The transfer of oxygen atoms in oxidation reduction reactions IV. The action of hydrogen peroxide with sulphite and thiosulphite and of oxygen, manganese dioxide and of permanganate with sulphite. Am. Chem. Soc. J. 74, 380-382.

Hegg, D.A. and Hobbs, P.V. (1978) Oxidation of sulphur dioxide in aqueous systems with particular reference to the atmosphere. Atmos. Env. 12, 241-253.

Heather, R.C. and Goodeve, C.F. (1934) The oxidation of sulphurous acid III. Catalysis by manganese sulphate. Trans. Far. Soc. 30, 1149-1156.

Hoffman, M.R. and Edwards, J.D. (1975) Kinetics of the oxidation of sulphite by hydrogen peroxide in acidic solution. J. Phys. Chem. 79, 2096-2098.

Hoigne, J. and Bader, H. (1976) The role of hydroxyl radical reactions in oxidation processes in aqueous solutions. Water. Res. 10, 373-386.

Hoigne, J., Bader, H., Haag, W.R. and Staehelin, J. (1984) Rate constants of reactions of ozone with organic and inorganic compounds in water, III inorganic compounds and radicals. Submitted to Water Research.

Johnstone, H.F. and Coughanour, D.R. (1958) Absorption of sulphur dioxide from air. Ind. Eng. Chem. 50, 1169-1172.

Junge, C. and Ryan, T.G. (1958) Study of SO_2 oxidation in solution and its role in atmospheric chemistry. Q. J. Roy. Met. Soc. 84, 46-55.

Karraker, D.G. (1963) The kinetics of the reaction between sulphurous acid and ferrous ion. J. Phys. Chem. 67, 871-874.

Larson, T., Horike, N. and Harrison, H. (1978). Oxidation of sulphur dioxide by oxygen and ozone in aqueous solution. A kinetic study with significance to atmospheric rate processes. Atmos. Env. 12, 1597-1611.

Latimer, W.M. (1952) Oxidation potentials. Prentice-Hall Inc, Englewood Cliffs, N.J.

Lee. Y.N., Senum, G.I. and Gaffney, J.S. (1984) Peroxyacetylnitrate (PAN) Stability, solubility, and reactivity - implications for tropospheric nitrogen cycles and precipitation chemistry. Atmos. Env. in the press.

Levvy and Penkett (1972) in Results of the Teeside investigations. Eggleton, A.E.J. and Atkins, D.H.F. Harwell Report AERE R 6983, HM Stationery Office.

Lister, M.W. and Rosenblum, P. (1961) Rate of reaction of hypochlorite ions with sulphite and iodide ions, Can. J. Chem. 41, 3013-3020.

Logan, J.A., Prather, M.J., Wofsy, S.C. and McElroy, M.B. (1981) Tropospheric Chemistry: A global perspective. J. Geophys. Res. 86, 7210-7254.

Maahs, H.G. (1983) Measurements of the oxidation rate of sulphur (IV) by ozone in aqueous solutions and their relevance to the SO_2 conversion in non urban tropospheric clouds. Atmos. Env. 17, 341-345.

Mader, P.M. (1958) Kinetics of the hydrogen peroxide sulphite reaction in alkaline solution. Am. Chem. Soc. J. 80, 2634-2639.

Martin, L.R. (1983) Kinetic studies of sulphite oxidation in aqueous solutions, in SO_2, NO and NO_2 oxidation mechanisms: Atmospheric considerations. Ed. Calvert, J.G. Butterworths, Boston-London.

Martin, L.R. and Damschen (1981) Oxidation of sulphur dioxide by hydrogen peroxide at low pH. Atmos. Env. 15, 1615-1621.

Matteson, M.J., Stober, W. and Luther, H. (1969) Kinetics of the oxidation of sulphur dioxide by aerosols of manganese sulphate. I. and E.C. Fundamentals, 6, 677-687.

McArdle, J.V. and Hoffman, M.R. (1984) Kinetics and mechanisms of the oxidation of aquated sulphur dioxide by hydrogen peroxide at low pH. J. Phys. Chem. in the press.

Miller, J.M. and de Pena, R. (1972) Contribution of scavenged sulphur dioxide to the sulphate content of rainwater. J. Geophys. Res. 30, 5905-5916.

Morgan, J.J. (1982) Factors governing the pH, availability of H^+ and oxidation capacity of rain. In "Atmospheric Chemistry" Ed. Goldberg, E.D., Dahlem Konferenz, Springer Verlag, Berlin.

Morris, J.C. (1978) The chemistry of aqueous chlorine in relation to water chlorination. In Water Chlorination, Environmental Impact and Health effects. Vol. 1, Ann Arbor.

Neytzell de Wilde, F.G. and Taverner, L. (1958) Experiments relating to the possible production of air oxidising acid leach liquor by autooxidation for the extraction of uranium. 2nd U.N. International Conf. on the peaceful uses of atomic energy. Proceedings 3, 303-317.

Penkett, S.A. (1972) Oxidation of SO_2 and other atmospheric gases by ozone in aqueous solution. Nature 240, 105-106.

Penkett, S.A. and Garland, J.A. (1974) Oxidation of sulphur dioxide in artificial fogs by ozone. Tellus 26, 284-290.

Penkett, S.A., Jones, B.M.R. and Brice, K.A. (1977) Rate of oxidation of sodium sulphite solutions by oxygen, ozone and hydrogen peroxide and its relevance to the formation of sulphate in cloud and rainwater. Harwell Report AERE - 8584. H.M. Stationery Office.

Penkett, S.A., Jones, B.M.R., Brice, K.A. and Eggleton, A.E.J. (1979a) The importance of Atmospheric Ozone and hydrogen peroxide in oxidising sulphur dioxide in cloud and rainwater. Atmos. Environ. 13, 123-137.

Penkett, S.A., Jones, B.M.R. and Eggleton, A.E.J. (1979b) A study of SO_2 oxidation in stored rainwater samples. Atmos. Env. 13, 139-147.

Penkett, S.A., Jones, B.M.R. and Semb, A. (1983) A study of the rate of SO_2 oxidation by chlorine in aqueous solution. Presented at the I.A.M.A.P. Conference of tropospheric sulphur chemistry, Oxford.

Perner, D., Ehhalt, D.H., Patz, H.W., Platt, U., Roth, E.P. and Voltz, A. (1976) OH radicals in the lower troposphere. Geophys. Res. Lett. 3, 466-468.

Reinders, W. and Vles, S.I. (1925) The catalytic oxidation of sulphites. Rec. Trav. Chem. 44, 249-268.

Ross, F. and A.B. (1977) Selected specific rates of reactions of transients from water in Aqueous Solution III Hydroxyl radical and perhydroxyl radical and their radical ions. NSRDS - NBS 59, US Dept. of Commerce.

Schroeter. L.C. (1963) Kinetics of air oxidation of sulphurous acid salts. J. Pharm. Sci. 52, 559-563.

Scott, W.D. and Hobbs, P.V. (1962) The formation of sulphate in water droplets. J. Atmos. Sci. 24, 54-57.

Sillen, L.A. (1964) Stability constant of metal ion complexes Section I - Inorganic Legends. Second Edition. Chem. Soc. (London), Special Publication No. 17.

Uri, N. (1952) Inorganic free radicals in solution. Chem. Rev. 50, 375-453.

Van den Heuvel, P.P. and Mason, B.J. (1963) The formation of ammonium sulphate in water droplets exposed to gaseous sulphur dioxide and ammonia. Q.J.R. Meteor. Soc. 89, 271-275.

Weiss, J. (1975) Investigations of the radical HO_2 in solution. Trans. Far. Soc. 31, 668-681.

Zika, R., Saltzman, E., Chameides, W.L. and Davis, D.D. (1982) H_2O_2 in rainwater collected in South Florida and the Bahama Islands. J. Geophys. Res. 87, 5015-5017.

TRANSITION METALS
AS POTENTIAL CATALYSTS IN ATMOSPHERIC OXIDATION PROCESSES

Rudi van Eldik
Institute for Physical Chemistry
J. W. Goethe-University
Am Niederurseler Hang
D-6000 Frankfurt/Main
F.R.G.

ABSTRACT

In addition to the photochemical free radical pathways involving O_3, OH, H_2O_2, HO_2, RO_2 and $O(^3P)$ as oxidizing agents for atmospheric oxidation processes, substantial evidence for transition metal catalyzed pathways has been reported. In these homogeneous aqueous-phase oxidation processes the initial step usually involves the formation of a metal complex with the species to be oxidized, which is the basis for the catalytic activity.

In this contribution the general behaviour of transition metal ions in aqueous solution is discussed with emphasis on: chemical forms, equilibria, solubility and stability of different oxidation states, complex formation ability, reaction modes and associated mechanisms, redox properties. The interaction of transition metal ions with dissolved gases such as CO_2, SO_2 and their conjugated bases, and the reactivity of the produced complexes including elimination, aquation, redox and photochemical reactions is treated in detail. The relevance of the described processes in atmospheric oxidation reactions is stressed.

1. INTRODUCTION

Since it is the objective of this ASI to treat the chemistry of multiphase atmospheric systems from various points of view to clarify different aspects of this very complex system, this contribution will focus on the possible role transition metal ions can play during atmospheric oxidation processes. Earlier contributions in this volume have emphasized the importance of atmospheric oxidation processes in the concept of acid rain. For this purpose we will restrict the discussion to the oxidation of S(IV), i.e. SO_2, HSO_3^- and SO_3^{2-}, and NO

NATO ASI Series, Vol. G6
Chemistry of Multiphase Atmospheric Systems
Edited by W. Jaeschke

and NO_2, during which SO_4^{2-}, HSO_4^-, HNO_2 and HNO_3 are produced and largely account for the acidity of rain.

The natural pH of rain based on the dissolution of $CO_2(g)$ through the equilibria

$$\begin{aligned} CO_2(g) &\rightleftharpoons CO_2(aq) \\ CO_2(aq) + H_2O &\rightleftharpoons HCO_3^- + H^+ \\ HCO_3^- &\rightleftharpoons CO_3^{2-} + H^+ \end{aligned} \tag{1}$$

is estimated to be 5.6 [1-3]. In contrast rain of pH between 3 and 4 has been recorded over large parts of eastern US, southeastern Canada and northern Europe[4]. This increase in acidity is directly related to the atmospheric oxidation processes mentioned above. These can occur photochemically in the gas phase involving free radicals like OH, HO_2, RO_2 and O, or in the aqueous (droplet) phase involving species such as O_2, O_3, H_2O_2 and the mentioned free radicals[1,4,5-7].

Earlier contributions by Pruppacher, Hoffman and Winkler in this volume have emphasized the important role of the aqueous phase in such oxidation processes. SO_2 oxidation occurs mainly in cloud water, and a good correlation between produced sulfate concentration and relative humidity exists. On the other hand, oxidation can also occur in the aqueous layer surrounding a solid particle, thus emphasizing the homogeneous nature of the oxidation process. Penkett included numerous examples of catalytic activities of transition metal ions during the aqueous phase oxidation of S(IV)[2,8]. In general, kinetic data for this system was reported to fluctuate tremendously[3], with an average rate constant of approx. 1×10^{-3} s^{-1} at 25°C for the oxidation by O_2. Addition of complexation agents such as EDTA[9] and 1,10-phenanthroline[10] reduces the rate constant to ca. 1×10^{-7} s^{-1}, demonstrating the strong influence of trace metal ions present in such systems. The principal catalysts in atmospheric oxidation processes are uncertain and catalytic mechanisms are virtually unstudied[8,11,12]. It is, therefore, our intention to center the following discussion around the role of the transition metal ion in such systems.

For those not familiar with some of the important aspects of transition metal chemistry, a short discussion on the nature of transition metal ions in aqueous solution will be presented. This is followed by a section describing some fundamental aspects concerning the interaction of dissolved gas molecules like CO_2, SO_2 and their conjugated bases with transition metal ions in aqueous solution. Finally, some atmospheric oxidation processes will be discussed in detail, and present and future activities will be emphasized.

2. SOLUTION CHEMISTRY OF TRANSITION METAL IONS

In considering the role of transition metal ions as homogeneous catalysts the following aspects should be treated[2]:

a. Abundance of transition metals in the atmosphere

b. Chemical forms in which the transition metals exist

c. Solubility limitations

d. Stability of different oxidation states

e. Complex formation ability

f. Availability of ligands for complexation

The abundance of transition metals in atmospheric aerosols on a global basis can be summarized as follows (data in $\mu g/m^3$)[2]: Fe 0.5 - 2.0; Cu 0.01 - 0.80; Zn 0.10 - 0.80; Ti 0.01 - 0.08; Mn 0.01 - 0.12; V 0.01 - 0.10; Ni 0.008 - 0.05; Cr 0.005 - 0.02; Cd 0.001 - 0.010; Co 0.0003 - 0.003; Sc 0.0001; Hg 0.0003 - 0.002; Ag 0.001 - 0.003. Although some of these are relatively low, typical rain water concentrations[3] are $(3.8 \pm 1.6) \times 10^{-6}$ and $(2.6 \pm 0.5) \times 10^{-7}$ M for Fe and Mn, respectively, with much higher concentrations (as high as 10^{-4} M) being recorded in the Los Angeles area. If one keeps in mind that very effective catalysis has been observed for $[Cu^{2+}]$ as low as 10^{-8} M, then the mentioned concentration ranges are adequate to catalyze many oxidation processes[13,14]. Potential catalysts are Fe, Cu, Mn, V and Ni[2], of which Mn and Fe have received considerable attention up to now. It follows that the chemistry of the first row transition metal ions is of importance in the further discussion. These ions usually occur in the 2+ and 3+ oxidation states in solution, with some exemptions like Cu which occurs in the 1+ and 2+ oxidation state.

In general metal ions do not exist as free ions in aqueous solution, but as hydrated or complexed species. Transition metal ions have vacant d orbitals which can via hybridization interact with vacant s and p orbitals to produce three types of complexes, viz. tetrahedral - sp^3 hybridization - coordination number 4; square planar - dsp^2 hybridization - coordination number 4; and octahedral - d^2sp^3 hybridization - coordination number 6. These vacant orbitals are filled by electron pairs donated by coordinating ligands, or in the absence of ligands by the coordinating oxygen atom of the solvent (water) molecules. Such aquated metal ions exhibit characteristic acid-base equilibria which for a M^{2+} species with coordination number 4 can be summarized as follows:

$$
\begin{aligned}
M(H_2O)_4{}^{2+} &\rightleftharpoons M(H_2O)_3OH^+ + H^+ \\
M(H_2O)_3OH^+ &\rightleftharpoons M(H_2O)_2(OH)_2 + H^+ \\
M(H_2O)_2(OH)_2 &\rightleftharpoons M(H_2O)(OH)_3{}^- + H^+ \\
M(H_2O)(OH)_3{}^- &\rightleftharpoons M(OH)_4{}^{2-} + H^+
\end{aligned}
\tag{2}
$$

Each of these aquo/hydroxo species will exhibit a characteristic solubility and complexation behaviour. Metal hydroxo species are in general weakly soluble and often precipitate as metal oxide species, viz.

$$M(H_2O)_2(OH)_2 \rightleftharpoons MO(\downarrow) + 3H_2O \qquad (3)$$

Aquated metal ions exhibit characteristic pK_a values which depend on the nature of the metal ion and the surrounding ligands. Water bound to a metal ion is much more acidic than uncoordinated (free) solvent water. Many aquo complexes have pK_a values between 6 and 8, compared to the pK_a value of 15.7 for free water. A number of examples are given to illustrate this point: $Co(NH_3)_5OH_2^{3+}$ 6.3 [15], 6.6 [16]; cis-$Rh(en)_2(OH_2)_2^{3+}$ 6.1 [17]; cis-$Rh(en)_2(OH_2)OH^{2+}$ 8.1 [17]; trans-$Rh(en)_2(OH_2)_2^{3+}$ 4.3 [18]; trans-$Rh(en)_2(OH_2)OH^{2+}$ 7.7 [18]; trans-$Rh(en)_2(X)OH_2^{2+}$ 6.1 (X = Cl), 6.4 (X = Br), 6.8 (X = I)[19]; $Fe(OH_2)_6^{3+}$ 2.7 [20], 2.8 [21,22]. These values demonstrate that the strength of the O-H bond decreases significantly during coordination of the oxygen atom to the metal ion. The increase of between 8 and 13 orders of magnitude in the value of K_a results from a corresponding increase in the deprotonation rate constant which is 2.5×10^{-5} s^{-1} for uncoordinated water, and 1 and 10^4 s^{-1} for complexes with pK_a values of 10 and 6, respectively. The reverse protonation (neutralization) reaction is assumed to be diffusion controlled with a rate constant of ca. 10^{10} M^{-1} s^{-1}. The rapid deprotonation of coordinated water results in the formation of metal hydroxy species, which is an essential aspect of many catalytic processes[23] (see further discussion). Similar decreases in pK_a values are observed for the coordination of other ligands such as ammines or carboxylic acids[24,25]. In general the pK_a value of a coordinated ligand is determined by the distance the acidic group is removed from the metal center, the nature of the central metal ion, the overall charge on the complex, and the surrounding non-participating ligands. However, these effects are of minor importance, the largest effect being caused by coordination itself.

Returning to the important aspect of solubility of metal aquo/hydroxo complex species, serious limitations have been predicted for the catalytic activity of different metal ions due to their low solubility[2]. The latter strongly depends on the pH of the aqueous phase, which could vary between 3 and 9 for typical atmospheric conditions. The solubility of a number of transition metal ions strongly decreases with increasing pH due to the production of metal hydroxy and metal oxide species[2]. For instance $Fe(OH)_3$ has a solubility of 6×10^{-2} M at pH 2 (due to the formation of $Fe(H_2O)_6^{3+}$) compared to a value of 6×10^{-8} M at pH 4 and 6×10^{-14} at pH 6 [2]. Nevertheless, significant catalytic activity could occur at very low concentrations as mentioned before.

In the presence of other ligands, metal aquo/hydroxo species will exhibit characteristic complexation behaviour. Ligands containing functional acid/base groups will exhibit a typical pH dependence in their reactivity with metal ions, deprotonated species being usually much more reactive than their protonated counterparts. In the case of a monodentate ligand typical complexation reactions would be

$$M(H_2O)_4{}^{2+} + L \rightleftharpoons M(H_2O)_3L^{2+} + H_2O \xrightarrow{L} M(H_2O)_2L_2{}^{2+} + H_2O, \text{ etc.} \quad (4)$$

compared to the following situation for a bidentate ligand

$$M(H_2O)_4{}^{2+} + L_2 \rightleftharpoons M(H_2O)_3L\text{-}L^{2+} + H_2O \xrightarrow{\text{ring-closure}} M(H_2O)_2L_2{}^{2+} + H_2O \quad (5)$$

Ligands likely to be found in atmospheric aerosols in order of decreasing abundance are[2] $SO_4{}^{2-}(HSO_4{}^-) > NO_3{}^- > Cl^- > NH_3 \sim CO_3{}^{2-}(HCO_3{}^-) \sim OH^- \sim SO_2 > H_2S(HS^-)$. The stability of the produced complexes depends on the basicity and nucleophilicite of the coordinating ligands. A relative nucleophilicity scale[26] developed on the basis of substitution reactions of a series of Pt(II) complexes suggests the following order:

	O-bonded	<	N-bonded	<	S-bonded	<	P-bonded
Ex:	H_2O, CH_3OH		NH_3, en		$SO_3{}^{2-}$, $(CH_3)_2S$		PEt_3, $P(Ph)_3$

The formation of a complex, which can be expressed by

$$M(OH_2)_x{}^{n+} + L \underset{k_d}{\overset{k_f}{\rightleftharpoons}} M(OH_2)_{x-1}L^{n+} + H_2O \quad (6)$$

where the formation constant $K = k_f/k_d$, is determined by the efficiency of k_f which strongly depends on the nucleophilicity of L.

Complex formation is further determined by the stability of the coordinated water molecules on the aquated metal ion. For the first row transition metal ions it has been observed that coordinated water is very labile for V^{2+}, Ni^{2+}, Co^{2+}, Fe^{2+}, Mn^{2+}, Cr^{2+} and Cu^{2+}, but significantly stabile for Co^{3+} and Cr^{3+} [27,28]. This is clearly seen in the water exchange rates of such ions

$$M(OH_2)_x{}^{n+} + H_2O^* \rightleftharpoons M(OH_2)_{x-1}{}^*OH_2{}^{n+} + H_2O \quad (7)$$

Typical examples are (rate constants at 25°C in s^{-1}): Mn^{2+} 3 x 10^7; Fe^{2+} 3 x 10^6; Co^{2+} 3 x

10^6; Ni^{2+} 3 x 10^4; Cu^{2+} 5 x 10^9; Fe^{3+} 1.6 x 10^2; $Fe(OH)^{2+}$ 1.4 x 10^5. It follows that Cu^{2+} has very labile coordinated water molecules, which could account for its effective catalytic action. On the other hand, these examples demonstrate that solvent exchange strongly depends on pH as shown for Fe^{3+} where deprotonation increases the exchange rate constant by three orders of magnitude[21]. The complex formation ability of a metal ion is often controlled by the water exchange step since other potential ligands can then compete with solvent water to bind to the metal center[29].

Substitution reactions like solvent exchange can occur according to one of the following mechanisms: (S = H_2O)

Associative (Sn2):

$$MS_4 + L \xrightarrow{k_A} MS_4L \xrightarrow{fast} MS_3L + S$$

$$\text{Rate} = k_A[MS_4][L] \qquad (8)$$

Interchange (I):

$$MS_4 + L \underset{}{\overset{K}{\rightleftharpoons}} [MS_4 \cdot L] \text{ (ion pair or encounter complex)}$$

$$[MS_4 \cdot L] \xrightarrow{k_I} MS_3L + S$$

$$\text{Rate} = k_I K[MS_4][L] / \{1 + K[L]\} \qquad (9)$$

Dissociative (Sn1):

$$MS_4 \xrightarrow{k_d} MS_3 + S$$

$$MS_3 \xrightarrow[\text{fast}]{L} MS_3L$$

$$\text{Rate} = k_d[MS_4] \qquad (10)$$

In the associative mechanism an intermediate of higher coordination number is produced and the rate of the process depends linearly on [L]. In contrast, the dissociative mechanism produces an intermediate of lower coordination number and the rate is independent of the concentration of L. The interchange mechanism is of vital importance in many reactions involving transition metal complexes. The entering and departing ligands interchange in an associative or dissociative way[30] during the rate-determining step, and the rate of the process usually exhibits a curved dependence on [L] changing from linear to independent on increasing [L]. Thus kinetic analyses enable a distinction between the limiting forms of the different substitution mechanisms.

Transition metal complexes can undergo various other reactions. For example they can aquate (the reverse steps in reactions (4) to (6)) to the corresponding aquo/hydroxo species and this process is usually acid and base catalyzed. In addition, many complexes undergo geometrical (cis/trans) or linkage isomerization processes. Typical examples of the latter are M-ONO to M-NO_2, or M-OSO_2 to M-SO_3, or vice versa in both cases. Another important reaction type is electron transfer processes which can occur intra- or intermolecular. In the case of atmospheric oxidation processes two possible mechanisms are of importance[31]

a) Outersphere electron transfer

The redox partners exchange an electron without the formation of a common bond

$$Ox_1 + Red_2 \rightleftharpoons Ox_1 \cdot Red_2 \quad \text{(ion-pair)}$$

$$\downarrow \text{E.T.}$$

$$Red_1 + Ox_2 \rightleftharpoons Red_1 \cdot Ox_2 \tag{11}$$

b) Innersphere electron transfer

A bond is formed between the redox partners and electron transfer occurs along the bridging atom (s)

$$Ox_1 + Red_2 \rightleftharpoons Ox_1 - Red_2$$

$$\downarrow \text{E.T.}$$

$$Red_1 + Ox_2 \rightleftharpoons Red_1 - Ox_2 \tag{12}$$

In the case of heterogeneous oxidation processes the metal surface can accomodate both reactants to catalyze the outersphere reaction path. Many electron transfer reactions are also initiated or catalyzed photochemically and these will be referred to in more detail.

To summarize, transition metal ions exist in the absence of other ligands as aquated complex ions in aqueous solution, and exhibit characteristic solvent exchange and complex formation abilities. Water molecules bound to metal centers are orders of magnitude more acidic than uncoordinated water and are capable to act as reactive centers in many processes of atmospheric importance, as will be seen in the forthcomming discussion.

3. REACTIONS INVOLVING COORDINATED LIGANDS

In this section we want to focus on the various ways dissolved gases and their conjugate bases can interact with metal aquo/hydroxo species. We will treat the mechanistic details of some model systems to emphasize the various reaction modes and corresponding mechanisms. In the previous section we showed that many metal aquo species are very labile, which would complicate the investigation of such systems. Therefore, the selected systems include less labile metal aquo species in which some coordination sites (four or five) have been blocked off by strong non-participating nucleophiles, thus simplifying the reaction system significantly. Furthermore, the $CO_2/HCO_3^-/CO_3^{2-}$ system was chosen as a model system since the chemistry involved is well understood and no complications due to competing oxidation processes as in the case of dissolved SO_2, exist.

3.1. Formation and reactivity of metal carbonato complexes

When $CO_2(g)$ is dissolved in aqueous solution the following reactions/equilibria set in[24]

$$CO_2(aq) + H_2O \underset{k_{-1}}{\overset{k_1}{\rightleftharpoons}} H_2CO_3 \overset{K_o}{\rightleftharpoons} HCO_3^- + H^+, \qquad K_1$$

$$HCO_3^- \rightleftharpoons CO_3^{2-} + H^+, \qquad K_2 \qquad (13)$$

Typical values at 25°C are: $k_1 = 3.7 \times 10^{-2}\ s^{-1}$ [32]; $k_{-1} = 18.3\ s^{-1}$ [33]; $pK_o = 3.7$ [34]; $pK_1 = p(K_o k_1/k_{-1}) = 6.4$; $pK_2 = 10.3$ [35]. Of these species $CO_2(aq)$, HCO_3^- and CO_3^{2-} can react with a metal aquo/hydroxo moiety to produce a stable metal carbonate complex. The reaction with carbonic acid is not considered to be important since less than 1 % of the dissolved $CO_2(g)$ is present in this form. Kinetic studies revealed that $CO_2(aq)$ is the most reactive species and can bind to a metal hydroxo species according to the general mechanism[24]

$$ML_5(OH_2)^{p+} \rightleftharpoons ML_5(OH)^{(p-1)+} + H^+, \qquad K_3$$

$$ML_5(OH)^{(p-1)+} + CO_2 \underset{k_{-2}}{\overset{k_2}{\rightleftharpoons}} ML_5(OCO_2H)^{(p-1)+}$$

$$ML_5(OCO_2H)^{(p-1)+} \rightleftharpoons ML_5(OCO_2)^{(p-2)+} + H^+, \qquad K_4 \qquad (14)$$

The produced monodentate carbonato species is stable and may, depending on the nature of the other ligands L, undergo a ring-closing reaction to produce a bidentate carbonato species

$$ML_5(OCO_2)^{(p-2)+} \xrightarrow{k_3} ML_4CO_3^{(p-2)+} + L \qquad (15)$$

Alternatively, carbonato complexes can be produced via the reaction of HCO_3^- and CO_3^{2-} with metal hydroxo species, but these reactions are orders of magnitude slower than the reaction with CO_2(aq) since they involve substitution at the metal center, i.e. M-OH bond breakage.

The kinetics of CO_2 uptake has been studied for numerous systems[24] employing stopped-flow measuring techniques. The corresponding rate constants are between 40 and 600 $M^{-1}\ s^{-1}$ for a series of octahedral transition metal complexes. These rate constants demonstrate how the reactivity of a M-OH species differs from a HOH species which reacts with CO_2(aq) at a rate constant of ca. $1 \times 10^{-3}\ M^{-1}\ s^{-1}$. No metal-oxygen bond breakage occurs during CO_2 uptake, and log k_2 increases linearly with pK_3, i.e. with increasing nucleophilicity of the coordinated hydroxy ion[19,24] (Fig. 1). Extrapolation to the pK_3 value of water, i.e. 15.7, results in a k_2 value of ca. $10^3\ M^{-1}\ s^{-1}$, which is indeed close to that reported for the reaction of OH^- with CO_2(aq). The advantage of a M-OH species, however, being the fact that such CO_2 uptake reactions occur at much lower pH values (usually between 7 and 9).

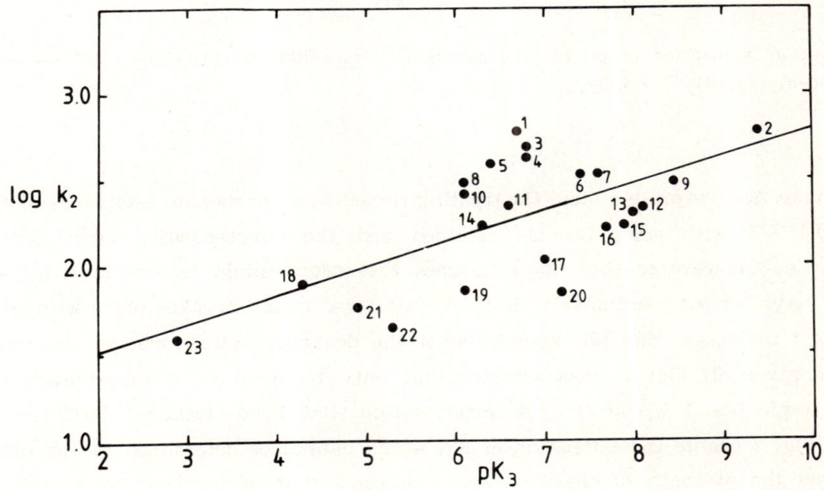

Fig. 1: Plot of log k_2 versus pK_3 for various CO_2 uptake reactions at 25°C [24]

(1) $Ir(NH_3)_5OH^{2+}$
(2) Cu(glygly)OH
(3) $Rh(NH_3)_5OH^{2+}$
(4) trans-$Rh(en)_2(I)OH^+$
(5) trans-$Rh(en)_2(Br)OH^+$
(6) trans-$Rh(en)_2(OCO_2)OH$
(7) trans-$Co(NH_3)_4(CN)OH^+$
(8) cis-$Co(en)_2(OH_2)OH^{2+}$
(9) $Zn(CR)OH^+$
(10) trans-$Rh(en)_2(Cl)OH^+$
(11) $Co(NH_3)_5OH^{2+}$
(12) cis-$Rh(en)_2(OH)_2^+$
(13) cis-$Co(cyclam)(OH)_2^+$
(14) S-$Co(tetren)OH^{2+}$
(15) $Co(tren)(OH)_2^+$
(16) trans-$Rh(en)_2(OH)_2^+$
(17) cis-$Cr(C_2O_4)_2(OH_2)OH^{2-}$
(18) trans-$Rh(en)_2(OH_2)OH^{2+}$
(19) cis-$Rh(en)_2(OH_2)OH^{2+}$
(20) trans-$Co(cyclam)(OH)_2^+$
(21) cis-$Co(cyclam)(OH_2)OH^{2+}$
(22) $Co(tren)(OH_2)OH^{2+}$
(23) trans-$Co(cyclam)(OH_2)OH^{2+}$

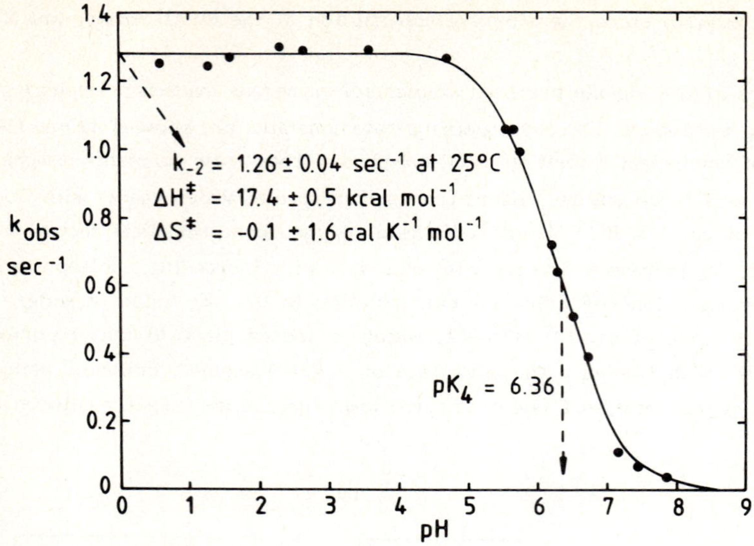

Fig. 2: k_{obs} as a function of pH for the reaction[19] trans-$Rh(en)_2(Cl)OCO_2$ + H^+ ⟶ trans-$Rh(en)_2(Cl)OH_2^{2+}$ + CO_2

CO_2 uptake is a reversible step. On treating monodentate carbonato complexes of the type $ML_5(OCO_2)^{(p-2)+}$ with acid, CO_2 is released and the corresponding aquo complex is produced, i.e. the reverse step (k_{-2}) becomes rate-determining. Isotope labelling experiments[36,37] have proved conclusively that acid-catalyzed aquation takes place without metal-oxygen bond breakage. The pH dependence of the decarboxylation process, for which an example is given in Fig. 2, demonstrates that only the protonated bicarbonato complex undergoes rapid (ca. 1 s^{-1} at 25°C) decarboxylation (O-C bond cleavage). Furthermore, the results in Fig. 2 enable the estimation of K_4, which cannot be determined in any other way and reflects the strength of the M-O and indirectly that of the O-C bond, such that a decrease in log k_{-2} with increasing pK_4 was observed[19,24]. Rapid scan spectrophotometric measurements have provided further evidence for the rapid protonation of the metal carbonato complex followed by rate-determining decarboxylation.

It follows that the reversible uptake of CO_2 can be presented as outlined in (14). The intimate nature of the CO_2 uptake/decarboxylation processes was determined from volume of acivation measurements[38], from which it follows that the processes can be vizualized in terms of the following transition states:

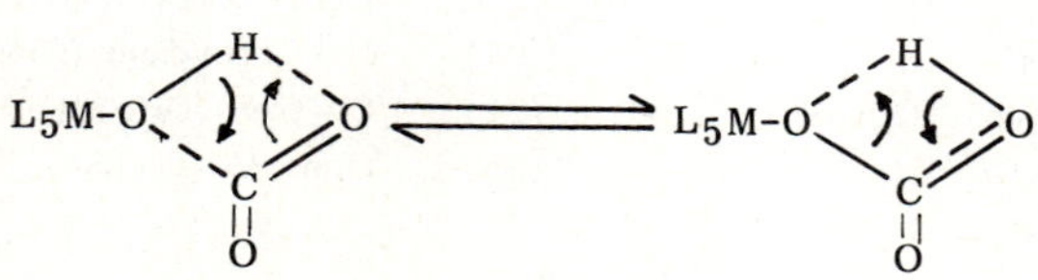

(16)

Up to now we have only considered the reactivity of a single metal-aquo moiety. In the case of a diaquo complex, the cis isomer can undergo ring-closure subsequent to the CO_2 uptake step (17). This is usually slow since it will involve metal-oxygen bond breakage on the metal center. Similarly, acid-catalyzed decarboxylation will involve a rate-determining ring-opening reaction (M-O bond breakage) followed by rapid decarboxylation (O-C bond breakage) as outlined in (18).

CO_2 uptake:

$$\underline{cis}\text{-}ML_4(OH_2)_2{}^{p+} \rightleftharpoons \underline{cis}\text{-}ML_4(OH_2)OH^{(p-1)+} + H^+$$

$$\underline{cis}\text{-}ML_4(OH_2)OH^{(p-1)+} + CO_2 \xrightarrow{\text{fast}} \underline{cis}\text{-}ML_4(OH_2)OCO_2{}^{(p-2)+} + H^+$$

$$\underline{cis}\text{-}ML_4(OH_2)OCO_2{}^{(p-2)+} \xrightarrow{\text{slow}} ML_4(CO_3)^{(p-2)+} + H_2O \qquad (17)$$

Decarboxylation:

$$ML_4(CO_3)^{(p-2)+} + H_3O^+ \xrightarrow{\text{slow}} \underline{cis}\text{-}ML_4(OH_2)OCO_2H^{(p-1)+}$$

$$\underline{cis}\text{-}ML_4(OH_2)OCO_2H^{(p-1)+} \xrightarrow{\text{fast}} \underline{cis}\text{-}ML_4(OH_2)OH^{(p-1)+} + CO_2 \qquad (18)$$

Finally, it should be kept in mind that these CO_2 uptake and decarboxylation reactions are presented here as model systems for the interaction of dissolved gases with metal aquo species. These reactions do suggest that SO_2 could react in a similar way, and the next section will be devoted to this possibility.

3.2 Formation and reactivity of sulfito complexes

Untill recently, work in this area mainly concerned the chemistry of S-bonded sulfito complexes and very few reference had been made to the possible formation of O-bonded sulfito species. SO_2(g) dissolves readily in water to produce sulfurous acid "H_2SO_3" which consists of dissolved and hydrated SO_2 [39,40]. The aqueous chemistry of dissolved SO_2 can be summarized as follows:

$$H_2O + SO_2(aq) \underset{k_{-4}}{\overset{k_4}{\rightleftharpoons}} HSO_3^- + H^+ \quad , \quad K_5$$

$$HSO_3^- \rightleftharpoons SO_3^{2-} + H^+ \quad , \quad K_6$$

$$2HSO_3^- \underset{k_{-5}}{\overset{k_5}{\rightleftharpoons}} S_2O_5^{2-} + H_2O \quad , \quad K_7 \qquad (19)$$

Typical values for the rate and overall equilibrium constants at 25°C are: $k_4 = 1 \times 10^8$ s^{-1} [41], 3.4×10^6 s^{-1} [42]; $k_{-4} = 2.5 \times 10^9$ M^{-1} s^{-1} [41], 2×10^8 M^{-1} s^{-1} [42]; $pK_5 = 1.9$ [15,43]; $pK_6 = 6.3$ [15,43]; $k_5 = 7 \times 10^2$ M^{-1} s^{-1} [41]; $k_{-5} = 1 \times 10^4$ s^{-1} [41]; $K_7 = 0.07$ [44], 0.088 M [45]. The magnitude of K_7 is such that solutions of $Na_2S_2O_5$ exist as $\geq$ 99 % HSO_3^- at $[Na_2S_2O_5] \leq 0.05$ M, conditions appropriate for the present discussion. It follows that SO_2(aq), HSO_3^- and SO_3^{2-} should be considered as potential reactive species in the presence of metal aquo/hydroxo complexes. Furthermore, it is important to note that k_4 and k_{-4} are orders of magnitude larger that k_1 and k_{-1} reported for the CO_2 system. Similarly, pK_5 and pK_6 are approx. 4 units smaller (more acidic) than pK_1 and pK_2 (see (13)).

Although some indirect evidence for the formation of O-bonded sulfito complexes has been reported before[46-51], our kinetic studies revealed the general tendency that metal aquo/hydroxo species can react very fast with SO_2(aq) to produce O-bonded sulfito complexes at 3 < pH < 8, and much slower (substitution controlled) with HSO_3^- or SO_3^{2-} at pH > 9 to produce S-bonded complexes. The O- and S-bonded isomers exhibit large spectral differences that greatly simplify the analysis of such systems.

The first study to report detailed kinetic data on the formation and stability of O-bonded sulfito complexes was performed on the $Co(NH_3)_5OH_2^{3+}/SO_2$ system[15]. Bubbling SO_2 gas or adding $Na_2S_2O_5$, $NaHSO_3$ or Na_2SO_3 to buffered solutions of $Co(NH_3)_5OH_2^{3+}$ at a pH between 4 and 7 resulted in an instantaneous colour change. Immediate acidification reproduced the $Co(NH_3)_5OH_2^{3+}$ species demonstrating the reversibility of the process. ^{17}O tracer experiments demonstrated[52] that no Co-O bond is broken during the formation and acidification reactions of $Co(NH_3)_5OSO_2^+$. A redox reaction was found to follow the formation step and Co^{2+} and SO_4^{2-} were identified as reaction products. Based on our experience with the corresponding CO_2 uptake process the following mechanism was suggested to account for the reported observations.

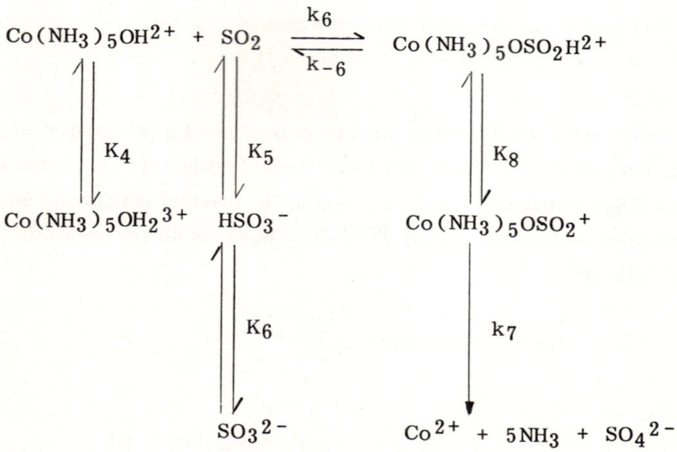

(20)

All participating hydrogen ions are omitted from the above scheme for simplicity reasons and the K's are defined as acid dissociation constants. The validity of this mechanism can be determined from an analysis of the kinetic data for the formation of the $Co(NH_3)_5OSO_2^+$ species, for which the following expression should hold:

$$k_{obs} = k_6 \left\{\frac{K_4}{K_4 + [H^+]}\right\} \left\{\frac{[H^+]^2 \; [\text{total S}]}{[H^+]^2 + K_5[H^+] + K_5K_6}\right\} \qquad (21)$$

k_{obs} as a function of pH should produce a bell shaped curve with a maximum around pH 4; a fall off in k_{obs} at higher pH due to a decrease in the $[SO_2(aq)]$, and a fall off at lower pH due to a decrease in the $[Co(NH_3)_5OH^{2+}]$. The experimentally observed pH dependence[15] partly fulfils the expected curve except for deviations at low pH where a sharp increase in k_{obs} with decreasing pH was observed. This could be ascribed to an interference of the reverse SO_2 elimination process under such conditions for which eq. (21) must be modified to

$$k_{obs} = k_6 \left(\frac{K_4}{K_4+[H^+]}\right) \left(\frac{[H^+]^2 \; [\text{total S}]}{[H^+]^2+K_5[H^+]+K_5K_6}\right) + k_{-6} \left(\frac{[H^+]}{K_8+[H^+]}\right) \qquad (22)$$

Typical values for the rate constants in (20) are: $k_6 = (2.6 \pm 0.6) \times 10^8$ M^{-1} s^{-1} and k_{-6} = 4×10^4 s^{-1} at 10°C; $k_7 = (2.5 \pm 0.4) \times 10^{-3}$ s^{-1} at 15°C. It is important to note that k_6 and k_{-6} are orders of magnitude larger than k_2 and k_{-2} for the corresponding CO_2 uptake/decarboxylation reactions. SO_2 uptake (k_6) is almost diffusion controlled and significantly faster than k_4 for the reaction of SO_2 with H_2O (see eq. (19)) when the latter is also expressed as a second order rate constant. This once again emphasizes the important role played by the central metal ion to which the hydroxy ligand is coordinated. In a subsequent study[53] it was found that both $Cr(NH_3)_5OH_2^{3+}$ and $Rh(NH_3)_5OH_2^{3+}$ react with $SO_2(aq)$ to produce O-bonded sulfito complexes, which on acidification decompose to the aquo complex. The corresponding values of k_6 are $(1.8 \pm 0.3) \times 10^8$ and $(2.9 \pm 0.6) \times 10^8$ M^{-1} s^{-1} at 25°C, respectively, which are slightly smaller than $(4.7 \pm 0.3) \times 10^8$ M^{-1} s^{-1} found for the Co(III) system at 25°C[15]. The $Cr(NH_3)_5OSO_2^+$ and $Rh(NH_3)_5OSO_2^+$ species undergo subsequent reactions during which O- to S-bonded isomerization and/or substitution to produce disulfito species, occurs.

The redox reaction of $Co(NH_3)_5OSO_2^+$ (k_7) is much slower than SO_2 uptake or elimination, but notably faster than that usually found for S-bonded sulfito complexes[54,55]. The observed kinetic data are in agreement with the expression:

$$k_{obs} = \frac{k_7 \; K_4 \; K_9 \; [SO_2]}{K_4[H^+] + K_4K_9[SO_2] + [H^+]^2} \qquad (23)$$

which is the rate law for the following suggested redox mechanism.

$$Co(NH_3)_5OH^{2+} + SO_2 \underset{}{\overset{K_9}{\rightleftharpoons}} Co(NH_3)_5OSO_2^{+} + H^{+}$$

$$Co(NH_3)_5OSO_2^{+} \xrightarrow{k_7} Co^{2+}(aq) + 5NH_3 + SO_3^{-}$$

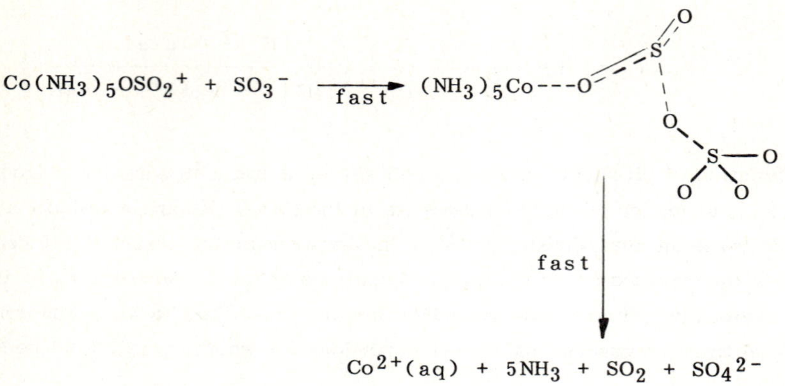

(24)

The outlined mechanism was selected on the basis of numerous experiments performed to elucidate its nature. No evidence for the formation of $S_4O_6^{2-}$ could be found[15]. The rate-determining step ($k_7 = (1.4 \pm 0.3) \times 10^{-2}\ s^{-1}$ at 25°C) involves direct oxidation of sulfite to SO_3^{-} and the formation of Co^{2+}. The activation parameters for k_7, however, suggest that the rate-determining step involves loss of coordinated NH_3, probably trans to the sulfito ligand. This prossibility is underlined by the later observations that trans-$Co(NH_3)_4(CN)OSO_2$ and $Co(tetren)OSO_2^{+}$ do not exhibit fast redox reactions[56-58] but slow linkage isomerization to produce the S-bonded isomers. It follows that the nature of the redox reaction strongly depends on the ligands occupying the other five positions on the metal ion.

Other studies on the behaviour of O-bonded Co(III) sulfito complexes involve ligands such as ethylenediamine and triaminotriethylamine[59,60]. These complexes undergo subsequent redox and substitution reactions depending on the pH of the solution. Similar results were recently reported for the reaction of $Pt(NH_3)_5OH^{3+}$ and $Pt(NH_3)_4(OH)_2^{2+}$ with SO_2 [61,62].

At this point we can draw some general conclusions. The above quoted studies revealed that metal hydroxy species can in general react rapidly with $SO_2(aq)$ to produce O-bonded sulfito species. Acidification of the latter produces unstable bisulfito complexes which rapidly loose SO_2 via O-S bond breakage. The O-bonded species can undergo fairly rapid electron transfer reactions to produce sulfate and reduced metal ions. In the absence of redox reactions, linkage isomerization and subsequent substitution reactions lead to the formation of S-bonded sulfito complexes. Some preliminary results[63] indicate that S-bonded sulfito complexes could isomerize photochemically to O-bonded species, which subsequently undergo rapid electron transfer.

The formation of O-bonded sulfito species via the rapid uptake of $SO_2(aq)$ does create the possibility of alternative reaction routes for metal ion catalyzed oxidation of S(IV). In principle one could think in terms of two reaction routes:

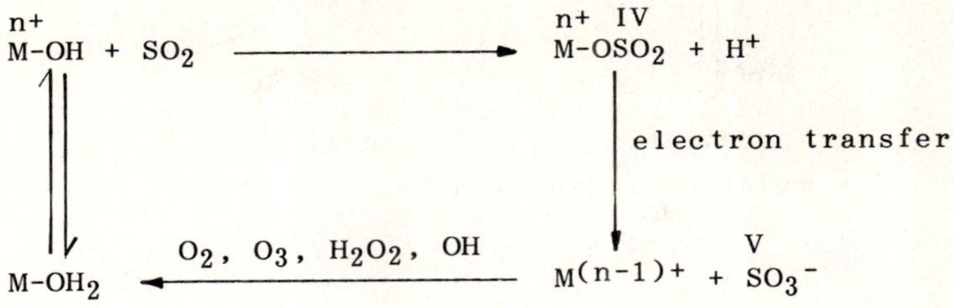

(25)

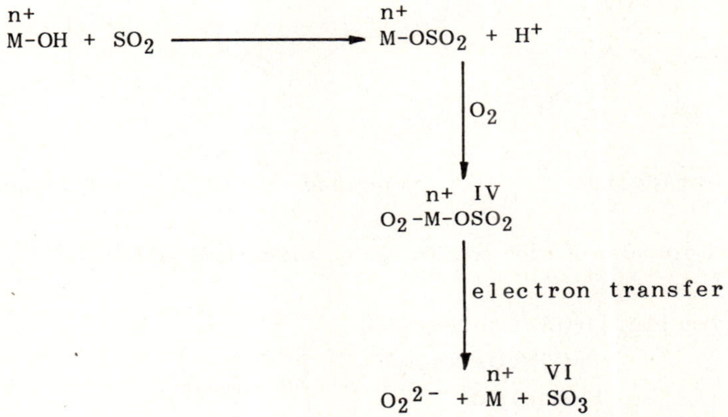

(26)

In the first case (25) the metal ion acts as initiator by oxidizing S^{IV} to the S^{V} radical SO_3^-, which can subsequently undergo further propagation reactions. To complete the cycle however the reduced $M^{(n-1)+}$ species must be oxidized by other available oxidants such as O_2, O_3, H_2O_2 or OH. In the second case the metal ion does not act as an oxidant but brings SO_2 and the oxidant such as O_2 together in a bridged species O_2-M-OSO_2 and transmits the electron from the sulfur center to the oxidant. Similar reactions could occur in systems where SO_2 is bound directly to the metal center, and not via an oxygen atom as in the above cases.

3.3. Formation and reactivity of metal SO_2 complexes

SO_2 is capable of coordinating directly to a transition metal center, and five different coordination geometries have been reported[64]:

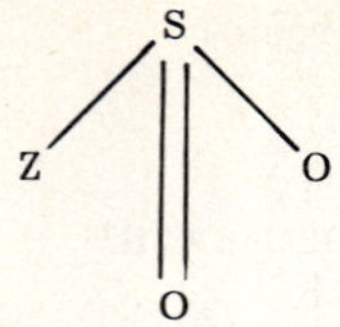

pyramidal
Z = metal or ligand

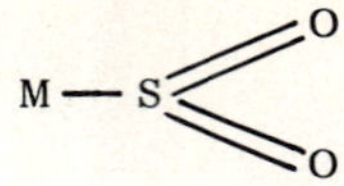

coplanar

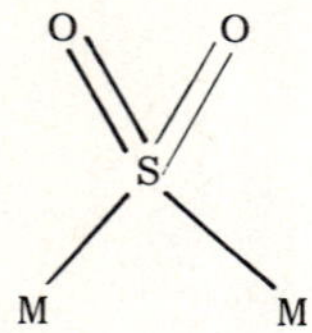

bridging

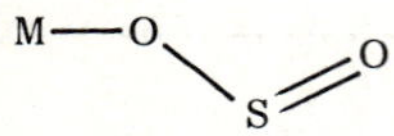

O-bonded

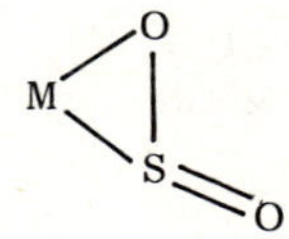

O,S-bonded

Typical examples of such species can be summarized as follows[64]

pyramidal: $Ir(Cl)(CO)(PPh_3)_2SO_2$
$Rh(Cl)(CO)(PPh_3)_2SO_2$
$Pt(PPh_3)_3SO_2$
$Pt(PPh_3)_2(SO_2)_2$
} Z = metal

$Pt(CH_3)(PPh_3)_2I{-}SO_2$
$Cu_2(I)_2(PPh_2Me)_4{\cdot}SO_2$
} Z = ligand

coplanar: $[Ru(Cl)(NH_3)_4SO_2]Cl$
$Mn(Cp)(CO)_2SO_2$
$Ni(PPh_3)_2(SO_2)_2$
$Ni(PPh_3)_3SO_2$

bridging: $[(Cp)Fe(CO)_2]_2SO_2$
$Fe_2(CO)_8SO_2$
$Pt_3(PPh_3)_3(SO_2)_3$
$[Ir(I)_2(CO)(PPh_3)]_2SO_2$

O-bonded: F_5SbOSO

O,S-bonded: $Rh(NO)(PPh_3)_2(\eta^2\text{-}SO_2)$
$Ru(Cl)(NO)(PPh_3)_2(\eta^2\text{-}SO_2)$
$Mo(CO)_3(phen)(\eta^2\text{-}SO_2)$

In many cases the crystal structure of such species has been obtained from X-ray diffraction analysis[65-67]. Such SO_2 complexes can in some cases react with oxygen to produce sulfato species:

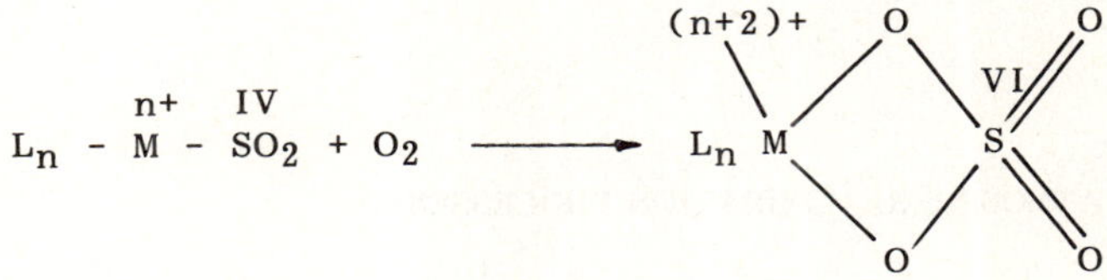

(27)

If we consider the sulfato species to formally have an S(VI) center, then the metal ion must be in an oxidized oxidation state in order to produce an overall similarly charged complex than the SO_2 complex, since the product contains a SO_4^{2-} ligand. This reaction was found to occur for $[Rh(Cl)(PPh_3)_2(SO_2)]_2$, $Rh(Cl)(PPh_2Me)_3(SO_2)$ and $Ir(SPh)(CO)(PPh_3)_2(SO_2)$, and in general for SO_2 containing five-coordinate Rh, Ir and Ru [64]. In these cases the SO_2 complex has a metal center in the +1 oxidation state such that the metal center is in the +3 oxidation state for the sulfato complexes produced in (27). Similar reactions have also been reported for other metal species. In general, reaction (27) does represent an alternative route for the metal catalyzed oxidation of SO_2. It is unknown whether an $O_2\text{-}M\text{-}SO_2$ type intermediate is formed at any stage during the reaction, similar to that suggested in (26).

3.4. Formation and reactivity of metal nitrito complexes

Very similar processes to those discussed in the previous section can play a role during the metal catalyzed oxidation of NO and NO_2 to HNO_2 and HNO_3. It is well-known that NO is an excellent ligand and can coordinate to almost any metal center. Electron transfer to the metal, or via the metal to another oxidant will result in the formation of NO^+. The latter is a well-known nitrosation reagent usually present in solutions of HNO_2 according to one of the equilibria[68]

$$H^+ + HNO_2 \rightleftharpoons H_2NO_2^+ \longrightarrow NO^+ + H_2O$$

$$\text{or} \quad HNO_2 + HNO_2 \rightleftharpoons \text{"}N_2O_3\text{"} + H_2O \qquad (28)$$

$$\text{where "}N_2O_3\text{"} \equiv NO^+ \; NO_2^-$$

NO^+ can coordinate to a metal hydroxy center without any metal-oxygen bond breakage to produce a nitrito complex M-ONO. On acidification of the latter, NO^+ is eliminated and the metal aquo species regenerated[69-71]. Alternatively, the O-bonded nitrito complexes isomerize to the corresponding N-bonded species[70-73] to produce stable nitrite complexes. The reaction sequence via NO^+ uptake is usually much faster than direct substitution on the metal center by NO_2^-, in agreement with the general tendency observed for CO_2 and SO_2 uptake processes. It is therefore expected that NO_2 or NO_2^+ will exhibit similar reactions with metal aquo/hydroxo species.

4. INVESTIGATED ATMOSPHERIC OXIDATION PROCESSES

The oxidation of S(IV) to S(VI) is surely the atmospheric oxidation process that has received the most attention of numerous investigators. The importance of this process, the possible catalytic role of metal species, the various suggested reaction mechanisms and the importance of this process during the acid rain phenomenon has been stressed in earlier chapters.

Both Fe(II) and Fe(III) have been reported to exhibit catalytic activity during the oxidation of $SO_2(aq)/HSO_3^-$ solutions. It was recently pointed out[74] that the Fe(II) catalyzed reaction exhibits a distinct induction period during which Fe(III) species are produced. Various investigators have studied the Fe(III) catalyzed reaction and Martin[75] recently indicated that a reasonable correlation is obtained for the available kinetic data. Fuzzi[76] reported a strong pH dependence of the process: a first order dependence on S(IV)at pH < 4 and a second order dependence at pH > 5. A maximum rate was found around pH 4 and ascribed to the formation of a $Fe(III)HSO_3^-$ complex species. Very similarly, the Fe(II) catalyzed reaction was suggested to proceed according to the following mechanism[77]:

<u>Equilibria</u>

$$Fe(II) + HSO_3^- \rightleftharpoons FeHSO_3^+$$
$$Fe(II) + 2H_2O \rightleftharpoons Fe(OH)_2 + 2H^+$$
$$Fe(II).H_2O^{2+} + O_2 \rightleftharpoons Fe(O_2)OH^+ + H^+$$

<u>Initiation</u>

$$FeHSO_3^+ + OH^- \longrightarrow Fe(OH)HSO_3 \qquad \text{slow}$$
$$FeHSO_3^+ + Fe(OH)_2 \longrightarrow Fe(OH)HSO_3 + FeOH^+ \qquad \text{slow}$$
$$Fe(OH)HSO_3 + O_2 \longrightarrow Fe^{III}OH^{2+} + OH^- + SO_4^-.$$
$$Fe(O_2)OH^+ + HSO_3^- \longrightarrow Fe^{III}OH^{2+} + OH^- + SO_4^-.$$
$$Fe^{III}OH^{2+} + HSO_3^- \rightleftharpoons Fe^{III}(H_2O)SO_3^+ \longrightarrow Fe(II).H_2O^{2+} + SO_3^-.$$

Propagation

Free-radical path

$$SO_4^{-}\cdot + HSO_3^{-} \longrightarrow HSO_4^{-} + SO_3^{-}\cdot$$
$$SO_3^{-}\cdot + O_2 \longrightarrow SO_5^{-}\cdot$$
$$SO_5^{-}\cdot + HSO_3^{-}\cdot \longrightarrow HSO_4^{-} + SO_4^{-}\cdot$$

Non-free-radical path

$$Fe^{III}SO_3^{+} + HSO_3^{-} \longrightarrow Fe^{III}(SO_3)HSO_3$$
$$Fe^{III}(SO_3)HSO_3 + O_2 \longrightarrow Fe^{III}SO_4^{+} + HSO_4^{-}$$

Termination

$$SO_5^{-}\cdot + \text{organics} \longrightarrow \text{inert products} \qquad (29)$$

The suggested mechanism is in agreement with the empirical rate law[74]

$$\frac{d}{dt}[S(VI)] = 10^{0.8-\sqrt{I}/1+\sqrt{I}} \frac{[Fe(II)]^{1.16}\ [HSO_3^{-}]^{0.61}}{[H^{+}]^{1.11}\ [SO_4^{2-}]^{0.43}} \qquad (30)$$

in which the rate is expressed as M h^{-1}. Fe(II) is oxidized to Fe(III) via a direct reaction with O_2 or via the formation of a $Fe(O_2)OH^{+}$ intermediate, during which $SO_4^{-}\cdot$ is produced. Throughout the suggested mechanism $FeHSO_3^{+}$ and $Fe(OH)HSO_3$ species are produced as intermediates.

Similar suggestions have been made for the Mn(II) catalyzed oxidation of S(IV). Various investigators have found a second order dependence on Mn(II) and have suggested the participation of a Mn_2^{4+} species. Martin[75] reported a fairly good correlation between the kinetic data of various groups. The recently suggested mechanism[77] consists of the following reactions:

Equilibria

$$Mn(II) + HSO_3^{-} \rightleftharpoons MnHSO_3^{+}$$
$$2Mn(II) \rightleftharpoons Mn_2^{4+}$$

Initiation

$$Mn_2^{4+} + HSO_3^{-} \longrightarrow Mn_2HSO_3^{3+} \qquad \text{slow}$$
$$Mn_2HSO_3^{3+} + O_2 \longrightarrow 2Mn(II) + OH\cdot + SO_4^{-}\cdot \qquad \text{fast}$$
$$OH\cdot + HSO_3^{-} \longrightarrow H_2O + SO_3^{-}\cdot$$

Propagation

Free-radical path

Free-radical path

$$SO_4^{-}\cdot + HSO_3^{-} \longrightarrow HSO_4^{-} + SO_3^{-}\cdot$$
$$SO_3^{-}\cdot + O_2 \longrightarrow SO_5^{-}\cdot$$
$$SO_5^{-}\cdot + HSO_3^{-} \longrightarrow SO_4^{-}\cdot + HSO_4^{-}$$

Non-free-radical path

$$MnHSO_3^{+} + HSO_3^{-} \longrightarrow Mn(HSO_3)_2$$
$$Mn(HSO_3)_2 + O_2 \longrightarrow Mn(II) + 2HSO_4^{-}$$

Termination

$$SO_5^{-}\cdot + \text{organics} \longrightarrow \text{inert products} \qquad (31)$$

This mechanism is in agreement with the empirical rate law[74]

$$\frac{d}{dt}[S(VI)] = 10^{6.4-4\sqrt{I}/1+\sqrt{I}}\,[Mn(II)]^2$$
$$(\text{in } M\ h^{-1}) \qquad + 10^{3.5-\sqrt{I}/1+\sqrt{I}}\,[Mn(II)][HSO_3^{-}] \qquad (32)$$

No evidence for the participation of Mn(III) species was reported. The formation of Mn(II) bisulfite species is of primary importance since these species react with O_2 to produce the S(V) and S(VI) species.

A comparison of the suggested mechanisms in (29) and (31) with the general trends reported in Section 3.2. immediately questions the validity of the assumed intermediate bisulfite complex species. The formation of O-bonded sulfito or SO_2 complexes should be considered as realistic reaction paths especially in the light of the fact that such species are known to react with O_2 as discussed before. Furthermore, our earlier studies have shown bisulfito complexes to be unstable, such that hydroxy bisulfito complexes should rather be considered to be aquo sulfito species.

Metal oxides such as Fe_2O_3 can also act as photochemical initiators by producing electrons and positively charged holes during irradiation. These processes are further described in the contribution by Guesten in this volume. In such a case the positive hole can oxidize SO_2 or HSO_3^{-} via a heterogeneous surface reaction.

5. PRESENT AND FURTURE ACTIVITIES

In order to determine the role played by SO_2 uptake reactions by metal hydroxo moieties to produce O-bonded sulfito complexes, or the direct addition of SO_2 to a metal center, during atmospheric oxidation processes, it is essential that the model systems investigated be extended to metal ions relevant to atmospheric chemistry. However, the reactivity patterns of aquated metal ions such as Fe(III) and Mn(II) are extremely complex due the number of possible reaction sites available and the lability of coordinated H_2O. We therefore suggest to study the behaviour of such metal ions in the presence of $SO_2(aq)/HSO_3^-$ under well controlled conditions, i.e. by blocking off 4 or 5 of the available coordination sites with non-participating ligands. In this way the reactivity of a single aquo/hydroxo moiety will be investigated prior to more complex situations.

In a similar way it is important to study the thermal and photochemical redox properties of species such as $M\text{-}OSO_2$, $M\text{-}SO_3$ and $M\text{-}SO_2$. The important role played by other oxidants such as O_2, O_3, H_2O_2 and OH should also be investigated in detail for such systems.

ACKNOWLEDGEMENTS

The author gratefully acknowledges financial support from the Deutsche Forschungsgemeinschaft and the Scientific Affairs Division of NATO under grant No RG 114.81.

REFERENCES

1. L.R. Ember, Acid pollutants: hitchhikers to ride the wind, Chem. Eng. News, Sept. 14, 20 (1981).
2. T.E. Graedel and C.J. Weschler, Chemistry with aqueous atmospheric aerosols and raindrops, Rev. Geophys. Space Phys. 19, 505 (1981).
3. G. Gravenhorst, S. Beilke, M. Betz and H.-W. Georgii, Sulfur dioxide absorbed in rain water, in "Effects of acid precipitation of terrestrial ecosystems", T.C. Hutchinson and M. Havas (Eds.), Plenum Publishing Co., 41 (1980).
4. J.J. Morgan, Factors governing the pH, availability of H^+, and oxidation capacity of rain, in "Atmospheric Chemistry", E.D. Goldberg (Ed.), Springer-Verlag, 17 (1982).
5. W.L. Chameides and D.D. Davis, Chemistry in the troposphere, Chem. Eng. News, Oct. 4, 39 (1982).
6. H. Niki, Homogeneous gas phase oxidation processes in the troposphere, in "Atmospheric Chemistry", E.D. Goldberg (Ed.), Springer-Verlag, 301 (1982.)
7. R.J. Cicerone, Tropospheric gases, aerosols and photochemical reactions group report, in "Atmospheric Chemistry", E.D. Goldberg (Ed.), Springer-Verlag, 357 (1982).
8. T.E. Graedel, Aqueous chemistry in the atmosphere group report, in "Atmospheric Chemistry", E.D. Goldberg (Ed.), Springer-Verlag, 93 (1982).
9. A. Huss, P.K. Lim and C.A. Eckert, On the uncatalyzed oxidation of sulfur(IV) in aqueous solutions, J. Am. Chem. Soc. 1OO, 6252 (1978).
10. P.K. Lim, A. Huss and C.A. Eckert, Oxidation of aqueous sulfur dioxide. 3. The effects of chelating agents and phenolic antioxidants, J. Phys. Chem. 86, 4233 (1982).
11. S.A. Penkett, Laboratory studies of S(IV) to S(VI), this volume.
12. J.H. Overton, V.P. Aneja and J.L. Durham, Production of sulfate in rain and raindrops in polluted atmospheres, Atmos. Environ. 13, 355 (1979).
13. Research priorities and criteria to establish factors which govern precipitation chemistry - Workshop report, p 3 - 10, 1982 (in press).
14. F.E. Brinckman, G.J. Olson and W.P. Iverson, in "Atmospheric Chemistry", E.D. Goldberg (Ed.), Springer-Verlag, 231 (1982).
15. R. van Eldik and G.M. Harris, Kinetics and mechanism of the formation, acid-catalyzed decomposition and intramolecular redox reaction of oxygen-bonded sulfitopentaamminecobalt(III) ions in aqueous solution, Inorg. Chem. 19, 880 (1980).
16. E. Chaffee, T.P. Dasgupta and G.M. Harris, Kinetics and mechanism of aquation and formation reactions of carbonato complexes. 5. Carbon dioxide uptake by hydroxopentaamminecobalt(III)ion to form carbonatopentaamminecobalt(III) ion, J. Am. Chem. Soc. 95, 4169 (1973).

17. D.A. Palmer, R. van Eldik, H. Kelm and G.M. Harris, Kinetics and mechanism of aquation and formation reactions of carbonato complexes. 17. Carbon dioxide uptake by and decarboxylation of the cis-bis(ethylenediamine)rhodium(III) system in aqueous solution, Inorg. Chem. 19, 1009 (1980).

18. R. van Eldik, D.A. Palmer and G.M. Harris, Kinetics and mechanism of aquation and formation reactions of carbonato complexes. 18. Carbon dioxide uptake by and decarboxylation of the trans-bis(ethylenediamine)rhodium(III) system in aqueous solution, Inorg. Chem. 19, 3673 (1980).

19. R. van Eldik, D.A. Palmer, H. Kelm and G.M. Harris, Kinetics and mechanism of aquation and formation reactions of carbonato complexes. 20. Carbon dioxide uptake by and decarboxylation of trans-halogenobis(ethylenediamine)rhodium(III) complexes in aqueous solution, Inorg. Chem. 19, 3679 (1980).

20. A. Jost, Fast reactions in solutions at high pressures. 3. The effect of pressure on the reactions of Fe(III) ions with thiocyanate in water up to 2 kbar at 25°C, Ber. Bunsenges. Phys. Chem. 80, 316 (1976).

21. T.W. Swaddle and A.E. Merbach, High-pressure oxygen-17 fourier transform nuclear magnetic resonance spectroscopy. Mechanism of water exchange on iron(III) in acidic aqueous solution, Inorg. Chem. 20, 4212 (1981).

22. R. Doss, R. van Eldik and H. Kelm, A reinvestigation of the effect of pressure on the reaction of Fe(III) with thiocyanate ions in aqueous solution, Ber. Bunsenges. Phys. Chem. 86, 925 (1982).

23. C.J. Hipp and D.H. Busch, Ligand reactions - the effect of metal complexes on chemical processes, in "Coordination Chemistry Vol. 2", A.E. Martell (Ed.), ACS Monograph 174, 221 (1978).

24. D.A. Palmer and R. van Eldik, The Chemistry of metal carbonato and carbon dioxide complexes, Chem. Rev. 83, 651 (1983).

25. R. van Eldik, Reactions involving coordinated ligands, in "Adv. in inorganic and bioinorganic mechanisms" Vol. 3, A.G. Sykes (Ed.), Academic Press.

26. M.L. Tobe, "Mechanisms of inorganic reactions", Nelson, 51 (1972).

27. D.W. Margerum, G.R. Cayley, D.C. Weatherburn and G.K. Pagenkopf, Kinetics and mechanisms of complex formation and ligand exchange, in "Coordination Chemistry Vol. 2", A.E. Martell (Ed.), ACS Monograph 174, 1 (1978).

28. H. Kruger, Techniques for the kinetic study of fast reactions in solution, Chem. Soc. Rev. 11, 227 (1982).

29. A.E. Merbach, The elucidation of solvent exchange mechanisms by high-pressure NMR studies, Pure Appl. Chem. 54, 1479 (1982).

30. R. van Eldik and H. Kelm, Interpretation of the volume of activation of inorganic reactions in solution, Rev. Phys. Chem. Jpn. 50, 185 (1980).

31. D.E. Pennington, Oxidation-reduction reactions of coordination complexes, in "Coordination Chemistry Vol. 2", A.E. Martell (Ed.), ACS Monograph 174, 476 (1978).

32. K.S. Johnson, Carbon dioxide hydration and dehydration kinetics in seawater, Limnol. Oceanogr. 27, 849 (1982).

33. R. van Eldik and D.A. Palmer, Effect of pressure on the kinetics of the dehydration of cabonic acid and the hydrolysis of CO_2 in aqueous solution, J. Sol. Chem. 11, 339 (1982).

34. K.F. Wissbrun, D.M. French and A. Patterson, The true ionization constant of carbonic acid in aqueous solution from 5 to 45°C, J. Phys. Chem. 58, 693 (1954).

35. M. Paabo and R.G. Bates, Dissociation of deuteriocarbonate ion in deuterium oxide from 5 to 50°C, J. Phys. Chem. 73, 3014 (1969).

36. J.P. Hunt, A.C. Rutenberg and H. Taube, Mechanism of aquation of carbonatopenta-amminecobaltic ion in acid solution, J. Am. Chem. Soc. 74, 268 (1952).

37. C.A. Bunton and D.R. Llewellyn, Tracer studies in the aquation and hydrolysis of cobalt complexes, J. Chem. Soc. 1692 (1953).

38. U. Spitzer, R. van Eldik and H. Kelm, Mechanistic information from the effect of pressure on the formation and acid-catalyzed aquation reactions of carbonatopenta-amminecobalt(III), -rhodium(III) and -iridium(III) ions in aqueous solution, Inorg. Chem., 21, 2821 (1982).

39. F.A. Cotton and G. Wilkinson, "Inorganic Chemistry", 4th Ed., John Wiley, 540 (1980).

40. M. Schmidt in "Sulfur in organic and inorganic chemistry Vol. 2", A. Senning (Ed.), Marcel Dekker, 84 (1972).

41. R.H. Betts and R.H. Voss, The kinetics of oxygen exchange between the sulfite ion and water, Can. J. Chem. 48, 2035 (1970).

42. M. Eigen, K. Kustin and G. Maass, The rate of hydration of SO_2 in aqueous solution, Z. Phys. Chem. N.F. 3O, 130 (1961).

43. Gmelin's Handbuch der Anorganischen Chemie, Verlag Chemie, No. 9, Part B2, 466 (1960).

44. D.W.A. Bourne, T. Higuchi and I.H. Pitman, Chemical equilibriums in solutions of bisulfite salts, J. Pharm. Sci. 63, 865 (1974).

45. R.E. Conrick, T.M. Tam and E. von Deuster, Equilibrium constant for the dimerization of bisulfite ion to form $S_2O_5^{2-}$, Inorg. Chem. 21, 103 (1982).

46. R.S. Murray, D.R. Stranks and J.K. Yandell, The unusual reactivity of sulfite ions and sulfito ligands, J. Chem. Soc. Chem. Commun., 604 (1969).

47. S.M. Farrell and R.S. Murray, An intermediate in the reaction between trans-aquabis(ethylenediamine)sulfitocobalt(III) and sulfite ion in aqueous solution, J. Chem. Soc. Dalton, 322 (1977).

48. M.A. Thacker and W.C.E. Higginson, Kinetics of hydrolysis of di-μ-hydroxo-bis(nitrilotriacetatocobaltate(III)) to diaquo(nitrilotriacetato)cobalt(III), and some reactions of the latter with non-metallic substrates in aqueous solution, J. Chem. Soc. Dalton, 704 (1975).

49. D.W. Carlyle and E.L. King, The influence of sulfite ion upon the rate of aquation of various complexes of chromium(III) ion. The stability of sulfitochromium(III) ion, Inorg. Chem. 9, 2333 (1970).

50. Gy. Bazsa and H. Diebler, Kinetics of the rapid interaction of chromium(III) with iodate and sulfite, React. Kinet. Catal. Lett. 2, 217 (1975).

51. A.D. James and R.S. Murray, Some reactions of pentacyanosulfitoferrate(II) with oxidising agents, J. Chem. Soc. Dalton, 319 (1977).

52. R. van Eldik, J. von Jouanne and H. Kelm, Mechanistic information from ^{17}O-NMR measurements. Evidence for the existence of oxygen-bonded (sulfito)pentaammine-cobalt(III) in weakly acidic aqueous solution, Inorg. Chem. 21, 2818 (1982).

53. R. van Eldik, Kinetic data for SO_2 uptake by Rh(III) and Cr(III) hydroxo complexes in aqueous solution, Inorg. Chim. Acta 42, 49 (1980).

54. M.A. Thacker, K.L. Scott, M.E. Simpson, R.S. Murray and C.E. Higginson, Redox decomposition of trans-tetraammineaquosulfitocobalt(III) in aqueous solution, J. Chem. Soc. Dalton, 647 (1974).

55. K.L. Scott, Preparation and characterization of cis- and trans-tetraammine-bis(sulfito)cobaltate(III) anions and the kinetics of redox decomposition in aqueous acid solution, J. Chem. Soc. Dalton, 1486 (1974).

56. R. van Eldik, unpublished results.

57. J. Kraft and R. van Eldik, Inorg. Chem. in press.

58. A.C. Dash, A.A. El-Awady and G.M. Harris, Kinetics and mechanism of the reactions of sulfito complexes in aqueous solution. 3. Formation, acid-catalyzed decomposition, and intramolecular isomerization of oxygen-bonded (αβS)-sulfitotetraethylenepent-amine)cobalt(III) ion and the hydrolysis of its sulfur-bonded analogue, Inorg. Chem. 20, 3160 (1981).

59. A.A. El-Awady and G.M. Harris, Kinetics and mechanism of the reactions of sulfito complexes in aqueous solution. 2. Formation and aquation of aquo(O-sulfito)(2,2',2"-triaminotriethylamine)cobalt(III) ion, Inorg. Chem. 20, 1660 (1981).

60. A.A. El-Awady and G.M. Harris, Kinetics and mechanism of the reactions of sulfito complexes in aqueous solution. 4. Intramolecular electron-transfer and sulfito ligand addition reactions of aquo(0-sulfito)(2,2',2"-triaminotriethylamine)cobalt(III) ion, Inorg. Chem. 20, 4251 (1981).

61. K.C. Koshy and G.M. Harris, Kinetics and mechanism of the reactions of sulfite complexes in aqueous solution. 5. Formation, isomerization and sulfite addition reactions of oxygen-bonded (sulfito)pentaammineplatinum(IV) ion and the subsequent intramolecular redox reactions of the sulfur-bonded intermediate, cis-di(sulfito)-tetraammineplatinum(IV) ion, Inorg. Chem., 22, 2947 (1983).

62. K.C. Koshy and G.M. Harris, prepared for publication.

63. U. Spitzer and R. van Eldik, Kinetics and mechanisms of the formation, substitution and aquation reactions of sulfur-bonded (sulfito)amminecobalt(III) complexes in aqueous solution, Inorg. Chem. 21, 4008 (1982).

64. G.J. Kubas, Diagnostic features of transition-metal-SO_2 coordination geometries, Inorg. Chem. 18, 182 (1979).

65. D.C. Moody and R.R. Ryan, Pyramidal metal-sulfur dioxide coordination. Chemistry of $Pt(SO_2)_2(PPh_3)_2$ and synthesis and molecular structure of $Pt_3(SO_2)_3(PPh_3)_3.C_7H_8.SO_2$, Inorg. Chem. 16, 1052 (1977).

66. M. Cowie and S.K. Dwight, Binuclear rhodium complexes. Their chemistry with sulfur dioxide and the structure of $[Rh_2Cl_2(\mu-SO_2)((C_6H_5)_2PCH_2P(C_6H_5)_2)_2]$, Inorg. Chem. 19, 209 (1980).

67. C.E. Briant, D.G. Evans and D.M. Mingos, General synthetic route for SO_2 cluster compounds of platinum and the structural characterization of $Pt_5(\mu-CO)_2(\mu-SO_2)_3(CO)(PPh_3)_4$, J. Chem. Soc. Chem. Commun., 1144 (1982).

68. G. Stedman, Reaction mechanisms of inorganic nitrogen compounds, Adv. Inorg. Chem. Radiochem. 22, 143 (1979).

69. D.A. Buckingham, I.I. Olsen, A.M. Sargeson and H. Satrapa, The mechanism of substitution reactions of pentaamminecobalt(III) complexes. Product distributions in the induced aquation of some $Co(NH_3)_5X^{2+}$ ions in the presence of added anions, Inorg. Chem. 6, 1027 (1967).

70. R.G. Pearson, P.M. Henry, J.G. Bergmann and F. Basolo, Mechanism of substitution reactions of complex ions. 6. Formation of nitrito- and nitrocobalt(III) complexes. O-nitrosation, J. Am. Chem. Soc. 76, 5920 (1954).

71. R.K. Murmann and H. Taube, The mechanism of the formation and rearrangement of nitritocobalt(III) ammines, J. Am. Chem. Soc. 78, 4886 (1956).

72. M. Mares, D.A. Palmer and H. Kelm, Activation volumes for the linkage isomerization reactions of nitritopentaammine complexes of Co(III), Rh(III) and Ir(III) in aqueous solution, Inorg. Chim. Acta 27, 153 (1978).

73. H. Ghazi-Bajat, R. van Eldik and H. Kelm, Kinetics and mechanism of the formation of nitritopentaminecobalt(III) in aqueous acidic solution, Inorg. Chim. Acta 60, 81 (1982).

74. A. Huss, P.K. Lim and C.A. Eckert, Oxidation of aqueous sulfur dioxide. 1. Homogeneous Mn(II) and Fe(II) catalysis at low pH, J. Phys. Chem. 86, 4224 (1982).

75. R.B. Martin, Kinetic studies of sulfite oxidation in aqueous solution, Aerospace Corporation, 1982.

76. S. Fuzzi, Study of iron(III) catalyzed sulfur dioxide oxidation in aqueous solution over a wide range of pH, Atmos. Environ. 12, 1439 (1978).

77. A. Huss, P.K. Lim and C.A. Eckert, Oxidation of aqueous sulfur dioxide. 2. High-pressure studies and proposed reaction mechanisms, J. Phys. Chem. 86, 4229 (1982).

PHOTOCATALYTIC DEGRADATION OF ATMOSPHERIC POLLUTANTS ON THE SURFACE OF METAL OXIDES

Hans Guesten

Kernforschungszentrum Karlsruhe
Institut fuer Radiochemie
Postfach 3640
D-7500 Karlsruhe
F.R.G.

ABSTRACT

Evidence has been accumulating in laboratory studies of an enhanced photocatalytic degradation or oxidation of atmospheric pollutants on solid surfaces of inorganic aerosol particles of natural and anthropogenic origin. The soil-generated inorganic aerosols as well as fly ash - the major contributor of anthropogenic aerosols - are mainly composed of metal oxides. Organic pollutants such as polynuclear aromatic hydrocarbons and pesticides are associated with aerosol particles in the low micron range which have tropospheric lifetimes of several days. Examples of photoassisted heterogeneous degradation reactions of atmospheric pollutants such as halogenated organic compounds, sulphur dioxide and polynuclear aromatic hydrocarbons are discussed. The possible mechanisms of heterogeneous catalytic photoreactions is discussed with reference to the presence of metal oxides with semiconduction properties in tropospheric aerosols. Photoelectrons released near the surface are transferred to the adsorbed pollutant and co-adsorbed oxygen molecules making possible chemical oxidation or degradation processes. With the knowledge of the quantum yield of the heterogeneous photodegradation process the tropospheric residence time of pollutants adsorbed on airborne particulate matter can be estimated. Possible oxidation mechanisms are discussed in regard to the surface characteristics of the atmospheric aerosols such as surface area, acid or base character, and the photogeneration of active sites.

NATO ASI Series, Vol. G6
Chemistry of Multiphase Atmospheric Systems
Edited by W. Jaeschke

INTRODUCTION

During the past decade it has become clear in atmospheric chemistry that a number of homogeneous gas-phase reactions of photolytically generated oxidising radicals play a key role in the degradation of atmospheric pollutants. In particular, the OH radical is now considered as the "vacuum cleaner" of the troposphere; it determines the amount of anthropogenic pollutants which are transported to the stratosphere. During the last years evidence has been accumulated in laboratory studies of an enhanced photocatalytic degradation of atmospheric pollutants on solid surfaces. These active surfaces are constituted mainly by land areas and by atmospheric particulate matter. Furthermore, sunlight provides an abundant supply of potential chemical energy. Especially the tropospheric aerosols constitute an ensemble of particles of different shapes, sizes, and chemical compositions with large surface-to-volume ratios. Thus, heterogeneous reactions between gas-phase pollutants and tropospheric aerosols, summarized under the title of this conference, "Chemistry of Multiphase Atmospheric Systems", may play an important role in the degradation of trace gases of natural as well as anthropogenic origins. Often, pollutants and aerosol particles are emitted from the same point source at high concentrations, for example, fly ash and SO_2 from fossil-fuel fired power plants. Thus, in urban regions the contribution of multiphase chemical reactions in competition with homogeneous gas-phase reactions might be dominating. In the presence of solar light the tropospheric aerosols may be capable of acting as photochemical catalysts inducing chemical reactions through photoactivated surfaces.

In this paper we shall discuss the results of such laboratory studies on the photodegradation of atmospheric pollutants adsorbed on inorganic aerosol particles of natural and anthropogenic origin. In general, it is the class of soil-generated inorganic aerosols which are often composed of metal oxides. We do not consider here possible contributions to heterogeneous photocatalytic reactions by tropospheric aerosols such as sea salt, organic aerosols or aerosol particles whose origin stems from gas-to-particle conversion.

NATURAL SOURCES OF INORGANIC AEROSOLS

The annual quantity of the overall global aerosol content has been estimated at about 1100 megatonnes (Mt) for those of natural origin (Almquist, 1974). Figures of 1250 Mt and 800 - 2200 Mt have been estimated by other authors (Nazarov et al. and Budyko and Davitaya, quoted by Grigoryev and Kondratyev, 1981). Besides sea salt aerosols with an estimated 1000 Mt yr^{-1} for particles less than 40 μm in diameter (Eriksson 1959), wind erosion of soil is recognized as the major source of inorganic aerosols on a global scale with an estimated value of 500 Mt yr^{-1} (Goldberg, quoted by Peterson and Junge, 1971). Desert areas, which constitute 18 % of the land area on the earth, are obviously the main source. The Sahara contributes to this figure about 260 Mt yr^{-1} across the west coast of Africa (Jaenicke and Schuetz, 1978) and 70 Mt via east Mediterranean trajectories (Ganor and Mamane, 1982). Recent results indicate that the above given figures for the global aerosol content have been severely underestimated. The total Sahara dust transported via the four cardinal points amounts to 750 - 800 Mt yr^{-1} (D'Almeida, 1983). Thus, the source strength of soil-generated aerosols exceeds the generally accepted number of 220 Mt yr^{-1}, as compiled by Natusch and Bauer (1978), by far (see table 1). The contribution from soil greatly exceeds the amount of aerosols injected into the atmosphere by volcanoes, forest wildfires and meteorites.

ANTHROPOGENIC SOURCES OF INORGANIC AEROSOLS

Current estimates of the magnitude of anthropogenic emissions are very uncertain and range between 5 - 30 % of the natural production (Jaenicke, 1980). Almquist (1974) estimated the annual quantity to 200 Mt for emissions of anthropogenic origin. Inorganic aerosols of anthropogenic sources originate from combustion, incineration, smelting, refining, corro-

Table 1: Global summary of the source strengths of atmospheric particulate matter of inorganic origin (Natusch and Bauer, 1978)

Source	Strength (Mt/yr) Natural	Anthropogenic
Fly ash from coal	-	39.7
Iron and steel industry emissions	-	9.9
Non-fossil fuel (wood, mill wastes)	-	8.8
Petroleum combustion	-	2.2
Incineration	-	4.4
Agricultural emission	-	11
Cement manufacture	-	7.7
Miscellaneous	-	17.6
Sea salt	1102	-
Soil dust	220	-
Volcanic particles	4.4	-
Forest fires	3.3	-
Total	1329.7	101

sion, wear etc. Major contributors of anthropogenic aerosols are coal-fired power plants.

The annual output of airborne fly ash from coal-fired power plants can be estimated by the following consideration. The worldwide source strength of SO_2 from coal-fired power plants is approximately 80 Mt yr^{-1}. If we take a realistic ratio of the mass of SO_2 to the emitted mass of coal fly ash particles of 10 : 1 ($\sim$100 mg fly ash to $\sim$1 g SO_2 per m^3), we obtain an annual worldwide emission of about 8 Mt of coal fly ash. This is certainly a conservative figure since many countries do not retain the fly ash by electrostatic precipitators. Indeed, older data give 40 Mt yr^{-1} of airborne coal fly ash (Natusch and Bauer, 1978). A summary of the source strength of inorganic atmospheric aerosols of natural and anthropogenic origins is compiled in table 1. These figures, however, need a revision, as indicated in the preceding subsection.

SIZE AND CHEMICAL COMPOSITION OF INORGANIC TROPOSPHERIC AEROSOLS

In view of a photocatalysed heterogeneous degradation of trace gases and pollutants in the troposphere the particle size distribution and the chemical composition is more important than the mere mass of the tropospheric aerosols. Gilette et al. (1974) and Schuetz and Jaenicke (1974) suggest that the major mass of the soil-generated aerosols is in the range of 2 to 100 um. Other authors assume that the total soil-derived particle mass with $d > 1$ µm contributes to 95 percent (Duce, 1978). Thus, the main mass of the natural aerosols of inorganic composition in the troposphere has a particle size of $d > 1$ µm. The aerosol size distribution of fly ash, the major contributor to the aerosols of anthropogenic origin, ranges between 0.1 to 5 µm (Dlugi and Guesten, 1983). Larger particles are effectively retained by electrostatic precipitators (Vandegrift et al., 1973). If we take into account the vertical distribution of the particle number density of 5×10^7 particles in a one cm^2 atmospheric column (Peterson, 1976) and a mean diameter of only 1 µm for the tropospheric aerosols, the global tropospheric aerosols provide a total surface area of about 8×10^8 km^2. This figure corresponds to an area nearly twice as large as the surface of our globe. The total surface area calculated is certainly a very conservative figure because it is calculated for a non-porous sphere of only 1 µm diameter. Only sand and fly ash may exhibit non-porous surfaces. Other tropospheric aerosols of crustal dust such as clay or loess are of porous structure. Furthermore, the particle number density in a one cm^2 atmospheric column (Peterson, 1976) was chosen for normal atmospheric conditions not taking into account the burden of particles from anthropogenic origin. Total number densities of airborne particles in cities can be as high as 300,000 per cm^3. Over continents, number densities range from several hundred per cm^3 to about 15,000, depending on the proximity to centers of population and on meteorological conditions. Since the particle density with height decreases rapidly, only the first kilometers provide the main surface area for possible catalytic degradation reactions of atmospheric pollutants and trace gases. Most important, since the surface is a quadratic function of the particle radius and 1 μm is probably too low a number for the mean diameter of the tropospheric aerosols of inorganic composition, the real surface area provided by these tropospheric aerosols may be 5 to 10 times that of our globe. With a tropospheric lifetime of one week assumed for these atmospheric aerosols, this total surface area is renewed some 50 times in the course of a year.

It is of interest to contrast the surface areas available for heterogeneous reactions and the concentration of an atmospheric pollutant. If we again take the emitted mass of SO_2 to the emitted mass of particles in a power plant plume of 1 g SO_2 to 100 mg of fly ash per m^3, the fly ash with a surface area of about 5 m^2/g provides a surface area of 0.5 m^2/m^3. If we further assume that all the SO_2 molecules in one m^3 readily adsorb to this surface area in a monolayer, the adsorption isotherm at room temperature predicts that about 4×10^{19} SO_2 molecules can adsorb to the 0.5 m^2 surface area of fly ash. This figure is out of proportion to the actual concentration of about 8×10^{21} molecules of SO_2 present in the stack emission per m^3. Thus, if substantial chemical conversions of SO_2 occur on the fly ash particles, this reaction can only occur via a heterogeneous catalytic reaction.

Unfortunately, we know much less about the inorganic chemistry of the atmospheric aerosols. In the majority of the papers elemental analysis is reported and the actual chemical forms are not identified in which these elements are present. The most important contribution to natural aerosols stems from chemical weathering. The dissolution of aluminosilicates as the major component of the weathering of feldspar, kaolinite and montmorillonites etc. results in the release of cations and silicic acid which is enriched with elements abundant in the lithic material. Many of the minor and trace elements are likely present as metal oxides. Eruptions of volcanoes inject into the atmosphere large amounts of particles composed of metal oxides. For example, the chemical analysis of airborne ash samples of the Mount St. Helens eruption revealed about 65 - 70 % of SiO_2, 15 - 18 % Al_2O_3, 2.5 - 4.6 % Fe_2O_3, 2.3 - 4.8 % CaO, 0.5 - 1.5 % MgO and 0.37 - 0,64 % TiO_2 (Rose and Hoffmann, 1982). Volatile minor and trace elements are enriched in the metal oxides and transported in the eruptive cloud through the atmosphere. It is not surprising that fly ash which is produced by a wide variety of high-temperature industrial processes, is composed of a large number of inorganic oxides. A typical coal fly ash might have the following composition (Seinfeld, 1975): ~40 % SiO_2, ~20 % Al_2O_3, ~15 % Fe_2O_3, ~10 % CaO and ~15 % of other metal oxides. More important under the aspect of environmental concern and catalytic chemical reactions is that metals such as Mn, Cr, Ni, Ti, V etc. are enriched in the surface of the aluminosilicate matrix (Dlugi and Guesten, 1983; Stinespring and Stewart, 1981; Linton et al., 1977).

In summary, the lower troposphere contains an immense number of adsorptive and catalytic aerosol particles which provide a tremendous surface area for heterogeneous and photocatalytic reactions of atmospheric pollutants. Since the atmospheric lifetime of these aerosols of natural and anthropogenic origins discussed above is on the order of several days to about one week, the probability of multiphase chemical reactions is especially high in urban areas where pollutants of low vapour pressure get readily adsorbed onto the surface of the particulate matter and are transported with this matter through the atmosphere. Examples of surface-associated pollutants are sulphur dioxide in a power plant plume, and adsorbed organic molecules such as polynuclear aromatic hydrocarbons on fly ash.

HETEROGENEOUS PHOTOCHEMISTRY IN THE TROPOSPHERE

In general, a number of different heterogeneous photoreactions may occur in the troposphere.

a) Direct photoexcitation of the adsorbed pollutant P:

$$P_{ads.} \xrightarrow[> 300\ nm]{h\nu} P^*_{ads.}$$

$$P^*_{ads} \longrightarrow product(s)$$

Like in photo-induced homogeneous gas-phase reactions, electronic excitation of the atmospheric pollutant P can only be achieved when the pollutant absorbs light beyond the tropospheric cut-off of the solar light of 300 nm. It is interesting to note that organic molecules tend to exhibit bathochromic shifts of the light absorption in the adsorbed state compared to their gas-phase or solution absorption spectra (Nicholls and Leermakers, 1971). Thus, often electronic excitation of an adsorbed organic pollutant might lead to degradation or oxidation reactions which would not occur in the gas-phase. Examples of such bathochromic shifts of surface-adsorbed industrial pollutants have been given by Parlar and Korte (1981). A useful new laboratory technique for studying the absorption spectra of molecules in the adsorbed state is photoacoustic spectroscopy.

b) Catalytic conversion of an adsorbed pollutant P by photoexcitation of the solid surface S:

$$S \xrightarrow{h\nu} S^*$$

$$S^* + P \longrightarrow product(s)$$

In this type of heterogeneous photocatalysis degradation or reaction of surface-adsorbed pollutants can even be achieved at wavelengths well beyond the gas or liquid phase or solid state absorption threshold of the pollutant. The induced photoreaction might be non-catalytic, i.e. stoichiometric or catalytic. Repeatedly photoexcited solid surfaces mobilize reactive species which react with the adsorbed pollutant.

c) Surface reaction of an electronically excited pollutant:

$$P^* + S \longrightarrow PS$$

Electronically excited pollutant molecules in the gas-phase react with the surface of an atmospheric aerosol particle. Though there are no examples of such a heterogeneous photoassisted reaction it might well happen in the atmosphere. A possible example might be the reaction of electronically excited SO_2 with organic particulate matter.

Though a sizeable number of papers deal with the environmental aspects of a heterogeneous catalytic photodegradation of atmospheric pollutants the lack of extrapolation of the laboratory experiments to the troposphere is evident in particular. Many heterogeneous photochemical reactions of pollutants studied are essentially qualitative and kinetic information is not reported. A series of heterogeneous photochemical degradation studies on pesticides and chlorinated organic industrial chemicals focus on the reaction and/or degradation products (Glotfelty, 1978). Often the wavelength quality and intensity of the light source, the absorptivity and concentration of the pollutant as well as the efficiency of light conversion to products are not reported. The descriptors "rapid" or "slow" are virtually meaningless for tropospheric extrapolation when describing a simulated laboratory experiment involving sunlight simulation. The major difficulty arises from the fact that quantum yields for solid-state reactions are extremely difficult to measure, especially when actual environmental concentrations are used. Only with the knowledge of the quantum yield for the photodegradation of an adsorbed pollutant the mean photodegradation lifetime in the troposphere can be estimated with the equation

$$k_1 = 1/\tau = \int_{300\ \mathrm{nm}}^{450\ \mathrm{nm}} I(\nu, z) \cdot \sigma(\nu) \cdot Q d\nu. \qquad (1)$$

The tropospheric lifetime is the product of the solar flux intensity I at the zenith angle z over the absorbing wavelength ν of the pollutant, its photoabsorption cross-section σ in this range of wavelengths and the quantum yield Q of heterogeneous photodegradation. While in text books on photochemistry the procedure for measuring quantum yields in solution and in the gas-phase is described in details, there are no general procedures indicated in the literature for solid-state reactions. The main drawback is the determination of the number of light quanta absorbed by the adsorbed pollutant. Only recently an approach to this problem has been published which allows an estimation of the quantum yield for a photoreaction in the adsorbed state (Blau and Guesten, 1982).

THE PHYSICAL PICTURE OF AGED TROPOSPHERIC AEROSOLS

At this point some considerations should be mentioned of the physical state of the inorganic aerosols and the interfaces at which the multiphase atmospheric chemistry occurs. In general, heterogeneous photocatalytic reactions might occur by solid-gas or solid-liquid interaction. Current trends tend to consider the aged tropospheric aerosols as an aqueous aerosol with a water content of 30 - 50 wt % (Graedel and Weschler, 1981). Around an insoluble core an aqueous solution of about 30 % of the total weight at humidities above 40 % envelopes the surface of the inorganic aerosol particle (Whitby et al., 1972). The solution may contain dissolved inorganic and organic compounds. The organic molecules are preferentially oriented toward the core with their polar part. Since molecular oxygen and transition metal ions are abundant in the aqueous solution, some of the chemical reactions at the solid-liquid interface might eventually take place by homogeneous catalytic or redox reactions. Furthermore, the soluble part of the inorganic aerosol particle largely determines the pH of the aqueous solution around the core. In dry air, however, no water has been detected on the surface of desert aerosols (Ho et al., 1974). Thus, depending on the relative humidity the multiphase chemistry at the surface of tropospheric aerosol particles can occur by heterogeneous gas-solid or liquid-solid reactions.

Another view deserves discussion, in particular regarding organic pollutants on inorganic aerosols. In urban areas the organic pollutants and the inorganic aerosol particles are often emitted together from high-temperature sources. Due to their low vapour pressure, the organic molecules with high molecular weights are readily adsorbed by the inorganic aerosol particles soon after leaving the stack. An example is the adsorption of polynuclear aromatic hydrocarbons on fly ash (Schure and Natusch, 1982). Thus, in this type of solid-gas phase adsorption processes at activated surfaces the organic molecule is tightly bound. It is likely that by the subsequent condensation of water the adsorbed organic molecule is not dissolved unless it has very polar groups or substituents.

PHOTOASSISTED HETEROGENEOUS DEGRADATION OF ATMOSPHERIC POLLUTANTS RELEVANT TO TROPOSPHERIC CHEMISTRY

a) Halogenated organic compounds

To our knowledge one of the first speculations on a heterogeneous degradation of atmospheric trace gases on the dry heated Sahara desert can be found in a paper on budget considerations regarding atmospheric N_2O (Schuetz et al., 1970). In the discussions on the ozone-chlorofluorocarbon problem (Rowland and Molina, 1975) it was suggested that photodissociation of these inert halocarbons on desert areas could provide an effective sink in the troposphere (Alyea et al., 1978). Laboratory experiments by Ausloos et al. (1977) have demonstrated the photodecomposition of chlorofluorocarbons on sand surfaces at wavelengths well beyond the gas-phase absorption threshold of these molecules. It was further found that the rate of photodecomposition greatly depends on the nature of the solid and the concentration of surface water (Gaeb et al., 1980). Heterogeneous catalytic degradation of the chlorofluoromethanes on zinc oxide surfaces revealed a spontaneous dark reaction on special activated surfaces while the photolytically induced degradation was also possible on "untreated" surfaces. With increasing exposure time increasing concentrations of chloride anions are formed on the surface of ZnO (Filby et al., 1981). The initial rate of chloride formation is slightly dependent on the chlorofluoromethane pressure in the range of 10^{-2} to 400 Torr only. This proves that the heterogeneous photocatalysed degradation of the chlorofluoromethanes is a real surface-induced reaction. A reaction mechanism was proposed in which indications of an electron capture reaction of the chlorofluorocarbons were given either directly by electrons generated in the valence-conduction band of ZnO by light or via the chemisorbed O_2^- or O^- species (Filby et al., 1981).

$$ZnO \xrightarrow[> 300\ nm]{\Delta,\ h\nu} (ZnO^+,\ e^-)$$

$$(ZnO^+, e^-) + O_2 \longrightarrow ZnO(O_2^-)$$

$$(ZnO^+, e^-) + CFCl_3 \longrightarrow ZnO + CFCl_3^-$$

$$ZnO(O_2^-) + CFCl_3 \longrightarrow ZnO(O_2) + CFCl_3^-$$

$$CFCl_3^- \longrightarrow Cl^- + \cdot CFCl_2$$

It is likely that the presence of semiconducting metal oxides in sand, such as mixed oxides of iron and titanium, is responsible for the photodegradation of the halogenated organic compounds. Other chlorinated hydrocarbons, in particular CCl_4, hexachlorobenzene, pentachlorophenol as well as several pesticides such as DDT (1,1,1-trichloro-2,2-bis(p-chlorophenyl)ethane), undergo photomineralization to CO_2 when irradiated with long-wavelength UV light (> 300 nm) in the presence of sand (Gaeb et al., 1975, 1980). Indications of a heterogeneous catalytic photodegradation of halocarbons over the Sahara desert can be taken from the paper of Pierotti et al. (1978). The atmospheric concentrations of CCl_4 and methylchloroform (CH_3CCl_3) are significantly lower over the Sahara while no measurable difference with respect to background air was detectable for the chlorofluoromethanes. This is in agreement with laboratory experiments (Ausloos et al., 1977). Under identical irradiation conditions the relative photolysis rates for CCl_4, $CFCl_3$, and CF_2Cl_2 were 100 : 10 : 3. Due to the very slow vertical diffusion of the chlorofluorocarbons a slow tropospheric removal rate of less than 1 % per year would have a major effect on the ozone depletion in the stratosphere. Thus, even a minor contribution of tropospheric heterogeneous photodegradation of the chlorofluorocarbons reducing their average residence time by a rather small amount would have a large effect on current estimates of ozone depletion in the stratosphere. The analytical techniques now available do not allow to measure such small differences in chlorofluorocarbon concentration due to degradation in the troposphere. Obviously, heterogeneous photocatalysis is not restricted to halocarbons. In general, photomineralization and oxidation of organic material catalysed by sand have been globally regarded as a significant source of CO_2 in the atmosphere. It has been estimated that nearly 35 % of the net rate of CO_2 production from the annual combustion of fossil fuels is produced via heterogeneous photocatalysis of adsorbed organic matter (Sancier and Wise, 1981). It is of interest that the non-detection of organic compounds on Mars has been explained by a heterogeneous photocatalytic oxidation mechanism operative under UV irradiation (Chun et al., 1978).

b) Sulphur dioxide

In recent years considerable attention has been paid to the study of heterogeneous reactions of SO_2 and the contribution of these reactions to atmospheric sulphate aerosol burdens (Judeikis et al., 1978; Dlugi, 1982). The results from these studies suggest that freshly emitted aerosols can be quite effective in converting SO_2 to particulate sulphate. As a consequence of these laboratory studies, transformation studies of SO_2 in power plant plumes have been performed since power plants are by far the major anthropogenic source of SO_2 (Newman, 1981; Forrest et al., 1981). It was observed that the oxidation rate seldom exceeded 5 % hr^{-1} with essentially all the oxidation occurring within the first few kilometers of the source of emission. This was explained by a heterogeneous catalytic oxidation of SO_2 on the fly ash particles in which the rate of the reaction decreased exponentially and ultimately approached zero. The significant increase in reaction capacity at high relative humidities suggests that the catalytic oxidation reaction takes place in adsorbed water films. The reason for the capacity-limited nature of the heterogeneous catalytic reaction was found in lowering of the surface pH by the acid sulphates formed. Since SO_2 is considerably less soluble in acid solutions than in neutral or basic solutions, the rate of the reaction decreases. Moreover, in the course of the reaction the surface of the fly ash is covered more and more with a layer of sulphate and/or sulphuric acid which prevents diffusion of SO_2 into the liquid phase of the aerosol. However, it is interesting to note that conversion rates of SO_2 to sulphate are substantially lower during the night (< 1 % hr^{-1}). Midday and afternoon rates range from ~1 to 4 % hr^{-1} (Forrest et al., 1981). This effect has been explained by additional conversion of SO_2 through homogeneous photochemical reactions. Smog chamber experiments on the heterogeneous catalytic SO_2 oxidation on fly ashes from different power plants, however, support the contribution of a heterogeneous photocatalysis (Dlugi and Guesten, 1983). In figure 1 the mass of sulphate produced per unit of mass of two fly ashes during the catalytic SO_2 oxidation at a relative humidity of 80 % is plotted versus the irradiation time. It has been demonstrated that the reaction capacity ($g_{SO_4^{2-}}$ per $g_{aerosol}$), calculated from the disappearance of SO_2 in the smog chamber in an overall first-order reaction, is comparable to the increasing sulphate concentration on the fly ash sample.

As a result of irradiation by simulated solar light, the reaction capacity of the fly ash designated K I increases by about 20 %. Under the same reaction conditions fly ash K II does not show this photocatalytically induced enhancement of the reaction capacity. Since all the experimental details in the reactions were the same, except for the fly ashes used, it is

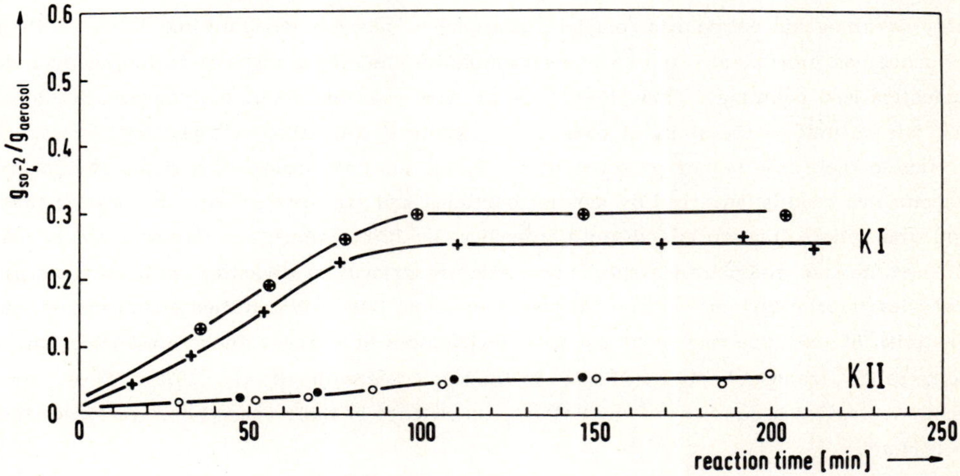

Figure 1: The mass of sulphate produced per mass of fly ash during the catalytic SO_2 (4.5 mg $SO_2 \cdot m^{-3}$) oxidation at a relative humidity of 80 %.
fly ash K I: +-+-+ without and ⊕-⊕-⊕ with UV irradiation
fly ash K II: o-o-o, ●-●-● with and without UV irradiation

plausible to assume that in the first case the fly ash K I acts as a photocatalyst and in the second case the fly ash K II does not exhibit this photocatalytic activity. Indeed, analytical data and other physical properties suggest that the two fly ashes are of different morphological structures (Dlugi and Guesten, 1983).

Projecting these results to a power plant plume leads to the assumption that depending on the type of fly ash a heterogeneous photocatalytic oxidation is operative in addition to heterogeneous catalytic SO_2 oxidation. Due to the capacity-limited nature of these catalytic reactions both mechanisms promote SO_2 oxidation only within the first kilometers behind the stack.

Much less is known about the heterogeneous reactions of the oxides of nitrogen. The heterogeneous reaction of NO_2 on a variety of materials likely to be found in urban aerosols and in the environment results in nitrate and nitrite formation on the surface of the aerosols studied. The catalytic reaction ceases when the surface is saturated with the reaction products. To our knowledge a possible photocatalytic heterogeneous decomposition of NO_2 is briefly mentioned in one paper only (Judeikis et al., 1979).

c) Polynuclear aromatic hydrocarbons

It is now well established that atmospheric particulate matter of urban and industrial origins is loaded with a class of cancerogenic organic compounds known as polynuclear aromatic hydrocarbons (PAH) (National Academy of Science, 1972), Furthermore, the PAHs are the only carcinogenic compounds found in atmospheric aerosols (Gelboin and Ts'O, 1978). This evidence has incriminated the PAHs responsible for the strong increase in lung cancer in all industrialized countries. Thus, there is a growing concern about the sources, the amount, the fate as well as the sinks of these carcinogenic hydrocarbons (Hites, 1981; Suess, 1976).

Due to their low vapour pressure virtually all the PAHs released into the atmosphere as vapour are readily adsorbed by aerosol particles and transported with this matter through the troposphere (Cautreels and van Cauwenberghe, 1978; Schure and Natusch, 1982). Studies of particle size distributions show that PAHs are primarily associated with aerosols in the very low micron and sub-micron ranges (Pierce and Katz, 1975; Butler and Crossley, 1979). Particles of this size may have atmospheric lifetimes of a week. There are indications of a long-range transport of the PAHs in Europe (Bjorseth et al., 1979). Thus, for an ecotoxicological assessment of the PAHs the knowledge of their atmospheric residence time is important.

During the last years evidence has been accumulated in laboratory studies of an enhanced photocatalytic degradation of PAHs on solid surfaces (Lane and Katz, 1977; Fox and Olive, 1979; Lotz et al., 1979). Indirectly, evidence for a photochemically operative degradation mechanism was obtained by comparing the summer/winter ratio of the airborne PAH concentration in a municipal garden and in a nearby railroad underpass (Guesten and Heinrich, 1979). Recently, quantitative data have been reported from laboratory experiments with simulated solar light concering the photodegradation of PAHs on silica gel as a model aerosol. Using equation 1, the measured quantum yields of the photodegradation of six PAHs from the EPA Priority Pollutant List allow to estimate the mean tropospheric lifetime of a PAH in Central Europe (Blau and Guesten, 1982).

While the quantum yields of the photodegradation of the two isomeric PAHs anthracene and phenanthrene differ by a factor of 10^3, the distances covered from the source differ by a factor of more than 10^4. This illustrates that beyond the quantum yield the absorption cross-section (see Equation 1) over the range of the solar spectrum plays an important role for the rate of photodegradation of a pollutant molecule.

The large differences in the rate of the heterogeneous photodegradation of the PAHs might explain why in geographic regions subject to atmospheric long-range transport the local phenanthrene concentration is higher by a factor of up to 20 compared to that of anthracene. When the air-mass trajectories from England or Central Europe are directed to Norway, the airborne phenanthrene concentration is higher by a factor of 10 to 20 than that of anthracene (Bjorseth et al., 1979). Norwegian drinking water resources contain much more phenanthrene than anthracene though their emission concentrations are roughly the same (Olufsen, 1980). While the phenanthrene to anthracene ratio in the Ruhr valley near Essen is 6:1, ratios of 25:1 have been assayed in the remote Sauerland area (Hettche, 1972). Accordingly, the atmospheric lifetime of the much discussed cancerogenic benzo(a)pyrene in Central Europe is about 10 hours in the summer.

The question is whether silica gel is a good model aerosol for other mineral components in the environment. Although SiO_2 and aluminosilicates are by far the major components of the mineral components of natural (Reiter et al., 1978) and anthropogenic aerosols (Seinfeld, 1975), the surface enrichment of volatile trace elements can alter the catalytic activity of the bulk silicate matrices. Fly ash derives from the diverse mineral matter in the feed coal. They are altered by passing through the combustion process. The photodegradation of anthracene and phenanthrene adsorbed on fly ash is shown in figure 2. When the incident light is only weakly adsorbed by the PAH, the rate of photo-de gradation can be described by a first order rate expression

$$-\frac{d[PAH]}{dt} = K \cdot Q' \cdot I_a \, [PAH], \qquad (2)$$

i.e. the photodegradation rate at a given PAH concentration is proportional to the product of the relative quantum yield Q', the absorbed light energy I_a and a proportional factor K used to correct the relative quantum yield to give a real quantum yield (Blau and Guesten, 1982). Thus, at identical experimental concentration and a controlled incident light intensity inside the rotating bed photoreactor with the same geometry in all experiments, the semi-log plots of decreasing PAH concentration versus irradiation time are linear. The half-lives of the photodegradation of PAH adsorbed on various adsorbents are representative of the quantum yields of photodegradation.

Under identical irradiation conditions the photodegradation of phenanthrene and anthracene greatly increases from silica gel to fly ash designated KI. Whereas the half-lives of phenanthrene are on the order of hours, those of anthracene on silica gel and fly ash KI are counted by minutes only (Dlugi and Guesten, 1983). The acceleration of the catalytic photodegradation of anthracene and phenanthrene is most likely explained by a mechanism of catalytic photooxidation on the surface of the fly ash.

As a logical extension of our studies of the photodegradation of PAHs in the adsorbed state (Blau and Guesten, 1982; Dlugi and Guesten, 1983), the photodegradation of phenanthrene on various metal oxides and fly ashes was investigated (Guesten et al., 1984). Phenanthrene was selected as a model PAH because of its relative photochemical stability which allows to better differentiate the photodegradation rates on the various adsorbents studied. In figure 3 the semi-log plots have been traced for the photodegradation of phenanthrene on various metal oxides, sea sand and fly ash. The concentration of phenanthrene was 0.6 mg/g aerosol (3.3×10^{-6} Mol g^{-1}). Taking into account a surface of 33 A^2 for the phenanthrene molecule (Fournier and Pezerat, 1982) one can estimate for all the adsorbents investigated that phenanthrene is covered at sub-monolayer concentrations. The

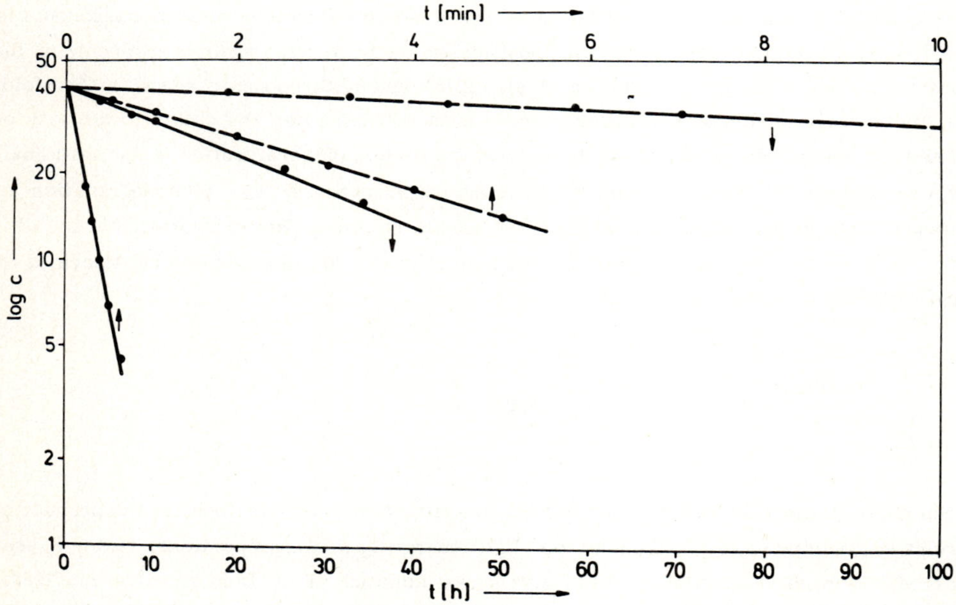

Figure 2: Heterogeneous photocatalytic degradation of anthracene (upper abszissa) and phenanthrene (lower abszissa) adsorbed on silica gel (---) and fly ash KI (——).

environmental concentration ratio of an adsorbed PAH to the mass of the airborne particulate matter is approximately 1 : 10^5 to 1 : 5 x 10^5 (Pierce and Katz, 1975; Miguel and Friedlander, 1978). Though our concentration of phenanthrene per gramme of adsorbent is about 50 to 100 times higher than actual environmental concentrations of PAHs it was demonstrated that the quantum yield of the photodegradation of PAHs on silical gel is independent of the concentration in the range of 10^{-7} to 10^{-5} Mol·g^{-1} (Blau and Guesten, 1982).

The half-lives of phenanthrene in figure 3 vary from 260 hours on silica gel to less than one hour on rutile. Obviously, there are two classes of metal oxides, one on which the photodegradation of phenanthrene is very slow and one with rapid photodegradation. The former class consists of metal oxides (SiO_2, Al_2O_3, SnO_2, MgO) which do not absorb the 300 - 400 nm simulated solar light of the photoreactor. In the latter class of metal oxides (TiO_2, ZnO, Fe_2O_3) the solid can be excited by light in the near-surface region. While in the former class of metal oxides the adsorbed phenanthrene is directly excited by photons to its electronic states, photons of band-gap energy excite electron-hole pairs at the surface in the latter class of metal oxides with semi-conduction properties. The half-life of the photodegradation of phenanthrene on fly ash KI is about 25 h. This value lies between the two classes of metal oxides. Thus, on the basis of the results in figure 3, one may conclude that the photodegradation of phenanthrene on a very pure aluminosilicate would be very slow, i.e. comparable to the photodegradation of phenanthrene on silica gel. The more rapid photodegradation of phenanthrene on fly ash KI and sea sand is then due to the presence of semi-conducting metal oxides at the surface of the aluminosilicate matrix.

Table 2: Mean tropospheric lifetime τ and distance R of PAHs transported from sources in Central Europe in the summer *)

PAH	τ [h]	R [km]
Anthracene	0.02	0.3
Phenanthrene	1100	15000
Benzo(a)anthracene	2.5	35
Benzo(b)chrysene	6.5	90
Fluoranthene	110	1600
Benzo(a)fluoranthene	0.25	3.3
Benzo(b)fluoranthene	80	1100
Benzo(j)fluoranthene	3	40
Benzo(k)fluoranthene	14	200
Benzo(a)pyrene	10	140

*) (Blau and Güsten, 1982)

The results on the photocatalytic degradation of atmospheric pollutants described in the last chapter can be explained best by the different reaction mechanisms generalized in one of the previous paragraphs. For example, the slow photodegradation of phenanthrene on the non-light absorbing metal oxides proceeds via direct electronic excitation of the adsorbed phenanthrene molecule. The large difference in the photostability of the two isomeric PAHs anthracene and phenanthrene (see table 2) is the result of the individual excited state behaviour of the two PAHs. At room temperature the excited singlet state of anthracene in the adsorbed state compared to phenanthrene is 31 nanoseconds to only 2 nanoseconds (Oelkrug et al., 1979). Since the photoreaction of anthracene in the adsorbed state is primarily a photooxidation (Fox and Olive, 1979; Kortuem and Braun, 1960) the long-lived excited singlet state of anthracene is more susceptible to a reaction with molecular oxygen than that of phenanthrene. Another explanation is that formation of the triplet state of phenanthrene is greatly enhanced while in adsorbed anthracene exclusively the singlet state is populated (Oelkrug et al., 1979). Thus, on metal oxides whose band-gap between the valence and conduction bands lies in the far UV and cannot be reached by solar light, the adsorbed pollutant reacts photochemically only after direct absorption of solar light.

If, however, the band-gap of a metal oxide is small enough to be reached by solar light,

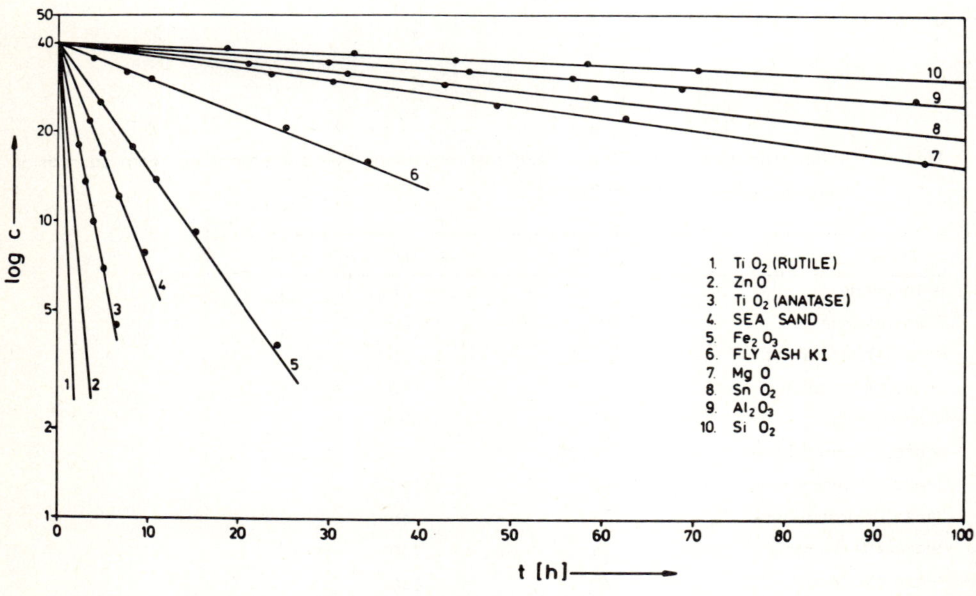

Figure 3: Semi-logarithmic plots of the catalytic photodegradation of phenanthrene on various metal oxides, sea sand and fly ash KI.

light absorption causes the transfer of electrons to the higher energy conduction band. These electrons released near the surface of the metal oxide can be transferred to the adsorbed pollutant or co-adsorbed oxygen molecules which subsequently induce oxidation processes. Arguments have been advanced in support of the idea that mainly oxygen is the active species in the photodegradation process (Howe, 1982; Bickley, 1978). It has been shown, for example, that anthracene adsorbed on Al_2O_3 is oxidized by light to anthraquinone (Kortuem and Braun, 1960). Since the absorption cross-section of the band-gap of metal oxides with semi-conduction properties such as TiO_2, Fe_2O_3 or ZnO, etc., lies in the near UV and is much larger than the absorption cross-section of the adsorbed organic molecule at sub-monolayer concentration, the photodegradation process is much more effective (see fig. 3). Thus, the principal function of these metal oxides (MO_2) is to act as a source or a sink for photogenerated electrons. Depending on the chemical nature of the adsorbed pollutant, either electron transfer from the conduction band to the pollutant molecule or electron transfer from the adsorbed pollutant to the solid in an electron hole might occur (see fig. 4).

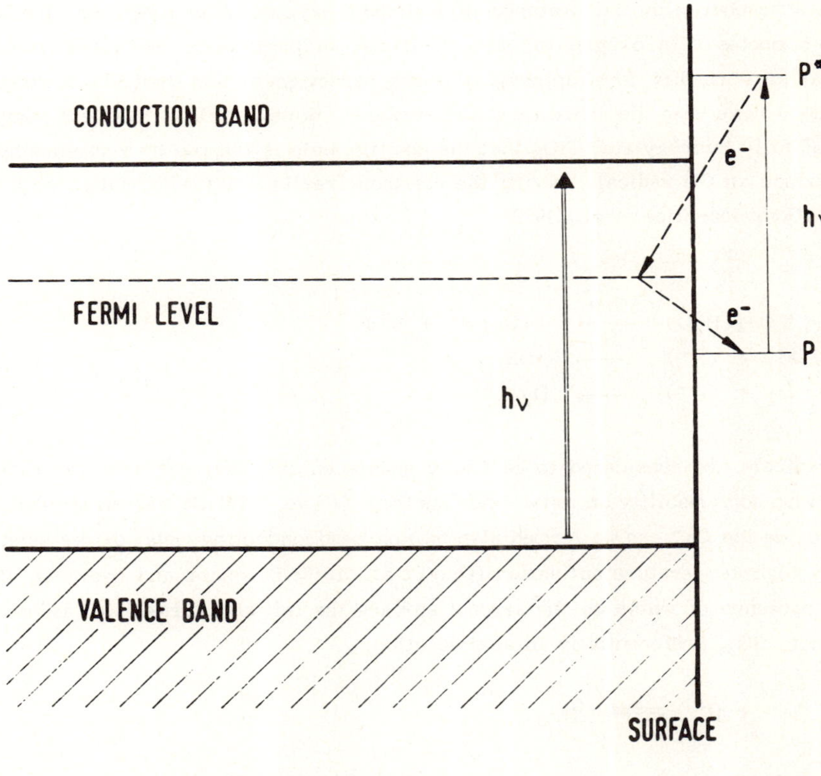

Figure 4: Schematic description of the light absorption by a semi-conducting solid and possible mechanisms for the reaction of an adsorbed pollutant molecule P in its ground state P_o and electronically excited state P^*.

The trapping of charge carriers by pollutant molecules P at the surface prevents the recombination of the electron (e^-) and the hole (h^+).

$$MO_2 \xrightarrow{h\nu} MO_2\ (e^- + h^+)$$
$$e^- + h^+ \longrightarrow \text{heat or } h\nu'$$
$$e^- + P \longrightarrow P^-$$
$$h^+ + P \longrightarrow P^+$$

Thus, one would expect in the first step of surface site-pollutant interaction the formation of radical anions or radical cations. The last process of radical cation formation is favoured by an electronically excited P* of an adsorbed pollutant. In general, the energy relationship of the surface levels of the adsorbed pollutant to the conduction and valence band are very important. In particular, the imperfect surface of the oxides with semi-conduction properties and its doping with altervalent ions determines the height of the Fermi level.

In the atmosphere, in the presence of abundant oxygen, it is likey that the adsorbed pollutant competes with oxygen for the electrons. In particular, molecules with a high electron affinity readily form anions. All three paramagnetic ion radicals of oxygen (O^-. O_2^- and O_3^-) have been identified on oxide surfaces (Howe, 1982). It has been suggested in the case of fully hydroxylated TiO_2 that the positive hole is trapped by a surface hydroxide ion to produce an OH radical, leaving the electron free to react with molecular oxygen to form O_2^- (Gonzalez-Elipe et al., 1979):

$$TiO_2(H_2O) \xrightarrow{h\nu} TiO_2(e^- + h^+)$$
$$OH^- + (h^+) \longrightarrow \cdot OH$$
$$O_2 + (e^-) \longrightarrow O_2^-.$$

The OH radicals are considered to be freely mobile on the TiO_2 surface. The radical O_2^- possesses also some mobility on metal oxide surfaces (Howe, 1982). It must be pointed out that the hyperoxide ion O_2^- can be formed also on non-semiconducting metal oxides such as MgO or CaO by high-temperature activation (Ono et al., 1979; Derouane and Indovina, 1972).

In the presence of water on the aerosol surface the initially formed O_2^- is immediately protonated to HO_2, preferentially in acid solution:

$$O_2^- + H^+ \rightleftharpoons HO_2.$$

In this respect, the hyperoxide ion is the base of the hydroperoxi radical. Thus, following the picture of Graedel and Weschler (1981), the chemistry within aqueous atmospheric aerosols of inorganic origin enters the radical chemistry in aqueous solution.

Hence, the production of OH and HO_2 radicals in the liquid phase around the inorganic core of an atmospheric aerosol should be possible via heterogeneous photocatalysis. We can further assume that the hydroperoxi radical accepts a second electron to form HO_2^- which is desorbed as H_2O_2 into the liquid phase of the aerosol particle. In presence of ferrous ions the well-known Fenton reaction (Walling, 1975) might produce OH radicals by the decomposition of H_2O_2 through reductive cleavage:

$$HO_2 + e^- \longrightarrow HO_2^- \xrightarrow{H^+} H_2O_2$$
$$H_2O_2 + Fe_2^+ \longrightarrow OH^- + \cdot OH + Fe^{3+}.$$

Thus, in the steady state of the liquid phase around the inorganic aerosol predominantly the OH radical might be responsible for oxidation reactions of adsorbed pollutants. Room temperature rate constants for reactions of OH radicals with organic compounds can be estimated for the gas phase and for liquid water (Guesten et al., 1981).

SURFACE SITE - POLLUTANT INTERACTION

It was pointed out above that in the first step of the surface site-pollutant interaction the formation of radical cations of anions is assumed. Depending on the electron affinity of the adsorbed pollutant, the pollutant molecule competes with the co-adsorbed oxygen for the electrons generated via heterogeneous photocatalysis. For large organic molecules such as the PAHs, cationic or anionic radicals have been identified on the surface of metal oxides by ESR spectroscopy. However, there is confusion about the nature of the radical species formed on the surface of silica gel or aluminum oxide (Flockhart et al., 1983). It has been thought that the radical cation of perylene can be formed simultaneously with the co-adsorbed O_2^- (Hall and Dollish, 1968). These assumptions are based on the non-uniformity of the surface of composed metal oxides. The local variations in the structure of atmospheric aerosols provide a non-uniform distribution of surface groups, defects, impurities and different types of crystal planes. Furthermore, depending on the chemical composition and doping with altervalent ions the active surface sites of the metal oxides exhibit acid-base character. In aluminosilicates, for example, which are among the common minerals in nature (and in fly ash), the different surface sites may act as Lewis acid favouring the formation of radical cations of the adsorbed pollutant molecule or as Lewis bases favouring the formation of radical anions by transferring electrons to the adsorbed pollutant molecule.

Whether or not the adsorbed pollutant tends to accept electrons depends on its electron affinity. Accordingly, the ionisation potential IP can often be correlated with the chemical reactivity of a molecule. It was pointed out by Kotzias et al. (1981) that, at least qualitatively a correlation might exist of the yield of CO_2 in the photomineralization of organic compounds on silica gel with the ionisation potentials of the organic compound. In figure 5 the yields of CO_2 are plotted versus the ionisation potentials of the organic compounds photodegraded on silica gel (left ordinate). By kinetic transformation of the CO_2 yields into reaction rate constants one obtains a fairly good linear correlation with IP (right ordinate). Thus, one can possibly predict the rate of the heterogeneous photodegradation of atmospheric pollutants from their ionisation potentials. Gas-phase ionisation potentials are conveniently measured today by photoelectron spectroscopy. The linear correlation is not necessarily valid for airborne particulate matter. As has been shown in figs. 2 and 3 the rate of the photodegradation of PAHs on fly ash is greatly enhanced as compared to silica gel. Moreover, great differences in the catalytic and photocatalytic activities of fly ashes from different coal-fired power plants have been demonstrated to exist by two reactions of environmental significance, namely by the example of the heterogeneous sulphur dioxide oxidation of SO_2 and the photodegradation of PAHs (Dlugi and Guesten, 1983; Guesten et al., 1984). While one fly ash gives rise to a fairly fast photocatalytic oxidation of SO_2 as well as degradation of PAHs, the fly ash from another coal-fired power plant showed no photocatalytic activity at all in the two reactions mentioned above. Even the very photolabile

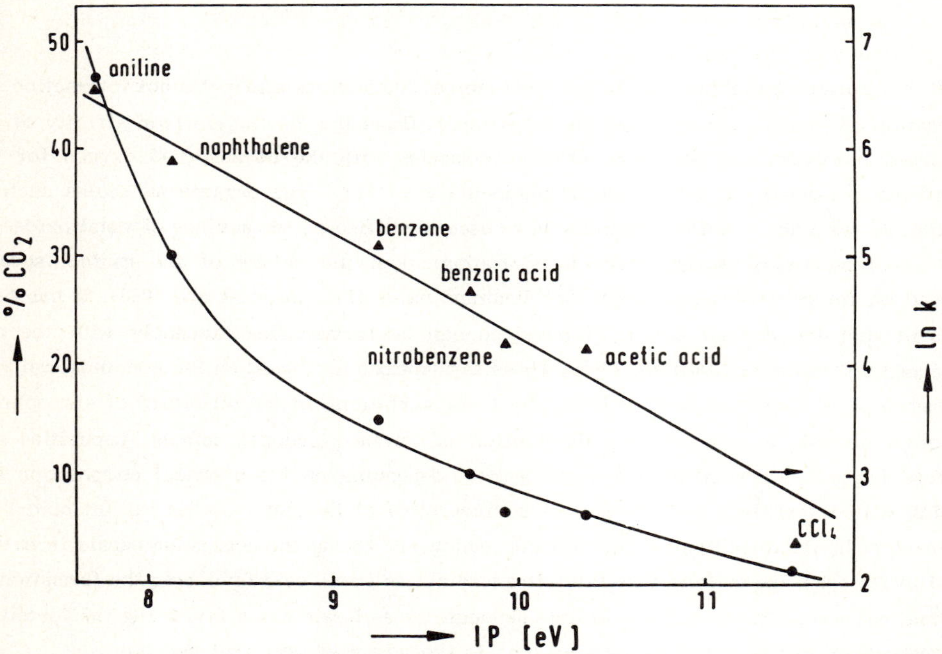

Figure 5: Dependence of the CO_2 yield of the photomineralization of organic compounds adsorbed on silica gel with their corresponding ionisation potentials (left ordinate) and after kinetic transformation to reaction rate constants (right ordinate) (CO_2 yields are taken from Kotzias et al., 1981).

anthracene was fully recovered after an attempt of more than four hours to achieve photodegradation on a fly ash designated K II. It is obvious that differing surface characteristics are responsible for the photocatalytic properties of the fly ashes. At present, only speculations can be made on the differing properties of the two fly ashes. Since, however, the fly ashes derive from the diverse lithic matter in the feed coal, it is not surprising that the surface and catalytic properties of the fly ashes are highly variable.

CONCLUSIONS

1. Substantial evidence is available that heterogeneous photocatalytic reactions of atmospheric pollutants can play an important role in the oxidation and degradation of these pollutants in the troposphere. Extrapolation of laboratory data on heterogeneous photocatalysis with a view to estimating the residence times of adsorbed pollutants in the troposphere are difficult to conduct.
2. Depending on the origin of the natural or anthropogenic aerosols, the adsorbed pollutants are either subject to a fairly rapid photodegradation by solar light or may be stabilized to withstand photolytic degradation. In the latter case the adsorbed pollutants undergo a long-range transport in the lower troposphere during which they might react with other gaseous atmospheric pollutants or reactive natural trace gases.
3. To clarify the different catalytic and photocatalytic activities of the various types of fly ash detailed surface analytical investigations are necessary.
4. Since many inorganic aerosol particles in the troposphere are of natural (sand, etc.) or anthropogenic (fly ash, etc.) origin and contain surface-enriched metal oxides with semi-conduction properties in the aluminosilicate matrices, the use of silica gel as a model aerosol is of little value in photodegradation tests performed to predict the tropospheric lifetime of chemicals.

ACKNOWLEDGEMENT

The comments and discussion on various aspects of the work by Drs. T. Cvitaš, R. Dlugi, G. Heinrich and C. Zetzsch have been most helpful.

REFERENCES

Almquist E. (1974) An analysis of global air pollution. Ambio 3, 161-167.

Alyea F.N., Cunnold D.M. and Prinn R.G. (1978) Meteorological constraints on tropospheric halocarbon and nitrous oxide destructions by siliceous land surfaces. Atmos. Environ. 12, 1009-1011.

Ausloos P., Rebbert R.E. and Glasgow L. (1977) Photodecomposition of chloromethanes adsorbed on silica surfaces. J. Res. Natl. Bur. Stand. 82, 1 - 8.

Bickley R.I. (1978) Photo-induced reactivity at oxide surfaces. Chem. Phys. Solids Their Surf. 7, 118 - 156.

Bjorseth A., Lunde G. and Lindskog A. (1979) Long-range transport of polynuclear aromatic hydrocarbons. Atmos. Environ. 13, 45 - 53.

Blau L. and Guesten H. (1982) Quantum yields of the photodecomposition of polynuclear aromatic hydrocarbons adsorbed on silica gel, in "Polynuclear Aromatic Hydrocarbons: Physical and Biochemical Chemistry". M. Cooke, A.J. Dennis and G.L. Fisher, eds., pp. 133 - 144, Battelle Press, Columbus (Ohio) and Springer-Verlag, New York - Heidelberg - Berlin.

Butler J.D. and Crossley P. (1979) An appraisal of relative airborne sub-urban concentrations of polynuclear aromatic hydrocarbons monitored indoors and outdoors. Sci. Total. Environ. 11, 53 - 58.

Cautreels W. and van Cauwenberghe K. (1978) Experiments on the distribution of organic pollutants between airborne particulate matter and the corresponding gas phase. Atmos. Environ. 12, 1133 - 1141.

Chun S.F.S., Pang K.D., Cutts J.A. and Ajello J.M. (1978) Photocatalytic oxidation of organic compounds on Mars. Nature 274, 875 - 876.

D'Almeida G.A. (1983) Der Staubtransport aus der Sahara. Dissertation, University of Mainz, Federal Republic of Germany.

Derouane E.G. and Indovina V. (1972) ESR of O_2^- species adsorbed on thermally activated MgO powders. Chem. Phys. Lett. 14, 455 - 459

Dlugi R. (1982) Untersuchungen zur Bildung schwefelhaltiger Aerosolteilchen: Die Bedeutung der heterogenen SO_2-Oxidation. Report-No. KfK-3329, Kernforschungszentrum Karlsruhe, Federal Republic of Germany.

Dlugi R. and Guesten H. (1983) The catalytic and photocatalytic activity of coal fly ashes. Atmos. Environ. 17, 1765 - 1771.

Duce R. A. (1978) Speculations on the budget of particulate and vapor phase non-methane organic carbon in the global troposphere. Pageoph. 116, 244 - 273.

Eriksson E. (1959) The yearly circulation of chloride and sulphur in nature: meteorological, geochemical, and pedological implications, Part 1. Tellus 11, 375 - 403.

Filby W. G., Mintas M. and Guesten H, (1981) Heterogeneous catalytic degradation of chlorofluoromethanes on zinc oxide surfaces. Ber. Bunsenges. Phys. Chem. 85, 189 - 192.

Flockhart B.D., Sesay I.M. and Pink R.C. (1983) Electron transfer at alumina surfaces. Part 7.- Nature of the radical species formed from perylene on the surface of an activated alumina - cation or anion? J. Chem. Soc. Faraday Trans. I, 79, 1009 - 1015.

Forrest J., Garber R.W. and Newman L. (1981) Conversion rates in power plant plumes based on filter pack data: the coal-fired Cumberland plume. Atmos. Environ. 15, 2273 - 2282.

Fournier J. and Pezerat H. (1982) Etude du mode d'adsorption des hydrocarbures polycycliques aromatiques sur les amiantes. Cas du phenanthrene. J. chim. phys. chim. biol. 79, 589 - 596.

Fox M.A. and Olive S. (1979) Photooxidation of anthracene on atmospheric particulate matter. Science 205, 582 - 583.

Gaeb S., Schmitzer J., Turner W.V. and Korte F. (1980) Mineralization of CCl_4 and CCl_2F_2 on solid surfaces. Z. Naturf. 35b, 946 - 952.

Gaeb S., Nitz S., Parlar H. and Korte F. (1975) Beitraege zur oekologischen Chemie. CV. Photomineralisation of certain aromatic xenobiotica. Chemosphere 4, 251 - 256.

Ganor E. and Mamane Y. (1982) Transport of Saharan dust across the Eastern Mediterranean. Atmos. Environ. 16, 581 - 587.

Gelboin H.V. and Ts'O, P.O.P. (1978) Polycyclic Hydrocarbons and Cancer, Vol. 1. Acad. Press, New York - San Francisco - London.

Gilette D.A., Blifford I.H. and Fryrear D.W. (1974) The influence of wind velocity on the size distribution of aerosols generated by wind erosion of soils. J. Geophys. Res. 79, 4068 - 4075.

Glotfelty D.E. (1978) The atmosphere as a sink for applied pesticides. J. Air. Pollut. Control Assoc. 28, 917 - 921.

Gonzalez-Elipe A.R., Munuera G. and Soria J. (1979) Photo-adsorption and photo-desorption of oxygen on highly hydroxylated TiO_2 surfaces. J. Chem. Soc. Faraday Trans. I, 75, 748 - 761.

Graedel T.E. and Weschler C.J. (1981) Chemistry within aqueous atmospheric aerosols and raindrops. Rev. Geophys. Space Phys. 19, 505 - 539.

Grigoryev A.A. and Kondratyev K.Ja.: Atmospheric dust observed from space. WMO Bulletin January 1981, p.3.

Guesten H. and Heinrich G. (1979) Polycyclic aromatic hydrocarbons in the lower atmosphere of Karlsruhe, in:"Polynuclear Aromatic Hydrocarbons", P.W. Jones and P.Leber, eds., pp. 357 - 370, Ann Arbor Science Publ. Inc., Ann Arbor, Mich..

Guesten H, Filby W.G. and Schoof S. (1981) Prediction of hydroxyl radical reaction rates with organic compounds in the gas phase. Atmos. Environ. 15, 1763 - 1765.

Guesten H., Božičević Z. and Klasinc L. (1984) Photodegradation of phenanthrene on metal oxides and fly ashes. Staub-Reinh. Luft, 44, pp 448-451.

Hall W.K. and Dollish F.R. (1968) The chemistry of radical ion formation on aluminum silicates. J. Colloid Interface Sci. 26, 261 - 265.

Hettche O. (1972) Polyzyklische Aromaten als schaedliche Fremdstoffe in der Aussenluft, VDI-Zeitschrift 114, 346 - 347.

Hites R.A. (1981) Sources and fates of atmospheric polycyclic aromatic hydrocarbons, in "Atmospheric Aerosols. Source/air quality relationships". E.S. Macias and P.K. Hopke, eds., ACS Symp. Ser. 167, 187 - 196.

Ho W., Hidy G.M. and Govan R.M. (1974) Microwave measurements of the liquid water content of atmospheric aerosols. J. Appl. Meteorol. 13, 871 - 879.

Howe R.F. (1982) EPR spectroscopy in surface chemistry: Recent developments. Advan. Colloid Interface Sci. 18, 1 - 55,

Jaenicke R. and Schuetz L. (1978) Comprehensive study of physical and chemical properties of the surface aerosols in the Cape Verde Island region. J. Geophys. Res. 83, 3583 - 3599.

Jaenicke R. (1980) Atmospheric aerosols and global climate. J. Aerosol Sci. 11, 577 - 588.

Judeikis H.S., Stewart T.B. and Wren A.G. (1978) Laboratory studies of heterogeneous reactions of SO_2. Atmos. Environ. 12, 1633 - 1641.

Judeikis H.S., Siegel S., Stewart T.B., Hedgpeth H.R. and Wren A.G. (1979) Laboratory studies of heterogeneous reactions of oxides of nitrogen, in "Nitrogeneous Air Pollutants. Chemical and Biological Implications", D. Grosjean, ed., pp. 83 - 109, Ann Arbor Science Publ. Inc., Ann Arbor, Mich..

Kortuem G. and Braun W. (1960) Photochemische Reaktionen des Anthracens in adsorbiertem Zustand. Liebigs Ann. Chem. 632, 104 - 115.

Kotzias D., Nitz S. and Korte F. (1981) Lichtinduzierter Totalabbau organischer Molekuele adsorbiert and Kieselgel. Chemosphere 12, 415 -427.

Lane D.A. and Katz M. (1977) The photomodification of benzo(a)pyrene, benzo(b)fluoranthene, and benzo(k)fluoranthene under simulated atmospheric conditions. Advan. Environ. Sci. Technol. 9, 137 - 154.

Linton R.W., Williams P., Evans Jr., C.A. and Natusch D.F.S. (1977) Determination of the surface predominance of toxic elements in airborne particles by ion microprobe mass spectrometry and Auger electron spectrometry. Anal. Chem. 49, 1514 - 1521.

Lotz F., Nitz S. and Korte F. (1979) Photomineralisierung adsorbierter organischer Chemikalien im Mikromasstab. Chemosphere 10, 763 - 768.

Miguel A.H. and Friedlander S.K. (1978) Distribution of benzo(a)pyrene and coronene with respect to particle size in Pasadena aerosols in the submicron range. Atmos. Environ. 12, 2407 - 2413.

Natusch D.F.S. and Bauer C.F. (1978) Collection and analysis of trace elements in the atmosphere, in: Air Pollution Control, Part III, pp. 217 - 315, W. Strauss, ed., J. Wiley and Sons, New York - Chichester - Brisbane - Toronto.

Newman L. (1981) Atmospheric oxidation of sulfur dioxide: A review as viewed from power plant plumes and smelter plume studies. Atmos. Environ. 15, 2231 - 2239.

Nicholls C.H. and Leermakers P.A. (1971) Photochemical and spectroscopic properties of organic molecules in adsorbed or other perturbing polar environments. Advan. Photochem. 8, 315 - 336.

Oelkrug D., Plauschinat M. and Kessler R.W. (1979) Radiative and non-radiative transitions in aromatic molecules chemisorbed on γ -alumina. J. Luminesc. 18/19, 434 - 438.

Olufsen B. (1980) Polynuclear aromatic hydrocarbons in Norwegian drinking water resources, in "Polynuclear Aromatic Hydrocarbons: Chemistry and Biological Effects". A. Bjorseth and A.J. Dennis, eds., pp. 333 - 343, Battelle Press, Columbus, Ohio.

Ono Y., Takagiwa H. and Fukuzumi S. (1979) On the formation of O_2^- ion on calcium oxide. Z. phys. Chem. 115, 51 - 54.

Parlar H. and Korte F. (1981) Wie effizient ist der photoinduzierte Abbau organischer Umweltchemikalien in heterogener Phase? Chem. Ztg. 105, 127 - 134.

Particulate Polycyclic Organic Matter. National Academy of Science, Washington, D.C., 1972.

Peterson J.T. and Junge C.E. (1971) Source of particulate matter in the atmosphere, in "Man's Impact on Climate", edited by W. Mattews, W. Kellog and G.D. Robinson, 584 pp., MIT Press, Cambridge, Mass.

Peterson J.T. (1976) Calculated actinic fluxes (290 - 700 nm) for air pollution photochemistry applications. EPA Report 600/4-76-025, June 1976.

Pierce R.C. and Katz M. (1975) Dependence of polynuclear aromatic hydrocarbon content on size distribution of atmospheric aerosols. Environ. Sci. Technol. 9, 347 - 353.

Pierotti D., Rasmussen L.E. and Rasmussen R.A. (1978) The Sahara as a possible sink for trace gases. Geophys. Res. Letters 5, 1001 - 1004.

Reiter R., Sladkovic, R. and Poetzl K. (1978) Chemische Komponenten des Reinluftaerosols in Abhaengigkeit von Luftmassencharakter und meteorologischen Bedingungen. Ber. Bunsenges. Phys. Chem. 82, 1188 - 1193.

Rose W.I. and Hoffmann M.F. (1982) The May 18, 1980, Eruption of Mount St. Helens: The nature of the eruption, with an atmospheric perspective, in: Atmospheric Effects and Potential Climatic Impact of the 1980 Eruptions of Mount St. Helens. NASA Conf. Publ. 2240, 1 - 10.

Rowland F.S. and Molina M.J. (1975) Chlorofluoromethanes in the environment. Rev. Geophys. Space Phys. 13, 1 - 35.

Sancier K.M. and Wise H. (1981) Photoassisted oxidation of organic material catalyzed by sand. Atmos. Environ. 15, 639 - 640.

Schuetz L. and Jaenicke R. (1974) Particle number and mass distribution above 10^{-4} cm radius in sand and aerosol of the Sahara desert. J. Appl. Meteorol. 13, 863 - 870.

Schuetz K., Junge C., Beck R. and Albrecht B. (1970) Studies of atmospheric N_2O. J. Geophys. Res. 75, 2230 - 2246.

Schure M.R. and Natusch D.F.S. (1982) The effect of temperature on the association of POM with airborne particles, in "Polynuclear Aromatic Hydrocarbons: Physical and Biochemical Chemistry". M. Cooke, A.J. Dennis and G.L. Fisher, eds., pp. 713 - 724, Battelle Press, Columbus (Ohio) and Springer-Verlag, New York - Heidelberg - Berlin.

Seinfeld J.H. (1975) Air Pollution. Physical and chemical fundamentals. McGraw-Hill, Inc.

Stinespring C.D. and Stewart G.W. (1981) Surface enrichment of aluminosilicate minerals and coal combustion ash particles. Atmos. Environ. 15, 307 - 313.

Suess M.J. (1976) The environmental load and cycle of polynuclear aromatic hydrocarbons. Sci. Total Environ. 6, 239 - 250.

Vandegrift A.E., Shannon L.J. and Gonnan P.G. (1973) Controlling fine particles. Chem. Eng. 80, No. 14, 107 - 114.

Walling C. (1975) Fenton's reagent revisited. Acc. Chem. Res. 8, 125 - 131.

Whitby K.T., Husar R.B. and Liu B.X.H. (1972) The aerosol size distribution of Los Angeles smog. J. Colloid Interface Sci. 39, 177-204.

Part V

THE ROLE OF MULTIPHASE ATMOSPHERIC CHEMISTRY IN SOURCE-RECEPTOR CONSIDERATIONS

ACID RAIN IN NORTH AMERICA: CONCEPTS AND STRATEGIES

Volker A. Mohnen
Atmosperic Sciences Research Center
SUNYA, Albany, NY

Jack Calvert
National Center for Atmospheric Research
Boulder, CO

Jerre Wilson
United States Military Academy
West Point, NY
U.S.A.

ABSTRACT

As theoretical and observational studies of precipitation chemistry continue to expand, so does our understanding of the relevant facts governing this phenomenon. This paper addresses our current understanding of key processes involved in acid deposition as they bear upon the development of an acid rain control strategy. Possible strategies and the impact of these courses of action are also discussed.

INTRODUCTION

Acid rain occurs at many places in North America, but exhibits a significant maxima in the northeastern United States and adjacent Canadian provinces. Figures 1 - 3 show the pH contour map and the precipitation weighted concentration of sulfate and nitrate ions derived from the NADP and CANSAP networks from 1979 through the end of 1981 (Semonin, personal communication, 1983).

NATO ASI Series, Vol. G6
Chemistry of Multiphase Atmospheric Systems
Edited by W. Jaeschke

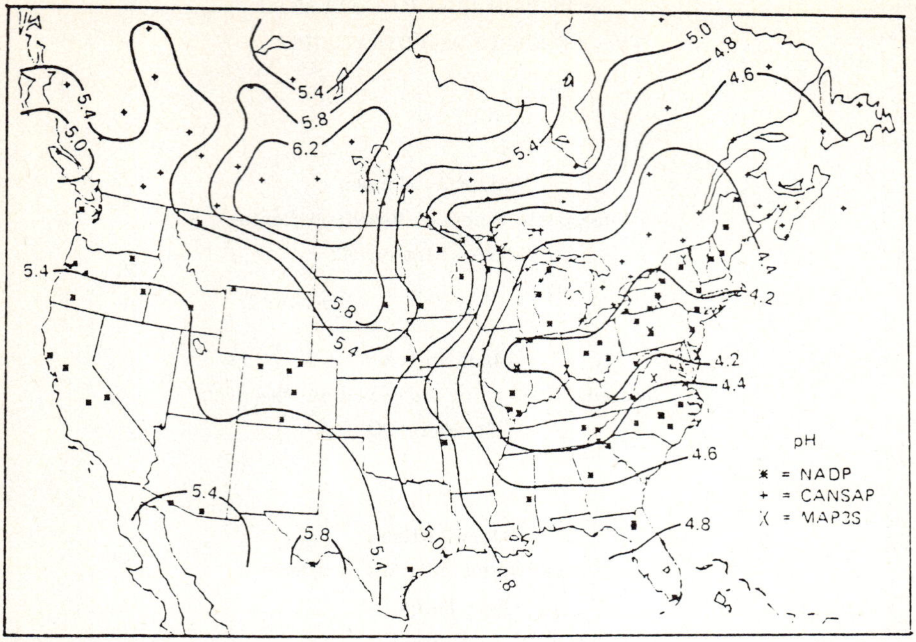

Fig. 1 pH contour map of North America. (Semonin, personal communication, 1983)

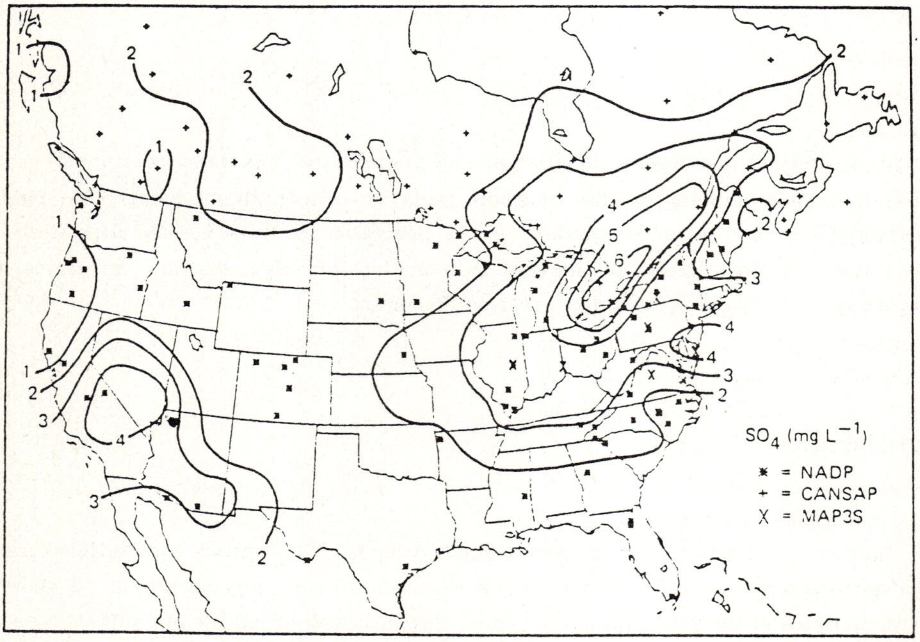

Fig. 2 Precipitation weighted sulfate concentration contour map for North America.

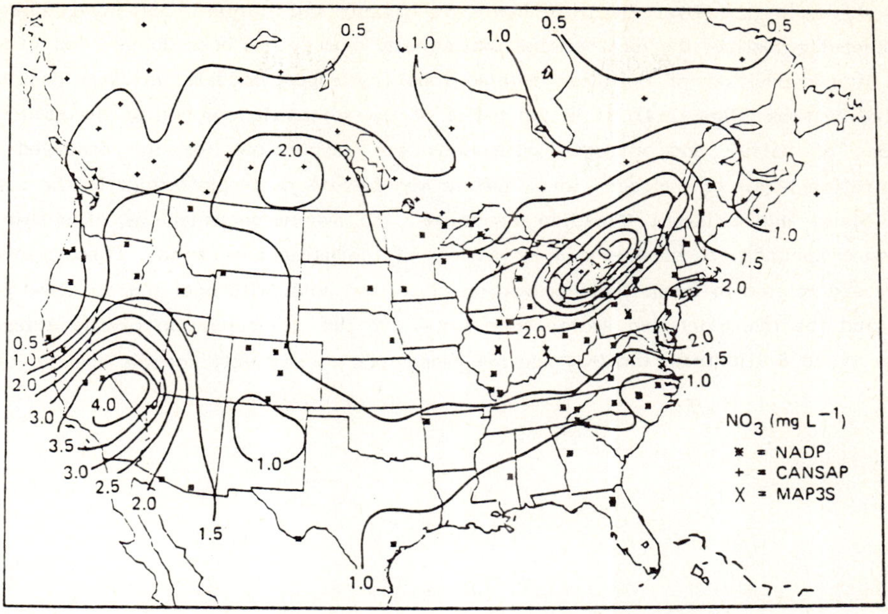

Fig. 3 Precipitation weighted nitrade concentration contour map for North America.

There are large regions in North America containing lakes that are potentially sensitive, based on bedrock geology, to acidification by acid precipitation (Fig. 4, from Galloway, personal communication). The places marked by a star indicate lakes located within such sensitive areas and with clearly demonstrated "damages". The solid circles also are locators for lakes within sensitive regions but for which no "damage" was found. The main difference between these two aquatic systems sensitive to acidic deposition is that one set of lakes received precipitation with pH levels well below 4.6 while the precipitation acidity received by the other lakes was well above the 4.6 level.

Identification of aquatic systems sensitive to acidic deposition should take into account not only the bedrock geology, but

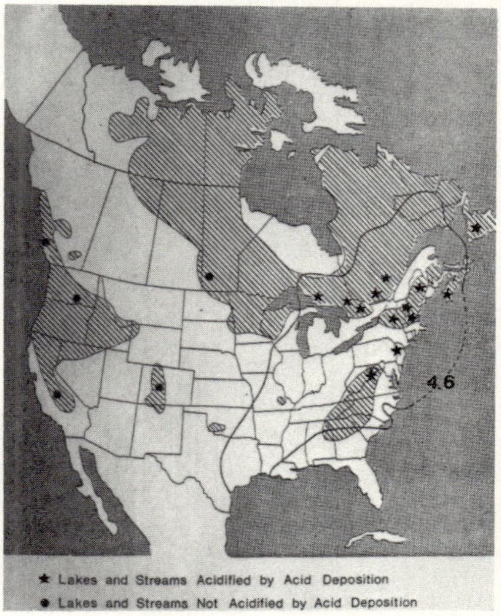

Fig. 4 Sensitivity map of North America based on bedrock geology.

include the vegetation cover, the litter layer, the organic and inorganic soil layer. However, as a general guideline, the fact remains that aquatic damage has been documented in areas where high deposition of pollution related ions (hydrogen, sulfate, nitrate) occurs on sensitive systems (Figs. 1-4). It is the belief of the scientific community that man-made emissions of sulfur and nitrogen compounds are responsible for the observed high concentration levels of hydrogen ions, sulfate and nitrate in precipitation. In the eastern United States and adjacent Canadian provinces, the man-made emissions of sulfur and nitrogen compounds are in excess of 90 percent of the natural emissions. These man-made sources are related to fossil fuel combustion of oil and coal, with coal and oil-fired power plants and the transportation sector accounting for the major portion of the emissions. Figures 5 and 6 illustrate the SO_2 and NO_x emissions respectively for the eastern United States.

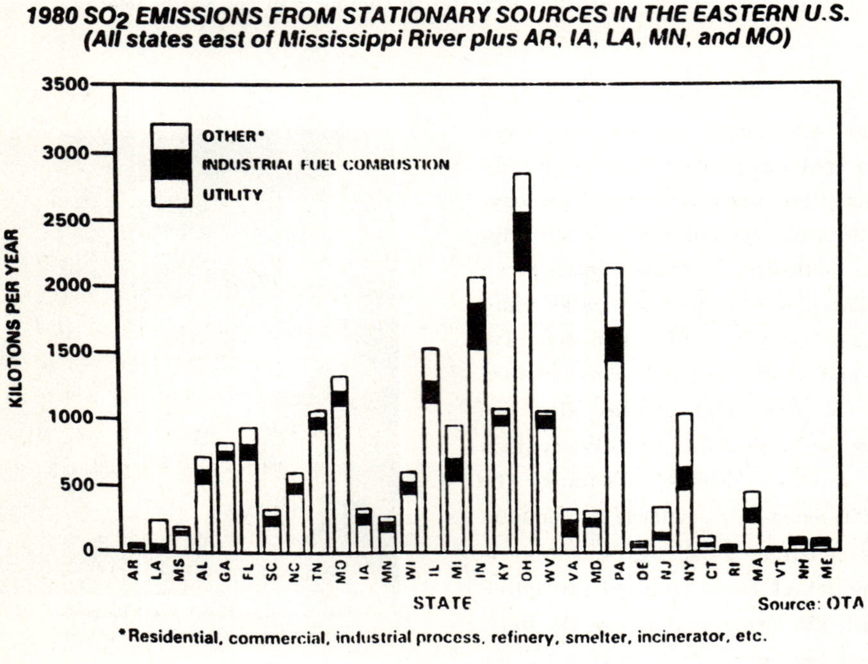

Fig. 5

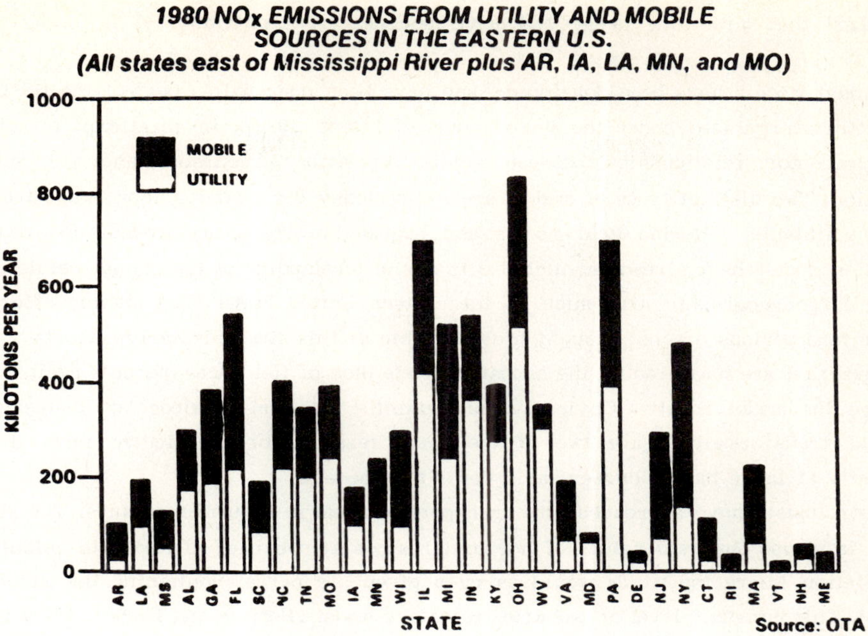

Fig. 6

TRANSPORT

One of the basic observations about acid rain is the approximate coincidence of the region of highest deposition with the areas of greatest sulfur dioxide and nitrogen oxide emissions. This circumstance represents some of the strongest evidence linking acid deposition with man-made emissions of SO_X and NO_X. These two pollutants deposit directly on the surface of the earth as dry acid deposition and are transformed in the atmosphere, at least partially, to sulfuric and nitric acid, which deposit in both dry and wet form. Considerable efforts have been underway to quantify the role of all of the precursors in the formation and deposition of acidic compounds. The speed and direction of the winds at the level of emission release are the major influence in the movement of these pollutants. How far they are moved depends upon whether the given pollutants are gases or particles, how they are transported vertically during their lifetime, what chemical reactions they undergo, and how efficient they are removed (scavenged) from the air. The consensus in the scientific community is that the long range transport of pollutants leads to acid deposition. It is also recognized that both long range and local sources may contribute to the acidity at a receptor.

Regional models have been developed--and have been operated as part of the US-Canada joint efforts organized under the Memorandum of Intent (1980)--in an attempt to establish source-receptor relationships between acidic deposition precursors and wet and dry deposition. Two distinct types of models are in use today for studying long range transport of air pollutants: Eulerian grid models and Lagrangian trajectory models. Existing acid deposition models have already contributed to our understanding of the annual wet deposition over a large geographic area such as the eastern United States. But similar efforts for shorter time periods are not thought to be feasible at this time. Similarly, efforts to model dry deposition are hampered by the almost complete lack of field measurements against which to judge the model results. Obviously, uncertainties in our knowledge of transport and chemical transformations are two of the basic reasons for the limited faith that the community at large has in current acid deposition models.

Meteorologists have approached the transport problem in a number of ways. The simplest method is to use observed values of horizontal winds at specified altitudes to calculate by interpolation where the winds would carry a given air parcel containing the material of interest. This constant-level or isobaric trajectory model (Heffter and Ferber, 1977) ignores the vertical component of the wind, so the method does not provide true trajectories in many circumstances, such as near frontal zones or over mountain ranges. On the basis of thermodynamic arguments, it is expected that vertical motions of air parcels should adhere rather closely to constant-entropy surfaces in the atmosphere, and from this isentropic trajectory models have evolved. However, the isentropic model still does not account for vertical motions in clouds or when isentropic surfaces intersect the terrain. Although curved-trajectory simulations can produce rather reliable results in simple meteorological situations, they are frought with uncertainty when conditions become complex, such as near frontal systems. For example compared with the interior of the United States, a much larger portion of the average annual precipitation in the northeast is associated with cyclonic disturbances. Thus the areas of eastern North America regarded as being sensitive to acid deposition receive precipitation from storm systems that are difficult to "model" and with origins at almost any latitude in the interior or the west. In other words, it is precisely in storm environments that it is most critical to trace pollutants with accuracy, and in those situations, current models are weakest.

Verification of any transport model is critical to its acceptance both as a description of scientific understanding and as a tool for analyzing policy choices. Methods include constant altitude balloons (Pack, et al., 1978) and chemical tracers. In the past, the first method did not yield conclusive answers over longer transport distances, and the latter method has only very recently been successfully employed in the field of long range transport (Captex 1983). While techniques for computing trajectories have advanced markedly over the past decade, they still are plagued by a host of additional uncertainties: most notably are the sparsity of meteorological data, both in time (upper-air soundings are made only twice daily) and space (stations are widely separated in the U.S.). Models that are built on basic physical and chemical processes and that can test hypotheses and guide the design and assessment of field measurement programs with the eventual goal of predicting acid deposition rates and source-receptor relationships and of providing reliable estimates of the effects of emission control strategies are desired (NCAR: Regional Acid Deposition: Design and Management Plan for a comprehensive modeling system, 1983).

A "Credible" model should consider as a minimum the following transport and mixing parameters:

HORIZONTAL MOTIONS

- Diffusion
- Windshear
  - nocturnal jets
  - frontal systems
- Orographic effects
- Mixing layer depth
- Above boundary layer

VERTICAL MOTIONS

- Buoyant bubbles
- Cloud venting
  - updrafts & downdrafts
  - injection into free troposphere or stratosphere
- Orographic effects

It is not possible at this time to unequivocally associate an uncertainty value to current acid deposition models resulting from their simplified treatment of transport processes.

Comparison of the results of eight different models have been made in predicting the

Table I. Comparison of eight different model predictions for percentage of sulfate derived from different State emissions which are deposited at an Adirondack receptor[a]

Source State	Percentage of sulfate derived from State(s)	
	Mean	Range
PA	18.3	13.2-23.6
OH	18.6	14.8-26.7
NY	14.1	9.3-24.2
IN	8.0	5.1- 9.5
WV	6.4	4.9- 8.9
MD/DE/NJ/DC	4.1	2.9- 6.4
MA/CT/RI	3.2	1.3- 5.8
MI	6.6	3.9- 9.0
IL	4.5	2.3- 6.5
KY	3.2	1.7- 4.4
Other[b]	14.4	5.5-32.4

[a]Table is based on the data summarized by Fay (1983).

[b]Other 31 states east of the Mississippi River.

percentage of sulfate derived from different state emissions which are deposited at an Adirondack receptor (Fay, 1983). The results of this comparison are given in Table I. Emissions from the Canadian sources are not included in these considerations. Wide ranges in estimates, usually over a factor of two or three in magnitude, are evident; this does not lead to great confidence in the models. Furthermore the range of values exhibited in Table I cannot reflect the absolute error in each of these models which are of uncertain quality. However qualitative judgments with these models can be and have been made in assessing potential source areas which might impact sensitive receptor regions. For example, although the error limits are very large, the estimates of Table I suggest clearly that a large fraction of the total anthropogenic sulfate originating in the United States and deposited in the Adirondacks, is derived from only a few of the eastern States.

However, a healthy degree of skepticism should be applied regarding the usefulness of these current models in accurately predicting acid depositions over eastern North America.

In particular those models are considered less reliable when the time averaging period is shortened to an event by even basis or when the geographic region of interest (receptor) is small.

The deterministic models that are available by necessity employ approximations to the atmospheric transformation processes that are hypothesized to be important to acid deposition. Generally, the chemical transformations are treated parametrically in all existing models reflecting a lack of knowledge of specific mechanisms. For example, the models selected by the U.S./Canada Work Group #2 use characteristics to assess acid deposition as outlined in Table II (Source: From NAS, 1983).

It becomes apparent that none of the currently available models contains an aqueous phase submodule capable of realistically treating the SO_2 transformations. Furthermore, at this time none of the models contain any detailed nitrogen chemistry.

TRANSFORMATION

Considerable insight has been obtained over the past years into the chemical transformations that occur during transport. For example it is now known that certain cloud and precipitation elements play a major role in the overall conversion of sulfur and nitrogen compounds to sulfates and nitrates. A comparison of the theoretical rates of the various pathways for SO_2 and NO_x conversion to acids has been made in our accompanying article in this volume (Calvert et al., 1984), and we not review all of the evidence for the important reactants here. We will discuss the effects of this chemistry on the nature of the source-receptor relationships in this section. Of course both gas phase and liquid phase reactions result in acid generation in the troposphere. The HO-radical reactions with SO_2 and NO_2 are important gas phase processes which in theory provide rates of conversion of SO_2 conversion of about 16% per 24 h period in summer and about 3% per 24 h period in winter during cloud free conditions. NO_2 conversion rates by the gas phase processes are very much higher (factor of ten or more). The liquid phase (cloud) chemistry which leads to SO_2 (HSO_3^-) oxidation by H_2O_2, O_3, iron and manganese ion-catalysis, etc., can be much faster, 100% h^{-1} or greater, for conditions encountered in the cloudy troposphere.

It is not now possible to judge accurately the contribution of gas and liquid phase processes to the total acid deposition which occurs in sensitive regions. The relative importance varies, of course, depending on a variety of meteorological conditions such as the extent of cloud cover, relative humidity, presence and concentrations of various pollutants, intensity of the solar radiation, and the amount of precipitation.

Most researchers who have analyzed regional air quality data have assumed a simple linear mechanism for transformation of SO_2 into sulfate; that is, the rate of sulfuric acid (or sulfate) generation is directly proportional to the SO_2 gas phase concentration,

TABLE II Characteristics of Several Regional-Scale Air Pollution Models Used to Assess Acid Deposition

Characteristic	AES[a]	ASTRAP[a]	CAPITA[a]	ENAMAP-1[a]	MEP[a]	MOE[a]
Type	Lagrangian-box	Statistical-trajectory	Monte Carlo	Puff-trajectory	Lagrangian	Statistical
Output	Monthly SO_2 and $SO_4^=$ concentrations and dry and wet S depositions	Monthly SO_2 and $SO_4^=$ concentrations and dry depositions and bulk S wet deposition	Monthly SO_2 and $SO_4^=$ concentrations and dry and wet depositions	Monthly SO_2 and $SO_4^=$ concentrations and dry and wet depositions	Monthly concentration and dry and wet depositions of sulfur	Long-term SO_2 and $SO_4^=$ concentrations; annual dry and wet sulfur depositions
Input	Annual and seasonal SO_2 emissions; daily precipitation amounts; winds and temperatures twice daily at four heights	Annual and seasonal SO_2 emissions for six-layered grid; stack parameters for major sources; 6-h precipitation amounts; wind profiles twice daily	Annual SO_2 emissions; 6-h precipitation probabilities; twice daily upper-air wind profiles and three times daily surface winds	Annual SO_2 emissions; 3-h precipitation amounts; wind profiles twice daily	Annual and seasonal SO_2 emissions; 3-h precipitation amounts; 6-h surface pressures	Point sources; area sources treated as effective point sources; statistics of durations of wet and dry periods and average precipitation rate during wet periods; statistical treatment of winds
Number of cells in the grid	52 x 37	User specified	52 x 37	46 x 41	User specified	User specified
Grid size (km)	127 x 127	Receptor point locations	127 x 127	70 x 70	Point receptors	Point receptors
Analysis of precipitation (by preprocessor)	Objective analysis of daily precipitation amounts	Hourly data are summed to produce 6-h totals across a grid of about a 76-km spacing	No actual precipitation rates used; time averages of precipitation probabilities used for each grid square	Hourly U.S. and 6-h Canadian data are summed to produce 3-h totals	Objective analysis of 3-h amounts	Climatological lenghts of Eulerian and Lagrangian wet and dry periods; rate of 1 mm/h
Mixed layer	Monthly climatological heights	Diurnal pattern in cluding nocturnal surface-based inversion; maximum: 1000 (w) and 1800 (s)	Day: 800 (winter); 1200 (spring/fall); 1350 (summer); Night: 300	1150 (winter); 1300 (spring/fall); 1450 (summer)	Varies diurnally over model domain	1000
Oxidation rate for SO_2 (percent/h)	1.0	Varies diurnally: 0.2-5.5 (summer); avg. of 2.0; 0.1-1.5 (winter); avg. of 0.5	0.6 (winter); 1.2 (summer)	1.0	Seasonal and diurnal variation: mean of 1.0	1.0
Dry deposition velocity (cm/sec)						
SO_2	0.5	Vary diurnally: 0.45 (summer avg.); 0.25 (winter avg.) (SO_2 and SO_4^{2-} similar but not identical)	0.31 (winter); 1.20 (summer)	0.38 (winter); 0.48 (summer)	Seasonal and diurnal variation: mean of 0.75	0.5
$SO_4^=$	0.1		0.07 (winter); 0.15 (summer)	0.22 (winter); 0.28 (summer)	0.25	0.05
Wet removal rate (percent/h)						
SO_2	$3x10^6P_{24}(t)/H$	- -	0.66 PP (winter); 11.7 PP (summer)	28.0 $P_1(t)$	Dependent on pH and temperature	10.8
$SO_4^=$	$85x10^6P_{24}(t)/H$ where H is the mixed layer (mm)	- -	5.0 PP (winter); 29.0 PP (summer) where PP = probability of precipitation (%) each 6-h period	7.0 $P_1(t)$	Precipitation rate dependent: mean of 30 $P_1(t)^{0.5}$	36.0 (used in stochastic scavenging model with τ_d=56 h and τ_w=7 h (the avg. dry- and wet-period durations)
Bulk S	- -	Minimum of $100(P_6(t)/10)^{1/2}$ and 80	- -	- -	- -	- -

P_x (t)=liquid precipitation rate (mm/xh)

TABLE II (Continued, Part 2)

Characteristic	RCDM-3[a]	UMACID[a]	ELSTAR[b]	RTM-II[c]	STEM-I[d]	STEM-II[e]	SURADS[f]
Type	Analytical	Puff-trajectory	Lagrangian trajectory	Hybrid Lagrangian/ Eulerian	Time-dependent Eulerian	Time-dependent Eulerian	Eulerian
Output	Monthly concentration and dry and wet depositions of sulfur	Estimates of source contributions to downwind concentrations and contributions of upwind sources on receptors at 6-h time steps	Hourly concentrations of photochemical products and intermediates; dry deposition; budgets for SO_x, NO_x	3-h, daily, mounthly concentrations of SO_2, $SO_4^{=}$; wet and dry sulfur deposition	Hourly and daily SO_2 and $SO_4^{=}$ concentrations; instantaneous and accumulated SO_2 and $SO_4^{=}$ deposition; daily regional sulfur budget	Hourly and daily species concentrations; instantaneous and accumulated deposition rates	Hourly and daily SO_2, $SO_4^{=}$ concentrations; dry deposition rates; daily SO_2 budget
Input	Emissions and centroids of emissions for area sources; spatially averaged annual and seasonal wet and dry periods; monthly, seasonal, and annual resultant winds and persistence factors	Annual SO_2 emission rates; 3-h precipitations amounts; wind profiles twice daily	Emissions of SO_x, NO_x, HC; winds extrapolated hourly in all layers from surface conditions, temperature profiles	Daily or hourly emissions; meteorological data	Surface and upper-air winds; cloud cover and ceiling, surface roughness, evaporation rates, vertical temperature and humidity profiles	Surface and upper-air winds; cloud cover and ceiling, rainfall rate, surface roughness, evaporation rates, vertical temperature and humidity profiles	Initial SO_2 and sulfate updated hourly; hourly SO_2 emissions; estimates of NO_x and HC emissions; radar-based precipitation; 12-h synoptic wind field aloft at fine levels; hourly surface winds; temperature profile, sunlight, cloud cover
Number of cells in the grid	70 x 70	41 x 32	Variable: 100 x 100	52 x 46 x 2 (episodic); 26 x 23 x 2 (long term)	User specified	User specified	Variable: e.g., 20 - 23
Grid size (km)	80 x 80	80 x 80	5 x 5	40 x 40 (episodic); 80 x 80 (long term)	80 x 80	Regional scale	80 x 80
Analysis of precipitation (by preprocessor)	Spatially averaged precipitation amounts and the average durations of wet and dry periods	Hourly data summed for 3-h periods for each 80-km grid square	Not included	Objective analysis on hourly data to generate grid-by-grid rates	- -	- -	Cloud chemistry and scavenging algorithm in fine layers
Mixed layer	600 (winter); 1200 (summer); 1000 (annual)	Varies only with month	4 or 5 layers as function of maximum mixing height on test day	Linear interpolation of daily minimum and maximum; spatial interpolation by spline fit	Calculated, varies diurnally	Calculated, varies diurnally	Five layers to 1500 m; preprocessors calculate mass conservative winds by layer and estimate diurnally variable mixing in each layer
Oxidation rate for SO_2 (percent/h)	Chemical conversion time scale 2.4×10^5 sec (1 percent/h)	Winter: 1.4 (day); Summer 2.8 (day); 0.2 (night)	Variable; tied to photochemical model	Function of solar zenith angle and geographic location; northeastern U.S. in summer, varies from 0.1 to 2.0 with 0.78 avg.; northern Great Plains, 0.3 avg.	- -	- -	Constant (0.5 to 3.0) or diurnally variable as calcuted

TABLE II (Continued)

Characteristic	RCDM-3[a]	UMACID[a]	ELSTAR[b]	RTM-II[c]	STEM-I[d]	STEM-II[e]	SURADS[f]
Dry deposition velocity (cm/sec)	Weighted by the percent of dry time	Varies according to time after sunrise and land use					
SO_2	0.50	0.10 - 0.55 (winter); 0.10 - 0.82 (summer)	Varies with aerodynamic resistance	One-dimension diffusional model depends on land use and time of day	Calculated based on surface type, evaporation, and meteorological conditions	Calculated based on surface type, evaporation, and meteorological conditions	Diurnally variable; calculated
$SO_4^=$	0.05	0.05 - 0.28 (winter); 0.03 - 0.43 (summer)	0.55 times SO_2 rate				0.1 to 1.0 times SO_2 rate
Wet removal rate (percent/h)							
SO_2	- -	$5.0 \times 10^4 P_1(t)/H$	Not included	Depends on precipitation rate and cloud type	User-specified subroutine to calculate zeroth and first-order removal rate constants	Calculated based on liquid-water content of clouds, temperature, pH, rainfall intensity, drop size and number distribution, and chemical and physical properties of the absorbed species	Not included
$SO_4^=$	- -	$2.2 \times 10^5 P_1(t)^{-0.625}/H$ where H = mixing height (mm)					
Bulk S	$\wedge P_t \frac{T_d + T_w}{T_w}$, where T_d and T_w are the average duration of the dry and wet periods (h), respectively, and $\wedge$ is the scavenging coefficient equal to 0.34	- -					

$P_w(t)$ = liquid precipitation rate (mm/xh)

a U.S./Canada Work Group #2 (1982),
b Lloyd et al. (1979),
c Durran et al. (1979); Liu et al. (1982); and Stewart et al. (1983),
d Carmichael and Peters (1983),
e Carmichael et al. (1983),
f Mueller and Hidy (1982b),

Source: NAS, 1983

unaffected by other impurity concentrations such as NO_x, hydrocarbons, etc. Some simplified models of acid generation and transport such as that of Rodhe et al. (1981) have included gas phase reactions involving the termination of $HO-HO_2$ and $HO-RO_2$-radical chains by SO_2 ($HO + SO_2 \longrightarrow H_2SO_4$). These models have shown in theory a significant nonproportional reduction of acid deposition at a given receptor site for a given reduction in SO_2 emissions from a given source. However recent studies of Stockwell and Calvert (1983) show that the SO_2 reaction with HO does not terminate chains, and modeling results based upon the new gas phase chemistry and a highly simplified cloud chemistry show a nearly proportional response of acid deposition reduction at a given receptor site to an SO_2 cutback at a given emission source (Calvert and Mohnen, 1983).

Nonproportional response of acid deposition to sulfur dioxide cutback can in theory result if there is an oxidant limitation. Results from the ASRC field station at Whiteface Mountain, New York, strongly suggest that during the winter months H_2O_2 levels in clouds are very low, sometimes below 1 ppb(wt) in cloud water. Measurements during the summer months always showed H_2O_2 to be present and at levels as high as 1000 ppb at times. Thus H_2O_2 oxidant limitation to SO_2 (HSO_3^-) oxidation in clouds and rain water is seemingly unimportant in summer when about 65% of the annual sulfate in precipitation is deposited. In winter oxidation by H_2O_2 seems unlikely and what does occur must proceed largerly through O_3 and other available oxidants for HSO_3^-. Since acid deposition on an annual basis is dominated by the summertime deposition, and no oxidant limitation is evident then, a near proportional response of the acid formed by liquid phase oxidation to cutbacks in SO_2 emissions may result. As a consequence, the chemistry of gas and aqueous phase sulfur dioxide oxidation may not induce significant non-linearity into the source-receptor relationships; this conclusion is in accord with that recently formulated from available evidence by the recent MAS Committee on Atmospheric Transport and Transformation in Acid Precipitation (NAS, 1983).

The more detailed knowledge acquired in recent months on atmospheric transport and chemical transformations and the anticipated results from field experiments being considered by the Interagency Task Force on Acid Precipitation will make possible Eulerian modeling approach for establishing a source-receptor relationship. In the Eulerian modeling approach, the non-linear equations of atmospheric motions and chemical reactions were solved directly on the entire domain under consideration, e.g., the United States and southern Canada east of the Rocky Mountains. The spatial resolution of the Eulerian model, both horizontally and vertically, is set to treat adequately the transport processes. The chemical mechanism, which may consist of up to 100 reactions, usually requires small incremental time steps and consumes the majority of the computer resources. Within the confines of reasonable costs, the Eulerian model can accomodate the simulation of all processes relevant to acid deposition on a regional scale.

The National Center for Atmospheric Research (NCAR), under Interagency Agreement with the U.S. Environmental Protection Agency, has initiated the construction of a Regional Acid Deposition Model (RADM) in June 1983. The model will have a domain scale of about 2500 km and horizontal grid resolution of about 50 km. There will be 10 to 15 vertical levels, and chemistry will include all of the chemical species important to acid generation in the troposphere. The time table for the development of RADM is of the order of 3-5 years which includes the development of transport and chemistry modules and the establishment of an appropriate field data base for model validation.

It is hoped that the NCAR scientists working with the cooperation of the faculties of the universities in the University Corporation for Atmospheric Research, the governing body of NCAR, the research teams of the National Laboratories (Argonne, Battelle Pacific Northwest, Brookhaven), and several private research groups funded by the U.S. Environmental Agency, will provide a new scientifically sound model for both research planning and control strategy development. Of course other capable groups are also developing improved models for long-range transport and transformation and acid deposition, and the prospects for the availability of reliable models in the years ahead seem good.

STRATEGY

In the meantime, however, public pressure is increasingly demanding some federal action to reduce the deposition of acidifying substance onto ecologically sensitive receptor areas. The NAS (1983) report on acid deposition expressed reservation and conservatively stated: "We do not believe it is practical at this time to rely upon currently available models to distinguish among alternative control strategies".

However, decisions on almost all issues of public concern routinely made in light of uncertainties in knowledge. To no surprise, therefore, 12 bills have been introduced in the United States Congress to address the issue of "acid rain". They are in the form of either research only and lakeliming bills (Byrd-Rahall) or major SO_2 reduction bills that approximate a 50% reduction (Mitchell/Strafford and Waxman/Sikorski). The research bills are supported by the Administration, most coal interests, the utility industry, the mining industry and the mine workers. The SO_2 rollback bills are supported by mainly northeastern Members of Congress, the government of Canada, and environmental organizations. Some form of control is also supported by some members of the National Governors Association, which is expected to issue recommendations in late October.

Canada argues that over 50 percent of her acid deposition damage is caused by United States and has proposed, during formal negotiations, that both countries reduce their SO_2 emissions by 50 percent. Recent activities of the ECE suggest growing European interest in the establishment of national control programs. At a recent meeting in Geneva, the ECE adopted a proposal calling for a substantial reduction in SO_2 emissions by ECE members.

During June, three major scientific reports on acid rain were released. They concluded that over 90 percent of acid rain in the Northeast and Canada comes from man-made sources

(National Acid Precipitation Task Force), that over a very broad region and over a long period of time a reduction in sulfur dioxide emissions will result in a proportional sulfur deposition reduction but that current information does not allow us to predict the effect of those reductions on sensitive receptors (National Academy of Sciences); and that some further reductions in emissions be undertaken now to avoid the potential for irreversible damage later (White House Science Advisor's Panel).

The fundamental question that is being decided by the United States Government at this time (November 1983) is whether or not further steps are warranted--for controlling SO_2 and NO_x emissions beyond those called for under the Clean Air Act--light of our present scientific understanding and available damage data. Options availabe are as follows:

1. Minimal change options

These options specially recognize that the science supporting the need for and scope of acid deposition controls is inadequate. They would recognize that the arbitrary regional SO_2 rollbacks contained in most congressional proposals are not based on sound science and are likely to be inefficient at solving the acid deposition problem.

A. Accelerated research and lake-liming

This approach would accelerate the acid rain research program under the National Acid Precipitation Task Force. Current plans call for assessments of current and anticipated future damage (1985), a methodology for evaluating costs and benefits of alternative control strategies (1987), and a final evaluation of optimal control strategies (1989). The results could enhance policy makers' ability to choose among control and mitigation options and greatly aid in efforts to effectively implement those decisions.

Option A can be taken separately or combined with one or all of the following approaches as a full package:

- Development of LIMB and other control technology which can be retrofitted to existing boilers and has the potential for 50 to 60 percent reduction of both NO_x and SO_2. This technology could be applied commercially by the mid-1990's.
- Authority to allow emissions trading between new and existing

sources.

The cost and stringency of new source performance standards are substantially greater than requirements on most existing sources. If new and existing sources were allowed to average their emissions under a "bubble", then the same emissions could be achieved at less cost initially. However, in the long run, emissions will not decline unless these new sources are forced to install more stringent controls when existing sources retire. Thus those cost savings will decline over time. The benefits would be an acceleration of emissions reduction and a reduction in the overall cost of SO_2 reductions prior to year 2010.

- State-wide cap on SO_2 emissions (does not apply to predominately gas-fired states).

Sulfur dioxide emissions in some eastern and midwestern states are projected to increase by one million tons from 1980 levels prior to the year 2000, with utility sources contributing to most of this increase. A cap could prevent these emission increases. A cap on state-level increases on SO_2 emissions would impose significantly lower annualized costs than other emission reduction options. However, individual sources and states might find a cap unduly burdensome if they anticipate significant growth.

B. Application of Current and augmented rule-making authorities

This option would provide EPA with additional legislative authority to:

- Improve the Clean Air Act's provisions for controlling interstate air pollution and
- Set regionally-variable secondary air quality standards, as opposed to uniform standards, for acid deposition and precursor pollutants and improve the ability to enforce the attainment of all secondary standards.

The present Clean Air Act requires the Administrator to periodically review, and where appropriate, revise existing national air quality standards and standards for new sources (NSPS). Revision of NSPS over time is virtually certain. Such changes could cause total emissions to decline more rapidly than present long-term projections which assume no increased stringency in the NSPS for major SO_2 emitters.

Use of this augmented Clean Air Act authority could produce substantial SO_2

reductions. This would use the State Implementation Plan process which allows states full authority on determining how sources will meet the required emission levels.

2. Major emissions reductions options

These options call for a major rollback in SO_2 emissions. Proponents of these approaches believe that in spite of the scientific uncertainty, the weight of the evidence suggests that action be taken now to control the dominant source of acid deposition--SO_2 emissions. Option C focuses on the near-term need to protect aquatic resources in the Northeast while learning more from accelerated research before deciding to expand a control program to other parts of the United States. Option D accepts the notion of a need to control SO_2 emissions on a broad regional scale to address concerns about both known damages in the Northeast and potential damages in other parts of the country.

Either of these options could prove to be very costly and could have significant impacts on high sulfur coal miners and utility rate payers in the affected states.

Financing schemes or subsidies to mitigate these impacts could take the form of a generation tax on fossil fuel electricity production, an emissions fee on each ton of SO_2 produced, transferable emissions permits, or direct federal budget outlays. Furthermore, prohibitions against fuel switching while protecting high sulfur coal miners would increase the cost of these options.

C. Experimental emission reduction option

This option would require by 1990 a reduction of three million tons of SO_2 below 1980 levels in a 13 state targeted area which includes New England, New York, Pennsylvania, Ohio, West Virginia, Maryland, Delaware and New Jersey. This option was designed to reduce acid deposition in the most sensitive regions, the Adirondacks and New England while providing an opportunity for further research before controls are imposed on more areas.

Emissions reductions would be allocated to the 13 states based on a formula which requires reductions proportional to the average emissions rate of each state in 1980. SO_2 emissions in the other 18 states, east of the Mississippi River, and outside the 13 stage region, would be capped at 1980 levels (allowing some growth in areas that are predominantly gas-fired).

These plans require a 25 percent reduction in the emissions of states in the 1000 km zone of influence of the Algoma-Muskoka region of Canada. As such, Canada should reduce SO_2 emissions in those areas of Canada within 1000 km of the Adirondacks by a comparable amount.

D. Large scale regional rollback over either: a) 31 states (Stafford/Mitchell), or b) 48 states (Waxman/Sikorski)

Option D-a would mandate an SO_2 emissions reduction target of 8 million tons below 1980 levels in 31 states by 1983-1995. Owners seeking to build new plants after 1995 would have to secure offsets. Each state would be allocated an emissions reduction through a formula that attempts to stimulate a least cost approach. Stafford/Mitchell use a formula based on a rate per Btu of fossil fuel consumed, as discussed in Option C above.

Option D-b would mandate a reduction of 14-million tons (up to 10-million tons of SO_2 from existing power plants with nor requirement to other growth from new sources and 4-million tons NO_x) over 48 states, with control technologies funded by a 1 mil/kilowatt hour fee on electricity generation in the 48 contiguous states.

SUMMARY

From a practical point of view there appears to be little doubt that the Adirondack Mountains and the northeastern U.S. in general have suffered the most from the deposition of acidifying substances. Figure 7 shows the annual precipitation weighted sulfate concentration ranging from 2.1 to 2.6 mg/liter for 1980. The annual amount of precipitation in this region is of the order of 110 cm which then yields average deposition values between 23 and 28 kg/ha.yr. Deposition values below 20 kg/ha.yr are desirable in order to avoid environmental degradation which would mandate an average deposition reduction in this region of the order of 25% below current levels.

An experimental program as described above under option C would avoid the doubtful task of establishing with some degree of certainty a source-receptor relationship before any emission reduction program can be initiated. Such a policy, if implemented, would allow research to catch up with the desire of policy makers and the general public at large to efficiently and swiftly reduce the deposition of acidifying substances in all ecologically sensitive regions, while keeping the overall economic impact of such steps at a minimum.

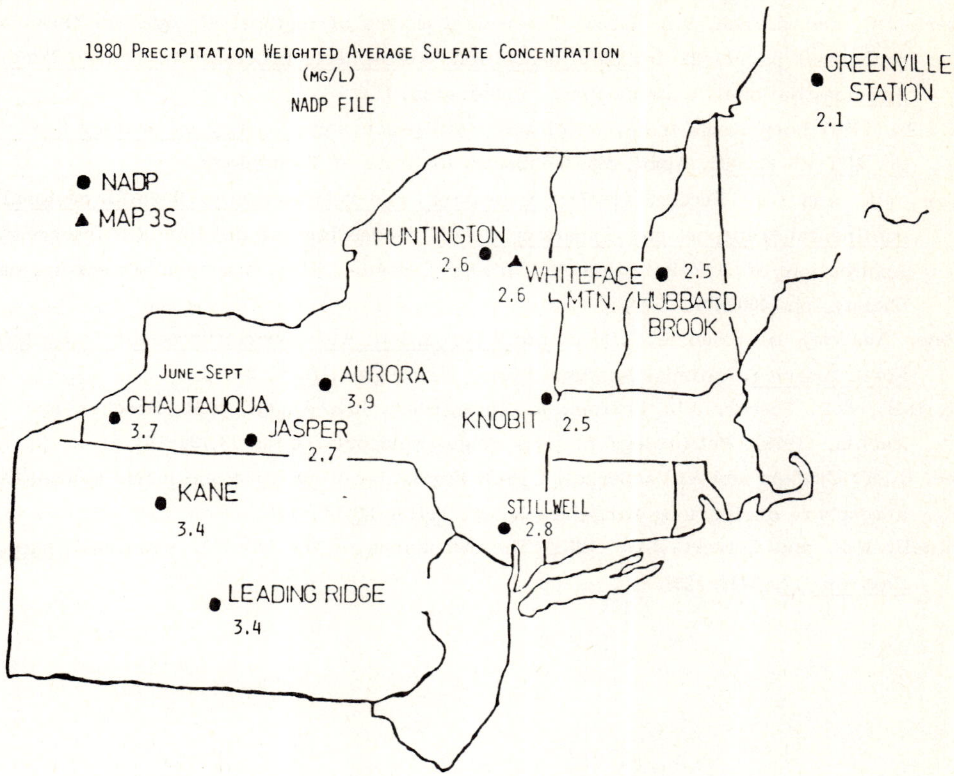

Fig. 7

REFERENCES

Calvert, J.G., Heikes, J.G., Stockwell, W.R., Mohnen, V.A., and Kerr, J.A. (1984) Some considerations of the important chemical processes in acid deposition, this volume.

Calvert, J.G. and Mohnen, V.A. (1983) Theoretical models of regional air quality, chemical processes, pp. 75-82 in: Acid deposition; Atmospheric processes in eastern North America, National Academy Press, Washington, D.C.

Fay, J.A. (1983) Long range transport of acid rain precursors, Energy Laboratory Report No. MIT-EL 83-005, April, Massachusetts Institute of Technology.

Heffter, J.L. and G.L. Ferber, (1977): Development and verification of the ARL regional-continental transport and dispersion model. Proceedings of the Joint Conference on Application of Air Pollution Meteorology, Boston, MA, American Meteorological Society, pp. 400-407.

National Academy of Sciences, (1983). Acid Deposition: Atmospheric Processes in Eastern North America. National Academy Press, Washington, D.C., 375 p.

Pack, D.H., G.F. Ferber, J.L. Heffter, K. Telegardad, J.K. Angell, W.H. Hoecker and L. Machta, (1978): Meteorology of long range transport. Tellus 33,132-141.

Rodhe, H., P. Crutzen and A. Vanderpol, (1981): Formation of sulfuric and nitric acid in the atmosphere during long range transport. Tellus 33,132-141.

Stockwell, W.R. and Calvert, J.G. (1983) The mechanism of the HO-SO_2 reaction. Atmos. Environ. 17, 2231-2237.

SOME CONSIDERATIONS OF THE IMPORTANT CHEMICAL PROCESSES IN ACID DEPOSITION

by Jack G. Calvert, Brian G. Heikes, and William R. Stockwell
ACAD Division, National Center for Atmospheric Research*
Boulder, CO, 80307
USA

Volker A. Mohnen
Atmospheric Science Research Center
State University New York, Albany

and

J. Alistair Kerr
Department of Chemistry
University of Birmingham, England

*The National Center for Atmospheric Research is sponsored by the National Science Foundation.

ABSTRACT:

The chemistry of the gas phase and that of the aqueous, cloud water, phase couple strongly to create many important pathways for acid generation in the troposphere. The theoretical rates of the alternative reactions forming sulfuric and nitric acids are estimated. The major gas phase routes to acids involve the HO-radical attack on SO_2 and NO_2. The gas phase generated species, HO, HO_2, NO_3, and N_2O_5 may be captured by and react in cloud droplets to provide important pathways for HSO_3^- oxidation to H_2SO_4, the NO_3 and N_2O_5 conversion to $HONO_2$, and the $HO_2(O_2^-)$ reactions to form H_2O_2, provided that the sticking coefficients for these reactive transients are greater than 10^{-3}. The molecules of O_3, H_2O_2, and $CH_3COO_2NO_2$, products of the gas phase reactions, when absorbed into cloud water along with SO_2, are important reactants for HSO_3^- oxidation to H_2SO_4. In general the rates of sulfuric acid generation in cloud water are very much larger than those for the gas phase;

NATO ASI Series, Vol. G6
Chemistry of Multiphase Atmospheric Systems
Edited by W. Jaeschke

these differences are rationalized in terms of the chemical and physical properties of these systems. The mechanisms and the rates of the gas phase, aqueous phase, and solid-photosensitized formation of the important HSO_3^- oxidant, hydrogen peroxide, are compared. Some of the major uncertainties in the chemistry of the acid development in the troposphere, important in the construction of acid rain models, are identified.

1. INTRODUCTION

Scientists now recognize that a large share of the acids deposited both in North America and in Europe originate from the oxidation of SO_2, NO, and NO_2 gases which are emitted largely from power plants, industrial operations, motor vehicles, and other fossil fuel combustion sources. Early signs of ecological damage have been observed in certain sensitive areas of the world which are deficient in soils with a good acid buffering capacity and which are recipients of a large input of acids through "acid rain" and/or dry deposition. Scientists throughout the world are actively working to assess the extent of damage which has occurred and that which can be expected in years ahead. Government leaders of many nations are attempting to evaluate alternative control strategies for acid deposition which can alleviate the existing and potential future problems.

The control of acid deposition is complicated by many factors which mirror our incomplete understanding of the subject. It is well established that the impurity acids and their precursor gases can be carried for many hundreds of kilometers through the natural motions of the tropospheric air masses. Current models of long-range air transport do not allow an accurate description of the extent to which emissions from a specific impurity source result in acid deposition at a given sensitive receptor site. Most models used today in control strategy development do not attempt to incorporate the extremely complex chemical and physical processes which control acid generation, transport, and deposition. They often employ average wind flows, fixed or diurnally varied rates of impurity conversion to acids and assumed rates of dry and wet deposition of acid precursors and acids. Such models are useful to the control strategists in deriving qualitative conclusions about the likely source regions which impact a sensitive area. However when such models are applied to estimate the acid deposition over large transport distances often involved in the acid deposition phenomenon in the eastern North America, the results can vary significantly from model to model (factors of 2-4 are not uncommon), and judgments based upon the results are of unpredictable quality. Uncertainties in air mass motions over long times and great distances account for a large part of the problem. However uncertainties also arise as a consequence of the sensitivity of such models to the assumed chemical conversion rates. Thus Golomb et al. (1983) found that among the four basic steps of the transformation process considered in his analysis (acid generation rates, dry deposition of the precursors, dry deposition of the product acids, and wet deposition of the acids), the rate of acid generation is the most important factor in determining rain acidity. They conclude that accurate modeling must include time- and space-variability of the rates of acid generation.

Models which provide a more quantitative description of the atmospheric processes which result in acid generation within the troposphere would appear to be valuable in the development of a scientifically sound and lasting strategy for the control of acid deposition in ecologically sensitive areas of the world. Several research groups, including ours at the National Center for Atmospheric Research, Boulder, Colorado, are now involved in the design of more scientifically accurate computer models of the many chemical and physical processes which result in acid generation. This task is extremely complex and demands a great variety of detailed inventory data as well as scientific information, some of which is not now available. For example, if we attempt to model acid development, transport, and deposition accurately in eastern North America, we must know with reasonable confidence the following information for the entire area east of the Mississippi River and with a spatial resolution of a few tens of kilometers:

a) The geographical distribution and composition of both the natural and the anthropogenic emissions of SO_2, NO_x, hydrocarbons, aldehydes, NH_3, and other acid neutralizing compounds, and other emissions which enter the chemical reactions which lead to acids or influence their generation;

b) the nature of the complex transport processes that distribute the emissions and their reaction products;

c) the mechanisms and the rates of the dry deposition of the initial impurity gases and the acids and other products of the chemical reactions which occur in the troposphere;

d) the nature and the rates of the various chemical pathways which lead to acid generation in the gas phase, the aqueous (cloud) phase, and on certain solid aerosols or aqueous suspensions of these aerosols in the troposphere;

e) the nature of the physical processes which occur in clouds and result in mixing and redistributing the pollutants in air masses, capturing impurity gases in the cloud water, promoting the rapid chemical production of acids, and the delivery of the acids, dissolved in the rain water, at the receptor site.

In our present considerations we will focus upon only one small aspect of this complex system, namely, the chemistry of the acid rain development. In particular we will attempt to identify those chemical processes which are most significant in acid formation and the quality of the information on these processes which is now at hand. In the following section 2 we will attempt to evaluate the significance of the potentially important sulfuric acid generating reactions which occur in the aqueous phase of clouds, rain water, and fogs. In section 3 we will examine the aqueous phase pathways to nitric acid formation. In section 4 we will review the current information on the rates of acid generation through gas phase chemical processes, and we will contrast these rates with those of the aqueous phase processes. The mechanism differences which occur with phase change (gas $\rightleftharpoons$ liquid) and the factors which favor the often higher rate of acid formation in the aqueous phase will be considered in section 5. In section 6 we will focus upon the significance of the mechanisms for the generation of the important oxidant, H_2O_2. The possible importance of other peroxy compounds as acid generators will be explored in section 7.

2. THE POTENTIALLY IMPORTANT AQUEOUS PHASE REACTIONS LEADING TO SULFURIC ACID FORMATION IN THE TROPOSPHERE

Various possible important oxidizing agents for SO_2 (HSO_3^-) in cloud and rain water have been identified in extensive research during the past few years. Recent evaluations of the chemistry of acid rain suggest that H_2O_2, O_3, and the Fe(III) and Mn(II) ion-catalyzed oxidation of SO_2 (HSO_3^-) are probably the most important species for conditions commonly encountered in the troposphere of eastern North America (Penkett et al., 1979; Martin, 1984; Hoffmann and Jacob, 1984; Schwartz, 1984; Brock and Durham, 1984; Calvert and Mohnen, 1983). Other processes which may be important for aqueous HSO_3^- oxidation are: a) gas-phase generated HO- and HO_2-radicals which diffuse to and are incorporated into cloud droplets (Chameides and Davis, 1982); b) the graphitic-carbon-catalyzed reaction (Chang et al., 1981); and c) the nitrite ion reaction (Oblath et al., 1981; Martin et al., 1981).

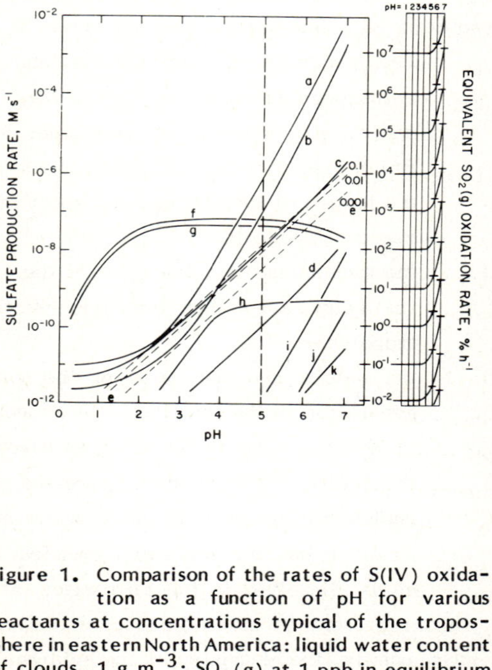

Figure 1. Comparison of the rates of S(IV) oxidation as a function of pH for various reactants at concentrations typical of the troposphere in eastern North America: liquid water content of clouds, 1 g m^{-3}; SO_2(g) at 1 ppb in equilibrium with cloud; curve
a, O_3 (-5°C); **b**, O_3 (15°C);
c, Fe^{3+} (10^{-5}M), Mn^{2+}(10^{-6}M)-catalyzed (15°C);
d, Fe^{3+} (10^{-7}M), Mn^{2+}(10^{-8}M)-catalyzed (15°C);
e, HO-radical captured from gas by cloud water with α = 0.1, 0.01, and 0.001;
f, H_2O_2 (-5°C); **g**, H_2O_2 (15°C);
h, graphitic carbon-catalyzed (15°C);
i, "uncatalyzed" oxidation by O_2 (15°C);
j, "uncatalyzed" oxidation by O_2 (-5°C);
k, nitrite ion (15°C).

We can compare in figure 1, adopted from the recent NCAR review (1983), the instantaneous rates of these reactions as estimated for typical impurity levels in a non-urban parcel of tropospheric air over eastern North America. For these considerations the gaseous SO_2 concentration is assumed to be 1 ppb, and this is in equilibrium with cloud water. The left-hand ordinate of this figure shows the aqueous sulfuric acid production rate in moles l^{-1} s^{-1}; this rate is independent of the liquid water content of the fog or cloud. The right-hand ordinate gives the equivalent gas phase SO_2 conversion rates (% h^{-1}) for a typical cloud with a liquid water content of 1 g m^{-3} of air; note the scale on this axis varies with pH of the solution for pH values greater than 5. For clouds or fogs with a lower liquid water content than that assumed here, a proportionally lower equivalent gas phase conversion rate can be estimated. The abscissa of the figure is the pH of the cloud water solution. Remember that pure water equilibrated with atmospheric CO_2 alone would have a pH of about 5.6. Charlson and Rodhe (1982) have pointed out that the pH of natural rain water might range from 4.5 to 5.6 due to the variability of the sulfur cycle alone, and in the absence of basic materials (such as NH_3 and $CaCO_3$), average pH values of about 5 are expected to occur in pristine locations. Thus the region of the figure 1 to the left of the vertical dashed line at pH = 5 is that of common interest in atmospheric considerations of acid rain development. Note that for most of the oxidants considered here, the rates are strongly pH dependent; for all but the reactant H_2O_2 the rates fall off sharply as the cloud water becomes more acidic (pH decreases); this results in large part from the decreased solubility of SO_2 with pH decrease.

First note in figure 1 that the uncatalyzed oxidation of S(IV) by oxygen alone is very slow; see curves i and j (for 15° and -5°C, respectively). We have used the data of Penkett et al. (1979) in estimating these rates, and they probably represent generous upper limits to the actual uncatalyzed rates. The results from the two temperatures span the range of values commonly encountered in cloud water. Even at the higher temperature for which the reaction is most rapid (curve i), less than 0.1% h^{-1} conversion rates are expected for cloud water with pH $\leq$ 5.6.

Theoretical rates of the iron- and manganese-catalyzed oxidation of SO_2 (HSO_3^-) are shown in curves c and d of figure 1. The synergestic effect of the combination of iron and manganese ions as reported by Martin (1984) is important in this case. The range of concentrations assumed here are: [Fe(III)] = 10^{-5}, [Mn(II)] = 10^{-6} M (curve c); [Fe(III)] = 10^{-7}, [Mn(II)] = 10^{-8} M (curve d); these are believed to be typical of those encountered in rainwater in eastern North America. At the highest concentrations chosen here, it can be seen that the rates of acid generation can be over 100% h^{-1} in cloud water at pH values $\geq$ 5. As the oxidation of HSO_3^- occurs in clouds and the acidity of the water climbs, the theoretical rate of acid generation falls; at pH values of about 5.0, 4.0, and 3.0, the rates are about 100%, 10%, and 1% h^{-1}, respectively. Hoffmann and Jacob (1984) found that at the much high levels of Fe and Mn ions observed in Los Angeles fog water (up to 200 and 10 μeq l^{-1}, respectively) the catalysis of HSO_3^- oxidation by these metal ions can dominate all other pathways of acid formation, especially during the nighttime hours.

The highest rates of HSO_3^- oxidation for the "typical" conditions picked for figure 1 are observed for the oxidant ozone in solutions at pH > 5; for our conditions the ozone in the cloud water is assumed to be at equilibrium with a typical level of gaseous ozone in the relatively clean troposphere, 50 ppb. The rates in this case were calculated from the rate constants of Maahs (1982); they are somewhat slower for 15°C (curve b) than for supercooled, liquid cloud water at -5°C (curve a). In this case the enhancement of the rate at the lower temperature arises from the increased Henry's law constant (increased solubility) for the reactants; this effect more than offsets the decrease in rate constant which occurs at the lower temperature. Again with the O_3 reactant the rate decreases as the cloud water becomes more acidic; for -5°C it is estimated to be: 10^4, 10^3, 10^2, 10, and 1% h^{-1} for pH values of about 5.1, 4.5, 4.0, 3.5, and 2.7, respectively.

The theoretical rates for the oxidant hydrogen peroxide, curves f (-5°C) and g (15°C) of figure 1, are also very high (several hundred percent per hour) and relatively insensitive to pH over the range of values encountered in cloud and rain water. In this case acid catalysis of the reaction compensates for the decrease in SO_2 solubility as the pH decreases. We have accepted the rate data of Martin (1984) for this system. In deriving figure 1 we have assumed that the aqueous concentration of H_2O_2 is 10^{-5} M, a value representative of ambient measurements in cloud and rain water in the non-urban atmosphere during the summer months. Theoretically the gas phase concentration of H_2O_2 which would have to exist in air free of liquid water in order to provide this cloud water concentration at equilibrium is only 0.3 ppb at 15°C. In section 6 we will see that for many conditions common to the troposphere, this quantity of H_2O_2 can develop readily in theory. It is clear that for our chosen conditions H_2O_2 can be the most important oxidizing agent for HSO_3^- in cloud water at pH values less than 4.5, a condition common to acid rain episodes.

Chameides and Davis (1982) have suggested a novel and seemingly important source of oxidant in cloud water. In theory a very significant contribution to HSO_3^- oxidation can arise from HO and HO_2-radicals formed in the gas phase reactions and transported to the cloud droplets; see the dashed curves labeled e in figure 1. The contribution of this process depends on the intensity of the solar radiation, the transmissivity of the cloud, and the properties of the air mass and cloud droplets. The values for the "standard model" of Chameides and Davis are assumed here: solar zenith angle, 30°; cloud transmissivity, 0.5; liquid water content of cloud, 1.5 g m^{-3}; temperature, 7°C, [O_3], 30 ppb, etc. The results shown are for an assumed 12 μm radius cloud droplet size and three different choices for the sticking coefficient α, 10^{-1}, 10^{-2}, and 10^{-3}; α is the fraction of the radicals striking the droplet surface which are captured by the droplet. The actual values of α for HO and HO_2 radicals are unknown, but it is a reasonable expectation that the value will lie within the range of those employed here. We can see from the data of figure 1 that for $\alpha > 10^{-3}$ in theory the rate of oxidation of HSO_3^- by HO-radicals in cloud droplets can be significant during the hours of peak sunlight intensity. Chameides and Davis have shown that these rates are lowered to about 37% of their initial values when the solar zenith angle is increased from 30 to 60°.

The results shown in figure 1 for this oxidation mode can provide only a qualitative measure of its significance for several reasons. Of course one should average such rate data over the solar cycle to provide a more realistic view of the representative 24-hr rates. Furthermore there are some technical problems in our use of the Chameides and Davis estimates of the [HO] in the cloud droplets. Thus they assume that the pressure of SO_2 is 55 ppt and show that a factor of 10 decrease in SO_2 results in an increase in [HO] by a factor of 1.6. The increase of the SO_2 level to 1000 ppt (1 ppb) in our examples considered in figure 1 should lower the (HO) and lower the estimated rates somewhat. In addition the present treatment does not take into account the presence of the several soluble organic species (CH_2O or $HOCH_2OH$, HCO_2H, etc.) which would likely be present in the actual air masses of interest in acid rain development. These species would compete with HSO_3^- for HO radicals and alter the [HO] as well. Also we have neglected (as well as Chameides and Davis) the possible role that HO_2 (O_2^-) species may play in oxidizing HSO_3^- in the cloud droplets. It seem reasonable to assume that HO_2 (and/or O_2^-) will be an effective oxidant for HSO_3^- although rate data related to this reaction are not now available. This contribution, if significant, will increase the rates in theory. In any case the results in figure 1 indicate qualitatively that the Chameides and Davis hypothesis of the gaseous free radical diffusion to and entrapment in cloud water, leads in theory to significant daytime rates of HSO_3^- oxidation in cloud water. Their hypothesis leads to another seemingly important source of HSO_3^- oxidation which results from the HO_2 transfer and subsequent generation of H_2O_2 in the cloud water which in turn will oxidize HSO_3^- quickly; we will consider this in more detail in section 6.

Through a comparison of the instantaneous rates shown in figure 1, it is clear that graphitic carbon (1 mg l^{-1} cloud water), curve j, and nitrite ion (in equilibrium with 0.1 ppb HONO (g)), curve k, are probably both unimportant oxidants for HSO_3^- in clouds. For the rates with carbon we have used the data of Chang et al. (1981); the rate data of Martin et al. (1981) were employed in the nitrite rate estimates. For the carbon catalysis to compete with H_2O_2 in solutions with pH values less than 6, its concentration must exceed 0.1 g l^{-1}, a impossibly high concentration for tropospheric cloud water. It is assumed that the nitrite ion in solution originates entirely from HONO(g), and we have taken this species to be present at the reasonable concentration of one-tenth of that measured in the highly polluted atmosphere of Los Angeles during the nighttime hours (Platt et al. 1980a).

3. AQUEOUS PHASE REACTIONS WHICH MAY GENERATE NITRIC ACID IN CLOUDS

Several reactant combinations which exist in the troposphere can in theory generate nitric acid in solution; $HONO$-O_3, $HONO$-H_2O_2, NO_2-NO_2-$H_2O(1)$, NO-NO_2-$H_2O(1)$ and others are possible sources. However recent research has shown that these reactions occur at negligibly slow rates for the conditions present in the typical tropospheric air mass; see Schwartz (1984); NCAR (1983). The most attractive source of $HONO_2$ in cloud chemistry involves the uptake of N_2O_5 and/or NO_3 by cloud water; see Heikes and Thompson (1981, 1983); NCAR (1983); Richards (1983). The reaction sequence is as follows:

$$O_3 + NO_2 \longrightarrow NO_3 + O_2 \tag{1}$$

$$NO_3 + NO_2 \rightleftharpoons N_2O_5 \tag{2}$$

$$N_2O_5 + H_2O(l) \longrightarrow 2H^+(aq) + 2NO_3^-(aq) \tag{3}$$

In theory it seems likely that NO_3 radicals will also be captured by cloud water and react in part with soluble organic matter, HSO_3^-, etc., to generate $HONO_2$ ($NO_3^-(aq)$, $H^+(aq)$) as well as oxidation products of the organic species, H_2SO_4, etc., in 4:

$$NO_3 + H_2O(aq) + \text{oxidizable impurities}(aq) \longrightarrow H^+(aq) + NO_3^-(aq) + \text{oxidized impurities}(aq) \tag{4}$$

Recent theoretical studies of Heikes and Thompson (1983) suggest that the rates of $HONO_2$ formation in clouds by reaction 1-4 can be very significant (hundreds of percent h^{-1}) provided that the sticking coefficients α for N_2O_5 and NO_3 on cloud droplets or moist aerosol particles are greater than 10^{-3}. The nighttime generation of $HONO_2$ and nitrate aerosol (NH_4NO_3) in Los Angeles (Richards, 1983) and the depletion of NO_3 (and by inference, its coupled N_2O_5 storage source) in nighttime atmospheres of high humidity or in the presence of clouds or rain (Platt et al., 1980b; Platt, 1983), provide indirect field evidence for these reactions.

Quantitative proof of the rates of these processes must await an evaluation of the N_2O_5 and NO_3 sticking coefficients on water droplets or moist particles. However the existing indirect evidence, both theoretical and experimental, points to the N_2O_5-NO_3 cloud scavenging mechanism as a most important source of nitric acid.

4. THE RATES OF ACID GENERATION IN GAS PHASE REACTIONS

Recent studies of the gas phase mechanisms of acid generation in the troposphere suggest a variety of contributing reaction pathways. Current analysis of potentially important gas phase reactions (Calvert and Stockwell, 1983b, 1984; Calvert and Mohnen, 1983) point to the dominance of SO_2 oxidation by HO-radicals (reaction 5) with much smaller contributions from $O(^3P)$-atoms (reaction 6), Criegee intermediates (RRCOO•), reactive products of the alkene-ozone reactions (reaction 7), and possible other species such as CH_3O_2.

$$HO + SO_2\ (+M) \longrightarrow HOSO_2\ (+M) \xrightarrow{(O_2)} SO_3 + HO_2 \tag{5}$$

$$O(^3P) + SO_2 (+M) \longrightarrow SO_3 (+M) \qquad (6)$$

$$RRCOO\cdot + SO_2 \longrightarrow RRCO + SO_3 \qquad (7)$$

$$SO_3 + H_2O \longrightarrow H_2SO_4 \qquad (8)$$

Note in figure 2 that studies of the apparent second order rate constant for the first step in 5 are in general accord with one another. Excluding the earliest results of Davis et al. (plus signs in figure 2) and Harris and Wayne (locus of estimates shown by the dashed curve), the recent data define reasonably well the values of k_5 as a function of air pressure (25°C). At one atmosphere pressure of air, $k_5 = 1.1 \times 10^{-12}$ cc molec^{-1} s^{-1} = 6.7×10^8 l mole^{-1} s^{-1} at 25°C; this reaction exhibits a negative activation energy so that at -5°C k_5 = 1.6×10^{-12} cc molec^{-1} s^{-1} = 9.8×10^8 l mole^{-1} s^{-1}.

Using a reasonably complete reaction mechanism to describe the various tropospheric gas phase reactions, Calvert and Stockwell (1983) have estimated by computer simulations the rates of acid generation for a typical clear summer day at 25°C and 40°N latitude, the season of the year during which acid deposition in the eastern North America is most important. They have assumed various pollution levels for several scenarios employed:

- A, a highly polluted tropospheric air mass typical of a Los Angeles smoggy day;
- B, highly polluted with hydrocarbons as in A but with NO_x at one-tenth of that used in A;
- C, highly polluted in hydrocarbons as in A but with NO_x at one-hundredth of that used in A;
- D, highly polluted in NO_x as in A but the hydrocarbons are at one-tenth of that used in A;
- E, highly polluted in NO_x as in A but hydrocarbons at one-hundredth that of A;
- F, impurities at 0.1-times those in A;
- G, impurities at 0.01-times those in A;
- H, impurities at 0.001-times those in A.

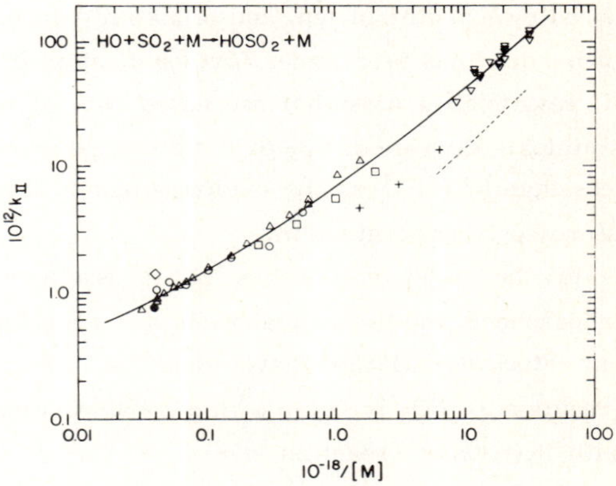

Figure 2. Plot of the reciprocal of the apparent second order rate constant, cm^3 molec^{-1} s^{-1}, for the reaction between HO and SO_2 versus the reciprocal of the concentration of air (molec cm^{-3}) for temperatures near 25°C; data shown are from Leu (1982), inverted triangles; Castleman and Tang (1976/77), triangles; Paraskevopoulos et al. (1983), open circles; Wine et al. (1984), squares; Cox and Sheppard (1980), diamonds; Davis et al. (1975), plus signs; Izumi et al. (1984), filled circles; Harris and Wayne (1975), locus of data, dashed curve.

Some of the results of the simulations for SO_2 oxidation rates are shown in figure 3 for which the relative humidity was taken as 50%. Note that the rates (% SO_2 oxidized h^{-1}) vary with the time of day and the level of pollution present. The lowest rates occur for cases E and D where the NO/hydrocarbons ratios at the start of the day are the highest used. The rate for the highly polluted atmosphere A with a typical NO_x and hydrocarbon composition, reaches a maximum of about 1.5% h^{-1}. This rate rises to 3.4% h^{-1} maximum in case F where the pollution levels of A were reduced by a factor of ten, and to 5.4% h^{-1} in case G where the dilution of case A by a factor of hundred has been made. Further dilution of the impurities by another factor of ten, case H, results in a somewhat decreased rate of SO_2 oxidation, 2.0% h^{-1} maximum. As we might anticipate the rate of SO_2 oxidation drops precipitously during the nighttime hours to values less than 0.1% h^{-1} as the oxidizing agents (largely HO-radicals generated photochemically) decrease in concentration.

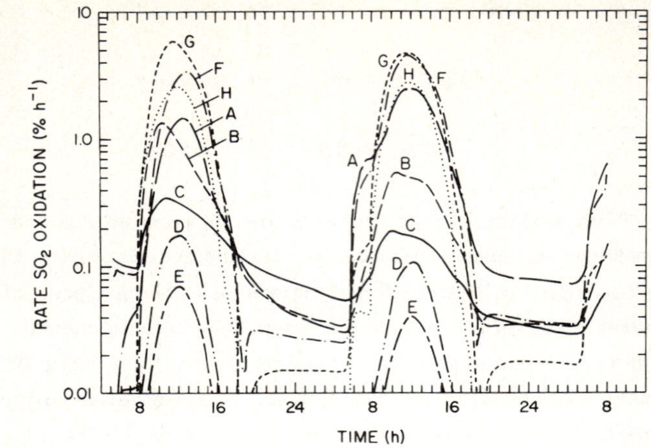

Figure 3. Theoretical gas phase oxidation rates for SO_2 for various polluted air scenarios; see text; after Calvert and Stockwell (1984); relative humidity, 50%(25°C).

In the case of the gas phase reactions leading to nitric acid generation in the troposphere, the theoretical rates are much higher than those for H_2SO_4 formation. Calvert and Stockwell (1983b) have identified the major sources as the HO-reaction with NO_2 (reaction 9), NO_3 reaction with aldehydes (reaction 10), and the homogeneous N_2O_5 reactions with H_2O vapor (reaction 11):

$$HO + NO_2\ (+M) \longrightarrow HONO_2\ (+M) \qquad (9)$$

$$NO_3 + RCHO \longrightarrow HONO_2 + RCO \qquad (10)$$

$$N_2O_5 + H_2O \longrightarrow 2HONO_2 \qquad (11)$$

The rate constant for the HO-NO_2 reaction 9 is about a factor of ten larger than the analogous SO_2 reaction 5; $k_9 = 1.0 \times 10^{-11}$ cc $molec^{-1}$ s^{-1} at 25°C and 1 atm of air (Anderson, 1980). Results of Calvert and Stockwell for rates of $HONO_2$ generation from gas phase reactions are shown in figure 4. Two different cases were considered in the simulations; for figure 4a, the value of $k_{11} = 3.4 \times 10^{-21}$ cc molec^{-1} s^{-1} was assumed; this was estimated from the time dependence of the atmospheric concentrations of NO_3, NO_2, O_3, and H_2O as measured by Platt et al. (1980b). There is some question as to the actual value of k_{11} since heterogeneous processes can lead to an overestimate of the rate constant for the homogeneous component of this reaction. Thus the value of $k_{11} = 0$ was chosen for the second set of simulations shown in figure 4b. Note that with either assumption for the value of k_{11}, the daylight rates of $HONO_2$ formation are nearly identical, and for these conditions reaction 9 dominates the available pathways for reaction. The rates are very much higher than those observed for SO_2. At nighttime the generation of $HONO_2$ through the N_2O_5-H_2O reaction 11 is most important, obviously, when the higher values of k_{11} are used (figure 4a). With $k_{11} = 0$ (figure 4b) the rates of $HONO_2$ formation are much lower at night but they are still significant as a result of the NO_3-aldehyde reaction 10.

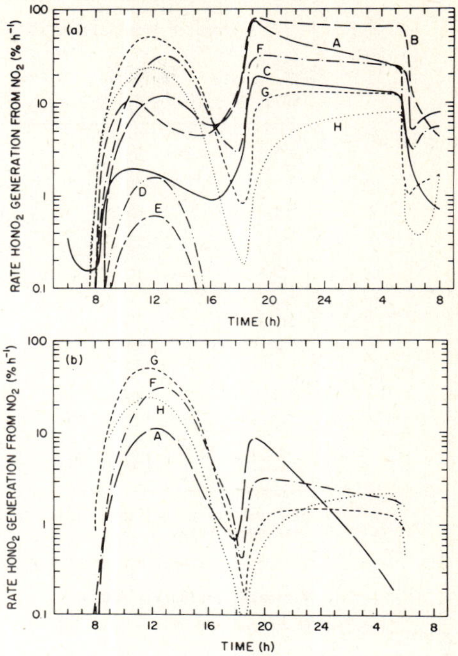

Figure 4. Theoretical gas phase oxidation rates for NO2 for various polluted air scenarios; see text;
a) simulation used $k_{11} = 3.4 \times 10^{-21}$;
b) $k_{11} = 0$ cm^3 molec^{-1} s^{-1};
after Calvert and Stockwell (1983).

A summary of the theoretical 24-hr average rates (% h^{-1}) for conversion of SO_2 to H_2SO_4 and NO_2 to $HONO_2$ is shown in Table 1 for conditions which reflect a clear, summertime troposphere with different degrees of pollutant contamination as outlined in the scenarios described previously. Observe that there is a marked dependence of the rates on the contamination of the air mass. The maximum theoretical rates of SO_2 oxidation occur with the relative clean air mass G, 1.0 and 1.25% h^{-1} for 50 and 100% relative humidity, respectively. These rates are far less than the instantaneous rates estimated in Section 2 for the cloud water reactions, but during a 24-hr cloud-free period they can result in from 24 to 30% conversion of the initial SO_2 to acid. The theoretical NO_2 to $HONO_2$ conversion rates are very significant for most of the air masses considered here; they are largest for air masses F and G, typical of those encountered in eastern North America during the long-range transport of acids in the troposphere (7-29% h^{-1}).

The true value of k_{11} remains unclear at this time; recently Tuazon et al. (1983) reported a new estimate of the constant from laboratory experiments: $k_{11} = (1.3 \pm 0.2) \times 10^{-21}$ cc molec^{-1}s^{-1}, a value intermediate to the two values used here. If one accepts this estimate for

Table 1. The theoretical gas phase rates for SO_2 and NO_2 conversion to acids

(Adopted from Calvert and Stockwell, 1983)

Polluted air scenario[a]	Relative humidity, %(25°C)	$R_{H_2SO_4}$ % h^{-1}	----- R_{HONO_2} ($\% h^{-1}$)--- (1)[b]	(2)[c]
A	10	0.50	---	---
	50	0.32	23	3.5
	100	0.33	---	4.7
B	50	0.30	36	---
C	50	0.15	8	---
D	50	0.03	0.3	---
E	50	0.01	0.1	---
F	50	0.67	19	6.7
	100	0.96	29	9.9
G	50	1.00	16	10.5
	100	1.25	19	12.2
H	50	0.54	8	5.5
	100	0.71	11	7.1

[a]See text for a qualitative description of the impurity concentrations for each case shown here.

[b]Estimated assuming $k_{11} = 3.4 \times 10^{-21}$ cc $molec^{-1}$ s^{-1} (Platt et al., 1980).

[c]Estimated taking $k_{11} = 0$.

simulations, integrated 24-hr rates for NO_2 conversion would become, about 60-80% of those shown for the R_{NO_2} column 1 of Table 1.

Clearly the homogeneous chemistry of the troposphere can lead to significant average rates of acid generation for the clear summertime months. The fractional contribution which such homogeneous processes make to the total acid which is deposited in sensitive areas is uncertain. If account is taken of the fractions of the time that the troposphere is relatively cloud-free and cloud-rich, then the qualitative conclusion which one can make is that the contribution from both gas phase and cloud reactions are significant in theory.

5. THE MECHANISM DIFFERENCES IN S(IV) OXIDATION WHICH OCCUR WITH PHASE CHANGE; THE FACTORS WHICH FAVOR THE HIGH RATES OF SULFURIC ACID GENERATION IN CLOUD WATER

One would not expect the rates of sulfur dioxide oxidation to be the same, or even similar, for the gas phase and liquid phase reactions, even if the concentration of the S(IV) and the oxidizing reactant were exactly the same in the two systems. There is a marked change in chemical structure and reactivity of the S(IV) as one dissolves in the gas phase form of sulfur dioxide (SO_2) in water; the major S(IV) species present in cloud water for tropospheric conditions is the bisulphite ion, HSO_3^-; a very little exists as sulfite ion (SO_3^{2-}) and the hydrate of SO_2 ($SO_2.H_2O$). It is not surprising that the very different chemical species, SO_2 and HSO_3^-(aq), react with a given oxidant with very different mechanisms and rate constants. This difference is obvious in Table 2 where a comparison is given of the apparent second order rate constants for the same oxidizing species (X) reacting with SO_2 and HSO_3^-(aq). In this presentation the rate of the gas phase reaction between X and SO_2 is given by, Rate of SO_2 oxidation = $[X][SO_2]k$, while for the solution phase, Rate of HSO_3^- oxidation = $[X][S(IV)]k$. Any dependence of the rate on the concentration of a third body M as in the gas phase reaction, $HO + SO_2 + M \longrightarrow HOSO_2 + M$, or the pH for the solution phase reactions as with the HSO_3^- oxidation by H_2O_2, has been incorporated into the rate constant k. It is clear from the comparison of Table 2 rate constant data that for a

Table 2. Comparison of apparent second order, gas phase and aqueous phase rate constants for SO_2 or S(IV) oxidation by a given reactant (X) at 25°C

	k^a, $1\ mole^{-1}\ s^{-1}$	
Reactant X	Gas phase [reference]	Aqueous phase [reference]
O_2	0.0	5×10^{-5}(pH = 4.7) [Penkett et al., 1979] For M^{++}-catalyzed, large [Martin, 1983]
O_3	$<5 \times 10^{-3}$ [Daubendiek, Calvert, 1975]	4×10^7 (pH = 5.6) [Maahs, 1982] 1.0×10^6(pH = 4.0) 1.4×10^4(pH = 2.0)
H_2O_2	<1.4 [Braslavsky, et al., 1976]	1.0×10^2(pH = 5.6) [Martin, 1983] 3.0×10^3(pH = 4.0) 4.0×10^5(pH = 2.0)
NO_3	<4.2 [Daubendiek, Calvert, 1975]	about $10^{6 \pm 1}$[authors' estimate]^b
HO	6.7×10^8 [Calvert, Stockwell, 1983b]	9.5×10^9 [Farhataziz, Ross, 1977]
HO_2 (O_2^-)	$<6 \times 10^2$ [Graham, et al., 1979]	Probably large [authors' estimate]
CH_3COO_2	$<4 \times 10^2$ [Cox, Derwent, 1976]	Probably large [authors' estimate]
$CH_3COO_2NO_2$	$<8 \times 10^{-3}$ [Pate, et al., 1976]	3×10^3 [Lee, et al., 1983]

[a]The rate constant is expressed as an apparent second order constant such that Rate = $[X][SO_2]k$ or Rate = $[X][S(IV)]k$.

[b]Based on analogy with other NO_3 and HO reactions in the solution phase.

given concentration of SO_2 and S(IV), each of the oxidizing species shown in the table reacts many magnitudes faster (10^6-times is typical) in the aqueous phase than in the gas phase. The smallest difference occurs with the most reactive species, the HO-radical, where the aqueous phase rate constant is only 14-times greater than that in the gas phase. In this case the solution phase reaction rate constant is very near the maximum theoretical limit, the diffusion controlled encounter rate constant for reactants in water solution. Clearly the mechanisms of reaction are very different for SO_2 and HSO_3^- (S(IV)), and the conversion of SO_2(g) to HSO_3^-(aq) when it dissolves in water is in large part responsible for the more rapid rates observed in cloud water systems.

The detailed pathways by which the SO_2 and HSO_3^-(aq) reactants lead to sulfuric acid on reaction with any of the reactants X in Table 2 are, of course, very different. For example consider the case of the HO-radical reactions. In the gas phase the reaction of HO with SO_2 proceeds by the addition of the HO to the SO_2 in reaction 5, and subsequent steps lead to H_2SO_4. For example the following elementary reactions may be important in the troposphere:

$$HO + SO_2\ (+M) \longrightarrow HOSO_2\ (+M) \tag{5}$$

$$HOSO_2 + O_2 \longrightarrow HO_2 + SO_3 \tag{12}$$

$$\longrightarrow HOSO_2O_2 \tag{13}$$

$$SO_3 + H_2O \longrightarrow H_2SO_4 \tag{8}$$

$$HOSO_2O_2 + NO \longrightarrow HOSO_2O + NO_2 \tag{14}$$

$$HOSO_2O + RH \longrightarrow H_2SO_4 + R \tag{15}$$

Recent rate studies (Stockwell and Calvert, 1983) suggest that reaction 12 may be favored over reaction 13. In contrast to this mechanism, that for the HO- radical reaction with SO_2 dissolved in water starts with the attack of HO on HSO_3^- in the aqueous phase. This may proceed by electron transfer or H-atom abstraction to form an intermediate S(V) species which ultimately leads to S(VI) and H_2SO_4 ($2H^+$(aq), SO_4^{2-}(aq)).

$$SO_2(g) + H_2O(aq) \rightleftharpoons H^+(aq) + HSO_3^-(aq) \tag{16}$$

$$HO(aq) + HSO_3^-(aq) \longrightarrow HO^-(aq) + HSO_3\cdot(aq) \tag{17}$$

$$\longrightarrow H_2O(aq) + SO_3^{\overline{\cdot}}(aq) \tag{18}$$

$$HSO_3^{\cdot}(aq) \rightleftharpoons H^+(aq) + SO_3^{\overline{\cdot}}(aq) \tag{19}$$

$$SO_3^{\bar{\cdot}}(aq) + O_2(aq) \longrightarrow SO_5^{\bar{\cdot}}(aq) \quad (20)$$

$$SO_5^{\bar{\cdot}}(aq) + HSO_3^-(aq) \longrightarrow SO_5^{2-}(aq) + HSO_3\cdot(aq) \quad (21)$$

$$SO_5^{2-}(aq) + HSO_3^-(aq) \longrightarrow 2SO_4^{2-}(aq) + H^+(aq) \quad (22)$$

Reaction 20 coupled with 21 represent the elements of a chain reaction which can repeat many times for each initial act of formation of $HSO_3\cdot(aq)$ or $SO_3^{\bar{\cdot}}(aq)$ in 17 and 18. Ultimately reaction 22 forms the oxidized sulfuric acid product.

As we have seen the gas phase and solution phase routes to H_2SO_4 involve different forms of S(IV) and no common intermediates, so one would certainly expect that the rates of the two processes would be very different. The gas and liquid phase mechanisms for the reaction of each of the reactants X of Table 2 with $SO_2(g)$ or S(IV) are also very different.

There is another important factor which contributes as well to the greater rates for the aqueous phase generation of H_2SO_4 compared to those of the gas phase. This results from the concentrating power of water which is reflected in the equilibrium distribution of a given reactant between the gas phase and the aqueous phase. The equilibrium ratio of the concentration of a given species X in the aqueous phase to that in the gas phase is given by 23 (Schwartz, 1984):

$$[X(aq)]/[X(g)] = H_X RT \quad (23)$$

where H_X is the Henry's law constant for species X in water, R is the gas constant, and T is the temperature (K). The value of RT at 25°C is 24.5 l-atm-mole^{-1}, so that when the value of $H_X > 0.041$ M atm^{-1} at 25°C, then at equilibrium the concentration of X in the liquid water will be greater than that in the gas phase. The Henry's law constants (M atm^{-1}, 25°C) for most of the reactants in table 2 are greater than 0.041 (e.g., SO_2, 1.3; H_2O_2, 1×10^5; HO_2, 10^3; $CH_3COO_2NO_2$, 3), so that we expect a concentration enrichment for these species in the liquid phase. It is highly probable that H_X values for HO, NO_3, and CH_3COO_2 are very much greater than 0.041 M atm^{-1} as well, although these quantities have not been determined to date. The lower values of H_X for O_2 (1.3×10^{-3}) and O_3 (1.25×10^{-2}) result in a somewhat lower concentration of this species in the water phase at equilibrium.

Obviously the phase of the reacting species does play an important role in establishing the mechanism and the rate of SO_2 ($HSO_3^-(aq)$) oxidation which occurs in the troposphere, and the several factors which we have discussed favor the higher rates in the aqueous phase.

6. THE INTERDEPENDENCE OF THE AQUEOUS AND GAS PHASE CHEMISTRY IN S(IV) OXIDATION IN CLOUD WATER; THE MECHANISM OF FORMATION OF OZONE AND HYDROGEN PEROXIDE, TWO FAVORED SOLUTION PHASE, $HSO_3^-(aq)$ OXIDANTS IN CLOUD WATER

One should recognize that the two seemingly most important oxidants for HSO_3^- in cloud water arise in large part from gas phase reactions. Ozone in the troposphere over North America originates from gas phase reactions. The ozone generation in the stratosphere depends on O-atom formation from O_2 photolysis followed by $O + O_2 + M \rightarrow O_3 + M$; stratospheric folding episodes insert some of this O_3 into the troposphere. The fraction of the tropospheric ozone which arises from stratospheric insertion remains uncertain. However it is generally accepted that a very large amount of the ozone which occurs in the troposphere was generated within the troposphere itself through reactions involving NO_x, hydrocarbons, and other hydrocarbon oxidation products (for examples see Demerjian et al, 1974; Calvert, 1978). The important elements in the O_3 generating scheme can be represented as follows:

$$NO_2 + h\nu(\lambda < 4300\ A) \rightarrow NO + O(^3P) \qquad (24)$$

$$O(^3P) + O_2 + M \rightarrow O_3 + M \qquad (25)$$

$$O_3 + NO \rightarrow NO_2 + O_2 \qquad (26)$$

$$O_3 + (\text{other impurities, light, etc.}) \rightarrow \text{other products} \qquad (27)$$

Ozone reacts with NO in 26 and other reactants (e.g., quanta of sunlight with wavelengths less than 3100 A, NO_2, alkenes, etc.) in the generalized reaction 27.

For typical tropospheric conditions the ozone concentration during the daylight hours is described reasonably well (usually within a factor of about two, Calvert and Stockwell, 1983a) by the relation 28:

$$[O_3] \simeq ([NO_2]/[NO])(k_{24}/k_{26}) \qquad (28)$$

The buildup of ozone during a given day depends largely on the conversion of NO to NO_2 with the accompanying increase in the $[NO_2]/[NO]$ ratio as well as the intensity of the sunlight within the NO_2 absorption band which determines the magnitude of the photodissociation rate constant k_{24}. The NO to NO_2 conversion results largely from the chain reactions of HO, HO_2, and alkylperoxy radicals with NO, hydrocarbons, aldehydes, CO, etc. For example reactions with formaldehyde lead to NO_2 generation in the following reaction sequence:

$$HO + CH_2O \rightarrow H_2O + HCO \qquad (29)$$

$$HCO + O_2 \longrightarrow HO_2 + CO \qquad (30)$$

$$HO_2 + NO \longrightarrow HO + NO_2 \qquad (31)$$

$$HO + CH_2O \text{ (or RH, CO, etc.)} \longrightarrow H_2O + HCO \text{ (}HO_2\text{, }RO_2\text{, etc.)} \qquad (32)$$

As these reactions occur in the troposphere, NO is converted to NO_2, the $[NO_2]/[NO]$ ratio increases, and for a given solar intensity, the $[O_3]$ rises in accord with relation 28. Thus a photochemically active mixture which is rich in hydrocarbons and NO_x, typical of polluted urban atmospheres, favors ozone generation during the daylight hours. It is uncommon to find less than 50 ppb of O_3 in the troposphere during a typical summer day over eastern North America, and concentrations of several hundred ppb are formed at times in stagnant, polluted air masses over or near large urban areas. During the winter months ozone levels are typically lower than those in summer. Ozone records at Whiteface Mountain in New York for the past ten years show that the $[O_3]$ maximizes at about 50-60 ppb in midsummer and falls to about 20-30 ppb in midwinter. Obviously the occurrence of the gas phase chemistry is important for the generation of ozone, and it is prerequisite to the oxidation of $HSO_3^-(aq)$ by ozone in cloud water.

Hydrogen peroxide can be formed in gas phase reactions; in theory other aqueous phase processes and solid-sensitizers can form it as well. It is interesting to consider here to what extent the gas phase and other reactions can in theory provide the quantities of H_2O_2 observed in cloud and rain water.

In the gas phase hydrogen peroxide arises from the interaction of hydroperoxy (HO_2) and hydrated hydroperoxy ($H_2O{\cdot}HO_2$) radicals:

$$2HO_2 \longrightarrow H_2O_2 + O_2 \qquad (33)$$

$$2HO_2 \rightleftharpoons H_2O_4^{\dagger} \xrightarrow{M} H_2O_2 + O_2 \qquad (34)$$

$$HO_2 + H_2O \rightleftharpoons H_2O{\cdot}HO_2 \qquad (35)$$

$$H_2O{\cdot}HO_2 + HO_2 \longrightarrow H_2O_2 + O_2 + H_2O \qquad (36)$$

$$H_2O{\cdot}HO_2 + HO_2 \rightleftharpoons H_2O_4{\cdot}H_2O^{\dagger} \xrightarrow{M} H_2O_2 + O_2 + H_2O \qquad (37)$$

The experimental rate laws for this system show that both direct second order, disproportionation channels 33 and 36 exist as well as pressure dependent pathways 34 and 37. Recent kinetic data evaluations (NASA, 1983) recommend the following rate constant values for the water-free system:

$k_{33} = 2.5 \times 10^{-13} e^{560/T}$ cm^3 molec^{-1} s^{-1}

$k_{34} = 1.7 \times 10^{-33}[M] e^{1000/T}$ cm^3 molec^{-1} s^{-1}

where [M] is the concentration of air (N_2, O_2) in molec cm^{-3}.

The effect of water on these reactions can be seen in figure 5 in which the results of many investigators are summarized. Note that the apparent second order rate constant for H_2O_2 formation shows a proportional increase with H_2O vapor concentration. The results of Sander et al. (1982) at 100 (triangles) and 700 torr (diamonds) of N_2 indicate that the water effect is independent of the total pressure of the nitrogen as well. The combined data of the many investigators shown in figure 5 are in reasonable agreement and give the following relation for the ratio of the apparent second order rate constants for H_2O_2 formation in the presence of H_2O (k_{H_2O}) and the absence of H_2O vapor (k):

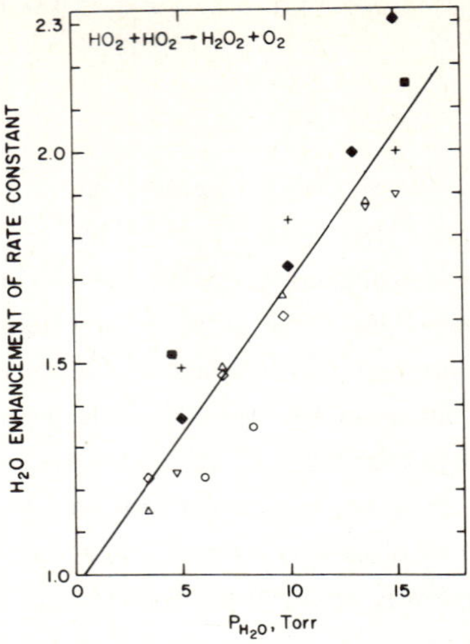

Figure 5. The water enhancement factor on the gas phase rate constant at 298 K for H_2O_2 formation versus pressure of H_2O; data shown are from: Sander et al. (1982), 100 torr air, triangles; 700 torr air, open diamonds; DeMore (1979), filled squares; Lii et al. (1981), inverted triangles; Hamilton (1975), plus signs; Hamilton and Lii, closed diamonds; Cox and Burrows (1979), circles; after Sander et al. (1982).

$$k_{H_2O}/k = 0.97 + 0.07 + (0.07 \pm 0.01) P_{H_2O}(\text{torr}).$$

The sensitivity of the rate constant for H_2O_2 generation to the [H_2O], results in a doubling of the rate constant with increase in H_2O 0 to 14 torr of H_2O. Thus the conditions of high humidity, common during the summer months in eastern North America, favor H_2O_2 generation.

Kircher and Sander (1984) have represented all of the H_2O_2 rate constant data in the following useful form:

$$k_{H_2O_2}(\text{overall}) = (k_{33} + k_{34})(1 + 1.4 \times 10^{-21} e^{2200/T}[H_2O]) \quad (38)$$

Note that the rate constants k_{33}, k_{34}, and $k_{H_2O_2}$(overall) have negative activation energies and increase with decrease in temperature. Thus in the lower temperatures and high humidities common to the upper troposphere during the summer months in eastern North America favor the gas phase generation of H_2O_2.

Some indication can be had in Table 3 of the typical concentrations of H_2O_2 found in

Table 3. Some representative estimates of $[H_2O_2]$ in rain, cloud, and fog water

Conditions	$[H_2O_2]$ ppm(wt)	$[H_2O_2]$ mole l^{-1}	Investigators
Clouds, New York(Long Island), Sept., Oct., 1982	1.0-2.9	$(2.9\text{-}8.6) \times 10^{-5}$	Daum, et al.(1983)
Clouds, Whiteface Mtn., N.Y., Summer, 1982	0.24-0.84	$(0.7\text{-}2.4) \times 10^{-5}$	Kadlecek, et al.(1983)
Clouds, Los Angeles, near coast, 1983	0.03-3.0	$(1\text{-}88) \times 10^{-6}$	Richards et al.(1983)
Fog, Los Angeles, urban, early morning, 1983	~0.001	$\sim 2.9 \times 10^{-8}$	Hoffmann(1983b)
Rain, S. Florida, 1982	0.34-2.4	$(1\text{-}7) \times 10^{-5}$	Zika, et al.(1982)

Table 4. The theoretical distribution of H_2O_2 between cloud water and the gas phase

Total H_2O_2 initially in the gas phase, ppb	Temp., K	$n_{H_2O_2(aq)} / n_{H_2O_2(total)}$	$[H_2O_2]_{aq}$ moles l^{-1} $\times 10^{-5}$	$[H_2O_2]_{aq}$ ppm(wt)
1	273	0.93	4.2	1.4
2	273	0.93	8.3	2.8
5	273	0.93	20.8	7.1
1	283	0.85	3.7	1.2
2	283	0.85	7.3	2.5
5	283	0.85	18.3	6.2
1	298	0.63	2.6	0.9
2	298	0.63	5.2	1.8
5	298	0.63	13.0	4.4

clouds and rain water during the late spring, summer, and fall seasons in the United States. These measurements should be free from artifact H_2O_2 formation which plagued early gas phase estimates employing bubblers (Heikes et al, 1982). Hydrogen peroxide concentrations of a few tenths to a few ppm (by wt) in cloud water are not uncommon in non-urban regions of the troposphere. The $[H_2O_2]$ in cloud and rain water observed at Whiteface Mountain in the summer of 1982 showed a strong directional variability, reflecting different air masses with different potentials for H_2O_2 generation. In Table 4 we have summarized the theoretical equilibrium relationship between the total $[H_2O_2]$ which must be generated in the gas phase in order to derive various concentrations of H_2O_2 in cloud water for various temperatures. A cloud water content of 1 g m^{-3} and the Henry's law data summarized by Martin and Damschen (1981) were used here. We see from these data that the mole fraction of H_2O_2 in the cloud water phase, $n_{H_2O_2(aq)}/n_{H_2O_2(total)}$, decreases from 0.93 to 0.63 as the temperature of the system rises from 273 to 298 K. Note that in order to generate liquid phase concentrations of H_2O_2 which are of the magnitude observed in clouds and rain water (table 3), we require less than about 2 ppb of gaseous H_2O_2 to feed the clouds. Compare this value of $[H_2O_2]$ with that expected theoretically from computer simulations of the gas phase chemistry of various polluted air masses (Calvert and Stockwell, 1983b). For several of the scenarios the results are shown for two different relative humidities (50 and 100%, 25°C). Particularly interesting are cases A, F, G, and H which represent compositions typical of a highly polluted air mass (A) with subsequent impurity levels at 0.1, 0.01, and 0.001-times that of A, respectively. A very unusual result is expected for case A in theory; H_2O_2 forms during the nighttime hours. The NO_3-aldehyde reactions are the source of the HO_2-radicals for this case (Calvert and Stockwell, 1983b). Note that the amounts of H_2O_2 formed during one day in cases F, G, and H are sufficient in theory to account entirely for those seen in clouds and rain water for relatively clean air masses. The very low H_2O_2 generation for the highly polluted air mass A during the first day of reaction is in accord with the Hoffmann (1983b) observation of about 0.001 pm (wt) H_2O_2 in Los Angeles fogs in the morning

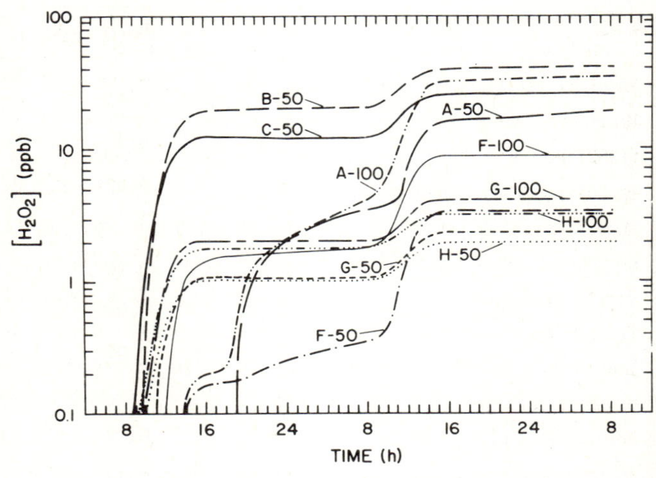

Figure 6. Theoretical H_2O_2 concentration-time profiles for several polluted troposphere scenarios; see text; the numbers following the letters designating the scenario are the relative humidities which are assumed (25°C); after Calvert and Stockwell (1983).

hours. This low yield of H_2O_2 in theory arises largely from the depletion of the HO_2-radical concentration by NO through its reaction 31:

$$HO_2 + NO \longrightarrow HO + NO_2 \quad (31)$$

Hydrogen peroxide formation is also very sensitive to the intensity of the solar energy which helps control the generation of HO and HO_2 radicals and to the temperature of the air mass.

Observations of H_2O_2 in cloud water or rain at Whiteface Mountain in the Adirondacks during the 1982 year show a marked lowering of the $[H_2O_2]$ during the winter months. The theoretical concentrations of the HO_2-radical are much lower for the winter months than those of the summer in eastern North America. This results for several reasons: the sunlight intensities and the length of the day are much less during the winter months, and the NO levels are commonly higher in winter than in summer, at least in the Ohio River Valley (Shaw and Paur, 1983). The rate of H_2O_2 formation through the gas phase reactions is proportional to the square of the HO_2-radical concentration. Thus lowering the $[HO_2]$ in winter to one-tenth of that in summer, a reasonable expectation, will lower the rate of H_2O_2 generation to one-hundredth of that in summer. Obviously when we compare the $[H_2O_2]$ in cloud water to that expected in theory we must remember that at least some of the H_2O_2 originally in the cloud water has reacted with $HSO_3^-(aq)$ and possibly other impurities present in the cloud water. It is possible that the amounts of H_2O_2 formed during the winter months results in its deficiency with respect to HSO_3^- in the cloud water. For such conditions the O_3 oxidant may be the most important reactant for HSO_3^- in clouds. This is obviously not the case for the summer months in the Adirondacks, when acid deposition is dominant.

Although it appears from the discussion presented here that the amounts of H_2O_2 observed in cloud water and rain in summer can be generated through the gas phase reactions alone, there are other sources which may also contribute significantly. For example, Zika et al. (1982) reported that the time dependence of $[H_2O_2]$ in rain during a storm in south Florida and the Bahamas did not follow that for SO_4^{2-} and NO_3^- in the rain water. They concluded that aqueous phase H_2O_2 generation may be important. Chameides and Davis (1982) have suggested from theoretical considerations that HO_2- radicals, generated in the gas phase, can be transported to the cloud water, and aqueous phase reactions of HO_2 (and O_2^- derived from HO_2) can account for a significant rate of H_2O_2 formation in clouds. Schwartz (1983a) has arrived at a similar conclusion from other theoretical consideration. It is instructive to estimate the maximum enhancement of the rate of H_2O_2 formation which one can expect from the cloud water processes. Assume HO_2-radicals are generated only in the gas phase and that in the presence of the clouds they are transferred to the liquid phase efficiently to establish the equilibrium, $HO_2(g) \rightleftharpoons HO_2(aq)$, between the gas and the cloud water phases. The Henry's law constant for HO_2 at 25°C has been estimated to be about 1.2×10^3 M atm^{-1} (Schwartz, 1983). Since the dissociative equilibrium 39 is established quickly with $K_{39} = 2.05 \times 10^{-5}$ M, the effective Henry's law constant is given by $H^*_{HO_2} = H_{HO_2}(1+K_{38}/[H^+(aq)])$.

In the aqueous phase H_2O_2 generation occurs through the two reactions 40 and 41:

$$HO_2(aq) \rightleftharpoons H^+ + O_2^-(aq) \tag{39}$$

$$HO_2(aq) + HO_2(aq) \longrightarrow H_2O_2(aq) + O_2(aq) \tag{40}$$

$$HO_2(aq) + O_2^-(aq) \longrightarrow HO_2^-(aq) + O_2(aq) \tag{41}$$

$$HO_2^-(aq) + H^+(aq) \longrightarrow H_2O_2(aq) \tag{42}$$

The values of k_{40} and k_{41} are 1.35×10^6 and 8.5×10^7 l mole^{-1} s^{-1}, respectively (Behar et al, 1970; Behar and Czapski, 1970).

With these assumptions it can be shown that the maximum ratio of the moles of H_2O_2 formed in the aqueous phase to that in the gas phase per cubic meter of air will be given by 43:

$$\frac{d(n_{H_2O_2(aq)})/dt}{d(n_{H_2O_2(g)})/dt} = \frac{(H_{HO_2}RT)^2\{k_{40} + (k_{41}K_{39}/[H^+(aq)]\}L}{k_{38}(10^6)} \tag{43}$$

For a liquid water content of the clouds (L) = 1 g m^{-3}, a pH of 5 for the cloud water, and taking R = 0.08205 l atm mole^{-1} deg^{-1}, the moles of H_2O_2 formed in the aqueous phase are a maximum of 30-times those formed in the gas phase. Of course an equilibrium between HO_2(g) and HO_2(aq) is not maintained as assumed here, but rather the rate of H_2O_2 formation in the cloud water is limited by diffusion mass transport of HO_2 to the liquid droplets (Schwartz, 1983a). Schwartz concluded as well that it is necessary to treat the dissolution of HO_2 as reversible, not as irreversible. Attempts to calculate the contribution from these liquid phase processes are restricted today since the results are a strong function of the assumed sticking coefficient of HO_2 on the cloud droplets. Chameides and Davis (1982) conclude that for $\alpha > 0.01$, normal cloud droplet size distributions and liquid water content of the clouds, the generation of H_2O_2 can be very significant in clouds during daylight hours. For example they estimate for $\alpha = 0.01$, average bulk rates of aqueous H_2O_2 generation of 5×10^{-6} to 1.5×10^{-6} M h^{-1} with assumed cloud transmissivities of 0.5 and 0.1, respectively. This is comparable to that expected in theory from gas phase processes. Obviously an accurate value of α for HO_2 will be necessary before one can establish a realistic estimate of the importance of this process; if α is as large as 0.01 then it probably is important.

Another multiphase process for H_2O_2 generation in the troposphere should be considered as well. This involves the reactions at the surface of certain light absorbing semiconductor solids suspended in cloud droplets. Solids such as ZnO, CdS, TiO_2, Fe_2O_3, etc., promote H_2O_2 generation in aerated, aqueous suspensions (for examples, see Hoffmann and Jacob, 1984; Rubin et al., 1952; Calvert et al., 1954). In solid photosensitizers of this sort, the promotion of a valence band electron to a conduction band, e^- (CB), through light absorption is accompanied by creation of "hole" or electron deficient region, (+); for a solid of ZnO-like ionic structure the process can be represented in the highly simplified fashion of 44:

$$M^{2+}O^{2-}(s) + h\nu \rightarrow (+) + e^-(CB) \qquad (44)$$

Since the extinction coefficient of solids such as ZnO is very high at wavelengths below 4000 A, the (+) is generated near the solid surface, is rather immobile, and can promote oxidation by accepting an electron by transfer from some adsorbed, oxidizable species. The e^-(CB) is relatively free to migrate throughout the crystal until it recombines with a hole, or it is trapped at an impurity site or by a reactant at the surface.

Mechanistic and photochemical studies in the ZnO(s), H_2O(l), O_2(g), suspensions with added $K_2C_2O_4$, $NaHCO_2$, or other soluble organic materials, provide several key observations which help define the reaction mechanism for H_2O_2 formation:

a) Studies using ^{18}O-labeled O_2(g) showed that the oxygen in the H_2O_2 product originated largely from the O_2-gas and not the water solvent (Calvert et al., 1954).

b) H_2O_2 formation is enhanced by the presence of soluble, oxidizable organic and inorganic compounds in the suspensions (Rubin et al., 1952).

c) The maximum quantum yield of H_2O_2, $\Phi = 0.50$, was observed at low light intensities, high O_2 concentrations (suspension equilibrated with O_2(g) at 1 atm) and high concentrations of added oxidizable substance (e.g., sodium formate at concentrations greater than 0.1 M).

d) For experiments in which only a very small extent of reaction occured, the equivalents of H_2O_2 formed were equal to those of oxidizable compound which were oxidized.

These results are consistent with the H_2O_2 generation following reduction of oxygen adsorbed on the solid through capture of conduction band electrons, and oxidation of the oxidizable material at holes. For example in the ZnO(s), O_2(g), H_2O(l), sodium formate system, the suggested mechanism is as follows:

$$\left.\begin{array}{lrcl}
 & O_2(ads) + e^-(CB) & \rightarrow & O_2^-(ads) \\
 & O_2^-(ads) & \rightleftharpoons & O_2^-(aq) \\
 & O_2^-(aq) + H_2O(aq) & \rightleftharpoons & HO_2(aq) + OH^-(aq) \\
 & 2HO_2(aq) & \rightarrow & H_2O_2(aq) + O_2(aq) \\
 & HO_2(aq) + O_2^-(aq) & \rightarrow & HO_2^-(aq) + O_2(aq) \\
 & HO_2^-(aq) + H_2O(aq) & \rightleftharpoons & H_2O2(aq) + OH^-(aq) \\
\text{and/or} & O_2^-(ads) + e^-(CB) & \rightarrow & O_2^{2-}(ads) \\
 & O_2^{2-}(ads) + 2H_2O(aq) & \rightarrow & H_2O_2(aq) + 2OH^-(aq)
\end{array}\right\} *$$

$$\left.\begin{array}{rcl}
HCO_2^-(ads) + (+) & \rightarrow & HCO_2(ads) + M^{2+}O^{2-}(s) \\
HCO_2(ads) + (+) & \rightarrow & H^+(aq) + CO_2 \cdot H_2O(aq)
\end{array}\right\} **$$

* O_2-reduction at the solid surface by e^-(CB)

** HCO_2^--oxidation at the solid surface by holes (+)

Since both H_2O_2 formation from O_2 and $CO_2 \cdot H_2O$ generation from HCO_2^- require a 2-e^- charge transfer per molecule, and only one conduction band electron together with one hole are released per quantum of absorbed light, it follows that the maximum theoretical quantum yield of H_2O_2 formation is 0.5 as observed experimentally. At high light intensities, low concentrations of added oxidizable molecules, or on buildup of H_2O_2 product, a lower quantum efficiency of H_2O_2 production is found. This is thought to result from the increased efficiencies of hole (+) and electron e^-(CB) combination as well as acceptance of electrons by H_2O_2 with subsequent decomposition:

$$e^-(CB) + (+) \longrightarrow M^{2+}O^{2-}(s)$$
$$H_2O_2(ads) + e^-(CB) \longrightarrow HO(aq) + OH^-(aq)$$
$$OH(aq) + H_2O_2(aq) \longrightarrow H_2O(aq) + HO_2(aq) \; . \; . \; . \; . \; .$$

It is interesting to speculate as to the possible significance of H_2O_2 generation rates from the solid phase photosensitized reactions in cloud water. The amount of ZnO, Fe_2O_3, TiO_2, and other recognized photosensitizers in cloud water is unknown today. It must be remembered that such oxides and sulfides would dissolve ultimately in cloud water as the water becomes more acidic, probably over a period of several hours, and hence the prolonged availability of the necessary solid phase for these reactions is somewhat questionable.

We can make a very qualitative estimate of the amounts of such photosensitizer solids required to produce significant amounts of H_2O_2 in cloud water using a very simplified model. We will assume that w_i (μg m^{-3} air) is the amount of the solid photosensitizer _i_ and that this is present in the cloud water; the density of the solid is ρ_i (g cm^{-3}); its average radius is r_i (μm); Φ_i is the quantum yield of H_2O_2 generation for the _i_th solid in the cloud droplet; and I_i is the number of quanta of sunlight (beyond the threshold for photosensitization) which is incident on the air mass (quanta cm^{-2} s^{-1}). The number of particles (n_i) of the ith solid present m^{-3} air is given approximately by 44:

$$n_i = w_i \times 10^{-6}/(4/3)\pi(r_i \times 10^{-4})^3 \rho_i \qquad (44)$$

Most inorganic photosensitizers have very high extinction coefficients for wavelengths below the threshold absorption for hole-electron generation, so light penetration to depths of more than a few hundred Ångstrom units is unlikely. Therefore as a first approximation we can assume that all of the incident light within this absorption region which is intercepted by the solid is absorbed and contributes to the process of electron-hole generation. Therefore the fraction (F_i) of the light incident on a one m^3 volume of air which is absorbed by the _i_th solid will be given approximately by 45:

$$F_i = n_i\pi(r_i \times 10^{-6})^2 \qquad (45)$$

Therefore the rate of H_2O_2 generation (molec s^{-1}) within the cloud droplets contained in one

m^3 of air (liquid water content, 1 g m^{-3}) will be given by 46:

$$R_i(\text{molec } s^{-1}\ m^{-3}\ \text{air}) = F_i I_i \Phi_i (10^4) \quad (46)$$

We may represent the H_2O_2 formation rate in mole l^{-1} h^{-1} and sum over all solid photosensitizers in the cloud water to derive the total rate of H_2O_2 formation from this process; relation 47 results

$$R_{H_2O_2}(\text{M } h^{-1} \text{ in cloud water}) = \sum_{i=1}^{n} R_i = \sum_{i=1}^{n} (W_i I_i \Phi_i)(4.5\text{x}10^{-20})/r_i \rho_i \quad (47)$$

For illustration let us assume that a single solid photosensitizer is present with an average radius r = 1 μm, ρ = 10 g cm^{-3}, and the solid has a threshold of 4000 A. A reasonable value for the tropospheric actinic irradiance I within the clouds for these conditions (2900 A < λ < 4000 A) is 7.0 x 10^{15} quanta cm^{-2} s^{-1} (solar zenith angle, 40°). With these assumed values equation 47 gives: $R_{H_2O_2}$(M h^{-1} in cloud water) = wΦ (3.2 x 10^{-5}). Reasonable values for Φ range from a maximum of 0.5 for ZnO-like suspensions with high concentrations of oxidizable species such as formate ion, oxalate ion, hydrated formaldehyde ($HOCH_2OH$) etc., to about 10^{-5} for hemotite (α-Fe_2O_3), Hoffmann (1983a). For H_2O_2 production by this process to be significant compared to other sources, requires rates of about 2 x 10^{-6} M h^{-1} in the cloud water. Thus the minimum amount of photosensitizer solid in the cloud water which would satisfy this expectation for the most favorable case (highest quantum yield) is 0.13 μg m^{-3} air and 6.5 mg m^{-3} for the least favorable case (lowest quantum yield). In summary it appears that solid phase photosensitized formation of H_2O_2 can be important only if reasonable efficient sensitizers are present. These may exist under some circumstances, but we need information on the nature and amounts of each solid species present in cloud and rain water before a definitive evaluation of this system can be made.

7. THE POTENTIAL ROLE OF HSO_3^-(aq) OXIDATION BY OTHER PEROXY COMPOUNDS WHICH MAY FORM IN THE TROPOSPHERE

Gas phase tropospheric reactions are expected to create a variety of potentially important oxidants in addition to O_3 and H_2O_2; peroxynitric acid (HO_2NO_2), peroxyacetyl nitrate ($CH_3COO_2NO_2$) and its higher homologues, peroxyacetic acid (CH_3COO_2H), methyl hydroperoxide (CH_3O_2H) and its higher homologues (RO_2H). A qualitative indication of the theoretical amounts of these species which are expected to form during a clear day (25°C) can be seen in Table 5 for various polluted air mass scenarios described previously. Fair amounts of CH_3O_2H and RO_2H are anticipated for the relatively clean atmospheres. Except for the very polluted atmosphere (case A), the compounds other than the hydroperoxides and peroxyacyl

Table 5. Theoretical concentrations of some oxidizing compounds formed in the troposphere by late afternoon in various polluted air scenarios (50% relative humidity, 25°C)[a]

Case[b]	HO_2NO_2	$CH_3COO_2NO_2$ (PAN)	CH_3COO_2H	CH_3O_2H	RO_2H
	Concentration, ppb				
A	5.6×10^{-2}	9.4	5.9×10^{-5}	2.0×10^{-3}	5.9×10^{-3}
B	3.4×10^{-2}	7.7	3.5×10^{-2}	1.2	3.0
C	2.9×10^{-3}	0.42	1.2×10^{-2}	0.52	2.4
D	6.5×10^{-7}	5.5×10^{-4}	2.2×10^{-10}	1.3×10^{-8}	6.9×10^{-8}
E	2.8×10^{-8}	1.5×10^{-5}	5.7×10^{-13}	9.7×10^{-11}	2.0×10^{-10}
F	4.5×10^{-3}	0.79	2.2×10^{-4}	8.1×10^{-3}	1.4×10^{-2}
G	2.5×10^{-4}	2.4×10^{-2}	7.2×10^{-4}	7.1×10^{-2}	4.7×10^{-2}
H	2.6×10^{-5}	1.2×10^{-3}	2.4×10^{-4}	0.12	1.7×10^{-2}

[a]Adapted from Calvert and Stockwell (1983b)

[b]Impurity concentrations are described in the text.

nitrate are relatively unimportant for the conditions chosen (25°C). In theory the thermally unstable product HO_2NO_2 would be of somewhat greater importance for the relatively common lower temperatures of the troposphere. Little is known about the actual tropospheric concentrations of these minor oxidizing compounds for any conditions. By far the most important of the oxidizing minor constituents in theory is the peroxyacetyl nitrate (PAN). The very dirty air masses A and B are the most efficient producers, but even the cleaner air masses F, G, and H with 0.1, 0.01, and 0.001-times the impurity of A, give significant PAN formation. This theoretical expectation is consistent with the

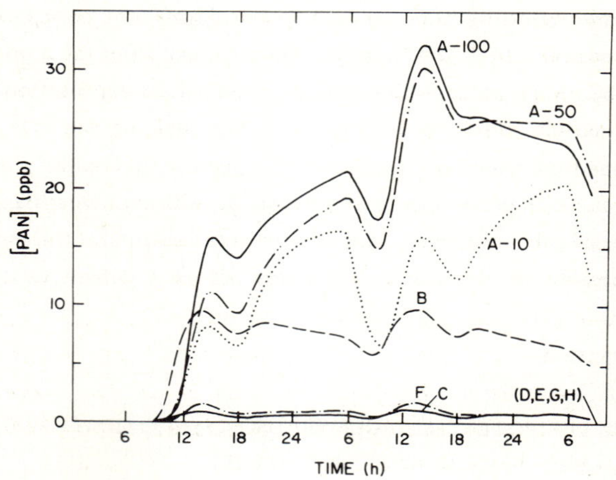

Figure 7. Theoretical peroxyacetyl nitrate (PAN) concentration-time profiles for several polluted troposphere scenarios; see the text; the numbers following the scenario A indicate the relative humidity which is assumed (25°C); for all other scenarios shown here, relative humidity, 50% (25°C); after Calvert and Stockwell (1983).

atmospheric observations of Penkett (1983) and Singh and Salas (1983) and others.

The simulations of Calvert and Stockwell (1983b) suggest an important yet unobserved result for PAN generation. The time dependence of PAN for various polluted air scenarios is shown in figure 7. Note that for the highly polluted air mass A we predict in theory that PAN will be formed at night as well as during the day. This unusual happening results in theory from the attack of NO_3 radicals an acetaldehyde (CH_3CHO):

$$O_3 + NO_2 \longrightarrow NO_3 + O_2$$
$$NO_3 + CH_3CHO \longrightarrow HONO_2 + CH_3CO$$
$$CH_3CO + O_2 \longrightarrow CH_3COO_2$$
$$CH_3COO_2 + NO_2 \longrightarrow CH_3COO_2NO_2 \quad (PAN)$$

At night the $[NO_3]$ maximizes as its destruction by sunlight becomes unimportant, and its reaction with NO ($NO_3 + NO \longrightarrow 2NO_2$) is minimized; O_3 oxidizes the NO efficiently for these conditions, and NO is not regenerated by NO_2 photolysis, a process important only during the daylight hours. The dominant reaction of the CH_3COO_2 radicals for these low [NO] conditions is that with NO_2 to form PAN.

Recently Lee et al. (1983) have reported the first measurements of the Henry's law constant for $CH_3COO_2NO_2$ (about 3 M atm^{-1} at 22°C) and its second order rate constant for reaction with HSO_3^- in aqueous solution at pH = 5.6 ($k \simeq 3 \times 10^3$ M^{-1} s^{-1} at 22°C). From these data we can estimate that although the fraction of PAN present at equilibrium with cloud water (1 g m^{-3} liquid water content) will be small (7×10^{-5}), the rate of oxidation of HSO_3^- for an air mass containing 5 ppb of SO_2 and 5 ppb of PAN would correspond to an equivalent gas phase conversion rate of about 2.7% h^{-1} at pH = 5.6. The dependence of the rate constant on pH has not been reported yet. If the constant is independent of pH, then the rates for HSO_3^- oxidation by PAN would become much less important as the pH of the cloud water becomes less and the solubility of the SO_2 decreases. Thus at pH = 4.5, the rate for reaction with 5 ppb each of SO_2 and PAN would fall to about 0.24% h^{-1}; at 1 ppb of PAN and SO_2 the rates are relatively unimportant: 0.53% h^{-1} for pH = 5.6 and 0.05% h^{-1} for pH = 4.5. Obviously the experimental determination of the pH dependence of this rate constant is necessary to evaluate properly the significance of PAN as an oxidant for HSO_3^- in cloud water.

8. CONCLUSIONS

We have seen that many complex interactions occur between the gas phase and the liquid phase (cloud water) chemistry which result in acid generation in the troposphere. Although many of the important processes have been identified and are reasonably well characterized, several others should be recognized as highly uncertain at this time. Amoung the seemingly important processes now poorly understood are the gas phase reactions between N_2O_5 and H_2O, the sticking coefficients for HO, HO_2, NO_3, and N_2O_5 on cloud droplets, and the actual composition, both the inorganic and organic components, of cloud and rain water; these quantities govern the rates of many potentially important cloud water reactions leading to acid generation. Obviously further quantitative studies of these very complex chemical systems will be required before one can have great confidence in the modeling of acid rain generation.

ACKNOWLEDGMENT

This work was partially supported by the U.S. Environmental Protection Agency under Interagency Agreements DW930237-01-0 and DW930144-01-0 but has not been subject to EPA review procedures.

REFERENCES

Anderson L. G. (1980) Absolute rate constants for the reaction of OH with NO_2 in N_2 and He from 225 to 389 K., J. Phys. Chem. 84, 2152-2155.

Behar D. and Czapski G. (1970) Ozonide and hydroperoxy radicals in flash photolysis of hydrogen peroxide. Israel J. Chem. 8, 690-708.

Behar D., Czapski G., Rabani J., Dorfman L. M., and Schwartz H. A. (1970) The acid dissociation constant and decay kinetics of the perhydroxyl radical. J. Chem. Phys. 74, 3209-3213.

Braslavsky S. E., Bottenheim J. W., Gunning H. E., and Strausz O. P. (1976) The H_2O_2-SO_2 gas phase reaction. Annual Report, Project C.75-3B, Hydrocarbon Research Center, University of Alberta.

Brock J. R., and Durham J. L. (1984) Aqueous aerosol as a reactant in the atmosphere. Chapter 5, pp. 209-249, in "SO_2, NO, and NO_2 oxida tion mechanisms; atmospheric considerations", Calvert J.G., Ed., Butterworth Publishing Co., Boston.

Calvert J. G. (1976) Test of the theory of ozone generation in the Los Angeles atmosphere. Environ. Sci. Technol. 10, 248-256.

Calvert J. G. and Mohnen V. (1983) The chemistry of acid formation, Appendix A, pp. 155-201, in "Acid deposition; atmospheric processes in eastern North America", National Academy Press, Washington, D. C.

Calvert J. G. and Stockwell W. R. (1983a) Deviations from the O_3-NO-NO_2 photostationary state in tropospheric chemistry. Can. J. Chem. 61, 983-992.

Calvert J. G. and Stockwell W. R. (1984) Mechanism and rates of the gas-phase oxidations of sulfur dioxide and nitrogen oxides in the atmosphere. Chapter 1, pp. 1-62, in "SO_2, NO, NO_2 oxidation mechanism, atmospheric considerations", Calvert J. G., Ed., Butterworth Publishing Bo., Boston.

Calvert J. G. and Stockwell W. R. (1983b) Acid generation in the troposphere by gas-phase chemistry. Environ. Sci. Technol. 17, 428A-443A.

Calvert J. G., Thuerer K., Rankin G. T., and MacNevin W. M. (1954) A study of the mechanism of the photochemical synthesis of hydrogen peroxide at zinc oxide surfaces. J. Amer. Chem. Soc. 76, 2575-2578.

Castleman A. W. Jr. and Tang I. N. (1976/77) Kinetics of the association reaction of SO_2 with hydroxyl radical, J. Photochem., 6, 349-354.

Chameides W. L. and Davis D. D. (1982) The free radical chemistry of cloud droplets and its impact upon the composition of rain. J. Geophys. Res. 87, 4863-4877.

Chang S. G., Toossi R., and Novakov T. (1981) The importance of soot particles and nitrous acid in oxidizing SO_2 in atmospheric aqueous droplets. Atmos. Environ. 15, 1287-1292.

Charlson R. J. and Rodhe H. (1982) Factors controlling the acidity of natural rainwater. Nature 295, 683-685.

Cox R. A. and Burrows J. P. (1979) Kinetics and mechanism of the disproportionation of HO_2 in the gas phase. J. Phys. Chem. 83, 2560-2568.

Cox R. A. and Sheppard D. (1980) Reactions of the OH radicals with gaseous sulfur compounds. Nature 284, 330-331.

Daubendiek R. L. and Calvert J. G. (1975) A study of the N_2O_5-SO_2-O_3 reaction system. Environ. Lett. 8, 103-116.

Daum P. H., Schwartz S. E., and Newman L. (1983) Studies of gas- and aqueous- phase composition of stratoform clouds, in Precipitation scaven ging, dry deposition, and resuspension, Vol. 1, Pruppacher et al., Ed., pp. 31-51.

Davis D. D., Ravishankara A. R., and Fischer S. (1979) SO_2 oxidation via the hydroxy radical: atmospheric fate of HSO_x radicals. Geophys. Res. Lett. 6, 113-116.

Demerjian K. L., Kerr J. A., and Calvert J. G. (1974) The mechanism of photochemical smog, chapter 1, Adv. Environ. Sci. Technol. 4, 1-256.

DeMore W. B. (1979) Reaction of HO_2 with O_3 and the effect of water vapor on HO_2 kinetics. J. Phys. Chem. 83, 1113-1118.

Farhataziz and Ross A. B. (1977) Selected specific rates of reaction of transients from water in aqueous solution. III. Hydroxyl radical and perhydroxyl radical and their radical ions. Rept. No. NSRDS-NBS-59, National Bureau of Standards, Washington, D. C.

Golomb D., Batterman S., Gruhl J., and Labys W. (1983) Sensitivity analysis of the kinetics of acid rain models. Atmos. Environ. 17, 645-653.

Graham R. A., Winer A. M., Atkinson R., and Pitts J. N. Jr. (1979) Rate constants for the reaction of HO_2 with HO_2, SO_2, CO, N_2O, trans-2-butene, and 2,3-dimethyl-2-butene at 300 K. J. Phys. Chem. 83, 1563-1567.

Hamilton E. J. Jr. (1975) Water vapor dependence of the kinetics of the self-reaction of HO_2 in the gas phase. J. Chem. Phys. 63, 3682-3683.

Hamilton E. J. Jr. and Lii, R.-R. (1977) The dependence on H_2O and on NH_3 of the kinetics of the self-reaction of HO_2 in the gas-phase formation of $HO_2 \cdot H_2O$ and $HO_2 \cdot NH_3$ complexes. Int. J. Chem. Kinet. 9, 875-885.

Harris G. W. and Wayne R. P. (1975) Reaction of hydroxyl radicals with NO, NO_2, and SO_2. J. Chem. Soc., Faraday Trans. I 71, 610-617.

Heikes B. G., Lazrus A. L., Kok G. L., Kunen S. M., Gandrud B. W., Gitlin S. N., and Sperry P. D. (1982) Evidence for aqueous phase hydrogen peroxide synthesis in the troposphere. J. Geophys. Res. 87, 3045-3051.

Heikes B., and Thompson A. M. (1981) Nitric acid formation mechanisms in-cloud: gas phase, photochemical and/or heterogeneous paths. EOS Trans., Amer. Geophys. Union, 62, 884.

Heikes B. G. and Thompson A. M. (1983) Effects of heterogeneous processes in NO_3, HONO, and HNO_3 chemistry in the troposphere. J. Geophys, Res., 88, 10883 - 10895

Hoffmann M. R. (1983a) Aqueous phase photo-assisted oxidation of S(IV) by oxygen on hemotite (α-Fe_2O_3(s)) at low pH. Paper presented in NATO Advanced Study Institute, "Chemistry of Multiphase Atmospheric Systems", Corfu, Greece, Sept. 26-Oct. 8.

Hoffmann M. R. (1983b) The chemistry of bisulphite aldehyde addition products in fog and cloudwater, paper presented at NATO Advanced Study Institute on Chemistry of Multiphase Atmospheric Systems, Corfu, Greece, Sept. 26-Oct. 8.

Hoffmann M. R. and Jacob D. J. (1984) Kinetics and mechanisms of the catalytic oxidation of dissolved sulfur dioxide in aqueous solution: an application to nighttime fog water chemistry, chapter 3, pp. 101-172, in "SO_2, NO, NO_2 oxidation mechanisms; atmospheric considerations," Calvert J. G., Ed., Butterworth Publishing Co., Boston.

Izumi K., Mizuochi M., Yoshioka M., Murano K., and Fukuyama T. (1984) A redetermination of the rate constant for the reaction of HO radicals with SO_2, Environ. Sci. Technol. 18, 116 - 118

Kadlecek J., McLaren S., Camarota N., Mohnen V. and Wilson J. (1983) Cloud water chemistry at Whiteface Mountain, in Precipitation scavenging, dry deposition and resuspension, Vol. 1. Precipitation Scavenging, pp. 103-114, Elsevier Science Publishing Co., N. Y.

Kircher C. C. and Sander S. P. (1984) Kinetics and mechanism of HO_2 and DO_2 disproportionations. J. Phys. Chem., 88, 2082 - 2091; the authors are grateful to Dr. Sander for use of this paper before publication.

Lee Y.-N., Senum G. I. and Gaffney J. S. (1983) Peroxyacetyl nitrate (PAN) stability, solubility, and reactivity-implications for tropospheric nitrogen cycles and precipitation chemistry, paper at CACGP Symposium on tropospheric chemistry with emphasis on sulfur and nitrogen cycles and the chemistry of clouds and precipitation, Oxford, England, Aug. 28-Sept. 3.

Leu M.-T. (1982) Rate constants for the reaction of OH with SO_2 at low pressures. J. Phys. Chem. 86, 4558-4562.

Lii R.-R., Sauer M. C. Jr., Gordon S. (1981) Temperature dependence of the gas- phase self-reaction of HO_2 in the presence of H_2O. J. Phys. Chem. 85, 2833-2834.

Maahs H. G. (1982) The importance of ozone in the oxidation of sulfur dioxide in nonurban tropospheric clouds, preprint volume, Second symposium on chemistry of the non-urban troposphere, Williamsburg, V., Amer. Meteor. Soc., Boston, Mass.

Martin L. R. (1984) Kinetic studies of sulfite oxidation in aqueous solution, Chapter 2, pp. 62-100, in "SO_2, NO, NO_2 oxidation mechanisms; atmospheric considerations", Calvert J. G., Ed., Butterworth Publishing Co., Boston.

Martin L. R. and Damschen D. E. (1981) Aqueous oxidation of sulfur dioxide by hydrogen peroxide at low pH. Atmos. Environ. 15, 1615-1621.

Martin L. R., Damschen D. E. and Judeikes H. S. (1981) Sulfur dioxide oxidation reactions in aqueous solutions, ATR/81(7700)-1, Aerospace Corporation, El Secundo, Calif., prepared for the U. S. Environmental Protection Agency, Contract No. 68-02-2702.

NASA (1983) Chemical kinetics and photochemical data for use in stratospheric modeling, Evaluation No. 6, JPL Publication 83-62.

NCAR (1983) Regional acid deposition: Models and physical processes, NCAR technical note 214+STR, August, National Center for Atmospheric Research, Boulder, Colorado.

Oblath S. B., Markowitz S. S., Novakov T., and Cheng S. G. (1981) Kinetics of the formation of hydroxylamine disulfonate by reaction of nitrite with sulfites. J. Phys. Chem. 85, 1017-1021.

Paraskevopoulos G., Singleton D. L., and Irvin R. S. (1983) Rates of OH radical reactions. The reaction HO + SO_2 + N_2. Chem. Phys. Lett. 100, 83-87.

Pate C. T., Atkinson R. and Pitts J. N. Jr. (1976) Rate constants for the gas phase reaction of peroxyacetyl nitrate with selected atmospheric constituents. J. Environ. Sci. Health - Envir. Sci. Eng. A11, 19-31.

Penkett S. A. (1983) PAN in the natural atmosphere. Nature 302, 293-294.

Penkett S. A., Jones B. M. R., Brice K. A., and Eggleton A. E. J. (1979) The importance of atmospheric ozone and hydrogen peroxide in oxidizing sulfur dioxide in cloud and rainwater. Atmos. Environ. 13, 123-137.

Platt U. (1983) The origin of nitrous and nitric acid in the atmosphere, presented at the NATO Advanced Study Institute on Chemistry of Multiphase Atmospheric Systems, Corfu, Greece, Sept. 26-Oct. 8.

Platt U., Perner D., Harris G. W., Winer A. M. and Pitts J. N. Jr. (1980a) Observations of nitrous acid in an urban atmosphere by differential optical absorption. Nature 285, 312-314.

Platt. U., Perner D., Winer A. M., Harris G. W., and Pitts J. N. Jr. (1980b) Detection of NO_3 in the polluted troposphere by differential optical absorption. Geophys. Res. Lett., 7, 89-92.

Richards L. W. (1983) Comments on the oxidation of NO_2 to nitrate - day and night. Atmos. Environ. 17, 397-402.

Richards L. W., Anderson J. A., Blumenthal D. L., McDonald J. A., Kok G. L., and Lazrus A. L. (1983) Hydrogen peroxide and S(IV) in Los Angeles cloud water. Atmos. Environ, 17, 911-914.

Rubin T. R., Calvert J. G., Rankin G. T. and MacNevin W. (1953) Photochemical synthesis of hydrogen peroxide at zinc oxide surfaces. J. Amer. Chem. Soc. 75, 2850-2853.

Sander S. P., Peterson M., Watson R. T., and Patrick R. (1982) Kinetic studies of the HO_2 + HO_2 and DO_2 + DO_2 reactions at 298 K. J. Phys. Chem. 86, 1236-1240.

Schwartz S. E. (1983) Gas- and aqueous-phase chemistry of HO_2 in liquid-water clouds. Paper presented American Chemical Society Meeting, Division of Environmental Chemistry, Washington D. C., September.

Schwartz S. E. (1984) Gas-aqueous reactions of sulfur and nitrogen oxides in liquid-water clouds. Chapter 4, pp. 173-208 in "SO_2, NO, NO_2 oxidation mechanisms; atmospheric consideration", Calvert J. G., Ed., Butterworth Publishing Co., Boston.

Shaw R. W. Jr. and Paur R. J. (1983) Measurements of sulfur gases and particles during sixteen months in the Ohio River Valley. Atmos. Environ. 17, 1431-1438.

Singh H. G. and Salas L. J. (1983) Peroxyacetyl nitrate in free troposphere. Nature 302, 326-328.

Stockwell W. R. and Calvert J. G. (1983) The mechanism of the HO-SO_2 reaction. Atmos. Environ. 17, 2231-2237.

Tuazon E., Atkinson A., Plum C. N., Winer A. M, and Pitts J. N. Jr. (1983) The reaction of gas phase N_2O_5 with water vapor. Geophys. Res. Lett. 10, 953-956.

Wine P.H., Semmes D. H., Thompson R. J., Gump C. A., Ravishankara A. R., Torabi A., and Nicovich J. M. (1984) Kinetics of the reaction HO + SO_2 + M $HOSO_2$ + M. Temperature and pressure dependence in the fall-off region, J. Phys. Chem., 88, 2095 - 2104.

Zika R., Saltzman E., Chameides W. L. and Davis D. D. (1982) H_2O_2 levels in rainwater collected in south Florida and the Bahama Islands. J. Geophys. Res. 87, 5015-5017.

A STUDY OF CLOUDWATER ACIDITY DOWNWIND OF URBAN AND POWER PLANT SOURCES

F.W. Lurmann, J.R. Young and G.M. Hidy
Environmental Research & Technology, Inc.
2625 Townsgate Road
Westlake Village, California 91361
U.S.A.

ABSTRACT

A modeling study is described which investigates the relationship of emissions to cloudwater acidity downwind of isolated urban and power plant emissions sources. The parameterized model employed in the study combines a gas-phase, aerosol-particle and aqueous-phase chemical mechanism with scavenging, mass transfer and meteorology in a simple Lagrangian box model framework. The effects of urban area and power plant emissions reduction scenarios on the chemical composition of cloudwater are examined for a specific environmental situation representing the eastern United States. The case chosen consists of a 24-hour simulation in which morning emissions are injected in fair weather followed by formation of a stratus layer (i.e., fog) which persists from sunset through the night. The results indicate significant nonlinear coupling between the sulfur oxide-nitrogen oxide-reactive hydrocarbon system which differs depending on the emissions rates among the species. The calculations suggest the consequences of the coupled chemical processes for emission changes.

1. INTRODUCTION

The role of sulfur oxides (SO_x) and nitrogen oxides (NO_x) emissions in determining the chemical composition of cloudwater and precipitation is a key issue in formulating control strategies to reduce acid deposition in environmentally sensitive areas. The combined effects of meteorology, gas-phase chemistry, scavenging, mass transfer and aqueous-phase chemistry on the distribution and composition of acidic species in the atmosphere make estimation of the impacts of emissions reductions on acid deposition very difficult. This difficulty arises because neither the individual processes, nor the interactions among the processes

NATO ASI Series, Vol. G6
Chemistry of Multiphase Atmospheric Systems
Edited by W. Jaeschke

are well understood at this time.

Numerous researchers have suggested that in-cloud pollutant scavenging processes in addition to below-cloud processes are important in determining precipitation acidity (e.g., Hegg and Hobbs 1981; Schwartz et al. 1983). Other researchers have suggested that direct deposition of acidic cloudwater in mountainous regions may contribute substantially to the total acidic deposition (Borys et al. 1983; Waldman et al. 1983). The relative importance of the two processes must certainly vary with environmental conditions such as pollutant loadings and their vertical distributions, cloud type, and precipitation rate, etc. Both processes should be included in comprehensive atmospheric models to address emission-deposition acidity relationships. However, since the time scale for pollutant interactions with cloudwater in stable nonprecipitating systems may be much longer than with cloudwater, rainwater, or ice in precipitating systems, we have chosen to focus our analysis on the former in this paper.

A parameterized model has been developed to investigate the interactions among the atmospheric processes which determine the chemical composition of water in warm cloud, nonprecipitating systems. The model is a simple Lagrangian box model with a combined gas/aerosol/aqueous phase chemical mechanism. The model formulation is biased towards detailed treatment of the chemistry and cloud droplet mass transfer rather than meteorology. The model employs a photochemical mechanism for the gas and aerosol species in the RHC*/NO_x/SO_x (*RHC = "reactive" hydrocarbons) system based on the Atkinson et al. (1982) scheme. The principal gases and aerosol particle species that interact with the aqueous phase are SO_2, SO_4^{2-}, HNO_3, NO_3^-, O_3, H_2O_2, HCHO, NH_3, NH_4^+, OH, HO_2 and H^+. Aerosol scavenging and mass transfer of gases to aqueous-phase droplets are parameterized using mean droplet diameters and a two-film resistance model for gas-liquid transfer. The aqueous chemical mechanism includes 20 equilibrium reactions; kinetic reactions for oxidation of S(IV) by H_2O_2, O_3, OH, N(III), NO_2 and O_2 catalyzed by iron, manganese and carbon, oxidation of nitrogen species; and kinetic reactions for H_2O_2 production from aqueous free radical species.

A single meteorological and air quality scenario and two sets of baseline emission strengths are considered in the analysis. The meteorology is intended to represent summer conditions in eastern North America with a stable, low-altitude stratus layer forming at sunset and lasting through the night. Two baseline emissions scenarios were selected to represent emissions from a hypothetical power plant complex and an urban area. The emissions are injected in the morning at the upwind part of the air mass trajectory. Emissions strengths were selected to reflect the relative contributions of power plants and urban sources to the overall emissions inventory for the eastern United States. Additional model runs were performed with reduced power plant and urban emissions to investigate the emissions-cloudwater acidity relationship for SO_x, NO_x and reactive hydrocarbon (RHC).

2. MODEL DESCRIPTION

The Lagrangian box model is perhaps the simplest framework in which to examine the relationships between the various atmospheric processes occurring between sources and receptors. This section describes the chemical mechanisms, scavenging algorithms, and approach to deposition employed in the Lagrangian box model.

Gas-Phase Chemistry

The model employs a slightly condensed version of the Atkinson et al. (1982) photochemical mechanism. The mechanism includes reactions and species of known importance to the photooxidation of atmospheric RHC, NO_x, and SO_x and resulting formation of ozone, nitrogen dioxide, nitric acid, PAN, sulfate and hydrogen peroxide. The mechanism has been tested extensively against smog chamber data and has been shown to predict observed time concentration profiles well ($\pm$20% for key species). The comparability of the mechanism's predictions for nitric acid and, especially, hydrogen peroxide are less certain (than for NO, NO_2, and O_3) because few reliable measurements are available from chamber studies.

The mechanism shown in Table 1 includes the inorganic and organic reactions of importance for mesoscale lower tropospheric applications. The reactive hydrocarbons initially present are represented by eight classes of compounds: $\geq C_4$ alkanes, ethylene, terminally double-bonded alkenes, internally double-bonded alkenes, monoalkylbenzenes, di/trialkylbenzenes, formaldehyde, and other aldehydes. Slow reacting compounds such as methane, ethane, propane and benzene are not included. The lumped alkane chemistry is based on average rate constants and product yields for n-butane, isobutane, n-pentane, isopentane, n-hexane, and n-heptane. Explicit chemistry for ethylene, propylene, and trans-2-butene is used to represent the three classes of alkenes. Chemistry for the two classes of aromatic compounds is represented by the explicit chemistry of toluene and average chemistry of the xylenes. Chemistry for the two classes of aldehydes is represented by the explicit chemistry of formaldehyde and acetaldehyde. Thus, the scheme allows for relatively detailed delineation of the different reaction pathways and oxidation rates for the major organic compounds found in the lower troposphere.

Several additional species have been incorporated into the mechanism for this application. Two "inert" gases, CO_2 and NH_3, have been added to the mechanism due to their importance in the aqueous phase. Additional aerosol species have been added; these include aerosol protons, ammonium, iron, manganese, carbonate and elemental carbon.

Table 1: THE CONDENSED ERT PHOTOCHEMICAL MECHANISM

	Reaction	Rate Constant (ppm min units)
Inorganics		
1.	$NO_2 + h\nu \xrightarrow{O_2} NO + O_3$	radiation dependent
2.	$NO + O_3 \longrightarrow NO_2 + O_2$	$1.0 \times 10^6\ T^{-1}\ e^{-1450/T}$
3.	$O_3 + h\nu \longrightarrow 2\ OH$	$\beta_1 k_1 \times 7.5 \times 10^{-6}\ [H_2O]$
4.	$OH + NO \xrightarrow{M} HONO$	$8.7 \times 10^8\ T^{-2}$
5.	$OH + NO_2 \xrightarrow{M} HNO_3$	$1.5 \times 10^9\ T^{-2}$
6.	$HONO + h\nu \longrightarrow OH + NO$	radiation dependent
7.	$HO_2 + NO \longrightarrow OH + NO_2$	$3.7 \times 10^6\ T^{-1}$
8.	$HO_2 + NO_2 \xrightarrow{M} HO_2NO_2$	$1.5 \times 10^8\ T^{-2}$
9.	$HO_2NO_2 \xrightarrow{M} HO_2 + NO_2$	$7.8 \times 10^{-15}\ e^{-10420/T}$
10.	$HO_2 + HO_2 \longrightarrow H_2O_2 + O_2$	$3.4 \times 10^4\ T^{-1}\ e^{1100/T}$
11.	$HO_2 + HO_2 + H_2O \longrightarrow H_2O_2 + O_2 + H_2O$	$5.8 \times 10^{-5}\ T^{-2}\ e^{5800/T}$
12.	$OH + CO \xrightarrow{O_2} HO_2$	$1.3 \times 10^5\ T^{-1}$
13.	$NO_2 + O_3 \longrightarrow NO_3$	$5.3 \times 10^4\ T^{-1}\ e^{-2450/T}$
14.	$NO + NO_3 \longrightarrow 2NO_2$	$8.4 \times 10^6\ T^{-1}$
15.	$NO_2 + NO_3 \xrightarrow{M} NO_2O_5$	$3.1 \times 10^7\ T^{-1}\ e^{-1100/T}$
16.	$N_2O_5 \xrightarrow{M} NO_2 + NO_3$	$3.5 \times 10^{18}\ e^{-12280/T}$
17.	$N_2O_5 + H_2O \longrightarrow 2\ HNO_3$	$5.7 \times 10^{-4}\ T^{-1}$
18.	$NO_3 + h\nu \longrightarrow 0.3\ NO + 0.7\ NO_2 + 0.7\ O_3$	radiation dependent
19.	$OH + O_3 \longrightarrow HO_2$	$7.0 \times 10^5\ T^{-1}\ e^{-940/T}$
20.	$HO_2 + O_3 \longrightarrow OH$	$4.8 \times 10^3\ T^{-1}\ e^{-580/T}$
20a.	$H_2O_2 + h\nu \longrightarrow 2OH$	radiation dependent
Formaldehyde		
21.	$HCHO + h\nu \xrightarrow{O_2} HO_2 + HO_2 + CO$	radiation dependent
22.	$HCHO + h\nu \longrightarrow CO + H_2$	radiation dependent
23.	$OH + HCHO \xrightarrow{O_2} HO_2 + CO$	$4.4 \times 10^6\ T^{-1}$
Acetaldehyde		
24.	$CH_3CHO + h\nu \xrightarrow{O_2} CH_3O_2 + HO_2 + CO$	radiation dependent
25.	$OH + CH_3CHO \xrightarrow{O_2} CH_3CO_3$	$3.0 \times 10^6\ T^{-1}\ e^{250/T}$
26.	$CH_3CO_3 + NO_2 \longrightarrow PAN$	$2.1 \times 10^6\ T^{-1}$
27.	$PAN \longrightarrow CH_3CO_3 + NO_2$	$1.2 \times 10^{18}\ e^{-13543/T}$
28.	$CH_3CO_3 + NO \xrightarrow{O_2} NO_2 + CH_3O_2$	$3.1 \times 10^6\ T^{-1}$
29.	$CH_3O_2 + NO \longrightarrow HCHO + HO_2 + NO_2$	$3.1 \times 10^6\ T^{-1}$

Table 1: (Continued)

	Reaction	Rate Constant (ppm min units)
Alkanes		
30.	OH + Alkane $\longrightarrow$ AO_2	$6.6 \times 10^6\ T^{-1}\ e^{-400/T}$
31.	AO_2 + NO $\longrightarrow$ −.8 NO + 1.7 NO_2 + .9 HO_2 + .4 CH_3CHO + .45 MEK + .3 CH_3COCH_3	$3.1 \times 10^6\ T^{-1}$
32.	OH + MEK $\xrightarrow{O_2}$ − NO + NO_2 + CH_3CHO + CH_3CO_3	$4.4 \times 10^6\ T^{-1}\ e^{-330/T}$
33.	MEK + $h\nu$ $\longrightarrow$ CH_3CO_3 + $C_2H_5O_2$	radiation dependent
34.	$C_2H_5O_2$ + NO $\longrightarrow$ NO_2 + CH_3CHO + HO_2	$3.1 \times 10^6\ T^{-1}$
Alkenes		
35.	OH + Ethene $\xrightarrow{O_2}$ 2HCHO + NO_2 − NO + HO_2	$9.7 \times 10^5\ T^{-1}\ e^{380/T}$
36.	OH + Propene $\xrightarrow{O_2}$ HCHO + CH_3CHO + HO_2 + NO_2 − NO	$1.8 \times 10^6\ T^{-1}\ e^{540/T}$
37.	OH + Butene $\xrightarrow{O_2}$ 1.8 CH_3CHO + 0.9 NO_2 + 0.9 HO_2 − NO	$5.0 \times 10^6\ T^{-1}\ e^{540/T}$
38.	O_3 + Ethene $\longrightarrow$ HCHO + 0.4 $\dot{C}H_2\dot{O}_2$ + 0.4 CO + 0.12 HO_2	$4.2 \times 10^3\ T^{-1}\ e^{-2560/T}$
39.	O_3 + Propene $\longrightarrow$ 0.5 HCHO + 0.5 CH_3CHO + 0.2 $\dot{C}H_2\dot{O}_2$ + 0.2 $CH_3\dot{C}HO\dot{O}$ + 0.3 CO + 0.2 HO_2 + 01 OH + 0.2 CH_3O_2	$3.1 \times 10^3\ T^{-1}\ e^{-1900/T}$
40.	O_3 + Butenes $\longrightarrow$ CH_3CHO + 0.4 $CH_3\dot{C}HO\dot{O}$ + 0.3 HO_2 + 0.2 OH + 0.45 CH_3O_2 + 0.2 CO	$3.3 \times 10^3\ T^{-1}\ e^{-1050/T}$
41.	$\dot{C}H_2\dot{O}_2$ + NO $\longrightarrow$ HCHO + NO_2	$3.1 \times 10^3\ T^{-1}$
42.	$\dot{C}H_2\dot{O}_2$ + NO_2 $\longrightarrow$ HCHO + NO_3	$3.1 \times 10^5\ T^{-1}$
43.	$\dot{C}H_2\dot{O}_2$ + H_2O $\longrightarrow$ Product	$1.5\ T^{-1}$
44.	$CH_3\dot{C}HO\dot{O}$ + NO $\longrightarrow$ CH_3CHO + NO_2	$3.1 \times 10^6\ T^{-1}$
45.	$CH_3\dot{C}HO\dot{O}$ + NO_2 $\longrightarrow$ CH_3CHO + NO_3	$3.1 \times 10^5\ T^{-1}$
46.	$CH_3\dot{C}HO\dot{O}$ + H_2O $\longrightarrow$ Product	$1.5\ T^{-1}$
Aromatics		
47.	OH + Toluene $\longrightarrow$ 0.15 ARO_2 + 0.20 Cresol + 0.20 HO_2 + 0.65 ADD	$2.7 \times 10^6\ T^{-1}$
48.	$\dot{O}H$ + Xylene $\longrightarrow$ 0.25 Cresol + 0.25 HO_2 + 0.75 ADD	$7.9 \times 10^6\ T^{-1}$

Table 1: (Continued)

Reaction	Rate Constant (ppm min units)
Aromatics (continued)	
49. ADD + NO $\longrightarrow$ 0.75 NO_2 + 0.75 HO_2 + 0.75 DIAL + α_1 $(CHO)_2$ + α_2 CH_3COCHO	$3.1 \times 10^6\ T^{-1}$
50. OH + DIAL $\longrightarrow$ EL	$1.3 \times 10^7\ T^{-1}$
51. EL + NO_2 $\longrightarrow$ $ELNO_2$	$2.1 \times 10^6\ T^{-1}$
52. $ELNO_2$ $\longrightarrow$ EL + NO_2	$1.2 \times 10^{18}\ e^{-13543/T}$
53. EL + NO $\longrightarrow$ 3 NO_2 - 2 NO + α_3 HO_2 + $\alpha_3 (CHO)_2$ + α_4 CH_3CO_3 + α_4 CH_3COCHO + α_3 CO	$3.1 \times 10^6\ T^{-1}$
54. OH + $(CHO)_2$ $\xrightarrow{O_2}$ HO_2 + CO	$8.8 \times 10^6\ T^{-1}$
55. $(CHO)_2$ + $h\nu$ $\longrightarrow$ HCHO + CO	radiation dependent
56. OH + CH_3COCHO $\xrightarrow{O_2}$ CH_3CO_3 + CO	$6.6 \times 10^6\ T^{-1}$
57. CH_3COCHO + $h\nu$ $\xrightarrow{O_2}$ CH_3CO + HO_2 + CO	radiation dependent
58. OH + Cresol $\longrightarrow$ - NO + .75 NO_2 + .75 HO_2 + .75 DIAL	$1.9 \times 10^7\ T^{-1}$
59. NO_3 + Cresol $\longrightarrow$ - NO_2 + HNO_3	$k = 6.6 \times 10^6\ T^{-1}$
SO_x	
60. SO_2 + OH $\xrightarrow{M}$ H_2SO_4 + HO_2	$2 \times 10^9\ T^{-2.5}$
61. SO_2 + $\dot{C}H_2\dot{O}_2$ $\longrightarrow$ HCHO + H_2SO_4	$3 \times 10^4\ T^{-1}$
62. SO_2 + $CH_3\dot{C}HOO$ $\longrightarrow$ CH_3CHO + H_2SO_4	$3 \times 10^4\ T^{-1}$

Notes

α_1 = .075 k_{48} [xylene]/(k_{47} [toluene] + k_{49} [xylene])

α_2 = .75 - α_1

α_3 = k_{47} [toluene] + .5 k_{48} [xylene]/(k_{47} [toluene] + k_{48} [xylene])

α_4 = 1 - α_3

HNO_3 is the total inorganic nitrate (e.g., the sum of HNO_3(g) and NH_4NO_3 aerosol).

ARO_2 = $C_6H_5 \cdot CH_2O$

DIAL = CHOCH=CHCHO

ADD = $C_6H_5(CH_3)(OH) \cdot O_2$

EL = CHOCH(OH)-C(O_2)HCHO

$ELNO_2$ = CHOCH(OH)-C($OONO_2$)HCHO

Aqueous-Phase Chemistry

In order to adequately describe the relevant chemistry within aqueous droplets, it is necessary to consider the system from both a kinetic and an equilibrium viewpoint. Many aqueous-phase reactions (e.g., acid-base and complex formation) occur at rates orders of magnitude faster than the most rapid gas-phase reactions. These very fast aqueous-phase reactions may be assumed to be in equilibrium (i.e., the forward and reverse reaction rates are equal). The equilibrium constant for the reaction then defines the ratio of products to reactants. Other less rapid reactions are treated kinetically in the same manner as the gas-phase reactions.

Modeling aqueous-phase equilibrium chemistry is accomplished using a formulation in terms of equilibrium constants. A set of nonlinear equations provided by mass action laws and mole balance equations are solved by an iterative procedure (Morel and Morgan 1972; Westall 1979). The equilibrium problem is defined by choosing a set of chemical compounds (components) from which all chemical species to be considered can be formed. Stoichiometrically balanced chemical reactions are written for the formation of each species from the set of components and the corresponding equilibrium constants for these reactions are evaluated. Total concentrations of each component, species formation reaction stoichiometries and equilibrium constants are the model inputs.

Formation reactions for the chemical species considered are listed in Table 2 along with Henry's Law reactions, which describe the equilibrium distribution of soluble gases between the gas and aqueous phases. The equilibrium constants for these reactions are necessary to accurately describe the mass-transfer process between the gas and aqueous phases.

Equilibrium constants for the reactions as a function of temperature are calculated using the equation

$$\log K(T) = A + B\left(\frac{1000}{T}\right) + C\left(\frac{1000}{T}\right)^2 \qquad (1)$$

where K is the equilibrium constant and T is temperature in °K. Values for the constant A, B and C which describe the temperature dependence of the equilibrium constants for the reactions considered are listed in Table 2, along with the log of the equilibrium constant at 25°C. The constants A, B and C were determined either by fitting measured equilibrium constant data at different temperatures or using the measured standard enthalpy of reaction, ΔH° in Kcal/mole, and measured equilibrium constant at 298°K with the van' Hoff equation,

$$\ln \frac{K_2}{K_1} = \frac{\Delta H^\circ}{R} \frac{1}{T_1} - \frac{1}{T_2} \qquad (2)$$

where R is the gas constant.

Table 2: SPECIES FORMATION REACTIONS, HENRY'S LAW REACTIONS AND TEMPERATURE DEPENDENCIES OF EQUILIBRIUM CONSTANTS USED IN DESCRIBING THE EQULILIBRIUM CHEMISTRY OF ATMOSPHERIC DROPLETS*

Species Formation Reactions	A**	B**	C**	Log (K)†
1. $SO_4^{2-} + H^+ \rightleftharpoons HSO_4^-$	5.949	-1.180	0.0	1.99
2. $HSO_3^- + H^+ \rightleftharpoons SO_2(aq)$	4.843	-0.874	0.0	1.91
3. $HSO_3^- \rightleftharpoons SO_3^{2-} + H^+$	-8.860	0.486	0.0	-7.23
4. $NO_2^- + H^+ \rightleftharpoons HONO$	1.457	0.546	0.0	3.29
5. $CO_2(aq) \rightleftharpoons HCO_3^- + H^+$	-14.252	5.194	-0.846	-6.35
6. $CO_2(aq) \rightleftharpoons CO_3^{2-} + 2H^+$	-28.055	8.059	-1.392	-16.68
7. $NH_4^+ \rightleftharpoons NH_3(aq) + H^+$	-0.086	-2.732	0.0	-9.25
8. $HO_2(aq) \rightleftharpoons O_2^- + H^+$	-4.9	0.0	0.0	-4.9
9. $H_2O \rightleftharpoons H^+ + OH^-$	-3.833	-3.031	0.0	-14.00

Henry's Law Reactions	A**	B**	C**	Log (K)†
1. $SO_2(g) \rightleftharpoons SO_2(aq)$	-4.648	1.411	0.0	0.085
2. $HONO(g) \rightleftharpoons HONO(aq)$	-5.235	2.062	0.0	1.68
3. $HNO_3(g) \rightleftharpoons H^+ + NO_3^-$	-6.173	3.781	0.0	6.51
4. $CO_2(g) \rightleftharpoons CO_2(aq)$	-4.657	0.756	0.0581	-1.47
5. $NH_3(g) \rightleftharpoons NH_3(aq)$	-3.501	1.604	0.0	1.88
6. $NO(g) \rightleftharpoons NO(aq)$	-2.710	0.0	0.0	-2.71
7. $NO_2(g) \rightleftharpoons NO_2(aq)$	-1.900	0.0	0.0	-1.90
8. $H_2O_2(g) \rightleftharpoons H_2O_2(aq)$	-5.153	2.987	0.0	4.87
9. $O_3(g) \rightleftharpoons O_3(aq)$	-5.158	1.042	0.0	-1.66
10. $HCHO(g) \rightleftharpoons CH_2(OH)_2(aq)$	-5.571	2.808	0.0	3.85

* All unspecified species and components shown above are aqueous-phase species and components.

** Values for the constants A, B, and C in Equation 1.

† Equilibrium constant at 25°C. Henry's Law constant units are molar/atm.

Table 4: RATE EXPRESSIONS FOR Fe AND Mn CATALYZED OXIDATION OF S(IV) BY O_2 VALID OVER VARIOUS pH RANGES AND S(IV) CONCENTRATIONS*

[S(IV)] \ pH	<4	>5
$> 10^{-4}$ M	$R = 4.7\ [Mn^{2+}]^2[H^+]^{-1} + \frac{0.82[Fe^{3+}][S(IV)(aq)]}{[H^+]}_T \times \left(1 + \frac{1.7 \times 10^3\ [Mn^{2+}]^{1.5}}{(6.3 \times 10^{-6})+[Fe^{3+}]}\right)$	$R = 4.7\ [Mn^{2+}]^2[H^+]^{-1} + 10^7[Fe^{3+}][S(IV)(aq)^2]_T$
$<10^{-6}$ M	$R = 5000\ [Mn^{2+}][HSO_3^-] + \frac{0.82\ [Fe^{3+}][S(IV)(aq)]}{[H^+]}_T$ or three times the above if both Fe and Mn are present	$R = 5000\ [Mn^{2+}][HSO_3^-]$

* Rates were interpolated for pH values between 4 and 5 and S(IV) concentrations between 10^{-6} and 10^{-4} M. Rate expressions are according to Martin (1982).

Table 5: AQUEOUS-PHASE FREE RADICAL CHEMICAL REACTIONS AND RATE CONSTANTS*

	Reaction	Rate Constants ($M^{-1}min^{-1}$)
1.	$H_2O_2 + h\nu \longrightarrow 2OH$	2.1×10^{-6} (min^{-1})
2.	$O_3 + h\nu \xrightarrow{H_2O} 2\ OH$	4.7×10^{-5} (min^{-1})
3.	$O_3 + OH \longrightarrow HO_2$	1.8×10^{11}
4.	$O_3 + HO_2 \longrightarrow OH + 2O_2$	2.4×10^{8}
5.	$OH + HO_2 \longrightarrow H_2O + O_2$	1.9×10^{11}
6.	$2HO_2 \longrightarrow H_2O_2 + O_2$	5.2×10^{7}
7.	$2OH \longrightarrow H_2O_2$	3.6×10^{11}
8.	$H_2O_2 + OH \longrightarrow HO_2 + H_2O$	1.6×10^{9}

* Selected from Graedel and Goldberg (1983).

concentrations. Film theory allows one to calculate the diffusion from the bulk gas to bulk liquid (or visa versa) based on the sum of diffusion resistances in each boundary layer. The first order ordinary differential equations describing concentrations in the bulk gas and aqueous phase are (see Appendix A for derivation):

$$\frac{dC^g}{dt} = \frac{3L\ (C^a/K - C^g)}{R\left(\frac{\delta g}{D_g} + \frac{\delta a}{kTKD_a}\right)} \tag{3}$$

$$\frac{dC^a}{dt} = \frac{3\,(KC^g - C^a)}{R\left(\frac{kTYK\delta g}{Dg} + \frac{\delta a}{Da}\right)} \tag{4}$$

where C^g = bulk gas concentration (ppm)

C^a = bulk aqueous concentration (moles/liter),

R = water droplet radius (m),

L = volume fraction liquid H_2O in air (dimensionless),

K = Henry's Law equilibrium constant (moles/liter-ppm),

D_g = gas diffusion coefficient in air (m^2/min),

D_a = aqueous diffusion coefficient in water (m^2/min),

δg, δa = gas and aqueous film thickness (m),

k = gas constant (8.2 x 10^4 liter-ppm/mole-degree), and

T = temperature (°K).

A potential problem using this parameterized approach is proper specification of the diffusion film thickness. The theory of convective and diffusive mass transfer to a moving sphere indicates the gas boundary layer thickness can be determined from the Sherwood number (N_{Sh}), the dimensionless ratio of convective to diffusive mass transfer:

$$\delta_g = \frac{2R}{N_{Sh}} \qquad (5)$$

The Sherwood number can be determined from the Froessling equation:

$$N_{Sh} = 2 + 0.6 N_{Re}^{0.5} N_{Sc}^{0.33} \qquad (6)$$

where N_{Re} and N_{Sc} are the Reynolds and Schmidt numbers, respectively. For the case of the small (~10 μm) stagnant droplet employed in the simulations reported here, the Sherwood number is 2, hence δg equals the droplet radius. There is no comparable theory to guide the determination of the aqueous-phase diffusion film thickness. Fortunately, modeling studies have been carried out which compare numerical solution of the convective diffusion equation to the film theory approximation (Walcek and Pruppacher 1983). These modeling results indicate the aqueous diffusion film thickness is a function of the concentration gradient. When the bulk aqueous concentration is nearly in equilibrium with the surface concentration, the aqueous film thickness is approximately one-tenth the droplet radius. In situations where the bulk aqueous-phase concentrations are far from equilibrium with the surface concentrations, smaller thicknesses (~ 0.01 R) are appropriate. This parameterization of δa was developed for falling droplets, however, in the absence of other information, it is used in the modeling here for all droplets (i.e., stagnant or moving). Aqueous phase film thicknesses of 0.01 R and 0.1 R are used for species with fast and slow aqueous kinetics reaction rates, respectively, since the faster reactions can create the largest concentration gradients over the droplet profile.

The gas-aqueous species mass transfer equations include the effects of molecular diffusion inside and outside of the droplet. They do not include a term for the potential effects of interfacial or surface resistance to mass transfer. Schwartz (1983) has found that the latter may be important for diffusion to small (~10 μm) droplets if the surface accommodation coefficient is low (< 0.01). Graedel et al. 1983 have suggested that organic films on natural droplet surfaces may be common and present substantial resistance to mass transfer across droplet surfaces. Since little is known concerning the accommodation coefficients for the species of interest on natural surfaces, the effects of interfacial resistance to mass transfer have been ignored.

Aerosol particle scavenging is treated in two steps in the model. At the onset of condensation, a fraction of the ambient aerosol is instantaneously scavenged by the aqueous phase. For small hygroscopic particles such as sulfate, nitrate and ammonium, this fraction is large (0.5 to 1.0) since these particles are ideal condensation nuclei. Initial scavenging fractions of 0.7 are used in the simulations reported herein. Aerosols are then continuously scavenged at a constant rate while droplets are present. The following mass transfer equations are used to describe aerosol scavenging:

$$\frac{dC^P}{dt} = - \varepsilon C^P \tag{7}$$

$$\frac{dC^a}{dt} = \frac{\varepsilon C^P}{LkT} \tag{8}$$

where C^P = aerosol particle concentration (ppm)
ε = aerosol particle scavenging rate (min^{-1})

and the remaining symbols are as defined previously. The continuous aerosol scavenging rate by cloudwater droplets can be estimated from theoretical models of precipitation scavenging at very low precipitation rates. For the cases reported here, a 20% hr^{-1} rate has been used based on the Scott (1982) model with a precipitation rate of 0.1 mm/hr.

These scavenging equations are also used for gaseous free radicals for which Henry's Law constants are unavailable. Scavenging rates for the free radicals can be calculated from collision frequency and accommodation coefficients as functions of liquid water content and drop radius (Fuchs and Sutugin 1971). There is very little theoretical or experimental work to guide the specification of accommodation coefficients (Sherwood et al. 1975). Reported values range from 19^{-9} to unity. An accommodation coefficient of 10^{-3} has been employed for the calculations reported here. At a liquid water content of 0.2 gm/m^3 and accommodation coefficient of 10^{-3}, the OH scavenging rate is approximately 1% per second. The HO_2 and

Table 6: MASS TRANSFER SPECIES

Method	Aerosol	Aqueous
IRREVERSIBLE	SO_4^{2-}	SO_4^{2-}
SCAVENGING	NH_4^+	NH_4^+
	H^+	H^+
	Fe(III)	Fe(III)
	Mn(II)	Mn(II)
	CO_3^{2-}	CO_3^{2-}
	Gases	
REVERSIBLE	SO_2	αSO_2
SCAVENGING		$+ (1-\alpha)HSO_3^-$
		$+ (1-\alpha)H^+$*
	CO_2	CO_2
	HNO_3	$NH_3^- + H^+$
	HNO_2	$NO_2^- + H^+$
	NH_3	$NH_4^+ + OH^-$
	NO	NO
	NO_2	NO_2
	H_2O_2	H_2O_2
	O_3	O_3
	HCHO	$CH_2(OH)_2$
IRREVERSIBLE	OH	OH
SCAVENGING	HO_2	HO_2
	N_2O_5	$NO_3^- + H^+$

* The coefficient α depends on the solution pH.

N_2O_5 scavenging rates are 28% and 60% lower than the rate for OH.

Mass transfer of the compounds shown in Table 6 are included in the model. The overall mass transfer rate of gases to small droplets (<10 μm diameter) can be determined by either the gas or aqueous phase diffusion resistance depending on the gas's solubility. The primary diffusion resistance for highly soluble gases such as nitric acid and hydrogen peroxide is in the gas phase and is generally small. For less soluble gases such as ozone, nitrogen dioxide, and sulfur dioxide (at low pH), the aqueous diffusion resistance is dominant, however, the total resistance is small. Table 7 shows the characteristic times for mass transfer of gas- and aqueous-phase species under the conditions employed in the simulations. The important point to note here is the time constants for restoration of aqueous species, derived from insoluble gases, to their equilibrium concentrations after kinetic reactions are very small. This means, based on the mass transfer equations above, the bulk concentrations of HSO_3^- is maintained very close to its equilibrium concentration even under conditions of fast kinetic destruction rates.

It should be noted that numerous gases dissociate rapidly in the aqueous phase. As the time scale for the dissociation is much shorter than a typical model time step, several species are mass transferred directly to their dissociation products (shown in Table 5) to more accurately predict aqueous speciation. In the case of S(IV), SO_2 gas is diffused to SO_2(aq), HSO_3^- and H^+, the relative distribution being a function of pH.

Continuous condensation and evaporation of cloudwater droplets is not considered in the model at this time. If nucleation scavenging of aerosols is extremely efficient and gases are

Table 7: CHARACTERISTIC TIMES FOR MASS TRANSFER

Henry's Law Constant (molar/atm)	τ_{gases} (seconds)	$\tau_{aqueous}$ (seconds)	Dominant Resistance
10^{-2}	8500	4×10^{-4}	Aqueous
1	88	4×10^{-4}	Aqueous
10^{2}	4	2×10^{-3}	Gas
10^{4}	3	0.15	Gas
10^{6}	3	15	Gas

$$\tau_{gases} = \frac{R}{3L}\left(\frac{\delta g}{Dg} + \frac{\delta a}{kTKDa}\right)$$

$$\tau_{aqueous} = \frac{R}{3}\left(\frac{kTK\delta g}{Dg} + \frac{\delta a}{Da}\right)$$

Evaluated for:
$R = 5\ \mu m$
$L = 2 \times 10^{-7}$
$T = 298°K$
$Dg = 0.13\ cm^2/sec$
$Da = 2 \times 10^{-5}\ cm^2/sec$
$\delta a = 0.1\ R$
$\delta g = R$
$k = 0.082\ \ell$ -atm/deg mole

always being diffused to droplets, the solutions with and without continuous condensation and evaporation should be very similar.

The model has the capability to completely evaporate cloudwater droplets at any point in time. Upon evaporation, the aqueous-phase pollutants are allocated to the gas and aerosol phases as shown in Table 6. The ammonium and protons are returned with the sulfates as $(NH_4)_xH_ySO_4$ subject to the constraint that x+y=2. If the available ammonium exceeds 2 times the sulfate, the excess maximum is allocated to gaseous ammonia.

Deposition

Modeling of atmospheric pollutant concentrations generally includes the effects of wet and dry deposition. Wet deposition is ignored in the simulations reported here since the analysis pertains only to small water droplets with negligible deposition velocities.

Table 8: DRY DEPOSITION VELOCITIES (cm/sec)

Species	Day	Night
HNO_3	4.0	0.8
H_2O_2	4.0	0.8
HCHO	1.5	0.30
CH_3CHO	1.5	0.30
SO_2	1.0	0.20
NH_3	1.0	0.20
NO_2	0.8	0.16
O_3	0.8	0.16
PAN	0.5	0.10
$ELNO_2$*	0.5	0.10
SO_4^{2-}(AR)**	0.2	0.04
H^+(AR)	0.2	0.04
NH_4^+(AR)	0.2	0.04

*A PAN-type compound formed from the oxidation of aromatic hydrocarbons.
**AR≡ Particulate aerosol.

Dry deposition of gases and aerosol particles is included in the model. Dry deposition of nitric acid and hydrogen peroxide are believed to be rapid, and need to be included in simulation of periods longer than 3-6 hours. Deposition of other gases and aerosols is included for consistency. Table 8 lists the day and nighttime deposition velocities used in the model. Deposition measurements suggest that 1.0 and 0.2 cm/sec are common daytime and nighttime deposition velocities for SO_2 on many surfaces (Sheih et al. 1979). Deposition velocities for other species have been scaled from the SO_2 deposition velocities.

3. ENVIRONMENTAL CONDITIONS

The chemical composition of cloudwater is expected to vary widely with geographic location, as does air quality and precipitation quality, owing to variations in meteorology and emissions. We have selected just one set of meteorological and air quality conditions, and two sets of baseline emission strengths for the analysis presented here. The meteorological and air quality conditions are intended to represent summer conditions in eastern North America. The two sets of baseline emission strengths were selected to represent emissions from a hypothetical power plant complex and an urban area. Additional model runs were performed with the power plant and urban emissions reduced by 50 and 30%, respectively, to explore the implications of acidity-emissions relationships.

The model is used here to simulate air quality and cloudwater composition for a hypothetical 1-day Lagrangian transport of a 30x30 km air parcel. Simulations are initiated at sunrise and either urban or power plant emissions are entrained as the air parcel moves at low wind speeds (~2 m/sec) over the emission sources from 0600 to 1000 hours. The air parcel is then assumed to move over rural regions void of significant emissions until near sunrise the following day. Liquid water condensation occurs in the evening (2000 hours) and persists throughout the night. This condition may reflect low stratus or fog formation due to gradual nighttime cooling of moist air. A constant liquid water content of 0.2 gm/m^3 and

Table 9: INITIAL POLLUTANT CONCENTRATIONS

Gases	Concentration (ppb)	
SO_2	3.0	
NO_x	2.0	
NO		0.5
NO_2		1.5
NO_y	0.2	
HNO_3		0.1
PAN		0.1
RHC	50.0 (ppbC)	
>C3 Alkanes		30.0
Ethylene		2.5
Terminal Alkenes		0.0
Internal Alkenes		0.0
Monoalkylbenzenes		10.0
Di/trialkylbenzenes		2.5
Formaldehyde		3.0
Other Aldehydes		2.0
O_3	50.0	
H_2O_2	0.1	
NH_3	7.0	
CO	200.0	
CO_2	320,000.0	
H_2O	based on 50% RH without clouds and 100% RH with clouds	
Aerosols	**Concentrations ($\mu g/m^3$)**	
SO_4^{2-}	5.0	
NH_4^+	5.0	
H^+	5.0	
Fe	0.5	
Mn	0.025	
Elemental Carbon	5.0	
Carbonate	0.9	

average droplet diameter of 10 μm, which are common for continental stratus (Pruppacher 1981), are used throughout this period.

Mixing heights and temperatures are assumed to follow typical diurnal patterns. The mixing height is assumed to rise from 250 meters at sunrise to a maximum of 1500 meters at 1400 hours. This condition is allowed to persist until 2000, at which time the mixing height is decreased to 250 meters. The temperature in the air parcel is assumed to increase gradually from 15°C at sunrise to a daytime maximum of 25°C and then drop to 15°C at night.

The model requires specification of initial surface layer pollutant concentrations and concentrations of pollutant aloft which are entrained as the mixing height rises. We have assumed surface layer and air aloft pollutant concentration are the same (i.e., not strongly influenced by upwind nighttime emissions or dry depositions). Table 9 lists the concentrations used to initiate the model simulations. For the essential gases, these include 3 ppb SO_2, 2 ppb NO_x, 0.2 ppb NO_y, 50 ppb O_3, 0.1 ppb H_2O_2, 7 ppb NH_3, and 50 ppbC reactive hydrocarbons. For the aerosol particles, we have assumed 5 $\mu g/m^3$ NH_4HSO_4, 0.5 $\mu g/m^3$ of iron, 0.025 $\mu g/m^3$ of manganese, 5 $\mu g/m^3$ of elemental carbon and 0.9 $\mu g/m^3$ of carbonate carbon. The SO_2, NO_x, RHC, O_3, and SO_4^{2-} concentrations are within the range commonly observed at rural locations in the northeastern United States (Mueller and Hidy, 1983; Martinez and Singh 1979; Lonneman 1979). Concentrations of other compounds are reasonable estimates. It

Table 10: SURE REGIONAL EMISSION INVENTORY (10^3 Tons/yr)

Type	SO_x	NO_x	RHC
Major Point	25	6.7	1
Minor Point and Area	5	11.3	17
Total	30	18	18

TRAJECTORY EMISSION RATES (10^3 kg/hr)

Scenario	SO_x	NO_x	RHC
Power Plant Complex	26	8	0
Urban Area	4	10	17
Natural	0	0.05	0

should be noted that the RHC composition (60% alkanes, 5% ethylene, 20% mono-alkylbenzenes, 5% di/tri-alkylbenzenes, 6% formaldehyde, and 4% other aldehydes on a carbon basis) represents a fairly aged and only moderately reactive mixture which is expected in rural regions. Similarly, it is assumed that the sulfate has had sufficient transport time to be exposed to ammonia sources and thus exists as ammonium bisulfate.

The emission rates for the simulations were selected to reflect the relative contributions of power plants and urban sources to the overall emissions inventory for the northeastern United States (Klemm and Brennan 1982). Table 10 lists the regional and specific trajectory emission rates. We have assumed that 3% of the SO_x is emitted as primary sulfate, that 7% of the NO_x is emitted as NO_2, and that the RHC emission composition is typical of that measured in urban areas (50% alkanes, 15% alkenes, 30% aromatics, and 5% aldehydes). In addition, NH_3/NH_4^+ emissions of 330 and 200 kg/hr are assumed in the urban and power plant complex scenarios, respectively. The split between NH_3 gas and NH_4^+ aerosol is set so as to represent the SO_4^{2-} emissions as NH_4HSO_4. It should be noted that the emission sources selected represent a very large power plant complex (~3,000 MW) and a moderate size (population ~ 300,000) urban area.

Natural NO_x emissions of 60 $\mu g/m^2$-hr were included in the simulation after passing the source region. This NO_x natural emission rate is close to the upper limit reported in recent investigations of NO emissions from soils (Galbally and Roy 1978; Lipschultz et al. 1981). These emissions are very small relative to the anthropogenic emissions included, however, they were included to prevent the model from computing unrealistically low NO_x concentration.

4. RESULTS AND DISCUSSION

The results of the modeling study are grouped into separate areas for discussion. A brief overview of the general characteristics and features common to all the simulations is presented first. This is followed by more detailed individual discussions of the baseline urban and power plant complex emissions scenarios and then the urban and power plant complex emissions reductions scenarios. Finally, implications of the results on acidic deposition control strategies and the assumptions and limitations inherent in the modeling approach are discussed.

The predicted concentrations of the key gas-, aerosol-, and aqueous-phase species as a function of time along the trajectory for the simulations with baseline emissions are shown in Figures 1 and 2 for the urban and power plant emission simulations, respectively. These figures show a sharp discontinuity in the concentration profiles for the soluble gases and aerosol particles at 2000 hours when liquid water is introduced into the system. Most of the important gas-phase chemistry occurs in the daytime and, therefore, prior to cloudwater formation. Much of the NO_x is oxidized to HNO_3 and PAN; ozone reaches a maximum; a significant amount of SO_2 is oxidized to sulfuric acid aerosol; and significant amounts of H_2O_2 are formed.

The acidity of cloudwater at the onset of condensation is fairly high, which reflects the moderate to high sulfate and nitric acid concentrations resulting from gas-phase oxidation reactions, Cloudwater acidity increases substantially during the first few minutes in the urban case and then increases slowly through the remainder of the night in the urban and power plant cases as a result of aqueous-phase oxidation of S(IV) to S(VI), sulfate aerosol scavenging, and N_2O_5 hydrolysis.

The oxidation of S(IV) to S(VI) in the cloudwater is the most important process contributing to acidification after condensation occurs. S(IV) oxidation by H_2O_2 is the dominant aqueous-phase S(VI) formation pathway, however, oxidation by O_2 catalyzed by Fe and Mn is also important in the power plant emissions cases. This result follows directly from two factors: (1) the abundance of H_2O_2 (especially in the urban emission cases), and (2) the absence of mass transfer limitations in these small drops. Oxidation of S(IV) by the other pathways are predicted to be negligible under the conditions examined here. These results are summarized in Table 11.

The chemical half-life of SO_2 gas in the presence of nighttime fogs is predicted to be 2 and 30 hours in the urban and power plant emission cases, respectively. These can be compared to SO_2 chemical half-lives in cloud-free conditions of hundreds of hours at night and 20-60 hours in the daylight. Thus, the chemical transformations occurring in cloudwater may strongly influence the overall oxidation of SO_x emissions in the atmosphere.

Oxidation of NO_x emissions occurs primarily in the gas phase during daylight hours and results in the formation of significant amounts of gaseous nitric acid, PAN, and, in the presence of NH_3, ammonium nitrate aerosol particles. Scavenging of the inorganic nitrate is

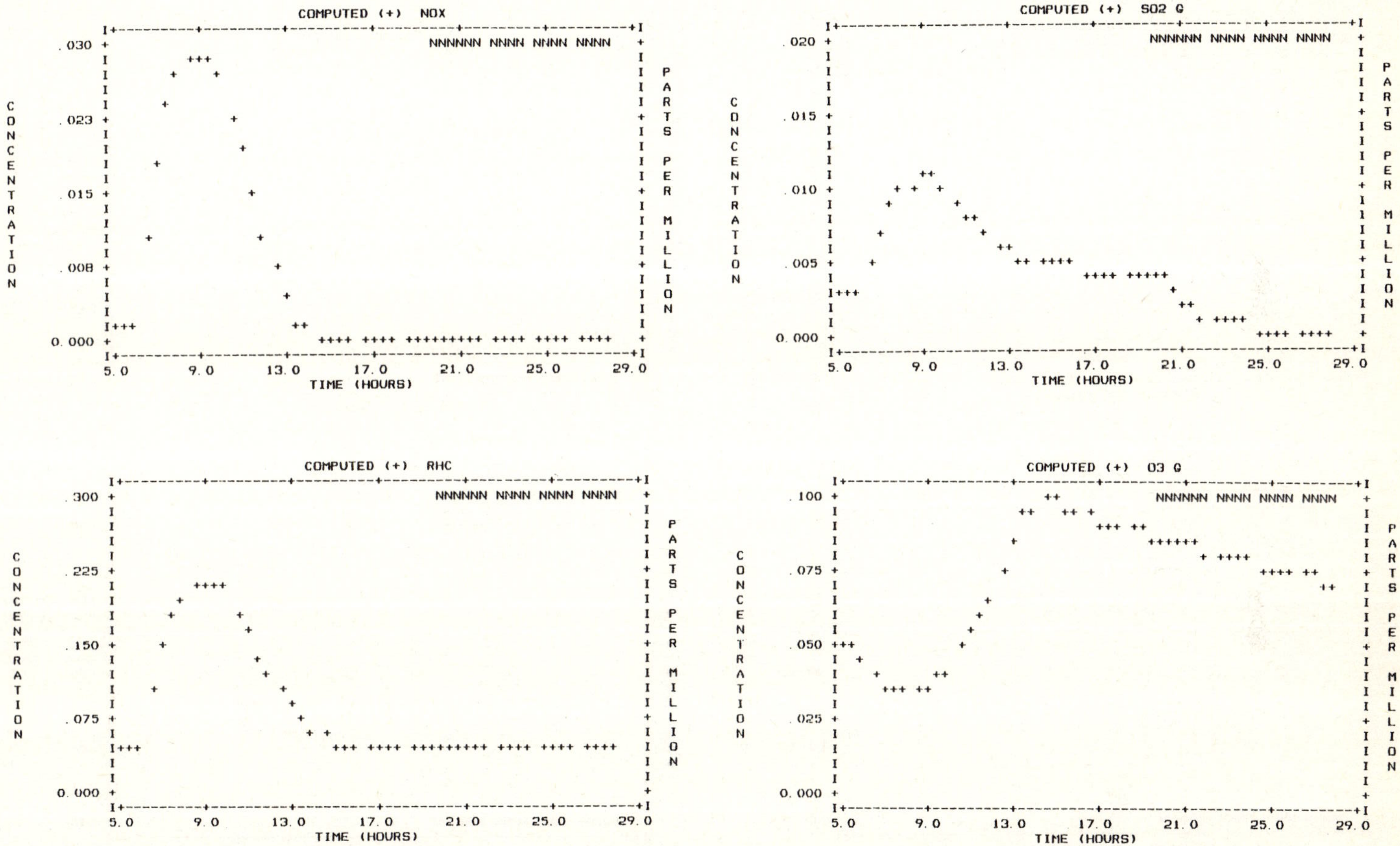

Fig. 1: Predicted concentrations of key gas-, aerosol- and aqueous-phase species as a function of time for the simulation with baseline urban emissions (N = night, G = gas, AR = aerosol particle, L = liquid).

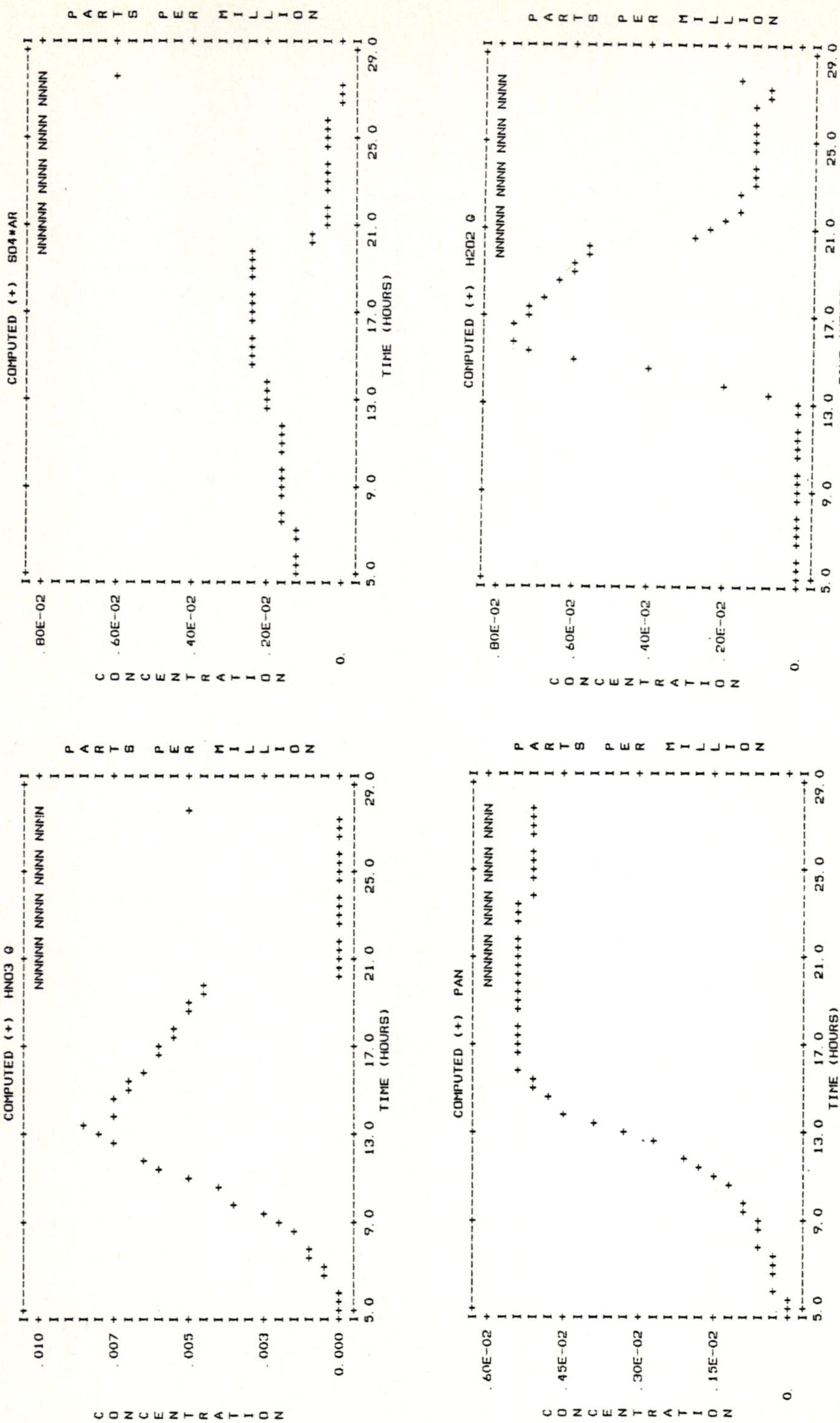

Fig. 1: Continued

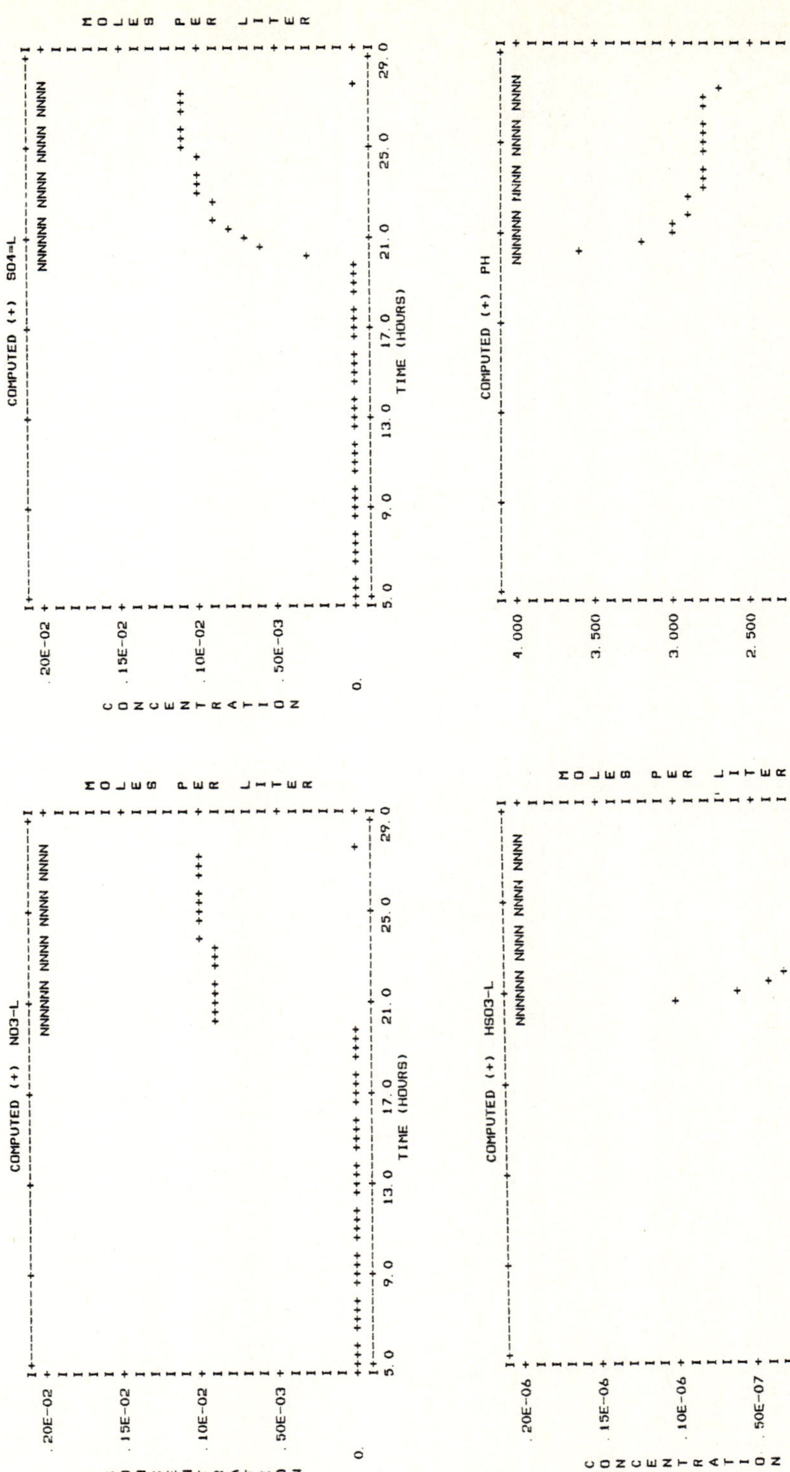

Fig. 1: Continued

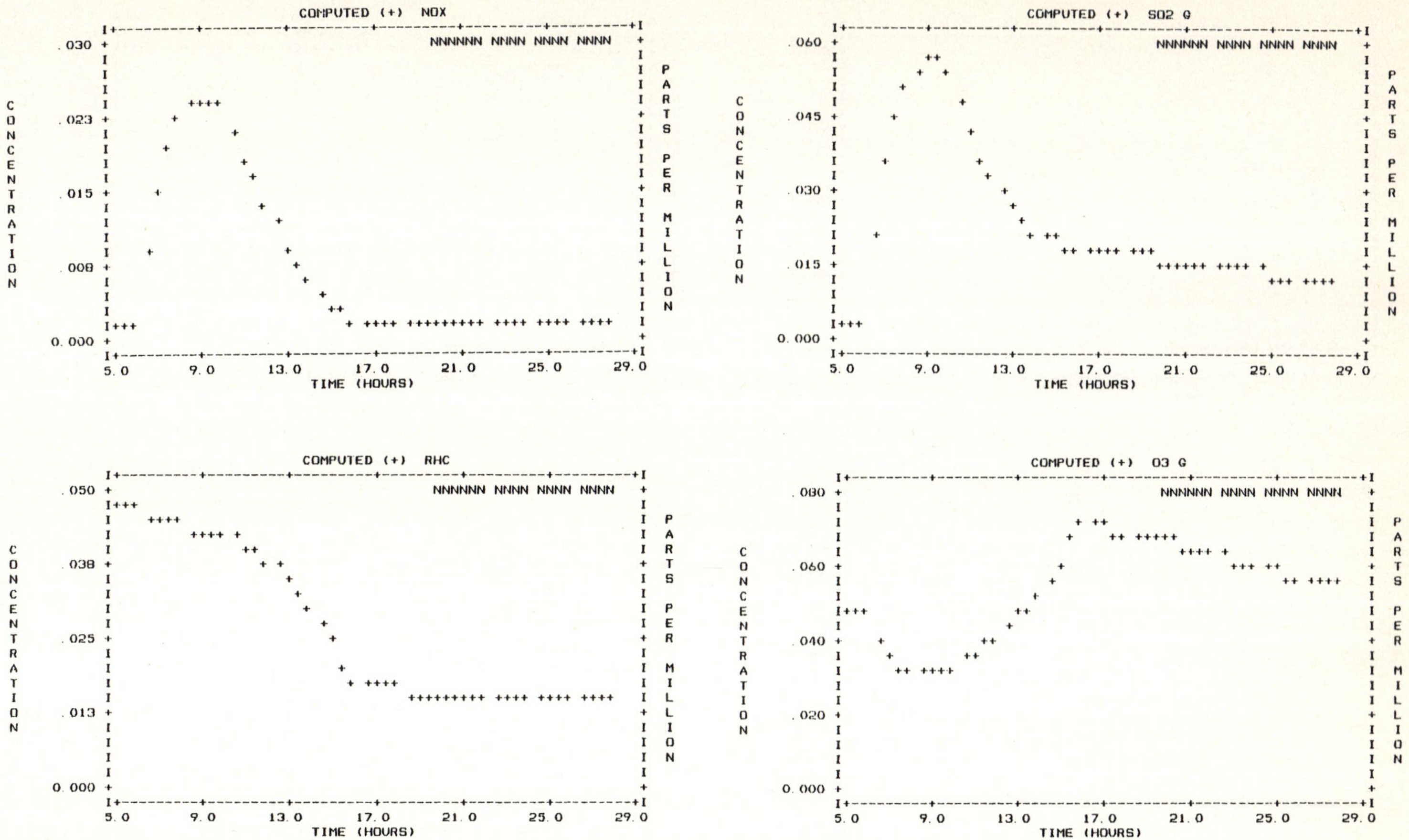

Fig. 2: Predicted concentrations of key gas-, aerosol- and aqueous-phase species as a function of time for the simulation with baseline power plant emissions (N = night, G = gas, AR = aerosol particle, L = liquid).

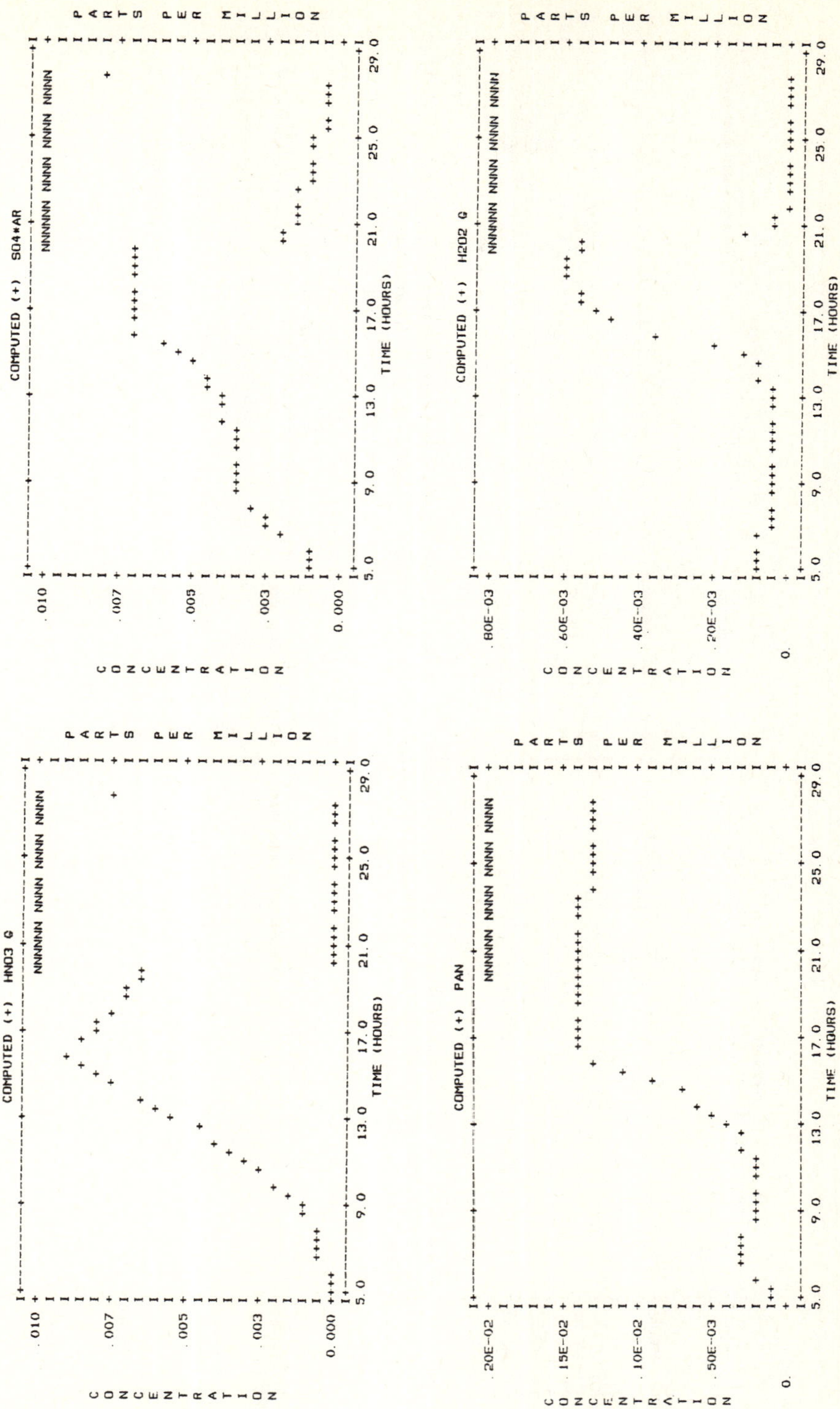

Fig. 2: Continued

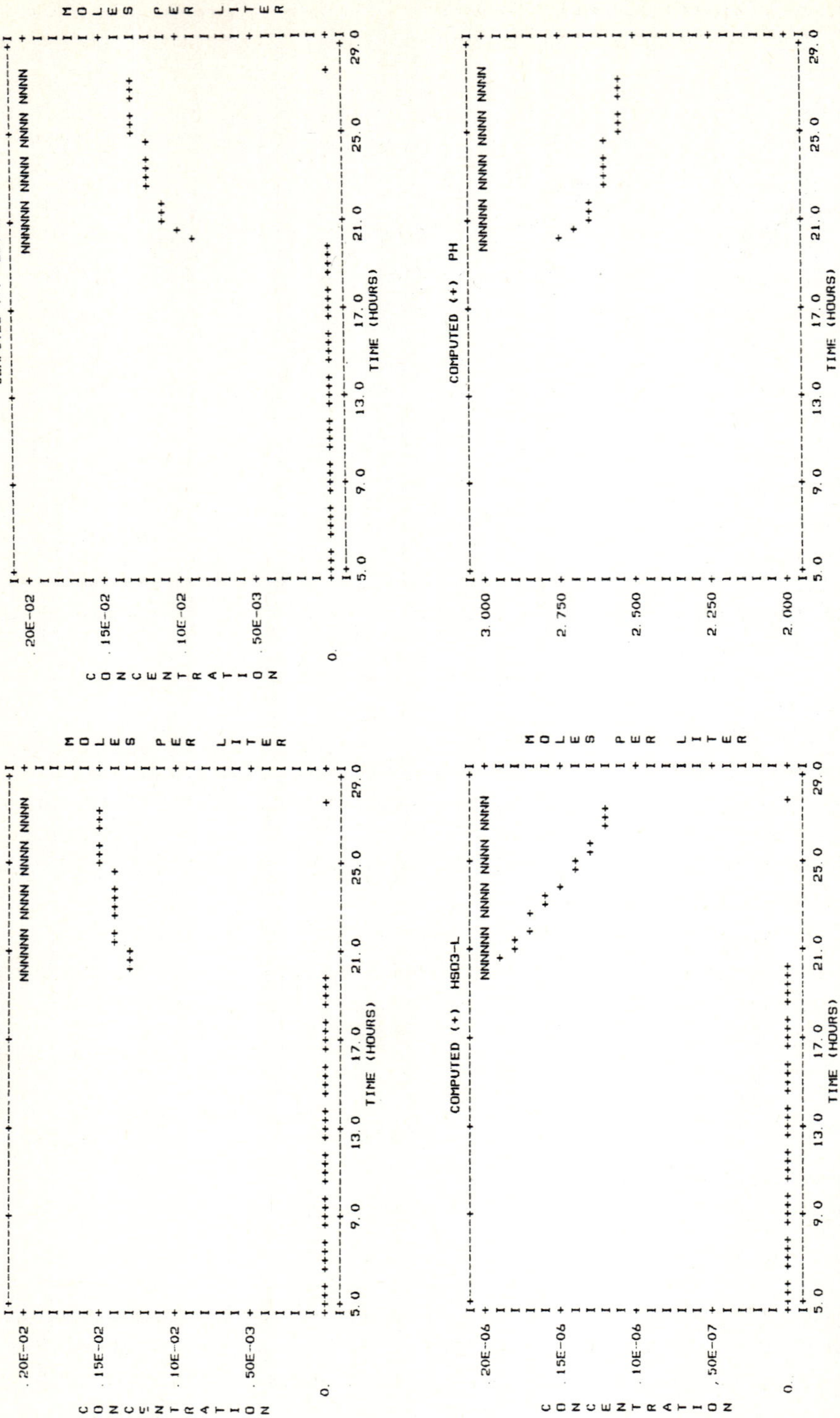

Fig. 2: Continued

Table 11: RELATIVE IMPORTANCE OF S(IV) OXIDATION PATHWAYS TO AQUEOUS S(VI) PRODUCTION (AS PERCENT)

Simulation	2030 hours			0330 hours		
	H_2O_2	Fe/Mn	Others(1)	H_2O_2	Fe/Mn	Others(1)
URBAN:						
Baseline	93.0	6.8	0.2	93.7	6.1	0.2
Reduced NO_x	96.2	3.7	0.1	87.7	11.7	0.6
Reduced RHC	92.0	7.9	0.1	80.6	13.1	0.3
Reduced NO_x, SO_x, RHC	97.4	2.4	0.2	96.8	3.0	0.2
POWER PLANT:						
Baseline	79.5	20.0	0.5	1.0	97.3	1.7
Reduced SO_x	70.8	28.4	0.8	1.3	97.0	1.7
Reduced NO_x	93.6	1.4	0.0	6.6	38.0	5.4
Reduced SO_x and NO_x	97.5	2.5	0.0	73.8	25.4	0.8

(1) Other pathways include O_3(aq), catalysis elemental carbon, OH(aq), NO_2(aq), and NO_2^-(aq).

efficient and contributes significantly (20-40%) to cloudwater acidity. As discussed above, low level, continuous NO emissions are included in the simulations. During the night these NO emissions are oxidized to NO_2, NO_3 and N_2O_5. Subsequent scavenging of the N_2O_5 by water droplets and hydrolysis accounts for additional nitrate formation during the cloud lifetime. Cloudwater nitrate levels are predicted to increase by up to 15% during the cloud lifetime as a result of the N_2O_5 hydrolysis pathway.

The formation of hydroxymethane sulfonic acid (HMSA) from reaction of HSO_3^- with dissolved formaldehyde does not play a significant role in cloudwater as either a source of strong acid or a sink for S(IV). Maximum concentrations predicted are generally less than 0.1 μM.

Urban Emissions Baseline Results

The results for the case with baseline urban emissions (Figure 1) show concentration levels and trends which are generally in agreement with observation in urban areas. NO_x and RHC concentrations in the urban area reach maxima of ~30 ppb and 220 ppbC, respectively. The ozone concentration increases from 50 ppb upwind of the urban area to 100 ppb downwind in the afternoon hours. Nitric acid (gas + aerosol) and PAN concentrations reach maxima of 8.0 and 5.0 ppb, respectively, with the nitric acid maximum occurring several hours before the PAN maximum. H_2O_2 concentrations reach a maximum of 8 ppb in the afternoon. Significant H_2O_2 production occurs after NO_x levels have been reduced and competition for HO_2 decreases. SO_2 concentrations reach a maximum of 10 ppb in the urban area and then decay to ~4 ppb in the late afternoon as a result of diffusion, dry deposition, and oxidation to SO_4^{2-} aerosol. Particulate sulfate concentrations gradually increase from 5 to 9 $\mu g/m^3$ between sunrise and the time water condensation occurs.

With the onset of liquid water, nucleation scavenging of aerosol particles containing sulfate, nitrate, ammonium, and acidity plus diffusion of gaseous nitric acid and ammonia result in fairly acidic cloudwater (pH 3.6) reflecting the preponderance of acids over bases in the system. During the first 30 minutes after condensation, cloudwater acidity increases and the pH drops to 3.2. The pH range predicted here is not unrealistically low based on observations of cloudwater downwind of eastern U.S. urban areas (Falconer and Falconer 1980; Castillo et al. 1983). Also, if one considers that precipitation may commonly dilute the cloudwater pollutant concentrations by a factor of 10 or more, the predicted acidity is not inconsistent with precipitation chemistry data. The cloudwater is predicted to become substantially more acidic during the lifetime of the stratus clouds. By the end of the aqueous-phase calculation (0330 hours), the cloudwater pH has decreased to 2.7. In the period between the onset of condensation and the end of the calculation, the $S(VI)/NO_3^-$ equivalents ratio increases from 0.8 to 2.3. This increase in cloudwater sulfate indicates the importance of the S(IV) oxidation and particulate SO_4^{2-} scavenging relative to nitric acid formation from N_2O_5 hydrolysis. While the $S(VI)/NO_3^-$ ratios are in agreement with those observed in precipitation in the eastern United States (1 to 3), the final pH in this simulation may be unrealistically low relative to rural observations. While the pH has been measured in urban areas (Munger et al. 1983), it seems low for the case where urban emissions have been transported for ~18 hours.

A number of factors may be responsible for potentially overpredicting the cloudwater acidity. First, the box model does not include the effects of lateral dispersion of the urban plume beyond the 30 km parcel width and hence may overpredict the gas and aerosol particle concentrations at the onset of cloud formation. Second, there are large uncertainties in the kinetics of $H_2O_2(g)$ formation and H_2O_2 dry deposition velocities (as well as HNO_3, SO_4^{2-}). Since it is the relatively high $H_2O_2(g)$ concentrations which cause continued acidification of the cloudwater through the aqueous-phase oxidation of HSO_3^- to sulfate by $H_2O_2(aq)$, the high acidity may be an artifact of high H_2O_2 yields from the $RHC/NO_x/SO_x$ chemical mechanism or underestimation of H_2O_2 dry deposition velocities. Third, potential mass transfer limitations at the droplet surface in the presence of organics have been ignored. The mass transfer of $SO_2(g)$ to aqueous S(IV) is essentially instantaneous in the calculations, If substantial interfacial resistance exists, for example, due to organic films as suggested by Graedel et al. (1982), the acidification by S(IV) oxidation may be significantly slower than predicted here. Hence, there are substantial uncertainties in the understanding of these processes which are reflected in the calculations.

Power Plant Emissions Baseline Results

The results for the case with the power plant complex emissions (Figure 2) differ from the urban case primarily because of the higher SO_x and lower RHC loadings. Gas-phase SO_2 concentrations reach a maximum of $\sim$60 ppb while the parcel is passing the emission sources and then decreases to 16 ppb by the late afternoon. NO_x levels reach a maximum of $\sim$25 ppb in the morning and decay to 1 ppb in the evening. RHC levels are quite low, 15-50 ppb, throughout this simulation. At these low HC/NO_x ratios, nitric acid formation is strongly favored relative to PAN formation and the model predicts 6.0 ppb of HNO_3 (gas + aerosol) and only 1.4 ppb of PAN at the onset of liquid water condensation. Significant O_3 scavenging by NO_x is predicted in the source region followed by slow O_3 production reaching $\sim$70 ppb by the late afternoon. H_2O_2 production is slower than in the urban emission case owing to slower NO_x oxidation at these low RHC concentrations. Late in the day H_2O_2 concentrations reach 0.6 ppb which is an order of magnitude lower than in the urban emissions case. Gas-phase sulfate aerosol formation is substantial; sulfate concentrations increase from 5 to 27 ug/m^3 between the early morning and late afternoon due to a combination of SO_2 oxidation and primary emissions. All of these results reflect the fact that oxidation of SO_x and NO_x, in a relative sense, is occurring more slowly in the power plant case than the urban case due to the lower RHC levels.

Predicted concentrations in cloudwater formed downwind of the power plant complex show much greater sulfuric acid contributions to acidity than nitric acid contributions, as is expected from their emissions. The initial pH of the cloudwater is 2.7 and gradually decreases to 2.5 by 0330 hours due to S(IV) oxidation, sulfate aerosol scavenging, and nitric acid formation via N_2O_5 hydrolysis at night. The first two processes are the most important in this case and contribute additional acidity in comparable amounts throughout the cloudwater lifetime. Cloudwater hydrogen ion concentrations increase by a factor 1.5 over the 6.5 hour lifetime of the cloud. The S(VI)/NO_3^- equivalents ratio increases from 1.4 to 1.9 during this period. S(IV) oxidation by H_2O_2 is initially more important in enhancing droplet acidity, however, after the H_2O_2 is depleted, oxidation occurs at a fairly slow rate via O_2 catalysis by Mn and Fe. Scavenging of particulate sulfate is thus a more important contributor to aqueous-phase sulfate production than oxidation processes in the later part of the cloud lifetime. In terms of the overall droplet acidity, scavenging of sulfuric acid and nitric acid formed in the gas phase is more important ($\sim$80% of acidity) than aqueous oxidation of S(IV) to S(VI) in the power plant emission case.

The reasonableness of the predicted low cloudwater pH values are of interest since few observations of cloudwater pH below 3 have been made in rural regions. In this case, the predicted pH is below 3 at the onset of water condensation indicating the predicted nitric acid gas and particulate sulfate concentrations are overestimated. Few measurements of gaseous nitric acid exist for rural locations so it is difficult to evaluate the reasonableness of the 6 ppb concentration predicted here. However, the particulate sulfate prediction can be compared with the EPRI SURE observations. The calculated 27 $\mu g/m^3$ is consistent with the 20-40 ug/m^3 observed during summer sulfate episodes in the eastern United States. The relatively high sulfate and nitrate concentrations suggest dispersion and dry deposition of these pollutants and/or their precursors may have been underestimated in the calculations. This characteristic of the results, however, does not detract from the usefulness of the results for examining coupling between the gas/particle/aqueous chemistries. It should just be recognized that the results apply to summer episodic conditions only. The situation in other seasons will be complicated by the addition of ice particles in the cloud, as well as other factors, including temperature and sunlight reductions.

Reduced Urban Emissions Results

The results for the baseline emissions runs and reduced emission runs are summarized in Tables 12 to 15. Table 12 lists the concentrations of gases and aerosols just prior to cloudwater formation (2000 hours) and the aqueous species concentrations 30 minutes after cloudwater formation (2030 hours). Table 13 lists the aqueous species concentrations 30 minutes prior to cloudwater evaporation (0330 hours, second day) and the gas and aerosol concentrations immediately after evaporation (0400 hours, second day). Tables 14 and 15 show the percent changes in concentrations in the runs with reduced emissions relative to the runs with baseline emissions at these four times. To simplify the interpretation of the results for cloudwater acidity, Tables 14 and 15 also include an acidity-emissions reduction benefit ratio. This ratio is defined as the ratio of the reduction of predicted cloudwater hydrogen ion concentration to the reduction in potential acidity of the emissions. The potential acidity of the emissions is calculated from the molar emission rates assuming oxidation of nitrogen and sulfur molecules generates one and two hydrogen ions, respectively. Results showing a positive ratio indicate there is a benefit to the emission reduction. Results showing a ratio of unity indicate the acidity-emission relationship is linear.

In the case with urban NO_x emissions reduced by 30% from their baseline rates, the predicted concentrations reflect the expected trend for a photochemical system operating in a higher HC/NO_x ratio regime than the baseline cases. The NO_x is oxidized more rapidly to NO_y and the composition of the NO_y shifts from 45% HNO_3 in the baseline case to 28% HNO_3 in the reduced NO_x case, since PAN formation becomes more favorable at higher relative organic loadings. A 30% NO_x emission reduction is predicted to reduce gas, aerosol and cloudwater

Table 12: SPECIES CONCENTRATIONS NEAR THE TIME OF CLOUDWATER FORMATION

		Urban Emissions Case				Power Plant Emissions Case			
		Baseline	Reduced NO_x	Reduced RHC	Reduced NO_x,RHC,SO_x	Baseline	Reduced SO_x	Reduced NO_x	Reduced SO_x and NO_x
Gas/Aerosol @ 2000 Hours									
NO_x	(ppb)	0.02	0.02	0.17	0.02	1.11	1.15	0.11	0.03
RHC	(ppbC)	40.9	44.7	31.4	32.6	16.0	16.1	13.8	13.5
HNO_3	(ppb)	4.44	2.17	5.65	2.83	6.33	6.33	2.51	2.50
PAN	(ppb)	5.52	5.54	4.31	4.65	1.44	1.41	1.94	1.97
O_3	(ppb)	86.2	69.4	89.1	73.1	66.1	64.8	63.8	62.2
H_2O_2	(ppb)	5.72	6.81	4.27	6.43	0.57	0.48	5.92	5.63
SO_2	(ppb)	4.07	4.17	3.98	3.36	16.5	9.12	15.2	8.39
SO_4^{2-}	($\mu g/m^3$)	9.35	8.97	9.74	8.54	27.4	16.8	33.9	20.4
Aqueous @ 2030 Hours									
S(VI)	(μeq)	1223	1160	1040	1020	2157	1422	3690	2321
NO_3^-	(μeq)	904	443	1153	577	1323	1323	513	510
NH_4^+	(μeq)	1313	1291	1346	1298	1292	1287	1224	1209
H^+	(μeq)	643	150	690	140	2019	1226	2819	1436
H_2O_2	(μmol)	374	337	280	240	18	23	209	295
pH		3.19	3.82	3.16	3.85	2.70	2.91	2.55	2.84

Table 13: SPECIES CONCENTRATIONS NEAR THE TIME OF CLOUDWATER EVAPORATION

		Urban Emissions Case				Power Plant Emissions Case			
		Baseline	Reduced NO_x	Reduced RHC	Reduced NO_x,RHC,SO_x	Baseline	Reduced SO_x	Reduced NO_x	Reduced SO_x and NO_x
Gas/Aerosol @ 0400 Hours									
NO_x	(ppb)	0.51	0.52	0.60	0.52	0.89	0.89	0.75	0.73
RHC	(ppbC)	40.4	44.5	30.4	32.5	15.4	15.6	13.4	13.2
HNO_3	(ppb)	4.95	2.55	5.97	3.20	7.28	7.15	2.92	2.80
PAN	(ppb)	4.97	5.02	3.82	4.22	1.27	1.24	1.71	1.75
O_3	(ppb)	70.8	57.0	74.6	70.0	54.2	53.4	52.4	51.8
H_2O_2	(ppb)	1.66	0.54	0.75	0.71	0.0	0.0	0.0	0.14
SO_2	(ppb)	0.16	0.05	0.58	0.12	12.1	6.38	7.22	2.29
SO_4^{2-}	($\mu g/m^3$)	23.7	24.5	21.4	20.5	32.0	20.9	56.6	40.5
Aqueous @ 0330 Hours									
S(VI)	(μeq)	2360	2431	2135	2255	2952	2097	5219	3883
NO_3^-	(μeq)	1045	536	1264	675	1539	1512	615	591
NH_4^+	(μeq)	1449	1443	1442	1444	1408	1383	1391	1356
H^+	(μeq)	1768	1357	1620	1322	2913	1905	4276	2888
H_2O_2	(μmol)	158	51	74	66	0.05	0.11	0.24	0.14
pH		2.75	2.86	2.79	2.88	2.54	2.72	2.37	2.54

Table 14: PERCENT CHANGE IN SPECIES CONCENTRATION FROM THE BASELINE CASE NEAR THE TIME OF CLOUDWATER FORMATION

	Urban Emissions			Power Plant Emissions		
	Reduced NO_x	Reduced RHC	Reduced NO_x,RHC,SO_x	Reduced SO_x	Reduced NO_x	Reduced SO_x and NO_x
Gas/Aerosol @ 2000 Hours						
NO_x	0	+750	0	+4	-90	-93
RHC	+9	-23	-20	+1	-14	-16
HNO_3	-51	+27	-36	0	-60	-61
PAN	0	-22	-16	-2	+35	+37
O_3	-19	+3	-15	-2	-3	-6
H_2O_2	+19	-25	+12	-16	+939	+888
SO_2	+2	-2	-17	-45	-8	-49
SO_4^{2-}	-4	+10	-9	-39	+24	-26
Aqueous @ 2030 Hours						
S(VI)	-5	-15	-17	-34	+71	+8
NO_3^-	-51	+28	-36	0	-61	-61
H_2O_2	-10	-25	-36	-6	+1060	+1540
H^+	-77	+7	-78	-39	+40	-29
Acidity-Emission Reduction Benefit Ratio	+4.06	--	+2.61	+0.95	-4.44	+0.58

Table 15: PERCENT CHANGE IN SPECIES CONCENTRATION FROM THE BASELINE CASE NEAR THE TIME OF CLOUDWATER EVAPORATION

	Urban Emissions			Power Plant Emissions		
	Reduced NO_x	Reduced RHC	Reduced NO_x,RHC,SO_x	Reduced SO_x	Reduced NO_x	Reduced SO_x and NO_x
Gas/Aerosol @ 0400 Hours						
NO_x	+2	+18	+2	+2	-14	-16
RHC	+10	-25	-20	+1	-13	-14
HNO_3	-48	+21	-35	-2	-60	-62
PAN	+1	-23	-15	-2	+35	+38
O_3	-19	+5	-1	-1	-3	-4
H_2O_2	-67	-55	-57	--	--	--
SO_2	-69	+260	-25	-88	-87	-96
SO_4^{2-}	+3	-10	-14	-35	+77	+27
Aqueous @ 0330 Hours						
S(VI)	+3	-10	-4	-32	+77	+32
NO_3^-	-49	+21	-35	-2	-60	-62
H_2O_2	-68	-53	-58	--	--	--
H^+	-23	-8	-25	-35	+47	-1
Acidity-Emission Reduction Benefit Ratio	+1.21	--	+0.84	+0.85	-5.22	+0.02

nitrate levels by ~50%. The NO_x reduction has only a minor effect on the gas/aerosol SO_x chemistry in this case. Particulate sulfate concentrations are 0.4 $\mu g/m^3$, or 4% lower than in the case with the baseline urban emissions just before the clouds are formed. Gaseous H_2O_2 concentrations are slightly higher than in the baseline case owing to the lower NO_x concentrations. With the lower initial nitric acid levels and slightly lower sulfate concentrations, the initial pH of the cloudwater is 5.8 which is much higher than in the baseline case. However, within minutes the pH drops to ~4 due to the combined effects of higher solubility of SO_2 and fast oxidation rates for O_3 and O_2 catalyzed by Fe and Mn at higher pH values. The cloudwater acidity is predicted to increase steadily throughout the cloud lifetime such that by 0300 the pH has decreased to 2.86 and essentially all of the SO_2 gas has been transformed to aqueous sulfuric acid. The S(VI)/NO_3^- equivalents ratio changes from 2.0 to 4.6 during the cloud lifetime. The benefit of this emission reduction is predicted to be large, however, the relative benefit decreases with cloudwater lifetime. The 30% reduction in NO_x emission represents a 19% reduction in potential acidity of the total emissions. The acidity-emissions reduction benefit ratio decreases from 4.1 to 1.2 during the cloud lifetime. The important result here is that this emission reduction strategy for urban emissions for the most part produces a greater than 1:1 reduction in cloudwater acidity.

The reduction of urban RHC emissions by 30% has just the opposite effect of the NO_x emissions reductions. The photochemical system operates at a somewhat lower HC/NO_x ratio regime where formation of HNO_3 relative to PAN from NO_x oxidation is favored and H_2O_2 production is somewhat slower than in the baseline case. Cloudwater nitrate levels are predicted to increase by 21 to 28%. Although slightly higher particulate sulfate concentrations are predicted prior to the time of cloud formation, slower aqueous oxidation of S(IV) to S(VI) occurs because of lower H_2O_2 concentrations and lower initial pH (3.35 vs 3.50). The cloudwater acidity is higher during the first few hours of the cloud lifetime than in the baseline case, however, by the end of the simulation, the acidity is slightly lower. The fact that the initial acidity is higher may have importance for precipitating systems where cloudwater lifetime is fairly short. In fact, if one considers that reduction of RHC emissions is the primary strategy used to reduce urban ozone concentrations in the United States for the last ten years, these results suggest this strategy may be working somewhat against the ancillary goal of reducing precipitation acidity (particularly that contributed by nitric acid and nitrate aerosols).

The results for the simulation with urban NO_x, RHC, and SO_x emissions reduced by 30% shows significant reductions of the acidic oxidation products. The HC/NO_x ratio regime is similar or slightly higher than the baseline case so the relative oxidation rates are similar. The oxidation of NO_x to NO_y is shifted somewhat more towards PAN than in the baseline case, however, the HNO_3 (gas + particulate) concentration just prior to cloudwater condensation is still reduced by 36%. The conversion of the available SO_2 gas to particulate SO_4^{2-} occurs slightly faster than in the baseline case, however, the particulate sulfate concentrations are still 10% lower than in the baseline case immediately before the liquid water is introduced into the system. The concentration profiles in the cloudwater are similar to the case with only reduced NO_x emissions. The predicted pH drops from 5.4 to ~3.8 within a few minutes due to rapid oxidation of S(IV). The acidity continues to increase throughout the cloudwater lifetime to a pH of 2.8 at 0330. The acidity-emissions reduction ratio is greater than unity for most of this period. However, comparison of this case to the case with only NO_x emissions reduced shows very little additional acidity reduction benefit is derived from the additional urban RHC and SO_x emission controls.

In summary, results from the simulations with urban emissions suggest cloudwater acidity should be quite high at transport distances corresponding to 18-24 hour travel time. The results with reduced emissions suggest controlling NO_x emissions would be an effective strategy in reducing cloudwater acidity. The control of RHC emissions alone is predicted to have a fairly minor effect on acidity of cloud droplets with long lifetimes. For droplets with short lifetimes (i.e., in precipitating systems), the control of RHC tends to increase droplet acidity and, therefore, may not be an effective strategy. Also, the concurrent control of urban NO_x, RHC, and SO_x results in cloudwater acidity reductions which are comparable to those predicted by controlling NO_x emissions alone.

Reduced Power Plant Emissions Results

The simulation with the power plant SO_x emissions reduced by 50% show ambient SO_2 gas and particulate sulfate reductions of 45 and 39%, respectively, just prior to the time of cloudwater formation. If the background SO_2 and SO_3^{2-} concentrations are subtracted from the predictions, the SO_2 and SO_4^{2-} reductions are very close to 50%, as is expected. The results show the concentrations of other species controlled by the $HC/NO_x/O_3$ chemistry are essentially the same as in the baseline case. The 50% SO_x emission reduction produces almost comparable reductions in cloudwater S(VI) and acidity. S(VI) and H^+ concentrations are reduced by 32 to 39% throughout the cloudwater lifetime. This emission reduction results in cloudwater with pH values in the range of 2.7 to 2.9 instead of 2.5 to 2.7 in the baseline case. The 50% SO_x emissions reduction is equivalent to a 41% reduction in potential acidity of the total emissions. The acidity-emission reductions benefit ratio is 0.85 to 0.95 which indicates that cloudwater acidity responds almost linearly to the emission reduction. This result is essentially in agreement with the arguments of NAS (1983) and Oppenheimer (1983) on the linear proportionality of the acidity-emission reduction relationship for So_2. This result is quite encouraging since this control strategy is the one most commonly proposed by policy makers.

The results for the simulation with power plant NO_x emissions reduced by 50% (with SO_x emissions unchanged) show several interesting features. First, as seen in the case with reduced urban NO_x emissions, the relative reduction in nitric acid concentrations exceed the relative reduction in NO_x emissions since the NO_y composition is shifted towards PAN at these higher HC/NO_x ratios. PAN concentrations are, in fact, predicted to be slightly higher with the 50% NO_x emission reduction. Second, H_2O_2 concentrations of $\sim$ 6 ppb are predicted in this case. These H_2O_2 concentrations are an order of magnitude greater than those predicted in the baseline case. Third, reduction of the NO_x emissions is predicted to enhance the gas-phase SO_2 oxidation rate noticeably. Particulate sulfate concentrations are predicted to increase by 6 $\mu g/m^3$ or 24% just prior to cloudwater formation.

These changes in the gas/particle chemistry affect the cloudwater chemical composition as expected. While the acidity of cloudwater at the time of condensation is similar (2.75) to the baseline case (i.e., the increased SO_4^{2-} offsets the reduced HNO_3), S(IV) oxidation proceeds about an order of magnitude faster in the case with lower NO_x emissions. Instead of 27% of the gas-phase SO_2 being lost during the cloud lifetime, 53% of the SO_2 gas is transformed to sulfuric acid in the cloudwater. This results in predictions showing very acidic cloudwater. After one-half hour of the cloud lifetime the cloudwater pH has decreased to 2.55 versus 2.70 in the baseline case. After 6.5 hours, the cloudwater pH has decreased to 2.37 versus 2.54 in the baseline case. Hence, the acidity-emissions reduction benefit ratio is negative for this control strategy, This occurs because sulfuric acid formation rates in the gas and aqueous phase are enhanced as NO_x emissions are reduced. The enhancement of SO_4^{2-} formation rates greatly offsets the reduction in NO_3^- under the conditions simulated here.

The results for the simulation with 50% reduction of SO_x and NO_x emissions from the power plants show the combined features of the cases with the emissions reduced individually. Both nitric acid and particulate sulfate concentrations are lower than in the baseline case at the time of cloudwater formation. However, the reduction in SO_4^{2-} concentrations is only 25%, rather than 39% predicted in the case with only SO_x emissions reduced. H_2O_2 gas concentrations are $\sim$5 ppb as in the case with only reduced NO_x emissions. The initial acidity of the cloudwater is 3.1, however, with the high H_2O_2 levels, S(IV) oxidation proceeds rapidly. The cloudwater pH decreases to 2.84 within one-half hour and to 2.54 by the end of the simulation. Thus, after 6.5 hours, the cloudwater acidity is comparable to the baseline case. While the acidity-emissions reduction benefit ratio is initially positive and significant, the ratio is very small by the end of the cloudwater lifetime.

Interpretation

The results for the simulations with baseline urban and power plant emissions indicate acidity in stable summer stratus clouds or fog can be quite high in the eastern United States given the available knowledge of chemical processes. Even though the liquid water content of these types of clouds is low, the clouds may act as efficient chemical reactors converting SO_2 gas to aqueous sulfuric acid. The efficiency of the clouds as chemical reactors appears do depend strongly on the gas-phase H_2O_2 production rates and the initial loading of acidic

2. SOLUTION CHEMISTRY OF TRANSITION METAL IONS

In considering the role of transition metal ions as homogeneous catalysts the following aspects should be treated[2]:

a. Abundance of transition metals in the atmosphere
b. Chemical forms in which the transition metals exist
c. Solubility limitations
d. Stability of different oxidation states
e. Complex formation ability
f. Availability of ligands for complexation

The abundance of transition metals in atmospheric aerosols on a global basis can be summarized as follows (data in ug/m^3)[2]: Fe 0.5 - 2.0; Cu 0.01 - 0.80; Zn 0.10 - 0.80; Ti 0.01 - 0.08; Mn 0.01 - 0.12; V 0.01 - 0.10; Ni 0.008 - 0.05; Cr 0.005 - 0.02; Cd 0.001 - 0.010; Co 0.0003 - 0.003; Sc 0.0001; Hg 0.0003 - 0.002; Ag 0.001 - 0.003. Although some of these are relatively low, typical rain water concentrations[3] are $(3.8 \pm 1.6) \times 10^{-6}$ and $(2.6 \pm 0.5) \times 10^{-7}$ M for Fe and Mn, respectively, with much higher concentrations (as high as 10^{-4} M) being recorded in the Los Angeles area. If one keeps in mind that very effective catalysis has been observed for $[Cu^{2+}]$ as low as 10^{-8} M, then the mentioned concentration ranges are adequate to catalyze many oxidation processes[13,14]. Potential catalysts are Fe, Cu, Mn, V and Ni[2], of which Mn and Fe have received considerable attention up to now. It follows that the chemistry of the first row transition metal ions is of importance in the further discussion. These ions usually occur in the 2+ and 3+ oxidation states in solution, with some exemptions like Cu which occurs in the 1+ and 2+ oxidation state.

In general metal ions do not exist as free ions in aqueous solution, but as hydrated or complexed species. Transition metal ions have vacant d orbitals which can via hybridization interact with vacant s and p orbitals to produce three types of complexes, viz. tetrahedral - sp^3 hybridization - coordination number 4; square planar - dsp^2 hybridization - coordination number 4; and octahedral - d^2sp^3 hybridization - coordination number 6. These vacant orbitals are filled by electron pairs donated by coordinating ligands, or in the absence of ligands by the coordinating oxygen atom of the solvent (water) molecules. Such aquated metal ions exhibit characteristic acid-base equilibria which for a M^{2+} species with coordination number 4 can be summarized as follows:

$$\begin{aligned} M(H_2O)_4{}^{2+} &\rightleftharpoons M(H_2O)_3OH^+ + H^+ \\ M(H_2O)_3OH^+ &\rightleftharpoons M(H_2O)_2(OH)_2 + H^+ \\ M(H_2O)_2(OH)_2 &\rightleftharpoons M(H_2O)(OH)_3{}^- + H^+ \\ M(H_2O)(OH)_3{}^- &\rightleftharpoons M(OH)_4{}^{2-} + H^+ \end{aligned} \qquad (2)$$

Each of these aquo/hydroxo species will exhibit a characteristic solubility and complexation behaviour. Metal hydroxo species are in general weakly soluble and often precipitate as

These assumptions and uncertainties dictate that the model results should not be overinterpreted. In particular, extrapolation of these results from cloudwater to precipitation is not feasible as the physics of precipitation formation and below-cloud scavenging processes have not been included in the model formulation. Further, extrapolation to longer transport times neglects the continuing roles of dilution and dry deposition in lowering pollutant concentrations and should not be attempted. In particular, the role of nitrate in cloudwater acidification is expected to decrease rapidly as the transport time scale increases beyond that analyzed here. Also, nonlinearities in the gas-phase SO_2 chemistry are expected to decrease with transport time.

5. CONCLUSIONS

State-of-the-science atmospheric chemistry and mass transfer have been incorporated into a relatively simple Lagrangian box modeling framework to predict relationships between emissions and cloudwater acidity in mesoscale transport situations. A limited number of simulations have been carried out for summer conditions in eastern North America for cases with urban and power plant emissions interacting with nonprecipitating cloud systems. Some salient results from the modeling study are as follows:

(1) Cloudwater acidity is predicted to be fairly high (pH 2.5 to 3.6) relative to acidity commonly observed in precipitation. These results are consistent with available cloudwater observations, however, ther predicted acidity tends to be high.

(2) Sulfuric acid is predicted to be the primary source of cloudwater acidity for droplets with long lifetimes. This characteristic is predicted not only for power plant cases where it is expected, but also for urban plumes where NO_x emissions typically exceed SO_x emissions by substantial amounts.

(3) The SO_x oxidation pathways contributing to droplet acidity differed substantially between the urban and power plant emission cases. Water droplet scavenging of SO_2 gas followed by aqueous oxidation to sulfuric acid via the H_2O_2 reaction was found to be most important acidification process for simulations with urban emissions. Gas-phase oxidation of SO_2 to particulate sulfate and subsequent scavenging by cloudwater droplets was found to be most important for cases with power plant emissions.

(4) The contribution of NO_x oxidation products to cloudwater acidity was significant. Scavenging of gaseous nitric acid and particulate NO_3^- accounted for 20 to 40% of the cloudwater acidity. While the majority of nitrate was formed by daytime gas-phase oxidation of NO_2 by OH, nighttime oxidation of NO_2 to N_2O_5 followed by hydrolysis in droplets accounted for up to 15% of the nitrate after 6 hours.

(5) For urban emission sources, the control of NO_x alone and the control of NO_x, RHC,

and SO_x concurrently were found to be effective in reducing acidity in cloudwater. The control of urban RHC emissions alone was not found to be effective for decreasing acidity.

(6) For power plant emission sources, the control of SO_x emissions was found to be effective in reducing cloudwater acidity in mesoscale transport situations. A 50% reduction in SO_x emissions was predicted to reduce cloudwater acidity by 35 to 39%. A 50% reduction in NO_x emissions was found to enhance both particulate sulfate and H_2O_2 concentrations which lead to increased cloudwater acidity. Concurrent reductions of SO_x and NO_x emissions from power plants were found to have little effect on cloudwater acidity because a reduction in HNO_3 levels was offset by higher gas- and aqueous-phase SO_2 to SO_4^{2-} oxidation rates.

The results presented here illustrate the potential for strong co-pollutant interactions in the atmosphere. These interactions require systematic study over a broad range of conditions to explore the possible consequences of different emission control strategies for oxidants, NO_x and SO_x. The calculations so far are highly idealized and apply to specific environmental conditions. Since significant nonlinear responses to emission changes were found, further study using more realistic simulations are warranted. The results discussed here for idealized cases should be interpreted with caution, and should not be extrapolated to long term averages, other seasons or different geographical settings. However, the methodology illustrated should provide incentive for more in-depth study of these complex chemical systems in relation to emission control policy.

APPENDIX A

Derivation of Gas-Aqueous Species Mass Transfer Equations

This appendix describes the derivation of the mass transfer equations for gas and aqueous species used in the model. Consider an air-water droplet mixture wherein chemical species exist in the gas and aqueous phases. Assume the water droplet spacing is large relative to the droplet size and, for simplicity, assume the drops are uniform in size. Rather than attempting to solve the parabolic diffusion equation for species concentrations at various radial distances inside and outside of the droplets, we employ a lumped-parameter approach to parameterize the diffusion process in terms of the bulk gas- and aqueous-phase concentrations.

Consider first the mass flux of gaseous species to water droplet. The flux (ϕ_g) is given by

$$\phi_g = -4\pi R^2 D_g \frac{dC^g}{dr} \qquad \text{(A-1)}$$

where

R = droplet radius (m);

C^g = gaseous species concentration (moles/m^3 here, subsequently ppm);

D_g = diffusion coefficient of gaseous species in air (m^2/sec); and

r = radial distance from droplet center.

Let C^g_s and C^g_∞ denote the gas concentrations at the droplet surface and far away from the droplet as illustrated in Figure A-1. Assume that the diffusion boundary layer (i.e., the region where $dC/dr \neq 0$) thickness (δ_g) is known, so that the concentration gradient can be approximated as

$$\frac{dC^g}{dr} = \frac{C^g_\infty - C^g_s}{\delta_g} \qquad \text{(A-2)}$$

The rate of change of the bulk gas concentration can then be calculated as the product of the mass flux to an individual droplet and the droplet number density in air. The droplet number density is given by $N_D = 3L/4\pi R^3$, where L is the dimensionless volume fraction of liquid water in air (cm^3 water/cm^3 air). The rate of change of the bulk gas-phase species concentration is then given by:

$$\frac{dC^g_\infty}{dt} = -\frac{3LD_g}{R\delta_g}(C^g_\infty - C^g_s) \qquad \text{(A-3)}$$

The film theory approximation is probably more appropriate for diffusion occurring outside the drop than inside. However, as a first approximation, assume the diffusion of species within the droplet can be treated similarly. As illustrated in Figure A-1, let C^a_o denote bulk aqueous species concentration and C^a_s denote the aqueous species surface concentration. The rate of change of the bulk concentration for any drop in the system is given by:

$$\frac{dC^a_o}{dt} = \frac{3D_a}{R\delta_a}(C^a_s - C^a_o) \qquad \text{(A-4)}$$

where

D_a = diffusion coefficient of aqueous species in water (m^2/sec), and

δ_a = aqueous diffusion film thickness (m).

At the air-water interface, the gas and aqueous species are in chemical equilibrium described by their Henry's Law constant K_H:

$$K_H = \frac{C^a_s}{C^g_s} \qquad \text{(A-5)}$$

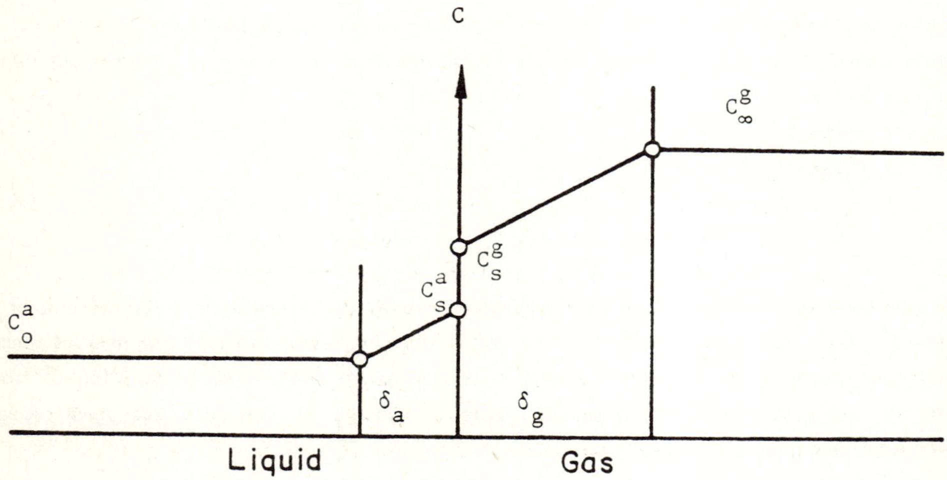

Fig. A-1: Schematic of the Liquid Gas Interface.

Also, since the mass of the chemical constituents is conserved in the mixed-phase system, the mass flux from the gas phase must equal the influx to the aqueous phase:

$$\frac{dC_{\infty}^{g}}{dt} = - kLT \frac{dC_{o}^{a}}{dt} \qquad \text{(A-6)}$$

where

k = universal gas constant, and

T = temperature.

Substituting A-3 and A-4 into A-6, and defining γ= kLT gives:

$$\frac{LD_g}{g} (C_{\infty}^{g} - C_{s}^{g}) = \frac{\gamma D_a}{\delta_a} (C_{s}^{a} = C_{o}^{a}) \qquad \text{(A-7)}$$

Substituting the Henry's Law relationship (Equation A-5) in (A-7) gives:

$$C_{s}^{g} = \frac{C_{s}^{a}}{K_H} = \frac{\frac{LD_g}{\delta_g} C_{\infty}^{g} + \frac{\gamma D_a}{\delta_a} C_{o}^{a}}{\frac{LD_a}{\delta_g} + \frac{\gamma D_a K_H}{\delta_a}} \qquad \text{(A-8)}$$

Substituting this expression for the surface concentrations into Equations A-3 and A-4 provide the mass transfer equations for gas and aqueous species in terms of their bulk concentrations only:

$$\frac{dC_{\infty}^{g}}{dt} = \frac{3L(C_{o}^{a}/K_H - C_{\infty}^{g})}{R(\frac{\delta_g}{D_g} + \frac{\delta_a}{kTD_a K_H})} \qquad \text{(A-9)}$$

$$\frac{dC_{o}^{a}}{dt} = \frac{3(K_H C_{\infty}^{g} - C_{o}^{a})}{\frac{kTK_H \delta_g}{D_g} + \frac{\delta_a}{D_a}} \qquad \text{(A-10)}$$

These equation possess the properties physically expected. At chemical equilibrium, C_{o}^{a} = $K_H C_{\infty}^{g}$, and no mass transfer occurs. When the system is perturbed from equilibrium, mass transfer occurs so as to bring the system closer to equilibrium. For the case where C_{o}^{a} is initially zero, the equation indicates that the amount of gas transferred to the aqueous phase into order to bring the system to equilibrium is proportional to the liquid water content as expected. Also, the denominator of these equations can be physically interpreted as the sum of gas and aqueous diffusion resistance to mass transfer. The units employed in the model for the parameters in Equations A-9 and A-10 are listed in Table A-1.

Table A-1: UNIT CONVENTIONS EMPLOYED IN THE MODEL

Parameter	Units
C^g_∞	Parts per million in air (ppm)
C^a_o	Moles/liter water
K_H	Moles/liter-ppm
R, δ_g, δ_a	Meters
D_g, D_a	Meters^2/minute
L	Dimensionless (m^3 water/m^3 air)
k	8.205 x 10^4 ppm-liters/moles-°K
T	Degrees Kelvin

REFERENCES

Atkinson, R., A.C. Lloyd and L. Winges 1982. An Updated Chemical Mechanism for Hydrocarbon/NO_x/SO_2 Photooxidation Suitable for Inclusion in Atmospheric Simulation Models, Atmos. Environ. 16, 1341-1355.

Barrie, L.A. 1978. An Improved Model of Reversible SO_2-Washout by Rain. Atmos. Environ. 12, 407-412

Benner, W.H., R. Brodzinsky and T. Novakov 1982. Oxidation of SO_2 in Droplets Which Contain Soot Particles. Atmos. Environ. 16, 1333-1339.

Borys, R.S., P.J. Demott, E.E. Hindman and D. Feng 1983. The Significance of Snow Crystal and Mountain-Surface Riming to the Removal of Atmospheric Trace Constituents from Cold Clouds. In: Precipitation Scavenging, Dry Deposition and Resuspension, Volume 1, Precipitation Scavenging, Pruppacher, Semonin and Slinn (editors) pp 171-180.

Castillo, R.A., J.E. Jiusto and E. McLaren 1983. The pH and Ionic Composition fo Stratiform Cloud Water. Atmos. Environ. 18, 1497-1506.

Chameides, W.L. and D.D. Davis 1982. The Free Radical Chemistry of Cloud Droplets and Its Impact Upon the Composition of Rain, J. Geophys. Res. 87, 4863-4877.

Damschen, D.E. and L.R. Martin 1983. Aqueous Aerosol Oxidation of Nitrous Acid by O_2, O_3 and H_2O_2, Atmos. Environ. in press.

Dasgupta, P.K., K. DeCesare and J.C. Ullrey 1980. Determination of Atmospheric Sulfur Dioxide without Tetrachloromercurate (II) and the Mechanism of the Schiff Reaction, Anal. Chem. 52, 1912-1922.

Falconer, R.E. and P.D. Falconer 1980. Determination of Cloud Water Acidity at a Mountain Observatory in the Adirondack Mountains of New York State, J. Geophys. Res. 85, 7465-7470.

Fuchs, N.A. and A.G. Sutugin 1971. High-dispersed Aerosols, in: International Reviews of Aerosol Physics and Chemistry, Vol. 2. G.M. Hidy and J.R. Brock (eds.) pp. 1-60, Pergamon, New York.

Galbally, I.E. and C.R. Roy 1978. Loss of Fixed Nitrogen From Soils by Nitric Oxide Exhalation. Nature 275, 734.

Graedel, T.E., P.S. Gill and C.J. Weschler 1983. Effects of Organic Surface Films on the Scavenging of Atmospheric Gases by Raindrops and Aerosol Particles. Precipitation Scavenging, Dry Deposition and Resuspension, Volume 1: Precipitation Scavenging. H. Pruppacher, R.G. Semonin and W.G.N. Slinn, Eds. Elsevier Publishing, pp 417-430.

Graedel, T.E. and K.I. Goldberg 1983. Kinetic studies of Raindrop Chemistry I. Inorganic and Organic Processes, J. Geophys. Res. in press.

Hegg, D.A. and P.V. Hobbs 1981. Cloud Water Chemistry and the Production of Sulfate in Clouds. Atmos. Environ. 14, 1597-1604.

Klemm, H.A. and R.J. Brennan 1981. Emissions Inventory for the SURE Region, EPRI Report Number EA-1913.

Lee, Y.N. and S.E. Schwartz 1982. Kinetics of Oxidation of Aqueous S(IV) by Nitrogen Dioxide. In: Precipitation Scavenging, Dry Deposition and Resuspension, Volume 1, Precipitation Scavenging. Pruppacher, Semonin, and Slinn (eds) pp 453-470.

Lipschultz, F., O.C. Zafiriou, S.C. Wofsy, M.B. McElroy, F.W. Valois, S.W. Watson 1981. Production of NO and N_2O by Soil Nitrifying Bacteria. Nature 294, 641.

Lonneman 1979. EPA ESRL, Unpublished data.

Maahs, H.G. 1983. Measurements of the Oxidation Rate of Sulfur (IV) by Ozone in Aqueous Solution and Their Relevance to SO_2 Conversion in Nonurban Tropospheric Clouds, Atmos. Environ. 17, 341-345.

Martin, L.R. and D.E. Damschen 1981. Aqueous Oxidation of Sulfur Dioxide by Hydrogen Peroxide at Low pH, Atmos. Environ. 15, 1615-1621.

Martin, L.R. 1982. Kinetic Studies of Sulfite Oxidation in Aqueous Solution, In: SO_2, NO and NO_2 Oxidation Mechanisms: Atmospheric Considerations, ed. J.G. Calvert, Ann Arbor Science, Woburn, MA. in press.

Martinez, J.R. and H.B. Singh 1979. Survey of the Role of NO_x in Nonurban Ozone Formation. EPA-450/4-79-039. U.S. Environmental Protection Agency, Research Triangle Park, NC.

McArdle, J.V. and M.R. Hoffmann 1983. Kinetics and Mechanism of the Oxidation of Aquated Sulfur Dioxide by Hydrogen Peroxide at Low pH, J. Phys. Chem. 87, 5425.

Morel, F. and J.J. Morgan 1972. A Numerical Method for Computing Equilibria in Aqueous Chemical Systems. Environ. Sci. and Technol. 6, 58-67.

Mueller, P.K. and G.M. Hidy 1983. The Sulfate Regional Experiment: Report of Findings, Volume 3. EPRI Report Number EA-1901.

Munger J.W., D.J. Jacob, J.M. Waldman and M.R. Hoffmann 1983. Fogwater Chemistry in an Urban Atmosphere, J. Geophys. Res. 88, 5109-5121.

National Academy of Sciences 1983. Acid Deposition: Atmospheric Processes in Eastern North America, A Review of current Scientific Understanding. National Academy Press, Washington. DC.

Oblath, S.B., S.S. Markowitz, T. Novakov and S.C. Chang 1981. Kinetics of the Formation of Hydroxylamine Disulfonate by Reaction of Nitrite with Sulfites. J. Phys. Chem. 85, 1017-1021.

Oppenheimer, M. 1983. The Relationship of Sulfur Emissions to Sulfate in Precipitation. Atmos. Environ. 17,451-460.

Pruppacher, H. R. 1981. "The Microstructure of Clouds and Precipitation" in Clouds and Their Formation, Optical Properties, and Effects. Academic Press, New York. P.V. Hobbs and A.Deepak Editors.

Richards, L.W., J.A. Anderson, D.L. Blumenthal, J.A. McDonald, G.L. Kok and A.L. Lazrus 1983. Hydrogen Peroxide and Sulfur (IV) in Los Angeles Cloud Water. Atmos. Environ. 17, 911.

Schwartz, S.E. 1982. Gas-Aqueous Reactions of Sulfur and Nitrogen Oxides in Liquid-Water Clouds; In: SO_2, NO and NO_2 Oxidation Mechanisms: Atmospheric Conditions, ed. J. G. Calvert, Ann Arbor Science, Woburn, MA. in press.

Schwartz, S.E. 1983. Mass Transfer Considerations Pertinent to Aqueous-Phase Reaction of Gases in Liquid-Water Clouds. Proceedings of the NATO Conference on Chemistry of Multiphase Atmospheric Systems, Corfu, Greece, Sept. 26-Oct. 8, 1983.

Schwartz, S.E., P.H. Daum, M.R. Hjelmfelt and L. Newman 1983. Cloudwater Acidity Measurements and Formation Mechanisms -- Experimental Design. In: Precipitation Scavenging, Dry Deposition and Resuspension, Volume 1, Precipitation Scavenging. Pruppacher, Semonin and Slinn (editors), pp 15- 30.

Scott, B.C. 1982. Theoretical Estimates of the Scavenging Coefficient for Soluble Aerosol Particles as a Function of Precipitation Type, Rate and Altitude. Atmos. Environ. 16, 1753-1762.

Sheih, C.M., M.L. Wesely and B.B. Hicks 1979. Estimated Dry Deposition Velocities of Sulfur Over the Eastern U.S. and Surrounding Region. Atmos. Environ. 13, 1361-1368.

Sherwood, T.K., R.L. Pigford, C.R. Wilke 1975. Mass Transfer. McGraw-Hill, New York.

Stockwell, W.R. and J.G. Calvert 1983. The Mechanism of the HO-SO_2 Reaction. Atmos. Environ. in press.

Walcek, C.J. and H.R. Pruppacher 1983. On the Scavenging of SO_2 by Falling Cloud and Rain Drops, I. A Theoretical Study of SO_2 Absorption and Desorption. J. Atmos. Chem. in press.

Waldman, J.M., J.W. Munger, D.J. Jacob and M.R. Hoffmann 1983. Urban Cloudwater and Its Potential for Pollutant Deposition in a Los Angeles Pine Forest. Presented before the Division of Environmental Chemistry, American Chemical Society. 186th National Meeting, Washington, DC.

Westall, J. 1979. MICROQL I. A Chemical Equilibrium Program in BASIC. Swiss Federal Institute of Technology (EAWAG), CH-8600 Duebendorf, Switzerland.

RECEPTOR MODELING BASED ON ELEMENTAL ANALYSIS

Glen E. Gordon
Department of Chemistry
University of Maryland
College Park, MD 20742
U.S.A.

ABSTRACT

Although the main species involved in the formation and deposition of acids are compounds of sulfur and nitrogen, one can learn little about the locations of sources of the acid precursors and their transport and dispersion by observing these species per se. Fortunately, considerable progress has been made on the development of receptor models in which concentration patterns for many elements are used to identify sources of airborne particles and trace the movements of air masses bearing those particles over long distances. On an urban scale, the method can be used to determine contributions of various classes of sources to observed loadings of primary particulate material. Global receptor modeling is being used to identify source areas for Arctic haze and springtime dust episodes in the North Pacific Ocean, both involving transport distances of several thousand km. Trace element signatures for smelter complexes have been identified at receptors more than 1000 km away. Long distance tracing of particles will provide data needed for improving airmass trajectory calculations, which are quite inaccurate for long times and distances. Regional scale receptor modeling in populated, continental areas is more difficult than for shorter or longer distances because of the additions of particles when the air mass passes over cities, power plants and industrial complexes. Although the state of the art is not as advanced as for urban and global scales, considerable progress is being made.

RECEPTOR MODELING ON AN URBAN SCALE

Air-pollution agencies commonly estimate the contributions of air pollutants from given sources to ambient concentrations by use of source-emissions inventories coupled with atmospheric dispersion models. Because of large uncertainties in both the emissions

NATO ASI Series, Vol. G6
Chemistry of Multiphase Atmospheric Systems
Edited by W. Jaeschke

inventories and dispersion models, many researchers have been developing receptor models for source attribution. The basic idea of receptor modeling is that of measuring concentrations of species characteristic of certain sources at ambient sampling sites, the "receptors". For example, most particulate lead in U.S. cities results from combustion of leaded gasoline by motor vehicles. If one knows the "profile" of the source, i.e., the amounts of all species emitted relative to the "marker" species, then the ambient contributions of those species from the source can be calculated unless there is preferential deposition or chemical reaction of some species relative to others between emission and reception.

Most receptor modeling has been based on atmospheric concentrations of elements borne by particles. Other species are under investigation for use in receptor modeling, but elements have the advantages that they are not destroyed or created by chemical reactions and the combination of x-ray fluorescence (XRF) and instrumental neutron activation analysis (INAA) allows one to measure concentrations of about forty elements in individual air-filter samples. Knowledge of concentrations of a very large number of species is desirable for receptor modeling; however, many elements on particles are strongly correlated with each other, so the number of independent variables is fewer than the number of elements measured.

Rarely does the entire atmospheric concentration of an element originate from a single class of source, so it is necessary to resolve the observed concentration pattern into its components. This is the approach of the most widely used receptor model, "chemical element balances" (CEBs) (Gordon, 1980; Friedlander, 1973) also called "chemical mass balances" (CMBs) (Cooper and Watson, 1980) in which the concentration $\underline{C_i}$, of element $\underline{i}$ in a sample is given as

$$C_i = \sum_j m_j x_{ij} \alpha_{ij} \tag{1}$$

where m_j = mass concentration contributed during the sampling period by the $\underline{j}$th source ($\mu g/m^3$)

x_{ij} = concentration (by weight) of element $\underline{i}$ in material from source j, and

α_{ij} is a correction term included for generality to account for any gain or loss of element $\underline{i}$ between the source and receptor, but in most treatments the α_{ij} values are set equal to unity.

To apply the CEB method to urban data, one must know the compositions of particles emitted by the important types of sources in the area, i.e., the $\underline{x_{ij}}$ values. For point sources, this can be done by collecting suspended particles from the stack and analyzing them for the elements of interest. Then the object is to determine the source strengths, i.e., the $\underline{m_j}$ values, that give the best fit to the observed concentration pattern, the $\underline{C_i}$ values. Note that Eq. (1) is a matrix equation in which the $\underline{C_i}$s and $\underline{m_j}$s are vectors and the $\underline{x_{ij}}$s form the source concentration matrix. The $\underline{m_j}$ values are determined by a least-squares fit to the observed concentration pattern for each sample, as the solution is usually over-determined by fitting concentrations of more elements than the number of sources assumed. Mathematics of the fitting procedures are discussed in detail by Watson (1979) and not covered here; however, it should be noted that the results depend sensitively on the weighting factors applied to the various elements used in the fit (e.g., Kowalczyk et al., 1982).

Usually one does not attempt to fit the observed concentrations of all measured elements. Some such as Cl, Br, Se and Hg are so volatile that they may leave or condense on particles between the source and receptor, i.e., $\alpha_{ij} \neq 1$ in Eq. (1). Concentrations of some very trace species may not be well known in source materials.

Table 1 shows the results of a typical CEB treatment of an urban data set, the study of Washington, D.C. particles by Kowalczyk et al. (1982). Concentrations of about 40 elements were predicted on the basis of a linear combination of compositions of particles from seven types of sources. Concentrations of 26 elements were fitted by the least-squares procedure, with the remaining elements being allowed to float freely. There had previously been much discussion about whether it was better to fit only a few elements that are strong markers of the sources assumed, leaving most elements floating to test the quality of the fit, or to fit all well-behaved elements to take full advantage of the information content of the data set (Winchester, 1980). For this data set and choice of source profiles, Gordon et al. (1981) demonstrated that the results were virtually identical for fits to any number from 9 to 26 elements if certain key elements were included: Na, Pb, Ca and V, whose concentrations largely determine the strengths of the marine, motor-vehicle, limestone and oil-combustion sources, respectively. If these elements were omitted, the fits were often quite unstable, as reflected, for example, by negative source strengths for some components.

As Washington, D.C. has almost no industrial activity, the sources assumed for the CEBs were of types common to most U.S. urban areas, except that residual oil combustion is mostly confined to near coastal areas. As in this case, it is usually necessary to include two crustal type sources: a soil component and limestone (e.g., Dzubay, 1980). The latter is needed to account for excess Ca from many sources such as gravel roads, cement plants and erosion of building and highways.

Concentrations of most elements are satisfactorily predicted by the CEBs. In some other cases, the concentrations are underpredicted, but for reasons that are understood. The coal, oil and refuse components were based on analysis of particles collected in hot stacks, whereas it is known that substantial fractions of some elements, e.g. Se (Andren et al., 1975), are in the vapor phase, but some of the vapor phase component probably becomes attached to particles in ambient air. Losses of volatile elements such as Cl and Br from particles in ambient air and possibly from filters after collection are common. Concentrations of elements such as In, Cs, Ga and W are so poorly known in the source materials that it would be surprising if they were correctly predicted. However, there are several elements for which the underprediction cannot be easily explained, including Ni, Cu and Cr. Nickel may be underpredicted because of too small a Ni/V ratio of the oil component. Despite precautions against it, abrasion of Cu from brushes of the pump motors might have contaminated some samples with Cu. The excess Cr may have come from the $K_2Cr_2O_7$ bacteriacide that is often used in the cooling towers on buildings, where most samples were collected. (Other sources

Table 1. Average Results of Chemical Element Balances of 130 Samples from Washington, D.C. Area, Summer, 1976[a]

Element (mass unit)	Predicted contributions (mass/m^3)							Total		Predicted/Observed
	Soil	Limestone	Coal	Oil	Refuse	Motor vehicle[c]	Marine	Predicted	Observed[b]	
Na (ng)[c]	43	0.8	8.3	12	35	–	201	300	300±20	1.00
Mg (ng)[c]	74	101	27	3.8	5.4	32	26	270	440±30	0.61
Al (ng)[c]	812	9	517	0.4	6.1	–	–	1340	1350±110	1.00
K (ng)[c]	154	6	67	0.4	47	13	7	295	400±20	0.74
Ca (ng)[c]	66	635	47	8.2	7.1	47	7.6	820	860±40	0.95
Sc (pg)[c]	150	2	180	0.2	0.6	–	–	330	330±30	1.00
Ti (ng)[c]	52	0.83	31	0.03	1.0	–	–	85	110±10	0.77
V (ng)[c]	1.1	0.04	1.6	23	0.01	–	–	26	25±2	1.04
Cr (ng)	0.8	0.02	0.8	0.06	0.21	–	–	2.0	14±2	0.14
Mn (ng)[c]	13	2.3	1.6	0.1	0.3	1.3	–	18	17±2[d]	1.06
Fe (ng)[c]	511	8.3	362	2.8	2.8	34	–	920	1000±60	0.92
Co (pg)[c]	220	0.2	250	160	3	50	–	680	830±80	0.82
Ni (ng)	0.5	0.04	1.0	4.0	0.07	0.34	–	6.0	17±2	0.35
Cu (ng)	0.2	0.01	2.2	0.9	0.7	3.0	–	7.1	17±2	0.42
Zn (ng)[c]	1.1	0.04	2.6	1.6	51	7.3	–	64	85±6	0.75
Ga (pg)[c]	380	8	460	0.1	–	–	–	850	1290±170	0.66
As (ng)[c]	0.06	–	3.1	0.03	0.1	–	–	3.3	3.3±0.2	1.00
Se (pg)	0.9	0.2	780	35	16	35	0.1	870	2500±200	0.35
Br (ng)	0.1	0.01	2.1	0.05	0.7	167	1.2	171	136±9	1.26
Rb (ng)[c]	1.2	–	0.6	–	–	–	0.002	1.8	2.1±0.2	0.86
Sr (ng)[c]	3.65	1.3	3.5	0.1	0.03	–	–	8.6	10±1	0.86
Ag (pg)[c]	0.7	0.1	–	6	230	–	–	240	200±10	1.2
Cd (ng)[c]	0.001	–	0.13	0.003	0.64	1.03	–	1.8	2.4±0.2	0.75
In (pg)	0.7	0.1	2.3	–	2.4	–	0.4	5.9	20±1	0.30
Sb (ng)[c]	0.008	–	0.13	0.007	0.89	0.60	–	1.6	2.1±0.2	0.76

Table 1. Continued

Element (mass unit)	Predicted contributions (mass/m^3)							Total		Predicted/Observed
	Soil	Limestone	Coal	Oil	Refuse	Motor vehicle[c]	Marine	Predicted	Observed[b]	
I (ng)	0.06	0.003	2.1	–	–	–	1.14	3.3	2.0±0.1	1.65
Cs (pg)[c]	28	1.1	39	0.5	2.5	–	–	70	170±50	0.41
Ba (ng)	7	0.02	4.7	2.0	0.3	6.4	–	21	19±2	1.10
La (pg)[c]	750	14	310	16	1.6	–	–	1.1	1.5±0.1	0.73
Ce (pg)[c]	980	24	620	14	7	–	–	1600	2000±200	0.80
Sm (pg)[c]	68	2.7	57	1.2	0.3	–	–	130	200±20	0.65
Eu (pg)[c]	15	0.4	14	0.5	0.2	–	–	30	30±3	1.00
Yb (pg)	37	1.1	30	0.4	0.9	–	–	70	34±3	2.0
Lu (pg)	6.7	0.4	8	0.1	0.4	–	–	16	5.6±0.6	2.9
Hf (pg)[c]	31	0.6	22	0.8	0.4	–	–	55	100±10	0.55
Ta (pg)[c]	41	0.8	8	1.1	1.8	–	–	52	36±4	1.44
W (pg)	14	1.3	38	0.4	7.2	–	–	61	240±20	0.25
Pb (ng)[c]	0.15	0.02	2	0.4	34	428	–	465	440±20	1.06
Th (pg)[c]	110	3.6	100	1.6	0.8	–	–	220	250±20	0.88

[a]Based on Kowalczyk _et al._ (1982).

[b]Average and standard deviation of the mean value.

[c]Element fitted by least-squares procedure.

[d]Observed concentration reduced by factor of 0.69 as described in text.

of K are so strong that its concentration would not be affected.) In any case, significant sources of some elements may have been left out of the CEBs.

One of the most interesting elements is Mn. Manganese is needed in the least-squares fit, as it is one of the few elements that distinguish between coal fly ash and soil. When particles of all sizes are collected on filters, as in Washington, the relative concentrations of most lithophile elements are similar for these two source materials. But coal and the suspended particles produced from it are deficient in Mn relative to soil. (Arsenic is also very useful, being highly enriched in coal emissions relative to soil particles.) The CEBs of Table 1 indicate that most Mn, indeed, arises from soil and the fit of Mn is good. However, the Mn values were multiplied by a factor of 0.69 before running the CEBs. Kowalczyk et al. (1982) found from cascade-impactor studies that only 69% of the Mn was borne by particles having the predominantly large particle size of crustal material (exhibited by elements such as Al). The remaining 31% was fine-particle Mn, whose source could not be identified.

Although concentrations of some trace elements were underpredicted by factors of two or more, the Washington-area CEBs represent a vast improvement over the CEBs attempted for Pasadena (Friedlander, 1973) or Chicago (Gatz, 1975) during the early 1970s, in which predicted concentrations of some trace elements were low by more than an order of magnitude. The improvement results from better knowledge of compositions of particles released by sources from careful analysis of particles from important sources such as coal-fired power plants (Klein et al., 1975; Gladney et al., 1976; Natusch et al., 1974), municipal incinerators (Greenberg et al., 1978a, 1978b) and oil-fired power plants (Mroz, 1976). These studies showed, for example, that suspended particles from coal-fired plants have much higher ratios of "enriched" elements such as As, Se and Sb to those of lithophile elements such as Al and Fe than did the input coal or the particles trapped in electrostatic precipitators. When components containing these higher levels of enriched elements were used in CEBs, a much better accounting of most trace elements was obtained.

Although CEBs are far from perfect, they are probably more reliable than the traditional emissions-inventory/dispersion-model method for determining the sources responsible for total mass of suspended particles in ambient air and the observed levels of elements. Using a simple box model of the Washington, D.C. atmosphere, Kowalczyk et al. (1982) found good agreement between their CEB results and traditional estimates of emissions available for power plants, incinerators and motor-vehicle emissions. The CEB approach also allows one to determine contributions from sources such as entrained soil dust, limestone and marine aerosol, which are difficult or impossible to handle by traditional methods. Soil dust is especially important if one is trying to achieve attainment for total suspended particulate matter (TSP), as it typically accounts for about half of TSP. Receptor models are beginning to be used in regulatory decisions. For example, the Oregon Department of Environmental Quality used the results of CEBs at several sites in Portland, OR to devise the most cost-effective way in which to reduce TSP in non-attainment areas of that city (Gore et al., 1981). The result was that TSP could be most cheaply reduced by control of dust entrainment on roads in industrial areas.

Despite the successes of the CEB or CMB method, it has some serious weakness. One must have an a priori knowledge of the types and compositions of components. One can infer that some sources have been missed only by a consistent underprediction of concentrations of some elements. Additional sources can be identified, if at all, only by trying other sources and observing changes in the overall fit. Furthermore, CEBs treat only primary particles and, unless $_{ij}$ terms other than unity are used, assume no modification of the particles between release and reception. Some authors have partially overcome the first limitation by including sulfate or $(NH_4)_2SO_4$ as a component in the CEBs (Dzubay, 1980; Watson, 1979), but one doesn't know a priori the concentrations of other elements that should be included along with the secondary sulfates. (Below I discuss a method for determining source profiles at the receptor that takes care of the second problem.)

Another receptor model, factor analysis, avoids some limitations of CEBs, but with some loss of other information. Factor analysis is a complex statistical procedure for examining large data sets for multiple correlations among elements that result from the fact that their concentrations depend on strengths of their common sources (Hopke et al., 1976; Harmon, 1976). The first step in factor analysis is a normalization of all concentrations according to equation:

$$z_{ij} = \frac{x_{ij} - \bar{x}_i,}{\sigma_i} \qquad (2)$$

where z_{ij} is the normalized concentration of element i in sample j,
x_{ij} is the actual concentration of element i in sample j,
$\bar{x}_i$ is the average concentration of element i, for the entire data set and
σ_i is the standard deviation of the concentration of element i.

In terms of the normalized concentrations, z_{ij}, the concentration of each element has a mean value $\bar{z}_i = 0$ and a variance, $\sigma_i^2 = 1.0$, This normalization has the effect of giving each element equal weighting in the factor analysis regardless of its average concentration. The objective of factor analysis is that of finding some minimum number of common factors whose collective variance can explain most of the observed variance of each element.

An example of factor analysis of 77 samples taken in Boston, for which data for 18 elements were sufficiently complete for inclusion, is listed in Table 2 (Hopke et al., 1976). Squares of the listed factor loadings, a_{ij}, indicate the fraction of the observed variance of element i that is accounted for by variance of the intensity of factor j. The sum of the a_{ij}^2 values for an element is the "communality", h^2, for that element, i.e., the fraction of its observed variance that is accounted for by variance of the common factors. The sum of the a_{ij}^2 values for factor j indicates the total amount of variance explained by that factor and the sum of these values, also the sum of the communalities, is the total variance of the system explained by all factors. As a result of the normalization of Eq. (2), each element has one unit of variance, so the total variance of the system is equal to the number of elements

Table 2. Results of Factor Analysis of Elemental Concentrations in the Boston Area Atmosphere[a]

Element	Factor Loadings (x100)						Communality
	Factor 1	Factor 2	Factor 3	Factor 4	Factor 5	Factor 6	(%)
Na	17	82	8	7	6	-2	71
Al	89	11	24	10	24	19	96
Cd	-1	84	-4	8	-3	2	71
Sc	90	9	13	9	25	26	97
V	31	4	72	25	6	20	73
Cr	76	-12	44	-11	9	-8	82
Mn	61	13	16	17	56	-1	75
Fe	90	8	25	1	22	17	96
Co	66	7	59	21	23	10	89
Zn	38	1	31	6	24	61	68
Se	17	-1	6	12	68	16	54
Br	13	12	13	59	16	4	43
Sb	45	-5	2	17	19	33	38
La	80	6	20	51	-2	10	96
Ce	70	9	16	47	11	20	80
Sm	85	10	19	42	1	15	97
Eu	83	21	14	14	15	26	87
Th	82	6	16	23	27	9	84
Variance explained	7.6	1.5	1.5	1.3	1.2	0.85	14.0

[a]Hopke *et al.* (1976).

included, 18 in Table 2. In this example, the six factors account for 14 units, or 78% of the total system variance. There is no universal agreement on deciding how many factors to include. The first factor always accounts for the largest amount of variance, the incremental amount dropping with each added factor. In principle, the incremental variance explained should remain fairly high until all important factors are included, then drop sharply for the next factor. But in practice for urban applications, they usually drop slowly, so the cut-off point is not obvious. Some investigators use a rule that one should cut off beyond the factors that explain one unit of variance, while others suggest that one should retain only those that can be interpreted physically.

As in most factor analyses of air-filter or coarse-particle data sets, the first factor in Table 2 represents soil dust. Elements commonly associated with soil dust, e.g., Al, Fe, Sc, etc., have heavy loadings on this factor. The second factor is loaded with Na, Cl and little else, indicating marine aerosol. The third factor is loaded with V, suggesting oil-fired power plants and the fourth factor, loaded strongly with Br, surely represents motor-vehicle emissions. (Minor loadings of some rare earths on this factor are not understood.) Unfortunately, Pb was not included in this data set, but when it is, it is always heavily loaded on the same factor as Br in urban areas(e.g., Dzubay, 1980). The sixth factor, loaded with Zn and Sb, is similar to the incinerator component of Table 1.

Factor 5 is loaded with modest amounts of Mn and Se and little else. Cascade-impactor studies in Boston revealed Se to be strongly associated with fine particles and Mn to be bimodally distributed as in Washington (Gladney et al., 1974). The Mn associated with Se in Factor 5 is probably the fine-particle component, with the coarse part loaded on the soil factor. Concentrations of sulfur were not measured in the Boston study by Hopke et al. (1976); however, in other studies noted below, Se and S are found to be strongly correlated with each other, so sulfate would probably be found with Se and Mn on Factor 5.

Factor analysis overcomes some disadvantages of CEBs, as no a priori assumptions about the number and compositions of the components are needed. Also, as illustrated below, it can handle secondary components such as sulfate. However, a major disadvantage is that the normalization of Eq. (2) makes it impossible to obtain concentration patterns of the components. Thus, one can determine the fraction of variance of an element accounted for by each component, but not the fraction of the observed concentration. Also, as factor analysis operates on sample-to-sample variance, one must have a large data set to use it, whereas CEBs are applied for each sample individually. Despite these limitations, factor analysis is a useful tool, as the results can guide other interpretive methods, e.g., it can indicate the number and types of components to use the CEB analyses.

By use of a different normalizing procedure, Alpert and Hopke (1981) developed the use of an approach that is a sort of hybrid between factor analysis and CEBs, called "target-transformation factor analysis", TTFA. With TTFA, one obtains compositions of the source profiles and, if desirable, can force the compositions of some of them to be of the form of known materials in the atmosphere. Table 3 lists results of TTFA applied to data obtained in the Regional Air Pollution Study (RAPS) of the St. Louis atmosphere (Stevens et al., 1978; Loo et al., 1978). The data of Table 3 resulted from TTFA of two months' of data obtained at one of ten RAPS sampling sites at which particles were collected in two size fractions with dichotomous, virtual impactors and analyzed by XRF. The results shown are for the fine-particles fraction with particle diameters less than about 3 μm. In part because of the restriction to fine particles, the soil component (which may also include some coal fly ash) is not the first factor resolved, but the third factor. The first factor is motor-vehicle emissions, as it accounts for much of the observed Pb and Br. The second factor is largely a secondary sulfate component. This component is partly regional, as sulfur concentrations are not much lower at rural stations near the city. The fourth component is unique to St. Louis - a pigment plant (now defunct) which released huge quantities of particles rich in Ti

Table 3. Mass Contributions (ng/m^3) Determined by Target-Transformation Factor Analysis of July and August, 1976 Data from Station 112 of the Regional Air Pollution Study (fine fraction)[a]

Element	Motor Vehicle	Sulfate	Flyash/ Soil	Pigment	Refuse	Total	
						Predicted	Observed[b]
Al	23	21	170	-	-	230	200±24
Si	-	36	420	-	9	470	450±59
S	0.1	4570	62	2	-	4630	4360±320
Cl	11	21	-	2	27	61	80±9
K	6	30	49	2.4	60	150	150±9
Ca	49	-	52	14	1.5	120	110±10
Ti	-	12	8	46	-	66	64±13
Mn	-	-	2.3	2.0	11	15	17±3
Fe	-	21	120	37	45	220	220±19
Ni	0.4	0.8	0.1	-	0.9	2.1	2.2±0.2
Cu	2.6	0.2	-	-	11	14	15±2
Zn	3.6	-	-	1.5	81	86	75±8
Se	0.4	2.1	-	0.1	0.2	2.8	2.7±0.2
Br	135	0.5	8.2	-	0.1	140	132±8
Sr	0.4	0.2	0.5	0.1	-	1.2	1.1±0.1
Ba	3.1	0.9	0.2	12	0.6	16	15±4
Pb	480	120	16	-	57	680	720±53

[a]Based on Alpert and Hopke (1981).

[b]Average and standard deviation of the mean value.

and Fe (Rheingrover, 1981). The fifth factor is somewhat similar to the refuse component of Table 1, containing a variety of minor and trace elements. However, as noted below, it apparently includes an admixture of trace elements from other sources in the vicinity.

As it yields compositions of the factors, TTFA is more useful than ordinary factor analysis. However, it suffers a common weakness, illustrated below, in that by seeking inter-element correlations it confuses sources of different types which are located close together. The co-location of sources can result in strong correlations of elements from the nearby sources as a function of wind direction. Unless the co-located sources have considerably different operating schedules, the correlations imposed by wind direction may be strong enough to make the two or more sources appear as a single factor in TTFA (Gordon, 1980; Chang et al., 1983).

A weak point in receptor modeling is our poor knowledge of compositions of particles from certain types of sources as received at ambient sampling stations. In an urban setting, the particles will usually have travelled several km before being collected and will have had opportunity for modification of various types, such as fallout of very large particles. If particles are released from high temperature sources, portions of volatile elements may initially be in the gas phase, but condense on plume or ambient particles before reaching the sampler. Recently, attempts have been made to compensate for these effects by "dilution source sampling": instead of collecting particles directly from the stack gases, one cools and dilutes the stack gases with air prior to collection of the particles (Houck et al., 1982) Although this method is an improvement over direct stack sampling, it may not entirely simulate the effects of transit time and interaction with ambient species.

Rheingrover and Gordon (1983) developed a method, called "wind-trajectory analysis" (WTA), for obtaining compositions of particles from various sources as received at the sampling sites. They searched the huge RAPS data set at each station for sampling periods during which fluctuations of wind direction were small (σ_θ< 20 deg) and concentrations of a given element were abnormally high. Histograms of the mean wind directions for sampling periods meeting these criteria often formed clusters about particular directions pointing towards major point sources of the element(s). Treatments at other stations often produced similar clusters for the same elements. Maps of these trajectories for a given element observed at the various stations often yielded convergence of trajectories from two or more stations at the location of the source. When the source was thus identified, compositions of particles released from it were determined by linear regressions of the concentrations of all elements in the trajectory samples versus that of the element associated with the source. For elements that were strongly correlated, the slope of the regression line yielded the concentration of the element relative to the one against which the correlation was run. In this way, compositions were determined for many sources: an iron works, an iron-and-steel complex, Zn and Pb smelters, two Cu plants, the pigment plant mentioned above, an incinerator, motor vehicles and a fertilizer plant.

In addition to the fact that the components are determined at the receptor site, the WTA method has many advantages. First, it avoids the considerable time and expense of measurements at the source; in fact, it does not require knowledge or cooperation of personnel at the source. The data are obtained at the same time and with the same equipment as the ambient samples. Indeed, if ambient sampling networks are designed with this application in mind, the source components are obtainable at no extra cost except for computer treatment of the data (Gordon et al., 1983). A further advantage of WTA is that, except for motor vehicles, the components observed were for a particular source, not just for a class of sources as in other receptor models.

Table 4 illustrates the kind of data obtainable from WTA applied to fine-particles samples heavily influenced by an incinerator in St. Louis in comparison with the refuse component obtained from TTFA of data from one station of the RAPS network (see Table 3). Concentrations of many elements, including Mn, Cu, and Zn, are considerably greater for the TTFA result than that obtained by WTA. This descrepancy illustrates the difficulty that TTFA has in resolving components from sources that are located close together. The

Table 4. Comparison of Compositions of Fine Particles Released by South Incinerator in St. Louis

Element	Concentration (ppm or % by weight)	
	Wind-trajectory analysis[a]	TTFA[b] (Station 112)
Cl (%)	2.8±1.0	2.2
K (%)	3.4±1.1	4.8
Cr (ppm)	300±140	-
Mn (%)	-	0.86
Fe (%)	1.7±0.9	3.6
Ni (ppm)	440±150	700
Cu (%)	0.11±0.05	0.87
Zn (%)	3.7±1.1	6.5
Se (ppm)	-	200
Br (ppm)	-	50
Cd (%)	0.15±0.06	-
Sn (%)	0.41±0.13	-
Sb (ppm)	300±140	-
Ba (ppm)	-	500
Pb (%)	6.3±2.0	4.6

[a]Rheingrover and Gordon (1983)

[b]Alpert and Hopke (1981)

incinerator is in the same general direction from the sampling site as an iron works, a Cu plant and a Zn smelter, so elements from these sources are sufficiently correlated with emissions from the incinerator to raise concentrations of elements associated with these sources in the incinerator component. By contrast, the sharp spatial resolution of WTA allows one to cleanly separate these components. Recently, Chang et al. (1983) showed that the resolution of WTA can be further enhanced by subjecting the samples influenced by a particular source to TTFA instead of the less powerful linear regressions.

One deficiency of WTA is that it has not identified trajectories associated with coal-fired power plants. Perhaps because of their tall stacks, the plumes are too dilute to be observed when they touch down at the sampling sites. Possibly the strong vertical mixing needed to bring the plumes down is always associated with such unstable winds that the sampling periods do not meet the criterion that $\sigma_{\Theta} < 20$ deg. However, the failure to observe coal-fired power plants may result from two features specific to the St. Louis data set. First, the samples were analyzed by XRF, which is not very sensitive for elements by which coal emissions could be best observed, especially Se and As. Second, the various chalcophile element plants (Cu, Pb, Zn) are such strong sources of these and other elements associated with coal-fired plants that plumes from the latter may be obscured.

The above discussion summarizes the major themes of present knowledge of the application of receptor models based on elemental concentrations to urban scale problems. Although these approaches continue to be improved and tested, receptor-modeling is probably already more accurate than traditional methods. It is impossible to prove this statement in a real urban setting, as we don't know the correct answers. In a test sponsored by EPA, the application of various receptor-model techniques to simulated urban data, for which the correct answer were known, produced quite satisfactory results (Currie et al., 1983). However, the application of these methods to real data from Houston, TX, yielded less satisfying results, largely because emissions from many sources in the Houston area were not well characterized (Dzubay et al., 1983). (By contrast, Dzubay et al. (1982) made a satisfactory apportionment of light scattering and absorption in Houston by use of receptor-model techniques.) The Environmental Protection Agency (EPA) is presently making a more careful test of urban receptor modeling. Samples collected at three sites in Philadelphia during the summer of 1982 along with samples from several important sources in the area are being analyzed by XRF and the fine fractions are being further analyzed by INAA in our laboratory, so data will be available for more than 50 elements and chemical species. These data will be circulated to the receptor-model community within the next several months for "round-robin" comparison of results of various receptor-model treatments. Sufficient emissions and meteorological data were collected to make possible careful comparisons of the results of receptor modeling with those of the traditional method.

GLOBAL SCALE RECEPTOR MODELING

There is a strong need to determine the sources and transport of particles on a global basis. Among other reasons, this knowledge is needed for prediction of long-term climatic effects of changes in anthropogenic and natural emissions of particles. Today we don't know the fraction of globally circulating particles that result from human activities, much less their residence time, light-scattering and -absorption properties and their effects on cloud formation.

Receptor modeling is beginning to be applied to particles collected in areas remote from strong local sources of particles. In some of the earliest work of this sort, Rahn (1981) used non-crustal ratios of Mn/V in an attempt to establish the source area for most of the particles observed in the North American Arctic. The term "non-crustal" means that one subtracts crustal contributions by assuming that all Al is derived from the crust:

$$(Mn)_{ncr} = (Mn)_{obs} - (Al)_{obs}(Mn/Al)_{crust};\ (V)_{ncr} = (V)_{obs} - (Al)_{obs}(V/Al)_{crust}, \qquad (3)$$

where "ncr", "obs" and "crust" indicate non-crustal, observed and crustal, respectively. Rahn used Mason's (1966) crustal ratios of Mn/Al = 0.0117 and V/Al = 0.00166. He found that the non-crustal Mn/V ratios in Europe and Asia are approximately 3, whereas those from the northeast U.S. are about 0.3, The differences are apparently due to the extensive combustion of V-containing oil in the U.S. and to some unknown large source(s) of Mn, probably in Central Asia, possibly involving metals industries. As the ratio for North American Arctic particles was about 1.5, they postulated that the particles mainly originate in Eurasia. However, as discussed below, Rahn et al. have more recently indicated that midwestern U.S. air contains particles with a ratio greater than one, so that the value from the Northeast is not representative of North America or even the U.S., weakening the argument regarding the source of Arctic particulate matter.

Cunningham and Zoller (1981) performed a multi-element CEB of remote area particles by resolving data obtained in studies of the pristine atmosphere of the South Pole throughout the 1970s (e.g., Maenhaut et al., 1979). Reasonable fits to the observed concentrations were obtained with components such as sulfate, marine aerosol and crustal material. One of the most interesting results was their finding it necessary to invoke the presence of a meteoritic component to account for appreciable fractions of the observed concentrations of Co, Fe and Mn. The amount of such material in the atmosphere is so small that it would be impossible to detect it via these elements at virtually any other sampling site on the Earth's surface. Cunningham (1979) used a simple model to calculate the meteoritic infall rate to the Earth on the basis of the ambient concentration of material at the South Pole, obtaining an estimate, 1600 tons/yr, at the low end of the range of estimates made by meteorite researchers. Studies are now in progress to confirm this result by observation of elements that are more sensitive indicators of meteoritic material, e.g., platinum-group elements (Tuncel, 1983).

Heidam (1981) performed factor analysis on data from three sites in Greenland obtained by XRF of air filters. He obtained three distinct factors for most of the data sets: a crustal component, a sulfate component containing some metal contributions and a metal component, probably resulting mainly from North American anthropogenic activities.

The most extensive data set for remote areas is now being published by Zoller, Parrington and coworkers. Since 1979, they have collected weekly air-filter samples at Mauna Loa Observatory (MLO) in Hawaii. This is an excellent site for such studies as it is in the Northern Hemisphere, where most continental land mass and the most extensive anthropogenic activities are located. However, the island is so distant from large land masses that it is not strongly affected by nearby sources. The Observatory is located on Mauna Loa mountain at an altitude of 3400 m, well above the boundary layer. Winds at MLO typically follow a diurnal pattern: warming of the slopes during the day causes upslope winds, whereas cooling at night leads to downslope winds. Parrington et al. (1983a) took advantage of these wind conditions by designing a sampling controller that used time of day and wind direction to turn on pumps for the "downslope" or "upslope" filters. When these conditions were not satisfied, or a condensation nuclei counter indicated the presence of contamination, a pump on an "interim" filter was activated. Downslope filters collected material from the free troposphere at an average altitude of about 4 km. Upslope winds brought material up from just above the boundary layer as well as some marine aerosol and particles generated on the island itself, including human activities in the city of Hilo.

The most spectacular result of the MLO studies is the observation of huge increases in the concentrations of airborne crustal material from about March to June in both upslope and downslope samples (Parrington et al., 1983b). Ten-day back-trajectories at 500 and 700 mb calculated by NOAA (Harris, 1982), although not highly accurate over areas of such sparse meteorological data, indicated that these events are associated with air masses coming out of the desert areas of China. Great dust storms occur in those regions, raising the TSP levels to at least 2 mg/m^3 in Beijing (Daisey et al., 1983; Winchester et al., 1981). During the Spring dust periods, the concentration of crustal material in downslope periods at MLO is raised to an average of 730 ng/m^3 vs. the "clean period" average of 50 ng/m^3 (Parrington and Zoller, 1983). Similar levels of Asian dust have been observed at Enewetak by Duce et al. (1983).

During the Asian dust incursions, the dust is so dominant relative to other components that the total sample composition can be used, after minor corrections for marine and sulfate particles, as an Asian dust component. Using such an approach, Parrington and Zoller (1983) performed CEBs of the average upslope and downslope concentration patterns for both clean and dust periods, obtaining the results listed in Table 5. Local soil and dust is derived from Hawaiian volcanic material which has a composition sufficiently different from that of Asian dust that the two types of crustal material can be resolved. Sulfur was the dominant element in the sulfate component and Na was used as an indicator of marine aerosol. Sulfate is the dominant component in clean downslope samples (and, being soluble, the dominant type of cloud condensation nuclei in the free troposphere over the Pacific). Even in clean periods, continental crustal material, mostly from Asia, perdominates over local basalt-like material, and little marine aerosol reaches MLO under downslope conditions. During dust periods, the continental dust contribution increases by more than an order of magnitude. The rise in local basalt concentration may be real or an artifact resulting from errors in resolving it from the much stronger continental component. As expected, upslope samples contain greater concentrations of marine aerosol and local basalt. However, even for them, continental material is dominant during both seasons, suggesting that Asian dust is transported as efficiently in the lower free troposphere as at higher altitudes. The increased sulfate concentrations during the dust periods results from the fact that much of the sulfate arises from biological sources, whose period of maximum productivity in the later Spring and Summer overlaps partially with the Asian dust period (Parrington et al., 1983c).

Some of the most direct "receptor-model" results from this extensive data set can be obtained simply by plotting concentrations of all elements versus time. Most plots display a low baseline concentration with small fluctuations during non-dust periods and large increases during dust periods, even for elements not normally considered to be from crustal sources, e.g., As. It is not known if these elements have naturally elevated concentrations in the material that becomes airborne during dust storms or if the dust is contaminated by anthropogenic emissions during transit over populated and industrial areas of Asia. Daisey et al. (1983) recently reported enormous concentrations of many elements in the atmosphere of Beijing during March and April, 1983. Although this was during the Asian dust season, they attributed many elements to local sources, especially coal combustion in individual housing units.

For many elements, Parrington et al. (1983c) observed concentration spikes far greater than the fluctuations of normal baseline values. The spikes for various elements, including members from among group V, Zn, As, Sb, Se, W, In and Au, often coincided with each other in time and many of them were coincident with sulfate spikes. Parrington et al. (1983c)

Table 5. Results of Chemical Element Balances of Downslope and Upslope Samples Collected at Mauna Loa Observatory[a]

Component	Mass Concentration (ng/m^3)			
	Downslope		Upslope	
	Clean period	Dust period	Clean period	Dust period
Sulfate	240	480	450	640
Marine	16	22	250	320
Continental crust	50	730	160	870
Local basalt	13	25	82	115

[a]Parrington and Zoller (1983a).

interpreted these spikes as resulting from short-term wind patterns that bring particles to MLO from strong anthropogenic sources. Volcanic eruptions could also be a source of the spikes, but none could be definitely associated with a specific eruption prior to the eruption of Kilauea on the island in January, 1983 (Zoller et al., 1983). Back trajectories for some of these periods indicated sources in Asia, but some of the strongest, with elevated concentrations of V, Zn, Co and Sb, seemed to originate from Mexico and or Central America. Trajectories for other spikes indicated the source region to be southwest U.S., in which copper smelters are a strong source of atmospheric trace elements. To test this hypothesis for the origin of the latter events, Parrington et al. (1983c) normalized the concentrations of elements displaying spikes to the concentration of Cu during the event and compared the results with similarly normalized average concentration of the same species from studies of the plumes of five copper smelters in Arizona (Small et al., 1981). As shown in Table 6, the normalized concentrations of sulfate and V are much greater at MLO than in the plumes, the former probably because little of the SO_2 would have been converted to sulfate close to the

Table 6. Excess Concentrations of Several Species Normalized to Copper in Plumes from Copper Smelters and in Two Samples at Mauna Loa Observatory

Species	Concentration Relative to Copper (x1000)		
	Copper smelter plumes[a]	Mauna Loa Samples (week of 6/27/81)[b] Downslope	Mauna Loa Samples (week of 6/27/81)[b] Upslope
SO_4^{2-}	24,000	1.1×10^6	1.9×10^6
V	2.5	230	180
Zn	1,600	3,000	2,300
As	450	930	620
Se	80	100	130
Ag	3.5	4.3	1.5
In	2.4	2.0	1.3
Sb	33	96	66
W	2.4	7.2	17
Au	0.045	0.12	0.39

[a]Small et al. (1981). Concentrations of 50 samples from plumes of five copper smelters after subtraction of non-plume background concentrations.

[b]Parrington et al. (1983c). Net concentration after subtraction of average concentration observed in June samples.

stacks. The trajectories leading to MLO would have a good chance of passing over the Los Angeles Basin in which much oil is refined and consumed, adding V to the atmosphere. Normalized concentrations of the remaining elements are in reasonable agreement, in support of the hypothesis that the spikes represent a signal from the U.S. smelters, although further work would be needed to confirm this. Rahn et al. (1983) previously reported observation of particles from Canadian nickel/copper smelters on filters collected in Rhode Island, some 1500 km to the southeast. According to their results, the best "signal" of particles from such smelters is In, which normally has a very low ambient concentration.

Parrington et al. (1983c) also performed very successful factor analyses in the MLO data set. As noted above, in factor analyses of urban data sets, the increment of total communality drops off slowly with increasing number of factors, reflecting the large number of sources of slightly different compositions from those of the dominant sources in urban areas. By contrast, there were usually sharp breaks in the increment of communality after four, five or six factors of the various groups of MLO samples. This is not surprising as, especially in the case of downslope samples, the material travels long distances before reaching MLO, so there is good opportunity for mixing of particles from many local sources in the source area. Except during the spike periods discussed above, the particulate matter is strongly dominated by continental crustal material and sulfate, with minor marine and local basalt contributions.

Analysis of data from downslope, clean period yielded five factors: a crustal factor, a sulfate-Se factor containing some Na (suggesting a marine association), a "secondary" marine factor lightly loaded with Na and heavily loaded with Br and I (which was probably gone through the vapor phase), a factor loaded with Zn, As and Sb that apparently results from anthropogenic activities probably in Asia, and a factor loaded with V, Co and Zn that apparently results from the episodes from Mexico and Central America noted above. The downslope dust-period factor analysis is dominated heavily by crustal material. The first factor accounts for 14 of the 25 unit of variance in the system and 14 elements have loadings 0.85 of greater on the factor. The second factor represents crustal material of unusual composition that came to MLO for an 8-week period during the Spring of 1980 (Parrington and Zoller, 1983). Concentrations of several elements, including As, La and Sm, were much higher than in the usual Asian dust. The remaining factors were an anthropogenic factor (Se, Zn, Sb), the secondary marine factor (Br and I) and the sulfate-Se factor.

In summary, great progress is being made in global receptor modeling, but much more work is needed. Samples of suspected source materials need to be analyzed and their compositions compared with those of particles observed at receptors, e.g., dust from probable source areas in Asia, particles from large cities in Asia, etc. If clear associations can be made between the sources and receptors, this information can be used to improve trajectory calculations. Other kinds of information could be helpful such as satellite pictures of the movement of Asian dust over the Pacific Ocean.

REGIONAL SCALE RECEPTOR MODELING IN CONTINENTAL AREAS

Little work has been done on receptor modeling on a regional scale (~1500 km) in continental areas, which is more difficult than receptor modeling on either urban or global scales. The former is dominated by local sources and the transport of particles to remote sites is usually over water or lightly populated areas that add little anthropogenic material to the air. By contrast, regional scale receptor modeling involves transport over distances great enough for considerable modification and fractionation of particles by deposition, in-cloud chemistry, etc., and additions of new material by cities, power plants and industries along the way.

In an initial attempt at regional scale receptor modeling, Rahn et al. (1982) used non-

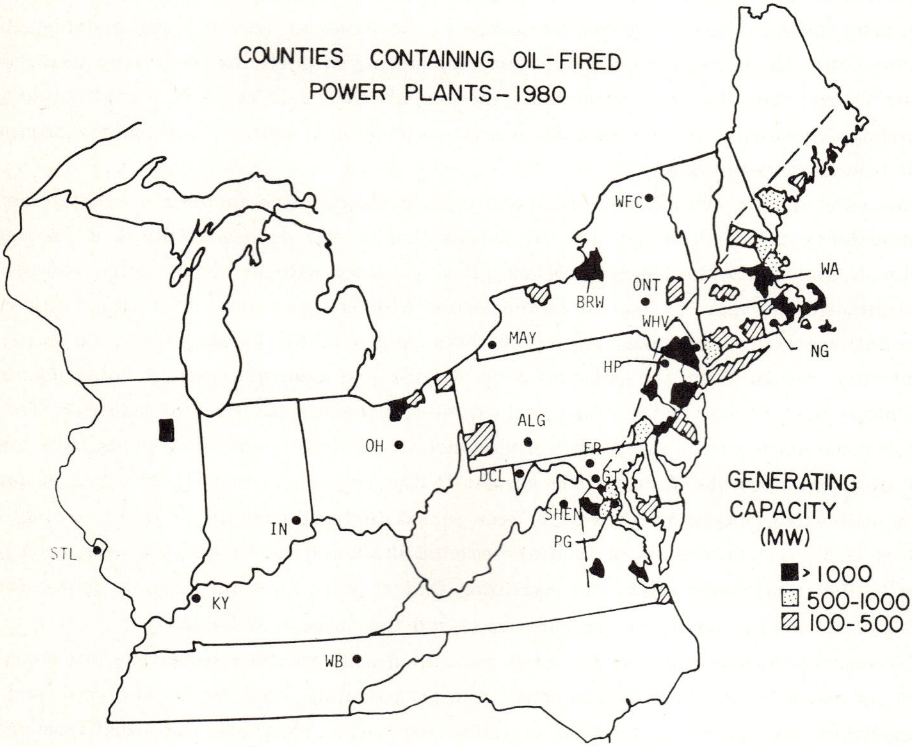

Figure 1. Map showing capacity of oil-burning power plants by county (U.S. Dept. of Energy, 1981) and rural and suburban sampling sites: STL = St. Louis area; Ohio River Valley sites in Kentucky (KY), Indiana (IN) and Ohio (OH), WB = Walker Branch, DCL = Deep Creek Lake, SHEN = Shenandoah Valley, ALG = Allegheny Mt., PG = Prince George's County, GT = Gaithersburg, FR = Frederick, MAY = Mayville, BRW = Brewerton, WFC = Whiteface Mt., ONT = Oneonta, WHV = West Haverstraw, HP = High Point, NG = Narragansett, WA = Watertown. Dashed line is "oil-combustion line."

crustal Mn/V ratios to determine the source areas of particles collected at sites in the East. The $(Mn/V)_{ncr}$ ratios were calculated from Eq. (2) except that Mason's crustal ratio of Mn/Al was multiplied by 0.7 to avoid occasional negative non-crustal Mn values, apparently resulting from Mn/Al ratios in eastern U.S. soils that are lower than the crustal average. At a sampling site at High Point, NJ (see Fig. 1), Rahn et al. found that $(Mn/V)_{ncr}$ ratios were normally greater than unity for air masses coming from the Midwest, whereas they were < 1 for air masses that had passed through the East. The difference in ratios was explained by the high V concentrations in the East resulting from the combustion of residual oil and unknown sources of Mn in the Midwest, probably the metal industries. The classic scenario for sulfate production in the northeast quadrant of the U.S., resulting from largescale experiments such as the Sulfur Regional Experiment (Mueller and Hidy, 1983) is that air masses often move up the Mississippi Valley and then the Ohio River Valley, collecting SO_2 from power plants along the way, which is slowly converted to H_2SO_4 and sulfate particles during transport to New York and New England. Indeed, Rahn et al. often found midwestern Mn/V ratios at High Point, NJ, during periods of high sulfate, in agreement with the classic picture. However, at Watertown, MA, and Narragansett, RI, high sulfate concentrations were normally associated with eastern ratios, i.e., $(Mn/V)_{ncr} < 1$, suggesting that much of the sulfate in New England is produced from sources of SO_2 in the same area, rather being transported from the Midwest. The implication was that the source area of precursors of acid precipitation in New England was that area itself.

Results obtained by Campbell (1977) of our laboratory are in general agreement with the $(Mn/V)_{ncr}$ ratios of Rahn et al. She collected particulate samples at a rural site near Frederick, MD, from June 1972 to December 1973. It would be difficult to calculate back-trajectories for the sampling periods, but sampling periods for which the local winds were predominantly from the northwest to southwest have 67% of their $(Mn/V)_{ncr}$ ratios greater than 1.0, whereas those with winds from the northeast to southeast have only 8% of their ratios greater than 1.

Despite this agreement, there are serious weaknesses in the use of Mn/V ratios to determine the previous histories of air masses arriving in the East. Vanadium concentrations are much greater in the East than the Midwest, but it is not clear that Mn concentrations are higher in the Midwest. In Table 7 are summarized the concentrations of Al, Mn and V and the $(Mn/V)_{ncr}$ ratios for available data at rural or suburban sites of the East and Midwest. The data are divided according to their location relative to the "oil-combustion line". From data on the capacities of oil-fired power plants (U.S. Department of Energy, 1981), I have constructed the map of oil-burning capacity by county (as data on a state-by-state basis do not have the necessary resolution) shown in Fig. 1. Nearly all counties are clustered along the Atlantic Coast or major rivers leading to it, as residual oil is so viscous that it must be transported by barge or tanker. A few counties along the shores of the Great Lakes also have significant oil-burning capacities, but the high density of oil-fired plants is east and south of a line that passes just to the west of Washington, D.C., west and north of Philadelphia and west of the Boston area. As shown in Table 7, V concentrations are typically about 3 ng/m^3

in the rural Midwest, rising to ~20 ng/m^3 near and east of the oil-combustion line.

When both coarse and fine particles are included, Mn concentrations in both the Midwest and East are in the range of 9 to 21 ng/m^3, with no obviously enhanced concentration in the Midwest. Recall from above, that unexplained fine particle Mn was observed in both Boston and Washington. On the basis of these limited data, it appears that the $(Mn/V)_{ncr}$ ratio is lower in the East simply because of the addition of V to air masses when they move into the East. Thus, the Mn/V ratio is a "negative" signal in that, if the ratio is high, the air mass has probably not been in the East, but once it is in the East, the addition of V by local sources will give the air an Eastern ratio regardless of its previous history. It is, thus, not surprising that Eastern ratios of Mn/V are usually observed at Narragansett and Watertown, as the air mass will often have travelled through areas such as Philadelphia and New York city before arriving at those sites. On the other hand, High Point, NJ, is west of the oil-combustion line, so the Mn/V ratio will often have a Midwestern value. Similar results have

Table 7. Atmospheric Concentrations of Al, V and Mn in Rural and Suburban Areas

Investigators	Location	No. Spls.	Years	Concentration (ng/m^3)			$\overline{Mn}_{ncr}/\overline{V}_{ncr}$
				Al	V	Mn	
West of Oil-Combustion Line:							
Loo *et al.* (1978)	St. Louis, MO area	3500	1976	765	3.0	21	8.5
Bondietti *et al.* (1976)	Walker Branch, TN	4	1975	725	3.0	12	3.4
Shaw and Paur (1983)	Union Cty., KY	400	1980/81	660	-	21	-
	Franklin Cty., IN	400	1980/81	520	-	>6	-
	Ashland Cty., OH	400	1980/81	425	-	12	-
Pierson *et al.* (1980)	Allegheny Mt., PA	35	1977	700	3	9	1.8
Husain *et al.* (1983)	Mayville, NY	63	1981	-	-	-	7.6[a]
	Brewerton, NY	51	1981	-	-	-	2.6[a]
	Whiteface Mt., NY	23	1981	-	-	-	1.9[a]
Stevens *et al.* (1983)	Shenandoah Valley, VA	32	1980	350	2	5.4	1.8
Near Oil-Combustion Line:							
Campbell (1977)	Frederick, MD	136	1972/73	860	14	21	1.1
Kowalczyk *et al.* (1982)	Gaithersburg, MD	8	1976	570	7	15	1.7
Husain *et al.* (1983)	Oneonta, NY	13	1981	-	-	-	0.60[a]
East of Oil-Combustion Line:							
Thurston *et al.* (1983)	Watertown, MA	322	1979/81	400	25	9.4	0.25
Rahn *et al.* (1982)	Narragansett, RI	233	1977/80	427	13	7.2	0.29
Husain *et al.* (1983)	West Haverstraw, NY	16	1981	-	-	-	0.74[a]
Muhlbaier (1978)	Prince George's Cty, MD	4	1976	1030	9.7	14	0.70
Gladney (1974)	Prince George's Cty, MD	77	1973	910	35	13.5	0.18

[a]Calculated without factor of 0.7 in Eq. (3) multiplying (Mn/Al) crustal ratio. This change could not be made as Husain *et al.* do not list absolute concentrations of elements.

recently been reported by Husain et al. (1983) based in studies throughout the State of New York. At sites in the western and northern areas of the state, high sulfate is often associated with Midwestern Mn/V ratios, but much less so in the eastern and southern areas.

The weaknesses of the use of Mn/V ratios does not mean that receptor modeling on a regional, continental scale is unsuccessful. Rather it means that a much wider range of elements and other species should be investigated for use in regional scale receptor models. It would, for example, be desirable to develop "positive" signals of midwestern air or of certain types of important sources such as coal-fired power plants that would persist even after addition of particles from other regions.

Another serious weakness of the approach of Rahn et al. (1982) is their use of air-filter samples without size segregation. Much of the Mn and many other elements arises from crustal material. To observe contributions from other types of sources, one must substract the crustal contributions based on ratios of the concentrations of the elements relative to those of elements arising mainly from crustal sources such as Al in Eq. (3). For Mn, in particular, the crustal contribution is often so close to the total observed that the net result is very sensitive to uncertainty of the assumed Mn/Al value of the crustal material (which varies somewhat from one area to another) and to errors of the analyses for Mn and Al. Similar problems would arise for all elements contributed mainly by the crust on air-filter samples. The errors could be greatly reduced and the signals enhanced if the samples were collected with a size-segregation device and only the fine fraction, say < 2.5-μm diam, analyzed. The enriched elements from most important sources are usually much more enriched in the fine particles than in the coarse ones, especially elements that pass through the vapor state in high temperature sources (Natusch et al., 1974).

In an attempt to develop a sound basis for receptor modeling on a regional basis in the northeast quadrant of the U.S., my group is performing INAA of samples collected by Environmental Protection Agency (EPA) personnel at four sites in the Midwest and East. In the Ohio River Valley Study (ORVS), Shaw and Paur (1983) took samples with dichotomous samplers at rural sites in Kentucky, Indiana and Ohio for 15 months starting in May, 1980, and have analyzed them by XRF. Stevens et al. (1983) collected similar data for a 30-day period at a site in the Shenandoah Valley of Virginia in July and August, 1980. As noted above, XRF is an excellent method for analyzing huge numbers of samples, but it is not very sensitive to trace elements, which may be the most sensitive indicators of certain regions or sources. By contrast, INAA is more expensive and time- consuming, but can be used to observe many potentially useful trace elements. Thus, we will use the results of XRF analyses and back-trajectories to select the most interesting samples from the ORVS data set for INAA to augment the XRF data to provide data for a total of about 40 elements. Because of their greater information content, we will mostly analyze fine fractions. As the number of samples available from Shenandoah Valley is much smaller, we are analyzing all of the fine fractions of them. By using back-trajectories to select samples, we hope to determine if there are characteristic concentration patterns depending on whether the air mass has come up the Mississippi and Ohio River Valleys, or down from Canada, or if it has passed through highly industrialized areas of Chicago/Gary, IN, Detroit, Cleveland, etc., or has come from rural areas of the South. Samson's (1981) similar sorting of previously available data sets, e.g., at Allegheny Mountain (Pierson et al., 1980), has demonstrated the value of using back-trajectories in this connection. We also plan to select samples on the basis of the amount of precipitation, if any, that occurred along the trajectories.

Stevens et al. (1983) performed factor analysis of chemical and XRF data from Shenandoah. They obtained four factors for the fine fraction: coal-fired power plants, soil, carbon and combustion of leaded gasoline. The factor attributed to coal-fired power plants contained heavy loadings of SO_4^{2-}, total S, mass, H^+, NH_4^+, NO_3^- and Se, and accounted for 40% of the variance of the 16 variables included in the analysis. The authors attach considerable significance to the association of Se with S and SO_4^{2-} on the first factor in attributing it to coal-fired plants. In this continental area, in which coal is a major source of atmospheric sulfur, the assignment of this factor is probably correct; however, because of their chemical similarities, S and Se are usually strongly correlated with each other, regardless of their source. As noted above, S an Se are strongly correlated at MLO, where much of both elements arise from natural sources, probably biological activity in the sea.

Semra Tuncel of our laboratory is subjecting the Shenandoah Valley fine fraction samples to INAA. Preliminary results from her partial analyses of these samples confirm the strong association of S and Se, using the more accurate Se data from INAA. She also finds substantial loadings of V, Sb, As and, surprisingly, Mn on the sulfate factor. The observation of Mn on this factor is surprising because Mn is depleted in coal relative to other crustal elements. There is some enrichment of Mn on fine particles in stacks of coal-fired power plants with respect to crustal elements in input coal, but even so, the crustal enrichment factor of Mn on fine particles from four coal-fired plants averaged only 0.52 ± 0.54 (Gordon et al., 1981). Another argument against Mn being from coal-fired power plants is the fact that Boston has unexplained fine Mn which was apparently associated with Se (see Factor 5 in Table 2) despite the fact that very little coal is burned there. However, it is possible that the fine Mn in Shenandoah Valley is from coal-fired plants, as it is not strongly enriched with respect to the crustal abundance pattern (e.g., Mason, 1966). The Mn/Al ratio of the fine particles is 0.103, whereas the crustal abundance ratio is 0.0117, so the enrichment with respect to the crust is 9.0. By contrast, the crustal enrichment factor of Se of fine particles in Shenandoah Valley is 1.02×10^5.

It is tempting to suggest that the Mn-SO_4^{2-} association arises from the catalytic conversion of SO_2 in cloud droplets containing Mn, a mechanism that has been discussed for years and was considered by other speakers at this Conference. Indeed, studies of the composition of rainfall (Lindberg, 1981; Penkett et al., 1979) have revealed strong correlations between Mn and SO_4^{2-}. However, Lindberg (1981) suggested that the correlations may result from oxidation of SO_2 in the collected water in the case of non-event sampling and, in the case of event sampling, from similar rain-volume, dilution effects on both Mn and SO_4^{2-}. The association of Mn with SO_4^{2-} may be a very important piece of information, but we do not yet know how to interpret it.

A recent step in receptor modeling on a regional and global scale has been the development of a tracer system involving ratios of concentrations of six elements to that of Se by Rahn and Lowenthal (1983). They take ratios to Se to eliminate the effects of dilution, with Se serving as a general indicator of pollution found at similar concentrations in many source areas. They suggest that other species could be used for this purpose, e.g., Pb or elemental carbon. The six elements whose ratios are used are As, Sb, Zn, In, noncrustal Mn and noncrustal V. These were selected from many elements investigated as being accurately measurable by INAA and having good discriminating power for particles from different source areas. The seven-element tracer system is an improvement over the use of noncrustal Mn/V ratios, but much more development and testing are needed. Some of the criteria for areas, e.g., the Interior U.S., were based on a few as four carefully selected samples. More statistics are needed to determine the ranges of the ratios for particles from certain areas and, with use of a wider range of species, patterns typical of areas more localized than "Interior U.S." can perhaps be developed.

INPROVEMENTS NOW IN PROGRESS

Results from the EPA-sponsored Philadelphia study of 1982 (discussed above) will soon become available and should be quite valuable for advancing the application of receptor modeling on an urban scale. During the Summer of 1983, EPA, the Department of Energy and their contractors conducted a field study at a rural site near Deep Creek Lake in western Maryland. This is a regional study in an area downwind from several coal-fired power plants. Samples were collected at the coal-fired plants predicted to have the greatest influence on the ambient site. Dilution sampling was done at the plant stacks and airborne sampling was conducted in the plumes. Samples are being analyzed by many techniques, including XRF, and will be analyzed by INAA in our laboratory. Each group has a Quality Assurance/Quality Control plan for documenting the reliability of the analyses, a big improvement over many older data sets, whose accuracies are often quite uncertain. Independent, but coordinated studies were performed simultaneously by Ford Motor Co. scientists at the nearby Allegheny Mountain, PA, site at which they conducted detailed studies in 1977 (Pierson et al., 1980). Among other things, the Deep Creek Lake experiments is designed to determine whether or not emissions from different coal-fired plants can be distinguished at the receptor site.

One of the potentially most important aspects of this study will be the analysis of particles collected on several stages of a micro-orifice impactor to investigate sub-0.1-μm particles (Ondov, 1983). Ondov et al. (1983) have shown that substantial amounts of moderately volatile trace elements are borne by these particles in stacks of coal-fired power plants. If appreciable amounts of these very fine particles survive coagulation into the accumulation mode, they may provide excellent signatures of emissions from individual coal-fired plants because of their enormous enrichments of many elements relative to the crustal abundance pattern. Also, the amounts of these particles released and their compositions may be quite sensitive to operating parameters of the boiler and the pollution-control equipment.

ADDITIONAL RECEPTOR-MODEL SIGNATURES

This review has focused on the use of particle-borne elements as source signatures, which has been developed more fully than other types of signatures. However, many other types of observations are needed and are being investigated for observing emissions from certain types of sources, especially those which emit mainly carbonaceous particles, e.g., diesel and nonleaded-gasoline engines, home oil and gas heating units, emissions from oil refineries and petrochemical plants. Much work is being done to develop the use of series of organic compounds, especially the polynuclear aromatic hydrocarbons (PAHs) for receptor-model application (Daisey, 1980). Although various types of sources release particles with distinctive PAH patterns, their use is complicated by reactions of the PAHs that change their pattern between the source and receptor (Friedlander, 1981). Differential thermal analysis of carbonaceous material is useful for distinguishing between elemental (or "graphitic") carbon and organic carbon (Johnson et al., 1981). The measurement of ^{14}C/total carbon ratios yields the fraction of carbon originating from modern sources as distinguished from fossil carbon (Cooper et al., 1981). Little use has been made of the vapor-phase species of volatile elements, but some such as Hg, Se and B could be useful markers of specific sources (Braman and Johnson, 1974; Fogg, 1983; Kitto et al., 1983).

One very promising area is the use of individual particles to identify sources. During the past decade, these have been rapid advances in analytical methods such as scanning electron microscopy coupled with XRF (SEM/XRF), secondary ion mass spectrometry (SIMS), and laser-microprobe mass analysis (LAMMA). Before these techniques can come into wide use, several problems must be solved. Some techniques depend on the vaporization and ionization efficiencies of the various elements, which are not well established for particles of various compositions. Many strengths and limitations of these methods were discussed by Van Grieken and Busek (this conference). Another problem has been the fact that particles had to be studied one at a time by a skilled microscopist. This is satisfactory for particles present in large quantities, but it does not provide statistics needed for rare types of particles. However, considerable progress had been made recently on the development of computer-assisted instruments that automatically compare characteristics of observed particles with those in a "library" of characteristics of particles from various types of sources (Johnson et al., 1983).

RECEPTOR METHOD FOR DEPOSITION STUDIES

Little use has been made of receptor-model methods in the study of deposition processes, although they could be very useful in the study of cloud and fog processes, dry deposition, etc. Most studies of these processes have involved measurements of major species such as H^+, NH_4^+, SO_4^{2-}, NO_3^-, etc. It is reactions of these species, of course, that lead to the production of acid or to its neutralization. However, such major species originate from many kinds of sources in many areas, so measurements of their concentrations give little information about sources and transport. By contrast, concentration patterns of elements and compounds and individual particles of characteristic types can often provide definitive information about sources and transport. While concentrations of many elements have been measured in some rainfall and dry deposition networks, these measurements are at the end of the line of so many processes that it is difficult, if not impossible to infer the prior history of the species. Furthermore, with only these "end-of-the-line" measurements it will be difficult to close the link between the history of air masses (e.g., via Rahn and Lowenthal's seven-element tracer system) and the sources of acid. Thus, much work is needed to develop the application of receptor-model techniques to the study of intermediate steps between release and deposition. For example, vertical profiles of concentrations of particle-borne elements should be made to learn more about vertical transport of material released from known sources. Concentrations and size-distributions of particles bearing certain elements should be measured near clouds, in interstitial cloud air, in cloud water and ice crystals, below clouds before and during rains, and in precipitation to obtain better information about the kinds of particles that become incorporated into droplets, which particle initiate ice formation, and to learn how much material s brought down by in-cloud processes vs. below-cloud scavenging. The advantage of using elemental concentration patters or other receptor-model techniques is that such characteristics can be used to identify the types of particles that most easily participate in such processes. We know, or can determine, sources of particles bearing certain elements, the size distribution of those particles, and the fractions of each element that dissolve if the particles are contacted with liquid water. [Muhlbaier (1978), for example, demonstrated many of these measurements in preliminary studies of clouds and precipition in Maryland.]

A great deal more could be done by receptor-model methods. For example, we have largely ignored the processing of particles by clouds during transport over long distances. Do particles retain their identities throughout processing? If they go through a cloud and acquire a water layer that dissolves soluble material and serve as a site for acid production, perhaps with subsequent neutralization, what happens if the cloud evaporates? Does the particle retain its composition except for the addition of SO_4^{2-}, NH_4^+, etc.? Do particles that participate in such processes grow significantly as a result of these processes? If one looks at particles near a major source and again 100s of km downwind, are there observable differences depending on whether or not there has been cloud formation along the way?

These are some of the interesting questions that could be addressed by use of receptor-model methods. Some could be done by use of bulk techniques, e.g., by INAA of material collected on the stages of cascade impactors to determine size distributions of particles bearing certain species. However, many would require analyses of individual particles, or of residue of individual cloud drops of ice crystals. For example, bulk techniques noted above have revealed correlations between Mn and sulfate, but are these species present on the same particles, or on different particles that just happen to be travelling in the same air mass? Perhaps individual particles studies could answer these kind of questions, e.g., by looking carefully at sulfate particles to see if they contain Mn, or carbon particles, etc., although most individual particle methods are not sensitive enough for observation of trace species.

SUMMARY

Receptor-model techniques will clearly be useful for determining the sources of particulate mass, elements and some other species on an urban scale. Already they are probably more accurate than traditional source-emissions/dispersion-modeling methods, although it is difficult to prove this point. Some local authorities have begun to use receptor models for making regulatory decisions; however, there is so much inertia associated with the use of traditional methods that it may take several years for the new approach to become widely used.

Receptor modeling on a global scale shows considerable promise on the basis of the few studies that have been done. Observations of smelter emissions 1000s of km away are very encouraging, as it would greatly aid in the testing and development of trajectory calculations if emissions from clearly identified sources could be observed at very distant locations. It would be even better if emissions from "pulse" type sources, e.g., volcanic eruptions, huge industrial explosions, etc., could be observed so that one could identify both times and places of emission. If global receptor modeling can be developed, it will have many fundamental and practical applications, including the determinations of the fraction of particles in global circulation that originate from human activity, with implications for the effect of human activity upon long-term global climate.

Regional scale receptor modeling is just beginning to be investigated. Although the method is very promising for helping us understand problems such as the sources and transport of acid and its precursors, it will probably be more difficult to establish than either urban or global scale modeling. The major problem is that transport on a regional scale usually occurs over populated areas which add new material to the air mass. However, in view of the potential value of the method, it is surely worthwhile to expend considerable effort to develop and test it.

The techniques used in receptor modeling should be applied to the study of vertical movement of emissions and processes that occur in clouds and fogs.

ACKNOWLEDGEMENTS

I thank my many colleagues, especially William Zoller, Josef Parrington, Robert Stevens and Thomas Dzubay, for valuable discussions about the subject of this paper as well as their sharing of new data. The work was supported in part by the National Science Foundation Atmospheric Chemistry Program under Grant No. ATM-82-19020.

REFERENCES

Alpert D. J. and Hopke P. K. (1981) A determination of the sources of airborne particles collected during the Regional Air Pollution Study. Atmos. Environ. 15, 675-687.

Andren A. W., Klein D. H. and Talmi Y. (1975) Selenium in coal-fired steam plant emissions. Environ. Sci. Technol. 9, 856-858.

Bondietti E. et al. (1976) Ecology and Analysis of Trace Contaminants (Edited by Van Hook and Shults) Oak Ridge National Laboratory Report No. ORNL/NSF/EATC-22.

Braman R. S. and Johnson D. L. (1974) Selective absorption tubes and emission technique for determination of ambient forms of mercury in air. Environ. Sci. Technol. 8, 996-1003.

Campbell (Thomae) S. (1977) Size and composition of atmospheric particles in rural areas near Washington, D.C. Ph.D. Thesis, University of Maryland.

Chang S.-N., Hopke P. K., Gordon G. E. and Rheingrover S. W. (1983) Target-transformation factor analysis of airborne particulate samples selected by wind-trajectory analysis (unpublished manuscript).

Cooper J. A., Currie L. A. and Klouda G. A. (1981) Assessment of contemporary carbon source contributions to urban air particulate levels using carbon-14 measurements. Environ. Sci. Technol. 15, 1045-1050.

Cooper J. A. and Watson J. G. Jr. (1980) Receptor oriented methods of air particulate source apportionment. J. Air. Pollut. Contr. Ass. 30, 1116- 1125.

Core J. E., Hanrahan P. L. and Cooper J. A. (1981) Air particulate control strategy development: a new approach using chemical mass balance methods. In Atmospheric Aerosol: Source/Air Quality Relationship (Edited by Macias and Hopke), pp. 107-124, ACS Symp. Series No. 167, American Chemical Society, Washington, D.C.

Cunningham W. C. (1979) The composition, sources and sinks of South Polar aerosols. Ph.D. Thesis, University of Maryland.

Cunningham W. C. and Zoller W. H. (1981) The chemical composition of remote area aerosols. J. Aerosol Sci. 12, 367-384.

Currie L. A., Gerlach R. W., Lewis C. W. et al. (1984) Interlaboratory comparison of source apportionment procedures: results for simulated data sets. Atmos. Environ. 18, pp 1517-1537.

Daisey J. M. (1980) Organic compounds in urban aerosols. Ann. N.Y. Acad. Sci. 338, 50-69.

Daisey J. M., Kneip T. J., Ming-xing W., Li-xin R. and Wei-xiu L. (1983) Organic and elemental composition of airborne particulate matter in Beijing, Spring 1981. Aerosol Sci. Technol. 2, 407-415.

Duce R. A., Arimoto R., Ray B. J., Unni C. K. and Harder P. J. (1983) Trace elements in marine aerosols from Enewetak Atoll: I. Concentrations, sources and temporal variability. J. Geophys. Res. 88, 5321-5342.

Dzubay T. G. (1980) Chemical elemental balance method applied to dichotomous sampler data. Ann. N.Y. Acad. Sci. 38, 126-144.

Dzubay T. G., Stevens R. K., Balfour W. D., Williamson H. J., Cooper J. A., Core J. E., DeCesar R. T., Crutcher E. R., Dattner S. L., Davis B. L., Heisler S. L., Shah J. J., Hopke P. K. and Johnson D. L. (1984) Interlaboratory comparison of receptor model results for Houston aerosol. Atmos. Environ. 18, pp. 1555-1566.

Dzubay T. G., Stevens R. K., Lewis C. W., Hern D., Courtney W.J., Tesch J. W. and Mason M. A. (1982) Visibility and aerosol composition in Houston, Texas. Environ. Sci. Technol. 16, 514-525.

Fogg T. R. (1983) Boron in the marine atmosphere. Ph.D. Thesis, Graduate School of Oceanography, University of Rhode Island.

Friedlander S. K. (1973) Chemical element balances and identification of air pollution sources. Environ. Sci. Technol. 7, 235-240.

Friedlander S. K. (1981) New developments in receptor modeling theory. In Atmospheric Aerosol: Source/Air Quality Relationships (Edited by Macias and Hopke), pp. 1-19. ACS Symp. Series No. 167, American Chemical Society, Washington, D.C.

Gatz D. G. (1975) Relative contributions of different sources of urban aerosols: application of a new estimation method to multiple sites in Chicago. Atmos. Environ. 9, 1-18.

Gladney E. S. (1974) Trace element emissions from coal-fired power plants: a study of the Chalk Point Electric Generating Station. Ph.D. Thesis, University of Maryland.

Gladney E. S., Small J. A., Gordon G. E. and Zoller W. H. (1976) Composition and size distribution of in-stack particulate material at a coal fired power plant. Atmos. Environ. 10, 1071-1077.

Gladney E. S., Zoller W. H., Jones A. G. and Gordon G. E. (1974) Composition and size distribution of atmospheric particulate matter in the Boston area. Environ. Sci. Technol. 8, 551-557.

Gordon G. E. (1980) Receptor models. Environ. Sci. Technol. 14, 792-800.

Gordon G.E., Pierson W. R., Daisey J. M., Lioy P. J., Cooper J. A., Watson J. G., Jr. and Cass G. R. (1984) Considerations for design of source apportionment studies. Atmos. Environ. 18, pp. 1567-1582.

Gordon G. E., Zoller W. H., Kowalczyk G. S. and Rheingrover S. W. (1981) Composition of source components needed for aerosol receptor models. In Atmospheric Aerosol: Source/Air Quality Relationship (Edited by Macias and Hopke), pp. 51-74, ACS Symp. Series No. 167, American Chemical Society, Washington, D.C.

Greenberg R. R., Gordon G.E., Zoller W. H., Jacko R. B., Neuendorf D. W. and Yost K. J. (1978a) Composition of particles from the Nicosia Municipal Incinerator. Environ. Sci. Technol. 12, 1329-1332.

Greenberg R. R., Zoller W. H. and Gordon G. E. (1978b) Composition and size distributions of particles released in refuse incineration. Environ. Sci. Technol. 12, 566-573.

Harmon H. H. (1976) Modern Factor Analysis (3rd edn. rev.) The University of Chicago Press, Chicago.

Harris J. M. (1982) The GMCC Atmospheric Trajectory Programs. NOAA Technical Memorandum ERL ARL-116, Boulder, CO.

Heidam N. Z. (1981) On the origin of the Arctic aerosols: a statistical approach. Atmos. Environ. 15, 1421-1427.

Hopke P. K., Gladney E. S., Gordon G. E., Zoller W. H. and Jones A. G. (1976) The use of multivariate analysis to identify sources of selected elements in the Boston urban aerosol. Atmos. Environ. 10, 1015-1025.

Houck J. E., Core J. E. and Cooper J. A. (1982) Source sampling for receptor oriented source apportionment. Nuclear Environmental Analysis, Inc., Beaverton, OR.

Husain L., Webber J. S., Canelli E., Detkiewicz V. A. and Halstead J. A. (1984) Mn/V as a tracer of aerosol sulfate transport. Atmosph. Environm. 18, pp.1059-1071.

Johnson D. L., Davis B. L., Dzubay T. G., Crutcher E. R., Courtney W. J., Jaklevic J. M., Thompson A. C. and Hopke P. K. (1984) Chemical and physical analyses of Houston aerosol for interlaboratory comparison of source apportionment procedures. Atmos. Environ. 18, pp.1539-1553.

Johnson R. L., Shah J. J., Cary R. A. and Huntzicker J.J. (1981) An automated thermal-optical method for the analysis of carbonaceous aerosol. In Atmospheric Aerosol: Source/Air Quality Relationship (Edited by Macias and Hopke), pp. 197-222. ACS Symp. Series No. 167, American Chemical Society, Washington, D.C.

Kitto M. E., Anderson D. L., Gordon G. E. and Ondov, J. M. (1983) Boron as a potential tracer for coal-fired power plant emissions. Presented at the American Chemical Society National Meeting, Washington, D.C., Aug.

Klein D. N., Andren A. W., Carter J. A., Emery J. F., Feldman C., Fulkerson W., Lyon W. S., Ogle J. C., Talmi Y., Van Hook R. I. and Bolton N. E. (1975) Pathways of thirty-seven elements through a coal-fired power plant. Environ. Sci. Technol. 9, 973-979.

Kowalczyk G. S., Gordon G. E. and Rheingrover S. W. (1982) Identification of atmospheric particulate sources in Washington, D.C., using chemical element balances. Environ. Sci. Technol. 16, 79-90.

Lindberg S. E. (1981) The relationship between manganese and sulfate ions in rain. Atmos. Environ. 15, 1749-1753.

Loo B. W., French W. R., Gatti R. C., Goulding F. S., Jaklevic J. M., Llacer J. and Thompson A. C. (1978) Large-scale measurements of airborne particulate sulfur. Atmos. Environ. 12, 759-771.

Maenhaut W., Zoller W. H., Duce R. A. and Hoffmann G. L. (1979) Concentration and size distribution of particulate trace elements in the South Polar atmosphere. J. Geophys. Res. 84, 2421-2431.

Mason B. (1966) Principles of Geochemistry (3rd. edn. rev.) John Wiley and Sons, Inc., New York.

Mroz E. J. (1976) The study of the elemental composition of particulate emissions from an oil-fired power plant. Ph.D. Thesis, University of Maryland.

Mueller P. K. and Hidy G. M. (1983) The sulfate regional experiment: report of findings. Electric Power Research Institute Report No. EA-1901.

Muhlbaier J. (1978) The chemistry of precipitation near the Chalk Point power plant. Ph.D. Thesis, University of Maryland.

Natusch D. F. S., Wallace J. R. and Evans C. A. (1974) Toxic trace elements: preferential concentration in respirable particles. Science 183, 202-204.

Ondov J. M. (1983) private communication. Martin-Marietta Laboratory, Baltimore, MD.

Ondov J. M., Biermann A. H., Heft R. E. and Koszykowski R. F. (1981) Elemental composition of atmospheric fine particles emitted from coal burned in a modern electric power plant equipped with a flue-gas desulfurization system. In Atmospheric Aerosol: Source/Air Quality Relationships (Edited by Macias and Hopke), pp. 173-186. ACS Symp. Series No. 167, American Chemical Society, Washington, D.C.

Parrington J. R. and Zoller W. H. (1983) Selective sampling of diurnal changes in aerosol composition at Mauna Loa, Hawaii. J. Geophys. Res. (in press).

Parrington J. R., Whittaker J. K. and Zoller W. H. (1983a) Instrumentation for and results of particulate sampling of diurnal winds at Mauno Loa Observatory, Hawaii. J. Appl. Meteor. (submitted).

Parrington J. R., Zoller W. H. and Aras N. K. (1983b) Asian dust: seasonal transport to the Hawaiian Islands, Science 220, 195-197.

Parrington J. R., Zoller W. H. and Gordon G. E. (1983c) Sources and composition of atmospheric particles of the free troposphere over the North Pacific. J. Geophys. Res. (submitted).

Penkett S. A., Jones B. M. R. and Eggleton A. E. J. (1979) A study of SO_2 oxidation in stored rainwater samples. Atmos. Environ. 13, 139-147.

Pierson W. R., Brachaczek W. W., Truex T. J., Butler J. W. and Korniski T. J. (1980) Ambient sulfate measurements on Allegheny Mountain and the question of atmospheric sulfate in the northeastern United States. Ann. N.Y. Acad. Sci. 338, 145-173.

Rahn K. A. (1981) Relative importances of North America and Eurasia as sources of Arctic aerosols. Atmos. Environ. 15, 1447-1455.

Rahn K. A. and Lowenthal D. H. (1984) Elemental tracers of distant regional pollution aerosols. Science 223, pp. 132-139.

Rahn K. A., Lowenthal D. H. and Lewis N. F. (1982) Elemental tracers and source areas of pollution aerosol in Narragansett, Rhode Island. Technical Report, Graduate School of Oceanography, University of Rhode Island.

Rahn K. A., Noelle L. F. and Lowenthal D. H. (1983) Elemental tracers of Canadian smelter aerosol transported into the northeastern United States. In Receptor Models Applied to Contemporary Pollution Problems (Edited by Dattner and Hopke) Air Pollution Control Assn., Pittsburgh, PA.

Rheingrover S. W. (1981) A statistical model for titanium pollution transport and dispersion in the atmosphere in St. Louis. J. Appl. Meteor. 20, 693-705

Rheingrover S. W. and Gordon G. E. (1983) Wind-trajectory method for determining compositions of particles from major air-pollution sources (unpublished manuscript).

Samson P. J. (1981) Trajectory analysis of summertime sulfate concentrations in the northeastern United States. J. Appl. Meteor. 19, 1382-1394.

Shaw R. W. and Paur R. J. (1983) Composition of aerosol particles collected at rural sites in the Ohio River Valley. Atmos. Environ. 17, 2031-2044.

Small M. A., Germani M. S., Small A. M., Zoller W. H. and Moyers J. L. (1981) Airborne plume study of emissions from the processing of copper ores in southwestern Arizona. Environ. Sci. Technol. 15, 293-299.

Stevens R. K., Dzubay T. G., Lewis C. W. and Shaw R. W. Jr. (1984) Source apportionment methods applied to the determination of ambient aerosols that affect visibility in forested areas. Atmos. Environ. 18, 261-272.

Stevens R. K., Dzubay T. G., Russwurm G. and Rickle D. (1978) Sampling and analysis of atmospheric sulfates and related species. Atmos. Environ. 12, 55-68.

Thurston G. D., Spengler J. D. and Samson P. J. (1983) An assessment of relationships between regional pollution transport and trace elements using wind trajectory analysis. In Receptor Models Applied to Contemporary Pollution Problems (Edited by Dattner and Hopke) Air Pollution Control Assn., Pittsburgh, PA.

Tuncel G. (1983) private communication, University of Maryland.

U.S. Department of Energy (1981) Energy Data Report: Inventory of Power Plants in the United States, 1980 Annual, No. DOE/EIA-0095 (80).

Watson J. G. Jr. (1979) Chemical element balance receptor model methodology for sources of fine and total suspended particulate matter in Portland, Oregon. Ph.D. Thesis, Oregon, Graduate Center.

Winchester J. W. (1980) Discussion. Ann. N.Y. Acad. Sci. 338, 190-191.

Winchester J. W., Mingxing W., Lixin R., Weixin L., Darzi M. and Leslie A. C. D. (1981) Aerosol composition in relation to air mass movements in North China. In Atmospheric Aerosol: Source/Air Quality Relationship (Edited by Macias and Hopke), pp. 287-302. ACS Symp. Series No. 167, American Chemical Society, Washington, D.C.

Zoller W. H., Parrington J. R. and Kotra J. P. (1983) Iridium enrichment in airborne particles from Kilauea volcano: January 1983. (1983), Science 222, 1118-1121.

UNCERTAINTIES IN QUANTIFYING SOURCE-RECEPTOR RELATIONSHIPS FOR ATMOSPHERIC ACIDS

Perry J. Samson, Jennie L. Moody, Jon Kahl and Gerald Keeler
Department of Atmospheric and Oceanic Science
The University of Michigan
Ann Arbor, Michigan 48109
U.S.A.

ABSTRACT

Computer simulation models which attempt to quantify the contribution of States and Provinces to atmospheric sulfur deposition at ecologically sensitive receptor locations are being developed by several investigators. There are many assumptions inherent to state-of-the-art acid deposition models which may limit their application for predicting changes due to emission control strategies. This paper addresses the variations in source-receptor relationships which one might expect due to (1) year-to-year variations in meteorological conditions and (2) spatially inhomogeneous gas-phase conversion of SO_2 to $SO_4^{=}$. The variations in source-receptor relationships predicted by a regional-scale acid deposition model using observed winds and precipitation for the years 1978 to 1980 are discussed. The variations in these relationships are found to vary substantially at some receptors from year-to-year. The greatest variations generally occur between receptors and nearby States.

Sulfur deposition modeling has, to date, assumed, that the transformation of sulfur dioxide (SO_2) to sulfate ($SO_4^{=}$) in the atmosphere is linearly proportional to the SO_2 concentration. To make an initial test of the possible effect of nonlinear transformation processes, this work assumes that the conversion is modified by the hydrocarbon to nitrogen oxide ratio (RHC/NO_x). The assumption is based on the results of sensitivity analyses performed with a simplified chemical mechanism for sulfate formation. The source-receptor relationships for ambient sulfate are shown to shift away from the industrial Midwest because of a presumed depression in the RHC/NO_x ratio in these regions.

NATO ASI Series, Vol. G6
Chemistry of Multiphase Atmospheric Systems
Edited by W. Jaeschke

INTRODUCTION

Our ability to forecast the environmental improvements to be gained as a result of reductions in sulfur dioxide emissions is limited by our ability to properly represent physical and chemical processes in the atmosphere. In recent years a number of investigators have developed acid (actually sulfur) deposition models which attempt to incorporate the emissions, transport, dispersion, chemical transformation, and deposition of pollutants over regional scales (> 500 km) where standard plume equations are no longer applicable. This paper investigates the variability in model predictions of source-receptor relationships. It will be shown that the "natural potential" for emissions in one region to impact another region can vary significantly, depending on both the assumptions which have been made in the modeling and the changes in meteorological conditions from year-to-year.

The Atmospheric Contributions to Interregional Deposition (ACID) model [1] is similar in design to other Lagrangian-type transport and deposition models [2-5], including first-order, linear chemical conversion and deposition processes. The ACID model, however, is unique in that it is receptor-oriented and is designed to assess the potential of upwind regions for contribution to the sulfur deposition at specific receptors. That is, the model calculates the potential deposition or concentration per unit of emissions (eg. kilograms per hectare of sulfur wet deposition per kilograms of sulfur emissions) based on the observed meteorological conditions and assumed chemical transformation processes. This allows an analysis of the variability in source-receptor relationships which is apart from any variations in emissions strength. The variability in the natural potential quantifies the confidence with which an emissions reduction program which will result in measurable decreases in sulfur deposition.

MODEL ASSUMPTIONS

The determination of the potential of upwind areas to contribute to the deposition at a receptor can be expressed in terms of probability. The "natural potential" for contribution to wet deposition, for example, is the probability of the sulfur precursors, sulfur dioxide (SO_2) and sulfate ($SO_4^{=}$), being transferred to the receptor multiplied by the probability that, once there, the material is deposited. Let $S_g(\vec{x},t)$ be the probability of sulfur as SO_2 arriving at a receptor located at $\vec{x}$ at time t and $S_p(\vec{x},t)$ be the same for sulfur as $SO_4^{=}$. Then, in terms of probability [6,7],

$$S_g(\vec{x},t) = \int_{-\infty}^{\infty} \int_{-\infty}^{\infty} \int_{t-\tau}^{t} Q(\vec{x},t|\vec{x}',t')\ D_g(\vec{x}',t')\ R(t|t')\ d\vec{x}'\ dt'$$

where $Q(\vec{x},t|\vec{x}',t')$ is the mass transfer function defining the probability of an air parcel located at $\vec{x}'$ at time t' arriving at receptor $\vec{x}$ at time t, $D_g(\vec{x}',t')$ is the probability of sulfur as SO_2 not being deposited upwind at $(\vec{x}',t')$ and $R(t|t')$ is the probability of sulfur as SO_2 not being reacted to $SO_4^=$ from time t' to time t. The equation is integrated over the domain from $t-\tau$ to t assuming the probability of mass transfer from beyond time $t-\tau$ to be zero.

The mass transfer function can be approximated from the computation of upwind (backward) trajectories. The axis of the computed trajectory is assumed to represent the highest probability of contributing to the receptor at a given time upwind. The distribution of potential contribution from regions located laterally from the trajectory will depend upon the degree of vertical mixing coupled with vertical wind velocity shear. As a first approximation it has been assumed that the potential mass transfer is normally distributed about the trajectory and the distribution increases linearly with time upwind of the receptor [8-10].

Thus far, this model has been specifically applied to the transport, transformation, and deposition of sulfur species. Consequently, the probability of chemical transformation of SO_2 to $SO_4^=$ species has been included as a linear, first-order process. This is accomplished by assuming linear transformation rates which vary both diurnally and annually from maximum rates in June during daylight hours (when photochemistry is greatest) to minimum rates during the nighttime period. The rate of change of S_g with time is described by

$$dS_g/dt = -k_t(t)S_g(t)$$

$$dS_p/dt = +k_t(t)S_g(t)$$

so the probability of not being converted can be computed as

$$R(t|t') = 1 - \int_{t-t'}^{t} k_t(t')\ S_g(t')\ dt'$$

This approximation is a gross simplification of a very complex system which involves both gas-phase transformation processes and in-cloud transformation. The justification for such a parameterization is based on field studies conducted in episodes of high ambient $SO_4^=$ concentrations [11]. Its extension to all meteorological conditions is open to debate, but it has been utilized in these models for lack of more complete information about the variations in the transformation process.

The dry deposition of SO_2 and $SO_4^=$ assumes linear rate coefficients over three hour periods. The dry deposition velocities vary diurnally and seasonally, having minima of 0.15 cm/sec for SO_2 and 0.05 cm/sec for $SO_4^=$ at nighttime regardless of season. The highest values are assumed to occur in the afternoon with 0.8 cm/sec and 0.3 cm/sec assumed for wintertime (January) SO_2 and $SO_4^=$ dry deposition velocities, respectively, increasing to 1.3 cm/sec and 0.45 cm/sec for summertime (July) SO_2 and $SO_4^=$ dry deposition velocities. Rates for other months are linearly interpolated between these values. The height of the layer through which the deposition is occurring is assumed to be the monthly average mixing height computed along the trajectory determined as the bottom of the first non-surface based stable ($\partial\Theta/\partial z \geq 0.005°C/m$) layer which meets the criteria $\Theta_{TOP} - \Theta_{BASE} \geq 2°C$.

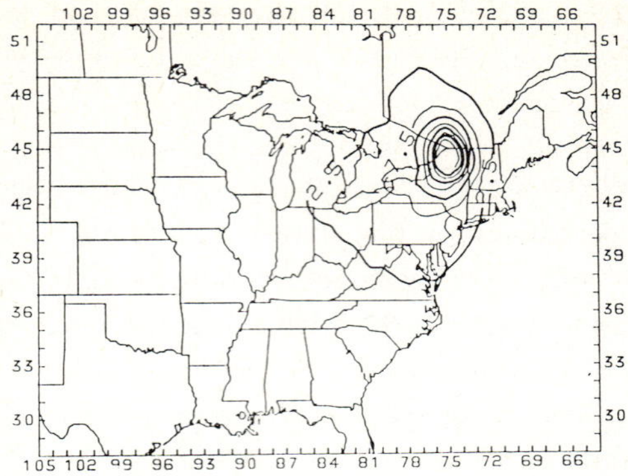

Figure 1. Calculated "natural potential" for contribution to sulfur wet deposition in the Adirondack Mountains for 1979 in units of $kg\text{-}ha^{-1}$ per 10^9 kg of sulfur emissions.

The wet scavenging of the sulfur compounds along the trajectories and at the receptor is also assumed to be linearly proportional to the species concentrations. The rate coefficients for wet deposition of SO_2, k_w, and ambient sulfate, κ_w, are determined by

$$k_w = \omega(SO_2)R/h$$

$$\kappa_w = 0.29\ R^{0.622}$$

where $\omega(SO_2)$ is the washout ratio for SO_2, R is the precipitation rate (mm/hr), and h is the mixing height (mm). The washout ratio is defined as the ratio of the aqueous concentration of pollutant to the ambient concentration in the atmosphere. the value of $\omega(SO_2)$ was chosen as a median value from the range described by Barrie [12]

$$\omega(SO_2) = 5 \times 10^4$$

and the coefficient for κ_w was derived

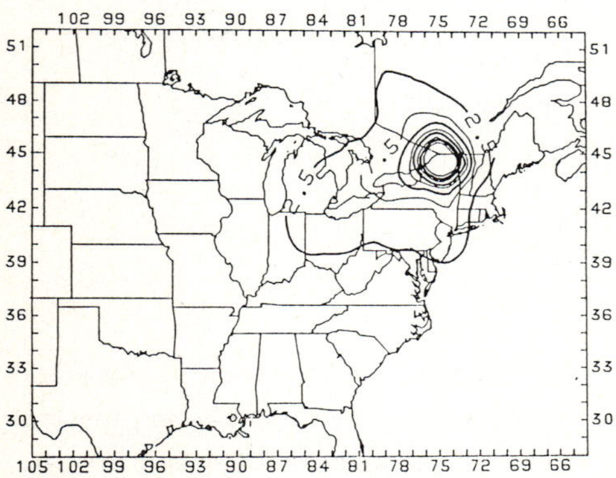

Figure 2. Calculated "natural potential" for contribution to sulfur wet deposition in the Adirondack Mountains for 1980 in units of $kg\text{-}ha^{-1}$ per 10^9 kg of sulfur emissions.

from a formulation [13,14] for clouds containing ice nuclei.

It is further assumed that all scavenged sulfur dioxide arrives at the surface as $SO_4^=$ through in-cloud and/or in-precipitation conversion. This removal mechanism does not distinguish between scavenging within the cloud (rainout) and scavenging beneath the cloud (washout) but rather lumps the two together as net removers of the sulfur mass in the transporting layer.

INTERANNUAL VARIABILITY

Figure 1 shows the "natural potential" for contribution to sulfur wet deposition in the Adirondack Mountains of upstate New York using 1979 meteorological data. It shows that the axis of highest natural potential was associated with winds from the southwest because these winds most often resulted in precipitation at the Whiteface Mountain receptor. It is also important to note that the area of highest "natural potential" is within New York State. That is, assuming the chemical conversion of SO_2 to $SO_4^=$ is linear, reductions within New York State would be more effective in reducing Adirondack wet deposition than would similar reductions upwind.

There are variations in the natural potential field from year to year owing to changing meteorological conditions. Figure 2 shows the same analysis as was presented in Figure 1 for 1980. The field is not skewed towards the southwest as much as it was in the 1979 calculation indicating a shift in the direction of the wind responsible for the precipitation events. Figure 3 is a plot of the percent difference in the two years (1979-1980) illustrating that central New York and Pennsylvania had less potential for contributing to sulfur wet deposition in the Adirondacks in 1980 while the potential for Ontario, Northern New York and Vermont increased.

Using 1978 emissions [15] the contributions to sulfur wet deposition have been calculated from the integration of natural potential fields with sulfur emissions

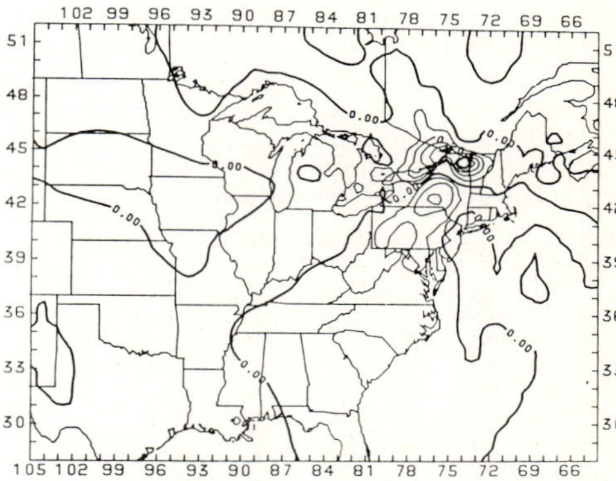

Figure 3. Percent difference between the Adirondack sulfur wet deposition natural potential fields for 1979 versus 1980. There is a negative area in the difference field in central Pennsylvania and New York indications a higher likelihood for those sources to contribute to sulfur wet deposition in 1980 than 1979.

fields over State and Provincial boundaries. Figure 4 shows the variation in source-receptor relationships between the five highest States and the Adirondacks. The contribution attributed to the State of Ohio varies from about 25% in 1978 to 19% in 1979. While this sample is too small to calculate the statistics of variability of source-receptor relationships, it demonstrates that even assuming linear conversion processes the reduction in sulfur wet deposition obtained from sulfur emissions reductions will depend upon the weather.

The variation in source-receptor relationships can be much larger than the ones shown for the Adirondacks. Figure 5 shows the calculated percent of absolute contribution from selected States to the MAP3S site at Charlottesville, Virginia.

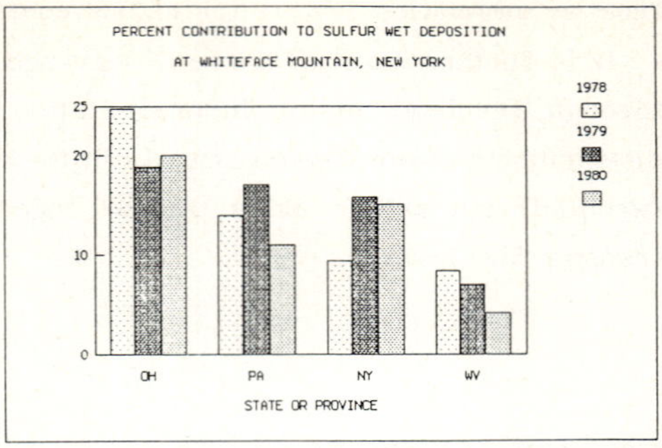

Figure 4. Calculated absolute contribution of four upwind States to sulfur wet deposition at Whiteface Mountain, New York over the period 1978 to 1980.

The calculated absolute contribution from upwind States to Charlottesville showed small changes in the percent contribution from the largest contributors, but relatively large changes in the contribution of nearby States. Thus the variations in source-receptor relationships for sulfur wet deposition resulting from year-to-year variations in meteorological conditions seem to be most pronounced for those States nearest the receptor.

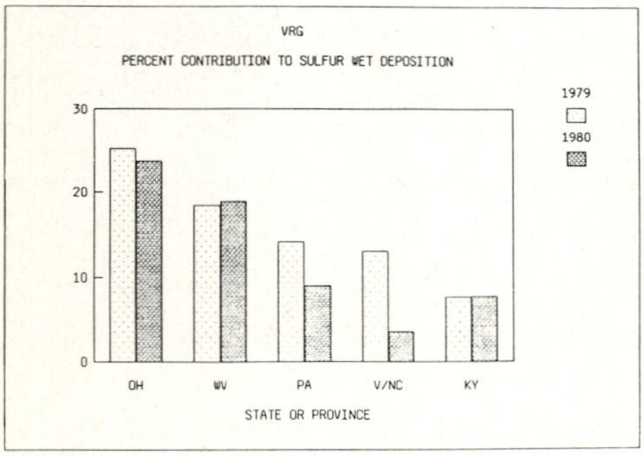

Figure 5. Calculated absolute contribution of five upwind States to sulfur wet deposition at Charlottesville, Virginia over the period 1979 to 1980.

PARAMETERIZATION OF GAS-PHASE NON-LINEARITY

Gas-phase tropospheric chemistry of sulfur dioxide is relatively well understood and documented [16]. In the probabalistic approach described herein the conversion probability is not constrained to be linear, in which case

$$dS_g/dt = -dS_p/dt = f(OH, rh, h\nu, \ldots; H_2O_2, O_3, pH, SO_2, \ldots)$$

The functional form of the right-hand side illustrates a partial list of those atmospheric conditions and chemical species which could result in spatially and temporally non-homogeneous transformation processes. The difficulty in adequately parameterizing this chemistry in regional-scale models arises in part from a lack of data on reactive hydrocarbon (RHC) and nitrogen oxide (NO_x) emissions and concentrations. In this work we have made an initial investigation of the response of the ACID model to gas-phase nonlinear chemistry to evaluate the magnitude of the perturbation this influence could make on source-receptor relationships.

A sensitivity analysis has been conducted on a modified chemical box model [17-19] describing the major components of regional-scale gas-phase chemistry to assess the systems response to varying RHC/NO_x. The results have been applied to the regional scale model in the form of a function which modifies the rate of sulfate formation. Based on this analyses it is assumed that local emissions of RHC and NO_x effect changes in the transformation rate of sulfur dioxide to sulfate.

The basic photochemical model used in this investigation was developed by Rodhe et al. [17]. The chemical reactions, listed in Table 1, approximate the gas-phase formation of oxidizing free radicals, and describe the subsequent formation of H_2SO_4 and HNO_3 from SO_2 and NO_x. Free radical production is accomplished through ozone and formaldehyde photolysis. The free radicals hydroxyl and hydroperoxy (OH, HO_2) are then available to participate in chain reactions with reactive hydrocarbons.

One of the products of the hydrocarbon reactions is an aldehyde (formaldehyde), which can then photolyze to produce a new free radical. The termination reactions for the free radical species are combined to form the oxidant hydrogen peroxide (H_2O_2) and two acid formation mechanisms, the homogeneous gas-phase reactions of SO_2 and NO_x with OH to form sulfuric acid (H_2SO_4) and nitric acid (HNO_3). Also included is a parameterization of the heterogeneous aqueous-phase reaction of SO_2 with H_2O_2 to form H_2SO_4. The relative importance of the aqueous oxidation pathway for NO_x is not known and therefore is not included.

The Rodhe et al. scheme takes into account only two hydrocarbon compounds, methane (CH_4), and ethylene (C_2H_4) with C_2H_2 representing the entire class of reactive hydrocarbons. As one modification of the mechanism, three separate hydrocarbon "species" have been used in this work. They are referred to as "very reactive", "reactive", and "less reactive". Ethylene is the reactive class and the two additional "species" are treated as pseudo-ethylene. They participate in the same reactions but proceed at 10^{-1} and 10^{+1} times the rate of the reactive class. This approximation has the effect of introducing time-varying reactivity to the hydrocarbon mix in an effort to more realistically represent the nature of hydrocarbon reactions in the atmosphere.

Table 1. Chemical reactions and photolysis, $j(sec^{-1})$, and reaction, k $(cm^3\ molecule^{-1}\ sec^{-1})$, rates assumed in this study.

$O_3 + h\nu \rightarrow O_2 + O(^1D)$	$j_1 = 4.1 \times 10^{-6}$
$O(^1D) + M \rightarrow O + M$	$k_2 = 5.0 \times 10^{-11}$
$O(^1D) + H_2O \rightarrow 2OH$	$k_3 = 2.3 \times 10^{-10}$
$CO + OH \rightarrow \cdots\cdots \rightarrow HO_2 + CO_2$	$k_4 = 3.0 \times 10^{-13}$
$2HO_2 \rightarrow H_2O_2 + O_2$	$k_5 = 2.5 \times 10^{-12}$
$HO_2 + OH \rightarrow H_2O + O_2$	$k_6 = 5.1 \times 10^{-11}$
$H_2O_2 + OH \rightarrow HO_2 + H_2O$	$k_7 = 6.9 \times 10^{-13}$
$H_2O_2 + h\nu \rightarrow 2OH$	$j_8 = 3.3 \times 10^{-6}$
$NO_2 + h\nu \rightarrow NO + O$	$j_9 = 4.0 \times 10^{-3}$
$NO + O_3 \rightarrow NO_2 + O_2$	$k_{10} = 1.4 \times 10^{-14}$
$NO + HO_2 \rightarrow OH + NO_2$	$k_{11} = 8.0 \times 10^{-12}$
$NO_2 + OH \rightarrow HNO_3$	$k_{12} = 8.0 \times 10^{-12}$
$SO_2 + OH \rightarrow \cdots\cdots \rightarrow H_2SO_4 + HO_2$	$k_{13} = 1.5 \times 10^{-12}$
$SO_2 + H_2O_2 + \text{"CLOUD"} \rightarrow H_2SO_4$	$k_{14} = 6.0 \times 10^{-16}$
$CH_4 + OH \rightarrow \cdots\cdots \rightarrow CH_2O + HO_2$	$k_{15} = 5.0 \times 10^{-15}$
$CH_2O + OH \rightarrow \cdots\cdots \rightarrow CO + HO_2$	$k_{16} = 1.2 \times 10^{-11}$
$CH_2O + h\nu \rightarrow \cdots\cdots \rightarrow CO + 2HO_2$	$j_{17} = 1.1 \times 10^{-5}$
$CH_2O + h\nu \rightarrow \cdots\cdots \rightarrow CO + H_2$	$j_{18} = 2.6 \times 10^{-5}$
$C_2H_4 + OH \rightarrow \cdots\cdots \rightarrow HO_2 + 2CH_2O$	$k_{19} = 9.0 \times 10^{-12}$

A second modification to the mechanism was proposed [18] to include the generation of HO_2 in equation (13) in Table 1. Additionally the rate of this reaction has been increased to its present value from the rate of 6 x 10^{-13} cm^3 $molecules^{-1}$ sec^{-1} proposed originally by Rodhe et al. This modification leads to a more linear relationship between changes in initial SO_2 concentrations and changes in $SO_4^=$ production than were found using the original Rodhe et al. mechanism wherein a direct nonlinear feedback was forced as OH levels were sensitive to SO_2 concentrations. Some recent evidence exists [19] supporting the contention that OH levels in reacting mixtures of hydrocarbons, NO_x and SO_2 are relatively insensitive to SO_2 concentrations.

The set of equations used in this model describes diurnally averaged chemical processes. Nocturnal chemical processes are not explicitly represented and the diurnal variability of photolysis (j) coefficients is neglected. The aqueous-phase transformation is given by a second order rate constant multiplied by the time dependent concentrations of H_2O_2 and SO_2. The rate constant was scaled so that an average of 50% of the oxidation of SO_2 occurs by this reaction path. This parameterization assumes the aqueous-phase reaction mechanism to be intrinsically linear. In reality the reactions leading to aqueous-phase oxidation are very complex and detailed chemical models have been devoted to aqueous processes [20]. Because of this simplification some potential sources of nonlinearity have been ignored which could either augment or compete with the contributions to nonlinearity determined in this study.

The sensitivity of $SO_4^=$ production to initial conditions is illustrated in Figure 6 by plotting the percent change in calculated $SO_4^=$ production from "episode" baseline production 24 hours downwind of the source region. In general, $SO_4^=$ production increases as RHC is increased and NO_x and RHC concentrations can cause substantial changes in the modeled $SO_4^=$ production from the baseline production rate.

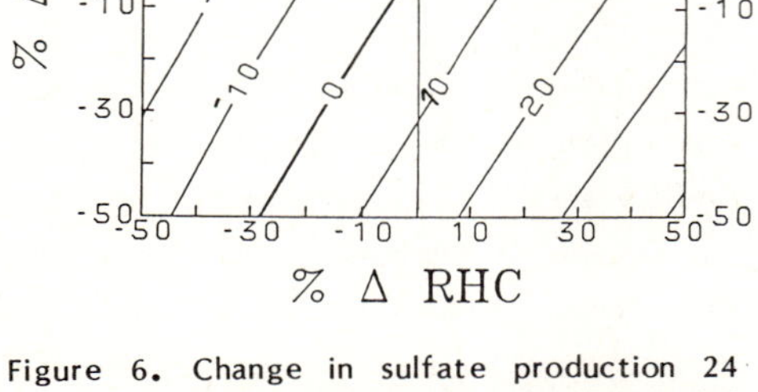

Figure 6. Change in sulfate production 24 hours downwind under "episode" conditions. Contours represent percentage change of sulfate production relative to a baseline value [20].

The sensitivity of $SO_4^=$ production to perturbations in initial RHC and NO_x concentrations can be quantified by calculating the sensitivity of the model to changes in the relative RHC/NO_x ratio. The sensitivity of the SO_2 transformation rate to changes in "reactivity", the ratio of $(RHC/NO_x)/(RHC/NO_x)_{baseline}$, is expressed as a function (R) which is the predicted fraction of baseline production for both "episode" and "typical" conditions. The function (R) can be described by a linear regression relationship of the generic form

$$(R) = \quad + \quad (R/R_{base})^{\gamma}$$

where R represents the ratio of RHC/NO_x and is the fractional sulfate production rate (i.e. the fractional change in sulfate production rate above the baseline rate). The regression coefficients β, λ and γ have values of -68.6, 69.6 and .006 [20]. The form of this equation is illustrated in Figure 7.

SOURCE-RECEPTOR RELATIONSHIPS INCLUDING GAS-PHASE CHEMISTRY

The ACID model was rund for a one year period (1978) using back trajectories from Whiteface Mountain, New York (44°23',73°56'). Figure 8 compares the percent of the annual ambient sulfate estimated to have originated in the four States calculated to contribute the most sulfur using a scenario with linear chemical reaction mechanism versus one with the modified method sensitive to RHC/NO_x. The abbreviations for State and Province names is listed in Table 2. The results show that the inclusion of the RHC/NO_x mechanism causes a systematic shift in culpability away from those States in the Ohio River Basin (where the ratio of emissions of RHC/NO_x is relatively low).

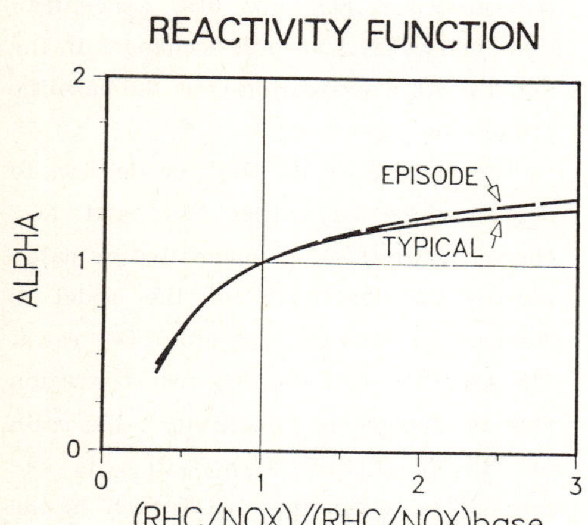

Figure 7. The gas-phase "reactivity" function (R) derived from the modified Rodhe et al. [17] model for episodic and more typical conditions as a function of the RHC/NO_x ratio of the underlying emissions.

While this analysis suggests that a fair amount of uncertainty exists in source-receptor relationships due to uncertainty in the importance on non-linear chemical mechanisms even over an annual basis, it must be stressed that this shift is only in response to the theoretical sensitivity of the atmosphere to gas-phase chemistry for ambient sulfate. It is not known, for example, what fraction of the sulfate observed in precipitation or the atmosphere is due to aqueous-phase production versus gas-phase production. Additionally, the sensitivity of the atmosphere to changes in RHC/NO_x may well have been overstated in this investigation because of the presence of natural sources of reactive hydrocarbons. Natural emissions, which were not included in our emissions inventory,

Table 2. Key to abbreviations for source regions.

SUD	- Sudbury, Ontario
SWO	- Southwestern Ontario
SEO	- Southeastern Ontario
NCQ	- North Central Quebec
NYK	- New York
OHO	- Ohio
ILL	- Illinois
PEN	- Pennsylvania
IND	- Indiana
MIC	- Michigan
WVG	- West Virginia

would smooth the variation in RHC/NO_x variations and produce smaller deviations from the base-line ratio used in the model. Nonetheless, non-linear chemical mechanisms which could produce systematic spatial biases in the transformation process must continue to be examined in order to properly evaluate the confidence of success for an emissions control strategy.

Figure 9 and 10 shows the calculated percent contribution of four States to annual sulfur dry and wet deposition, respectively. The results for wet deposition are far more robust than those shown previously for ambient sulfate. The effect of the reactivity function is more important for the wet deposition than for the dry, but still small. Ohio's contribution to sulfur dry deposition in the Adirondacks increases by about 3%, while the fraction from Sudbury drops by almost 4%.

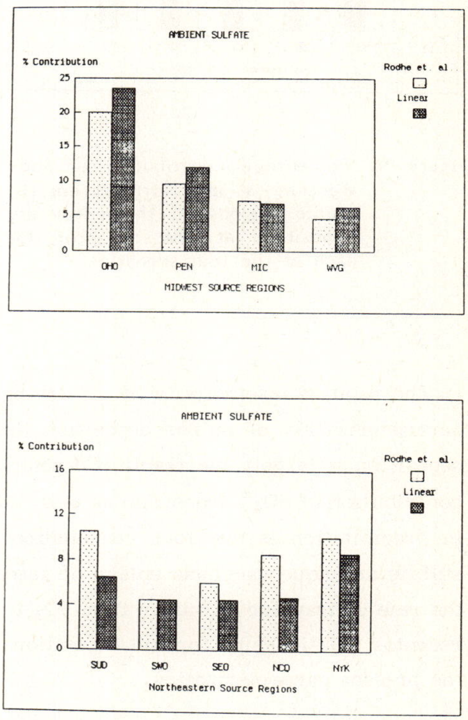

Figure 8. Percentage contribution of Midwestern (a) and Northeastern (b) source regions to the ambient sulfate at the receptor for each of the four scenarios.

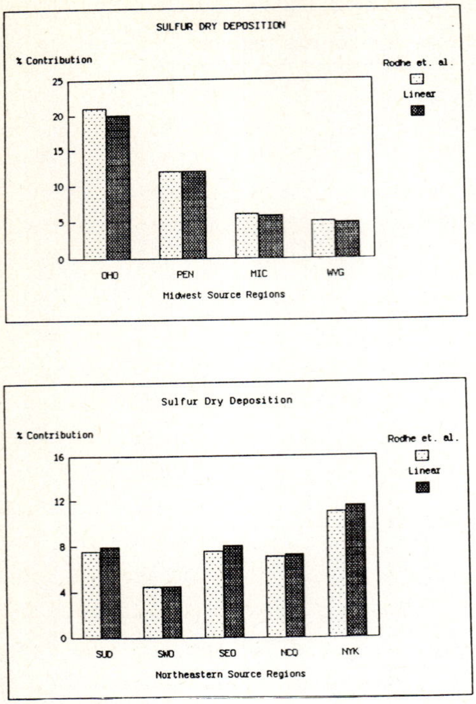

Figure 9. Percentage contribution of Midwestern (a) and Northeastern (b) source regions to the sulfur dry deposition at the receptor for each of the four scenarios.

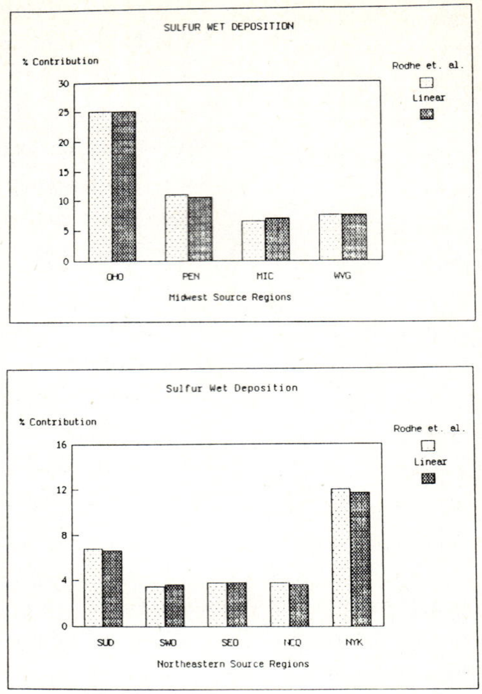

Figure 10. Percentage contribution of Midwestern (a) and Northeastern (b) source regions to the sulfur wet deposition at the receptor for each of the four scenarios.

The relative insensitivity of the dry and wet deposition to reactivity is due in part to the parameterization of sulfur deposition and scavenging used in the ACID model. The dry deposition is largely the result of deposition of SO_2, but the total deposition includes a small contribution of $SO_4^=$ deposition as well. Likewise, it is assumed that the scavenging of sulfur in precipitation is due to a combination of nucleation scavenging of particles containing sulfate and aqueous-phase uptake of gaseous SO_2. Thus a chemical mechanism which alters the rate of transfer of sulfur from SO_2 to $SO_4^=$ will not produce as large a shift in source-receptor relationships for wet deposition as it will for ambient sulfate concentrations with the present parameterization.

CONCLUSIONS

The receptor-oriented probabalistic modeling described in this paper has the ability to diagnose the spatial changes in source-receptor relationships for both varying meteorological conditions and for varying chemical transformation mechanisms. The modeling of year-to-year variations in source-receptor relationships for sulfur wet deposition has demonstrated that there can be substantial fluctuations in the natural potential for a given State or Province to contribute to a particular receptor, especially for source States near the receptor.

Additionally, the variations in source-receptor relationships for annual ambient sulfate concentrations have also been shown to be potentially sensitive to the inclusion of a simplified gas-phase mechanism in place of a spatially invariant linear conversion mechanism. The percentage of contribution computed to have originated in regions of low estimated RHC/NO_x emissions is less than that computed without the gas-phase reactivity function.

It is premature to make generalizations about the influence of non-linear atmospheric processes on source-receptor relationships for atmospheric sulfur species. This work represents an initial attempt to incorporate potentially complex nonlinear processes into a regional-scale sulfur deposition model. Future work will continue to examine the variability in model estimates due to the uncertainty of certain model assumptions and the variability of meteorology.

REFERENCES

1. P. J. Samson and M. J. Small, Modelling the Impacts of Total Acid Precipitation, Ann Arbor Science, MI (1983).
2. A. Eliassen and J. Saltbones, Atmospheric Environment 17, 1457 (1983).
3. J. D. Shannon, Atmospheric Environment 15, 689 (1981).
4. E. C. Voldner, M. P. Olson, K. Oikawa and M. Loiselle, J. Geophys. Res. 86, 5339 (1981)
5. C. M. Bhumralker, R. L. Mancuso, D. E. Wolf, W. B. Johnson and J. Pankrath, Tellus 33, 142 (1981).
6. R. G. Lamb and J. H. Seinfeld, Environ. Sci. Technol. 7, 253 (1973).
7. G. R. Cass, Atmospheric Environment 15, 1227 (1981).
8. P. J. Samson, J. Appl. Meteor. 19, 1382 (1980).
9. P. J. Samson and J. L. Moody, Air Pollution Modeling and its Application I, C.D. Wispelaere, Ed., Plenum Press, 43 (1981).
10. R. R. Draxler and A. D. Taylor, J. Appl. Meteor. 21, 367 (1982).
11. N. V. Gillani, R. B. Husar, J. D. Husar, D. E. Patterson and W. E. Wilson, Atmospheric Environment 12, 589 (1978).
12. L. A. Barrie, Atmospheric Environment 15, 31 (1981).
13. B. C. Scott, J. Appl. Meteor. 17, 1375 (1978).
14. D. J. McNaughton, APCA Specialty Conference on the Meteorology of Acid Deposition, P.J. Samson, Ed. (1983).
15. C. M. Benkovitz, Atmospheric Environment 16, 1551 (1982).
16. J. G. Calvert and W. R. Stockwell, Environ. Sci. Tech. 17, 428 (1983).
17. H. Rodhe, P. Crutzen and A. Vanderpol, Tellus 33, 132 (1981).
18. United States National Academy of Science, Acid Deposition Atmospheric Processes in eastern North America, NAS Press, Washington, D.C. (1983).
19. W. R. Stockwell and J. G. Calvert, Atmospheric Environment 17, 2231 (1983).
20. P. J. Samson and J. L. Moody, Atmospheric Environment (in press).
21. Chameides, W. L., D. D. Davis, J. Geophys. Res. 87, 4863 (1982).

MODES OF INTERACTION OF AIR POLLUTANTS WITH SPRUCE-FIR FORESTS: SUMMARY AND NEW MECHANISMS.

G.H. Kohlmaier, E.O. Siré, U. Fischbach, H. Broehl and M. Ploechl

Institute for Physical and Theoretical Chemistry
Johann Wolfgang Goethe-University
Niederurseler Hang
D-6000 Frankfurt am Main
F.R.G.

ABSTRACT

The longest standing hypothesis of interaction of acid precipitation (rain, snow, fog, particulate or gaseous compounds) with the vegetation is to be understood in a literal sense: deposition or interception of protons by the aboveground vegetation and soil leads to a negative stress for the plant metabolism. Both the leaves (needles) of the tree canopy as well as the soil-root system may be affected adversely by the excess acid.

Ulrich (Goettingen) has emphasized that a chronic intoxication of the soil may occur when more and more of the monovalent and bivalent nutrient cations (K^+, Ca^{++}, Mg^{++}) are replaced by protons on the cation exchange sites while at the same time Al^{+++} ions are liberated when the breakdown of the clay materials occurs.

On the other hand a drastically lowered Mg^{++} and Ca^{++} concentration of the spruce and fir needles has been observed which may be a consequence of the soil impoverishment but also an effect of direct acid leaching on the needles. Ozone and other photooxidants like PAN may support the latter effect in as much as the cuticula (wax) layer of the needle is destroyed and the cell membranes are made more permeable to ion exchange.

As a new mechanism it is suggested that the proton gradient across the thylacoid membrane of the chloroplast, responsible for the ATP production in the light reaction of photosynthesis, is reduced by acid infiltration or changed membrane permeability. Thus other parts of metabolism may be activated, since the ratio of ATP to $NADPH/H^+$ is changed.

It will be also discussed whether the primary and secondary products of anthropogenic emissions are interacting directly with the leaves, roots and soil of the system, or whether there may be a local chemical or photochemical interaction with some of the metabolic products of the coniferous trees, like terpenes, isoprene and ethylene, leading together with NO_x to the more toxic higher homologues of PAN as well as ozone.

NATO ASI Series, Vol. G6
Chemistry of Multiphase Atmospheric Systems
Edited by W. Jaeschke

INTRODUCTION

Damage of ecosystems caused by air pollution has been observed over long periods of time. About hundred years ago dieback of forests in the near neighbourhood of industrial facilities was reported and attributed to the fume gases, especially sulfur dioxide [1].

Though severe damages occured, they were restricted to small areas around the emission sources. Advancing industrialisation and increasing use of fossil fuels necessitate efforts to clean up highly polluted cities and conurbations, one such effort was the construction of higher chimneys to achieve more effective dispersion of pollutants. While this was done in Central Europe in the middle 1960s, few years later first reports appeared of ecosystem damage in Southern Sweden most probably resulting from air pollutants imported from neighbouring countries. The most obvious impact was the acidification of lakes, which was also observed in North America, leading to a decrease of 1-2 pH-units in the soft and poorly buffered waters of these regions [2]. This resulted in a decline and extinction of fish populations [3] as well as other communities of aquatic organisms [4].

In the middle 1970s, severe dieback of Silver fir (Abies alba) was observed in the southern parts of Germany, however, it was not possible to relate this clearly to air pollution effects. Within the last years other coniferous trees, especially spruce and pine, showed similar symptoms as observed in fir on rapidly increasing areas, and decidiuous trees were also injured, leading to a share of 34% of the total forested area of the FRG being affected [5]. Fig. 1 summarizes the time development of the observed forest dieback in the FRG, visualizing the explosive vehemence with which this new disease started.

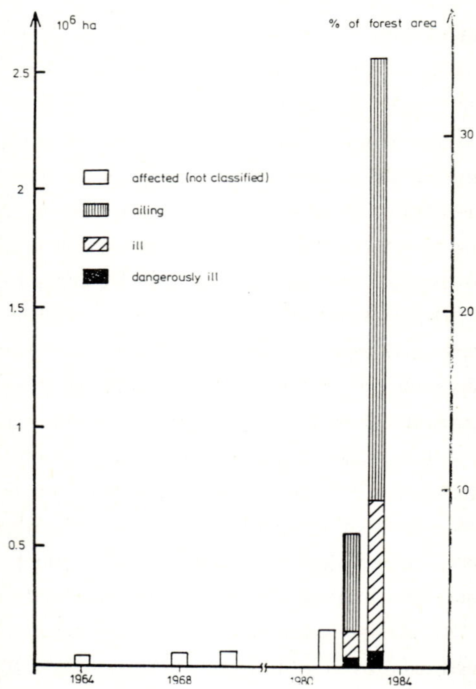

Fig. 1: Reported areal extent of tree disease in the Federal Republic of Germany for different years. The forest damage inventories 1982 and 1983 for the first time distinguish between three states of damage, as indicated in the figure. Earlier data are scarce and usually not classified.

BRIEF DESCRIPTION OF TREE DISEASE

Symptoms common to all affected species are damage of the fine root system and the mycorrhiza as well as leaf injury, expressed as decoloration and loss of older needles leading to a thinning of the lower crown in conifers or as deformation and necrosis on leaves and early leaffall in decidiuous trees. Growth disturbance is observed in fir as flattening of the upper crown and probably as a reduction in tree ring width in all species [6,7]. Lack of calcium and magnesium frequently occurs in conifer leaves and the water balance seems often to be disturbed [8]. Trees of all age-classes are affected, though older trees of fir and spruce seemed to be more susceptible. Damage usually increases with altitude up to a certain level and is most severe on southern and south-western slopes, but no clear correlation with soil conditions is observed [9]. Injured trees are scattered within the stands but frequently occur on exposed sites such as forest rims, recently total stands are affected in increasing amounts.

The time development of the disease is visualized for selected areas and tree species in Fig. 2. As can be seen, the transformation from a healthy to a seriously affected state may be very rapid especially for spruce, though there is some evidence that injury may have started earlier but was not recognized [10,11].

Several hypothesis about the primary factors responsible for these tree diseases have been proposed, including unusual climate/weather patterns, biological factors such as insect, pest, fungus, bacteria or virus infestation, or even an overaging of the forest. Though it could not be ruled out that some or all of these factors may play a role, today it is widely believed that air pollution is the main primary cause for the observed forest dieback [12], possibly accentuated or followed by other factors.

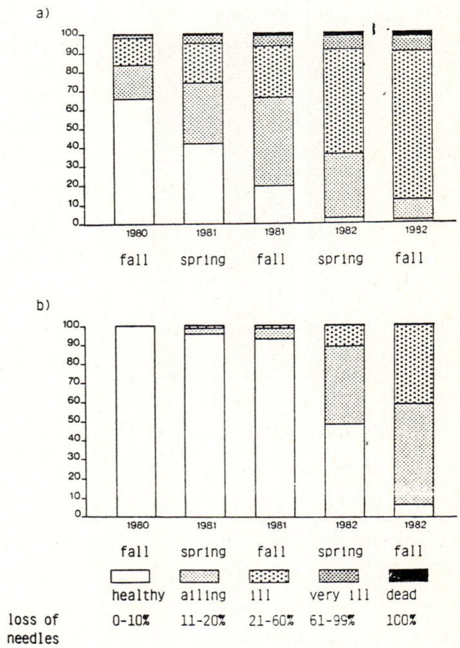

Fig. 2: Changes in the state of health of (a) silver fir (Abies alba) and (b) Norway spruce (Picea abies) for selected plots in the state of Baden-Wuerttemberg from fall 1980 to fall 1982, according to [10].

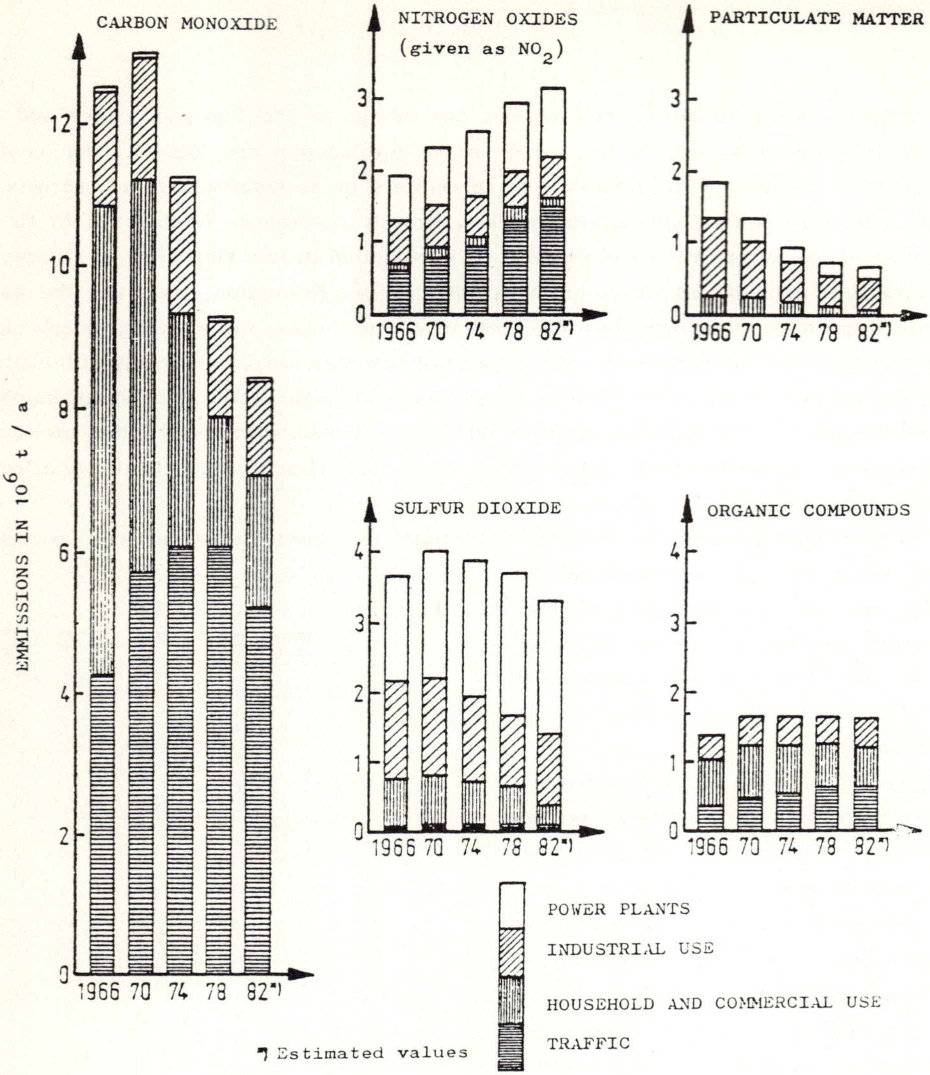

Fig. 3: Time development of the yearly emissions of selected air pollutants in the FRG, 1966 - 1982, according to [34].

EMISSION AND IMMISSION OF AIR POLLUTANTS

The first substances found to be toxic to plants were gaseous compounds like SO_2 and NO/NO_2, leading to acute injury when applied in concentrations in the order of few ppm. In the following times many other substances were identified as phytotoxicants, including ozone, organic compounds and heavy metals.

The emissions of these pollutants increased in Europe within the first half of this century to considerable amounts, leading to heavy pollution on large areas. Within the last decade, however, the emission rates of some pollutants, especially SO_2, remained nearly constant (but are supposed to increase in the coming decades) [13] or even declined, as in the case of particulate matter, while others, e.g. NO_x, continued to increase. Fig. 3 summarizes the emission trends for the main pollutants in the FRG for the years 1966-1982.

Much less is known about possible changes in atmospheric transport and the time development of immissions on different areas. From measurements of concentrations of SO_2, NO_x, sulfur compounds in particulate matter and H^+ in so-called clean-air regions in the FRG in general no clear trend appears, though an increase in NO_x and in some regions also in SO_2 has been proposed [14]. Fig. 4 shows the concentrations of some pollutants as measured by the German Federal Environmental Agency for a summer and a winter month (July and December, resp.) in 1982.

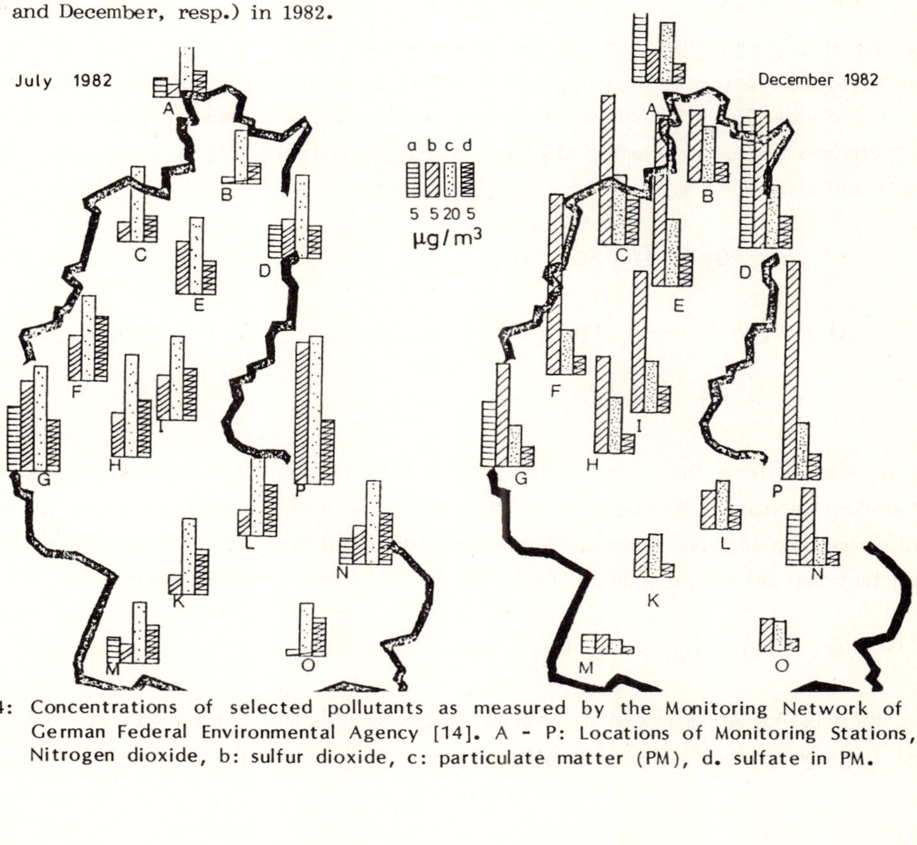

Fig. 4: Concentrations of selected pollutants as measured by the Monitoring Network of the German Federal Environmental Agency [14]. A - P: Locations of Monitoring Stations, a: Nitrogen dioxide, b: sulfur dioxide, c: particulate matter (PM), d. sulfate in PM.

Deposition of air pollutants widely differ in type and amount according to several factors. Besides the distance from the emission source, the ratio of wet and dry deposition depends on the characteristics of the substances as well as on climatic or meteorological conditions, especially on precipitation. While in general both are of comparable importance in northern mid-latitude regions [15], dry deposition may be an important process in heavily polluted areas as well as in ecosystems with high surface areas, as e.g. coniferous forests. Typical deposition rates of sulfur derived from measurement for the FRG lie between 5 and 6 g S m^{-2} a^{-1} with maximum values up to 10 g S m^{-2} a^{-1}. Dry deposition may account for up to two thirds of these values. Nitrogen deposition may typically amount to 0.4 to 1 g N m^{-2} a^{-1}, mostly as wet deposition. This may be compared with the mean value as calculated from the emission data: if out of the total sulfur emission in 1982 of about 3.6 10^6 t SO_2, or about 1.8 10^6 t S resp., a net of 2/3 is deposited over an area of 25 x 10^{10} m^2, about 4.8 g S m^{-2} a^{-1} are to be expected. Similarity with about 3.0 10^6 t of NO_2 or about 0.8 10^6 t N, and again a net deposition of 2/3, about 2.4 g m^{-2} a^{-1} are calculated as the average immission.

CHEMISTRY OF AIR POLLUTANT FORMATION

During atmospheric transport the emitted substances may undergo chemical and photochemical transformations in the gas phase as well as heterogeneous reaction at the gas/liquid or gas/solid interface of aerosols or water droplets within clouds or fog. One of these transformation systems connecting nitrogen and oxygen compounds is given by the following set of reactions:

$$NO_2 \xrightarrow{h\nu} NO + O$$

$$O + O_2 + M \longrightarrow O_3 + M$$

$$O_3 + NO \longrightarrow NO_2 + O_2$$

It is evident from the above system of coupled equations that tropospheric ozone is produced and consumed by NO_2/NO such that no excess ozone becomes available for other reactions as long as ozone is needed for the reoxidation of NO to NO_2. However, it is a well-known fact that NO can be oxidized by O_2 in the presence of carbon monoxide

$$NO + CO + O_2 \longrightarrow NO_2 + CO_2$$

or in the presence of saturated or olefinic hydrocarbons. The net reaction for butane or butene is given by the following equations:

$$C_4H_{10} + 2NO + 2O_2 \rightsquigarrow C_3H_7CHO + H_2O + 2NO_2$$

$$i\text{-}C_4H_8 + 2NO + 2O_2 \rightsquigarrow (H_3C)_2C{=}O + O{=}CH_2 + 2NO_2$$

When NO is oxidized according to the above reaction, ozone may accumulate to concentrations sufficient to act as phytotoxicant [16].

The aldehydes resulting from the reactions described above or emitted directly may in presence of OH-radicals and NO_2 be converted to peroxyacyl nitrates, the most common one being PAN. These photooxidants also may have toxic effects on plants, as will be discussed later.

The formation of acid from and other transformations of the emitted nitrogen and sulfur oxides as well as the reactions of these acids with ammonia from natural and anthropogenic sources leading to the corresponding ammonia salts are discussed in detail elsewhere in this volume.

It is commonly assumed that the primary and secondary pollutants enter the forest and grassland ecosystems either as gases like SO_2, NO_x, ozone or PAN, as particulate matter like soot or other aerosols, or dissolved in rain drops, fog droplets and snow containing aside from hydrogen ions NH_4^+, SO_3^{--}, SO_4^{--}, NO_2^-, NO_3^- and many others. A summary of the formation of primary and secondary pollutants is given in table 1.

However, we like to suggest that the primary and secondary pollutants may possibly undergo tertiary reaction within the airspace of the ecosystem and at the interface of the air and the plant-soil system, as outlined in table 2.

Each ecosystem exhibits specific source and sink functions for airborne components; ecosystems of temperate and higher latitudes show e.g. a seasonal cycle in the uptake and release of CO_2, oxygen and water vapor. There is also a large number of secondary components characterizing

Primary and secondary air pollutants

Type of emission or reaction	examples
A. Emission of primary pollutants from urban-industrial complexes and natural sources	SO_2, NO (little NO_2) HC = hydrocarbons such as alkenes, alkanes, aromates,. PM = particulate matter including soot, heavy metal oxides
B. (Photo)-Chemical formation of secondary pollutants in the atmosphere in homogeneous and heterogeneous reactions at the solid/liquid/gas interphase of aerosols and droplets	$NO \rightarrow NO_2 \rightarrow HNO_2, HNO_3$ $SO_2 \rightarrow H_2SO_3, H_2SO_4$ $NO_2 + h\nu \rightarrow NO + O$ $O + O_2 + M \rightarrow O_3 + M$ $O_3 + NO \rightarrow O_2 + NO_2$ $NH_3 + H_2SO_4 \rightarrow (NH_4)_2SO_4$ $NH_3 + HNO_3 \rightarrow NH_4NO_3$
A+B. Pollutants entering forest and grassland ecosystems as gases, particulate matter or dissolved in rain, fog and snow.	gases: SO_2, NO_x, O_3, PAN PM: soot, salt, oxides of heavy metals liquid: NH_4^+, H^+, SO_4^{--}, NO_3^-, NO_2^-

Tab. 1

the gas exchange of ecosystems which is not yet known in full detail. Table 3 shows some typical hydrocarbons as they are emitted from forest and grassland ecosystems.

It is estimated that forest ecosystems are large sources of such reactive and non-reactive hydrocarbons, which, however, because of the specific reaction patterns within the canopy and the soil have only a small chance to escape into the free atmosphere. Table 4 shows a summary of estimates on the emission rates of terpenes as given by different authors. As can be seen, the global emission rates of reactive hydrocarbons may vary between 175 and 830 Mt a^{-1}, a figure larger by a factor of up to ten compared to the anthropogenic emissions of this class of substances.

The stationary terpene concentration within or in the immediate neighbourhood of forest ecosystems do vary not only from system to system but also from day to night and from one season to another. Some results of concentration measurements are given in table 5.

Reactions of primary and secondary air pollutants within the air space of the ecosystem and the interphase of air and plant/soil system.

Type of emission or reaction	examples
A+B. Pollutants entering forest and grassland ecosystems as gases, particulate matter or dissolved in rain, fog and snow.	gases: SO_2, NO_x, O_3, PAN PM: soot, salt, oxides of heavy metals liquid: NH_4^+, H^+, SO_4^{--}, SO_3^-, NO_3^-, NO_2^-
C. Gaseous, liquid and particulate emissions from forest and grassland ecosystems	CO_2, CO, H_2O, O_2, CH_4, C_2H_4, isoprene, terpenes, resins and waxes, pollen, etc.
D. (Photo) Chemical formation of tertiary pollutants at the site of the forest/grassland ecosystem in the airspace sourrounding the plant/soil system	a) consumption of ozone: O_3 + ethene, isoprene, monoterpene $\rightarrow$ aldehydes, ketones, acids $\rightarrow$ $\xrightarrow{NO_2, h\nu}$ PAN and higher more poisenous homologues b) formation of ozone: $NO + R_1R_2C{=}CHR_3 + O_2 \rightarrow NO_2 + R_1R_2\dot{C}O \; O\dot{C}HR_3$ $NO_2 + h\nu \rightarrow NO + O$ $O + O_2 + M \rightarrow O_3 + M$
E. (Photo)Chemical formation of tertiary pollutants at the site of the forest/grassland ecosystem at the interphase of the air (water) /leaf surface	a) oxydation of waxes and resins of the cuticula of the leaf(needle) through O_3, H_2O_2 $\Rightarrow$ formation of aldehydes, ketones and acids destroying the leaf surface b) sulfonation of waxes, formation of detergent like compounds leading to micelles, amplifying other possible reactions

Tab. 2

Little is known on the reactive pathways of these hydrocarbons within the ecosystem. In the presence of primary and secondary air pollutants they may well undergo specific reactions leading to tertiary pollutants [17,18,19]. From work in Japan [20] we know that there is a negative correlation between advected ozone and terpene concentrations in forests. In this paper we will discuss two principle routes of terpene reactions:

1.) the reaction with ozone to form aldehydes and ketones by the intermediate of the Criegee radical

2.) the reaction with NO_x leading to the enhancement of ozone.

In general the reaction of olefinic hydrocarbons and ozone proceed via the primary ozonides to an aldehyde or ketone and a Criegee radical or ion.

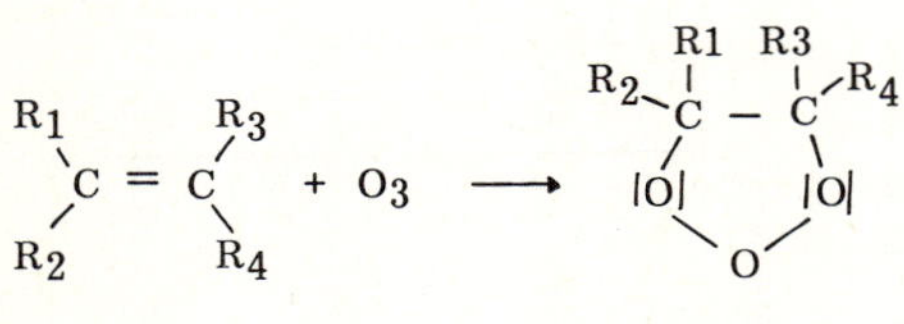

TABLE 3

Typical hydrocarbons naturally emitted from forest and grassland ecosystems (modified after [22])

Substances		Main Emitters
Alkanes	Ethane	grass
	Propane	coniferous trees, grass
	Butane	coniferous trees, grass
Alkenes	Ethene	all higher plants
	Isoprene	decidiuous trees
Aromatics	p-Cymene	coniferous and decidiuous trees
Acyclic Terpenes	Myrcene	alfalfa, coniferous trees
	Ocimene	alfalfa
	Citronellal	grass, coniferous trees
	Geraniol	grass
	Linalool	alfalfa, coniferous trees
Monocyclic Terpenes	Limonene	alfalfa, coniferous and decidiuous trees
	α-Pinene	decidiuous trees
	Terpenoline	coniferous trees
	β-Phellandrone	coniferous trees
	Isopulegol	grass
Bicyclic hydrocarbons	α-Pinene	coniferous and decidiuous trees
	β-Pinene	coniferous trees
	3-Carene	coniferous trees
	Camphene	coniferous trees
	Campher	coniferous trees
	1,8-Cineole	decidiuous trees

Tab. 4: Estimates of Emission Rates of Terpenes

Reactive Hydrocarbons (as $C_{10}H_{18}$)	global:	$175\ 10^6\ t\ a^{-1}$	
	local:	$1.2 - 10.5\ g\ m^{-2}\ a^{-1}$	[37]
		$= 333 - 2920\ \mu g\ m^{-2}\ h^{-1}$	
α-Pinene emissions (Pine forest, summer):		$1900\ \mu g\ m^{-2}\ h^{-1}$	[20]
extrapolated to 1 year:		$8.2\ g\ m^{-2}\ a^{-1}$	
Terpene emmissions (Mixed forest, fall)		$3120\ \mu g\ m^{-2}\ h^{-1}$	[17]
extrapolated to 1 year:		$13.7\ g\ m^{-2}\ a^{-1}$	
Terpene emissions, global:		$480\ 10^6\ t\ a^{-1}$	[38]
Isoprene emissions, global:		$350\ 10^6\ t\ a^{-1}$	

Tab. 5: Selected terpene concentrations in or in the near neighbourhood of forested areas.

Coniferous forest, eastern US (Terpenes and Hemiterpenes)	within the canopy	$50\ \mu g\ m^{-3}$	[18]
	outside the canopy	$3 - 32\ \mu g\ m^{-3}$	
Forested area in the Pacific North-west of the US (Terpenes)		$9.2\ \mu g\ m^{-3}$	[17]
Pine Forest, Japan (α-Pinene)	with O_3 - load	$0 - 3.8\ \mu g\ m^{-3}$	[20]
	at night	$15.8\ \mu g\ m^{-3}$	

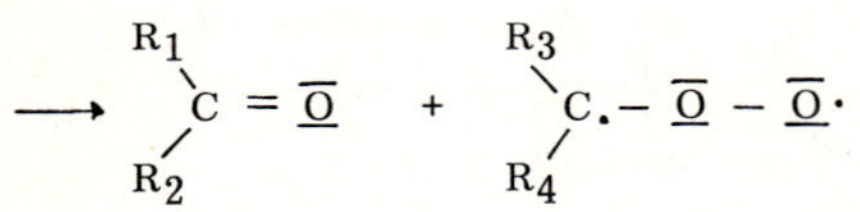

(I) (II)

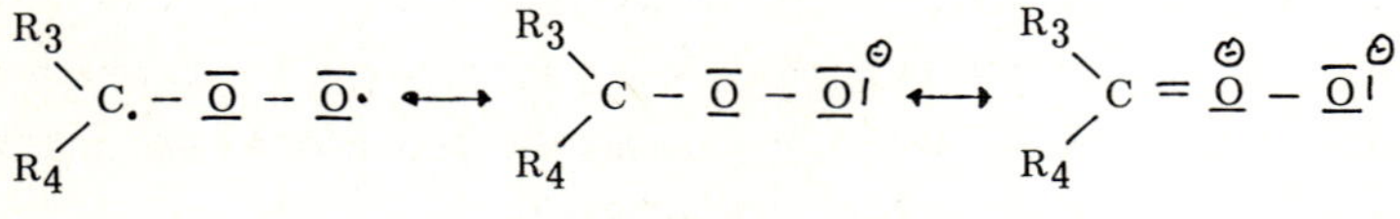

The Criegee compound (II) may polymerize according to

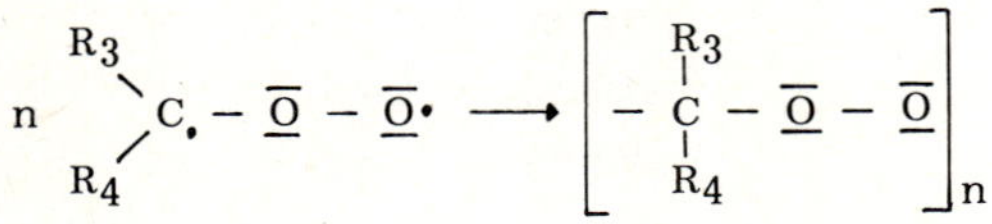

Such polymers may contribute to the aerosol formation which is known as "blue haze" over certain coniferous forest areas within the Unites States. The Criegee radical may also react with compound (I) to form a real ozonide:

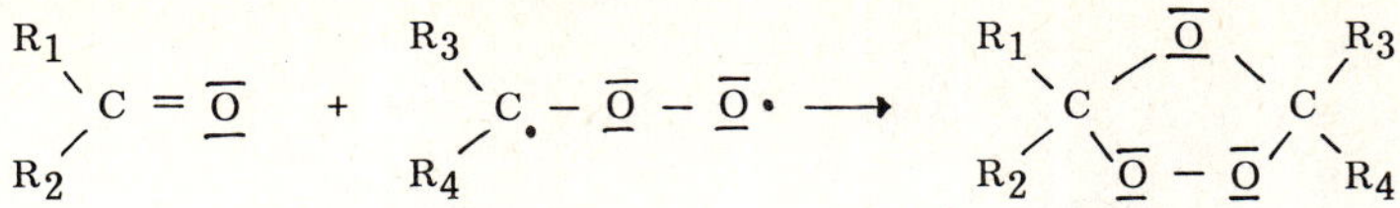

These compounds in turn are not very stable and decompose forming aldehydes, ketones or acids.

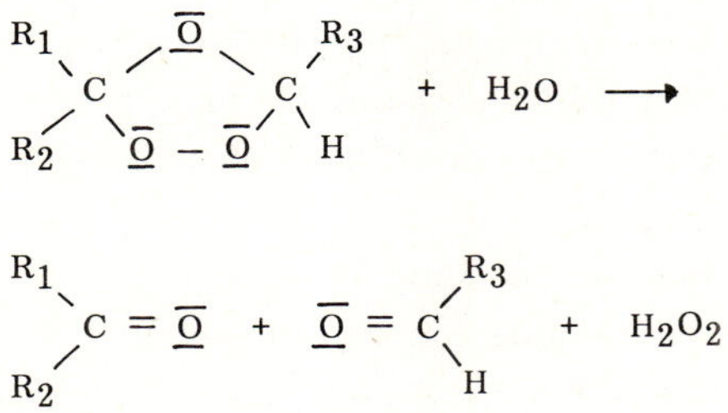

Within the hydrolysis hydrogen peroxide is formed which itself may react with the hydrocarbons from the ecosystem or the newly formed aldehydes to form acid according to:

In the presence of OH-radicals and NO_2 aldehydes may also be transformed to peroxiacyl nitrates:

$$R\text{-}CHO \xrightarrow[O_2,\ NO_2]{OH^\bullet} R\text{-}CO\text{-}OO\text{-}NO_2$$

If R stands for CH_3, then the above compound may be identified with PAN, however, if R stands for a higher homologue then according to [21] even more poisonous compounds may be formed. Graedel [22] has suggested that stable intermediates of the oxidative degradation of α- and β-Pinen may be the Pinon- and Norpinon aldehydes.

The reactive hydrocarbons of the ecosystem may also promote an oxidation of NO in the presence of molecular oxygen. Above we have already shown the net reaction for such a process involving e.g. butene. In general we may expect the following net reaction for an olefin

$$R_1R_2C{=}CR_3H \quad + \quad 2NO + 2O_2$$
$$\longrightarrow \quad R_1R_2C{=}O \quad + \quad R_3CHO \quad + \quad 2NO_2$$

or for the corresponding cyclic compound

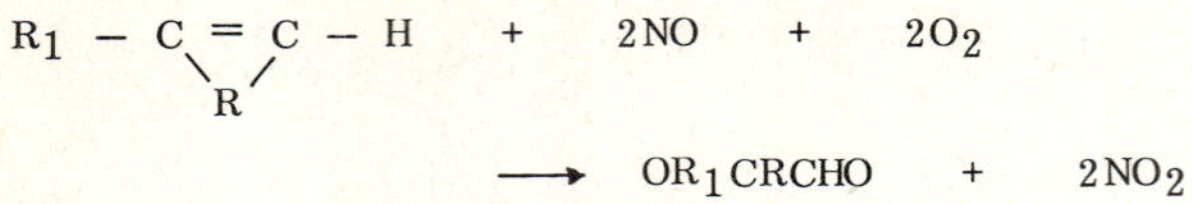

Heicklen [23] has suggested that the reaction may be initiated by OH-radicals followed by the formation of a peroxy-radical which transforms one oxygen atom to NO to form NO_2 while in a second sequence the molecule is broken forming the corresponding aldehyde and ketone and releasing OH. The newly formed NO_2 may photochemically release an oxygen atom, forming ozone which itself may react with the terpenes or their metabolites as described above.

Before entering the leaf through the stomata or the cuticula the primary, secondary and tertiary pollutants may even undergo further reactions on the leaf surface. There may be very important (photo)chemical processes on the waxy surface layer of the needle with ozone, SO_2 and other pollutants. It has been suggested that the waxy layer is perforated by reaction with ozone alleviating an ion exchange between hydrogen ions of the acid precipitation and magnesium, calcium and potassium ions of the leaf cells. We will refer to this process as acid leaching and will describe details in the chapter below. It has also been suggested that a light-assisted sulfonation may take place on the outer needle surface leading to detergent-like hydrocarbon compounds.

EFFECTS OF AIR POLLUTION ON PLANT METABOLISM

On the cellular level there are two paths of transporting gaseous pollutants into the inner parts of the leaves: transstomatal and transcuticular transport.

The greatest part of the gaseous substances (air polluting gases) enters through the stomata, which make up only 0.3 - 2.0% of the surface area, and only a very small amount via cuticula (see fig. 5).

Conditions which favour stomata opening usually result in increased uptake of e.g. SO_2, one of the main pollutants, and increased plant sensitivity.

Some mechanisms for closure and opening of stomata are:

1. The influence of light (photoactive -),
2. CO_2-concentration (chemoactive -), and
3. water changes (hydroactive mechanism) while temperature changes influence mechanism 1-3 (thermoactive mechanism).

Sulphur enters plants through roots as SO_4^{2-} and to the greatest amount through leaves via stomata as gaseous SO_2, which dissolves in the tissue (water of the cell walls) during formation of HSO_3^- or SO_3^{2-} according to the prevalent pH of the cell compartment. At a pH

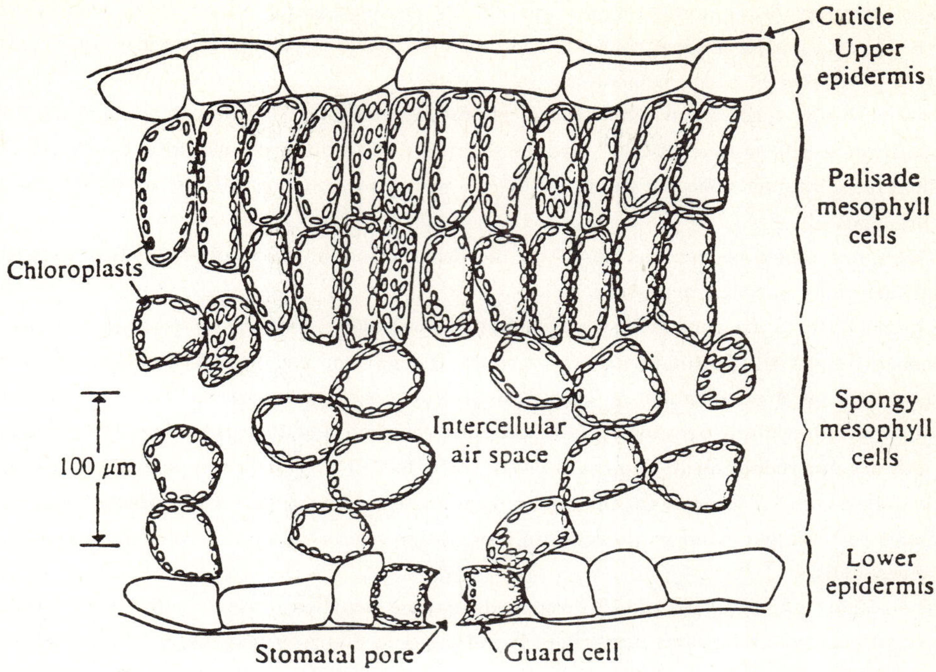

Schematic transverse section through a leaf, indicating the arrangement of various cell types.

Fig. 5: Legend as given in the figure.

of 7.2, which is normal for the cytoplasm of most plants, sulphur (IV) exists to about 50% in form of HSO_3^- and to another 50% in the form of SO_3^{2-}. An acidic solution favours HSO_3^- while in an alkaline solution SO_3^{2-} is favoured [25].

The sulphite ion may be partially oxidized to sulphate by sulphite oxidase, an enzyme which is present in plant mitochondria and needs Mg^{++} and cytochrome c for its activity. But it is not clear which share of the reaction is the result of enzymatic processes and how much takes place because of the redox conditions within the cell [26]. Small amounts of reducing sulphur components can be metabolized into organic compounds, mainly into sulphhydryl groups of sulphur containing amino acids without harmful effects [27]. Highest concentration of incorporated sulfur is found in the guard cells of the stomata, much less in epidermis and mesophyll [27]. An accumulation of S-compounds like SO_2, SO_4^{2-}, SO_3^{2-} and HSO_3^- exists predominantly at the edges and tips of leaves [28]. Under the influence of low SO_2 concentrations the stomata loose their ability to close and under higher concentrations over a long time range paralysis of the open stomata takes place [29].

The opening of the stomata will result in an increasing water loss. Besides this, additional SO_2 is able to penetrate into the inner parts of the leaf quite rapidly and to affect central processes of plant metabolism.

SO_2 pollution leads to an inhibition of photosynthetic CO_2-fixation. It is supposed that a competition exists between HCO_3^- and SO_3^{2-} for active sites on ribulose 1,5-diphosphate carboxylase. Higher concentrations of SO_3^{2-} result in non-competitive inhibition of this enzyme [25].

There are other mechanisms based on the destruction of chlorophyll by SO_2 fumigation which are very complex processes.

Chlorophyll is decomposed to phaeophytine and Mg^{2+} ions by SO_2 depending on the concentration of the pollutant and the time of fumigation, and besides phaeophytization in acid medium (at a pH of about 4) oxidation of chlorophyll takes place [25].

But the chlorophyll breakdown is a too slow reaction and therefore has little relevance for the drastic reduction in photosynthesis. The effect of SO_2 pollution on respiration is not very well known up to now. On the one hand there is reported that SO_2 reduces respiration [25] and on the other hand an increase in respiration even at very low concentrations can be observed [30].

The effects of sulfur IV and VI compounds may be modified by the simultaneous action of other pollutants. So stomatal movements in response to SO_2 may be greatly influenced in the presence of photooxidants of NO_x resulting in antagonistic or synergistic effects depending on concentration, type of plants, growth state and other external or internal factors. A summary of the present knowledge about these effects is given in [21].

Ozone and PAN (peroxiacetylnitrate) inhibit photosynthesis by oxidation of the SH-groups of disulfite-containing photoreducible chloroplast proteins. Besides this, photooxidants affect many parts of photosynthetic electron transport. Respiration may be inhibited by O_3 and PAN in the beginning as a consequence of oxidating the SH-groups of the enzymes involved [26].
After continued fumigation respiration usually increases because of further reactions leading to visible damage on the plant.

Aside from the gaseous pollutants (e.g. SO_2, NO_2, O_3), the plants are also penetrated by H^+-ions resulting from dissolution processes on the surface and the inner parts of the leaf and in the soil.

Leaf cells possess a buffering capacity which enables to neutralize protons through Ca^{++} and Mg^{++} ions on the ion exchange sites of the cell walls.

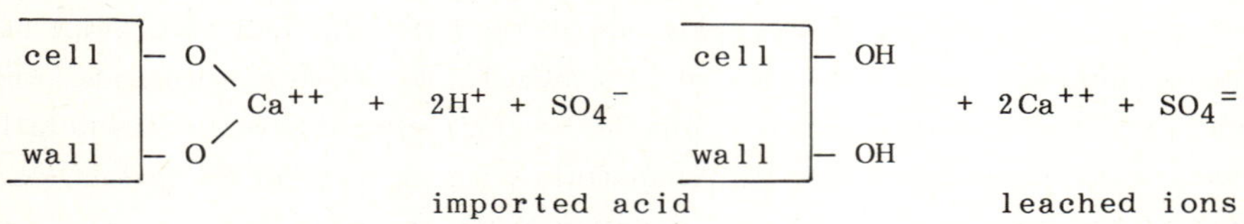

The cell acts as a biochemical pH-stat if the system has enough resiliance to overcome the acid stress.

Resiliance depends on a recharging of the exchange sites, using the basicity of Ca^{++} (Mg^{++}) imported from the soil with the transpiration stream through the xylem.

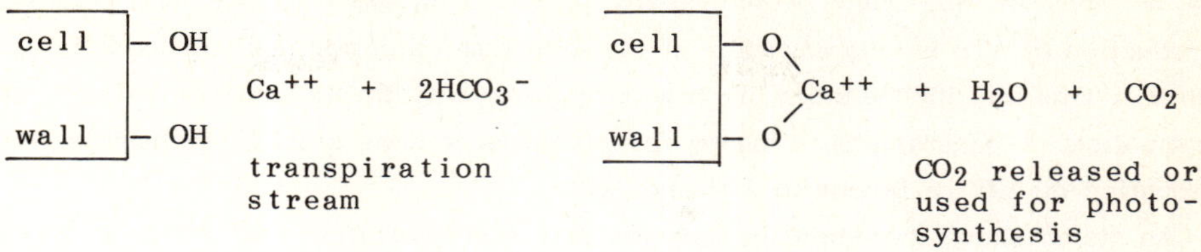

The removal of Ca^{++} (Mg^{++}) ions from the soil leads to soil acidification.

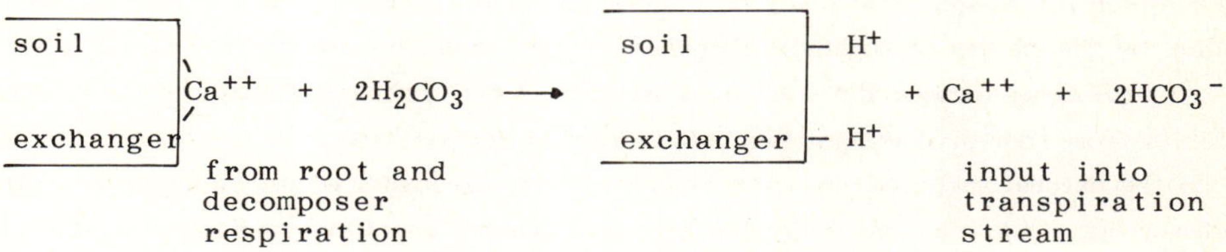

If Ca^{++} (Mg^{++}) ion resources are exhausted within the soil (through cronic acid stress) a re-exchange in the leaves is no longer possible. We expect a breakdown of pH-homeostasis in the leaf cells. At this stage a range of injuring reactions will be based on lowering the pH within the cellular space.

For example important pH dependent processes take place in the chloroplasts:

In the light reaction of photosynthesis water is decomposed photolytically into O_2, H^+ ions and electrons in the inner space of the thylakoid membrane (see fig. 6). According to

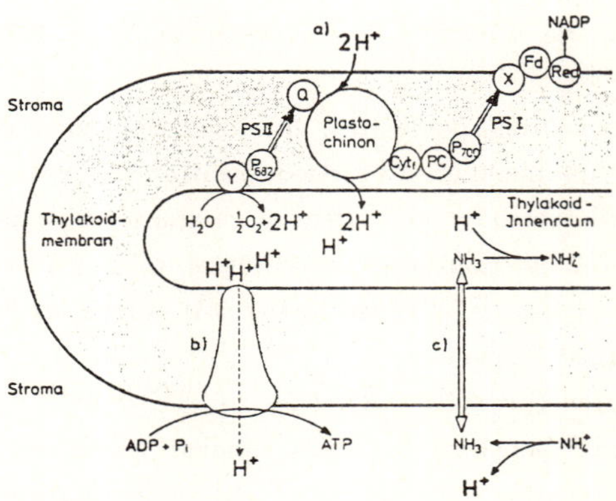

Fig. 6: Internal membrane system of chloroplast (thylakoid) in which the light reaction of photosynthesis takes place. Further explanation see text.

Mitchell [31] the resulting proton gradient between the inside and the outside of the thylacoid membrane drives the ATP production from ADP and phosphate. If the pH gradient is reduced by H^+ ions or NH_4^+ ions which infiltrate the cytoplasma it may be imagined that the production of ATP is stopped, while at the same time reduction equivalents $NADPH_2$ are still formed. This may induce other biochemical pathways within the primary and secondary plant metabolism. A general stress theory for plants as well as animals emphasizes the relative reduction of ATP in favour to ADP and AMP.

As stated above the acidic components of the air pollution also affect on the acid-base system of the soils.

Ulrich (1983) [32] has pointed out that soil may be devided into a series of buffer regions of a given pH range, which may be correlated mainly with the concentrations of Al ions which by themselves or together with Ca ions determine the toxicity of a soil.

At a pH range between 8.6 - 6.2 there exists a $CaCO_3$ buffer region where the proton input results from root or decomposition respiration is compensated.

If the natural soil environment of e.g. a coniferous forest is associated with a pH range between 6.2 - 3.5, then the following proton dependent reactions may be of significance.

1. Silicate buffer region (pH 6.2 - 5.0)

   weathering of feldspars (Na, K and Ca feldspars) to clay minerals of different Al/Si ratios depending on climatic and biotic conditions, is considered to be a slow process e.g.:

$$CaAl_2Si_2O_3 + 2H^+ + H_2O \longrightarrow Ca^{++} + Al_2Si_2O_5(OH)_4$$

   The reaction takes place under release of nutrient cations (e.g. Ca^{++}, K^+, ...).

2. Cation-exchange buffer region (pH 5.0 - 4.2)

   Cation-exchange processes as e.g. for the Kaolinite

$$[Al_2Si_2O_6(OH)_3]_mH_{m-n}\ Me + n\ H^+ = [Al_2Si_2O_6(OH)_3]_mH_m + Me^{n+}$$

   with Me^{n+} is equal to sodium, potassium, calcium, magnesium or to a small amount aluminium yielded by weathering or by decomposition of organic material.

   When the soil is penetrated by acid agens more and more of the exchangeable sites X_mMe may react to yield eventually a clay material where all metal ions are replaced by protons. If $CaCO_3$ or $MgCO_3$ is added to the soil (fertilization) the above process at least may be reversed in part.

3. Aluminium buffer region (pH 4.2 - 3.2)

   Following complete ion exchange the clay mineral itself may be disintegrated by additional H^+ ions yielding polymer oxyhydroxy-aluminium-ions which finally become monomer aluminium ions:

$$[Al_2Si_2O_5(OH)_4]_m \xrightarrow{H^+} [Al(OH)_{2.5}^{0.5+}]_m + \text{silica} \xrightarrow{H^+} Al(OH)_2^+ \xrightarrow{H^+} Al(OH)^{2+} \xrightarrow{H^+} Al^{3+}$$

4. With proceeding acidification (pH < 3) Mn- and Fe-oxyhydroxy-compounds exchange in the same way as aluminium (Fe-/Mn-buffer region).

Similar to reactions in the cell wall of leaves the acid input into the soils leads to a nutrient ion (K, Na, Mg, Ca, Fe, Mn) deletion which may lower the productivity of plants. Release of toxic ions like e.g. $Al(OH)_2^+$, $Al(OH)^{2+}$, Al^{3+} may be a limiting factor for growth and regeneration of most species. Huettermann (1983) [39] found that bacterial and soil animal activity as well as growth of roots can be restricted by aluminium.

Authors like Ulrich (1983) [34] have suggested that the ratio (Ca/Al) is critical for the toxicity of soil and therefore for the damage and finally death of the trees.

SUMMARY

Anthropogenically released air pollutants may affect ecosystems in different ways. Substances reaching the ecosystems may there undergo a variety of reactions with biogenically released compounds to enhance concentrations of specific pollutants or even form new, more phytotoxic substances then originally deposited. A summary of possible interactions of air pollutants and forest ecosystems, slightly modified taken from [36], is given in table 6.

This qualitative description can up to now only partially underset by the description of the involved processes on a physical, chemical or physiological level. Though great research efforts are needed to clarify the problem of air pollution effects, countermeasures should not be delayed because of the lack of exact knowledge about single processes, since it seems to be quite obvious that forest dieback will be caused or at least dramatically enhanced by air pollutants.

Table 6 Interaction of Air Pollution and Temperate Forest Ecosystems under Conditions of Low Air Contaminant Load, Designated Class I Interaction

Forest soil and vegetation: Activity and response	Ecosystem consequence and impact
1. Forest soils and vegetation release particulate and gaseous contaminants to the atmosphere	1. Atmospheric burden of contaminants from anthropogenic sources supplemented by forest additions—scale may be local, regional, or global
2. Forest soils and vegetation remove particulate and gaseous contaminants from the atmosphere	2. Air contaminants transferred from the atmosphere to the biosphere, forest ecosystems supplement natural removal mechanisms
3. No or minimal alteration of structure or metabolism of forest soils or vegetation	3. No adverse ecosystem change discernible, slight fertilization possible

Table 6 cont.

Interaction of Air Pollution and Temperate Forest Ecosystems under Conditions of Intermediate Air Contaminant Load, Designated Class II Interaction

Forest soil and vegetation: activity and response	Ecosystem consequence and impact
1. Forest tree reproduction, alteration or inhibition	1. Altered species composition
2. Forest nutrient cycling, alteration a. Reduced litter decomposition b. Increased plant leaching, soil leaching, and soil weathering c. Disturbance of microbial symbioses	2. Reduced growth, less biomass
3. Forest metabolism, alteration a. Decreased photosynthesis b. Increased respiration	3. Reduced growth, less biomass
4. Forest stress, alteration a. Phytophagous insects, increased or decreased activity b. Microbial pathogens, increased or decreased activity c. Foliar damage increased by direct air pollution influence	4. Altered ecosystem stress Increased or decreased insect infestations; Increased or decreased disease epidemics; Reduced growth, less biomass, altered species composition
5. Forest emmissions may be altered and react with air pollutants to enhance specific or form new phytotoxic substances	Enhanced chemical stress

Interaction of Air Pollution and Temperate Forest Ecosystems under Conditions of High Air Contaminant Load, Designated Class III Interaction

Forest soil and vegetation: Activity and response	Ecosystem consequence and impact
1. Severe morbility, excessive foliar damage	1. Dramatic change in species composition, reduced biomass, increased erodebility, nutrient attrition, altered microclimate and hydrology
2. Mortality	2. Forest simplification or destruction

REFERENCES

1. Schroeder, J.V., and C. Reuss, 1883. Die Beschaedigung der Vegetation durch Rauch und die Oberharzer Huettenrauchschaeden. Verlag v. Paul Parey, Berlin.
2. Wright, R.F., and Gjessing, E.T., 1976: Acid Precipitation: Changes in the Chemical Composition of Lakes. Ambio 5, No. 5-6, p. 219-223.
3. Schofield, C.F., 1976. Acid Precipitation: Effects on Fish. Ambio 5, No. 5-6, p. 228-230.
4. Hendrey, G.R., K. Baalsrud, T.S. Trauen, M. Laake and G. Racklum, 1976. Acid Precipitation: Some Hydrobiological Changes. Ambio 5, No. 5-6, p. 224-227.
5. Der Bundesminister fuer Ernaehrung, Landwirtschaft und Forsten, 1983. Neuartige Waldschaeden in der Bundesrepublik. Report of the Federal Ministry of Food, Agriculture and Forests on the occasion of the forest damage inventory 1983. Bonn, FRG.
6. P. Schuett, 1982. Das Krankheitsbild - verschiedene Baumarten, gleiche Symptome. Bild der Wissenschaft 12-82, p. 86-101.
7. Athari, S., and H. Kenneweg, 1983. Zur Problematik des "Waldsterbens aus der Sicht der Waldwachstumsforschung und der Waldinventur. Allg. Forst- und Jagd-Zeitung, 154. Jg., 6/7, p. 114-121.
8. Zoettle, H.W., and Mies, E., 1983. Die Fichtenerkrankung in Hochlagen des Suedschwarzwaldes. Allg. Forst- und Jagd-Zeitung, 154. Jg., 6/7, p. 110-114.
9. Der Bundesminister fuer Ernaehrung, Landwirtschaft und Forsten, 1982. Waldschaeden durch Luftverunreinigungen. Publications of the Ministry of Food, Agriculture and Forests, Series A: Applied Sciences, Vol. 273.
10. H. Schroeter, 1983. Entwicklung des Gesundheitszustandes von Tannen und Fichten auf Beobachtungsflaechen der FVA in Baden-Wuerttemberg. Allg. Forst- u. Jagd-Zeitung, 154. Jg., 6/7. p. 132-131.
11. Binns, W.O., and D.B. Redfern, 1983. Acid Rain and Forest Decline in W.Germany. Forest Commission Research and Development Paper 131, Forestry Research Station, Alice Holt Lodge, Wrecclesham, Farnham, Surrey GU104LH.
12. Ulrich, B., and J. Pankrath, 1983. Effects of Accumulation of Air Pollutants in Forest Ecosystems, D. Reidel Publishing Company, Dordrecht, London.
13. Highton, N.H., and M.J. Chadwick, 1982. The Effects on Changing Patterns of Energy Use on Sulfur Emission and Deposition in Europe. Ambio Vol. 11, No. 6, p. 324-329.
14. Umweltbundesamt, 1982. Monthly reports from the Monitoring Network of the Federal Environmental Agency, W.-Berlin.
15. Whelpdale, D. M., 1982. Wet and Dry Deposition. In: H.W. Georgii and W. Jaeschke (eds): Chemistry of the Unpolluted and Polluted Troposphere. D. Reidel Publ. Comp., Dordrecht: Holland.
16. Prinz, B., G.H.M. Krause, and H. Stratmann, 1982. Vorlaeufiger Bericht der Landesanstalt fuer Immissionsschutz ueber Untersuchungen zur Aufklaerung der Waldschaeden in der Bundesrepublik Deutschland.

17. Holdren, M.W., H.H. Westberg, and P.R. Zimmerman, 1979. Analysis of Monoterpene Hydrocarbons in Rural Atmospheres. J. Geophys. Res. 84, No. 68, p. 5083-5088.

18. Whitby, R.A., and P.E. Coffey, 1977. Measurements of Terpenes and Other Organics in an Adirondack Mountain Pine Forest. J. Geophys. Res. 82, No. 37, p. 5928-5934.

19. Pinker, R.T., and A. Delman, 1983. On the precipitation - vegetation - ozone connection in the Washington, D.C. area. Atmos. Environ. 17, 2, p. 221-226.

20. Yokouchi, Y., M. Okaniwa, Y. Ambe and K. Fuwa, 1983. Seasonal variation of Monoterpenes in the atmosphere of a Pine forest. Atmos. Environ. 17, 4,, p. 743-750.

21. Guderian, R., and R. Rabe, 1981. Effects of photochemical oxidants on plants. Final report of the European Communities, Directorate General for Environment, Consumer Protection and Nuclear Safety.

22. Graedel, T.E., 1979. Terpenoids in the Atmosphere. Rev. Geophys. Space, Phys. 17, 5, p. 937-947.

23. Heicklen, J., 1976. Atmospheric Chemistry. Academic Press.

24. Larcher, W., 1980. Physologica_ Plant Ecological. Springer-Verlag, Berlin, Heidelberg, New York, 2nd ed.

25. Malhotra, S.S., and D. Hocking, 1976. Biochemical and Cytological Effects of Sulphur Dioxide on Plant Metabolism. New Phytol. 76, p. 227-237.

26. Ziegler, I., 1973. The effect of air polluting gases on plant metabolism. In: Environmental Quality and Safety (Coulston F. and Korte F. (eds).).

27. Knabe, W., 1976. Effects of Sulfur Dioxide on Terrestrial Vegetation, Ambio 5 (5-6), p.p. 213-218.

28. Guderian, R., 1970a. Untersuchungen ueber quantitative Beziehungen zwischen dem Schwefelgehalt von Pflanzen und dem Schwefeldioxidgehalt der Luft. Z. Pflanzenkrankh. 77, 387.

29. Schaub, H., 1983. Private Communication.

30. Unsworth, M.H., 1981. Air pollution and plant productivity. In: Physiological Processes limiting Plant Productivity. (Johnson C.B. ed.), Butterworths, London, pp. 293-306.

31. Mitchell, P., 1966. Bot. Rev. 41, 445-502.

32. Ulrich, B., 1983. Gefahren fuer das Waldoekosystem durch Saure Niederschlaege, Sonderheft der Mitteilungen 1982 - Erweiterte Neuauflage 1983 der Landesanstalt fuer Oekologie Landwirtschaftsentwicklung und Forstplanung Nordrhein-Westfalen.

33. Huettermann, A., 1983. Immissionsschaeden im Bereich der Wurzeln von Waldbaeumen. LOELF. 1983, pp. 10a-14a.

34. Ulrich, B., 1983. Soil acidity and its relations to acid deposition. In: Effects of Accumulation of Air Pollutants in Forest Ecosystems (Ulrich, B., Pankrath, J. eds), D. Reidel Publishing Company, Dordrecht, London.

35. Lange, 1983 in: Hearing of the Hessian Parliament, 22.4.1983, UWA/10/5. LFA/10/4, WTA/10/3, Stenographic Report.

36. Smith, W.H., 1981. Air pollution and forests. Springer Verlag New York, Berlin, Heidelberg.

37. Rasmussen, R.A., 1972. What do the hydrocarbons from trees contribute to air pollution? J. Air Poll. Control Assoc., 22 (7), 537-543.

38. Zimmerman, P.R., R.B. Chatfield, J. Fishmann, P.J. Crutzen and P.L. Hanst, 1978. Estimates of the production of CO and H_2 from the Oxidation of hydrocarbon emission from vegetation. Geophys. Res. Lett., 6, 679-682.

INDEX

NATO ASI Series G

Vol. 1: Numerical Taxonomy. Edited by J. Felsenstein. X, 644 pages. 1983.

Vol. 2: Immunotoxicology. Edited by P.W. Mullen. VII, 161 pages. 1984.

Vol. 3: In Vitro Effects of Mineral Dusts. Edited by E.G. Beck and J. Bignon. XIII, 548 pages. 1985.

Vol. 4: Environmental Impact Assessment, Technology Assessment, and Risk Analysis. Edited by V.T. Covello, J.L. Mumpower, P.J.M. Stallen, and V.R.R. Uppuluri. X, 1068 pages. 1985.

Vol. 5: Genetic Differentiation and Dispersal in Plants. Edited by P. Jacquard, G. Heim, and J. Antonovics. XVII, 452 pages. 1985.

Vol. 6: Chemistry of Multiphase Atmospheric Systems. Edited by W. Jaeschke. XVI, 773 pages. 1986.